P9-AQE-915

MECHANICS OF MATERIALS

SIXTH EDITION

MECHANICS OF MATERIALS

Ferdinand P. Beer

Late of Lehigh University

E. Russell Johnston, Jr.

Late of University of Connecticut

John T. Dewolf

University of Connecticut

David F. Mazurek

United States Coast Guard Academy

Global Edition Adapted by Dr. Sanjeev Sanghi

Indian Institute of Technology Delhi, India

McGraw Hill

Connect
Learn
Succeed™

MECHANICS OF MATERIALS, SIXTH EDITION

Published by McGraw-Hill, a business unit of The McGraw-Hill Companies, Inc., 1221 Avenue of the Americas, New York, NY 10020. Copyright © 2012 by The McGraw-Hill Companies, Inc. All rights reserved. Printed in China. Previous editions © 2009, 2006, and 2002. No part of this publication may be reproduced or distributed in any form or by any means, or stored in a database or retrieval system, without the prior written consent of The McGraw-Hill Companies, Inc., including, but not limited to, in any network or other electronic storage or transmission, or broadcast for distance learning.

Some ancillaries, including electronic and print components, may not be available to Global Edition customers.

This book is printed on acid-free paper.

2 3 4 5 6 7 8 9 0 CTPS/CTPS 1 0 9 8 7 6 5 4 3 2

ISBN 978-0-07-131439-8
MHID 0-07-131439-3

All credits appearing on page or at the end of the book are considered to be an extension of the copyright page.

Cover image: © Steven Wauters; The cover photo shows the famous Holmenkollen Ski Jump in Oslo, Norway, a giant cantilevered structure clad in aluminum and glass that rises 58 meters in the air. Photo courtesy of JDS Architects.

superpos – consider everything left

AFD – consider everything to the right when cutting

www.mhhe.com

About the Authors

As publishers of the books written by Ferd Beer and Russ Johnston, we are often asked how did they happen to write the books together, with one of them at Lehigh and the other at the University of Connecticut.

The answer to this question is simple. Russ Johnston's first teaching appointment was in the Department of Civil Engineering and Mechanics at Lehigh University. There he met Ferd Beer, who had joined that department two years earlier and was in charge of the courses in mechanics. Born in France and educated in France and Switzerland (he held an M.S. degree from the Sorbonne and an Sc.D. degree in the field of theoretical mechanics from the University of Geneva), Ferd had come to the United States after serving in the French army during the early part of World War II and had taught for four years at Williams College in the Williams-MIT joint arts and engineering program. Born in Philadelphia, Russ had obtained a B.S. degree in civil engineering from the University of Delaware and an Sc.D. degree in the field of structural engineering from MIT.

Ferd was delighted to discover that the young man who had been hired chiefly to teach graduate structural engineering courses was not only willing but eager to help him reorganize the mechanics courses. Both believed that these courses should be taught from a few basic principles and that the various concepts involved would be best understood and remembered by the students if they were presented to them in a graphic way. Together they wrote lecture notes in statics and dynamics, to which they later added problems they felt would appeal to future engineers, and soon they produced the manuscript of the first edition of *Mechanics for Engineers*. The second edition of *Mechanics for Engineers* and the first edition of *Vector Mechanics for Engineers* found Russ Johnston at Worcester Polytechnic Institute and the next editions at the University of Connecticut. In the meantime, both Ferd and Russ had assumed administrative responsibilities in their departments, and both were involved in research, consulting, and supervising graduate students—Ferd in the area of stochastic processes and random vibrations, and Russ in the area of elastic stability and structural analysis and design. However, their interest in improving the teaching of the basic mechanics courses had not subsided, and they both taught sections of these courses as they kept revising their texts and began writing together the manuscript of the first edition of *Mechanics of Materials*.

Ferd and Russ's contributions to engineering education earned them a number of honors and awards. They were presented with the Western Electric Fund Award for excellence in the instruction of engineering students by their respective regional sections of the American Society for Engineering Education, and they both received the Distinguished Educator Award from the Mechanics Division of the same society. In 1991 Russ received the Outstanding Civil Engineer Award from the Connecticut Section of the American Society of Civil Engineers, and in 1995 Ferd was awarded an honorary Doctor of Engineering degree by Lehigh University.

John T. DeWolf, Professor of Civil Engineering at the University of Connecticut, joined the Beer and Johnston team as an author on the second edition of *Mechanics of Materials*. John holds a B.S. degree in civil engineering from the University of Hawaii, and M.E. and Ph.D. degrees in structural engineering from Cornell University. His research interests are in the area of elastic stability, bridge monitoring, and structural analysis and design. He is a registered Professional Engineer and a member of the Connecticut Board of Professional Engineers. He was selected as the University of Connecticut Teaching Fellow in 2006.

David F. Mazurek, Professor of Civil Engineering at the United States Coast Guard Academy, joined the team in the fourth edition. David holds a B.S. degree in ocean engineering and an M.S. degree in civil engineering from the Florida Institute of Technology, and a Ph.D. degree in civil engineering from the University of Connecticut. He is a registered Professional Engineer. He has served on the American Railway Engineering & Maintenance of Way Association's Committee 15—Steel Structures for the past seventeen years. Professional interests include bridge engineering, structural forensics, and blast-resistant design.

Dr. Sanjeev Sanghi is Professor in the Applied Mechanics department at Indian Institute of Technology, Delhi. He is also the Head of the Centre for Educational Technology at I.I.T. Delhi. He has been a recipient of the Ralph Boliagno award for most distinguished Teaching assistant at Cornell University and the Distinguished Teacher Award at I.I.T. Delhi. His research interests include single phase and multi phase turbulent flows, CFD and non-linear dynamics, and chaos. He has published research articles in The Journal of Fluid Mechanics, Physics of Fluids, Chaos, The Journal of Computational Physics, AIAA Journal, and other reputed journals.

Contents

3 Torsion 163

4 Pure Bending 243

8 Principal Stresses Under a Given Loading 535

9 Deflection of Beams 571

Preface

OBJECTIVES

The main objective of a basic mechanics course should be to develop in the engineering student the ability to analyze a given problem in a simple and logical manner and to apply to its solution a few fundamental and well-understood principles. This text is designed for the first course in mechanics of materials—or strength of materials—offered to engineering students in the sophomore or junior year. The authors hope that it will help instructors achieve this goal in that particular course in the same way that their other texts may have helped them in statics and dynamics.

GENERAL APPROACH

In this text the study of the mechanics of materials is based on the understanding of a few basic concepts and on the use of simplified models. This approach makes it possible to develop all the necessary formulas in a rational and logical manner, and to clearly indicate the conditions under which they can be safely applied to the analysis and design of actual engineering structures and machine components.

Free-body Diagrams Are Used Extensively. Throughout the text free-body diagrams are used to determine external or internal forces. The use of "picture equations" will also help the students understand the superposition of loadings and the resulting stresses and deformations.

Design Concepts Are Discussed Throughout the Text Whenever Appropriate. A discussion of the application of the factor of safety to design can be found in Chap. 1, where the concepts of both allowable stress design and load and resistance factor design are presented.

Optional Sections Offer Advanced or Specialty Topics. Topics such as residual stresses, torsion of noncircular and thin-walled members, bending of curved beams, shearing stresses in nonsymmetrical members, and failure criteria have been included in optional sections for use in courses of varying emphases. To preserve the integrity of the subject, these topics are presented in the proper sequence, wherever they logically belong. Thus, even when not covered in the course, they are highly visible and can be easily referred to by the students if needed in a later course or in engineering practice. For convenience all optional sections have been indicated by asterisks.

It is expected that students using this text will have completed a course in statics. However, Chap. 1 is designed to provide them with an opportunity to review the concepts learned in that course, while shear and bending-moment diagrams are covered in detail in Secs. 5.2 and 5.3. The properties of moments and centroids of areas are described in Appendix A; this material can be used to reinforce the discussion of the determination of normal and shearing stresses in beams (Chaps. 4, 5, and 6).

The first four chapters of the text are devoted to the analysis of the stresses and of the corresponding deformations in various structural members, considering successively axial loading, torsion, and pure bending. Each analysis is based on a few basic concepts, namely, the conditions of equilibrium of the forces exerted on the member, the relations existing between stress and strain in the material, and the conditions imposed by the supports and loading of the member. The study of each type of loading is complemented by a large number of examples, sample problems, and problems to be assigned, all designed to strengthen the students' understanding of the subject.

The concept of stress at a point is introduced in Chap. 1, where it is shown that an axial load can produce shearing stresses as well as normal stresses, depending upon the section considered. The fact that stresses depend upon the orientation of the surface on which they are computed is emphasized again in Chaps. 3 and 4 in the cases of torsion and pure bending. However, the discussion of computational techniques—such as Mohr's circle—used for the transformation of stress at a point is delayed until Chap. 7, after students have had the opportunity to solve problems involving a combination of the basic loadings and have discovered for themselves the need for such techniques.

The discussion in Chap. 2 of the relation between stress and strain in various materials includes fiber-reinforced composite materials. Also, the study of beams under transverse loads is covered in two separate chapters. Chapter 5 is devoted to the determination of the normal stresses in a beam and to the design of beams based on the allowable normal stress in the material used (Sec. 5.4). The chapter begins with a discussion of the shear and bending-moment diagrams (Secs. 5.2 and 5.3) and includes an optional section on the use of singularity functions for the determination of the shear and bending moment in a beam (Sec. 5.5). The chapter ends with an optional section on nonprismatic beams (Sec. 5.6).

Chapter 6 is devoted to the determination of shearing stresses in beams and thin-walled members under transverse loadings. The formula for the shear flow, $q = VQ/I$, is derived in the traditional way. More advanced aspects of the design of beams, such as the determination of the principal stresses at the junction of the flange and web of a W-beam, are in Chap. 8, an optional chapter that may be covered after the transformations of stresses have been discussed in Chap. 7. The design of transmission shafts is in that chapter for the same reason, as well as the determination of stresses under combined loadings that can now include the determination of the principal stresses, principal planes, and maximum shearing stress at a given point.

Statically indeterminate problems are first discussed in Chap. 2 and considered throughout the text for the various loading conditions encountered. Thus, students are presented at an early stage with a method of solution that combines the analysis of deformations with the conventional analysis of forces used in statics. In this way, they will have become thoroughly familiar with this fundamental method by the end of the course. In addition, this approach helps the students realize that stresses themselves are statically indeterminate and can be computed only by considering the corresponding distribution of strains.

The concept of plastic deformation is introduced in Chap. 2, where it is applied to the analysis of members under axial loading. Problems involving the plastic deformation of circular shafts and of prismatic beams are also considered in optional sections of Chaps. 3, 4, and 6. While some of this material can be omitted at the choice of the instructor, its inclusion in the body of the text will help students realize the limitations of the assumption of a linear stress-strain relation and serve to caution them against the inappropriate use of the elastic torsion and flexure formulas.

The determination of the deflection of beams is discussed in Chap. 9. The first part of the chapter is devoted to the integration method and to the method of superposition, with an optional section (Sec. 9.6) based on the use of singularity functions. (This section should be used only if Sec. 5.5 was covered earlier.) The second part of Chap. 9 is optional. It presents the moment-area method in two lessons.

Chapter 10 is devoted to columns and contains material on the design of steel, aluminum, and wood columns. Chapter 11 covers energy methods, including Castigliano's theorem.

PEDAGOGICAL FEATURES

Each chapter begins with an introductory section setting the purpose and goals of the chapter and describing in simple terms the material to be covered and its application to the solution of engineering problems.

Chapter Lessons. The body of the text has been divided into units, each consisting of one or several theory sections followed by sample problems and a large number of problems to be assigned. Each unit corresponds to a well-defined topic and generally can be covered in one lesson.

Examples and Sample Problems. The theory sections include many examples designed to illustrate the material being presented and facilitate its understanding. The sample problems are intended to show some of the applications of the theory to the solution of engineering problems. Since they have been set up in much the same form that students will use in solving the assigned problems, the sample problems serve the double purpose of amplifying the text and demonstrating the type of neat and orderly work that students should cultivate in their own solutions.

Homework Problem Sets.   Most of the problems are of a practical nature and should appeal to engineering students. They are primarily designed, however, to illustrate the material presented in the text and help the students understand the basic principles used in mechanics of materials. The problems have been grouped according to the portions of material they illustrate and have been arranged in order of increasing difficulty. Problems requiring special attention have been indicated by asterisks. Answers to problems are given at the end of the book, except for those with a number set in italics.

Chapter Review and Summary.   Each chapter ends with a review and summary of the material covered in the chapter. Notes in the margin have been included to help the students organize their review work, and cross references provided to help them find the portions of material requiring their special attention.

Review Problems.   A set of review problems is included at the end of each chapter. These problems provide students further opportunity to apply the most important concepts introduced in the chapter.

Computer Problems.   Computers make it possible for engineering students to solve a great number of challenging problems. A group of six or more problems designed to be solved with a computer can be found at the end of each chapter. These problems can be solved using any computer language that provides a basis for analytical calculations. Developing the algorithm required to solve a given problem will benefit the students in two different ways: (1) it will help them gain a better understanding of the mechanics principles involved; (2) it will provide them with an opportunity to apply the skills acquired in their computer programming course to the solution of a meaningful engineering problem. These problems can be solved using any computer language that provides a basis for analytical calculations.

Fundamentals of Engineering Examination.   Engineers who seek to be licensed as *Professional Engineers* must take two exams. The first exam, the *Fundamentals of Engineering Examination*, includes subject material from *Mechanics of Materials*. Appendix E lists the topics in *Mechanics of Materials* that are covered in this exam along with problems that can be solved to review this material.

SUPPLEMENTAL RESOURCES

Instructor's Solutions Manual.   The Instructor's and Solutions Manual that accompanies the sixth edition continues the tradition of exceptional accuracy and keeping solutions contained to a single page for easier reference. The manual also features tables designed to assist instructors in creating a schedule of assignments for their courses. The various topics covered in the text are listed in Table I, and a suggested number of periods to be spent on each topic is indicated. Table II provides a brief description of all groups of problems and a classification of the problems in each group according to the units used. Sample lesson schedules are also found within the manual.

McGRAW-HILL CREATE™

Craft your teaching resources to match the way you teach! With McGraw-Hill Create™, www.mcgrawhillcreate.com, you can easily rearrange chapters, combine material from other content sources, and quickly upload content you have written like your course syllabus or teaching notes. Arrange your book to fit your teaching style. Create even allows you to personalize your book's appearance by selecting the cover and adding your name, school, and course information. Order a Create book and you'll receive a complimentary print review copy in 3–5 business days or a complimentary electronic review copy (eComp) via email in minutes. Go to www.mcgrawhillcreate.com today and register to experience how McGraw-Hill Create empowers you to teach *your* students *your* way.

ADDITIONAL ONLINE RESOURCES

Mechanics of Materials 6e also features a companion website (www.mhhe.com/beerjohnston) for instructors. Included on the website are lecture PowerPoints, an image library, and animations. Via the website, instructors can also request access to C.O.S.M.O.S., a complete online solutions manual organization system that allows instructors to create custom homework, quizzes, and tests using end-of-chapter problems from the text. For access to this material, contact your sales representative for a user name and password.

ACKNOWLEDGMENTS

The authors thank the many companies that provided photographs for this edition. We also wish to recognize the determined efforts and patience of our photo researcher, Sabina Dowell.

Our special thanks go to Professor Dean Updike of the Department of Mechanical Engineering and Mechanics, Lehigh University, for his patience and cooperation as he checked the solutions and answers of all the problems in this edition.

We also gratefully acknowledge the help, comments, and suggestions offered by the many reviewers and users of previous editions of *Mechanics of Materials*.

John T. DeWolf
David F. Mazurek

List of Symbols

$A, B, C, \ldots$ Force reactions

L, M, G Points

A Area

b Distance, width

c Constant, distance, radius

C Centroid

C_w Constant of integration

C_P Column stability factor

d Distance, ultimate depth

D Diameter

e Distance, eccentricity distance

E Modulus of elasticity

f Frequency, motion

F Force

$F.S.$ Factor of safety

G Modulus of rigidity, shear modulus

h Distance, height

H Force

GPa, kPa Pascals

I Moment of inertia

J Product of inertia

P Polar moment of inertia

k Spring constant, shape factor, bulk modulus, constant

K Stress-concentration factor, flexural/spring constant

l Length, span

L Length, span

L_e Effective length

m Mass

M Couple

M, M_o Bending moment

M_D Bending moment, dead load (LRFD)

M_L Bending moment, live load (LRFD)

M_u Bending moment, ultimate load (LRFD)

n Number, ratio of modulus of elasticity, empirical direction

p Pressure

P Force, concentrated load

P_D Dead load (LRFD)

P_L Live load (LRFD)

P_u Ultimate load (LRFD)

q Shearing force per unit length, shear flow

Q Force

Q First moment of area

List of Symbols

a	Constant; distance
A, B, C, . . .	Forces; reactions
$A, B, C, \ldots$	Points
$A, \mathcal{Q}$	Area
b	Distance; width
c	Constant; distance; radius
C	Centroid
$C_1, C_2, \ldots$	Constants of integration
C_P	Column stability factor
d	Distance; diameter; depth
D	Diameter
e	Distance; eccentricity; dilatation
E	Modulus of elasticity
f	Frequency; function
F	Force
$F.S.$	Factor of safety
G	Modulus of rigidity; shear modulus
h	Distance; height
H	Force
H, J, K	Points
$I, I_x, \ldots$	Moment of inertia
$I_{xy}, \ldots$	Product of inertia
J	Polar moment of inertia
k	Spring constant; shape factor; bulk modulus; constant
K	Stress concentration factor; torsional spring constant
l	Length; span
L	Length; span
L_e	Effective length
m	Mass
M	Couple
$M, M_x, \ldots$	Bending moment
M_D	Bending moment, dead load (LRFD)
M_L	Bending moment, live load (LRFD)
M_U	Bending moment, ultimate load (LRFD)
n	Number; ratio of moduli of elasticity; normal direction
p	Pressure
P	Force; concentrated load
P_D	Dead load (LRFD)
P_L	Live load (LRFD)
P_U	Ultimate load (LRFD)
q	Shearing force per unit length; shear flow
Q	Force
Q	First moment of area

18

r	Radius; radius of gyration
$\mathbf{R}$	Force; reaction
R	Radius; modulus of rupture
s	Length
S	Elastic section modulus
t	Thickness; distance; tangential deviation
$\mathbf{T}$	Torque
T	Temperature
u, v	Rectangular coordinates
u	Strain-energy density
U	Strain energy; work
$\mathbf{v}$	Velocity
$\mathbf{V}$	Shearing force
V	Volume; shear
w	Width; distance; load per unit length
$\mathbf{W}, W$	Weight, load
x, y, z	Rectangular coordinates; distance; displacements; deflections
$\bar{x}, \bar{y}, \bar{z}$	Coordinates of centroid
Z	Plastic section modulus
α, β, γ	Angles
α	Coefficient of thermal expansion; influence coefficient
γ	Shearing strain; specific weight
γ_D	Load factor, dead load (LRFD)
γ_L	Load factor, live load (LRFD)
δ	Deformation; displacement
ϵ	Normal strain
θ	Angle; slope
λ	Direction cosine
ν	Poisson's ratio
ρ	Radius of curvature; distance; density
σ	Normal stress
τ	Shearing stress
ϕ	Angle; angle of twist; resistance factor
ω	Angular velocity

r	Radius of gyration
R	Force, reaction
R	Radius, modulus of rupture
s	Length
S	Plane section modulus
t	Thickness, distance, longitudinal deviation
T	Torque
T	Temperature
u	Rectangular coordinates
u	Surface energy density
U	Strain energy, work
v	Velocity
V	Shearing force
V	Volume, shear
w	Width, distance, loss per unit length
W, W	Weight, load
x, y, z	Rectangular coordinates, distances, displacements, deflections
$\bar{x}, \bar{y}$	Coordinates of centroid
Z	Plastic section modulus
α, β, γ	Angles
α	Coefficient of thermal expansion, influence coefficient
γ	Specific weight, specific weight
w_D	Dead load, dead load (DEAD)
w_L	Live load, live load (LIVE)
δ	Deformation, displacement
ϵ	Normal strain
θ	Angle, slope
λ	Direction cosine
ν	Poisson's ratio
ρ	Radius of curvature, distance, density
σ	Normal stress
τ	Shearing stress
ϕ	Angle, angle of twist, resistance factor
ω	Angular velocity

MECHANICS OF MATERIALS

This chapter is devoted to the study of the stresses occurring in many of the elements contained in these excavators, such as two-force members, axles, bolts, and pins.

Introduction—Concept of Stress

1.1   INTRODUCTION

The main objective of the study of the mechanics of materials is to provide the future engineer with the means of analyzing and designing various machines and load-bearing structures.

Both the analysis and the design of a given structure involve the determination of *stresses* and *deformations*. This first chapter is devoted to the concept of *stress*.

Section 1.2 is devoted to a short review of the basic methods of statics and to their application to the determination of the forces in the members of a simple structure consisting of pin-connected members. Section 1.3 will introduce you to the concept of *stress* in a member of a structure, and you will be shown how that stress can be determined from the *force* in the member. After a short discussion of engineering analysis and design (Sec. 1.4), you will consider successively the *normal stresses* in a member under axial loading (Sec. 1.5), the *shearing stresses* caused by the application of equal and opposite transverse forces (Sec. 1.6), and the *bearing stresses* created by bolts and pins in the members they connect (Sec. 1.7). These various concepts will be applied in Sec. 1.8 to the determination of the stresses in the members of the simple structure considered earlier in Sec. 1.2.

The first part of the chapter ends with a description of the method you should use in the solution of an assigned problem (Sec. 1.9) and with a discussion of the numerical accuracy appropriate in engineering calculations (Sec. 1.10).

In Sec. 1.11, where a two-force member under axial loading is considered again, it will be observed that the stresses on an *oblique* plane include both *normal* and *shearing* stresses, while in Sec. 1.12 you will note that *six components* are required to describe the state of stress at a point in a body under the most general loading conditions.

Finally, Sec. 1.13 will be devoted to the determination from test specimens of the *ultimate strength* of a given material and to the use of a *factor of safety* in the computation of the *allowable load* for a structural component made of that material.

1.2   A SHORT REVIEW OF THE METHODS OF STATICS

In this section you will review the basic methods of statics while determining the forces in the members of a simple structure.

Consider the structure shown in Fig. 1.1, which was designed to support a 30-kN load. It consists of a boom *AB* with a 30 × 50-mm rectangular cross section and of a rod *BC* with a 20-mm-diameter circular cross section. The boom and the rod are connected by a pin at *B* and are supported by pins and brackets at *A* and *C*, respectively. Our first step should be to draw a *free-body diagram* of the structure by detaching it from its supports at *A* and *C*, and showing the reactions that these supports exert on the structure (Fig. 1.2). Note that the sketch of the structure has been simplified by omitting all unnecessary details. Many of you may have recognized at this point that *AB* and *BC* are *two-force members*. For those of you who have not, we will pursue our analysis, ignoring that fact and assuming that the directions of the reactions at *A* and *C* are unknown. Each of these

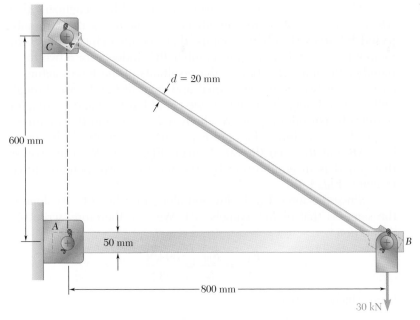

Fig. 1.1 Boom used to support a 30-kN load.

reactions, therefore, will be represented by two components, $\mathbf{A}_x$ and $\mathbf{A}_y$ at A, and $\mathbf{C}_x$ and $\mathbf{C}_y$ at C. We write the following three equilibrium equations:

$+\curvearrowleft \Sigma\, M_C = 0$: $A_x(0.6\text{ m}) - (30\text{ kN})(0.8\text{ m}) = 0$

$$A_x = +40\text{ kN} \tag{1.1}$$

$\xrightarrow{+}\ \Sigma\, F_x = 0$: $A_x + C_x = 0$

$$C_x = -A_x \qquad C_x = -40\text{ kN} \tag{1.2}$$

$+\uparrow \Sigma\, F_y = 0$: $A_y + C_y - 30\text{ kN} = 0$

$$A_y + C_y = +30\text{ kN} \tag{1.3}$$

We have found two of the four unknowns, but cannot determine the other two from these equations, and no additional independent equation can be obtained from the free-body diagram of the structure. We must now dismember the structure. Considering the free-body diagram of the boom AB (Fig. 1.3), we write the following equilibrium equation:

$+\curvearrowleft \Sigma\, M_B = 0$: $-A_y(0.8\text{ m}) = 0$ $A_y = 0$ (1.4)

Substituting for A_y from (1.4) into (1.3), we obtain $C_y = +30$ kN. Expressing the results obtained for the reactions at A and C in vector form, we have

$$\mathbf{A} = 40\text{ kN} \rightarrow \qquad \mathbf{C}_x = 40\text{ kN} \leftarrow,\ \mathbf{C}_y = 30\text{ kN} \uparrow$$

We note that the reaction at A is directed along the axis of the boom AB and causes compression in that member. Observing that the components C_x and C_y of the reaction at C are, respectively, proportional to the horizontal and vertical components of the distance from B to C, we conclude that the reaction at C is equal to 50 kN, is directed along the axis of the rod BC, and causes tension in that member.

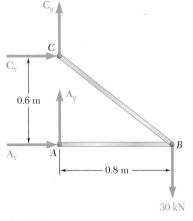

Fig. 1.2

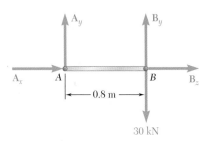

Fig. 1.3

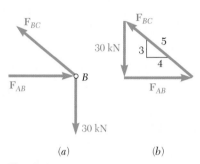

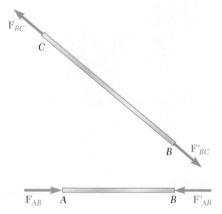

Fig. 1.4

Fig. 1.5

These results could have been anticipated by recognizing that *AB* and *BC* are two-force members, i.e., members that are subjected to forces at only two points, these points being *A* and *B* for member *AB*, and *B* and *C* for member *BC*. Indeed, for a two-force member the lines of action of the resultants of the forces acting at each of the two points are equal and opposite and pass through both points. Using this property, we could have obtained a simpler solution by considering the free-body diagram of pin *B*. The forces on pin *B* are the forces $\mathbf{F}_{AB}$ and $\mathbf{F}_{BC}$ exerted, respectively, by members *AB* and *BC*, and the 30-kN load (Fig. 1.4*a*). We can express that pin *B* is in equilibrium by drawing the corresponding force triangle (Fig. 1.4*b*).

Since the force $\mathbf{F}_{BC}$ is directed along member *BC*, its slope is the same as that of *BC*, namely, 3/4. We can, therefore, write the proportion

$$\frac{F_{AB}}{4} = \frac{F_{BC}}{5} = \frac{30 \text{ kN}}{3}$$

from which we obtain

$$F_{AB} = 40 \text{ kN} \qquad F_{BC} = 50 \text{ kN}$$

The forces $\mathbf{F}'_{AB}$ and $\mathbf{F}'_{BC}$ exerted by pin *B*, respectively, on boom *AB* and rod *BC* are equal and opposite to $\mathbf{F}_{AB}$ and $\mathbf{F}_{BC}$ (Fig. 1.5).

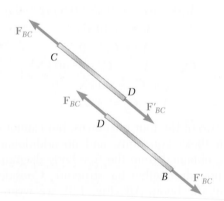

Fig. 1.6

Knowing the forces at the ends of each of the members, we can now determine the internal forces in these members. Passing a section at some arbitrary point *D* of rod *BC*, we obtain two portions *BD* and *CD* (Fig. 1.6). Since 50-kN forces must be applied at *D* to both portions of the rod to keep them in equilibrium, we conclude that an internal force of 50 kN is produced in rod *BC* when a 30-kN load is applied at *B*. We further check from the directions of the forces $\mathbf{F}_{BC}$ and $\mathbf{F}'_{BC}$ in Fig. 1.6 that the rod is in tension. A similar procedure would enable us to determine that the internal force in boom *AB* is 40 kN and that the boom is in compression.

1.3 STRESSES IN THE MEMBERS OF A STRUCTURE

While the results obtained in the preceding section represent a first and necessary step in the analysis of the given structure, they do not tell us whether the given load can be safely supported. Whether rod *BC*, for example, will break or not under this loading depends not only upon the value found for the internal force F_{BC}, but also upon the cross-sectional area of the rod and the material of which the rod is made. Indeed, the internal force F_{BC} actually represents the resultant of elementary forces distributed over the entire area *A* of the cross section (Fig. 1.7) and the average intensity of these distributed forces is equal to the force per unit area, F_{BC}/A, in the section. Whether or not the rod will break under the given loading clearly depends upon the ability of the material to withstand the corresponding value F_{BC}/A of the intensity of the distributed internal forces. It thus depends upon the force F_{BC}, the cross-sectional area *A*, and the material of the rod.

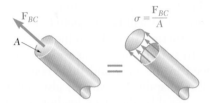

Fig. 1.7

The force per unit area, or intensity of the forces distributed over a given section, is called the *stress* on that section and is denoted by the Greek letter σ (sigma). The stress in a member of cross-sectional area *A* subjected to an axial load **P** (Fig. 1.8) is therefore obtained by dividing the magnitude *P* of the load by the area *A*:

$$\sigma = \frac{P}{A} \tag{1.5}$$

A positive sign will be used to indicate a tensile stress (member in tension) and a negative sign to indicate a compressive stress (member in compression).

Since SI metric units are used in this discussion, with *P* expressed in newtons (N) and *A* in square meters (m²), the stress σ will be expressed in N/m². This unit is called a *pascal* (Pa). However, one finds that the pascal is an exceedingly small quantity and that, in practice, multiples of this unit must be used, namely, the kilopascal (kPa), the megapascal (MPa), and the gigapascal (GPa).[†] We have

$$1\ \text{kPa} = 10^3\ \text{Pa} = 10^3\ \text{N/m}^2$$

$$1\ \text{MPa} = 10^6\ \text{Pa} = 10^6\ \text{N/m}^2$$

$$1\ \text{GPa} = 10^9\ \text{Pa} = 10^9\ \text{N/m}^2$$

[†]The principal SI units used in mechanics are listed in tables inside the front cover of this book.

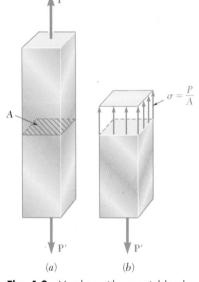

Fig. 1.8 Member with an axial load.

1.4 ANALYSIS AND DESIGN

Considering again the structure of Fig. 1.1, let us assume that rod BC is made of a steel with a maximum allowable stress $\sigma_{all} = 165$ MPa. Can rod BC safely support the load to which it will be subjected? The magnitude of the force F_{BC} in the rod was found earlier to be 50 kN. Recalling that the diameter of the rod is 20 mm, we use Eq. (1.5) to determine the stress created in the rod by the given loading. We have

$$P = F_{BC} = +50\,\text{kN} = +50 \times 10^3\,\text{N}$$

$$A = \pi r^2 = \pi\left(\frac{20\,\text{mm}}{2}\right)^2 = \pi(10 \times 10^{-3}\,\text{m})^2 = 314 \times 10^{-6}\,\text{m}^2$$

$$\sigma = \frac{P}{A} = \frac{+50 \times 10^3\,\text{N}}{314 \times 10^{-6}\,\text{m}^2} = +159 \times 10^6\,\text{Pa} = +159\,\text{MPa}$$

Since the value obtained for σ is smaller than the value σ_{all} of the allowable stress in the steel used, we conclude that rod BC can safely support the load to which it will be subjected. To be complete, our analysis of the given structure should also include the determination of the compressive stress in boom AB, as well as an investigation of the stresses produced in the pins and their bearings. This will be discussed later in this chapter. We should also determine whether the deformations produced by the given loading are acceptable. The study of deformations under axial loads will be the subject of Chap. 2. An additional consideration required for members in compression involves the *stability* of the member, i.e., its ability to support a given load without experiencing a sudden change in configuration. This will be discussed in Chap. 10.

The engineer's role is not limited to the analysis of existing structures and machines subjected to given loading conditions. Of even greater importance to the engineer is the *design* of new structures and machines, i.e., the selection of appropriate components to perform a given task. As an example of design, let us return to the structure of Fig. 1.1, and assume that aluminum with an allowable stress $\sigma_{all} = 100$ MPa is to be used. Since the force in rod BC will still be $P = F_{BC} = 50$ kN under the given loading, we must have, from Eq. (1.5),

$$\sigma_{all} = \frac{P}{A} \qquad A = \frac{P}{\sigma_{all}} = \frac{50 \times 10^3\,\text{N}}{100 \times 10^6\,\text{Pa}} = 500 \times 10^{-6}\,\text{m}^2$$

and, since $A = \pi r^2$,

$$r = \sqrt{\frac{A}{\pi}} = \sqrt{\frac{500 \times 10^{-6}\,\text{m}^2}{\pi}} = 12.62 \times 10^{-3}\,\text{m} = 12.62\,\text{mm}$$

$$d = 2r = 25.2\,\text{mm}$$

We conclude that an aluminum rod 26 mm or more in diameter will be adequate.

1.5 AXIAL LOADING; NORMAL STRESS

As we have already indicated, rod BC of the example considered in the preceding section is a two-force member and, therefore, the forces $\mathbf{F}_{BC}$ and $\mathbf{F}'_{BC}$ acting on its ends B and C (Fig. 1.5) are directed along the axis of the rod. We say that the rod is under *axial loading*. An actual example of structural members under axial loading is provided by the members of the bridge truss shown in Photo 1.1.

Photo 1.1 This bridge truss consists of two-force members that may be in tension or in compression.

Returning to rod BC of Fig. 1.5, we recall that the section we passed through the rod to determine the internal force in the rod and the corresponding stress was perpendicular to the axis of the rod; the internal force was therefore normal to the plane of the section (Fig. 1.7) and the corresponding stress is described as a *normal stress*. Thus, formula (1.5) gives us the *normal stress in a member under axial loading*:

$$\sigma = \frac{P}{A} \tag{1.5}$$

We should also note that, in formula (1.5), σ is obtained by dividing the magnitude P of the resultant of the internal forces distributed over the cross section by the area A of the cross section; it represents, therefore, the *average value* of the stress over the cross section, rather than the stress at a specific point of the cross section.

To define the stress at a given point Q of the cross section, we should consider a small area ΔA (Fig. 1.9). Dividing the magnitude of $\Delta \mathbf{F}$ by ΔA, we obtain the average value of the stress over ΔA. Letting ΔA approach zero, we obtain the stress at point Q:

$$\sigma = \lim_{\Delta A \to 0} \frac{\Delta F}{\Delta A} \tag{1.6}$$

Fig. 1.9

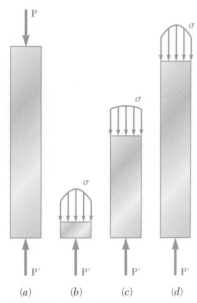

Fig. 1.10 Stress distributions at different sections along axially loaded member.

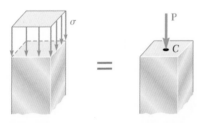

Fig. 1.11

In general, the value obtained for the stress σ at a given point Q of the section is different from the value of the average stress given by formula (1.5), and σ is found to vary across the section. In a slender rod subjected to equal and opposite concentrated loads **P** and **P′** (Fig. 1.10a), this variation is small in a section away from the points of application of the concentrated loads (Fig. 1.10c), but it is quite noticeable in the neighborhood of these points (Fig. 1.10b and d).

It follows from Eq. (1.6) that the magnitude of the resultant of the distributed internal forces is

$$\int dF = \int_A \sigma \, dA$$

But the conditions of equilibrium of each of the portions of rod shown in Fig. 1.10 require that this magnitude be equal to the magnitude P of the concentrated loads. We have, therefore,

$$P = \int dF = \int_A \sigma \, dA \qquad (1.7)$$

which means that the volume under each of the stress surfaces in Fig. 1.10 must be equal to the magnitude P of the loads. This, however, is the only information that we can derive from our knowledge of statics, regarding the distribution of normal stresses in the various sections of the rod. The actual distribution of stresses in any given section is *statically indeterminate*. To learn more about this distribution, it is necessary to consider the deformations resulting from the particular mode of application of the loads at the ends of the rod. This will be discussed further in Chap. 2.

In practice, it will be assumed that the distribution of normal stresses in an axially loaded member is uniform, except in the immediate vicinity of the points of application of the loads. The value σ of the stress is then equal to σ_{ave} and can be obtained from formula (1.5). However, we should realize that, when we assume a uniform distribution of stresses in the section, i.e., when we assume that the internal forces are uniformly distributed across the section, it follows from elementary statics† that the resultant **P** of the internal forces must be applied at the centroid C of the section (Fig. 1.11). This means that *a uniform distribution of stress is possible only if the line of action of the concentrated loads* **P** *and* **P′** *passes through the centroid of the section considered* (Fig. 1.12). This type of loading is called *centric loading* and will be assumed to take place in all straight two-force members found in trusses and pin-connected structures, such as the one considered in Fig. 1.1. However, if a two-force member is loaded axially, but *eccentrically* as shown in Fig. 1.13a, we find from the conditions of equilibrium of the portion of member shown in Fig. 1.13b that the internal forces in a given section must be

---

†See Ferdinand P. Beer and E. Russell Johnston, Jr., *Mechanics for Engineers,* 5th ed., McGraw-Hill, New York, 2008, or *Vector Mechanics for Engineers,* 9th ed., McGraw-Hill, New York, 2010, Secs. 5.2 and 5.3.

Fig. 1.12

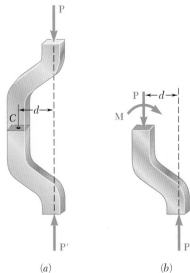

(a)   (b)

Fig. 1.13 Eccentric axial loading.

equivalent to a force **P** applied at the centroid of the section and a couple **M** of moment $M = Pd$. The distribution of forces—and, thus, the corresponding distribution of stresses—*cannot be uniform*. Nor can the distribution of stresses be symmetric as shown in Fig. 1.10. This point will be discussed in detail in Chap. 4.

1.6 SHEARING STRESS

The internal forces and the corresponding stresses discussed in Secs. 1.2 and 1.3 were normal to the section considered. A very different type of stress is obtained when transverse forces **P** and **P′** are applied to a member *AB* (Fig. 1.14). Passing a section at *C* between the points of application of the two forces (Fig. 1.15*a*), we obtain the diagram of portion *AC* shown in Fig. 1.15*b*. We conclude that internal forces must exist in the plane of the section, and that their resultant is equal to **P.** These elementary internal forces are called *shearing forces,* and the magnitude *P* of their resultant is the *shear* in the section. Dividing the shear *P* by the area *A* of the cross section, we

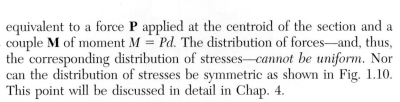

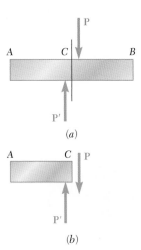

(a)

(b)

Fig. 1.15

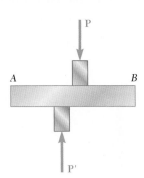

Fig. 1.14 Member with transverse loads.

obtain the *average shearing stress* in the section. Denoting the shearing stress by the Greek letter τ (tau), we write

$$\tau_{ave} = \frac{P}{A} \qquad (1.8)$$

It should be emphasized that the value obtained is an average value of the shearing stress over the entire section. Contrary to what we said earlier for normal stresses, the distribution of shearing stresses across the section *cannot* be assumed uniform. As you will see in Chap. 6, the actual value τ of the shearing stress varies from zero at the surface of the member to a maximum value τ_{max} that may be much larger than the average value τ_{ave}.

Photo 1.2 Cutaway view of a connection with a bolt in shear.

Shearing stresses are commonly found in bolts, pins, and rivets used to connect various structural members and machine components (Photo 1.2). Consider the two plates A and B, which are connected by a bolt CD (Fig. 1.16). If the plates are subjected to tension forces of magnitude F, stresses will develop in the section of bolt corresponding to the plane EE'. Drawing the diagrams of the bolt and of the portion located above the plane EE' (Fig. 1.17), we conclude that the shear P in the section is equal to F. The average shearing stress in the section is obtained, according to formula (1.8), by dividing the shear $P = F$ by the area A of the cross section:

$$\tau_{ave} = \frac{P}{A} = \frac{F}{A} \qquad (1.9)$$

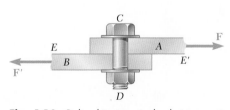

Fig. 1.16 Bolt subject to single shear.

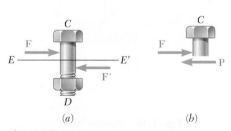

Fig. 1.17

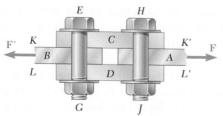

Fig. 1.18 Bolts subject to double shear.

The bolt we have just considered is said to be in *single shear.* Different loading situations may arise, however. For example, if splice plates C and D are used to connect plates A and B (Fig. 1.18), shear will take place in bolt HJ in each of the two planes KK' and LL' (and similarly in bolt EG). The bolts are said to be in *double shear.* To determine the average shearing stress in each plane, we draw free-body diagrams of bolt HJ and of the portion of bolt located between the two planes (Fig. 1.19). Observing that the shear P in each of the sections is $P = F/2$, we conclude that the average shearing stress is

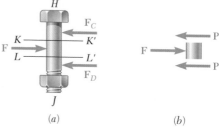

Fig. 1.19

$$\tau_{\text{ave}} = \frac{P}{A} = \frac{F/2}{A} = \frac{F}{2A} \qquad (1.10)$$

1.7 BEARING STRESS IN CONNECTIONS

Bolts, pins, and rivets create stresses in the members they connect, along the *bearing surface,* or surface of contact. For example, consider again the two plates A and B connected by a bolt CD that we have discussed in the preceding section (Fig. 1.16). The bolt exerts on plate A a force $\mathbf{P}$ equal and opposite to the force $\mathbf{F}$ exerted by the plate on the bolt (Fig. 1.20). The force $\mathbf{P}$ represents the resultant of elementary forces distributed on the inside surface of a half-cylinder of diameter d and of length t equal to the thickness of the plate. Since the distribution of these forces—and of the corresponding stresses—is quite complicated, one uses in practice an average nominal value σ_b of the stress, called the *bearing stress,* obtained by dividing the load P by the area of the rectangle representing the projection of the bolt on the plate section (Fig. 1.21). Since this area is equal to td, where t is the plate thickness and d the diameter of the bolt, we have

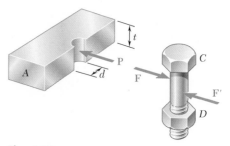

Fig. 1.20

$$\sigma_b = \frac{P}{A} = \frac{P}{td} \qquad (1.11)$$

Fig. 1.21

1.8 APPLICATION TO THE ANALYSIS AND DESIGN OF SIMPLE STRUCTURES

We are now in a position to determine the stresses in the members and connections of various simple two-dimensional structures and, thus, to design such structures.

As an example, let us return to the structure of Fig. 1.1 that we have already considered in Sec. 1.2 and let us specify the supports and connections at A, B, and C. As shown in Fig. 1.22, the 20-mm-diameter rod BC has flat ends of 20×40-mm rectangular cross section, while boom AB has a 30×50-mm rectangular cross section and is fitted with a clevis at end B. Both members are connected at B by a pin from which the 30-kN load is suspended by means of a U-shaped bracket. Boom AB is supported at A by a pin fitted into a double bracket, while rod BC is connected at C to a single bracket. All pins are 25 mm in diameter.

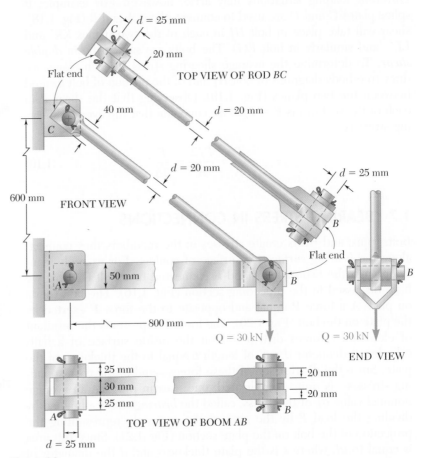

Fig. 1.22

a. Determination of the Normal Stress in Boom AB and Rod BC.

As we found in Secs. 1.2 and 1.4, the force in rod BC is $F_{BC} = 50$ kN (tension) and the area of its circular cross section is $A = 314 \times 10^{-6}$ m^2; the corresponding average normal stress is $\sigma_{BC} = +159$ MPa. However, the flat parts of the rod are also under tension and at the narrowest section, where a hole is located, we have

$$A = (20 \text{ mm})(40 \text{ mm} - 25 \text{ mm}) = 300 \times 10^{-6} \text{ m}^2$$

The corresponding average value of the stress, therefore, is

$$(\sigma_{BC})_{end} = \frac{P}{A} = \frac{50 \times 10^3 \text{ N}}{300 \times 10^{-6} \text{ m}^2} = 167 \text{ MPa}$$

Note that this is an *average value*; close to the hole, the stress will actually reach a much larger value, as you will see in Sec. 2.18. It is clear that, under an increasing load, the rod will fail near one of the holes rather than in its cylindrical portion; its design, therefore, could be improved by increasing the width or the thickness of the flat ends of the rod.

Turning now our attention to boom AB, we recall from Sec. 1.2 that the force in the boom is $F_{AB} = 40$ kN (compression). Since the area of the boom's rectangular cross section is $A = 30$ mm $\times$ 50 mm $= 1.5 \times 10^{-3}$ m², the average value of the normal stress in the main part of the rod, between pins A and B, is

$$\sigma_{AB} = -\frac{40 \times 10^3 \text{ N}}{1.5 \times 10^{-3} \text{ m}^2} = -26.7 \times 10^6 \text{ Pa} = -26.7 \text{ MPa}$$

Note that the sections of minimum area at A and B are not under stress, since the boom is in compression, and, therefore, *pushes* on the pins (instead of *pulling* on the pins as rod BC does).

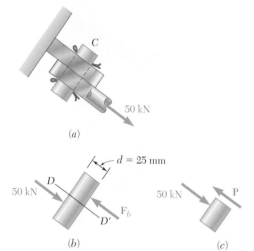

(a)

(b)

(c)

Fig. 1.23

b. Determination of the Shearing Stress in Various Connections.

To determine the shearing stress in a connection such as a bolt, pin, or rivet, we first clearly show the forces exerted by the various members it connects. Thus, in the case of pin C of our example (Fig. 1.23a), we draw Fig. 1.23b, showing the 50-kN force exerted by member BC on the pin, and the equal and opposite force exerted by the bracket. Drawing now the diagram of the portion of the pin located below the plane DD' where shearing stresses occur (Fig. 1.23c), we conclude that the shear in that plane is $P = 50$ kN. Since the cross-sectional area of the pin is

$$A = \pi r^2 = \pi \left(\frac{25 \text{ mm}}{2}\right)^2 = \pi (12.5 \times 10^{-3} \text{ m})^2 = 491 \times 10^{-6} \text{ m}^2$$

we find that the average value of the shearing stress in the pin at C is

$$\tau_{ave} = \frac{P}{A} = \frac{50 \times 10^3 \text{ N}}{491 \times 10^{-6} \text{ m}^2} = 102 \text{ MPa}$$

Considering now the pin at A (Fig. 1.24), we note that it is in double shear. Drawing the free-body diagrams of the pin and of the portion of pin located between the planes DD' and EE' where shearing stresses occur, we conclude that $P = 20$ kN and that

$$\tau_{ave} = \frac{P}{A} = \frac{20 \text{ kN}}{491 \times 10^{-6} \text{ m}^2} = 40.7 \text{ MPa}$$

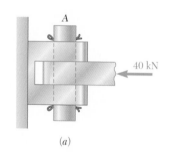

(a)

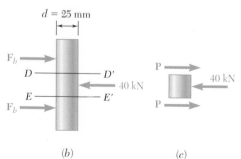

(b)

(c)

Fig. 1.24

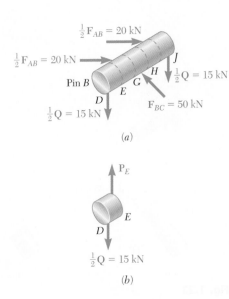

$\frac{1}{2}F_{AB} = 20$ kN

$\frac{1}{2}F_{AB} = 20$ kN

Pin B

$\frac{1}{2}Q = 15$ kN

$\frac{1}{2}Q = 15$ kN

$F_{BC} = 50$ kN

(a)

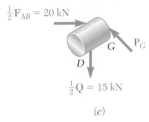

P_E

$\frac{1}{2}Q = 15$ kN

(b)

$\frac{1}{2}F_{AB} = 20$ kN

P_G

$\frac{1}{2}Q = 15$ kN

(c)

Fig. 1.25

Considering the pin at B (Fig. 1.25a), we note that the pin may be divided into five portions which are acted upon by forces exerted by the boom, rod, and bracket. Considering successively the portions DE (Fig. 1.25b) and DG (Fig. 1.25c), we conclude that the shear in section E is $P_E = 15$ kN, while the shear in section G is $P_G = 25$ kN. Since the loading of the pin is symmetric, we conclude that the maximum value of the shear in pin B is $P_G = 25$ kN, and that the largest shearing stresses occur in sections G and H, where

$$\tau_{\text{ave}} = \frac{P_G}{A} = \frac{25 \text{ kN}}{491 \times 10^{-6} \text{ m}^2} = 50.9 \text{ MPa}$$

c. Determination of the Bearing Stresses. To determine the nominal bearing stress at A in member AB, we use formula (1.11) of Sec. 1.7. From Fig. 1.22, we have $t = 30$ mm and $d = 25$ mm. Recalling that $P = F_{AB} = 40$ kN, we have

$$\sigma_b = \frac{P}{td} = \frac{40 \text{ kN}}{(30 \text{ mm})(25 \text{ mm})} = 53.3 \text{ MPa}$$

To obtain the bearing stress in the bracket at A, we use $t = 2(25 \text{ mm}) = 50$ mm and $d = 25$ mm:

$$\sigma_b = \frac{P}{td} = \frac{40 \text{ kN}}{(50 \text{ mm})(25 \text{ mm})} = 32.0 \text{ MPa}$$

The bearing stresses at B in member AB, at B and C in member BC, and in the bracket at C are found in a similar way.

1.9 METHOD OF PROBLEM SOLUTION

You should approach a problem in mechanics of materials as you would approach an actual engineering situation. By drawing on your own experience and intuition, you will find it easier to understand and formulate the problem. Once the problem has been clearly stated, however, there is no place in its solution for your particular fancy. Your solution must be based on the fundamental principles of statics and on the principles you will learn in this course. Every step you take must be justified on that basis, leaving no room for your "intuition." After an answer has been obtained, it should be checked. Here again, you may call upon your common sense and personal experience. If not completely satisfied with the result obtained, you should carefully check your formulation of the problem, the validity of the methods used in its solution, and the accuracy of your computations.

The *statement* of the problem should be clear and precise. It should contain the given data and indicate what information is required. A simplified drawing showing all essential quantities involved should be included. The solution of most of the problems you will encounter will necessitate that you first determine the *reactions at supports* and *internal forces and couples*. This will require the drawing of one or several *free-body diagrams,* as was done in Sec. 1.2, from which you will write *equilibrium equations.* These equations can be solved for the unknown forces, from which the required *stresses* and *deformations* will be computed.

After the answer has been obtained, it should be *carefully checked.* Mistakes in *reasoning* can often be detected by carrying the units through your computations and checking the units obtained for the answer. For example, in the design of the rod discussed in Sec. 1.4, we found, after carrying the units through our computations, that the required diameter of the rod was expressed in millimeters, which is the correct unit for a dimension; if another unit had been found, we would have known that some mistake had been made.

Errors in *computation* will usually be found by substituting the numerical values obtained into an equation which has not yet been used and verifying that the equation is satisfied. The importance of correct computations in engineering cannot be overemphasized.

1.10 NUMERICAL ACCURACY

The accuracy of the solution of a problem depends upon two items: (1) the accuracy of the given data and (2) the accuracy of the computations performed.

The solution cannot be more accurate than the less accurate of these two items. For example, if the loading of a beam is known to be 75,000 N with a possible error of 100 N either way, the relative error which measures the degree of accuracy of the data is

$$\frac{100 \text{ N}}{75,000 \text{ N}} = 0.0013 = 0.13\%$$

In computing the reaction at one of the beam supports, it would then be meaningless to record it as 14,322 N. The accuracy of the solution cannot be greater than 0.13 percent, no matter how accurate the computations are, and the possible error in the answer may be as large as $(0.13/100)(14,322 \text{ N}) \approx 20 \text{ N}$. The answer should be properly recorded as 14,320 ± 20 N.

In engineering problems, the data are seldom known with an accuracy greater than 0.2 percent. It is therefore seldom justified to write the answers to such problems with an accuracy greater than 0.2 percent. A practical rule is to use four figures to record numbers beginning with a 1 and three figures in all other cases. Unless otherwise indicated, the data given in a problem should be assumed known with a comparable degree of accuracy. A force of 40 N, for example, should be read 40.0 N, and a force of 15 N should be read 15.00 N.

Pocket calculators and computers are widely used by practicing engineers and engineering students. The speed and accuracy of these devices facilitate the numerical computations in the solution of many problems. However, students should not record more significant figures than can be justified merely because they are easily obtained. As noted, an accuracy greater than 0.2 percent is seldom necessary or meaningful in the solution of practical engineering problems.

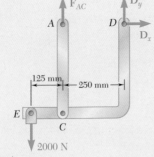

SAMPLE PROBLEM 1.1

In the hanger shown, the upper portion of link ABC is 10 mm thick and the lower portions are each 6 mm thick. Epoxy resin is used to bond the upper and lower portions together at B. The pin at A is of 10-mm diameter while a 6-mm-diameter pin is used at C. Determine (a) the shearing stress in pin A, (b) the shearing stress in pin C, (c) the largest normal stress in link ABC, (d) the average shearing stress on the bonded surfaces at B, (e) the bearing stress in the link at C.

SOLUTION

Free Body: Entire Hanger. Since the link ABC is a two-force member, the reaction at A is vertical; the reaction at D is represented by its components $\mathbf{D}_x$ and $\mathbf{D}_y$. We write

$$+\curvearrowleft \ \Sigma M_D = 0: \quad (2000 \text{ N})(375 \text{ mm}) - F_{AC}(250 \text{ mm}) = 0$$
$$F_{AC} = +3000 \text{ N} \qquad F_{AC} = 3000 \text{ N} \qquad \textit{tension}$$

a. Shearing Stress in Pin A. Since this 10-mm-diameter pin is in single shear, we write

$$\tau_A = \frac{F_{AC}}{A} = \frac{3000 \text{ N}}{\frac{1}{4}\pi(10 \text{ mm})^2} \qquad \tau_A = 38.2 \text{ MPa} \ \blacktriangleleft$$

b. Shearing Stress in Pin C. Since this 6-mm-diameter pin is in double shear, we write

$$\tau_C = \frac{\frac{1}{2}F_{AC}}{A} = \frac{1500 \text{ N}}{\frac{1}{4}\pi(6 \text{ mm})^2} \qquad \tau_C = 53.1 \text{ MPa} \ \blacktriangleleft$$

c. Largest Normal Stress in Link ABC. The largest stress is found where the area is smallest; this occurs at the cross section at A where the 10-mm hole is located. We have

$$\sigma_A = \frac{F_{AC}}{A_{\text{net}}} = \frac{3000 \text{ N}}{(10 \text{ mm})(30 \text{ mm} - 10 \text{ mm})} = \frac{3000 \text{ N}}{200 \text{ mm}^2} \qquad \sigma_C = 15 \text{ MPa} \ \blacktriangleleft$$

d. Average Shearing Stress at B. We note that bonding exists on both sides of the upper portion of the link and that the shear force on each side is $F_1 = (3000 \text{ N})/2 = 1500 \text{ N}$. The average shearing stress on each surface is thus

$$\tau_B = \frac{F_1}{A} = \frac{1500 \text{ N}}{(30 \text{ mm})(45 \text{ mm})} \qquad \tau_B = 1.111 \text{ MPa} \ \blacktriangleleft$$

e. Bearing Stress in Link at C. For each portion of the link, $F_1 = 1500 \text{ N}$ and the nominal bearing area is $(6 \text{ mm})(6 \text{ mm}) = 36 \text{ mm}^2$.

$$\sigma_b = \frac{F_1}{A} = \frac{1500 \text{ N}}{36 \text{ mm}^2} \qquad \sigma_b = 41.7 \text{ MPa} \ \blacktriangleleft$$

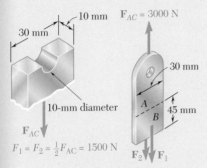

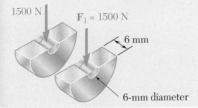

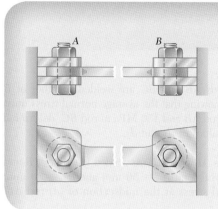

SAMPLE PROBLEM 1.2

The steel tie bar shown is to be designed to carry a tension force of magnitude $P = 120$ kN when bolted between double brackets at A and B. The bar will be fabricated from 20-mm-thick plate stock. For the grade of steel to be used, the maximum allowable stresses are: $\sigma = 175$ MPa, $\tau = 100$ MPa, $\sigma_b = 350$ MPa. Design the tie bar by determining the required values of (a) the diameter d of the bolt, (b) the dimension b at each end of the bar, (c) the dimension h of the bar.

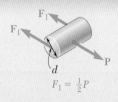

SOLUTION

a. Diameter of the Bolt. Since the bolt is in double shear, $F_1 = \frac{1}{2}P = 60$ kN.

$$\tau = \frac{F_1}{A} = \frac{60 \text{ kN}}{\frac{1}{4}\pi\, d^2} \qquad 100 \text{ MPa} = \frac{60 \text{ kN}}{\frac{1}{4}\pi\, d^2} \qquad d = 27.6 \text{ mm}$$

We will use $\quad d = 28$ mm ◄

At this point we check the bearing stress between the 20-mm-thick plate and the 28-mm-diameter bolt.

$$\tau_b = \frac{P}{td} = \frac{120 \text{ kN}}{(0.020 \text{ m})(0.028 \text{ m})} = 214 \text{ MPa} < 350 \text{ MPa} \qquad \text{OK}$$

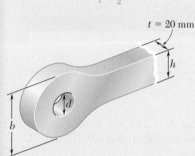

b. Dimension b at Each End of the Bar. We consider one of the end portions of the bar. Recalling that the thickness of the steel plate is $t = 20$ mm and that the average tensile stress must not exceed 175 MPa, we write

$$\sigma = \frac{\frac{1}{2}P}{ta} \qquad 175 \text{ MPa} = \frac{60 \text{ kN}}{(0.02 \text{ m})a} \qquad a = 17.14 \text{ mm}$$

$$b = d + 2a = 28 \text{ mm} + 2(17.14 \text{ mm}) \qquad b = 62.3 \text{ mm} \quad ◄$$

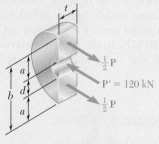

c. Dimension h of the Bar. Recalling that the thickness of the steel plate is $t = 20$ mm, we have

$$\sigma = \frac{P}{th} \qquad 175 \text{ MPa} = \frac{120 \text{ kN}}{(0.020 \text{ m})h} \qquad h = 34.3 \text{ mm}$$

We will use $\quad h = 35$ mm ◄

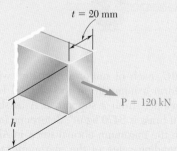

PROBLEMS

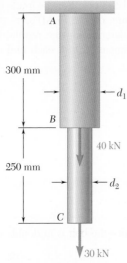

Fig. P1.1 and P1.2

1.1 Two solid cylindrical rods AB and BC are welded together at B and loaded as shown. Knowing that the average normal stress must not exceed 175 MPa in rod AB and 150 MPa in rod BC, determine the smallest allowable values of d_1 and d_2.

1.2 Two solid cylindrical rods AB and BC are welded together at B and loaded as shown. Knowing that $d_1 = 50$ mm and $d_2 = 30$ mm, find the average normal stress at the midsection of (a) rod AB, (b) rod BC.

1.3 Two solid cylindrical rods AB and BC are welded together at B and loaded as shown. Determine the magnitude of the force $\mathbf{P}$ for which the tensile stress in rod AB has the same magnitude as the compressive stress in rod BC.

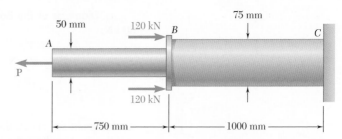

Fig. P1.3

1.4 In Prob. 1.3, knowing that $P = 160$ kN, determine the average normal stress at the midsection of (a) rod AB, (b) rod BC.

1.5 Two steel plates are to be held together by means of 16-mm-diameter high-strength steel bolts fitting snugly inside cylindrical brass spacers. Knowing that the average normal stress must not exceed 200 MPa in the bolts and 130 MPa in the spacers, determine the outer diameter of the spacers that yields the most economical and safe design.

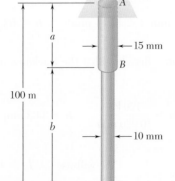

Fig. P1.6

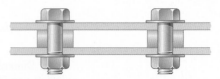

Fig. P1.5

1.6 Two brass rods AB and BC, each of uniform diameter, will be brazed together at B to form a nonuniform rod of total length 100 m which will be suspended from a support at A as shown. Knowing that the density of brass is 8470 kg/m^3, determine (a) the length of rod AB for which the maximum normal stress in ABC is minimum, (b) the corresponding value of the maximum normal stress.

1.7 Each of the four vertical links has an 8 × 36-mm uniform rectangular cross section and each of the four pins has a 16-mm diameter. Determine the maximum value of the average normal stress in the links connecting (*a*) points *B* and *D*, (*b*) points *C* and *E*.

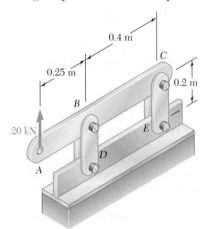

Fig. P1.7

1.8 Knowing that link *DE* is 3 mm thick and 25 mm wide, determine the normal stress in the central portion of that link when (*a*) $\theta = 0$, (*b*) $\theta = 90°$.

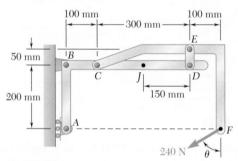

Fig. P1.8

1.9 Knowing that the central portion of the link *BD* has a uniform cross-sectional area of 800 mm², determine the magnitude of the load **P** for which the normal stress in that portion of *BD* is 50 MPa.

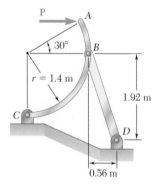

Fig. P1.9

1.10 Three forces, each of magnitude $P = 4$ kN, are applied to the mechanism shown. Determine the cross-sectional area of the uniform portion of rod *BE* for which the normal stress in that portion is +100 MPa.

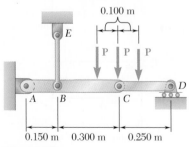

Fig. P1.10

1.11 The rigid bar *EFG* is supported by the truss system shown. Knowing that the member *CG* is a solid circular rod of 18-mm diameter, determine the normal stress in *CG*.

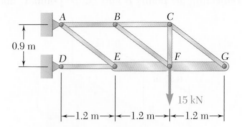

0.9 m

15 kN

|←1.2 m→|←1.2 m→|←1.2 m→|

Fig. P1.11 and P1.12

1.12 The rigid bar *EFG* is supported by the truss system shown. Determine the cross-sectional area of member *AE* for which the normal stress in the member is 105 MPa.

1.13 An aircraft tow bar is positioned by means of a single hydraulic cylinder connected by a 25-mm-diameter steel rod to two identical arm-and-wheel units *DEF*. The mass of the entire tow bar is 200 kg, and its center of gravity is located at *G*. For the position shown, determine the normal stress in the rod.

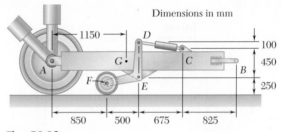

Dimensions in mm

1150   100
             450
             250

850   500   675   825

Fig. P1.13

1.14 A couple **M** of magnitude 1500 N · m is applied to the crank of an engine. For the position shown, determine (*a*) the force **P** required to hold the engine system in equilibrium, (*b*) the average normal stress in the connecting rod *BC*, which has a 450-mm² uniform cross section.

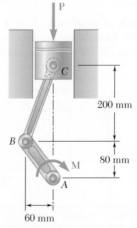

P

C

200 mm

B

80 mm

M

A

60 mm

Fig. P1.14

1.15 When the force **P** reached 8 kN, the wooden specimen shown failed in shear along the surface indicated by the dashed line. Determine the average shearing stress along that surface at the time of failure.

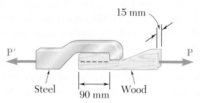

Steel   90 mm   Wood

Fig. P1.15

1.16 Two wooden planks, each 12 mm thick and 225 mm wide, are joined by the dry mortise joint shown. Knowing that the wood used shears off along its grain when the average shearing stress reaches 8 MPa, determine the magnitude P of the axial load that will cause the joint to fail.

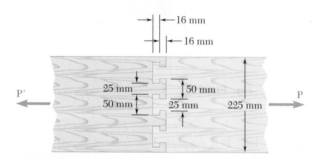

Fig. *P1.16*

1.17 A load **P** is applied to a steel rod supported as shown by an aluminum plate into which a 12-mm-diameter hole has been drilled. Knowing that the shearing stress must not exceed 180 MPa in the steel rod and 70 MPa in the aluminum plate, determine the largest load **P** that can be applied to the rod.

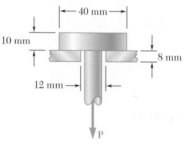

Fig. P1.17

1.18 Two wooden planks, each 22 mm thick and 160 mm wide, are joined by the glued mortise joint shown. Knowing that the joint will fail when the average shearing stress in the glue reaches 820 kPa, determine the smallest allowable length d of the cuts if the joint is to withstand an axial load of magnitude $P = 7.6$ kN.

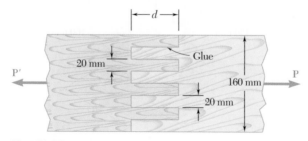

Fig. P1.18

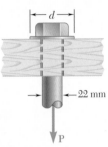

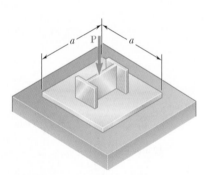

Fig. P1.21

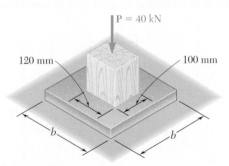

1.19 The load **P** applied to a steel rod is distributed to a timber support by an annular washer. The diameter of the rod is 22 mm and the inner diameter of the washer is 25 mm, which is slightly larger than the diameter of the hole. Determine the smallest allowable outer diameter d of the washer, knowing that the axial normal stress in the steel rod is 35 MPa and that the average bearing stress between the washer and the timber must not exceed 5 MPa.

Fig. P1.19

1.20 The axial force in the column supporting the timber beam shown is $P = 75$ kN. Determine the smallest allowable length L of the bearing plate if the bearing stress in the timber is not to exceed 3.0 MPa.

Fig. P1.20

1.21 An axial load **P** is supported by a short W200 × 59 column of cross-sectional area $A = 7650$ mm^2 and is distributed to a concrete foundation by a square plate as shown. Knowing that the average normal stress in the column must not exceed 200 MPa and that the bearing stress on the concrete foundation must not exceed 20 MPa, determine the side a of the plate that will provide the most economical and safe design.

1.22 A 40-kN axial load is applied to a short wooden post that is supported by a concrete footing resting on undisturbed soil. Determine (a) the maximum bearing stress on the concrete footing, (b) the size of the footing for which the average bearing stress in the soil is 145 kPa.

Fig. P1.22

1.23 Two identical linkage-and-hydraulic-cylinder systems control the position of the forks of a fork-lift truck. The load supported by the one system shown is 6 kN. Knowing that the thickness of member BD is 16 mm, determine (*a*) the average shearing stress in the 12-mm-diameter pin at B, (*b*) the bearing stress at B in member BD.

1.24 Knowing that $\theta = 40°$ and $P = 9$ kN, determine (*a*) the smallest allowable diameter of the pin at B if the average shearing stress in the pin is not to exceed 120 MPa, (*b*) the corresponding average bearing stress in member AB at B, (*c*) the corresponding average bearing stress in each of the support brackets at B.

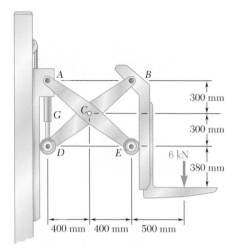

Fig. P1.23

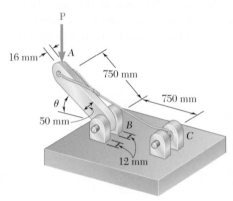

Fig. *P1.24* and *P1.25*

1.25 Determine the largest load **P** that can be applied at A when $\theta = 60°$, knowing that the average shearing stress in the 10-mm-diameter pin at B must not exceed 120 MPa and that the average bearing stress in member AB and in the bracket at B must not exceed 90 MPa.

1.26 Link AB, of width $b = 50$ mm and thickness $t = 6$ mm, is used to support the end of a horizontal beam. Knowing that the average normal stress in the link is -140 MPa, and that the average shearing stress in each of the two pins is 80 MPa, determine (*a*) the diameter d of the pins, (*b*) the average bearing stress in the link.

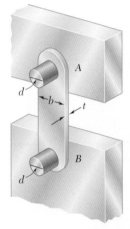

Fig. P1.26

1.27 For the assembly and loading of Prob. 1.7, determine (*a*) the average shearing stress in the pin at B, (*b*) the average bearing stress at B in member BD, (*c*) the average bearing stress at B in member ABC, knowing that this member has a 10 × 50-mm uniform rectangular cross section.

1.28 A 12-mm-diameter steel rod AB is fitted to a round hole near end C of the wooden member CD. For the loading shown, determine (*a*) the maximum average normal stress in the wood, (*b*) the distance b for which the average shearing stress is 620 kPa on the surfaces indicated by the dashed lines, (*c*) the average bearing stress on the wood.

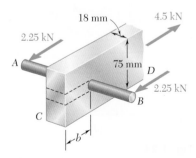

Fig. P1.28

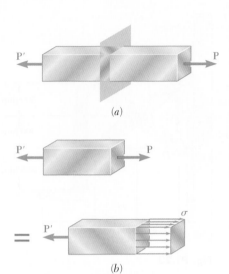

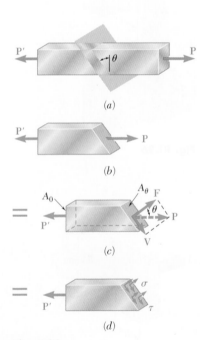

Fig. 1.26 Axial forces.

Fig. 1.28

1.11 STRESS ON AN OBLIQUE PLANE UNDER AXIAL LOADING

In the preceding sections, axial forces exerted on a two-force member (Fig. 1.26a) were found to cause normal stresses in that member (Fig. 1.26b), while transverse forces exerted on bolts and pins (Fig. 1.27a) were found to cause shearing stresses in those connections (Fig. 1.27b). The reason such a relation was observed between axial forces and normal stresses on one hand, and transverse forces and shearing stresses on the other, was because stresses were being determined only on planes perpendicular to the axis of the member or connection. As you will see in this section, axial forces cause both normal and shearing stresses on planes which are not perpendicular to the axis of the member. Similarly, transverse forces exerted on a bolt or a pin cause both normal and shearing stresses on planes which are not perpendicular to the axis of the bolt or pin.

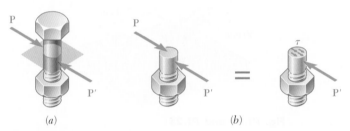

Fig. 1.27 Transverse forces.

Consider the two-force member of Fig. 1.26, which is subjected to axial forces **P** and **P′**. If we pass a section forming an angle θ with a normal plane (Fig. 1.28a) and draw the free-body diagram of the portion of member located to the left of that section (Fig. 1.28b), we find from the equilibrium conditions of the free body that the distributed forces acting on the section must be equivalent to the force **P**.

Resolving **P** into components **F** and **V**, respectively normal and tangential to the section (Fig. 1.28c), we have

$$F = P \cos \theta \qquad V = P \sin \theta \qquad (1.12)$$

The force **F** represents the resultant of normal forces distributed over the section, and the force **V** the resultant of shearing forces (Fig. 1.28d). The average values of the corresponding normal and shearing stresses are obtained by dividing, respectively, F and V by the area A_θ of the section:

$$\sigma = \frac{F}{A_\theta} \qquad \tau = \frac{V}{A_\theta} \qquad (1.13)$$

Substituting for F and V from (1.12) into (1.13), and observing from Fig. 1.28c that $A_0 = A_\theta \cos \theta$, or $A_\theta = A_0/\cos \theta$, where A_0 denotes the area of a section perpendicular to the axis of the member, we obtain

$$\sigma = \frac{P \cos \theta}{A_0/\cos \theta} \qquad \tau = \frac{P \sin \theta}{A_0/\cos \theta}$$

or

$$\sigma = \frac{P}{A_0} \cos^2 \theta \qquad \tau = \frac{P}{A_0} \sin \theta \cos \theta \qquad (1.14)$$

We note from the first of Eqs. (1.14) that the normal stress σ is maximum when $\theta = 0$, i.e., when the plane of the section is perpendicular to the axis of the member, and that it approaches zero as θ approaches 90°. We check that the value of σ when $\theta = 0$ is

$$\sigma_m = \frac{P}{A_0} \qquad (1.15)$$

as we found earlier in Sec. 1.3. The second of Eqs. (1.14) shows that the shearing stress τ is zero for $\theta = 0$ and $\theta = 90°$, and that for $\theta = 45°$ it reaches its maximum value

$$\tau_m = \frac{P}{A_0} \sin 45° \cos 45° = \frac{P}{2A_0} \qquad (1.16)$$

The first of Eqs. (1.14) indicates that, when $\theta = 45°$, the normal stress σ' is also equal to $P/2A_0$:

$$\sigma' = \frac{P}{A_0} \cos^2 45° = \frac{P}{2A_0} \qquad (1.17)$$

The results obtained in Eqs. (1.15), (1.16), and (1.17) are shown graphically in Fig. 1.29. We note that the same loading may produce either a normal stress $\sigma_m = P/A_0$ and no shearing stress (Fig. 1.29b), or a normal and a shearing stress of the same magnitude $\sigma' = \tau_m = P/2A_0$ (Fig. 1.29c and d), depending upon the orientation of the section.

(a) Axial loading

(b) Stresses for $\theta = 0$

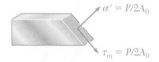

(c) Stresses for $\theta = 45°$

(d) Stresses for $\theta = -45°$

Fig. 1.29

1.12 STRESS UNDER GENERAL LOADING CONDITIONS; COMPONENTS OF STRESS

The examples of the previous sections were limited to members under axial loading and connections under transverse loading. Most structural members and machine components are under more involved loading conditions.

Consider a body subjected to several loads $\mathbf{P}_1$, $\mathbf{P}_2$, etc. (Fig. 1.30). To understand the stress condition created by these loads at some point Q within the body, we shall first pass a section through Q, using a plane parallel to the yz plane. The portion of the body to the left of the section is subjected to some of the original loads, and to normal and shearing forces distributed over the section. We shall denote by $\Delta \mathbf{F}^x$ and $\Delta \mathbf{V}^x$, respectively, the normal and the shearing

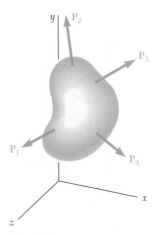

Fig. 1.30

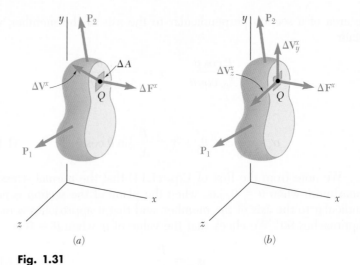

Fig. 1.31

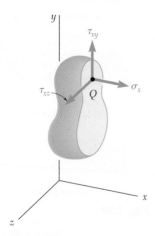

Fig. 1.32

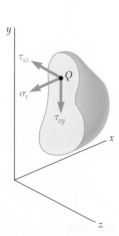

Fig. 1.33

forces acting on a small area ΔA surrounding point Q (Fig. 1.31a). Note that the superscript x is used to indicate that the forces $\Delta \mathbf{F}^x$ and $\Delta \mathbf{V}^x$ act on a surface perpendicular to the x axis. While the normal force $\Delta \mathbf{F}^x$ has a well-defined direction, the shearing force $\Delta \mathbf{V}^x$ may have any direction in the plane of the section. We therefore resolve $\Delta \mathbf{V}^x$ into two component forces, $\Delta \mathbf{V}_y^x$ and $\Delta \mathbf{V}_z^x$, in directions parallel to the y and z axes, respectively (Fig. 1.31b). Dividing now the magnitude of each force by the area ΔA, and letting ΔA approach zero, we define the three stress components shown in Fig. 1.32:

$$\sigma_x = \lim_{\Delta A \to 0} \frac{\Delta F^x}{\Delta A}$$

$$\tau_{xy} = \lim_{\Delta A \to 0} \frac{\Delta V_y^x}{\Delta A} \qquad \tau_{xz} = \lim_{\Delta A \to 0} \frac{\Delta V_z^x}{\Delta A}$$

(1.18)

We note that the first subscript in σ_x, τ_{xy}, and τ_{xz} is used to indicate that the stresses under consideration are exerted *on a surface perpendicular to the x axis*. The second subscript in τ_{xy} and τ_{xz} identifies *the direction of the component*. The normal stress σ_x is positive if the corresponding arrow points in the positive x direction, i.e., if the body is in tension, and negative otherwise. Similarly, the shearing stress components τ_{xy} and τ_{xz} are positive if the corresponding arrows point, respectively, in the positive y and z directions.

The above analysis may also be carried out by considering the portion of body located to the right of the vertical plane through Q (Fig. 1.33). The same magnitudes, but opposite directions, are obtained for the normal and shearing forces $\Delta \mathbf{F}^x$, $\Delta \mathbf{V}_y^x$, and $\Delta \mathbf{V}_z^x$. Therefore, the same values are also obtained for the corresponding stress components, but since the section in Fig. 1.33 now faces the *negative x axis*, a positive sign for σ_x will indicate that the corresponding arrow points *in the negative x direction*. Similarly, positive signs for τ_{xy} and τ_{xz} will indicate that the corresponding arrows point, respectively, in the negative y and z directions, as shown in Fig. 1.33.

Passing a section through Q parallel to the zx plane, we define in the same manner the stress components, σ_y, τ_{yz}, and τ_{yx}. Finally, a section through Q parallel to the xy plane yields the components σ_z, τ_{zx}, and τ_{zy}.

To facilitate the visualization of the stress condition at point Q, we shall consider a small cube of side a centered at Q and the stresses exerted on each of the six faces of the cube (Fig. 1.34). The stress components shown in the figure are σ_x, σ_y, and σ_z, which represent the normal stress on faces respectively perpendicular to the x, y, and z axes, and the six shearing stress components τ_{xy}, τ_{xz}, etc. We recall that, according to the definition of the shearing stress components, τ_{xy} represents the y component of the shearing stress exerted on the face perpendicular to the x axis, while τ_{yx} represents the x component of the shearing stress exerted on the face perpendicular to the y axis. Note that only three faces of the cube are actually visible in Fig. 1.34, and that equal and opposite stress components act on the hidden faces. While the stresses acting on the faces of the cube differ slightly from the stresses at Q, the error involved is small and vanishes as side a of the cube approaches zero.

Important relations among the shearing stress components will now be derived. Let us consider the free-body diagram of the small cube centered at point Q (Fig. 1.35). The normal and shearing forces acting on the various faces of the cube are obtained by multiplying the corresponding stress components by the area ΔA of each face. We first write the following three equilibrium equations:

$$\Sigma F_x = 0 \qquad \Sigma F_y = 0 \qquad \Sigma F_z = 0 \qquad (1.19)$$

Since forces equal and opposite to the forces actually shown in Fig. 1.35 are acting on the hidden faces of the cube, it is clear that Eqs. (1.19) are satisfied. Considering now the moments of the forces about axes x', y', and z' drawn from Q in directions respectively parallel to the x, y, and z axes, we write the three additional equations

$$\Sigma M_{x'} = 0 \qquad \Sigma M_{y'} = 0 \qquad \Sigma M_{z'} = 0 \qquad (1.20)$$

Using a projection on the $x'y'$ plane (Fig. 1.36), we note that the only forces with moments about the z axis different from zero are the shearing forces. These forces form two couples, one of counterclockwise (positive) moment $(\tau_{xy} \Delta A)a$, the other of clockwise (negative) moment $-(\tau_{yx} \Delta A)a$. The last of the three Eqs. (1.20) yields, therefore,

$$+\curvearrowleft \Sigma M_z = 0: \qquad (\tau_{xy} \Delta A)a - (\tau_{yx} \Delta A)a = 0$$

from which we conclude that

$$\tau_{xy} = \tau_{yx} \qquad (1.21)$$

The relation obtained shows that the y component of the shearing stress exerted on a face perpendicular to the x axis is equal to the x

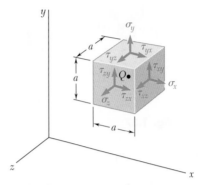

Fig. 1.34

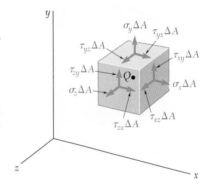

Fig. 1.35

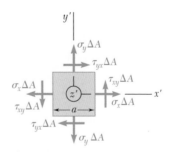

Fig. 1.36

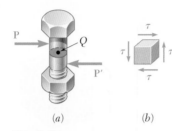

Fig. 1.37

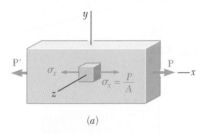

(a)

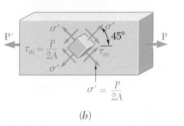

(b)

Fig. 1.38

component of the shearing stress exerted on a face perpendicular to the y axis. From the remaining two equations (1.20), we derive in a similar manner the relations

$$\tau_{yz} = \tau_{zy} \qquad \tau_{zx} = \tau_{xz} \qquad (1.22)$$

We conclude from Eqs. (1.21) and (1.22) that only six stress components are required to define the condition of stress at a given point Q, instead of nine as originally assumed. These six components are σ_x, σ_y, σ_z, τ_{xy}, τ_{yz}, and τ_{zx}. We also note that, at a given point, *shear cannot take place in one plane only;* an equal shearing stress must be exerted on another plane perpendicular to the first one. For example, considering again the bolt of Fig. 1.27 and a small cube at the center Q of the bolt (Fig. 1.37a), we find that shearing stresses of equal magnitude must be exerted on the two horizontal faces of the cube and on the two faces that are perpendicular to the forces **P** and **P'** (Fig. 1.37b).

Before concluding our discussion of stress components, let us consider again the case of a member under axial loading. If we consider a small cube with faces respectively parallel to the faces of the member and recall the results obtained in Sec. 1.11, we find that the conditions of stress in the member may be described as shown in Fig. 1.38a; the only stresses are normal stresses σ_x exerted on the faces of the cube which are perpendicular to the x axis. However, if the small cube is rotated by 45° about the z axis so that its new orientation matches the orientation of the sections considered in Fig. 1.29c and d, we conclude that normal and shearing stresses of equal magnitude are exerted on four faces of the cube (Fig. 1.38b). We thus observe that the same loading condition may lead to different interpretations of the stress situation at a given point, depending upon the orientation of the element considered. More will be said about this in Chap 7.

1.13 DESIGN CONSIDERATIONS

In the preceding sections you learned to determine the stresses in rods, bolts, and pins under simple loading conditions. In later chapters you will learn to determine stresses in more complex situations. In engineering applications, however, the determination of stresses is seldom an end in itself. Rather, the knowledge of stresses is used by engineers to assist in their most important task, namely, the design of structures and machines that will safely and economically perform a specified function.

a. Determination of the Ultimate Strength of a Material. An important element to be considered by a designer is how the material that has been selected will behave under a load. For a given material, this is determined by performing specific tests on prepared samples of the material. For example, a test specimen of steel may be prepared and placed in a laboratory testing machine to be subjected to a known centric axial tensile force, as described in Sec. 2.3. As the magnitude of the force is increased, various changes in the specimen are measured, for example, changes in its length and its diameter.

Eventually the largest force which may be applied to the specimen is reached, and the specimen either breaks or begins to carry less load. This largest force is called the *ultimate load* for the test specimen and is denoted by P_U. Since the applied load is centric, we may divide the ultimate load by the original cross-sectional area of the rod to obtain the *ultimate normal stress* of the material used. This stress, also known as the *ultimate strength in tension* of the material, is

$$\sigma_U = \frac{P_U}{A} \tag{1.23}$$

Several test procedures are available to determine the *ultimate shearing stress*, or *ultimate strength in shear*, of a material. The one most commonly used involves the twisting of a circular tube (Sec. 3.5). A more direct, if less accurate, procedure consists in clamping a rectangular or round bar in a shear tool (Fig. 1.39) and applying an increasing load P until the ultimate load P_U for single shear is obtained. If the free end of the specimen rests on both of the hardened dies (Fig. 1.40), the ultimate load for double shear is obtained. In either case, the ultimate shearing stress τ_U is obtained by dividing the ultimate load by the total area over which shear has taken place. We recall that, in the case of single shear, this area is the cross-sectional area A of the specimen, while in double shear it is equal to twice the cross-sectional area.

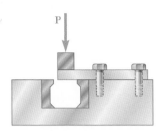

Fig. 1.39 Single shear test.

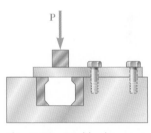

Fig. 1.40 Double shear test.

b. Allowable Load and Allowable Stress; Factor of Safety.

The maximum load that a structural member or a machine component will be allowed to carry under normal conditions of utilization is considerably smaller than the ultimate load. This smaller load is referred to as the *allowable load* and, sometimes, as the *working load* or *design load*. Thus, only a fraction of the ultimate-load capacity of the member is utilized when the allowable load is applied. The remaining portion of the load-carrying capacity of the member is kept in reserve to assure its safe performance. The ratio of the ultimate load to the allowable load is used to define the *factor of safety*.† We have

$$\text{Factor of safety} = F.S. = \frac{\text{ultimate load}}{\text{allowable load}} \tag{1.24}$$

An alternative definition of the factor of safety is based on the use of stresses:

$$\text{Factor of safety} = F.S. = \frac{\text{ultimate stress}}{\text{allowable stress}} \tag{1.25}$$

The two expressions given for the factor of safety in Eqs. (1.24) and (1.25) are identical when a linear relationship exists between the load and the stress. In most engineering applications, however, this relationship ceases to be linear as the load approaches its ultimate value, and the factor of safety obtained from Eq. (1.25) does not provide a

---

†In some fields of engineering, notably aeronautical engineering, the *margin of safety* is used in place of the factor of safety. The margin of safety is defined as the factor of safety minus one; i.e., margin of safety = *F.S.* − 1.00.

true assessment of the safety of a given design. Nevertheless, the *allowable-stress method* of design, based on the use of Eq. (1.25), is widely used.

c. Selection of an Appropriate Factor of Safety.

The selection of the factor of safety to be used for various applications is one of the most important engineering tasks. On the one hand, if a factor of safety is chosen too small, the possibility of failure becomes unacceptably large; on the other hand, if a factor of safety is chosen unnecessarily large, the result is an uneconomical or nonfunctional design. The choice of the factor of safety that is appropriate for a given design application requires engineering judgment based on many considerations, such as the following:

1. *Variations that may occur in the properties of the member under consideration.* The composition, strength, and dimensions of the member are all subject to small variations during manufacture. In addition, material properties may be altered and residual stresses introduced through heating or deformation that may occur during manufacture, storage, transportation, or construction.

2. *The number of loadings that may be expected during the life of the structure or machine.* For most materials the ultimate stress decreases as the number of load applications is increased. This phenomenon is known as *fatigue* and, if ignored, may result in sudden failure (see Sec. 2.7).

3. *The type of loadings that are planned for in the design, or that may occur in the future.* Very few loadings are known with complete accuracy—most design loadings are engineering estimates. In addition, future alterations or changes in usage may introduce changes in the actual loading. Larger factors of safety are also required for dynamic, cyclic, or impulsive loadings.

4. *The type of failure that may occur.* Brittle materials fail suddenly, usually with no prior indication that collapse is imminent. On the other hand, ductile materials, such as structural steel, normally undergo a substantial deformation called *yielding* before failing, thus providing a warning that overloading exists. However, most buckling or stability failures are sudden, whether the material is brittle or not. When the possibility of sudden failure exists, a larger factor of safety should be used than when failure is preceded by obvious warning signs.

5. *Uncertainty due to methods of analysis.* All design methods are based on certain simplifying assumptions which result in calculated stresses being approximations of actual stresses.

6. *Deterioration that may occur in the future because of poor maintenance or because of unpreventable natural causes.* A larger factor of safety is necessary in locations where conditions such as corrosion and decay are difficult to control or even to discover.

7. *The importance of a given member to the integrity of the whole structure.* Bracing and secondary members may in many cases be designed with a factor of safety lower than that used for primary members.

In addition to the these considerations, there is the additional consideration concerning the risk to life and property that a failure would produce. Where a failure would produce no risk to life and only minimal risk to property, the use of a smaller factor of safety can be considered. Finally, there is the practical consideration that, unless a careful design with a nonexcessive factor of safety is used, a structure or machine might not perform its design function. For example, high factors of safety may have an unacceptable effect on the weight of an aircraft.

For the majority of structural and machine applications, factors of safety are specified by design specifications or building codes written by committees of experienced engineers working with professional societies, with industries, or with federal, state, or city agencies. Examples of such design specifications and building codes are

1. *Steel:* American Institute of Steel Construction, Specification for Structural Steel Buildings
2. *Concrete:* American Concrete Institute, Building Code Requirement for Structural Concrete
3. *Timber:* American Forest and Paper Association, National Design Specification for Wood Construction
4. *Highway bridges:* American Association of State Highway Officials, Standard Specifications for Highway Bridges

*d. Load and Resistance Factor Design.   As we saw previously, the allowable-stress method requires that all the uncertainties associated with the design of a structure or machine element be grouped into a single factor of safety. An alternative method of design, which is gaining acceptance chiefly among structural engineers, makes it possible through the use of three different factors to distinguish between the uncertainties associated with the structure itself and those associated with the load it is designed to support. This method, referred to as *Load and Resistance Factor Design (LRFD)*, further allows the designer to distinguish between uncertainties associated with the *live load*, P_L, i.e., with the load to be supported by the structure, and the *dead load*, P_D, i.e., with the weight of the portion of structure contributing to the total load.

When this method of design is used, the *ultimate load*, P_U, of the structure, i.e., the load at which the structure ceases to be useful, should first be determined. The proposed design is then acceptable if the following inequality is satisfied:

$$\gamma_D P_D + \gamma_L P_L \leq \phi P_U \qquad (1.26)$$

The coefficient ϕ is referred to as the *resistance factor;* it accounts for the uncertainties associated with the structure itself and will normally be less than 1. The coefficients γ_D and γ_L are referred to as the *load factors;* they account for the uncertainties associated, respectively, with the dead and live load and will normally be greater than 1, with γ_L generally larger than γ_D. While a few examples or assigned problems using LRFD are included in this chapter and in Chaps. 5 and 10, the allowable-stress method of design will be used in this text.

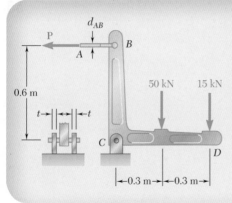

0.6 m

d_{AB}

P

A

B

50 kN   15 kN

t t

C

D

←0.3 m→←0.3 m→

SAMPLE PROBLEM 1.3

Two forces are applied to the bracket *BCD* as shown. (*a*) Knowing that the control rod *AB* is to be made of a steel having an ultimate normal stress of 600 MPa, determine the diameter of the rod for which the factor of safety with respect to failure will be 3.3. (*b*) The pin at *C* is to be made of a steel having an ultimate shearing stress of 350 MPa. Determine the diameter of the pin *C* for which the factor of safety with respect to shear will also be 3.3. (*c*) Determine the required thickness of the bracket supports at *C* knowing that the allowable bearing stress of the steel used is 300 MPa.

SOLUTION

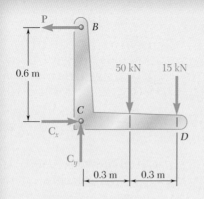

P

B

0.6 m

50 kN   15 kN

C

C_x

D

C_y

0.3 m   0.3 m

Free Body: Entire Bracket.   The reaction at *C* is represented by its components $\mathbf{C}_x$ and $\mathbf{C}_y$.

$+\curvearrowleft \Sigma M_C = 0$: $P(0.6 \text{ m}) - (50 \text{ kN})(0.3 \text{ m}) - (15 \text{ kN})(0.6 \text{ m}) = 0$ $P = 40 \text{ kN}$

$\Sigma F_x = 0$: $C_x = 40 \text{ k}$

$\Sigma F_y = 0$: $C_y = 65 \text{ kN}$ $C = \sqrt{C_x^2 + C_y^2} = 76.3 \text{ kN}$

a. Control Rod *AB*.   Since the factor of safety is to be 3.3, the allowable stress is

$$\sigma_{\text{all}} = \frac{\sigma_U}{F.S.} = \frac{600 \text{ MPa}}{3.3} = 181.8 \text{ MPa}$$

For $P = 40$ kN the cross-sectional area required is

$$A_{\text{req}} = \frac{P}{\sigma_{\text{all}}} = \frac{40 \text{ kN}}{181.8 \text{ MPa}} = 220 \times 10^{-6} \text{ m}^2$$

$$A_{\text{req}} = \frac{\pi}{4} d_{AB}^2 = 220 \times 10^{-6} \text{ m}^2 \qquad d_{AB} = 16.74 \text{ mm} \quad \blacktriangleleft$$

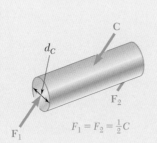

C

d_C

F_2

F_1

$F_1 = F_2 = \frac{1}{2}C$

b. Shear in Pin *C*.   For a factor of safety of 3.3, we have

$$\tau_{\text{all}} = \frac{\tau_U}{F.S.} = \frac{350 \text{ MPa}}{3.3} = 106.1 \text{ MPa}$$

Since the pin is in double shear, we write

$$A_{\text{req}} = \frac{C/2}{\tau_{\text{all}}} = \frac{(76.3 \text{ kN})/2}{106.1 \text{ MPa}} = 360 \text{ mm}^2$$

$$A_{\text{req}} = \frac{\pi}{4} d_C^2 = 360 \text{ mm}^2 \qquad d_C = 21.4 \text{ mm} \qquad \text{Use: } d_C = 22 \text{ mm} \quad \blacktriangleleft$$

The next larger size pin available is of 22-mm diameter and should be used.

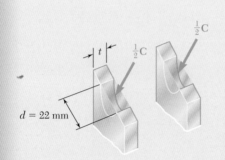

$\frac{1}{2}C$

t

$\frac{1}{2}C$

$\frac{1}{2}C$

$d = 22$ mm

c. Bearing at *C*.   Using $d = 22$ mm, the nominal bearing area of each bracket is $22t$. Since the force carried by each bracket is $C/2$ and the allowable bearing stress is 300 MPa, we write

$$A_{\text{req}} = \frac{C/2}{\sigma_{\text{all}}} = \frac{(76.3 \text{ kN})/2}{300 \text{ MPa}} = 127.2 \text{ mm}^2$$

Thus $22t = 127.2$ $t = 5.78$ mm   Use: $t = 6$ mm $\blacktriangleleft$

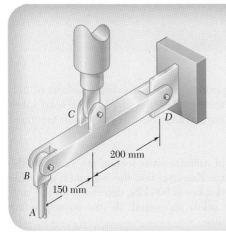

SAMPLE PROBLEM 1.4

The rigid beam BCD is attached by bolts to a control rod at B, to a hydraulic cylinder at C, and to a fixed support at D. The diameters of the bolts used are: $d_B = d_D = 10$ mm, $d_C = 12$ mm. Each bolt acts in double shear and is made from a steel for which the ultimate shearing stress is $\tau_U = 280$ MPa. The control rod AB has a diameter $d_A = 11$ mm and is made of a steel for which the ultimate tensile stress is $\sigma_U = 420$ MPa. If the minimum factor of safety is to be 3.0 for the entire unit, determine the largest upward force which may be applied by the hydraulic cylinder at C.

SOLUTION

The factor of safety with respect to failure must be 3.0 or more in each of the three bolts and in the control rod. These four independent criteria will be considered separately.

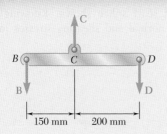

Free Body: Beam BCD. We first determine the force at C in terms of the force at B and in terms of the force at D.

$+\curvearrowleft \Sigma M_D = 0$: $B(350 \text{ mm}) - C(200 \text{ mm}) = 0$ $C = 1.750B$ (1)

$+\curvearrowleft \Sigma M_B = 0$: $-D(350 \text{ mm}) + C(150 \text{ mm}) = 0$ $C = 2.33D$ (2)

Control Rod. For a factor of safety of 3.0 we have

$$\sigma_{\text{all}} = \frac{\sigma_U}{F.S.} = \frac{420 \text{ MPa}}{3.0} = 140 \text{ MPa}$$

The allowable force in the control rod is

$$B = \sigma_{\text{all}}(A) = (140 \text{ MPa})\tfrac{1}{4}\pi(11 \times 10^{-3} \text{ m})^2 = 13.30 \text{ kN}$$

Using Eq. (1) we find the largest permitted value of C:

$$C = 1.750B = 1.750(13.3 \text{ kN}) \qquad C = 23.3 \text{ kN} \quad \blacktriangleleft$$

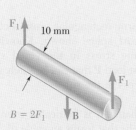

Bolt at B. $\tau_{\text{all}} = \tau_U/F.S. = (280 \text{ MPa})/3 = 93.33$ MPa. Since the bolt is in double shear, the allowable magnitude of the force **B** exerted on the bolt is

$$B = 2F_1 = 2(\tau_{\text{all}}A) = 2(93.33 \text{ MPa})(\tfrac{1}{4}\pi)(10 \times 10^{-3} \text{ m})^2 = 14.66 \text{ kN}$$

From Eq. (1): $C = 1.750B = 1.750(14.66 \text{ kN})$ $C = 25.7 \text{ kN}$ $\blacktriangleleft$

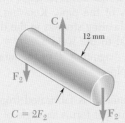

Bolt at D. Since this bolt is the same as bolt B, the allowable force is $D = B = 14.66$ kN. From Eq. (2):

$$C = 2.33D = 2.33(14.66 \text{ kN}) \qquad C = 34.2 \text{ kN} \quad \blacktriangleleft$$

Bolt at C. We again have $\tau_{\text{all}} = 93.33$ MPa and write

$$C = 2F_1 = 2(\tau_{\text{all}}A) = 2(93.33 \text{ MPa})(\tfrac{1}{4}\pi)(12 \times 10^{-3} \text{ m})^2 \qquad C = 21.1 \text{ kN} \quad \blacktriangleleft$$

Summary. We have found separately four maximum allowable values of the force C. In order to satisfy all these criteria we must choose the smallest value, namely: $C = 21.1 \text{ kN}$ $\blacktriangleleft$

PROBLEMS

Fig. P1.29 and P1.30

1.29 The 6-kN load **P** is supported by two wooden members of uniform cross section that are joined by the simple glued scarf splice shown. Determine the normal and shearing stresses in the glued splice.

1.30 Two wooden members of uniform cross section are joined by the simple scarf splice shown. Knowing that the maximum allowable tensile stress in the glued splice is 500 kPa, determine (*a*) the largest load **P** that can be safely supported, (*b*) the corresponding shearing stress in the splice.

1.31 Two wooden members of uniform rectangular cross section are joined by the simple glued scarf splice shown. Knowing that $P = 11$ kN, determine the normal and shearing stresses in the glued splice.

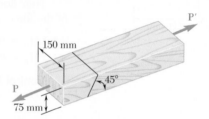

Fig. P1.31 and P1.32

1.32 Two wooden members of uniform rectangular cross section are joined by the simple glued scarf splice shown. Knowing that the maximum allowable shearing stress in the glued splice is 620 kPa, determine (*a*) the largest load **P** that can be safely applied, (*b*) the corresponding tensile stress in the splice.

1.33 A steel pipe of 300-mm outer diameter is fabricated from 6-mm-thick plate by welding along a helix that forms an angle of 25° with a plane perpendicular to the axis of the pipe. Knowing that the maximum allowable normal and shearing stresses in the directions respectively normal and tangential to the weld are $\sigma = 50$ MPa and $\tau = 30$ MPa, determine the magnitude P of the largest axial force that can be applied to the pipe.

1.34 A steel pipe of 300-mm outer diameter is fabricated from 6-mm-thick plate by welding along a helix that forms an angle of 25° with a plane perpendicular to the axis of the pipe. Knowing that a 250-kN axial force **P** is applied to the pipe, determine the normal and shearing stresses in directions respectively normal and tangential to the weld.

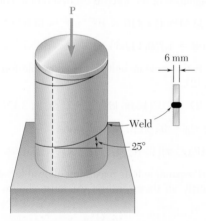

Fig. *P1.33* and *P1.34*

1.35 A 1060-kN load **P** is applied to the granite block shown. Determine the resulting maximum value of (a) the normal stress, (b) the shearing stress. Specify the orientation of that plane on which each of these maximum values occurs.

1.36 A centric load **P** is applied to the granite block shown. Knowing that the resulting maximum value of the shearing stress in the block is 18 MPa, determine (a) the magnitude of **P**, (b) the orientation of the surface on which the maximum shearing stress occurs, (c) the normal stress exerted on that surface, (d) the maximum value of the normal stress in the block.

1.37 Link BC is 6 mm thick, has a width $w = 25$ mm, and is made of a steel with a 480-MPa ultimate strength in tension. What is the safety factor used if the structure shown was designed to support a 16-kN load **P**?

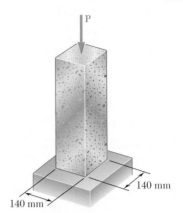

140 mm

140 mm

Fig. P1.35 and P1.36

1.38 Link BC is 6 mm thick and is made of a steel with a 450-MPa ultimate strength in tension. What should be its width w if the structure shown is being designed to support a 20-kN load **P** with a factor of safety of 3?

1.39 Members AB and AC of the truss shown consist of bars of square cross section made of the same alloy. It is known that a 20-mm square bar of the same alloy was tested to failure and that an ultimate load of 120 kN was recorded. If bar AB has a 15-mm square cross section, determine (a) the factor of safety for bar AB, (b) the dimensions of the cross section of bar AC if it is to have the same factor of safety as bar AB.

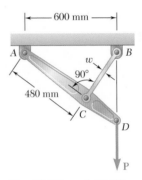

600 mm

A   w   B
        90°
480 mm
        /C
            D

P

Fig. P1.37 and *P1.38*

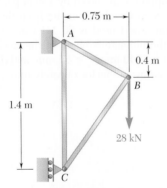

0.75 m

A

0.4 m

B

1.4 m

28 kN

C

Fig. P1.39 and P1.40

1.40 Members AB and AC of the truss shown consist of bars of square cross section made of the same alloy. It is known that a 20-mm square bar of the same alloy was tested to failure and that an ultimate load of 120 kN was recorded. If a factor of safety of 3.2 is to be achieved for bars, determine the required dimensions of the cross section of (a) bar AB, (b) bar AC.

1.41 Link AB is to be made of a steel for which the ultimate normal stress is 450 MPa. Determine the cross-sectional area of AB for which the factor of safety will be 3.50. Assume that the link will be adequately reinforced around the pins at A and B.

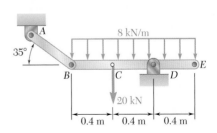

A   8 kN/m
35°
B| C   D   E

20 kN

0.4 m   0.4 m   0.4 m

Fig. P1.41

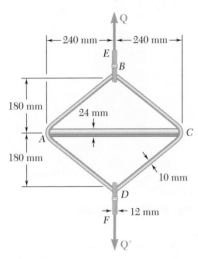

Fig. P1.42

1.42 A steel loop *ABCD* of length 1.2 m and of 10-mm diameter is placed as shown around a 24-mm-diameter aluminum rod *AC*. Cables *BE* and *DF*, each of 12-mm diameter, are used to apply the load **Q.** Knowing that the ultimate strength of the steel used for the loop and the cables is 480 MPa and that the ultimate strength of the aluminum used for the rod is 260 MPa, determine the largest load **Q** that can be applied if an overall factor of safety of 3 is desired.

1.43 Two wooden members shown, which support a 16-kN load, are joined by plywood splices fully glued on the surfaces in contact. The ultimate shearing stress in the glue is 2.5 MPa and the clearance between the members is 6 mm. Determine the required length *L* of each splice if a factor of safety of 2.75 is to be achieved.

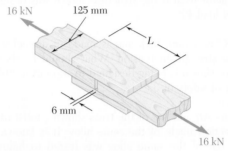

Fig. P1.43 and P1.44

1.44 For the joint and loading of Prob. 1.43, determine the factor of safety, knowing that the length of each splice is *L* = 180 mm.

1.45 A load **P** is supported as shown by a steel pin that has been inserted in a short wooden member hanging from the ceiling. The ultimate strength of the wood used is 60 MPa in tension and 7.5 MPa in shear, while the ultimate strength of the steel is 145 MPa in shear. Knowing that *b* = 40 mm, *c* = 55 mm, and *d* = 12 mm, determine the load **P** if an overall factor of safety of 3.2 is desired.

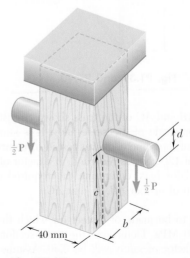

Fig. P1.45

1.46 For the support of Prob. 1.45, knowing that the diameter of the pin is $d = 16$ mm and that the magnitude of the load is $P = 20$ kN, determine (a) the factor of safety for the pin, (b) the required values of b and c if the factor of safety for the wooden member is the same as that found in part a for the pin.

1.47 Three steel bolts are to be used to attach the steel plate shown to a wooden beam. Knowing that the plate will support a 110-kN load, that the ultimate shearing stress for the steel used is 360 MPa, and that a factor of safety of 3.35 is desired, determine the required diameter of the bolts.

1.48 Three 18-mm-diameter steel bolts are to be used to attach the steel plate shown to a wooden beam. Knowing that the plate will support a 110-kN load and that the ultimate shearing stress for the steel used is 360 MPa, determine the factor of safety for this design.

Fig. P1.47 and P1.48

1.49 A steel plate 8 mm thick is embedded in a horizontal concrete slab and is used to anchor a high-strength vertical cable as shown. The diameter of the hole in the plate is 20 mm, the ultimate strength of the steel used is 250 MPa, and the ultimate bonding stress between plate and concrete is 2 MPa. Knowing that a factor of safety of 3.60 is desired when $P = 10$ kN, determine (a) the required width a of the plate, (b) the minimum depth b to which a plate of that width should be embedded in the concrete slab. (Neglect the normal stresses between the concrete and the bottom edge of the plate.)

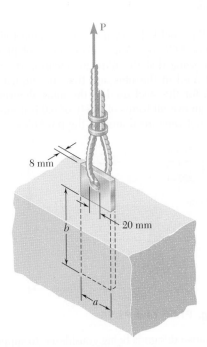

Fig. P1.49 and P1.50

1.50 Determine the factor of safety for the cable anchor in Prob. 1.49 when $P = 14$ kN, knowing that $a = 50$ mm and $b = 190$ mm.

1.51 In the steel structure shown, a 6-mm-diameter pin is used at *C* and 10-mm-diameter pins are used at *B* and *D*. The ultimate shearing stress is 150 MPa at all connections, and the ultimate normal stress is 400 MPa in link *BD*. Knowing that a factor of safety of 3.0 is desired, determine the largest load **P** that can be applied at *A*. Note that link *BD* is not reinforced around the pin holes.

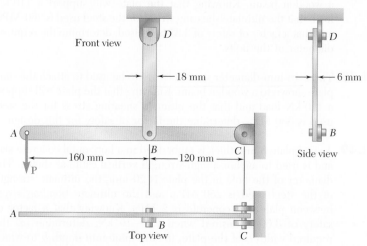

Fig. P1.51

1.52 Solve Prob. 1.51, assuming that the structure has been redesigned to use 12-mm-diameter pins at *B* and *D* and no other change has been made.

1.53 Each of the steel links *AB* and *CD* is connected to a support and to member *BCE* by 25-mm-diameter steel pins acting in single shear. Knowing that the ultimate shearing stress is 210 MPa for the steel used in the pins and that the ultimate normal stress is 490 MPa for the steel used in the links, determine the allowable load **P** if an overall factor of safety of 3.0 is desired. (Note that the links are not reinforced around the pin holes.)

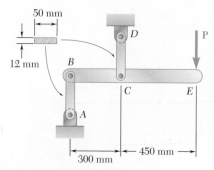

Fig. P1.53 and P1.54

1.54 An alternative design is being considered to support member *BCE* of Prob. 1.53, in which link *CD* will be replaced by two links, each of 6 × 50-mm cross section, causing the pins at *C* and *D* to be in double shear. Assuming that all other specifications remain unchanged, determine the allowable load **P** if an overall factor of safety of 3.0 is desired.

1.55 In the structure shown, an 8-mm-diameter pin is used at A, and 12-mm-diameter pins are used at B and D. Knowing that the ultimate shearing stress is 100 MPa at all connections and that the ultimate normal stress is 250 MPa in each of the two links joining B and D, determine the allowable load **P** if an overall factor of safety of 3.0 is desired.

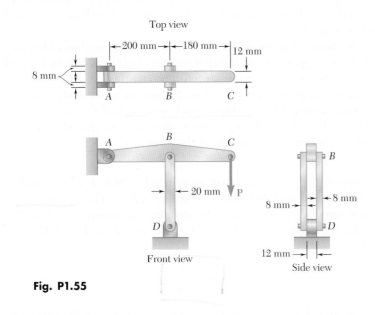

Fig. P1.55

1.56 In an alternative design for the structure of Prob. 1.55, a pin of 10-mm diameter is to be used at A. Assuming that all other specifications remain unchanged, determine the allowable load **P** if an overall factor of safety of 3.0 is desired.

***1.57** The Load and Resistance Factor Design method is to be used to select the two cables that will raise and lower a platform supporting two window washers. The platform has a mass of 72 kg and each of the window washers is assumed to have a mass of 88 kg with equipment. Since these workers are free to move on the platform, 75 percent of their total weight and the weight of their equipment will be used as the design live load of each cable. (*a*) Assuming a resistance factor $\phi = 0.85$ and load factors $\gamma_D = 1.2$ and $\gamma_L = 1.5$, determine the required minimum ultimate load of one cable. (*b*) What is the conventional factor of safety for the selected cables?

***1.58** A 40-kg platform is attached to the end B of a 50-kg wooden beam AB, which is supported as shown by a pin at A and by a slender steel rod BC with a 12-kN ultimate load. (*a*) Using the Load and Resistance Factor Design method with a resistance factor $\phi = 0.90$ and load factors $\gamma_D = 1.25$ and $\gamma_L = 1.6$, determine the largest load that can be safely placed on the platform. (*b*) What is the corresponding conventional factor of safety for rod BC?

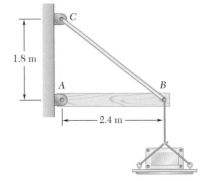

Fig. P1.57

Fig. P1.58

REVIEW AND SUMMARY

This chapter was devoted to the concept of stress and to an introduction to the methods used for the analysis and design of machines and load-bearing structures.

Section 1.2 presented a short review of the methods of statics and of their application to the determination of the reactions exerted by its supports on a simple structure consisting of pin-connected members. Emphasis was placed on the use of a *free-body diagram* to obtain equilibrium equations which were solved for the unknown reactions. Free-body diagrams were also used to find the internal forces in the various members of the structure.

Axial loading. Normal stress

Fig. 1.41

The concept of *stress* was first introduced in Sec. 1.3 by considering a two-force member under an *axial loading*. The <u>normal stress</u> in that member was obtained by dividing the magnitude P of the load by the cross-sectional area A of the member (Fig. 1.41). We wrote

$$\sigma = \frac{P}{A} \tag{1.5}$$

Section 1.4 was devoted to a short discussion of the two principal tasks of an engineer, namely, the *analysis* and the *design* of structures and machines.

As noted in Sec. 1.5, the value of σ obtained from Eq. (1.5) represents the *average stress* over the section rather than the stress at a specific point Q of the section. Considering a small area ΔA surrounding Q and the magnitude ΔF of the force exerted on ΔA, we defined the stress at point Q as

$$\sigma = \lim_{\Delta A \to 0} \frac{\Delta F}{\Delta A} \tag{1.6}$$

In general, the value obtained for the stress σ at point Q is different from the value of the average stress given by formula (1.5) and is found to vary across the section. However, this variation is small in any section away from the points of application of the loads. In practice, therefore, the distribution of the normal stresses in an axially loaded member is assumed to be *uniform*, except in the immediate vicinity of the points of application of the loads.

However, for the distribution of stresses to be uniform in a given section, it is necessary that the line of action of the loads **P** and **P'** pass through the centroid C of the section. Such a loading is called a *centric* axial loading. In the case of an *eccentric* axial loading, the distribution of stresses is *not* uniform. Stresses in members subjected to an eccentric axial loading will be discussed in Chap. 4.

Don't forget to subtract area of bolt if it's in tension

When equal and opposite *transverse forces* **P** and **P'** of magnitude P are applied to a member AB (Fig. 1.42), *shearing stresses* τ are created over any section located between the points of application of the two forces [Sec 1.6]. These stresses vary greatly across the section and their distribution *cannot* be assumed uniform. However, dividing the magnitude P—referred to as the *shear* in the section—by the cross-sectional area A, we defined the *average shearing stress* over the section:

$$\tau_{\text{ave}} = \frac{P}{A} \tag{1.8}$$

Shearing stresses are found in bolts, pins, or rivets connecting two structural members or machine components. For example, in the case of bolt CD (Fig. 1.43), which is in *single shear*, we wrote

$$\tau_{\text{ave}} = \frac{P}{A} = \frac{F}{A} \tag{1.9}$$

while, in the case of bolts EG and HJ (Fig. 1.44), which are both in *double shear*, we had

$$\tau_{\text{ave}} = \frac{P}{A} = \frac{F/2}{A} = \frac{F}{2A} \tag{1.10}$$

Bolts, pins, and rivets also create stresses in the members they connect, along the *bearing surface*, or surface of contact [Sec. 1.7]. The bolt CD of Fig. 1.43, for example, creates stresses on the semicylindrical surface of plate A with which it is in contact (Fig. 1.45). Since the distribution of these stresses is quite complicated, one uses in practice an average nominal value σ_b of the stress, called *bearing stress*, obtained by dividing the load P by the area of the rectangle representing the projection of the bolt on the plate section. Denoting by t the thickness of the plate and by d the diameter of the bolt, we wrote

$$\sigma_b = \frac{P}{A} = \frac{P}{td} \tag{1.11}$$

In Sec. 1.8, we applied the concept introduced in the previous sections to the analysis of a simple structure consisting of two pin-connected members supporting a given load. We determined successively the normal stresses in the two members, paying special attention to their narrowest sections, the shearing stresses in the various pins, and the bearing stress at each connection.

The method you should use in solving a problem in mechanics of materials was described in Sec. 1.9. Your solution should begin with a clear and precise *statement* of the problem. You will then draw one or several *free-body diagrams* that you will use to write *equilibrium equations*. These equations will be solved for *unknown forces*, from which the required *stresses* and *deformations* can be computed. Once the answer has been obtained, it should be *carefully checked*.

Transverse forces. Shearing stress

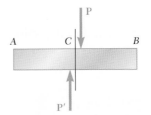

Fig. 1.42

Single and double shear

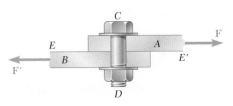

Fig. 1.43

Bearing stress

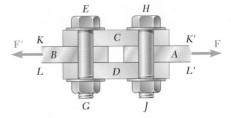

Fig. 1.44

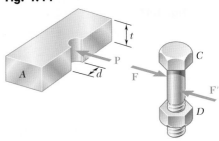

Fig. 1.45

Method of Solution

The first part of the chapter ended with a discussion of *numerical accuracy* in engineering, which stressed the fact that the accuracy of an answer can never be greater than the accuracy of the given data [Sec. 1.10].

Stresses on an oblique section

Fig. 1.46

$$A' = \frac{A}{\cos\theta}$$

In Sec. 1.11, we considered the stresses created on an *oblique section* in a two-force member under axial loading. We found that both *normal* and *shearing* stresses occurred in such a situation. Denoting by θ the angle formed by the section with a normal plane (Fig. 1.46) and by A_0 the area of a section perpendicular to the axis of the member, we derived the following expressions for the normal stress σ and the shearing stress τ on the oblique section:

$$\sigma = \frac{P}{A_0}\cos^2\theta \qquad \tau = \frac{P}{A_0}\sin\theta\cos\theta \qquad (1.14)$$

We observed from these formulas that the normal stress is maximum and equal to $\sigma_m = P/A_0$ for $\theta = 0$, while the shearing stress is maximum and equal to $\tau_m = P/2A_0$ for $\theta = 45°$. We also noted that $\tau = 0$ when $\theta = 0$, while $\sigma = P/2A_0$ when $\theta = 45°$.

Stress under general loading

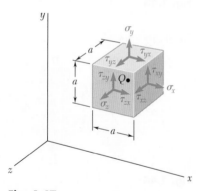

Fig. 1.47

Next, we discussed the state of stress at a point Q in a body under the most general loading condition [Sec. 1.12]. Considering a small cube centered at Q (Fig. 1.47), we denoted by σ_x the normal stress exerted on a face of the cube perpendicular to the x axis, and by τ_{xy} and τ_{xz}, respectively, the y and z components of the shearing stress exerted on the same face of the cube. Repeating this procedure for the other two faces of the cube and observing that $\tau_{xy} = \tau_{yx}$, $\tau_{yz} = \tau_{zy}$, and $\tau_{zx} = \tau_{xz}$, we concluded that *six stress components* are required to define the state of stress at a given point Q, namely, σ_x, σ_y, σ_z, τ_{xy}, τ_{yz}, and τ_{zx}.

Section 1.13 was devoted to a discussion of the various concepts used in the design of engineering structures. The *ultimate load* of a given structural member or machine component is the load at which the member or component is expected to fail; it is computed from the *ultimate stress* or *ultimate strength* of the material used, as determined by a laboratory test on a specimen of that material. The ultimate load should be considerably larger than the *allowable load*, i.e., the load that the member or component will be allowed to carry under normal conditions. The ratio of the ultimate load to the allowable load is defined as the *factor of safety*:

Factor of safety

$$\text{Factor of safety} = F.S. = \frac{\text{ultimate load}}{\text{allowable load}} \qquad (1.24)$$

The determination of the factor of safety that should be used in the design of a given structure depends upon a number of considerations, some of which were listed in this section.

Load and Resistance Factor Design

Section 1.13 ended with the discussion of an alternative approach to design, known as *Load and Resistance Factor Design*, which allows the engineer to distinguish between the uncertainties associated with the structure and those associated with the load.

1.59 A strain gage located at C on the surface of bone AB indicates that the average normal stress in the bone is 3.80 MPa when the bone is subjected to two 1200-N forces as shown. Assuming the cross section of the bone at C to be annular and knowing that its outer diameter is 25 mm, determine the inner diameter of the bone's cross section at C.

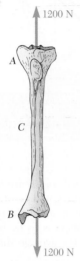

1200 N

A

C

B

1200 N

Fig. P1.59

1.60 Two horizontal 20-kN forces are applied to pin B of the assembly shown. Knowing that a pin of 20-mm diameter is used at each connection, determine the maximum value of the average normal stress (*a*) in link AB, (*b*) in link BC.

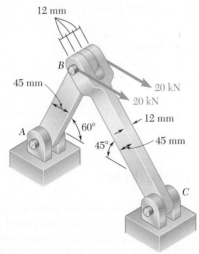

12 mm

B

45 mm

20 kN
20 kN

A

60°

12 mm

45 mm

45°

C

Fig. P1.60

1.61 For the assembly and loading of Prob. 1.60, determine (*a*) the average shearing stress in the pin at C, (*b*) the average bearing stress at C in member BC, (*c*) the average bearing stress at B in member BC.

1.62 In the marine crane shown, link *CD* is known to have a uniform cross section of 50×150 mm. For the loading shown, determine the normal stress in the central portion of that link.

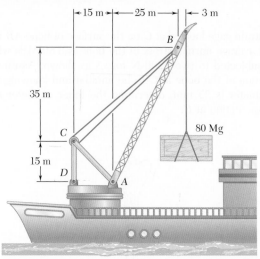

Fig. P1.62

1.63 The wooden members *A* and *B* are to be joined by plywood splice plates that will be fully glued on the surfaces in contact. As part of the design of the joint, and knowing that the clearance between the ends of the members is to be 6 mm, determine the smallest allowable length *L* if the average shearing stress in the glue is not to exceed 700 kPa.

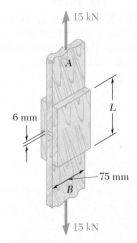

Fig. P1.63

1.64 Two wooden members of uniform rectangular cross section of sides $a = 100$ mm and $b = 60$ mm are joined by a simple glued joint as shown. Knowing that the ultimate stresses for the joint are $\sigma_U = 1.26$ MPa in tension and $\tau_U = 1.50$ MPa in shear and that $P = 6$ kN, determine the factor of safety for the joint when (*a*) $\alpha = 20°$, (*b*) $\alpha = 35°$, (*c*) $\alpha = 45°$. For each of these values of α, also determine whether the joint will fail in tension or in shear if *P* is increased until rupture occurs.

Fig. P1.64

1.65 Member *ABC*, which is supported by a pin and bracket at *C* and a cable *BD*, was designed to support the 16-kN load **P** as shown. Knowing that the ultimate load for cable *BD* is 100 kN, determine the factor of safety with respect to cable failure.

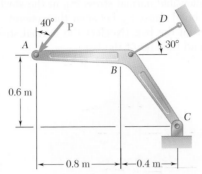

Fig. P1.65

1.66 Each of the two vertical links *CF* connecting the two horizontal members *AD* and *EG* has a uniform rectangular cross section 10 mm thick and 40 mm wide, and is made of a steel with an ultimate strength in tension of 400 MPa. The pins at *C* and *F* each have a 20-mm diameter and are made of a steel with an ultimate strength in shear of 150 MPa. Determine the overall factor of safety for the links *CF* and the pins connecting them to the horizontal members.

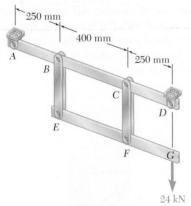

Fig. P1.66

1.67 Knowing that a force **P** of magnitude 750 N is applied to the pedal shown, determine (*a*) the diameter of the pin at *C* for which the average shearing stress in the pin is 40 MPa, (*b*) the corresponding bearing stress in the pedal at *C*, (*c*) the corresponding bearing stress in each support bracket at *C*.

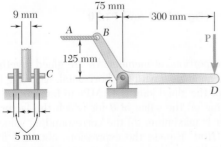

Fig. P1.67

1.68 A force **P** is applied as shown to a steel reinforcing bar that has been embedded in a block of concrete. Determine the smallest length L for which the full allowable normal stress in the bar can be developed. Express the result in terms of the diameter d of the bar, the allowable normal stress σ_{all} in the steel, and the average allowable bond stress τ_{all} between the concrete and the cylindrical surface of the bar. (Neglect the normal stresses between the concrete and the end of the bar.)

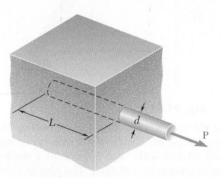

Fig. P1.68

1.69 The two portions of member AB are glued together along a plane forming an angle θ with the horizontal. Knowing that the ultimate stress for the glued joint is 17 MPa in tension and 9 MPa in shear, determine the range of values of θ for which the factor of safety of the members is at least 3.0.

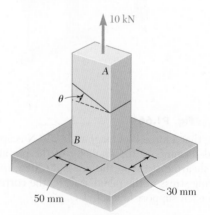

Fig. P1.69 and P1.70

1.70 The two portions of member AB are glued together along a plane forming an angle θ with the horizontal. Knowing that the ultimate stress for the glued joint is 17 MPa in tension and 9 MPa in shear, determine (a) the value of θ for which the factor of safety of the member is maximum, (b) the corresponding value of the factor of safety. (*Hint:* Equate the expressions obtained for the factors of safety with respect to normal stress and shear.)

COMPUTER PROBLEMS

The following problems are designed to be solved with a computer.

1.C1 A solid steel rod consisting of n cylindrical elements welded together is subjected to the loading shown. The diameter of element i is denoted by d_i and the load applied to its lower end by $\mathbf{P}_i$, with the magnitude P_i of this load being assumed positive if $\mathbf{P}_i$ is directed downward as shown and negative otherwise. (a) Write a computer program that can be used with SI units to determine the average stress in each element of the rod. (b) Use this program to solve Probs. 1.2 and 1.4.

Fig. P1.C1

1.C2 A 20-kN load is applied as shown to the horizontal member ABC. Member ABC has a 10×50-mm uniform rectangular cross section and is supported by four vertical links, each of 8×36-mm uniform rectangular cross section. Each of the four pins at A, B, C, and D has the same diameter d and is in double shear. (a) Write a computer program to calculate for values of d from 10 to 30 mm, using 1-mm increments, (1) the maximum value of the average normal stress in the links connecting pins B and D, (2) the average normal stress in the links connecting pins C and E, (3) the average shearing stress in pin B, (4) the average shearing stress in pin C, (5) the average bearing stress at B in member ABC, (6) the average bearing stress at C in member ABC. (b) Check your program by comparing the values obtained for $d = 16$ mm with the answers given for Probs. 1.7 and 1.27. (c) Use this program to find the permissible values of the diameter d of the pins, knowing that the allowable values of the normal, shearing, and bearing stresses for the steel used are, respectively, 150 MPa, 90 MPa, and 230 MPa. (d) Solve part c, assuming that the thickness of member ABC has been reduced from 10 to 8 mm.

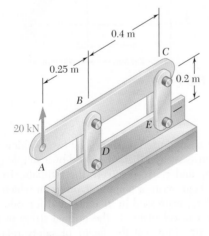

Fig. P1.C2

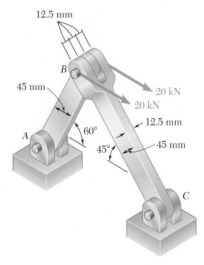

Fig. P1.C3

1.C3 Two horizontal 20-kN forces are applied to pin B of the assembly shown. Each of the three pins at A, B, and C has the same diameter d and is in double shear. (a) Write a computer program to calculate for values of d from 12.5 to 37.5 mm, using 1.25-mm increments, (1) the maximum value of the average normal stress in member AB, (2) the average normal stress in member BC, (3) the average shearing stress in pin A, (4) the average shearing stress in pin C, (5) the average bearing stress at A in member AB, (6) the average bearing stress at C in member BC, (7) the average bearing stress at B in member BC. (b) Check your program by comparing the values obtained for $d = 20$ mm with the answers given for Probs. 1.60 and 1.61. (c) Use this program to find the permissible values of the diameter d of the pins, knowing that the allowable values of the normal, shearing, and bearing stresses for the steel used are, respectively, 150 MPa, 90 MPa, and 250 MPa. (d) Solve part c, assuming that a new design is being investigated in which the thickness and width of the two members are changed, respectively, from 12.5 to 8 mm and from 45 to 60 mm.

1.C4 A 16-kN force **P** forming an angle α with the vertical is applied as shown to member ABC, which is supported by a pin and bracket at C and by a cable BD forming an angle β with the horizontal. (a) Knowing that the ultimate load of the cable is 100 kN, write a computer program to construct a table of the values of the factor of safety of the cable for values of α and β from 0 to 45°, using increments in α and β corresponding to 0.1 increments in tan α and tan β. (b) Check that for any given value of α, the maximum value of the factor of safety is obtained for $\beta = 38.66°$ and explain why. (c) Determine the smallest possible value of the factor of safety for $\beta = 38.66°$, as well as the corresponding value of α, and explain the result obtained.

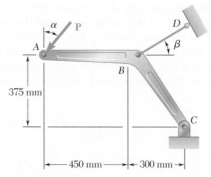

Fig. P1.C4

Fig. P1.C5

1.C5 A load **P** is supported as shown by two wooden members of uniform rectangular cross section that are joined by a simple glued scarf splice. (a) Denoting by σ_U and τ_U, respectively, the ultimate strength of the joint in tension and in shear, write a computer program which, for given values of a, b, P, σ_U, and τ_U, expressed in SI units, and for values of α from 5 to 85° at 5° intervals, can calculate (1) the normal stress in the joint, (2) the shearing stress in the joint, (3) the factor of safety relative to failure in tension, (4) the factor of safety relative to failure in shear, (5) the overall factor of safety for the glued joint. (b) Apply this program, using the dimensions and loading of the members of Probs. 1.29 and 1.31, knowing that $\sigma_U = 1.03$ MPa and $\tau_U = 1.47$ MPa for the glue used in Prob. 1.29, and that $\sigma_U = 1.26$ MPa and $\tau_U = 1.50$ MPa for the glue used in Prob. 1.31. (c) Verify in each of these two cases that the shearing stress is maximum for $\alpha = 45°$.

1.C6 Member ABC is supported by a pin and bracket at A, and by two links that are pin-connected to the member at B and to a fixed support at D. (*a*) Write a computer program to calculate the allowable load P_{all} for any given values of (1) the diameter d_1 of the pin at A, (2) the common diameter d_2 of the pins at B and D, (3) the ultimate normal stress σ_U in each of the two links, (4) the ultimate shearing stress τ_U in each of the three pins, (5) the desired overall factor of safety $F.S$. Your program should also indicate which of the following three stresses is critical: the normal stress in the links, the shearing stress in the pin at A, or the shearing stress in the pins at B and D (*b* and *c*). Check your program by using the data of Probs. 1.55 and 1.56, respectively, and comparing the answers obtained for P_{all} with those given in the text. (*d*) Use your program to determine the allowable load P_{all}, as well as which of the stresses is critical, when $d_1 = d_2 = 15$ mm, $\sigma_U = 110$ MPa for aluminum links, $\tau_U = 100$ MPa for steel pins, and $F.S. = 3.2$.

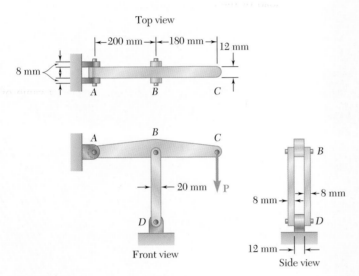

Fig. P1.C6

This chapter is devoted to the study of deformations occurring in structural components subjected to axial loading. The change in length of the diagonal stays was carefully accounted for in the design of this cable-stayed bridge.

Stress and Strain—Axial Loading

2.1 INTRODUCTION

In Chap. 1 we analyzed the stresses created in various members and connections by the loads applied to a structure or machine. We also learned to design simple members and connections so that they would not fail under specified loading conditions. Another important aspect of the analysis and design of structures relates to the *deformations* caused by the loads applied to a structure. Clearly, it is important to avoid deformations so large that they may prevent the structure from fulfilling the purpose for which it was intended. But the analysis of deformations may also help us in the determination of stresses. Indeed, it is not always possible to determine the forces in the members of a structure by applying only the principles of statics. This is because statics is based on the assumption of undeformable, rigid structures. By considering engineering structures as *deformable* and analyzing the deformations in their various members, it will be possible for us to compute forces that are *statically indeterminate*, i.e., indeterminate within the framework of statics. Also, as we indicated in Sec. 1.5, the distribution of stresses in a given member is statically indeterminate, even when the force in that member is known. To determine the actual distribution of stresses within a member, it is thus necessary to analyze the deformations that take place in that member. In this chapter, you will consider the deformations of a structural member such as a rod, bar, or plate under *axial loading*.

First, the *normal strain* ϵ in a member will be defined as the *deformation of the member per unit length*. Plotting the stress σ versus the strain ϵ as the load applied to the member is increased will yield a *stress-strain diagram* for the material used. From such a diagram we can determine some important properties of the material, such as its *modulus of elasticity*, and whether the material is *ductile* or *brittle* (Secs. 2.2 to 2.5). You will also see in Sec. 2.5 that, while the behavior of most materials is independent of the direction in which the load is applied, the response of fiber-reinforced composite materials depends upon the direction of the load.

From the stress-strain diagram, we can also determine whether the strains in the specimen will disappear after the load has been removed—in which case the material is said to behave *elastically*—or whether *a permanent set* or *plastic deformation* will result (Sec. 2.6).

Section 2.7 is devoted to the phenomenon of *fatigue*, which causes structural or machine components to fail after a very large number of repeated loadings, even though the stresses remain in the elastic range.

The first part of the chapter ends with Sec. 2.8, which is devoted to the determination of the deformation of various types of members under various conditions of axial loading.

In Secs. 2.9 and 2.10, *statically indeterminate problems* will be considered, i.e., problems in which the reactions and the internal forces *cannot* be determined from statics alone. The equilibrium equations derived from the free-body diagram of the member under consideration must be complemented by relations involving deformations; these relations will be obtained from the geometry of the problem.

In Secs. 2.11 to 2.15, additional constants associated with isotropic materials—i.e., materials with mechanical characteristics independent of direction—will be introduced. They include *Poisson's ratio,* which relates lateral and axial strain, the *bulk modulus,* which characterizes the change in volume of a material under hydrostatic pressure, and the *modulus of rigidity,* which relates the components of the shearing stress and shearing strain. Stress-strain relationships for an isotropic material under a multiaxial loading will also be derived.

In Sec. 2.16, stress-strain relationships involving several distinct values of the modulus of elasticity, Poisson's ratio, and the modulus of rigidity, will be developed for fiber-reinforced composite materials under a multiaxial loading. While these materials are not isotropic, they usually display special properties, known as *orthotropic* properties, which facilitate their study.

In the text material described so far, stresses are assumed uniformly distributed in any given cross section; they are also assumed to remain within the elastic range. The validity of the first assumption is discussed in Sec. 2.17, while *stress concentrations* near circular holes and fillets in flat bars are considered in Sec. 2.18. Sections 2.19 and 2.20 are devoted to the discussion of stresses and deformations in members made of a ductile material when the yield point of the material is exceeded. As you will see, permanent *plastic deformations* and *residual stresses* result from such loading conditions.

2.2 NORMAL STRAIN UNDER AXIAL LOADING

Let us consider a rod BC, of length L and uniform cross-sectional area A, which is suspended from B (Fig. 2.1a). If we apply a load **P** to end C, the rod elongates (Fig. 2.1b). Plotting the magnitude P of the load against the deformation δ (Greek letter delta), we obtain a certain load-deformation diagram (Fig. 2.2). While this diagram contains information useful to the analysis of the rod under consideration, it cannot be used directly to predict the deformation of a rod of the same material but of different dimensions. Indeed, we observe that, if a deformation δ is produced in rod BC by a load **P**, a load 2**P** is required to cause the same deformation in a rod $B'C'$ of the same length L, but of cross-sectional area $2A$ (Fig. 2.3). We note that, in both cases, the value of the stress is the same: $\sigma = P/A$. On the other hand, a load **P** applied to a rod $B''C''$, of the same

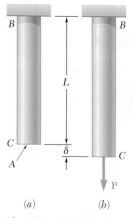

Fig. 2.1 Deformation of axially-loaded rod.

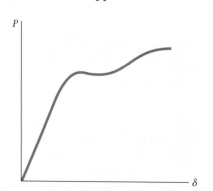

Fig. 2.2 Load-deformation diagram.

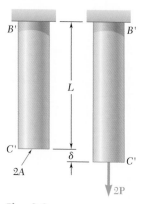

Fig. 2.3

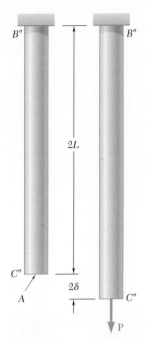

Fig. 2.4

cross-sectional area A, but of length $2L$, causes a deformation 2δ in that rod (Fig. 2.4), i.e., a deformation twice as large as the deformation δ it produces in rod BC. But in both cases the ratio of the deformation over the length of the rod is the same; it is equal to δ/L. This observation brings us to introduce the concept of *strain:* We define the *normal strain* in a rod under axial loading as the *deformation per unit length* of that rod. Denoting the normal strain by ϵ (Greek letter epsilon), we write

$$\epsilon = \frac{\delta}{L} \tag{2.1}$$

Plotting the stress $\sigma = P/A$ against the strain $\epsilon = \delta/L$, we obtain a curve that is characteristic of the properties of the material and does not depend upon the dimensions of the particular specimen used. This curve is called a *stress-strain diagram* and will be discussed in detail in Sec. 2.3.

Since the rod BC considered in the preceding discussion had a uniform cross section of area A, the normal stress σ could be assumed to have a constant value P/A throughout the rod. Thus, it was appropriate to define the strain ϵ as the ratio of the total deformation δ over the total length L of the rod. In the case of a member of variable cross-sectional area A, however, the normal stress $\sigma = P/A$ varies along the member, and it is necessary to define the strain at a given point Q by considering a small element of undeformed length Δx (Fig. 2.5). Denoting by $\Delta\delta$ the deformation of the element under the given loading, we define the *normal strain at point Q* as

$$\epsilon = \lim_{\Delta x \to 0} \frac{\Delta\delta}{\Delta x} = \frac{d\delta}{dx} \tag{2.2}$$

Since deformation and length are expressed in the same units, the normal strain ϵ obtained by dividing δ by L (or $d\delta$ by dx) is a *dimensionless quantity.* Thus, the same numerical value is obtained for the normal strain in a given member, whether SI metric units or U.S. customary units are used. Consider, for instance, a bar of length

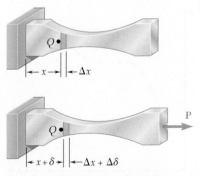

Fig. 2.5 Deformation of axially-loaded member of variable cross-sectional area.

$L = 0.600$ m and uniform cross section, which undergoes a deformation $\delta = 150 \times 10^{-6}$ m. The corresponding strain is

$$\epsilon = \frac{\delta}{L} = \frac{150 \times 10^{-6} \text{ m}}{0.600 \text{ m}} = 250 \times 10^{-6} \text{ m/m} = 250 \times 10^{-6}$$

Note that the deformation could have been expressed in micrometers: $\delta = 150 \ \mu$m. We would then have written

$$\epsilon = \frac{\delta}{L} = \frac{150 \ \mu\text{m}}{0.600 \text{ m}} = 250 \ \mu\text{m/m} = 250 \ \mu$$

and read the answer as "250 micros."

2.3 STRESS-STRAIN DIAGRAM

We saw in Sec. 2.2 that the diagram representing the relation between stress and strain in a given material is an important characteristic of the material. To obtain the stress-strain diagram of a material, one usually conducts a *tensile test* on a specimen of the material. One type of specimen commonly used is shown in Photo 2.1. The cross-sectional area of the cylindrical central portion of the specimen has been accurately determined and two gage marks have been inscribed on that portion at a distance L_0 from each other. The distance L_0 is known as the *gage length* of the specimen.

The test specimen is then placed in a testing machine (Photo 2.2), which is used to apply a centric load **P.** As the load **P** increases, the

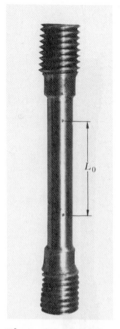

Photo 2.1 Typical tensile-test specimen.

Photo 2.2 This machine is used to test tensile-test specimens, such as those shown in this chapter.

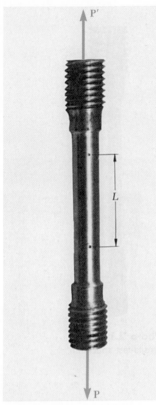

Photo 2.3 Test specimen with tensile load.

distance L between the two gage marks also increases (Photo 2.3). The distance L is measured with a dial gage, and the elongation $\delta = L - L_0$ is recorded for each value of P. A second dial gage is often used simultaneously to measure and record the change in diameter of the specimen. From each pair of readings P and δ, the stress σ is computed by dividing P by the original cross-sectional area A_0 of the specimen, and the strain ϵ by dividing the elongation δ by the original distance L_0 between the two gage marks. The stress-strain diagram may then be obtained by plotting ϵ as an abscissa and σ as an ordinate.

Stress-strain diagrams of various materials vary widely, and different tensile tests conducted on the same material may yield different results, depending upon the temperature of the specimen and the speed of loading. It is possible, however, to distinguish some common characteristics among the stress-strain diagrams of various groups of materials and to divide materials into two broad categories on the basis of these characteristics, namely, the *ductile* materials and the *brittle* materials.

Ductile materials, which comprise structural steel, as well as many alloys of other metals, are characterized by their ability to *yield* at normal temperatures. As the specimen is subjected to an increasing load, its length first increases linearly with the load and at a very slow rate. Thus, the initial portion of the stress-strain diagram is a straight line with a steep slope (Fig. 2.6). However, after a critical value σ_Y of the stress has been reached, the specimen undergoes a large deformation with a relatively small increase in the applied load. This deformation is caused by slippage of the material along oblique surfaces and is due, therefore, primarily to shearing stresses. As we can note from the stress-strain diagrams of two typical ductile materials (Fig. 2.6), the elongation of the specimen after it has started to yield can be 200 times as large as its deformation before yield. After a certain maximum value of the load has been reached, the diameter of a portion of the specimen begins to decrease, because of local instability (Photo 2.4a). This phenomenon is known as *necking*. After necking has begun, somewhat lower loads are sufficient to keep the specimen elongating further, until it finally ruptures (Photo 2.4b). We note that rupture occurs along a cone-shaped surface that forms an angle of approximately 45° with the original surface of the specimen. This indicates that shear is primarily responsible for the failure of ductile materials, and confirms the fact that, under an axial load,

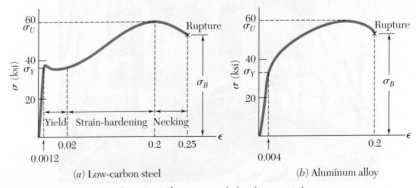

(a) Low-carbon steel   (b) Aluminum alloy

Fig. 2.6 Stress-strain diagrams of two typical ductile materials.

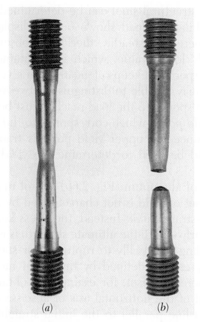

Photo 2.4 Tested specimen of a ductile material.

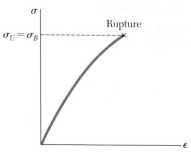

Fig. 2.7 Stress-strain diagram for a typical brittle material.

shearing stresses are largest on surfaces forming an angle of 45° with the load (cf. Sec. 1.11). The stress σ_Y at which yield is initiated is called the *yield strength* of the material, the stress σ_U corresponding to the maximum load applied to the specimen is known as the *ultimate strength,* and the stress σ_B corresponding to rupture is called the *breaking strength.*

Brittle materials, which comprise cast iron, glass, and stone, are characterized by the fact that rupture occurs without any noticeable prior change in the rate of elongation (Fig. 2.7). Thus, for brittle materials, there is no difference between the ultimate strength and the breaking strength. Also, the strain at the time of rupture is much smaller for brittle than for ductile materials. From Photo 2.5, we note the absence of any necking of the specimen in the case of a brittle material, and observe that rupture occurs along a surface perpendicular to the load. We conclude from this observation that normal stresses are primarily responsible for the failure of brittle materials.†

The stress-strain diagrams of Fig. 2.6 show that structural steel and aluminum, while both ductile, have different yield characteristics. In the case of structural steel (Fig. 2.6a), the stress remains constant over a large range of values of the strain after the onset of yield. Later the stress must be increased to keep elongating the specimen, until the maximum value σ_U has been reached. This is due to a property of the material known as strain-hardening. The

†The tensile tests described in this section were assumed to be conducted at normal temperatures. However, a material that is ductile at normal temperatures may display the characteristics of a brittle material at very low temperatures, while a normally brittle material may behave in a ductile fashion at very high temperatures. At temperatures other than normal, therefore, one should refer to *a material in a ductile state* or to *a material in a brittle state,* rather than to a ductile or brittle material.

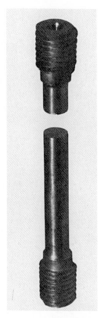

Photo 2.5 Tested specimen of a brittle material.

yield strength of structural steel can be determined during the tensile test by watching the load shown on the display of the testing machine. After increasing steadily, the load is observed to suddenly drop to a slightly lower value, which is maintained for a certain period while the specimen keeps elongating. In a very carefully conducted test, one may be able to distinguish between the *upper yield point*, which corresponds to the load reached just before yield starts, and the *lower yield point*, which corresponds to the load required to maintain yield. Since the upper yield point is transient, the lower yield point should be used to determine the yield strength of the material.

In the case of aluminum (Fig. 2.6*b*) and of many other ductile materials, the onset of yield is not characterized by a horizontal portion of the stress-strain curve. Instead, the stress keeps increasing—although not linearly—until the ultimate strength is reached. Necking then begins, leading eventually to rupture. For such materials, the yield strength σ_Y can be defined by the offset method. The yield strength at 0.2 percent offset, for example, is obtained by drawing through the point of the horizontal axis of abscissa $\epsilon = 0.2$ percent (or $\epsilon = 0.002$), a line parallel to the initial straight-line portion of the stress-strain diagram (Fig. 2.8). The stress σ_Y corresponding to the point Y obtained in this fashion is defined as the yield strength at 0.2 percent offset.

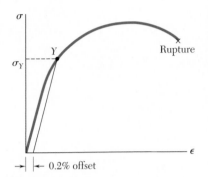

Fig. 2.8 Determination of yield strength by offset method.

A standard measure of the ductility of a material is its *percent elongation*, which is defined as

$$\text{Percent elongation} = 100 \frac{L_B - L_0}{L_0}$$

where L_0 and L_B denote, respectively, the initial length of the tensile test specimen and its final length at rupture. The specified minimum elongation for a 50-mm gage length for commonly used steels with yield strengths up to 350 MPa is 21 percent. We note that this means that the average strain at rupture should be at least 0.21.

Another measure of ductility which is sometimes used is the *percent reduction in area*, defined as

$$\text{Percent reduction in area} = 100 \frac{A_0 - A_B}{A_0}$$

where A_0 and A_B denote, respectively, the initial cross-sectional area of the specimen and its minimum cross-sectional area at rupture. For structural steel, percent reductions in area of 60 to 70 percent are common.

Thus far, we have discussed only tensile tests. If a specimen made of a ductile material were loaded in compression instead of tension, the stress-strain curve obtained would be essentially the same through its initial straight-line portion and through the beginning of the portion corresponding to yield and strain-hardening. Particularly noteworthy is the fact that for a given steel, the yield strength is the same in both tension and compression. For larger values of the strain, the tension and compression stress-strain curves diverge, and it should be noted that necking cannot occur in compression.

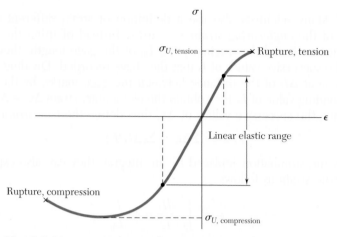

Fig. 2.9 Stress-strain diagram for concrete.

For most brittle materials, one finds that the ultimate strength in compression is much larger than the ultimate strength in tension. This is due to the presence of flaws, such as microscopic cracks or cavities, which tend to weaken the material in tension, while not appreciably affecting its resistance to compressive failure.

An example of brittle material with different properties in tension and compression is provided by *concrete*, whose stress-strain diagram is shown in Fig. 2.9. On the tension side of the diagram, we first observe a linear elastic range in which the strain is proportional to the stress. After the yield point has been reached, the strain increases faster than the stress until rupture occurs. The behavior of the material in compression is different. First, the linear elastic range is significantly larger. Second, rupture does not occur as the stress reaches its maximum value. Instead, the stress decreases in magnitude while the strain keeps increasing until rupture occurs. Note that the modulus of elasticity, which is represented by the slope of the stress-strain curve in its linear portion, is the same in tension and compression. This is true of most brittle materials.

*2.4 TRUE STRESS AND TRUE STRAIN

We recall that the stress plotted in the diagrams of Figs. 2.6 and 2.7 was obtained by dividing the load P by the cross-sectional area A_0 of the specimen measured before any deformation had taken place. Since the cross-sectional area of the specimen decreases as P increases, the stress plotted in our diagrams does not represent the actual stress in the specimen. The difference between the *engineering stress* $\sigma = P/A_0$ that we have computed and the *true stress* $\sigma_t = P/A$ obtained by dividing P by the cross-sectional area A of the deformed specimen becomes apparent in ductile materials after yield has started. While the engineering stress σ, which is directly proportional to the load P, decreases with P during the necking phase, the true stress σ_t, which is proportional to P but also inversely proportional to A, is observed to keep increasing until rupture of the specimen occurs.

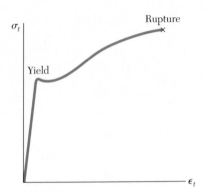

Fig. 2.10 True stress versus true strain for a typical ductile material.

Many scientists also use a definition of strain different from that of the *engineering strain* $\epsilon = \delta/L_0$. Instead of using the total elongation δ and the original value L_0 of the gage length, they use all the successive values of L that they have recorded. Dividing each increment ΔL of the distance between the gage marks, by the corresponding value of L, they obtain the elementary strain $\Delta\epsilon = \Delta L/L$. Adding the successive values of $\Delta\epsilon$, they define the *true strain* ϵ_t:

$$\epsilon_t = \Sigma\Delta\epsilon = \Sigma(\Delta L/L)$$

With the summation replaced by an integral, they can also express the true strain as follows:

$$\epsilon_t = \int_{L_0}^{L} \frac{dL}{L} = \ln\frac{L}{L_0} \tag{2.3}$$

The diagram obtained by plotting true stress versus true strain (Fig. 2.10) reflects more accurately the behavior of the material. As we have already noted, there is no decrease in true stress during the necking phase. Also, the results obtained from tensile and from compressive tests will yield essentially the same plot when true stress and true strain are used. This is not the case for large values of the strain when the engineering stress is plotted versus the engineering strain. However, engineers, whose responsibility is to determine whether a load P will produce an acceptable stress and an acceptable deformation in a given member, will want to use a diagram based on the engineering stress $\sigma = P/A_0$ and the engineering strain $\epsilon = \delta/L_0$, since these expressions involve data that are available to them, namely the cross-sectional area A_0 and the length L_0 of the member in its undeformed state.

2.5 HOOKE'S LAW; MODULUS OF ELASTICITY

Most engineering structures are designed to undergo relatively small deformations, involving only the straight-line portion of the corresponding stress-strain diagram. For that initial portion of the diagram (Fig. 2.6), the stress σ is directly proportional to the strain ϵ, and we can write

$$\sigma = E\epsilon \tag{2.4}$$

This relation is known as *Hooke's law,* after Robert Hooke (1635–1703), an English scientist and one of the early founders of applied mechanics. The coefficient E is called the *modulus of elasticity* of the material involved, or also *Young's modulus,* after the English scientist Thomas Young (1773–1829). Since the strain ϵ is a dimensionless quantity, the modulus E is expressed in the same units as the stress σ, namely in pascals or one of its multiples if SI units are used.

The largest value of the stress for which Hooke's law can be used for a given material is known as the *proportional limit* of that material. In the case of ductile materials possessing a well-defined yield point, as in Fig. 2.6a, the proportional limit almost coincides with the yield point. For other materials, the proportional limit cannot be defined as easily, since it is difficult to determine with accuracy the value of the stress σ for which the relation between σ and ϵ ceases to be linear. But from this very difficulty we can conclude for such materials that using Hooke's law for values of the stress slightly larger than the actual proportional limit will not result in any significant error.

Some of the physical properties of structural metals, such as strength, ductility, and corrosion resistance, can be greatly affected by alloying, heat treatment, and the manufacturing process used. For example, we note from the stress-strain diagrams of pure iron and of three different grades of steel (Fig. 2.11) that large variations in the yield strength, ultimate strength, and final strain (ductility) exist among these four metals. All of them, however, possess the same modulus of elasticity; in other words, their "stiffness," or ability to resist a deformation within the linear range, is the same. Therefore, if a high-strength steel is substituted for a lower-strength steel in a given structure, and if all dimensions are kept the same, the structure will have an increased load-carrying capacity, but its stiffness will remain unchanged.

For each of the materials considered so far, the relation between normal stress and normal strain, $\sigma = E\epsilon$, is independent of the direction of loading. This is because the mechanical properties of each material, including its modulus of elasticity E, are independent of the direction considered. Such materials are said to be *isotropic*. Materials whose properties depend upon the direction considered are said to be *anisotropic*.

An important class of anisotropic materials consists of *fiber-reinforced composite materials*. These composite materials are obtained by embedding fibers of a strong, stiff material into a weaker, softer material, referred to as a *matrix*. Typical materials used as fibers are graphite, glass, and polymers, while various types of resins are used as a matrix. Figure 2.12 shows a layer, or *lamina*, of a composite material consisting of a large number of parallel fibers embedded in a matrix. An axial load applied to the lamina along the x axis, that is, in a direction parallel to the fibers, will create a normal stress σ_x in the lamina and a corresponding normal strain ϵ_x which will satisfy Hooke's law as the load is increased and as long as the elastic limit of the lamina is not exceeded. Similarly, an axial load applied along the y axis, that is, in a direction perpendicular to the lamina, will create a normal stress σ_y and a normal strain ϵ_y satisfying Hooke's law, and an axial load applied along the z axis will create a normal stress σ_z and a normal strain ϵ_z which again satisfy Hooke's law. However, the moduli of elasticity E_x, E_y, and E_z corresponding, respectively, to each of the above loadings will be different. Because the fibers are parallel to the x axis, the lamina will offer a much stronger resistance to a loading directed along the x axis than to a loading directed along the y or z axis, and E_x will be much larger than either E_y or E_z.

A flat *laminate* is obtained by superposing a number of layers or laminas. If the laminate is to be subjected only to an axial load causing tension, the fibers in all layers should have the same orientation as the load in order to obtain the greatest possible strength. But if the laminate may be in compression, the matrix material may not be sufficiently strong to prevent the fibers from kinking or buckling. The lateral stability of the laminate may then be increased by positioning

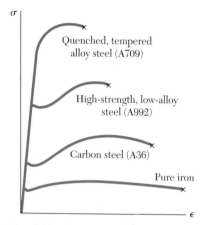

Fig. 2.11   Stress-strain diagrams for iron and different grades of steel.

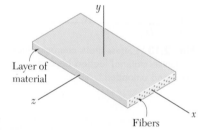

Fig. 2.12   Layer of fiber-reinforced composite material.

some of the layers so that their fibers will be perpendicular to the load. Positioning some layers so that their fibers are oriented at 30°, 45°, or 60° to the load may also be used to increase the resistance of the laminate to in-plane shear. Fiber-reinforced composite materials will be further discussed in Sec. 2.16, where their behavior under multiaxial loadings will be considered.

2.6 ELASTIC VERSUS PLASTIC BEHAVIOR OF A MATERIAL

If the strains caused in a test specimen by the application of a given load disappear when the load is removed, the material is said to behave *elastically*. The largest value of the stress for which the material behaves elastically is called the *elastic limit* of the material.

If the material has a well-defined yield point as in Fig. 2.6*a*, the elastic limit, the proportional limit (Sec. 2.5), and the yield point are essentially equal. In other words, the material behaves elastically and linearly as long as the stress is kept below the yield point. If the yield point is reached, however, yield takes place as described in Sec. 2.3 and, when the load is removed, the stress and strain decrease in a linear fashion, along a line *CD* parallel to the straight-line portion *AB* of the loading curve (Fig. 2.13). The fact that ϵ does not return to zero after the load has been removed indicates that a *permanent set* or *plastic deformation* of the material has taken place. For most materials, the plastic deformation depends not only upon the maximum value reached by the stress, but also upon the time elapsed before the load is removed. The stress-dependent part of the plastic deformation is referred to as *slip*, and the time-dependent part—which is also influenced by the temperature—as *creep*.

When a material does not possess a well-defined yield point, the elastic limit cannot be determined with precision. However, assuming the elastic limit equal to the yield strength as defined by the offset method (Sec. 2.3) results in only a small error. Indeed, referring to Fig. 2.8, we note that the straight line used to determine point *Y* also represents the unloading curve after a maximum stress σ_Y has been reached. While the material does not behave truly elastically, the resulting plastic strain is as small as the selected offset.

If, after being loaded and unloaded (Fig. 2.14), the test specimen is loaded again, the new loading curve will closely follow the earlier unloading curve until it almost reaches point *C*; it will then bend to the right and connect with the curved portion of the original stress-strain diagram. We note that the straight-line portion of the new loading curve is longer than the corresponding portion of the initial one. Thus, the proportional limit and the elastic limit have increased as a result of the strain-hardening that occurred during the earlier loading of the specimen. However, since the point of rupture *R* remains unchanged, the ductility of the specimen, which should now be measured from point *D*, has decreased.

We have assumed in our discussion that the specimen was loaded twice in the same direction, i.e., that both loads were tensile loads. Let us now consider the case when the second load is applied in a direction opposite to that of the first one. We assume that the

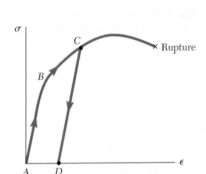

Fig. 2.13 Stress-strain characteristics of ductile material loaded beyond yield and unloaded.

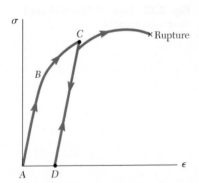

Fig. 2.14 Stress-strain characteristics of ductile material reloaded after prior yielding.

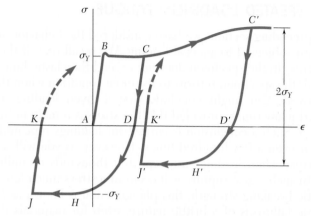

Fig. 2.15 Stress-strain characteristics for mild steel subjected to reverse loading.

material is mild steel, for which the yield strength is the same in tension and in compression. The initial load is tensile and is applied until point C has been reached on the stress-strain diagram (Fig. 2.15). After unloading (point D), a compressive load is applied, causing the material to reach point H, where the stress is equal to $-\sigma_Y$. We note that portion DH of the stress-strain diagram is curved and does not show any clearly defined yield point. This is referred to as the *Bauschinger effect.* As the compressive load is maintained, the material yields along line HJ.

If the load is removed after point J has been reached, the stress returns to zero along line JK, and we note that the slope of JK is equal to the modulus of elasticity E. The resulting permanent set AK may be positive, negative, or zero, depending upon the lengths of the segments BC and HJ. If a tensile load is applied again to the test specimen, the portion of the stress-strain diagram beginning at K (dashed line) will curve up and to the right until the yield stress σ_Y has been reached.

If the initial loading is large enough to cause strain-hardening of the material (point C'), unloading takes place along line $C'D'$. As the reverse load is applied, the stress becomes compressive, reaching its maximum value at H' and maintaining it as the material yields along line $H'J'$. We note that while the maximum value of the compressive stress is less than σ_Y, the total change in stress between C' and H' is still equal to $2\sigma_Y$.

If point K or K' coincides with the origin A of the diagram, the permanent set is equal to zero, and the specimen may appear to have returned to its original condition. However, internal changes will have taken place and, while the same loading sequence may be repeated, the specimen will rupture without any warning after relatively few repetitions. This indicates that the excessive plastic deformations to which the specimen was subjected have caused a radical change in the characteristics of the material. Reverse loadings into the plastic range, therefore, are seldom allowed, and only under carefully controlled conditions. Such situations occur in the straightening of damaged material and in the final alignment of a structure or machine.

2.7 REPEATED LOADINGS; FATIGUE

In the preceding sections we have considered the behavior of a test specimen subjected to an axial loading. We recall that, if the maximum stress in the specimen does not exceed the elastic limit of the material, the specimen returns to its initial condition when the load is removed. You might conclude that a given loading may be repeated many times, provided that the stresses remain in the elastic range. Such a conclusion is correct for loadings repeated a few dozen or even a few hundred times. However, as you will see, it is not correct when loadings are repeated thousands or millions of times. In such cases, rupture will occur at a stress much lower than the static breaking strength; this phenomenon is known as *fatigue*. A fatigue failure is of a brittle nature, even for materials that are normally ductile.

Fatigue must be considered in the design of all structural and machine components that are subjected to repeated or to fluctuating loads. The number of loading cycles that may be expected during the useful life of a component varies greatly. For example, a beam supporting an industrial crane may be loaded as many as two million times in 25 years (about 300 loadings per working day), an automobile crankshaft will be loaded about half a billion times if the automobile is driven 200,000 miles, and an individual turbine blade may be loaded several hundred billion times during its lifetime.

Some loadings are of a fluctuating nature. For example, the passage of traffic over a bridge will cause stress levels that will fluctuate about the stress level due to the weight of the bridge. A more severe condition occurs when a complete reversal of the load occurs during the loading cycle. The stresses in the axle of a railroad car, for example, are completely reversed after each half-revolution of the wheel.

The number of loading cycles required to cause the failure of a specimen through repeated successive loadings and reverse loadings may be determined experimentally for any given maximum stress level. If a series of tests is conducted, using different maximum stress levels, the resulting data may be plotted as a σ-n curve. For each test, the maximum stress σ is plotted as an ordinate and the number of cycles n as an abscissa; because of the large number of cycles required for rupture, the cycles n are plotted on a logarithmic scale.

A typical σ-n curve for steel is shown in Fig. 2.16. We note that, if the applied maximum stress is high, relatively few cycles are required to cause rupture. As the magnitude of the maximum stress is reduced, the number of cycles required to cause rupture increases, until a stress, known as the *endurance limit*, is reached. The endurance limit is the stress for which failure does not occur, even for an indefinitely large number of loading cycles. For a low-carbon steel, such as structural steel, the endurance limit is about one-half of the ultimate strength of the steel.

For nonferrous metals, such as aluminum and copper, a typical σ-n curve (Fig. 2.16) shows that the stress at failure continues to

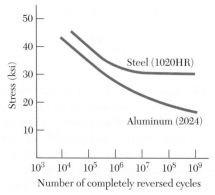

Fig. 2.16 Typical σ-n curves.

decrease as the number of loading cycles is increased. For such metals, one defines the *fatigue limit* as the stress corresponding to failure after a specified number of loading cycles, such as 500 million.

Examination of test specimens, of shafts, of springs, and of other components that have failed in fatigue shows that the failure was initiated at a microscopic crack or at some similar imperfection. At each loading, the crack was very slightly enlarged. During successive loading cycles, the crack propagated through the material until the amount of undamaged material was insufficient to carry the maximum load, and an abrupt, brittle failure occurred. Because fatigue failure may be initiated at any crack or imperfection, the surface condition of a specimen has an important effect on the value of the endurance limit obtained in testing. The endurance limit for machined and polished specimens is higher than for rolled or forged components, or for components that are corroded. In applications in or near seawater, or in other applications where corrosion is expected, a reduction of up to 50 percent in the endurance limit can be expected.

2.8 DEFORMATIONS OF MEMBERS UNDER AXIAL LOADING

Consider a homogeneous rod BC of length L and uniform cross section of area A subjected to a centric axial load $\mathbf{P}$ (Fig. 2.17). If the resulting axial stress $\sigma = P/A$ does not exceed the proportional limit of the material, we may apply Hooke's law and write

$$\sigma = E\epsilon \tag{2.4}$$

from which it follows that

$$\epsilon = \frac{\sigma}{E} = \frac{P}{AE} \tag{2.5}$$

Recalling that the strain ϵ was defined in Sec. 2.2 as $\epsilon = \delta/L$, we have

$$\delta = \epsilon L \tag{2.6}$$

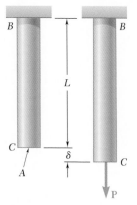

Fig. 2.17 Deformation of axially loaded rod.

and, substituting for ϵ from (2.5) into (2.6):

$$\delta = \frac{PL}{AE} \tag{2.7}$$

Equation (2.7) may be used only if the rod is homogeneous (constant E), has a uniform cross section of area A, and is loaded at its ends. If the rod is loaded at other points, or if it consists of several portions of various cross sections and possibly of different materials, we must divide it into component parts that satisfy individually the required conditions for the application of formula (2.7). Denoting, respectively, by P_i, L_i, A_i, and E_i the internal force, length, cross-sectional area, and modulus of elasticity corresponding to part i, we express the deformation of the entire rod as

$$\delta = \sum_i \frac{P_i L_i}{A_i E_i} \tag{2.8}$$

We recall from Sec. 2.2 that, in the case of a member of variable cross section (Fig. 2.18), the strain ϵ depends upon the position of the point Q where it is computed and is defined as $\epsilon = d\delta/dx$. Solving for $d\delta$ and substituting for ϵ from Eq. (2.5), we express the deformation of an element of length dx as

$$d\delta = \epsilon \, dx = \frac{P \, dx}{AE}$$

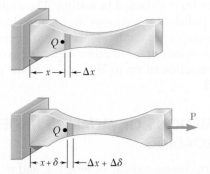

Fig. 2.18 Deformation of axially loaded member of variable cross-sectional area.

The total deformation δ of the member is obtained by integrating this expression over the length L of the member:

$$\delta = \int_0^L \frac{P \, dx}{AE} \tag{2.9}$$

Formula (2.9) should be used in place of (2.7), not only when the cross-sectional area A is a function of x, but also when the internal force P depends upon x, as is the case for a rod hanging under its own weight.

Determine the deformation of the steel rod shown in Fig. 2.19a under the given loads ($E = 200$ GPa).

EXAMPLE 2.01

We divide the rod into three component parts shown in Fig. 2.19b and write

$$L_1 = L_2 = 300 \text{ mm} \qquad L_3 = 400 \text{ mm}$$
$$A_1 = A_2 = 580 \text{ mm}^2 \qquad A_3 = 190 \text{ mm}^2$$

To find the internal forces P_1, P_2, and P_3, we must pass sections through each of the component parts, drawing each time the free-body diagram of the portion of rod located to the right of the section (Fig. 2.19c). Expressing that each of the free bodies is in equilibrium, we obtain successively

$$P_1 = 240 \text{ kN} = 240 \times 10^3 \text{ N}$$
$$P_2 = -60 \text{ kN} = -60 \times 10^3 \text{ N}$$
$$P_3 = 120 \text{ kN} = 120 \times 10^3 \text{ N}$$

Carrying the values obtained into Eq. (2.8), we have

$$\delta = \sum_i \frac{P_i L_i}{A_i E_i} = \frac{1}{E}\left(\frac{P_1 L_1}{A_1} + \frac{P_2 L_2}{A_2} + \frac{P_3 L_3}{A_3} \right)$$

$$= \frac{1}{200 \times 10^9}\left[\frac{(240 \times 10^3)(0.3)}{580 \times 10^{-6}} \right.$$
$$\left. + \frac{(-60 \times 10^3)(0.3)}{580 \times 10^{-6}} + \frac{(120 \times 10^3)(0.4)}{190 \times 10^{-6}} \right]$$

$$\delta = \frac{0.3457 \times 10^9}{200 \times 10^9} = 1.729 \times 10^{-3} \text{ m} = 1.729 \text{ mm}$$

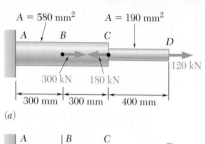

(a)

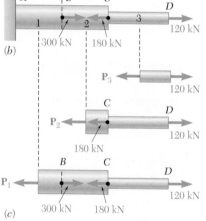

(b)

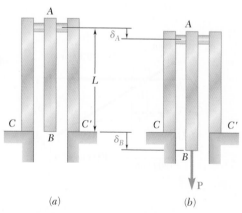

(c)

Fig. 2.19

The rod *BC* of Fig. 2.17, which was used to derive formula (2.7), and the rod *AD* of Fig. 2.19, discussed in Example 2.01, both had one end attached to a fixed support. In each case, therefore, the deformation δ of the rod was equal to the displacement of its free end. When both ends of a rod move, however, the deformation of the rod is measured by the *relative displacement* of one end of the rod with respect to the other. Consider, for instance, the assembly shown in Fig. 2.20a, which consists of three elastic bars of length L connected by a rigid pin at A. If a load **P** is applied at B (Fig. 2.20b), each of the three bars will deform. Since the bars AC and AC′ are attached to fixed supports at C and C′, their common deformation is measured by the displacement δ_A of point A. On the other hand, since both ends of bar AB move, the deformation of AB is measured by the difference between the displacements δ_A and δ_B of points A and B, i.e., by the relative displacement of B with respect to A. Denoting this relative displacement by $\delta_{B/A}$, we write

$$\delta_{B/A} = \delta_B - \delta_A = \frac{PL}{AE} \qquad (2.10)$$

where A is the cross-sectional area of AB and E is its modulus of elasticity.

Fig. 2.20 Example of relative end displacement, as exhibited by the middle bar.

89

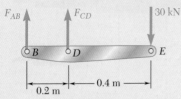

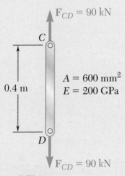

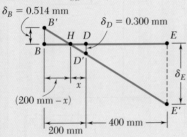

SAMPLE PROBLEM 2.1

The rigid bar BDE is supported by two links AB and CD. Link AB is made of aluminum ($E = 70$ GPa) and has a cross-sectional area of 500 mm²; link CD is made of steel ($E = 200$ GPa) and has a cross-sectional area of 600 mm². For the 30-kN force shown, determine the deflection (a) of B, (b) of D, (c) of E.

SOLUTION

Free Body: Bar BDE

$$+\gamma\Sigma M_B = 0: \qquad -(30\text{ kN})(0.6\text{ m}) + F_{CD}(0.2\text{ m}) = 0$$

$$F_{CD} = +90\text{ kN} \qquad F_{CD} = 90\text{ kN} \quad \textit{tension}$$

$$+\gamma\Sigma M_D = 0: \qquad -(30\text{ kN})(0.4\text{ m}) - F_{AB}(0.2\text{ m}) = 0$$

$$F_{AB} = -60\text{ kN} \qquad F_{AB} = 60\text{ kN} \quad \textit{compression}$$

a. Deflection of B. Since the internal force in link AB is compressive, we have $P = -60$ kN

$$\delta_B = \frac{PL}{AE} = \frac{(-60 \times 10^3\text{ N})(0.3\text{ m})}{(500 \times 10^{-6}\text{ m}^2)(70 \times 10^9\text{ Pa})} = -514 \times 10^{-6}\text{ m}$$

The negative sign indicates a contraction of member AB, and, thus, an upward deflection of end B:

$$\delta_B = 0.514\text{ mm} \uparrow \quad \blacktriangleleft$$

b. Deflection of D. Since in rod CD, $P = 90$ kN, we write

$$\delta_D = \frac{PL}{AE} = \frac{(90 \times 10^3\text{ N})(0.4\text{ m})}{(600 \times 10^{-6}\text{ m}^2)(200 \times 10^9\text{ Pa})}$$

$$= 300 \times 10^{-6}\text{ m} \qquad \delta_D = 0.300\text{ mm} \downarrow \quad \blacktriangleleft$$

c. Deflection of E. We denote by B' and D' the displaced positions of points B and D. Since the bar BDE is rigid, points B', D', and E' lie in a straight line and we write

$$\frac{BB'}{DD'} = \frac{BH}{HD} \qquad \frac{0.514\text{ mm}}{0.300\text{ mm}} = \frac{(200\text{ mm}) - x}{x} \qquad x = 73.7\text{ mm}$$

$$\frac{EE'}{DD'} = \frac{HE}{HD} \qquad \frac{\delta_E}{0.300\text{ mm}} = \frac{(400\text{ mm}) + (73.7\text{ mm})}{73.7\text{ mm}}$$

$$\delta_E = 1.928\text{ mm} \downarrow \quad \blacktriangleleft$$

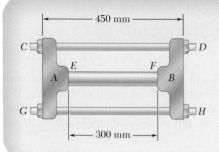

450 mm

300 mm

The rigid castings A and B are connected by two 18-mm-diameter steel bolts CD and GH and are in contact with the ends of a 38-mm-diameter aluminum rod EF. Each bolt is single-threaded with a pitch of 2.5 mm, and after being snugly fitted, the nuts at D and H are both tightened one-quarter of a turn. Knowing that E is 200 GPa for steel and 70 GPa for aluminum, determine the normal stress in the rod.

SOLUTION

Because the millimeter will be used as the unit for length, E will be expressed in Newtons per square millimeter (N/mm²). Note that $1 \text{ GPa} = 10^3 \text{ N/mm}^2$.

Deformations

Bolts CD and GH. Tightening the nuts causes tension in the bolts. Because of symmetry, both are subjected to the same internal force P_b and undergo the same deformation δ_b. We have

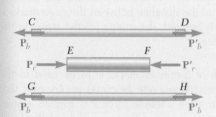

$$\delta_b = +\frac{P_b L_b}{A_b E_b} = +\frac{P_b(450 \text{ mm})}{\frac{1}{4}\pi(18 \text{ mm})^2(200 \times 10^3 \text{ N/mm}^2)} = +8.842 \times 10^{-6} P_b \quad (1)$$

Rod EF. The rod is in compression. Denoting by P_r the magnitude of the force in the rod and by δ_r the deformation of the rod, we write

$$\delta_r = -\frac{P_r L_r}{A_r E_r} = -\frac{P_r(300 \text{ mm})}{\frac{1}{4}\pi(38 \text{ mm})^2(70 \times 10^3 \text{ N/mm}^2)} = -3.779 \times 10^{-6} P_r \quad (2)$$

Displacement of D Relative to B. Tightening the nuts one-quarter of a turn causes ends D and H of the bolts to undergo a displacement of $\frac{1}{4}(2.5 \text{ mm})$ *relative to casting B*. Considering end D, we write

$$\delta_{D/B} = \tfrac{1}{4}(2.5 \text{ mm}) = 0.625 \text{ mm} \quad (3)$$

But $\delta_{D/B} = \delta_D - \delta_B$, where δ_D and δ_B represent the displacements of D and B. If we assume that casting A is held in a fixed position while the nuts at D and H are being tightened, these displacements are equal to the deformations of the bolts and of the rod, respectively. We have, therefore,

$$\delta_{D/B} = \delta_b - \delta_r \quad (4)$$

Substituting from (1), (2), and (3) into (4), we obtain

$$0.625 \text{ mm} = 8.842 \times 10^{-6} P_b + 3.779 \times 10^{-6} P_r \quad (5)$$

Free Body: Casting B

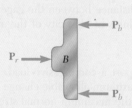

$$\xrightarrow{+} \Sigma F = 0: \qquad P_r - 2P_b = 0 \qquad P_r = 2P_b \quad (6)$$

Forces in Bolts and Rod Substituting for P_r from (6) into (5), we have

$$0.625 \text{ mm} = 8.842 \times 10^{-6} P_b + 3.779 \times 10^{-6}(2P_b)$$

$$P_b = 38.1 \times 10^3 \text{ N} = 38.1 \text{ kN}$$

$$P_r = 2P_b = 2(38.1 \text{ kN}) = 76.2 \text{ kN}$$

Stress in Rod

$$\sigma_r = \frac{P_r}{A_r} = \frac{76.2 \text{ kN}}{\frac{1}{4}\pi(38 \text{ mm})^2} \qquad \sigma_r = 67.2 \text{ MPa} \blacktriangleleft$$

PROBLEMS

2.1 An 80-m-long wire of 5-mm diameter is made of a steel with $E = 200$ GPa and an ultimate tensile strength of 400 MPa. If a factor of safety of 3.2 is desired, determine (a) the largest allowable tension in the wire, (b) the corresponding elongation of the wire.

2.2 A control rod made of yellow brass must not stretch more than 3 mm when the tension in the wire is 4 kN. Knowing that $E = 105$ GPa and that the maximum allowable normal stress is 180 MPa, determine (a) the smallest diameter that can be selected for the rod, (b) the corresponding maximum length of the rod.

2.3 Two gage marks are placed exactly 250 mm apart on a 12-mm-diameter aluminum rod with $E = 73$ GPa and an ultimate strength of 140 MPa. Knowing that the distance between the gage marks is 250.28 mm after a load is applied, determine (a) the stress in the rod, (b) the factor of safety.

2.4 An 18-m-long steel wire of 5-mm diameter is to be used in the manufacture of a prestressed concrete beam. It is observed that the wire stretches 45 mm when a tensile force **P** is applied. Knowing that $E = 200$ GPa, determine (a) the magnitude of the force **P**, (b) the corresponding normal stress in the wire.

2.5 A polystyrene rod of length 300 mm and diameter 12 mm is subjected to a 3-kN tensile load. Knowing that $E = 3.1$ GPa, determine (a) the elongation of the rod, (b) the normal stress in the rod.

2.6 A nylon thread is subjected to a 8.5-N tension force. Knowing that $E = 3.3$ GPa and that the length of the thread increases by 1.1 percent, determine (a) the diameter of the thread, (b) the stress in the thread.

2.7 Two gage marks are placed exactly 250 mm apart on a 12-mm-diameter aluminum rod. Knowing that, with an axial load of 6000 N acting on the rod, the distance between the gage marks is 250.18 mm, determine the modulus of elasticity of the aluminum used in the rod.

2.8 A cast-iron tube is used to support a compressive load. Knowing that $E = 69$ GPa and that the maximum allowable change in length is 0.025 percent, determine (a) the maximum normal stress in the tube, (b) the minimum wall thickness for a load of 7.2 kN if the outside diameter of the tube is 50 mm.

2.9 An aluminum control rod must stretch 2 mm when a 2-kN tensile load is applied to it. Knowing that $\sigma_{\text{all}} = 154$ MPa and $E = 70$ GPa, determine the smallest diameter and shortest length which may be selected for the rod.

2.10 A square yellow-brass bar must not stretch more than 2.5 mm when it is subjected to a tensile load. Knowing that $E = 105$ GPa and that the allowable tensile strength is 180 MPa, determine (a) the maximum allowable length of the bar, (b) the required dimensions of the cross section if the tensile load is 40 kN.

2.11 A 4-m-long steel rod must not stretch more than 3 mm and the normal stress must not exceed 150 MPa when the rod is subjected to a 10-kN axial load. Knowing that $E = 200$ GPa, determine the required diameter of the rod.

2.12 A nylon thread is to be subjected to a 10-N tension. Knowing that $E = 3.2$ GPa, that the maximum allowable normal stress is 40 MPa, and that the length of the thread must not increase by more than 1 percent, determine the required diameter of the thread.

2.13 The 4-mm-diameter cable BC is made of a steel with $E = 200$ GPa. Knowing that the maximum stress in the cable must not exceed 190 MPa and that the elongation of the cable must not exceed 6 mm, find the maximum load **P** that can be applied as shown.

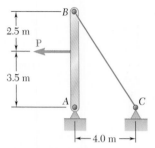

Fig. P2.13

2.14 Rod BD is made of steel ($E = 200$ GPa) and is used to brace the axially compressed member ABC. The maximum force that can be developed in member BD is 0.02P. If the stress must not exceed 126 MPa and the maximum change in length of BD must not exceed 0.001 times the length of ABC, determine the smallest-diameter rod that can be used for member BD.

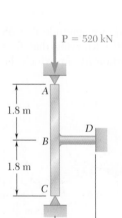

Fig. P2.14

2.15 A 1.2-m section of aluminum pipe of cross-sectional area 1100 mm² rests on a fixed support at A. The 15-mm-diameter steel rod BC hangs from a rigid bar that rests on the top of the pipe at B. Knowing that the modulus of elasticity is 200 GPa for steel and 72 GPa for aluminum, determine the deflection of point C when a 60-kN force is applied at C.

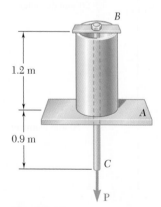

Fig. P2.15

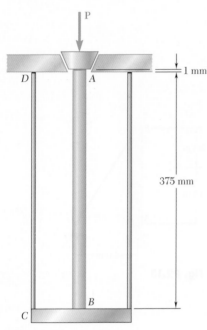

Fig. P2.16

2.16 The brass tube AB ($E = 105$ GPa) has a cross-sectional area of 140 mm^2 and is fitted with a plug at A. The tube is attached at B to a rigid plate that is itself attached at C to the bottom of an aluminum cylinder ($E = 72$ GPa) with a cross-sectional area of 250 mm^2. The cylinder is then hung from a support at D. In order to close the cylinder, the plug must move down through 1 mm. Determine the force **P** that must be applied to the cylinder.

2.17 A 250-mm-long aluminum tube ($E = 70$ GPa) of 36-mm outer diameter and 28-mm inner diameter can be closed at both ends by means of single-threaded screw-on covers of 1.5-mm pitch. With one cover screwed on tight, a solid brass rod ($E = 105$ GPa) of 25-mm diameter is placed inside the tube and the second cover is screwed on. Since the rod is slightly longer than the tube, it is observed that the cover must be forced against the rod by rotating it one-quarter of a turn before it can be tightly closed. Determine (a) the average normal stress in the tube and in the rod, (b) the deformations of the tube and of the rod.

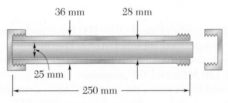

Fig. P2.17

2.18 A single axial load of magnitude $P = 58$ kN is applied at end C of the brass rod ABC. Knowing that $E = 105$ GPa, determine the diameter d of portion BC for which the deflection of point C will be 3 mm.

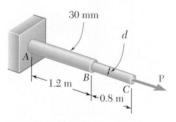

Fig. P2.18

2.19 Both portions of the rod ABC are made of an aluminum for which $E = 70$ GPa. Knowing that the magnitude of **P** is 4 kN, determine (a) the value of **Q** so that the deflection at A is zero, (b) the corresponding deflection of B.

2.20 The rod ABC is made of an aluminum for which $E = 70$ GPa. Knowing that $P = 6$ kN and $Q = 42$ kN, determine the deflection of (a) point A, (b) point B.

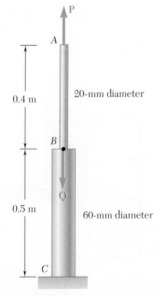

Fig. P2.19 and P2.20

2.21 Members *AB* and *BC* are made of steel (*E* = 200 GPa) with cross-sectional areas of 516 mm² and 412 mm², respectively. For the loading shown, determine the elongation of (*a*) member *AB*, (*b*) member *BC*.

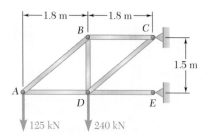

Fig. P2.21

2.22 The steel frame (*E* = 200 GPa) shown has a diagonal brace *BD* with an area of 1920 mm². Determine the largest allowable load **P** if the change in length of member *BD* is not to exceed 1.6 mm.

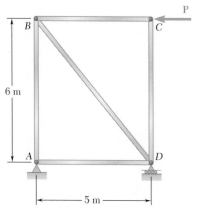

Fig. P2.22

2.23 For the steel truss (*E* = 200 GPa) and loading shown, determine the deformations of members *AB* and *AD*, knowing that their cross-sectional areas are 2400 mm² and 1800 mm², respectively.

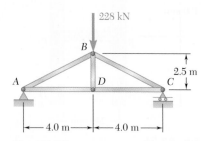

Fig. P2.23

2.24 Members *AB* and *CD* are 30-mm-diameter steel rods, and members *BC* and *AD* are 22-mm-diameter steel rods. When the turnbuckle is tightened, the diagonal member *AC* is put in tension. Knowing that *E* = 200 GPa and *h* = 1.2 m, determine the largest allowable tension in *AC* so that the deformations in members *AB* and *CD* do not exceed 1 mm.

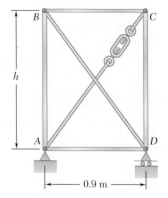

Fig. P2.24

2.25 Each of the links AB and CD is made of aluminum ($E = 75$ GPa) and has a cross-sectional area of 258 mm². Knowing that they support the rigid member BC, determine the deflection of point E.

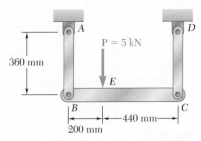

Fig. P2.25

2.26 Members ABC and DEF are joined with steel links ($E = 200$ GPa). Each of the links is made of a pair of 25 × 35-mm plates. Determine the change in length of (a) member BE, (b) member CF.

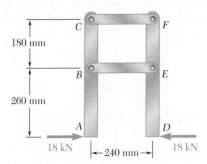

Fig. P2.26

2.27 Link BD is made of brass ($E = 105$ GPa) and has a cross-sectional area of 240 mm². Link CE is made of aluminum ($E = 72$ GPa) and has a cross-sectional area of 300 mm². Knowing that they support rigid member ABC, determine the maximum force **P** that can be applied vertically at point A if the deflection of A is not to exceed 0.35 mm.

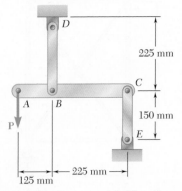

Fig. P2.27

2.28 Each of the four vertical links connecting the two rigid horizontal members is made of aluminum ($E = 70$ GPa) and has a uniform rectangular cross section of 10×40 mm. For the loading shown, determine the deflection of (a) point E, (b) point F, (c) point G.

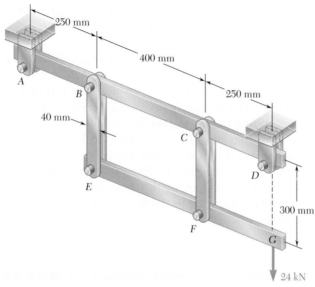

Fig. P2.28

2.29 Determine the deflection of the apex A of a homogenous circular cone of height h, density ρ, and modulus of elasticity E, due to its own weight.

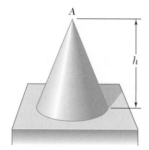

Fig. P2.29

2.30 A homogenous cable of length L and uniform cross section is suspended from one end. (a) Denoting by ρ the density (mass per unit volume) of the cable and by E its modulus of elasticity, determine the elongation of the cable due to its own weight. (b) Show that the same elongation would be obtained if the cable were horizontal and if a force equal to half of its weight were applied at each end.

2.31 The volume of a tensile specimen is essentially constant while plastic deformation occurs. If the initial diameter of the specimen is d_1, show that when the diameter is d, the true strain is $\epsilon_t = 2 \ln(d_1/d)$.

2.32 Denoting by ϵ the "engineering strain" in a tensile specimen, show that the true strain is $\epsilon_t = \ln(1 + \epsilon)$.

2.9   STATICALLY INDETERMINATE PROBLEMS

In the problems considered in the preceding section, we could always use free-body diagrams and equilibrium equations to determine the internal forces produced in the various portions of a member under given loading conditions. The values obtained for the internal forces were then entered into Eq. (2.8) or (2.9) to obtain the deformation of the member.

There are many problems, however, in which the internal forces cannot be determined from statics alone. In fact, in most of these problems the reactions themselves—which are external forces— cannot be determined by simply drawing a free-body diagram of the member and writing the corresponding equilibrium equations. The equilibrium equations must be complemented by relations involving deformations obtained by considering the geometry of the problem. Because statics is not sufficient to determine either the reactions or the internal forces, problems of this type are said to be *statically indeterminate*. The following examples will show how to handle this type of problem.

EXAMPLE 2.02

A rod of length L, cross-sectional area A_1, and modulus of elasticity E_1, has been placed inside a tube of the same length L, but of cross-sectional area A_2 and modulus of elasticity E_2 (Fig. 2.21a). What is the deformation of the rod and tube when a force **P** is exerted on a rigid end plate as shown?

Denoting by P_1 and P_2, respectively, the axial forces in the rod and in the tube, we draw free-body diagrams of all three elements (Fig. 2.21b, c, d). Only the last of the diagrams yields any significant information, namely:

$$P_1 + P_2 = P \tag{2.11}$$

Clearly, one equation is not sufficient to determine the two unknown internal forces P_1 and P_2. The problem is statically indeterminate.

However, the geometry of the problem shows that the deformations δ_1 and δ_2 of the rod and tube must be equal. Recalling Eq. (2.7), we write

$$\delta_1 = \frac{P_1 L}{A_1 E_1} \qquad \delta_2 = \frac{P_2 L}{A_2 E_2} \tag{2.12}$$

Equating the deformations δ_1 and δ_2, we obtain:

$$\frac{P_1}{A_1 E_1} = \frac{P_2}{A_2 E_2} \tag{2.13}$$

Equations (2.11) and (2.13) can be solved simultaneously for P_1 and P_2:

$$P_1 = \frac{A_1 E_1 P}{A_1 E_1 + A_2 E_2} \qquad P_2 = \frac{A_2 E_2 P}{A_1 E_1 + A_2 E_2}$$

Either of Eqs. (2.12) can then be used to determine the common deformation of the rod and tube.

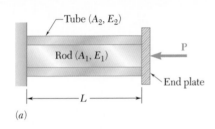

Tube (A_2, E_2)
Rod (A_1, E_1)
P
End plate
L

(a)

P_1 P_1'

(b)

P_2 P_2'

(c)

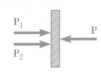

P_1
P
P_2

(d)

Fig. 2.21

EXAMPLE 2.03

A bar AB of length L and uniform cross section is attached to rigid supports at A and B before being loaded. What are the stresses in portions AC and BC due to the application of a load P at point C (Fig. 2.22a)?

Drawing the free-body diagram of the bar (Fig. 2.22b), we obtain the equilibrium equation

$$R_A + R_B = P \tag{2.14}$$

Since this equation is not sufficient to determine the two unknown reactions R_A and R_B, the problem is statically indeterminate.

However, the reactions may be determined if we observe from the geometry that the total elongation δ of the bar must be zero. Denoting by δ_1 and δ_2, respectively, the elongations of the portions AC and BC, we write

$$\delta = \delta_1 + \delta_2 = 0$$

or, expressing δ_1 and δ_2 in terms of the corresponding internal forces P_1 and P_2:

$$\delta = \frac{P_1 L_1}{AE} + \frac{P_2 L_2}{AE} = 0 \tag{2.15}$$

But we note from the free-body diagrams shown respectively in parts b and c of Fig. 2.23 that $P_1 = R_A$ and $P_2 = -R_B$. Carrying these values into (2.15), we write

$$R_A L_1 - R_B L_2 = 0 \tag{2.16}$$

Equations (2.14) and (2.16) can be solved simultaneously for R_A and R_B; we obtain $R_A = PL_2/L$ and $R_B = PL_1/L$. The desired stresses σ_1 in AC and σ_2 in BC are obtained by dividing, respectively, $P_1 = R_A$ and $P_2 = -R_B$ by the cross-sectional area of the bar:

$$\sigma_1 = \frac{PL_2}{AL} \qquad \sigma_2 = -\frac{PL_1}{AL}$$

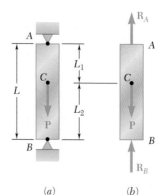

Fig. 2.22

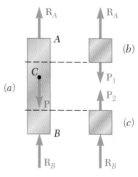

Fig. 2.23

Superposition Method. We observe that a structure is statically indeterminate whenever it is held by more supports than are required to maintain its equilibrium. This results in more unknown reactions than available equilibrium equations. It is often found convenient to designate one of the reactions as *redundant* and to eliminate the corresponding support. Since the stated conditions of the problem cannot be arbitrarily changed, the redundant reaction must be maintained in the solution. But it will be treated as an *unknown load* that, together with the other loads, must produce deformations that are compatible with the original constraints. The actual solution of the problem is carried out by considering separately the deformations caused by the given loads and by the redundant reaction, and by adding—or *superposing*—the results obtained.†

†The general conditions under which the combined effect of several loads can be obtained in this way are discussed in Sec. 2.12.

EXAMPLE 2.04

Determine the reactions at A and B for the steel bar and loading shown in Fig. 2.24, assuming a close fit at both supports before the loads are applied.

We consider the reaction at B as redundant and release the bar from that support. The reaction $\mathbf{R}_B$ is now considered as an unknown load (Fig. 2.25a) and will be determined from the condition that the deformation δ of the rod must be equal to zero. The solution is carried out by considering separately the deformation δ_L caused by the given loads (Fig. 2.25b) and the deformation δ_R due to the redundant reaction $\mathbf{R}_B$ (Fig. 2.25c).

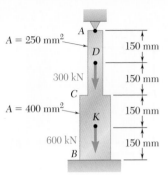

Fig. 2.24

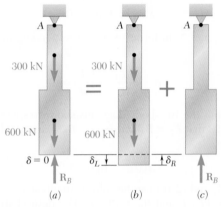

Fig. 2.25

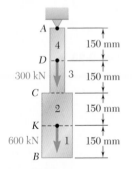

Fig. 2.26

The deformation δ_L is obtained from Eq. (2.8) after the bar has been divided into four portions, as shown in Fig. 2.26. Following the same procedure as in Example 2.01, we write

$$P_1 = 0 \qquad P_2 = P_3 = 600 \times 10^3 \text{ N} \qquad P_4 = 900 \times 10^3 \text{ N}$$
$$A_1 = A_2 = 400 \times 10^{-6} \text{ m}^2 \qquad A_3 = A_4 = 250 \times 10^{-6} \text{ m}^2$$
$$L_1 = L_2 = L_3 = L_4 = 0.150 \text{ m}$$

Substituting these values into Eq. (2.8), we obtain

$$\delta_L = \sum_{i=1}^{4} \frac{P_i L_i}{A_i E} = \left(0 + \frac{600 \times 10^3 \text{ N}}{400 \times 10^{-6} \text{ m}^2} \right.$$
$$\left. + \frac{600 \times 10^3 \text{ N}}{250 \times 10^{-6} \text{ m}^2} + \frac{900 \times 10^3 \text{ N}}{250 \times 10^{-6} \text{ m}^2} \right) \frac{0.150 \text{ m}}{E}$$
$$\delta_L = \frac{1.125 \times 10^9}{E} \tag{2.17}$$

Considering now the deformation δ_R due to the redundant reaction R_B, we divide the bar into two portions, as shown in Fig. 2.27, and write

$$P_1 = P_2 = -R_B$$
$$A_1 = 400 \times 10^{-6} \text{ m}^2 \qquad A_2 = 250 \times 10^{-6} \text{ m}^2$$
$$L_1 = L_2 = 0.300 \text{ m}$$

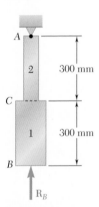

Fig. 2.27

Substituting these values into Eq. (2.8), we obtain

$$\delta_R = \frac{P_1 L_1}{A_1 E} + \frac{P_2 L_2}{A_2 E} = -\frac{(1.95 \times 10^3) R_B}{E} \qquad (2.18)$$

Expressing that the total deformation δ of the bar must be zero, we write

$$\delta = \delta_L + \delta_R = 0 \qquad (2.19)$$

and, substituting for δ_L and δ_R from (2.17) and (2.18) into (2.19),

$$\delta = \frac{1.125 \times 10^9}{E} - \frac{(1.95 \times 10^3) R_B}{E} = 0$$

Solving for R_B, we have

$$R_B = 577 \times 10^3 \text{ N} = 577 \text{ kN}$$

The reaction R_A at the upper support is obtained from the free-body diagram of the bar (Fig. 2.28). We write

$$+\uparrow \Sigma F_y = 0: \qquad R_A - 300 \text{ kN} - 600 \text{ kN} + R_B = 0$$
$$R_A = 900 \text{ kN} - R_B = 900 \text{ kN} - 577 \text{ kN} = 323 \text{ kN}$$

Once the reactions have been determined, the stresses and strains in the bar can easily be obtained. It should be noted that, while the total deformation of the bar is zero, each of its component parts *does deform* under the given loading and restraining conditions.

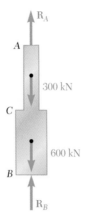

Fig. 2.28

EXAMPLE 2.05

Determine the reactions at A and B for the steel bar and loading of Example 2.04, assuming now that a 4.50-mm clearance exists between the bar and the ground before the loads are applied (Fig. 2.29). Assume $E = 200$ GPa.

We follow the same procedure as in Example 2.04. Considering the reaction at B as redundant, we compute the deformations δ_L and δ_R caused, respectively, by the given loads and by the redundant reaction $\mathbf{R}_B$. However, in this case the total deformation is not zero, but $\delta = 4.5$ mm. We write therefore

$$\delta = \delta_L + \delta_R = 4.5 \times 10^{-3} \text{ m} \qquad (2.20)$$

Substituting for δ_L and δ_R from (2.17) and (2.18) into (2.20), and recalling that $E = 200$ GPa $= 200 \times 10^9$ Pa, we have

$$\delta = \frac{1.125 \times 10^9}{200 \times 10^9} - \frac{(1.95 \times 10^3) R_B}{200 \times 10^9} = 4.5 \times 10^{-3} \text{ m}$$

Solving for R_B, we obtain

$$R_B = 115.4 \times 10^3 \text{ N} = 115.4 \text{ kN}$$

The reaction at A is obtained from the free-body diagram of the bar (Fig. 2.28):

$$+\uparrow \Sigma F_y = 0: \qquad R_A - 300 \text{ kN} - 600 \text{ kN} + R_B = 0$$
$$R_A = 900 \text{ kN} - R_B = 900 \text{ kN} - 115.4 \text{ kN} = 785 \text{ kN}$$

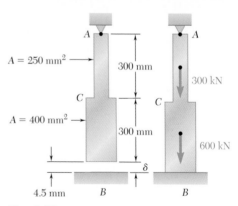

Fig. 2.29

2.10 PROBLEMS INVOLVING TEMPERATURE CHANGES

All of the members and structures that we have considered so far were assumed to remain at the same temperature while they were being loaded. We are now going to consider various situations involving changes in temperature.

Let us first consider a homogeneous rod AB of uniform cross section, which rests freely on a smooth horizontal surface (Fig. 2.30a). If the temperature of the rod is raised by ΔT, we observe that the rod elongates by an amount δ_T which is proportional to both the temperature change ΔT and the length L of the rod (Fig. 2.30b). We have

$$\delta_T = \alpha(\Delta T)L \tag{2.21}$$

where α is a constant characteristic of the material, called the *coefficient of thermal expansion*. Since δ_T and L are both expressed in units of length, α represents a quantity *per degree C*, if the temperature change is expressed in degrees Celsius.

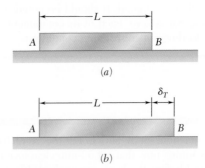

Fig. 2.30 Elongation of rod due to temperature increase.

With the deformation δ_T must be associated a strain $\epsilon_T = \delta_T/L$. Recalling Eq. (2.21), we conclude that

$$\epsilon_T = \alpha\Delta T \tag{2.22}$$

The strain ϵ_T is referred to as a *thermal strain,* since it is caused by the change in temperature of the rod. In the case we are considering here, there is *no stress associated with the strain* ϵ_T.

Let us now assume that the same rod AB of length L is placed between two fixed supports at a distance L from each other (Fig. 2.31a). Again, there is neither stress nor strain in this initial condition. If we raise the temperature by ΔT, the rod cannot elongate because of the restraints imposed on its ends; the elongation δ_T of the rod is thus zero. Since the rod is homogeneous and of uniform cross section, the strain ϵ_T at any point is $\epsilon_T = \delta_T/L$ and, thus, also zero. However, the supports will exert equal and opposite forces **P** and **P'** on the rod after the temperature has been raised, to keep it

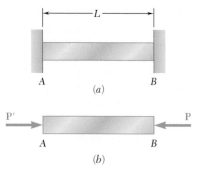

Fig. 2.31 Rod with ends restrained against thermal expansion.

from elongating (Fig. 2.31*b*). It thus follows that a state of stress (with no corresponding strain) is created in the rod.

As we prepare to determine the stress σ created by the temperature change ΔT, we observe that the problem we have to solve is statically indeterminate. Therefore, we should first compute the magnitude P of the reactions at the supports from the condition that the elongation of the rod is zero. Using the superposition method described in Sec. 2.9, we detach the rod from its support B (Fig. 2.32*a*) and let it elongate freely as it undergoes the temperature change ΔT (Fig. 2.32*b*). According to formula (2.21), the corresponding elongation is

$$\delta_T = \alpha(\Delta T)L$$

Applying now to end B the force $\mathbf{P}$ representing the redundant reaction, and recalling formula (2.7), we obtain a second deformation (Fig. 2.32*c*)

$$\delta_P = \frac{PL}{AE}$$

Expressing that the total deformation δ must be zero, we have

$$\delta = \delta_T + \delta_P = \alpha(\Delta T)L + \frac{PL}{AE} = 0$$

from which we conclude that

$$P = -AE\alpha(\Delta T)$$

and that the stress in the rod due to the temperature change ΔT is

$$\sigma = \frac{P}{A} = -E\alpha(\Delta T) \tag{2.23}$$

It should be kept in mind that the result we have obtained here and our earlier remark regarding the absence of any strain in the rod *apply only in the case of a homogeneous rod of uniform cross section.* Any other problem involving a restrained structure undergoing a change in temperature must be analyzed on its own merits. However, the same general approach can be used, i.e., we can consider separately the deformation due to the temperature change and the deformation due to the redundant reaction and superpose the solutions obtained.

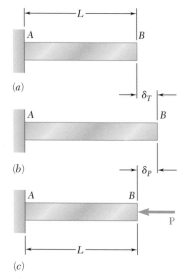

Fig. 2.32 Superposition method applied to rod restrained against thermal expansion.

EXAMPLE 2.06

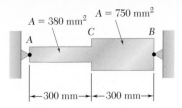

$A = 380 \text{ mm}^2$ $A = 750 \text{ mm}^2$

C B

A

|←—300 mm—→|←—300 mm—→|

Fig. 2.33

Determine the values of the stress in portions AC and CB of the steel bar shown (Fig. 2.33) when the temperature of the bar is $-45°C$ knowing that a close fit exists at both of the rigid supports when the temperature is $+24°C$. Use the values $E = 200$ GPa and $\alpha = 11.7 \times 10^{-6}/°C$ for steel.

We first determine the reactions at the supports. Since the problem is statically indeterminate, we detach the bar from its support at B and let it undergo the temperature change

$$\Delta T = (-45°C) - (24°C) = -69°C$$

The corresponding deformation (Fig. 2.34b) is

$$\delta_T = \alpha(\Delta T)L = (11.7 \times 10^{-6}/°C)(-69°C)(600 \text{ mm})$$
$$= -0.484 \text{ mm}$$

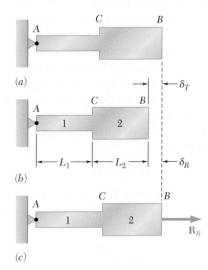

Fig. 2.34

Applying now the unknown force $\mathbf{R}_B$ at end B (Fig. 2.34c), we use Eq. (2.8) to express the corresponding deformation δ_R. Substituting

$$L_1 = L_2 = 300 \text{ mm}$$
$$A_1 = 380 \text{ mm}^2 \qquad A_2 = 750 \text{ mm}^2$$
$$P_1 = P_2 = R_B \qquad E = 200 \text{ GPa}$$

into Eq. (2.8), we write

$$\delta_R = \frac{P_1 L_1}{A_1 E} + \frac{P_2 L_2}{A_2 E}$$

$$= \frac{R_B}{200 \text{ GPa}}\left(\frac{300 \text{ mm}}{380 \text{ mm}^2} + \frac{300 \text{ mm}}{750 \text{ mm}^2}\right)$$

$$= (5.95 \times 10^{-6} \text{ mm/N})R_B$$

Expressing that the total deformation of the bar must be zero as a result of the imposed constraints, we write

$$\delta = \delta_T + \delta_R = 0$$
$$= -0.484 \text{ mm} + (5.95 \times 10^{-6} \text{ mm/N})R_B = 0$$

from which we obtain

$$R_B = 81.34 \times 10^3 \, \text{N} = 81.34 \, \text{kN}$$

The reaction at A is equal and opposite.

Noting that the forces in the two portions of the bar are $P_1 = P_2 = 81.34$ kN, we obtain the following values of the stress in portions AC and CB of the bar:

$$\sigma_1 = \frac{P_1}{A_1} = \frac{81.34 \, \text{kN}}{380 \, \text{mm}^2} = 214.1 \, \text{MPa}$$

$$\sigma_2 = \frac{P_2}{A_2} = \frac{81.34 \, \text{kN}}{750 \, \text{mm}^2} = 108.5 \, \text{MPa}$$

We cannot emphasize too strongly the fact that, while the *total deformation* of the bar must be zero, the deformations of the portions AC and CB *are not zero*. A solution of the problem based on the assumption that these deformations are zero would therefore be wrong. Neither can the values of the strain in AC or CB be assumed equal to zero. To amplify this point, let us determine the strain ϵ_{AC} in portion AC of the bar. The strain ϵ_{AC} can be divided into two component parts; one is the thermal strain ϵ_T produced in the unrestrained bar by the temperature change ΔT (Fig. 2.34b). From Eq. (2.22) we write

$$\epsilon_T = \alpha \, \Delta T = (11.7 \times 10^{-6}/°\text{C})(-69°\text{C})$$
$$= -807.3 \times 10^{-6} \, \text{mm/mm}$$

The other component of ϵ_{AC} is associated with the stress σ_1 due to the force $\mathbf{R}_B$ applied to the bar (Fig. 2.34c). From Hooke's law, we express this component of the strain as

$$\frac{\sigma_1}{E} = \frac{+214.1 \, \text{MPa}}{200 \, \text{GPa}} = +1070.5 \times 10^{-6} \, \text{mm/mm}$$

Adding the two components of the strain in AC, we obtain

$$\epsilon_{AC} = \epsilon_T + \frac{\sigma_1}{E} = -807.3 \times 10^{-6} + 1070.5 \times 10^{-6}$$
$$= +263.2 \times 10^{-6} \, \text{mm/mm}$$

A similar computation yields the strain in portion CB of the bar:

$$\epsilon_{CB} = \epsilon_T + \frac{\sigma_2}{E} = -807.3 \times 10^{-6} + 542.5 \times 10^{-6}$$
$$= -264.8 \times 10^{-6} \, \text{mm/mm}$$

The deformations δ_{AC} and δ_{CB} of the two portions of the bar are expressed respectively as

$$\delta_{AC} = \epsilon_{AC}(AC) = (263.2 \times 10^{-6})(300 \, \text{mm})$$
$$= 0.079 \, \text{mm}$$
$$\delta_{CB} = \epsilon_{CB}(CB) = (-264.8 \times 10^{-6})(300 \, \text{mm})$$
$$= -0.079 \, \text{mm}$$

We thus check that, while the sum $\delta = \delta_{AC} + \delta_{CB}$ of the two deformations is zero, neither of the deformations is zero.

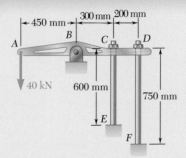

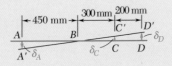

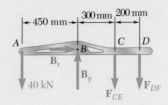

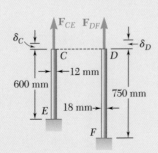

SAMPLE PROBLEM 2.3

The 12-mm-diameter rod CE and the 18-mm-diameter rod DF are attached to the rigid bar $ABCD$ as shown. Knowing that the rods are made of aluminum and using $E = 70$ GPa determine (a) the force in each rod caused by the loading shown, (b) the corresponding deflection of point A.

SOLUTION

Statics. Considering the free body of bar $ABCD$, we note that the reaction at B and the forces exerted by the rods are indeterminate. However, using statics, we may write

$$+\curvearrowleft \Sigma M_B = 0: \qquad (40 \text{ kN})(450 \text{ mm}) - F_{CE}(300 \text{ mm}) - F_{DF}(500 \text{ mm}) = 0$$

$$300F_{CE} + 500F_{DF} = 18000 \qquad (1)$$

Geometry. After application of the 40-kN load, the position of the bar is $A'BC'D'$. From the similar triangles BAA', BCC', and BDD' we have

$$\frac{\delta_C}{300 \text{ mm}} = \frac{\delta_D}{500 \text{ mm}} \qquad \delta_C = 0.6\delta_D \qquad (2)$$

$$\frac{\delta_A}{450 \text{ mm}} = \frac{\delta_D}{500 \text{ mm}} \qquad \delta_A = 0.9\delta_D \qquad (3)$$

Deformations. Using Eq. (2.7), we have

$$\delta_C = \frac{F_{CE}L_{CE}}{A_{CE}E} \qquad \delta_D = \frac{F_{DF}L_{DF}}{A_{DF}E}$$

Substituting for δ_C and δ_D into (2), we write

$$\delta_C = 0.6\delta_D \qquad \frac{F_{CE}L_{CE}}{A_{CE}E} = 0.6 \frac{F_{DF}L_{DF}}{A_{DF}E}$$

$$F_{CE} = 0.6 \frac{L_{DF}}{L_{CE}} \frac{A_{CE}}{A_{DF}} F_{DF} = 0.6 \left(\frac{750 \text{ mm}}{600 \text{ mm}}\right)\left[\frac{\frac{1}{4}\pi(12 \text{ mm})^2}{\frac{1}{4}\pi(18 \text{ mm})^2}\right] F_{DF} \quad F_{CE} = 0.333F_{DF}$$

Force in Each Rod. Substituting for F_{CE} into (1) and recalling that all forces have been expressed in kN, we have

$$300(0.333F_{DF}) + 500F_{DF} = 18000 \qquad\qquad F_{DF} = 30 \text{ kN} \quad \blacktriangleleft$$

$$F_{CE} = 0.333F_{DF} = 0.333(30 \text{ kN}) \qquad\qquad F_{CE} = 10 \text{ kN} \quad \blacktriangleleft$$

Deflections. The deflection of point D is

$$\delta_D = \frac{F_{DF}L_{DF}}{A_{DF}E} = \frac{(30 \times 10^3 \text{ N})(750 \text{ mm})}{\frac{1}{4}\pi(18)^2(70 \text{ GPa})} \qquad \delta_D = 1.263 \text{ mm}$$

Using (3), we write

$$\delta_A = 0.9\delta_D = 0.9(1.26 \text{ mm}) \qquad\qquad \delta_A = 1.137 \text{ mm} \quad \blacktriangleleft$$

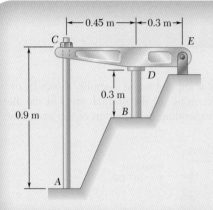

SAMPLE PROBLEM 2.4

The rigid bar CDE is attached to a pin support at E and rests on the 30-mm-diameter brass cylinder BD. A 22-mm-diameter steel rod AC passes through a hole in the bar and is secured by a nut which is snugly fitted when the temperature of the entire assembly is 20°C. The temperature of the brass cylinder is then raised to 50°C while the steel rod remains at 20°C. Assuming that no stresses were present before the temperature change, determine the stress in the cylinder.

Rod AC: Steel	Cylinder BD: Brass
$E = 200$ GPa	$E = 105$ GPa
$\alpha = 11.7 \times 10^{-6}/°C$	$\alpha = 20.9 \times 10^{-6}/°C$

SOLUTION

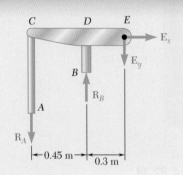

Statics. Considering the free body of the entire assembly, we write

$$+\!\!\uparrow\!\Sigma M_E = 0: \qquad R_A(0.75 \text{ m}) - R_B(0.3 \text{ m}) = 0 \qquad R_A = 0.4R_B \qquad (1)$$

Deformations. We use the method of superposition, considering $\mathbf{R}_B$ as redundant. With the support at B removed, the temperature rise of the cylinder causes point B to move down through δ_T. The reaction $\mathbf{R}_B$ must cause a deflection δ_1 equal to δ_T so that the final deflection of B will be zero (Fig. 3).

Deflection δ_T. Because of a temperature rise of $50° - 20° = 30°$C, the length of the brass cylinder increases by δ_T.

$$\delta_T = L(\Delta T)\alpha = (0.3 \text{ m})(30°\text{C})(20.9 \times 10^{-6}/°\text{C}) = 188.1 \times 10^{-6} \text{ m} \downarrow$$

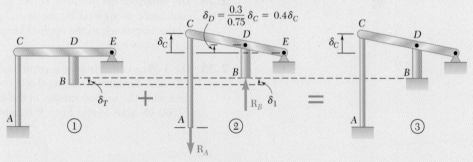

Deflection δ_1. We note that $\delta_D = 0.4\delta_C$ and $\delta_1 = \delta_D + \delta_{B/D}$.

$$\delta_C = \frac{R_A L}{AE} = \frac{R_A(0.9 \text{ m})}{\frac{1}{4}\pi(0.022 \text{ m})^2(200 \text{ GPa})} = 11.84 \times 10^{-9}R_A \uparrow$$

$$\delta_D = 0.40\delta_C = 0.4(11.84 \times 10^{-9}R_A) = 4.74 \times 10^{-9}R_A \uparrow$$

$$\delta_{B/D} = \frac{R_B L}{AE} = \frac{R_B(0.3 \text{ m})}{\frac{1}{4}\pi(0.03 \text{ m})^2(105 \text{ GPa})} = 4.04 \times 10^{-9}R_B \uparrow$$

We recall from (1) that $R_A = 0.4R_B$ and write

$$\delta_1 = \delta_D + \delta_{B/D} = [4.74(0.4R_B) + 4.04R_B]10^{-9} = 5.94 \times 10^{-9}R_B \uparrow$$

But $\delta_T = \delta_1$: $\qquad 188.1 \times 10^{-6} \text{ m} = 5.94 \times 10^{-9}R_B \qquad R_B = 31.7 \text{ kN}$

Stress in Cylinder: $\qquad \sigma_B = \frac{R_B}{A} = \frac{31.7 \text{ kN}}{\frac{1}{4}\pi(0.03 \text{ m})^2} \qquad \sigma_B = 44.8 \text{ MPa} \quad \blacktriangleleft$

PROBLEMS

2.33 An axial force of 200 kN is applied to the assembly shown by means of rigid end plates. Determine (*a*) the normal stress in the aluminum shell, (*b*) the corresponding deformation of the assembly.

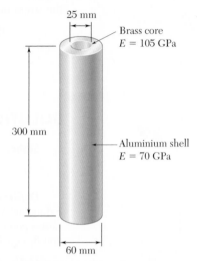

25 mm

Brass core
E = 105 GPa

300 mm

Aluminium shell
E = 70 GPa

60 mm

Fig. P2.33 and P2.34

2.34 The length of the assembly shown decreases by 0.40 mm when an axial force is applied by means of rigid end plates. Determine (*a*) the magnitude of the applied force, (*b*) the corresponding stress in the brass core.

2.35 The 1.35-m concrete post is reinforced with six steel bars, each with a 28-mm diameter. Knowing that E_s = 200 GPa and E_c = 29 GPa, determine the normal stresses in the steel and in the concrete when a 1560-kN axial centric force **P** is applied to the post.

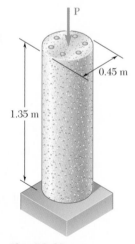

P

0.45 m

1.35 m

Fig. P2.35

2.36 An axial centric force of magnitude $P = 450$ kN is applied to the composite block shown by means of a rigid end plate. Knowing that $h = 10$ mm, determine the normal stress in (a) the brass core, (b) the aluminum plates.

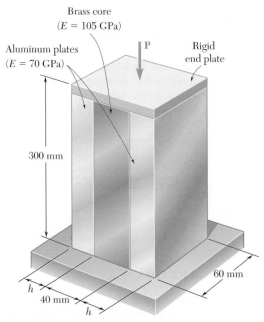

Brass core
($E = 105$ GPa)

Aluminum plates
($E = 70$ GPa)

P

Rigid
end plate

300 mm

60 mm

h

40 mm

h

Fig. P2.36 and P2.37

2.37 For the composite block shown in Prob. 2.36, determine (a) the value of h if the portion of the load carried by the aluminum plates is half the portion of the load carried by the brass core, (b) the total load if the stress in the brass is 80 MPa.

2.38 Compressive centric forces of 160 kN are applied at both ends of the assembly shown by means of rigid end plates. Knowing that $E_s = 200$ GPa and $E_a = 70$ GPa, determine (a) the normal stresses in the steel core and the aluminum shell, (b) the deformation of the assembly.

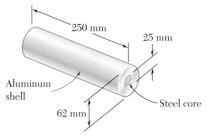

250 mm

25 mm

Aluminum
shell

62 mm

Steel core

Fig. P2.38

2.39 Three wires are used to suspend the plate shown. Aluminum wires of 3-mm diameter are used at A and B while a steel wire of 2-mm diameter is used at C. Knowing that the allowable stress for aluminum ($E_a = 70$ GPa) is 98 MPa and that the allowable stress for steel ($E_s = 200$ GPa) is 126 MPa, determine the maximum load **P** that can be applied.

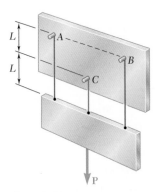

L

A

B

L

C

P

Fig. P2.39

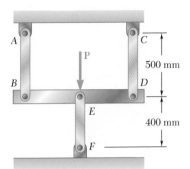

Fig. P2.40

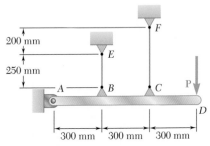

Fig. P2.43

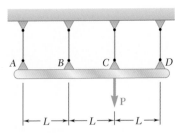

Fig. P2.44

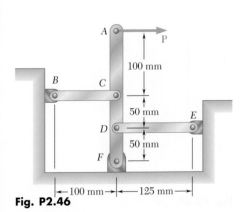

Fig. P2.46

2.40 Three steel rods (E = 200 GPa) support a 36-kN load **P**. Each of the rods AB and CD has a 200-mm^2 cross-sectional area and rod EF has a 625-mm^2 cross-sectional area. Neglecting the deformation of rod BED, determine (a) the change in length of rod EF, (b) the stress in each rod.

2.41 A steel tube (E = 200 GPa) with a 30-mm outer diameter and a 3-mm thickness is placed in a vise that is adjusted so that its jaws just touch the ends of the tube without exerting any pressure on them. The two forces shown are then applied to the tube. After these forces are applied, the vise is adjusted to decrease the distance between its jaws by 0.2 mm. Determine (a) the forces exerted by the vise on the tube at A and D, (b) the change in length of the portion BC of the tube.

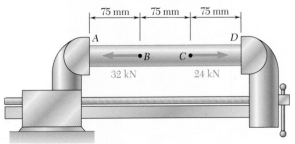

Fig. P2.41 and P2.42

2.42 Solve Prob. 2.41 assuming that after the forces have been applied, the vise is adjusted to increase the distance between its jaws by 0.1 mm.

2.43 The rigid bar $ABCD$ is suspended from four identical wires. Determine the tension in each wire caused by the load **P** shown.

2.44 The rigid bar AD is supported by two steel wires of 1.5-mm diameter (E = 200 GPa) and a pin and bracket at A. Knowing that the wires were initially taut, determine (a) the additional tension in each wire when a 1.0-kN load **P** is applied at D, (b) the corresponding deflection of point D.

2.45 The steel rods BE and CD each have a 16-mm diameter (E = 200 GPa); the ends of the rods are single-threaded with a pitch of 2.5 mm. Knowing that after being snugly fitted, the nut at C is tightened one full turn, determine (a) the tension in rod CD, (b) the deflection of point C of the rigid member ABC.

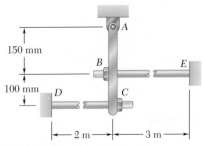

Fig. P2.45

2.46 Links BC and DE are both made of steel (E = 200 GPa) and are 12 mm wide and 6 mm thick. Determine (a) the force in each link when a 2.4-kN force **P** is applied to the rigid member AF shown, (b) the corresponding deflection of point A.

2.47 A 1.2-m concrete post is reinforced by four steel bars, each of 18-mm diameter. Knowing that $E_s = 200$ GPa, $\alpha_s = 11.7 \times 10^{-6}/°C$, $E_c = 25$ GPa, and $\alpha_c = 9.9 \times 10^{-6}/°C$, determine the normal stresses induced in the steel and in the concrete by a temperature rise of 27°C.

2.48 The assembly shown consists of an aluminum shell ($E_a = 73$ GPa, $\alpha_a = 23.2 \times 10^{-6}/°C$) fully bonded to a steel core ($E_s = 200$ GPa, $\alpha_s = 11.7 \times 10^{-6}/°C$) and is unstressed. Determine (a) the largest allowable change in temperature if the stress in the aluminum shell is not to exceed 40 MPa, (b) the corresponding change in length of the assembly.

2.49 The aluminum shell is fully bonded to the brass core and the assembly is unstressed at a temperature of 15°C. Considering only axial deformations, determine the stress in the aluminum when the temperature reaches 195°C.

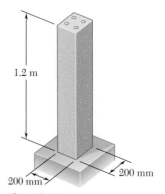

1.2 m

200 mm

200 mm

200 mm

Fig. P2.47

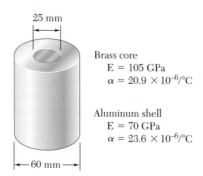

25 mm

Brass core
E = 105 GPa
$\alpha = 20.9 \times 10^{-6}/°C$

Aluminum shell
E = 70 GPa
$\alpha = 23.6 \times 10^{-6}/°C$

60 mm

Fig. P2.49

200 mm

18 mm

Aluminum shell

30 mm

Steel core

Fig. P2.48

2.50 Solve Prob. 2.49, assuming that the core is made of steel ($E_s = 200$ GPa, $\alpha_s = 11.7 \times 10^{-6}/°C$) instead of brass.

2.51 A rod consisting of two cylindrical portions AB and BC is restrained at both ends. Portion AB is made of steel ($E_s = 200$ GPa, $\alpha_s = 11.7 \times 10^{-6}/°C$) and portion BC is made of brass ($E_b = 105$ GPa, $\alpha_b = 20.9 \times 10^{-6}/°C$). Knowing that the rod is initially unstressed, determine the compressive force induced in ABC when there is a temperature rise of 50°C.

2.52 A steel railroad track ($E_s = 200$ GPa, $\alpha_s = 11.7 \times 10^{-6}/°C$) was laid out at a temperature of 6°C. Determine the normal stress in the rails when the temperature reaches 48°C, assuming that the rails (a) are welded to form a continuous track, (b) are 10 m long with 3-mm gaps between them.

2.53 A rod consisting of two cylindrical portions AB and BC is restrained at both ends. Portion AB is made of brass ($E_b = 105$ GPa, $\alpha_b = 20.9 \times 10^{-6}/°C$) and portion BC is made of aluminum ($E_a = 72$ GPa, $\alpha_a = 23.9 \times 10^{-6}/°C$). Knowing that the rod is initially unstressed, determine (a) the normal stresses induced in portions AB and BC by a temperature rise of 42°C, (b) the corresponding deflection of point B.

2.54 Solve Prob. 2.53, assuming that portion AB of the composite rod is made of steel and portion BC is made of brass.

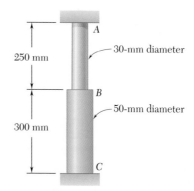

A

30-mm diameter

250 mm

B

50-mm diameter

300 mm

C

Fig. P2.51

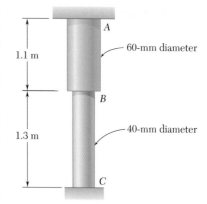

A

60-mm diameter

1.1 m

B

40-mm diameter

1.3 m

C

Fig. P2.53 and P2.54

2.55 A brass link ($E_b = 105$ GPa, $\alpha_b = 20.9 \times 10^{-6}$/°C) and a steel rod ($E_s = 200$ GPa, $\alpha_s = 11.7 \times 10^{-6}$/°C) have the dimensions shown at a temperature of 20°C. The steel rod is cooled until it fits freely into the link. The temperature of the whole assembly is then raised to 45°C. Determine (a) the final normal stress in the steel rod, (b) the final length of the steel rod.

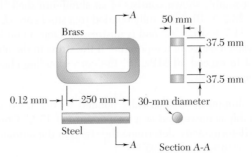

Fig. P2.55

2.56 Two steel bars ($E_s = 200$ GPa and $\alpha_s = 11.7 \times 10^{-6}$/°C) are used to reinforce a brass bar ($E_b = 105$ GPa, $\alpha_b = 20.9 \times 10^{-6}$/°C) that is subjected to a load $P = 25$ kN. When the steel bars were fabricated, the distance between the centers of the holes that were to fit on the pins was made 0.5 mm smaller than the 2 m needed. The steel bars were then placed in an oven to increase their length so that they would just fit on the pins. Following fabrication, the temperature in the steel bars dropped back to room temperature. Determine (a) the increase in temperature that was required to fit the steel bars on the pins, (b) the stress in the brass bar after the load is applied to it.

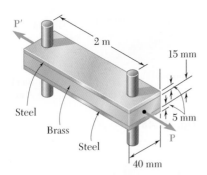

Fig. P2.56

2.57 Determine the maximum load P that can be applied to the brass bar of Prob. 2.56 if the allowable stress in the steel bars is 30 MPa and the allowable stress in the brass bar is 25 MPa.

2.58 Knowing that a 0.5-mm gap exists when the temperature is 24°C, determine (a) the temperature at which the normal stress in the aluminum bar will be equal to −75 MPa, (b) the corresponding exact length of the aluminum bar.

2.59 Determine (a) the compressive force in the bars shown after a temperature rise of 82°C, (b) the corresponding change in length of the bronze bar.

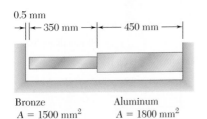

Bronze
$A = 1500$ mm^2
$E = 105$ GPa
$\alpha = 21.6 \times 10^{-6}$/°C

Aluminum
$A = 1800$ mm^2
$E = 73$ GPa
$\alpha = 23.2 \times 10^{-6}$/°C

Fig. P2.58 and P2.59

2.60 At room temperature (20°C) a 0.5-mm gap exists between the ends of the rods shown. At a later time when the temperature has reached 140°C, determine (a) the normal stress in the aluminum rod, (b) the change in length of the aluminum rod.

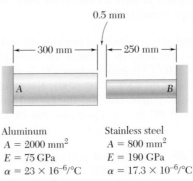

Aluminum
$A = 2000$ mm^2
$E = 75$ GPa
$\alpha = 23 \times 16^{-6}$/°C

Stainless steel
$A = 800$ mm^2
$E = 190$ GPa
$\alpha = 17.3 \times 10^{-6}$/°C

Fig. P2.60

2.11 POISSON'S RATIO

We saw in the earlier part of this chapter that, when a homogeneous slender bar is axially loaded, the resulting stress and strain satisfy Hooke's law, as long as the elastic limit of the material is not exceeded. Assuming that the load **P** is directed along the x axis (Fig. 2.35a), we have $\sigma_x = P/A$, where A is the cross-sectional area of the bar, and, from Hooke's law,

$$\epsilon_x = \sigma_x/E \tag{2.24}$$

where E is the modulus of elasticity of the material.

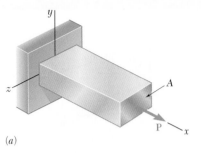

(a)

We also note that the normal stresses on faces respectively perpendicular to the y and z axes are zero: $\sigma_y = \sigma_z = 0$ (Fig. 2.35b). It would be tempting to conclude that the corresponding strains ϵ_y and ϵ_z are also zero. This, however, *is not the case*. In all engineering materials, the elongation produced by an axial tensile force **P** in the direction of the force is accompanied by a contraction in any transverse direction (Fig. 2.36).† In this section and the following sections (Secs. 2.12 through 2.15), all materials considered will be assumed to be both *homogeneous* and *isotropic*, i.e., their mechanical properties will be assumed independent of both *position* and *direction*. It follows that the strain must have the same value for any transverse direction. Therefore, for the loading shown in Fig. 2.35 we must have $\epsilon_y = \epsilon_z$. This common value is referred to as the *lateral strain*. An important constant for a given material is its *Poisson's ratio*, named after the French mathematician Siméon Denis Poisson (1781–1840) and denoted by the Greek letter ν (nu). It is defined as

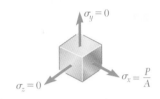

(b)

Fig. 2.35 Stresses in an axially-loaded bar.

$$\nu = -\frac{\text{lateral strain}}{\text{axial strain}} \tag{2.25}$$

or

$$\nu = -\frac{\epsilon_y}{\epsilon_x} = -\frac{\epsilon_z}{\epsilon_x} \tag{2.26}$$

for the loading condition represented in Fig. 2.35. Note the use of a minus sign in the equations to obtain a positive value for ν, the axial and lateral strains having opposite signs for all engineering materials.‡ Solving Eq. (2.26) for ϵ_y and ϵ_z, and recalling (2.24), we write the following relations, which fully describe the condition of strain under an axial load applied in a direction parallel to the x axis:

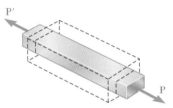

Fig. 2.36 Transverse contraction of bar under axial tensile force.

$$\epsilon_x = \frac{\sigma_x}{E} \qquad \epsilon_y = \epsilon_z = -\frac{\nu\sigma_x}{E} \tag{2.27}$$

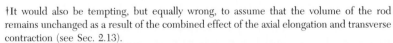

---

†It would also be tempting, but equally wrong, to assume that the volume of the rod remains unchanged as a result of the combined effect of the axial elongation and transverse contraction (see Sec. 2.13).

‡However, some experimental materials, such as polymer foams, expand laterally when stretched. Since the axial and lateral strains have then the same sign, the Poisson's ratio of these materials is negative. (*See* Roderic Lakes, "Foam Structures with a Negative Poisson's Ratio," *Science*, 27 February 1987, Volume 235, pp. 1038–1040.)

EXAMPLE 2.07

A 500-mm-long, 16-mm-diameter rod made of a homogenous, isotropic material is observed to increase in length by 300 μm, and to decrease in diameter by 2.4 μm when subjected to an axial 12-kN load. Determine the modulus of elasticity and Poisson's ratio of the material.

The cross-sectional area of the rod is

$$A = \pi r^2 = \pi(8 \times 10^{-3}\,\text{m})^2 = 201 \times 10^{-6}\,\text{m}^2$$

Choosing the x axis along the axis of the rod (Fig. 2.37), we write

$$\sigma_x = \frac{P}{A} = \frac{12 \times 10^3\,\text{N}}{201 \times 10^{-6}\,\text{m}^2} = 59.7\,\text{MPa}$$

$$\epsilon_x = \frac{\delta_x}{L} = \frac{300\,\mu\text{m}}{500\,\text{mm}} = 600 \times 10^{-6}$$

$$\epsilon_y = \frac{\delta_y}{d} = \frac{-2.4\,\mu\text{m}}{16\,\text{mm}} = -150 \times 10^{-6}$$

From Hooke's law, $\sigma_x = E\epsilon_x$, we obtain

$$E = \frac{\sigma_x}{\epsilon_x} = \frac{59.7\,\text{MPa}}{600 \times 10^{-6}} = 99.5\,\text{GPa}$$

and, from Eq. (2.26),

$$\nu = -\frac{\epsilon_y}{\epsilon_x} = -\frac{-150 \times 10^{-6}}{600 \times 10^{-6}} = 0.25$$

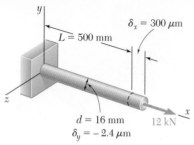

$\delta_x = 300\,\mu$m
$L = 500$ mm
$d = 16$ mm
12 kN
$\delta_y = -2.4\,\mu$m

Fig. 2.37

2.12 MULTIAXIAL LOADING; GENERALIZED HOOKE'S LAW

All the examples considered so far in this chapter have dealt with slender members subjected to axial loads, i.e., to forces directed along a single axis. Choosing this axis as the x axis, and denoting by P the internal force at a given location, the corresponding stress components were found to be $\sigma_x = P/A$, $\sigma_y = 0$, and $\sigma_z = 0$.

Let us now consider structural elements subjected to loads acting in the directions of the three coordinate axes and producing normal stresses σ_x, σ_y, and σ_z which are all different from zero (Fig. 2.38). This condition is referred to as a *multiaxial loading*. Note that this is not the general stress condition described in Sec. 1.12, since no shearing stresses are included among the stresses shown in Fig. 2.38.

Consider an element of an isotropic material in the shape of a cube (Fig. 2.39a). We can assume the side of the cube to be equal to unity, since it is always possible to select the side of the cube as a unit of length. Under the given multiaxial loading, the element will deform into a *rectangular parallelepiped* of sides equal, respectively, to $1 + \epsilon_x$, $1 + \epsilon_y$, and $1 + \epsilon_z$, where ϵ_x, ϵ_y, and ϵ_z denote the values of the normal strain in the directions of the three coordinate axes (Fig. 2.39b). You should note that, as a result of the deformations of

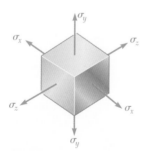

Fig. 2.38 Stress state for multiaxial loading.

the other elements of the material, the element under consideration could also undergo a translation, but we are concerned here only with the *actual deformation* of the element, and not with any possible superimposed rigid-body displacement.

In order to express the strain components ϵ_x, ϵ_y, ϵ_z in terms of the stress components σ_x, σ_y, σ_z, we will consider separately the effect of each stress component and combine the results obtained. The approach we propose here will be used repeatedly in this text, and is based on the *principle of superposition*. This principle states that the effect of a given combined loading on a structure can be obtained by *determining separately the effects of the various loads and combining the results obtained*, provided that the following conditions are satisfied:

1. Each effect is linearly related to the load that produces it.
2. The deformation resulting from any given load is small and does not affect the conditions of application of the other loads.

In the case of a multiaxial loading, the first condition will be satisfied if the stresses do not exceed the proportional limit of the material, and the second condition will also be satisfied if the stress on any given face does not cause deformations of the other faces that are large enough to affect the computation of the stresses on those faces.

Considering first the effect of the stress component σ_x, we recall from Sec. 2.11 that σ_x causes a strain equal to σ_x/E in the x direction, and strains equal to $-\nu\sigma_x/E$ in each of the y and z directions. Similarly, the stress component σ_y, if applied separately, will cause a strain σ_y/E in the y direction and strains $-\nu\sigma_y/E$ in the other two directions. Finally, the stress component σ_z causes a strain σ_z/E in the z direction and strains $-\nu\sigma_z/E$ in the x and y directions. Combining the results obtained, we conclude that the components of strain corresponding to the given multiaxial loading are

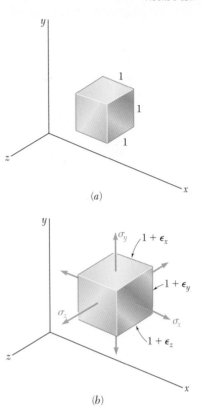

(a)

(b)

Fig. 2.39 Deformation of cube under multiaxial loading.

$$\epsilon_x = +\frac{\sigma_x}{E} - \frac{\nu\sigma_y}{E} - \frac{\nu\sigma_z}{E}$$

$$\epsilon_y = -\frac{\nu\sigma_x}{E} + \frac{\sigma_y}{E} - \frac{\nu\sigma_z}{E} \qquad (2.28)$$

$$\epsilon_z = -\frac{\nu\sigma_x}{E} - \frac{\nu\sigma_y}{E} + \frac{\sigma_z}{E}$$

The relations (2.28) are referred to as the *generalized Hooke's law for the multiaxial loading of a homogeneous isotropic material*. As we indicated earlier, the results obtained are valid only as long as the stresses do not exceed the proportional limit, and as long as the deformations involved remain small. We also recall that a positive value for a stress component signifies tension, and a negative value compression. Similarly, a positive value for a strain component indicates expansion in the corresponding direction, and a negative value contraction.

EXAMPLE 2.08

The steel block shown (Fig. 2.40) is subjected to a uniform pressure on all its faces. Knowing that the change in length of edge AB is -30×10^{-3} mm, determine (a) the change in length of the other two edges, (b) the pressure p applied to the faces of the block. Assume $E = 200$ GPa and $\nu = 0.29$.

(a) **Change in Length of Other Edges.** Substituting $\sigma_x = \sigma_y = \sigma_z = -p$ into the relations (2.28), we find that the three strain components have the common value

$$\epsilon_x = \epsilon_y = \epsilon_z = -\frac{p}{E}(1 - 2\nu) \tag{2.29}$$

Since

$$\epsilon_x = \delta_x / AB = (-30 \times 10^{-3} \text{ mm})/(100 \text{ mm})$$
$$= -300 \times 10^{-6} \text{ mm/mm}$$

we obtain

$$\epsilon_y = \epsilon_z = \epsilon_x = -300 \times 10^{-6} \text{ mm/mm}$$

from which it follows that

$$\delta_y = \epsilon_y(BC) = (-300 \times 10^{-6})(50 \text{ mm}) = -15 \times 10^{-3} \text{ mm}$$
$$\delta_z = \epsilon_z(BD) = (-300 \times 10^{-6})(75 \text{ mm}) = -22.5 \times 10^{-3} \text{ mm}$$

(b) **Pressure.** Solving Eq. (2.29) for p, we write

$$p = -\frac{E\epsilon_x}{1 - 2\nu} = -\frac{(200 \text{ GPa})(-300 \times 10^{-6})}{1 - 0.58}$$
$$p = 142.9 \text{ MPa}$$

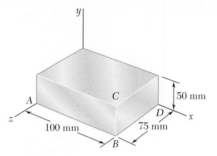

Fig. 2.40

*2.13 DILATATION; BULK MODULUS

In this section you will examine the effect of the normal stresses σ_x, σ_y, and σ_z on the volume of an element of isotropic material. Consider the element shown in Fig. 2.39. In its unstressed state, it is in the shape of a cube of unit volume; and under the stresses σ_x, σ_y, σ_z, it deforms into a rectangular parallelepiped of volume

$$v = (1 + \epsilon_x)(1 + \epsilon_y)(1 + \epsilon_z)$$

Since the strains ϵ_x, ϵ_y, ϵ_z are much smaller than unity, their products will be even smaller and may be omitted in the expansion of the product. We have, therefore,

$$v = 1 + \epsilon_x + \epsilon_y + \epsilon_z$$

Denoting by e the change in volume of our element, we write

$$e = v - 1 = 1 + \epsilon_x + \epsilon_y + \epsilon_z - 1$$

or

$$e = \epsilon_x + \epsilon_y + \epsilon_z \tag{2.30}$$

Since the element had originally a unit volume, the quantity e represents *the change in volume per unit volume;* it is referred to as the *dilatation* of the material. Substituting for ϵ_x, ϵ_y, and ϵ_z from Eqs. (2.28) into (2.30), we write

$$e = \frac{\sigma_x + \sigma_y + \sigma_z}{E} - \frac{2\nu(\sigma_x + \sigma_y + \sigma_z)}{E}$$

$$e = \frac{1 - 2\nu}{E}(\sigma_x + \sigma_y + \sigma_z) \qquad (2.31)\dagger$$

A case of special interest is that of a body subjected to a uniform hydrostatic pressure p. Each of the stress components is then equal to $-p$ and Eq. (2.31) yields

$$e = -\frac{3(1 - 2\nu)}{E}p \qquad (2.32)$$

Introducing the constant

$$k = \frac{E}{3(1 - 2\nu)} \qquad (2.33)$$

we write Eq. (2.32) in the form

$$e = -\frac{p}{k} \qquad (2.34)$$

The constant k is known as the *bulk modulus* or *modulus of compression* of the material. It is expressed in the same units as the modulus of elasticity E, that is, in pascals.

Observation and common sense indicate that a stable material subjected to a hydrostatic pressure can only *decrease* in volume; thus the dilatation e in Eq. (2.34) is negative, from which it follows that the bulk modulus k is a positive quantity. Referring to Eq. (2.33), we conclude that $1 - 2\nu > 0$, or $\nu < \frac{1}{2}$. On the other hand, we recall from Sec. 2.11 that ν is positive for all engineering materials. We thus conclude that, for any engineering material,

$$0 < \nu < \tfrac{1}{2} \qquad (2.35)$$

We note that an ideal material having a value of v equal to zero could be stretched in one direction without any lateral contraction. On the other hand, an ideal material for which $\nu = \frac{1}{2}$, and thus $k = \infty$, would be perfectly incompressible ($e = 0$). Referring to Eq. (2.31) we also note that, since $\nu < \frac{1}{2}$ in the elastic range, stretching an engineering material in one direction, for example in the x direction ($\sigma_x > 0$, $\sigma_y = \sigma_z = 0$), will result in an increase of its volume ($e > 0$).‡

---

†Since the dilatation e represents a change in volume, it must be independent of the orientation of the element considered. It then follows from Eqs. (2.30) and (2.31) that the quantities $\epsilon_x + \epsilon_y + \epsilon_z$ and $\sigma_x + \sigma_y + \sigma_z$ are also independent of the orientation of the element. This property will be verified in Chap. 7.

‡However, in the plastic range, the volume of the material remains nearly constant.

EXAMPLE 2.09

Determine the change in volume ΔV of the steel block shown in Fig. 2.40, when it is subjected to the hydrostatic pressure $p = 180$ MPa. Use $E = 200$ GPa and $\nu = 0.29$.

From Eq. (2.33), we determine the bulk modulus of steel,

$$k = \frac{E}{3(1 - 2\nu)} = \frac{200 \text{ GPa}}{3(1 - 0.58)} = 158.7 \text{ GPa}$$

and, from Eq. (2.34), the dilatation,

$$e = -\frac{p}{k} = -\frac{180 \text{ MPa}}{158.7 \text{ GPa}} = -1.134 \times 10^{-3}$$

Since the volume V of the block in its unstressed state is

$$V = (80 \text{ mm})(40 \text{ mm})(60 \text{ mm}) = 192 \times 10^3 \text{ mm}^3$$

and since e represents the change in volume per unit volume, $e = \Delta V/V$, we have

$$\Delta V = eV = (-1.134 \times 10^{-3})(192 \times 10^3 \text{ mm}^3)$$

$$\Delta V = -218 \text{ mm}^3$$

2.14 SHEARING STRAIN

When we derived in Sec. 2.12 the relations (2.28) between normal stresses and normal strains in a homogeneous isotropic material, we assumed that no shearing stresses were involved. In the more general stress situation represented in Fig. 2.41, shearing stresses τ_{xy}, τ_{yz}, and τ_{zx} will be present (as well, of course, as the corresponding shearing stresses τ_{yx}, τ_{zy}, and τ_{xz}). These stresses have no direct effect on the normal strains and, as long as all the deformations involved remain small, they will not affect the derivation nor the validity of the relations (2.28). The shearing stresses, however, will tend to deform a cubic element of material into an *oblique* parallelepiped.

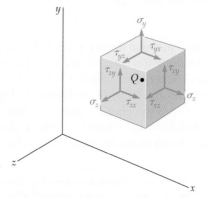

Fig. 2.41 General state of stress.

Consider first a cubic element of side one (Fig. 2.42) subjected to no other stresses than the shearing stresses τ_{xy} and τ_{yx} applied to faces of the element respectively perpendicular to the x and y axes. (We recall from Sec. 1.12 that $\tau_{xy} = \tau_{yx}$.) The element is observed to deform into a rhomboid of sides equal to one (Fig. 2.43). Two of the angles formed by the four faces under stress are reduced from $\frac{\pi}{2}$ to $\frac{\pi}{2} - \gamma_{xy}$, while the other two are increased from $\frac{\pi}{2}$ to $\frac{\pi}{2} + \gamma_{xy}$. The small angle γ_{xy} (expressed in radians) defines the *shearing strain* corresponding to the x and y directions. When the deformation involves a *reduction* of the angle formed by the two faces oriented respectively toward the positive x and y axes (as shown in Fig. 2.43), the shearing strain γ_{xy} is said to be *positive;* otherwise, it is said to be negative.

We should note that, as a result of the deformations of the other elements of the material, the element under consideration can also undergo an overall rotation. However, as was the case in our study of normal strains, we are concerned here only with the *actual deformation* of the element, and not with any possible superimposed rigid-body displacement.†

Plotting successive values of τ_{xy} against the corresponding values of γ_{xy}, we obtain the shearing stress-strain diagram for the material under consideration. This can be accomplished by carrying out a torsion test, as you will see in Chap. 3. The diagram obtained is similar to the normal stress-strain diagram obtained for the same material from the tensile test described earlier in this chapter. However, the values obtained for the yield strength, ultimate strength, etc., of a given material are only about half as large in shear as they are in tension. As was the case for normal stresses and strains, the initial portion of the shearing stress-strain diagram is a straight line. For values of the shearing stress that do not exceed the proportional

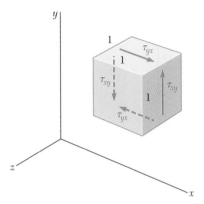

Fig. 2.42 Cubic element subjected to shearing stresses.

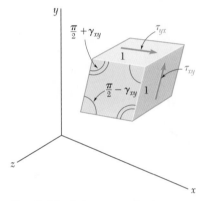

Fig. 2.43 Deformation of cubic element due to shearing stresses.

†In defining the strain γ_{xy}, some authors arbitrarily assume that the actual deformation of the element is accompanied by a rigid-body rotation such that the horizontal faces of the element do not rotate. The strain γ_{xy} is then represented by the angle through which the other two faces have rotated (Fig. 2.44). Others assume a rigid-body rotation such that the horizontal faces rotate through $\frac{1}{2}\gamma_{xy}$ counterclockwise and the vertical faces through $\frac{1}{2}\gamma_{xy}$ clockwise (Fig. 2.45). Since both assumptions are unnecessary and may lead to confusion, we prefer in this text to associate the shearing strain γ_{xy} with the *change in the angle* formed by the two faces, rather than with the *rotation of a given face* under restrictive conditions.

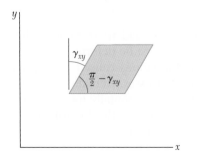

Fig. 2.44

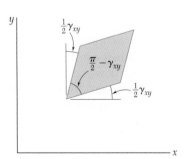

Fig. 2.45

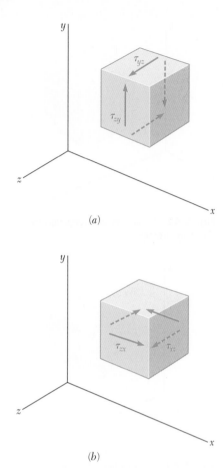

(a)

(b)

Fig. 2.46

limit in shear, we can therefore write for any homogeneous isotropic material,

$$\tau_{xy} = G\gamma_{xy} \tag{2.36}$$

This relation is known as *Hooke's law for shearing stress and strain,* and the constant G is called the *modulus of rigidity* or *shear modulus* of the material. Since the strain γ_{xy} was defined as an angle in radians, it is dimensionless, and the modulus G is expressed in the same units as τ_{xy}, that is, in pascals. The modulus of rigidity G of any given material is less than one-half, but more than one-third of the modulus of elasticity E of that material.†

Considering now a small element of material subjected to shearing stresses τ_{yz} and τ_{zy} (Fig. 2.46a), we define the shearing strain γ_{yz} as the change in the angle formed by the faces under stress. The shearing strain γ_{zx} is defined in a similar way by considering an element subjected to shearing stresses τ_{zx} and τ_{xz} (Fig. 2.46b). For values of the stress that do not exceed the proportional limit, we can write the two additional relations

$$\tau_{yz} = G\gamma_{yz} \qquad \tau_{zx} = G\gamma_{zx} \tag{2.37}$$

where the constant G is the same as in Eq. (2.36).

For the general stress condition represented in Fig. 2.41, and as long as none of the stresses involved exceeds the corresponding proportional limit, we can apply the principle of superposition and combine the results obtained in this section and in Sec. 2.12. We obtain the following group of equations representing the generalized Hooke's law for a homogeneous isotropic material under the most general stress condition.

$$\epsilon_x = +\frac{\sigma_x}{E} - \frac{\nu\sigma_y}{E} - \frac{\nu\sigma_z}{E}$$

$$\epsilon_y = -\frac{\nu\sigma_x}{E} + \frac{\sigma_y}{E} - \frac{\nu\sigma_z}{E} \tag{2.38}$$

$$\epsilon_z = -\frac{\nu\sigma_x}{E} - \frac{\nu\sigma_y}{E} + \frac{\sigma_z}{E}$$

$$\gamma_{xy} = \frac{\tau_{xy}}{G} \qquad \gamma_{yz} = \frac{\tau_{yz}}{G} \qquad \gamma_{zx} = \frac{\tau_{zx}}{G}$$

An examination of Eqs. (2.38) might lead us to believe that three distinct constants, E, ν, and G, must first be determined experimentally, if we are to predict the deformations caused in a given material by an arbitrary combination of stresses. Actually, only two of these constants need be determined experimentally for any given material. As you will see in the next section, the third constant can then be obtained through a very simple computation.

†See Prob. 2.91.

A rectangular block of a material with a modulus of rigidity $G = 630$ MPa is bonded to two rigid horizontal plates. The lower plate is fixed, while the upper plate is subjected to a horizontal force $\mathbf{P}$ (Fig. 2.47). Knowing that the upper plate moves through 1 mm under the action of the force, determine (a) the average shearing strain in the material, (b) the force $\mathbf{P}$ exerted on the upper plate.

EXAMPLE 2.10

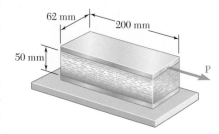

Fig. 2.47

(a) **Shearing Strain.** We select coordinate axes centered at the midpoint C of edge AB and directed as shown (Fig. 2.48). According to its definition, the shearing strain γ_{xy} is equal to the angle formed by the vertical and the line CF joining the midpoints of edges AB and DE. Noting that this is a very small angle and recalling that it should be expressed in radians, we write

$$\gamma_{xy} \approx \tan \gamma_{xy} = \frac{1 \text{ mm}}{50 \text{ mm}} \qquad \gamma_{xy} = 0.020 \text{ rad}$$

(b) **Force Exerted on Upper Plate.** We first determine the shearing stress τ_{xy} in the material. Using Hooke's law for shearing stress and strain, we have

$$\tau_{xy} = G\gamma_{xy} = (630 \text{ MPa})(0.020 \text{ rad}) = 12.6 \text{ MPa}$$

The force exerted on the upper plate is thus

$$P = \tau_{xy} A = (12.6 \text{ MPa})(200 \text{ mm})(62 \text{ mm}) = 156.2 \times 10^3 \text{ N}$$

$$P = 156.2 \text{ kN}$$

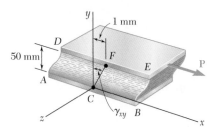

Fig. 2.48

2.15 FURTHER DISCUSSION OF DEFORMATIONS UNDER AXIAL LOADING; RELATION AMONG E, ν, AND G

We saw in Sec. 2.11 that a slender bar subjected to an axial tensile load $\mathbf{P}$ directed along the x axis will elongate in the x direction and contract in both of the transverse y and z directions. If ϵ_x denotes the axial strain, the lateral strain is expressed as $\epsilon_y = \epsilon_z = -\nu\epsilon_x$, where ν is Poisson's ratio. Thus, an element in the shape of a cube of side equal to one and oriented as shown in Fig. 2.49a will deform into a rectangular parallelepiped of sides $1 + \epsilon_x$, $1 - \nu\epsilon_x$, and $1 - \nu\epsilon_x$. (Note that only one face of the element is shown in the figure.) On the other hand, if the element is oriented at $45°$ to the axis of the load (Fig. 2.49b), the face shown in the figure is observed to deform into a rhombus. We conclude that the axial load $\mathbf{P}$ causes in this element a shearing strain γ' equal to the amount by which each of the angles shown in Fig. 2.49b increases or decreases.†

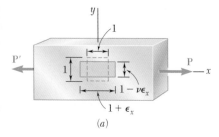

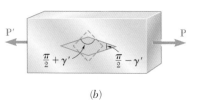

Fig. 2.49 Representations of strain in an axially-loaded bar.

†Note that the load $\mathbf{P}$ also produces normal strains in the element shown in Fig. 2.49b (see Prob. 2.73).

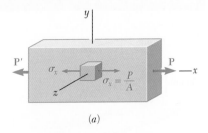

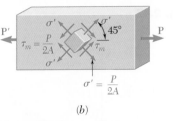

(b)

Fig. 1.38 (repeated)

The fact that shearing strains, as well as normal strains, result from an axial loading should not come to us as a surprise, since we already observed at the end of Sec. 1.12 that an axial load **P** causes normal and shearing stresses of equal magnitude on four of the faces of an element oriented at 45° to the axis of the member. This was illustrated in Fig. 1.38, which, for convenience, has been repeated here. It was also shown in Sec. 1.11 that the shearing stress is maximum on a plane forming an angle of 45° with the axis of the load. It follows from Hooke's law for shearing stress and strain that the shearing strain γ' associated with the element of Fig. 2.49b is also maximum: $\gamma' = \gamma_m$.

While a more detailed study of the transformations of strain will be postponed until Chap. 7, we will derive in this section a relation between the maximum shearing strain $\gamma' = \gamma_m$ associated with the element of Fig. 2.49b and the normal strain ϵ_x in the direction of the load. Let us consider for this purpose the prismatic element obtained by intersecting the cubic element of Fig. 2.49a by a diagonal plane (Fig. 2.50a and b). Referring to Fig. 2.49a, we conclude that this new element will deform into the element shown in Fig. 2.50c, which has horizontal and vertical sides respectively equal to $1 + \epsilon_x$ and $1 - \nu\epsilon_x$. But the angle formed by the oblique and horizontal faces of the element of Fig. 2.50b is precisely half of one of the right angles of the cubic element considered in Fig. 2.49b. The angle β into which this angle deforms must therefore be equal to half of $\pi/2 - \gamma_m$. We write

$$\beta = \frac{\pi}{4} - \frac{\gamma_m}{2}$$

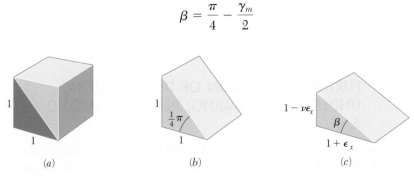

Fig. 2.50

Applying the formula for the tangent of the difference of two angles, we obtain

$$\tan \beta = \frac{\tan \dfrac{\pi}{4} - \tan \dfrac{\gamma_m}{2}}{1 + \tan \dfrac{\pi}{4} \tan \dfrac{\gamma_m}{2}} = \frac{1 - \tan \dfrac{\gamma_m}{2}}{1 + \tan \dfrac{\gamma_m}{2}}$$

or, since $\gamma_m/2$ is a very small angle,

$$\tan \beta = \frac{1 - \dfrac{\gamma_m}{2}}{1 + \dfrac{\gamma_m}{2}} \tag{2.39}$$

But, from Fig. 2.50c, we observe that

$$\tan \beta = \frac{1 - \nu \epsilon_x}{1 + \epsilon_x} \tag{2.40}$$

Equating the right-hand members of (2.39) and (2.40), and solving for γ_m, we write

$$\gamma_m = \frac{(1 + \nu)\epsilon_x}{1 + \dfrac{1 - \nu}{2}\epsilon_x}$$

Since $\epsilon_x \ll 1$, the denominator in the expression obtained can be assumed equal to one; we have, therefore,

$$\gamma_m = (1 + \nu)\epsilon_x \tag{2.41}$$

which is the desired relation between the maximum shearing strain γ_m and the axial strain ϵ_x.

To obtain a relation among the constants E, ν, and G, we recall that, by Hooke's law, $\gamma_m = \tau_m/G$, and that, for an axial loading, $\epsilon_x = \sigma_x/E$. Equation (2.41) can therefore be written as

$$\frac{\tau_m}{G} = (1 + \nu)\frac{\sigma_x}{E}$$

or

$$\frac{E}{G} = (1 + \nu)\frac{\sigma_x}{\tau_m} \tag{2.42}$$

We now recall from Fig. 1.38 that $\sigma_x = P/A$ and $\tau_m = P/2A$, where A is the cross-sectional area of the member. It thus follows that $\sigma_x/\tau_m = 2$. Substituting this value into (2.42) and dividing both members by 2, we obtain the relation

$$\frac{E}{2G} = 1 + \nu \tag{2.43}$$

which can be used to determine one of the constants E, ν, or G from the other two. For example, solving Eq. (2.43) for G, we write

$$G = \frac{E}{2(1 + \nu)} \tag{2.43'}$$

*2.16 STRESS-STRAIN RELATIONSHIPS FOR FIBER-REINFORCED COMPOSITE MATERIALS

Fiber-reinforced composite materials were briefly discussed in Sec. 2.5. It was shown at that time that these materials are obtained by embedding fibers of a strong, stiff material into a weaker, softer material, referred to as a *matrix*. It was also shown that the relationship between the normal stress and the corresponding normal strain created in a lamina, or layer, of a composite material depends upon the direction in which the load is applied. Different moduli of elasticity, E_x, E_y, and E_z, are therefore required to describe the relationship between normal stress and normal strain, according to whether the

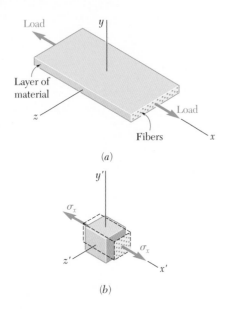

Fig. 2.51 Fiber-reinforced composite material under uniaxial tensile load.

load is applied in a direction parallel to the fibers, in a direction perpendicular to the layer, or in a transverse direction.

Let us consider again the layer of composite material discussed in Sec. 2.5 and let us subject it to a uniaxial tensile load parallel to its fibers, i.e., in the x direction (Fig. 2.51a). To simplify our analysis, it will be assumed that the properties of the fibers and of the matrix have been combined, or "smeared," into a fictitious equivalent homogeneous material possessing these combined properties. We now consider a small element of that layer of smeared material (Fig. 2.51b). We denote by σ_x the corresponding normal stress and observe that $\sigma_y = \sigma_z = 0$. As indicated earlier in Sec. 2.5, the corresponding normal strain in the x direction is $\epsilon_x = \sigma_x/E_x$, where E_x is the modulus of elasticity of the composite material in the x direction. As we saw for isotropic materials, the elongation of the material in the x direction is accompanied by contractions in the y and z directions. These contractions depend upon the placement of the fibers in the matrix and will generally be different. It follows that the lateral strains ϵ_y and ϵ_z will also be different, and so will the corresponding Poisson's ratios:

$$\nu_{xy} = -\frac{\epsilon_y}{\epsilon_x} \qquad \text{and} \qquad \nu_{xz} = -\frac{\epsilon_z}{\epsilon_x} \tag{2.44}$$

Note that the first subscript in each of the Poisson's ratios ν_{xy} and ν_{xz} in Eqs. (2.44) refers to the direction of the load, and the second to the direction of the contraction.

It follows from the above that, in the case of the *multiaxial loading* of a layer of a composite material, equations similar to Eqs. (2.28) of Sec. 2.12 can be used to describe the stress-strain relationship. In the present case, however, three different values of the modulus of elasticity and six different values of Poisson's ratio will be involved. We write

$$\epsilon_x = \frac{\sigma_x}{E_x} - \frac{\nu_{yx}\sigma_y}{E_y} - \frac{\nu_{zx}\sigma_z}{E_z}$$

$$\epsilon_y = -\frac{\nu_{xy}\sigma_x}{E_x} + \frac{\sigma_y}{E_y} - \frac{\nu_{zy}\sigma_z}{E_z} \tag{2.45}$$

$$\epsilon_z = -\frac{\nu_{xz}\sigma_x}{E_x} - \frac{\nu_{yz}\sigma_y}{E_y} + \frac{\sigma_z}{E_z}$$

Equations (2.45) may be considered as defining the transformation of stress into strain for the given layer. It follows from a general property of such transformations that the coefficients of the stress components are symmetric, i.e., that

$$\frac{\nu_{xy}}{E_x} = \frac{\nu_{yx}}{E_y} \qquad \frac{\nu_{yz}}{E_y} = \frac{\nu_{zy}}{E_z} \qquad \frac{\nu_{zx}}{E_z} = \frac{\nu_{xz}}{E_x} \tag{2.46}$$

These equations show that, while different, the Poisson's ratios ν_{xy} and ν_{yx} are not independent; either of them can be obtained from the other if the corresponding values of the modulus of elasticity are known. The same is true of ν_{yz} and ν_{zy}, and of ν_{zx} and ν_{xz}.

Consider now the effect of the presence of shearing stresses on the faces of a small element of smeared layer. As pointed out in

Sec. 2.14 in the case of isotropic materials, these stresses come in pairs of equal and opposite vectors applied to opposite sides of the given element and have no effect on the normal strains. Thus, Eqs. (2.45) remain valid. The shearing stresses, however, will create shearing strains which are defined by equations similar to the last three of the equations (2.38) of Sec. 2.14, except that three different values of the modulus of rigidity, G_{xy}, G_{yz}, and G_{zx}, must now be used. We have

$$\gamma_{xy} = \frac{\tau_{xy}}{G_{xy}} \qquad \gamma_{yz} = \frac{\tau_{yz}}{G_{yz}} \qquad \gamma_{zx} = \frac{\tau_{zx}}{G_{zx}} \qquad (2.47)$$

The fact that the three components of strain ϵ_x, ϵ_y, and ϵ_z can be expressed in terms of the normal stresses only and do not depend upon any shearing stresses characterizes *orthotropic materials* and distinguishes them from other anisotropic materials.

As we saw in Sec. 2.5, a flat *laminate* is obtained by superposing a number of layers or laminas. If the fibers in all layers are given the same orientation to better withstand an axial tensile load, the laminate itself will be orthotropic. If the lateral stability of the laminate is increased by positioning some of its layers so that their fibers are at a right angle to the fibers of the other layers, the resulting laminate will also be orthotropic. On the other hand, if any of the layers of a laminate are positioned so that their fibers are neither parallel nor perpendicular to the fibers of other layers, the lamina, generally, will not be orthotropic.†

EXAMPLE 2.11

A 60-mm cube is made from layers of graphite epoxy with fibers aligned in the x direction. The cube is subjected to a compressive load of 140 kN in the x direction. The properties of the composite material are: $E_x = 155.0$ GPa, $E_y = 12.10$ GPa, $E_z = 12.10$ GPa, $\nu_{xy} = 0.248$, $\nu_{xz} = 0.248$, and $\nu_{yz} = 0.458$. Determine the changes in the cube dimensions, knowing that (*a*) the cube is free to expand in the y and z directions (Fig. 2.52); (*b*) the cube is free to expand in the z direction, but is restrained from expanding in the y direction by two fixed frictionless plates (Fig. 2.53).

(a) Free in *y* and *z* Directions. We first determine the stress σ_x in the direction of loading. We have

$$\sigma_x = \frac{P}{A} = \frac{-140 \times 10^3 \text{ N}}{(0.060 \text{ m})(0.060 \text{ m})} = -38.89 \text{ MPa}$$

Since the cube is not loaded or restrained in the y and z directions, we have $\sigma_y = \sigma_z = 0$. Thus, the right-hand members of Eqs. (2.45) reduce to their first terms. Substituting the given data into these equations, we write

$$\epsilon_x = \frac{\sigma_x}{E_x} = \frac{-38.89 \text{ MPa}}{155.0 \text{ GPa}} = -250.9 \times 10^{-6}$$

$$\epsilon_y = -\frac{\nu_{xy}\sigma_x}{E_x} = -\frac{(0.248)(-38.89 \text{ MPa})}{155.0 \text{ GPa}} = +62.22 \times 10^{-6}$$

$$\epsilon_z = -\frac{\nu_{xz}\sigma_x}{E_x} = -\frac{(0.248)(-38.69 \text{ MPa})}{155.0 \text{ GPa}} = +62.22 \times 10^{-6}$$

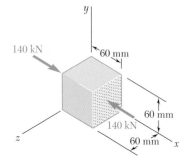

Fig. 2.52

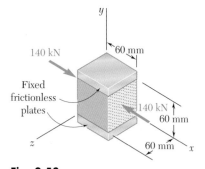

Fig. 2.53

---

†For more information on fiber-reinforced composite materials, see Hyer, M. W., *Stress Analysis of Fiber-Reinforced Composite Materials*, McGraw-Hill, New York, 1998.

The changes in the cube dimensions are obtained by multiplying the corresponding strains by the length $L = 0.060$ m of the side of the cube:

$$\delta_x = \epsilon_x L = (-250.9 \times 10^{-6})(0.060 \text{ m}) = -15.05 \ \mu\text{m}$$
$$\delta_y = \epsilon_y L = (+62.2 \times 10^{-6})(0.060 \text{ m}) = +3.73 \ \mu\text{m}$$
$$\delta_z = \epsilon_z L = (+62.2 \times 10^{-6})(0.060 \text{ m}) = +3.73 \ \mu\text{m}$$

(b) Free in z Direction, Restrained in y Direction. The stress in the x direction is the same as in part a, namely, $\sigma_x = -38.89$ MPa. Since the cube is free to expand in the z direction as in part a, we again have $\sigma_z = 0$. But since the cube is now restrained in the y direction, we should expect a stress σ_y different from zero. On the other hand, since the cube cannot expand in the y direction, we must have $\delta_y = 0$ and, thus, $\epsilon_y = \delta_y/L = 0$. Making $\sigma_z = 0$ and $\epsilon_y = 0$ in the second of Eqs. (2.45), solving that equation for σ_y, and substituting the given data, we have

$$\sigma_y = \left(\frac{E_y}{E_x}\right)\nu_{xy}\sigma_x = \left(\frac{12.10}{155.0}\right)(0.248)(-38.89 \text{ MPa})$$
$$= -752.9 \text{ kPa}$$

Now that the three components of stress have been determined, we can use the first and last of Eqs. (2.45) to compute the strain components ϵ_x and ϵ_z. But the first of these equations contains Poisson's ratio ν_{yx} and, as we saw earlier, this ratio is not equal to the ratio ν_{xy} which was among the given data. To find ν_{yx} we use the first of Eqs. (2.46) and write

$$\nu_{yx} = \left(\frac{E_y}{E_x}\right)\nu_{xy} = \left(\frac{12.10}{155.0}\right)(0.248) = 0.01936$$

Making $\sigma_z = 0$ in the first and third of Eqs. (2.45) and substituting in these equations the given values of E_x, E_y, ν_{xz}, and ν_{yz}, as well as the values obtained for σ_x, σ_y, and ν_{yx}, we have

$$\epsilon_x = \frac{\sigma_x}{E_x} - \frac{\nu_{yx}\sigma_y}{E_y} = \frac{-38.89 \text{ MPa}}{155.0 \text{ GPa}} - \frac{(0.01936)(-752.9 \text{ kPa})}{12.10 \text{ GPa}}$$
$$= -249.7 \times 10^{-6}$$

$$\epsilon_z = -\frac{\nu_{xz}\sigma_x}{E_x} - \frac{\nu_{yz}\sigma_y}{E_y} = -\frac{(0.248)(-38.89 \text{ MPa})}{155.0 \text{ GPa}} - \frac{(0.458)(-752.9 \text{ kPa})}{12.10 \text{ GPa}}$$
$$= +90.72 \times 10^{-6}$$

The changes in the cube dimensions are obtained by multiplying the corresponding strains by the length $L = 0.060$ m of the side of the cube:

$$\delta_x = \epsilon_x L = (-249.7 \times 10^{-6})(0.060 \text{ m}) = -14.98 \ \mu\text{m}$$
$$\delta_y = \epsilon_y L = (0)(0.060 \text{ m}) = 0$$
$$\delta_z = \epsilon_z L = (+90.72 \times 10^{-6})(0.060 \text{ m}) = +5.44 \ \mu\text{m}$$

Comparing the results of parts a and b, we note that the difference between the values obtained for the deformation δ_x in the direction of the fibers is negligible. However, the difference between the values obtained for the lateral deformation δ_z is not negligible. This deformation is clearly larger when the cube is restrained from deforming in the y direction.

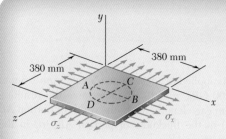

SAMPLE PROBLEM 2.5

A circle of diameter $d = 225$ mm is scribed on an unstressed aluminum plate of thickness $t = 18$ mm. Forces acting in the plane of the plate later cause normal stresses $\sigma_x = 84$ MPa and $\sigma_z = 140$ MPa. For $E = 70$ GPa and $\nu = \frac{1}{3}$, determine the change in (a) the length of diameter AB, (b) the length of diameter CD, (c) the thickness of the plate, (d) the volume of the plate.

SOLUTION

Hooke's Law. We note that $\sigma_y = 0$. Using Eqs. (2.28) we find the strain in each of the coordinate directions.

$$\epsilon_x = +\frac{\sigma_x}{E} - \frac{\nu\sigma_y}{E} - \frac{\nu\sigma_z}{E}$$

$$= \frac{1}{70 \text{ GPa}}\left[(84 \text{ MPa}) - 0 - \frac{1}{3}(140 \text{ MPa})\right] = +0.533 \times 10^{-3} \text{ mm/mm}$$

$$\epsilon_y = -\frac{\nu\sigma_x}{E} + \frac{\sigma_y}{E} - \frac{\nu\sigma_z}{E}$$

$$= \frac{1}{70 \text{ GPa}}\left[-\frac{1}{3}(84 \text{ MPa}) + 0 - \frac{1}{3}(140 \text{ MPa})\right] = -1.067 \times 10^{-3} \text{ mm/mm}$$

$$\epsilon_z = -\frac{\nu\sigma_x}{E} - \frac{\nu\sigma_y}{E} + \frac{\sigma_z}{E}$$

$$= \frac{1}{70 \text{ GPa}}\left[-\frac{1}{3}(84 \text{ MPa}) - 0 + (140 \text{ MPa})\right] = +1.600 \times 10^{-3} \text{ mm/mm}$$

a. Diameter AB. The change in length is $\delta_{B/A} = \epsilon_x d$.

$$\delta_{B/A} = \epsilon_x d = (+0.533 \times 10^{-3} \text{ mm/mm})(225 \text{ mm})$$

$$\delta_{B/A} = +0.12 \text{ mm} \quad \blacktriangleleft$$

b. Diameter CD.

$$\delta_{C/D} = \epsilon_z d = (+1.600 \times 10^{-3} \text{ mm/mm})(225 \text{ mm})$$

$$\delta_{C/D} = +0.36 \text{ mm} \quad \blacktriangleleft$$

c. Thickness. Recalling that $t = 18$ mm, we have

$$\delta_t = \epsilon_y t = (-1.067 \times 10^{-3} \text{ mm/mm})(18 \text{ mm})$$

$$\delta_t = -0.0192 \text{ mm} \quad \blacktriangleleft$$

d. Volume of the Plate. Using Eq. (2.30), we write

$$e = \epsilon_x + \epsilon_y + \epsilon_z = (+0.533 - 1.067 + 1.600)10^{-3} = +1.066 \times 10^{-3}$$

$$\Delta V = eV = +1.066 \times 10^{-3}[(380 \text{ mm})(380 \text{ mm})(18 \text{ mm})]$$

$$\Delta V = +2770 \text{ mm}^3 \quad \blacktriangleleft$$

PROBLEMS

2.61 A 2.75-kN tensile load is applied to a test coupon made from 1.6-mm flat steel plate ($E = 200$ GPa, $\nu = 0.30$). Determine the resulting change (a) in the 50-mm gage length, (b) in the width of portion AB of the test coupon, (c) in the thickness of portion AB, (d) in the cross-sectional area of portion AB.

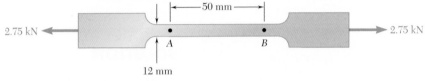

Fig. P2.61

Fig. P2.62

2.62 In a standard tensile test, an aluminum rod of 20-mm diameter is subjected to a tension force of $P = 30$ kN. Knowing that $\nu = 0.35$ and $E = 70$ GPa, determine (a) the elongation of the rod in a 150-mm gage length, (b) the change in diameter of the rod.

2.63 A 20-mm-diameter rod made of an experimental plastic is subjected to a tensile force of magnitude $P = 6$ kN. Knowing that an elongation of 14 mm and a decrease in diameter of 0.85 mm are observed in a 150-mm length, determine the modulus of elasticity, the modulus of rigidity, and Poisson's ratio for the material.

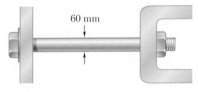

Fig. P2.64

2.64 The change in diameter of a large steel bolt is carefully measured as the nut is tightened. Knowing that $E = 200$ GPa and $\nu = 0.30$, determine the internal force in the bolt, if the diameter is observed to decrease by 13 μm.

2.65 A 2.5-m length of a steel pipe of 300-mm outer diameter and 15-mm wall thickness is used as a column to carry a 700-kN centric axial load. Knowing that $E = 200$ GPa and $\nu = 0.30$, determine (a) the change in length of the pipe, (b) the change in its outer diameter, (c) the change in its wall thickness.

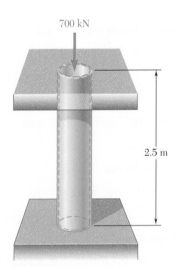

Fig. P2.65

2.66 A 30-mm square was scribed on the side of a large steel pressure vessel. After pressurization the biaxial stress condition at the square is as shown. For $E = 200$ GPa and $\nu = 0.30$, determine the change in length of (a) side AB, (b) side BC, (c) diagonal AC.

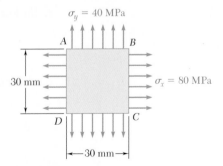

Fig. P2.66

2.67 The block shown is made of a magnesium alloy for which $E = 45$ GPa and $\nu = 0.35$. Knowing that $\sigma_x = -180$ MPa, determine (a) the magnitude of σ_y for which the change in the height of the block will be zero, (b) the corresponding change in the area of the face $ABCD$, (c) the corresponding change in the volume of the block.

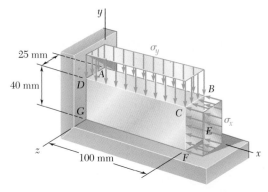

Fig. P2.67

2.68 An aluminum plate ($E = 74$ GPa, $\nu = 0.33$) is subjected to a centric axial load that causes a normal stress σ. Knowing that, before loading, a line of slope 2:1 is scribed on the plate, determine the slope of the line when $\sigma = 125$ MPa.

Fig. P2.68

2.69 The aluminum rod AD is fitted with a jacket that is used to apply a hydrostatic pressure of 42 MPa to the 300-mm portion BC of the rod. Knowing that $E = 70$ GPa and $\nu = 0.36$, determine (a) the change in the total length AD, (b) the change in diameter at the middle of the rod.

2.70 For the rod of Prob. 2.69, determine the forces that should be applied to the ends A and D of the rod (a) if the axial strain in portion BC of the rod is to remain zero as the hydrostatic pressure is applied, (b) if the total length AD of the rod is to remain unchanged.

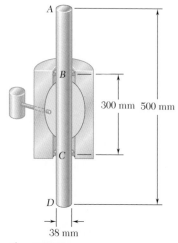

Fig. P2.69

2.71 In many situations it is known that the normal stress in a given direction is zero. For example, $\sigma_z = 0$ in the case of the thin plate shown. For this case, which is known as *plane stress*, show that if the strains ϵ_x and ϵ_y have been determined experimentally, we can express σ_x, σ_y, and ϵ_z as follows:

$$\sigma_x = E\frac{\epsilon_x + \nu\epsilon_y}{1 - \nu^2}$$

$$\sigma_y = E\frac{\epsilon_y + \nu\epsilon_x}{1 - \nu^2}$$

$$\epsilon_z = -\frac{\nu}{1 - \nu}(\epsilon_x + \epsilon_y)$$

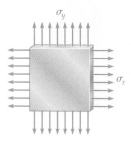

Fig. P2.71

2.72 In many situations physical constraints prevent strain from occurring in a given direction. For example, $\epsilon_z = 0$ in the case shown, where longitudinal movement of the long prism is prevented at every point. Plane sections perpendicular to the longitudinal axis remain plane and the same distance apart. Show that for this situation, which is known as *plane strain*, we can express σ_z, ϵ_x, and ϵ_y as follows:

$$\sigma_z = \nu(\sigma_x + \sigma_y)$$

$$\epsilon_x = \frac{1}{E}[(1 - \nu^2)\sigma_x - \nu(1 + \nu)\sigma_y]$$

$$\epsilon_y = \frac{1}{E}[(1 - \nu^2)\sigma_y - \nu(1 + \nu)\sigma_x]$$

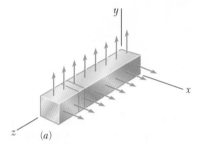

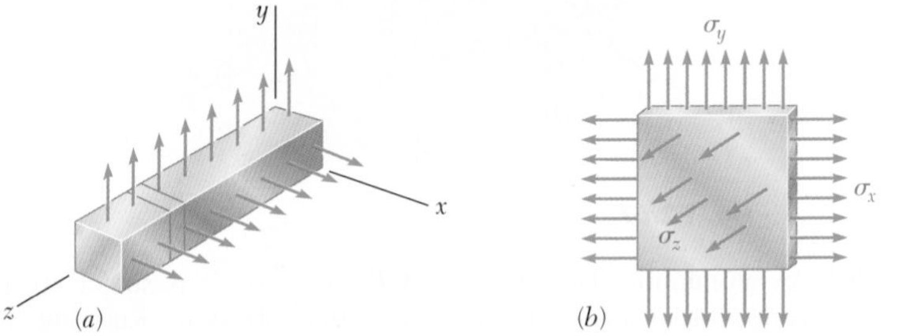

(a) (b)

Fig. P2.72

2.73 For a member under axial loading, express the normal strain ϵ' in a direction forming an angle of 45° with the axis of the load in terms of the axial strain ϵ_x by (a) comparing the hypotenuses of the triangles shown in Fig. 2.50, which represent respectively an element before and after deformation, (b) using the values of the corresponding stresses σ' and σ_x shown in Fig. 1.38, and the generalized Hooke's law.

2.74 The homogeneous plate *ABCD* is subjected to a biaxial loading as shown. It is known that $\sigma_z = \sigma_0$ and that the change in length of the plate in the x direction must be zero, that is, $\epsilon_x = 0$. Denoting by E the modulus of elasticity and by ν Poisson's ratio, determine (a) the required magnitude of σ_x, (b) the ratio σ_0/ϵ_z.

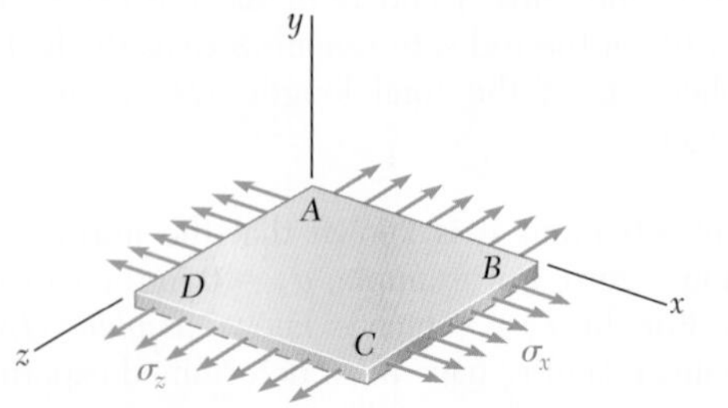

Fig. P2.74

2.75 An elastomeric bearing ($G = 0.9$ MPa) is used to support a bridge girder as shown to provide flexibility during earthquakes. The beam must not displace more than 10 mm when a 22-kN lateral load is applied as shown. Knowing that the maximum allowable shearing stress is 420 kPa, determine (a) the smallest allowable dimension b, (b) the smallest required thickness a.

Fig. P2.75

2.76 Two blocks of rubber, each of width $w = 60$ mm, are bonded to rigid supports and to the movable plate AB. Knowing that a force of magnitude $P = 19$ kN causes a deflection $\delta = 3$ mm, determine the modulus of rigidity of the rubber used.

2.77 The plastic block shown is bonded to a fixed base and to a horizontal rigid plate to which a force **P** is applied. Knowing that for the plastic used $G = 385$ MPa, determine the deflection of the plate when $P = 36$ kN.

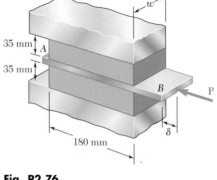

Fig. P2.76

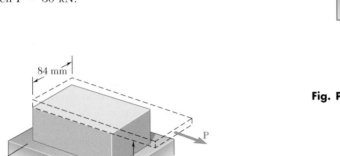

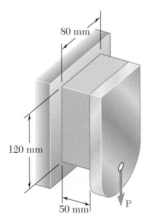

Fig. P2.77

2.78 A vibration isolation unit consists of two blocks of hard rubber bonded to plate AB and to rigid supports as shown. For the type and grade of rubber used $\tau_{all} = 1.54$ MPa and $G = 12.6$ MPa. Knowing that a centric vertical force of magnitude $P = 13$ kN must cause a 2.5-mm vertical deflection of the plate AB, determine the smallest allowable dimensions a and b of the block.

2.79 The plastic block shown is bonded to a rigid support and to a vertical plate to which a 240-kN load **P** is applied. Knowing that for the plastic used $G = 1050$ MPa, determine the deflection of the plate.

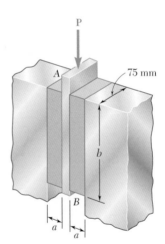

Fig. P2.78

Fig. P2.79

2.80 What load **P** should be applied to the plate of Prob. 2.79 to produce a 1.5-mm deflection?

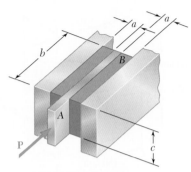

Fig. P2.81 and P2.82

2.81 Two blocks of rubber with a modulus of rigidity $G = 12$ MPa are bonded to rigid supports and to a plate AB. Knowing that $c = 100$ mm and $P = 45$ kN, determine the smallest allowable dimensions a and b of the blocks if the shearing stress in the rubber is not to exceed 1.4 MPa and the deflection of the plate is to be at least 5 mm.

2.82 Two blocks of rubber with a modulus of rigidity $G = 10$ MPa are bonded to rigid supports and to a plate AB. Knowing that $b = 200$ mm and $c = 125$ mm, determine the largest allowable load P and the smallest allowable thickness a of the blocks if the shearing stress in the rubber is not to exceed 1.5 MPa and the deflection of the plate is to be at least 6 mm.

*2.83 Determine the dilatation e and the change in volume of the 200-mm length of the rod shown if (a) the rod is made of steel with $E = 200$ GPa and $\nu = 0.30$, (b) the rod is made of aluminum with $E = 70$ GPa and $\nu = 0.35$.

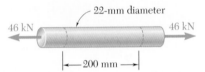

Fig. P2.83

*2.84 Determine the change in volume of the 50-mm gage length segment AB in Prob. 2.61 (a) by computing the dilatation of the material, (b) by subtracting the original volume of portion AB from its final volume.

*2.85 A 150-mm-diameter solid steel sphere is lowered into the ocean to a point where the pressure is 50 MPa (about 5 km below the surface). Knowing that $E = 200$ GPa and $\nu = 0.30$, determine (a) the decrease in diameter of the sphere, (b) the decrease in volume of the sphere, (c) the percent increase in the density of the sphere.

*2.86 (a) For the axial loading shown, determine the change in height and the change in volume of the brass cylinder shown. (b) Solve part a, assuming that the loading is hydrostatic with $\sigma_x = \sigma_y = \sigma_z = -70$ MPa.

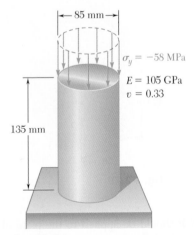

Fig. P2.86

*2.87 A vibration isolation support consists of a rod A of radius $R_1 = 10$ mm and a tube B of inner radius $R_2 = 25$ mm bonded to an 80-mm-long hollow rubber cylinder with a modulus of rigidity $G = 12$ MPa. Determine the largest allowable force **P** that can be applied to rod A if its deflection is not to exceed 2.50 mm.

*2.88 A vibration isolation support consists of a rod A of radius R_1 and a tube B of inner radius R_2 bonded to an 80-mm-long hollow rubber cylinder with a modulus of rigidity $G = 10.93$ MPa. Determine the required value of the ratio R_2/R_1 if a 10-kN force **P** is to cause a 2-mm deflection of rod A.

*2.89 The material constants E, G, k, and ν are related by Eqs. (2.33) and (2.43). Show that any one of the constants may be expressed in terms of any other two constants. For example, show that (a) $k = GE/(9G - 3E)$ and (b) $\nu = (3k - 2G)/(6k + 2G)$.

*2.90 Show that for any given material, the ratio G/E of the modulus of rigidity over the modulus of elasticity is always less than $\frac{1}{2}$ but more than $\frac{1}{3}$. [*Hint:* Refer to Eq. (2.43) and to Sec. 2.13.]

*2.91 A composite cube with 40-mm sides and the properties shown is made with glass polymer fibers aligned in the x direction. The cube is constrained against deformations in the y and z directions and is subjected to a tensile load of 65 kN in the x direction. Determine (a) the change in the length of the cube in the x direction, (b) the stresses σ_x, σ_y, and σ_z.

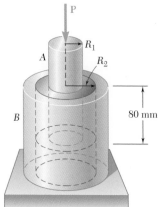

Fig. P2.87 and P2.88

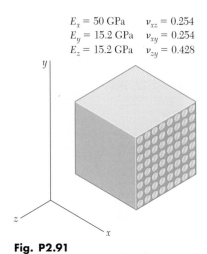

$$E_x = 50 \text{ GPa} \qquad \nu_{xz} = 0.254$$
$$E_y = 15.2 \text{ GPa} \qquad \nu_{xy} = 0.254$$
$$E_z = 15.2 \text{ GPa} \qquad \nu_{zy} = 0.428$$

Fig. P2.91

*2.92 The composite cube of Prob. 2.91 is constrained against deformation in the z direction and elongated in the x direction by 0.035 mm due to a tensile load in the x direction. Determine (a) the stresses σ_x, σ_y, and σ_z, (b) the change in the dimension in the y direction.

2.17 STRESS AND STRAIN DISTRIBUTION UNDER AXIAL LOADING; SAINT-VENANT'S PRINCIPLE

We have assumed so far that, in an axially loaded member, the normal stresses are uniformly distributed in any section perpendicular to the axis of the member. As we saw in Sec. 1.5, such an assumption may be quite in error in the immediate vicinity of the points of application of the loads. However, the determination of the actual stresses in a given section of the member requires the solution of a statically indeterminate problem.

In Sec. 2.9, you saw that statically indeterminate problems involving the determination of *forces* can be solved by considering the *deformations* caused by these forces. It is thus reasonable to conclude that the determination of the *stresses* in a member requires the analysis of the *strains* produced by the stresses in the member. This is essentially the approach found in advanced textbooks, where the mathematical theory of elasticity is used to determine the distribution of stresses corresponding to various modes of application of the loads at the ends of the member. Given the more limited mathematical means at our disposal, our analysis of stresses will be restricted to the particular case when two rigid plates are used to transmit the loads to a member made of a homogeneous isotropic material (Fig. 2.54).

If the loads are applied at the center of each plate,† the plates will move toward each other without rotating, causing the member to get shorter, while increasing in width and thickness. It is reasonable to assume that the member will remain straight, that plane sections will remain plane, and that all elements of the member will deform in the same way, since such an assumption is clearly compatible with the given end conditions. This is illustrated in Fig. 2.55,

Fig. 2.54 Axial load applied by rigid plates to a member.

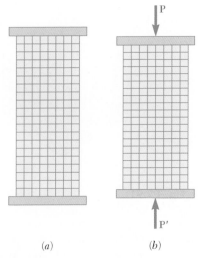

(a)　　　　(b)

Fig. 2.55 Axial load applied by rigid plates to rubber model.

†More precisely, the common line of action of the loads should pass through the centroid of the cross section (cf. Sec. 1.5).

which shows a rubber model before and after loading.† Now, if all elements deform in the same way, the distribution of strains throughout the member must be uniform. In other words, the axial strain ϵ_y and the lateral strain $\epsilon_x = -\nu\epsilon_y$ are constant. But, if the stresses do not exceed the proportional limit, Hooke's law applies and we may write $\sigma_y = E\epsilon_y$, from which it follows that the normal stress σ_y is also constant. Thus, the distribution of stresses is uniform throughout the member and, at any point,

$$\sigma_y = (\sigma_y)_{\text{ave}} = \frac{P}{A}$$

On the other hand, if the loads are concentrated, as illustrated in Fig. 2.56, the elements in the immediate vicinity of the points of application of the loads are subjected to very large stresses, while other elements near the ends of the member are unaffected by the loading. This may be verified by observing that strong deformations, and thus large strains and large stresses, occur near the points of application of the loads, while no deformation takes place at the corners. As we consider elements farther and farther from the ends, however, we note a progressive equalization of the deformations involved, and thus a more nearly uniform distribution of the strains and stresses across a section of the member. This is further illustrated in Fig. 2.57, which shows the result of the calculation by advanced mathematical methods of the distribution of stresses across various sections of a thin rectangular plate subjected to concentrated loads. We note that at a distance b from either end, where b is the width of the plate, the stress distribution is nearly uniform across the section, and the value of the stress σ_y at any point of that section can be assumed equal to the average value P/A. Thus, at a distance equal to, or greater than, the width of the member, the distribution of stresses across a given section is the same, whether the member is loaded as shown in Fig. 2.54 or Fig. 2.56. In other words, except in the immediate vicinity of the points of application of the loads, the stress distribution may be assumed independent of the actual mode of application of the loads. This statement, which applies not only to axial loadings, but to practically any type of load, is known as *Saint-Venant's principle,* after the French mathematician and engineer Adhémar Barré de Saint-Venant (1797–1886).

While Saint-Venant's principle makes it possible to replace a given loading by a simpler one for the purpose of computing the stresses in a structural member, you should keep in mind two important points when applying this principle:

1. The actual loading and the loading used to compute the stresses must be *statically equivalent.*
2. Stresses cannot be computed in this manner in the immediate vicinity of the points of application of the loads. Advanced theoretical or experimental methods must be used to determine the distribution of stresses in these areas.

†Note that for long, slender members, another configuration is possible, and indeed will prevail, if the load is sufficiently large; the member *buckles* and assumes a curved shape. This will be discussed in Chap. 10.

Fig. 2.56 Concentrated axial load applied to rubber model.

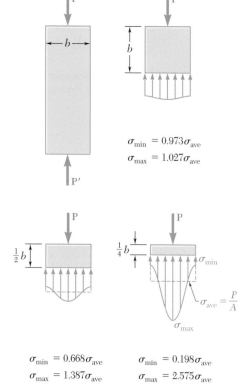

$\sigma_{\text{min}} = 0.973\sigma_{\text{ave}}$
$\sigma_{\text{max}} = 1.027\sigma_{\text{ave}}$

$\sigma_{\text{ave}} = \frac{P}{A}$

$\sigma_{\text{min}} = 0.668\sigma_{\text{ave}}$
$\sigma_{\text{max}} = 1.387\sigma_{\text{ave}}$

$\sigma_{\text{min}} = 0.198\sigma_{\text{ave}}$
$\sigma_{\text{max}} = 2.575\sigma_{\text{ave}}$

Fig. 2.57 Stress distributions in a plate under concentrated axial loads.

You should also observe that the plates used to obtain a uniform stress distribution in the member of Fig. 2.55 must allow the member to freely expand laterally. Thus, the plates cannot be rigidly attached to the member; you must assume them to be just in contact with the member, and smooth enough not to impede the lateral expansion of the member. While such end conditions can actually be achieved for a member in compression, they cannot be physically realized in the case of a member in tension. It does not matter, however, whether or not an actual fixture can be realized and used to load a member so that the distribution of stresses in the member is uniform. The important thing is to *imagine a model* that will allow such a distribution of stresses, and to keep this model in mind so that you may later compare it with the actual loading conditions.

2.18 STRESS CONCENTRATIONS

As you saw in the preceding section, the stresses near the points of application of concentrated loads can reach values much larger than the average value of the stress in the member. When a structural member contains a discontinuity, such as a hole or a sudden change in cross section, high localized stresses can also occur near the discontinuity. Figures 2.58 and 2.59 show the distribution of stresses in critical sections corresponding to two such situations. Figure 2.58 refers to a flat bar with a *circular hole* and shows the stress distribution in a section passing through the center of the hole. Figure 2.59 refers to a flat bar consisting of two portions of different widths connected by *fillets;* it shows the stress distribution in the narrowest part of the connection, where the highest stresses occur.

These results were obtained experimentally through the use of a photoelastic method. Fortunately for the engineer who has to design a given member and cannot afford to carry out such an analysis, the results obtained are independent of the size of the member and of the material used; they depend only upon the ratios of the geometric parameters involved, i.e., upon the ratio r/d in the case of a circular hole, and upon the ratios r/d and D/d in the case of fillets. Furthermore, the designer is more interested in the *maximum value* of the stress in a given section, than in the actual distribution of stresses in that section, since the main concern is to determine *whether* the allowable stress will be exceeded under a given loading, and not *where* this value will be exceeded. For this reason, one defines the ratio

$$K = \frac{\sigma_{\max}}{\sigma_{\text{ave}}} \qquad (2.48)$$

of the maximum stress over the average stress computed in the critical (narrowest) section of the discontinuity. This ratio is referred to as the *stress-concentration factor* of the given discontinuity. Stress-concentration factors can be computed once and for all in terms of the ratios of the geometric parameters involved, and the results obtained can be expressed in the form of tables or of graphs, as shown in Fig. 2.60. To determine the maximum stress occurring near a

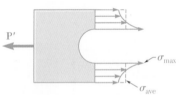

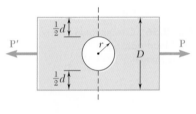

Fig. 2.58 Stress distribution near circular hole in flat bar under axial loading.

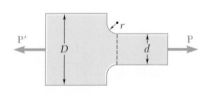

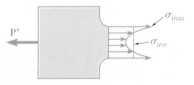

Fig. 2.59 Stress distribution near fillets in flat bar under axial loading.

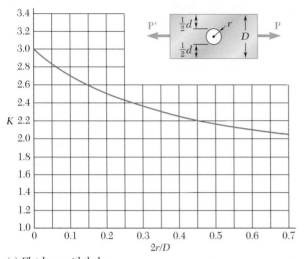

(a) Flat bars with holes

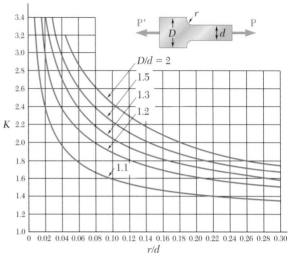

(b) Flat bars with fillets

Fig. 2.60 Stress concentration factors for flat bars under axial loading.[†]

Note that the average stress must be computed across the narrowest section: $\sigma_{ave} = P/td$, where t is the thickness of the bar.

discontinuity in a given member subjected to a given axial load P, the designer needs only to compute the average stress $\sigma_{ave} = P/A$ in the critical section, and multiply the result obtained by the appropriate value of the stress-concentration factor K. You should note, however, that this procedure is valid only as long as σ_{max} does not exceed the proportional limit of the material, since the values of K plotted in Fig. 2.60 were obtained by assuming a linear relation between stress and strain.

EXAMPLE 2.12

Determine the largest axial load **P** that can be safely supported by a flat steel bar consisting of two portions, both 10 mm thick and, respectively, 40 and 60 mm wide, connected by fillets of radius $r = 8$ mm. Assume an allowable normal stress of 165 MPa.

We first compute the ratios

$$\frac{D}{d} = \frac{60 \text{ mm}}{40 \text{ mm}} = 1.50 \qquad \frac{r}{d} = \frac{8 \text{ mm}}{40 \text{ mm}} = 0.20$$

Using the curve in Fig. 2.60b corresponding to $D/d = 1.50$, we find that the value of the stress-concentration factor corresponding to $r/d = 0.20$ is

$$K = 1.82$$

Carrying this value into Eq. (2.48) and solving for σ_{ave}, we have

$$\sigma_{ave} = \frac{\sigma_{max}}{1.82}$$

But σ_{max} cannot exceed the allowable stress $\sigma_{all} = 165$ MPa. Substituting this value for σ_{max}, we find that the average stress in the narrower portion ($d = 40$ mm) of the bar should not exceed the value

$$\sigma_{ave} = \frac{165 \text{ MPa}}{1.82} = 90.7 \text{ MPa}$$

Recalling that $\sigma_{ave} = P/A$, we have

$$P = A\sigma_{ave} = (40 \text{ mm})(10 \text{ mm})(90.7 \text{ MPa}) = 36.3 \times 10^3 \text{ N}$$
$$P = 36.3 \text{ kN}$$

[†]W. D. Pilkey, *Peterson's Stress Concentration Factors*, 2nd ed., John Wiley & Sons, New York, 1997.

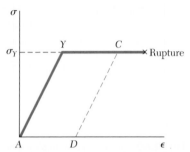

Fig. 2.61 Stress-strain diagram for an idealized elastoplastic material.

2.19 PLASTIC DEFORMATIONS

The results obtained in the preceding sections were based on the assumption of a linear stress-strain relationship. In other words, we assumed that the proportional limit of the material was never exceeded. This is a reasonable assumption in the case of brittle materials, which rupture without yielding. In the case of ductile materials, however, this assumption implies that the yield strength of the material is not exceeded. The deformations will then remain within the elastic range and the structural member under consideration will regain its original shape after all loads have been removed. If, on the other hand, the stresses in any part of the member exceed the yield strength of the material, plastic deformations occur and most of the results obtained in earlier sections cease to be valid. A more involved analysis, based on a nonlinear stress-strain relationship, must then be carried out.

While an analysis taking into account the actual stress-strain relationship is beyond the scope of this text, we gain considerable insight into plastic behavior by considering an idealized *elastoplastic material* for which the stress-strain diagram consists of the two straight-line segments shown in Fig. 2.61. We may note that the stress-strain diagram for mild steel in the elastic and plastic ranges is similar to this idealization. As long as the stress σ is less than the yield strength σ_Y, the material behaves elastically and obeys Hooke's law, $\sigma = E\epsilon$. When σ reaches the value σ_Y, the material starts yielding and keeps deforming plastically under a constant load. If the load is removed, unloading takes place along a straight-line segment CD parallel to the initial portion AY of the loading curve. The segment AD of the horizontal axis represents the strain corresponding to the permanent set or plastic deformation resulting from the loading and unloading of the specimen. While no actual material behaves exactly as shown in Fig. 2.61, this stress-strain diagram will prove useful in discussing the plastic deformations of ductile materials such as mild steel.

EXAMPLE 2.13

A rod of length $L = 500$ mm and cross-sectional area $A = 60$ mm² is made of an elastoplastic material having a modulus of elasticity $E = 200$ GPa in its elastic range and a yield point $\sigma_Y = 300$ MPa. The rod is subjected to an axial load until it is stretched 7 mm and the load is then removed. What is the resulting permanent set?

Referring to the diagram of Fig. 2.61, we find that the maximum strain, represented by the abscissa of point C, is

$$\epsilon_C = \frac{\delta_C}{L} = \frac{7 \text{ mm}}{500 \text{ mm}} = 14 \times 10^{-3}$$

On the other hand, the yield strain, represented by the abscissa of point Y, is

$$\epsilon_Y = \frac{\sigma_Y}{E} = \frac{300 \times 10^6 \text{ Pa}}{200 \times 10^9 \text{ Pa}} = 1.5 \times 10^{-3}$$

The strain after unloading is represented by the abscissa ϵ_D of point D. We note from Fig. 2.61 that

$$\epsilon_D = AD = YC = \epsilon_C - \epsilon_Y$$
$$= 14 \times 10^{-3} - 1.5 \times 10^{-3} = 12.5 \times 10^{-3}$$

The permanent set is the deformation δ_D corresponding to the strain ϵ_D. We have

$$\delta_D = \epsilon_D L = (12.5 \times 10^{-3})(500 \text{ mm}) = 6.25 \text{ mm}$$

EXAMPLE 2.14

A 0.75-m-long cylindrical rod of cross-sectional area $A_r = 48$ mm² is placed inside a tube of the same length and of cross-sectional area $A_t = 62$ mm². The ends of the rod and tube are attached to a rigid support on one side, and to a rigid plate on the other, as shown in the longitudinal section of Fig. 2.62. The rod and tube are both assumed to be elastoplastic, with moduli of elasticity $E_r = 210$ GPa and $E_t = 105$ GPa and yield strengths $(\sigma_r)_Y = 250$ MPa and $(\sigma_t)_Y = 310$ MPa. Draw the load-deflection diagram of the rod-tube assembly when a load **P** is applied to the plate as shown.

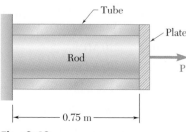

Fig. 2.62

We first determine the internal force and the elongation of the rod as it begins to yield:

$$(P_r)_Y = (\sigma_r)_Y A_r = (250 \times 10^6 \text{ Pa})(48 \times 10^{-6} \text{ m}^2) = 12 \text{ kN}$$

$$(\delta_r)_Y = (\epsilon_r)_Y L = \frac{(\sigma_r)_Y}{E_r} L = \frac{250 \times 10^6 \text{ Pa}}{210 \times 10^9 \text{ Pa}}(0.75 \text{ m})$$
$$= 0.893 \times 10^{-3} \text{ m} = 0.893 \text{ mm}$$

Since the material is elastoplastic, the force-elongation diagram *of the rod alone* consists of an oblique straight line and of a horizontal straight line, as shown in Fig. 2.63a. Following the same procedure for the tube, we have

$$(P_t)_Y = (\sigma_t)_Y A_t = (310 \times 10^6 \text{ Pa})(62 \times 10^{-6} \text{ m}^2) = 19.22 \text{ kN}$$

$$(\delta_t)_Y = (\epsilon_t)_Y L = \frac{(\sigma_t)_Y}{E_t} L = \frac{310 \times 10^6 \text{ Pa}}{105 \times 10^9 \text{ Pa}}(0.75 \text{ m})$$
$$= 2.21 \times 10^{-3} \text{ m} = 2.21 \text{ mm}$$

The load-deflection diagram of the tube alone is shown in Fig. 2.63b. Observing that the load and deflection of the rod-tube combination are, respectively,

$$P = P_r + P_t \qquad \delta = \delta_r = \delta_t$$

we draw the required load-deflection diagram by adding the ordinates of the diagrams obtained for the rod and for the tube (Fig. 2.63c). Points Y_r and Y_t correspond to the onset of yield in the rod and in the tube, respectively.

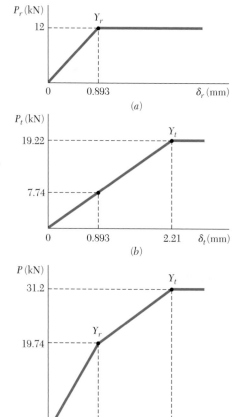

Fig. 2.63

139

EXAMPLE 2.15

If the load **P** applied to the rod-tube assembly of Example 2.14 is increased from zero to 25 kN and decreased back to zero, determine (a) the maximum elongation of the assembly, (b) the permanent set after the load has been removed.

(a) Maximum Elongation. Referring to Fig. 2.63c, we observe that the load $P_{max} = 25$ kN corresponds to a point located on the segment Y_rY_t of the load-deflection diagram of the assembly. Thus, the rod has reached the plastic range, with $P_r = (P_r)_Y = 12$ kN and $\sigma_r = (\sigma_r)_Y = 250$ MPa, while the tube is still in the elastic range, with

$$P_t = P - P_r = 25 \text{ kN} - 12 \text{ kN} = 13 \text{ kN}$$

$$\sigma_t = \frac{P_t}{A_t} = \frac{13 \text{ kN}}{62 \text{ mm}^2} = 210 \text{ MPa}$$

$$\delta_t = \epsilon_t L = \frac{\sigma_t}{E_t} L = \frac{210 \text{ MPa}}{105 \text{ GPa}}(0.75 \text{ m}) = 1.5 \text{ mm}$$

The maximum elongation of the assembly, therefore, is

$$\delta_{max} = \delta_t = 1.5 \text{ mm}$$

(b) Permanent Set. As the load **P** decreases from 25 kN to zero, the internal forces P_r and P_t both decrease along a straight line, as shown in Fig. 2.64a and b, respectively. The force P_r decreases along line CD parallel to the initial portion of the loading curve, while the force P_t decreases along the original loading curve, since the yield stress was not exceeded in the tube. Their sum P, therefore, will decrease along a line CE parallel to the portion $0Y_r$ of the load-deflection curve of the assembly (Fig. 2.64c). Referring to Fig. 2.63c, we find that the slope of $0Y_r$, and thus of CE, is

$$m = \frac{19.74 \text{ kN}}{0.893 \text{ mm}} = 22.1 \text{ kN/mm}$$

The segment of line FE in Fig. 2.64c represents the deformation δ' of the assembly during the unloading phase, and the segment 0E the permanent set δ_p after the load **P** has been removed. From triangle CEF we have

$$\delta' = -\frac{P_{max}}{m} = -\frac{25 \text{ kN}}{22.1 \text{ kN/mm}} = -1.131 \text{ mm}$$

The permanent set is thus

$$\delta_P = \delta_{max} + \delta' = 1.5 \text{ mm} - 1.131 \text{ mm}$$
$$= 0.369 \text{ mm}$$

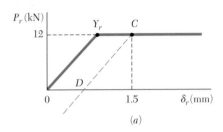

(a)

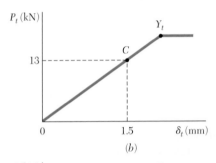

(b)

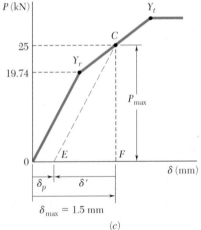

(c)

Fig. 2.64

We recall that the discussion of stress concentrations of Sec. 2.18 was carried out under the assumption of a linear stress-strain relationship. The stress distributions shown in Figs. 2.58 and 2.59, and the values of the stress-concentration factors plotted in Fig. 2.60 cannot be used, therefore, when plastic deformations take place, i.e., when the value of σ_{max} obtained from these figures exceeds the yield strength σ_Y.

Let us consider again the flat bar with a circular hole of Fig. 2.58, and let us assume that the material is elastoplastic, i.e., that its stress-strain diagram is as shown in Fig. 2.61. As long as no plastic deformation takes place, the distribution of stresses is as indicated in Sec. 2.18 (Fig. 2.65a). We observe that the area under the stress-distribution curve represents the integral $\int \sigma \, dA$, which is equal to the load P.

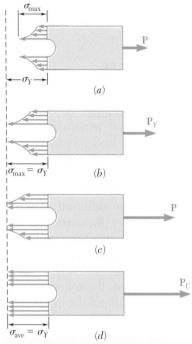

Fig. 2.65 Distribution of stresses in elastoplastic material under increasing load.

Thus this area, and the value of σ_{max}, must increase as the load P increases. As long as $\sigma_{max} \leq \sigma_Y$, all the successive stress distributions obtained as P increases will have the shape shown in Fig. 2.58 and repeated in Fig. 2.65a. However, as P is increased beyond the value P_Y corresponding to $\sigma_{max} = \sigma_Y$ (Fig. 2.65b), the stress-distribution curve must flatten in the vicinity of the hole (Fig. 2.65c), since the stress in the material considered cannot exceed the value σ_Y. This indicates that the material is yielding in the vicinity of the hole. As the load P is further increased, the plastic zone where yield takes place keeps expanding, until it reaches the edges of the plate (Fig. 2.65d). At that point, the distribution of stresses across the plate is uniform, $\sigma = \sigma_Y$, and the corresponding value $P = P_U$ of the load is the largest that can be applied to the bar without causing rupture.

It is interesting to compare the maximum value P_Y of the load that can be applied if no permanent deformation is to be produced in the bar, with the value P_U that will cause rupture. Recalling the definition of the average stress, $\sigma_{ave} = P/A$, where A is the net cross-sectional area, and the definition of the stress concentration factor, $K = \sigma_{max}/\sigma_{ave}$, we write

$$P = \sigma_{ave} A = \frac{\sigma_{max} A}{K} \tag{2.49}$$

for any value of σ_{max} that does not exceed σ_Y. When $\sigma_{max} = \sigma_Y$ (Fig. 2.65b), we have $P = P_Y$, and Eq. (2.49) yields

$$P_Y = \frac{\sigma_Y A}{K} \tag{2.50}$$

On the other hand, when $P = P_U$ (Fig. 2.65d), we have $\sigma_{ave} = \sigma_Y$ and

$$P_U = \sigma_Y A \tag{2.51}$$

Comparing Eqs. (2.50) and (2.51), we conclude that

$$P_Y = \frac{P_U}{K} \tag{2.52}$$

*2.20 RESIDUAL STRESSES

In Example 2.13 of the preceding section, we considered a rod that was stretched beyond the yield point. As the load was removed, the rod did not regain its original length; it had been permanently deformed. However, after the load was removed, all stresses disappeared. You should not assume that this will always be the case. Indeed, when only some of the parts of an indeterminate structure undergo plastic deformations, as in Example 2.15, or when different parts of the structure undergo different plastic deformations, the stresses in the various parts of the structure will not, in general, return to zero after the load has been removed. Stresses, called *residual stresses*, will remain in the various parts of the structure.

While the computation of the residual stresses in an actual structure can be quite involved, the following example will provide you with a general understanding of the method to be used for their determination.

EXAMPLE 2.16

Determine the residual stresses in the rod and tube of Examples 2.14 and 2.15 after the load **P** has been increased from zero to 25 kN and decreased back to zero.

We observe from the diagrams of Fig. 2.66 that after the load **P** has returned to zero, the internal forces P_r and P_t are *not* equal to zero. Their values have been indicated by point E in parts a and b, respectively, of Fig. 2.66. It follows that the corresponding stresses are not equal to zero either after the assembly has been unloaded. To determine these residual stresses, we shall determine the reverse stresses σ'_r and σ'_t caused by the unloading and add them to the maximum stresses $\sigma_r = 250$ MPa and $\sigma_t = 210$ MPa found in part a of Example 2.15.

The strain caused by the unloading is the same in the rod and in the tube. It is equal to δ'/L where δ' is the deformation of the assembly during unloading, which was found in Example 2.15. We have

$$\epsilon' = \frac{\delta'}{L} = \frac{-1.131 \text{ mm}}{750 \text{ mm}} = -1.508 \times 10^{-3} \text{ mm/mm}$$

The corresponding reverse stresses in the rod and tube are

$$\sigma'_r = \epsilon' E_r = (-1.508 \times 10^{-3})(210 \times 10^9 \text{ Pa}) = -316.7 \text{ MPa}$$

$$\sigma'_t = \epsilon' E_t = (-1.508 \times 10^{-3})(105 \times 10^9 \text{ Pa}) = -158.34 \text{ MPa}$$

The residual stresses are found by superposing the stresses due to loading and the reverse stresses due to unloading. We have

$$(\sigma_r)_{\text{res}} = \sigma_r + \sigma'_r = 250 \text{ MPa} - 316.7 \text{ MPa} = -66.7 \text{ MPa}$$

$$(\sigma_t)_{\text{res}} = \sigma_t + \sigma'_t = 210 \text{ MPa} - 158.34 \text{ MPa} = 51.7 \text{ MPa}$$

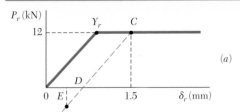

(a)

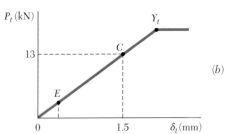

(b)

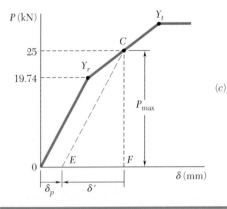

(c)

Fig. 2.66

Plastic deformations caused by temperature changes can also result in residual stresses. For example, consider a small plug that is to be welded to a large plate. For discussion purposes the plug will be considered as a small rod AB that is to be welded across a small hole in the plate (Fig. 2.67). During the welding process the temperature of the rod will be raised to over 1000°C, at which temperature its modulus of elasticity, and hence its stiffness and stress, will be almost zero. Since the plate is large, its temperature will not be increased significantly above room temperature (20°C). Thus, when the welding is completed, we have rod AB at $T = 1000$°C, with no stress, attached to the plate, which is at 20°C.

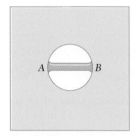

Fig. 2.67 Small rod welded to a large plate.

As the rod cools, its modulus of elasticity increases and, at about 500°C, will approach its normal value of about 200 GPa. As the temperature of the rod decreases further, we have a situation similar to that considered in Sec. 2.10 and illustrated in Fig. 2.31. Solving Eq. (2.23) for ΔT and making σ equal to the yield strength, $\sigma_Y = 300$ MPa, of average steel, and $\alpha = 12 \times 10^{-6}$/°C, we find the temperature change that will cause the rod to yield:

$$\Delta T = -\frac{\sigma}{E\alpha} = -\frac{300 \text{ MPa}}{(200 \text{ GPa})(12 \times 10^{-6}/°\text{C})} = -125°\text{C}$$

This means that the rod will start yielding at about 375°C and will keep yielding at a fairly constant stress level, as it cools down to room temperature. As a result of the welding operation, a residual stress approximately equal to the yield strength of the steel used is thus created in the plug and in the weld.

Residual stresses also occur as a result of the cooling of metals that have been cast or hot rolled. In these cases, the outer layers cool more rapidly than the inner core. This causes the outer layers to reacquire their stiffness (E returns to its normal value) faster than the inner core. When the entire specimen has returned to room temperature, the inner core will have contracted more than the outer layers. The result is residual longitudinal tensile stresses in the inner core and residual compressive stresses in the outer layers.

Residual stresses due to welding, casting, and hot rolling can be quite large (of the order of magnitude of the yield strength). These stresses can be removed, when necessary, by reheating the entire specimen to about 600°C, and then allowing it to cool slowly over a period of 12 to 24 hours.

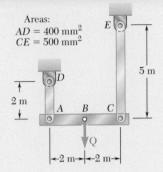

Areas:
$AD = 400$ mm^2
$CE = 500$ mm^2

2 m

5 m

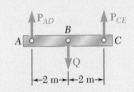

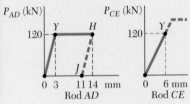

Load-deflection diagrams

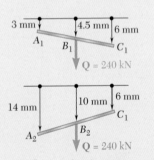

(a) Deflections for $\delta_B = 10$ mm ◀

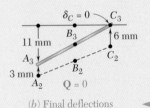

(b) Final deflections ◀

SAMPLE PROBLEM 2.6

The rigid beam ABC is suspended from two steel rods as shown and is initially horizontal. The midpoint B of the beam is deflected 10 mm downward by the slow application of the force $\mathbf{Q}$, after which the force is slowly removed. Knowing that the steel used for the rods is elastoplastic with $E = 200$ GPa and $\sigma_Y = 300$ MPa, determine (a) the required maximum value of Q and the corresponding position of the beam, (b) the final position of the beam.

SOLUTION

Statics. Since $\mathbf{Q}$ is applied at the midpoint of the beam, we have

$$P_{AD} = P_{CE} \qquad \text{and} \qquad Q = 2P_{AD}$$

Elastic Action. The maximum value of Q and the maximum elastic deflection of point A occur when $\sigma = \sigma_Y$ in rod AD.

$$(P_{AD})_{max} = \sigma_Y A = (300 \text{ MPa})(400 \text{ mm}^2) = 120 \text{ kN}$$

$$Q_{max} = 2(P_{AD})_{max} = 2(120 \text{ kN}) \qquad Q_{max} = 240 \text{ kN} \quad ◀$$

$$\delta_{A_1} = \epsilon L = \frac{\sigma_Y}{E} L = \left(\frac{300 \text{ MPa}}{200 \text{ GPa}}\right)(2 \text{ m}) = 3 \text{ mm}$$

Since $P_{CE} = P_{AD} = 120$ kN, the stress in rod CE is

$$\sigma_{CE} = \frac{P_{CE}}{A} = \frac{120 \text{ kN}}{500 \text{ mm}^2} = 240 \text{ MPa}$$

The corresponding deflection of point C is

$$\delta_{C_1} = \epsilon L = \frac{\sigma_{CE}}{E} L = \left(\frac{240 \text{ MPa}}{200 \text{ GPa}}\right)(5 \text{ m}) = 6 \text{ mm}$$

The corresponding deflection of point B is

$$\delta_{B_1} = \tfrac{1}{2}(\delta_{A_1} + \delta_{C_1}) = \tfrac{1}{2}(3 \text{ mm} + 6 \text{ mm}) = 4.5 \text{ mm}$$

Since we must have $\delta_B = 10$ mm, we conclude that plastic deformation will occur.

Plastic Deformation. For $Q = 240$ kN, plastic deformation occurs in rod AD, where $\sigma_{AD} = \sigma_Y = 300$ MPa. Since the stress in rod CE is within the elastic range, δ_C remains equal to 6 mm. The deflection δ_A for which $\delta_B = 10$ mm is obtained by writing

$$\delta_{B_2} = 10 \text{ mm} = \tfrac{1}{2}(\delta_{A_2} + 6 \text{ mm}) \qquad \delta_{A_2} = 14 \text{ mm}$$

Unloading. As force $\mathbf{Q}$ is slowly removed, the force P_{AD} decreases along line HJ parallel to the initial portion of the load-deflection diagram of rod AD. The final deflection of point A is

$$\delta_{A_3} = 14 \text{ mm} - 3 \text{ mm} = 11 \text{ mm}$$

Since the stress in rod CE remained within the elastic range, we note that the final deflection of point C is zero.

PROBLEMS

2.93 Knowing that $P = 40$ kN, determine the maximum stress when (a) $r = 12$ mm, (b) $r = 15$ mm.

2.94 Knowing that, for the plate shown, the allowable stress is 110 MPa, determine the maximum allowable value of P when (a) $r = 10$ mm, (b) $r = 18$ mm.

2.95 Knowing that the hole has a diameter of 9 mm, determine (a) the radius r_f of the fillets for which the same maximum stress occurs at the hole A and at the fillets, (b) the corresponding maximum allowable load **P** if the allowable stress is 100 MPa.

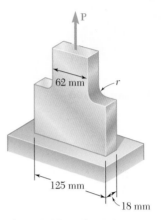

Fig. P2.93 and P2.94

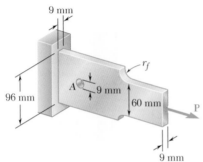

Fig. P2.95

2.96 For $P = 100$ kN, determine the minimum plate thickness t required if the allowable stress is 125 MPa.

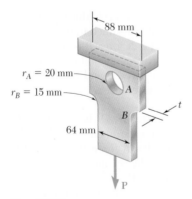

Fig. P2.96

2.97 The aluminum test specimen shown is subjected to two equal and opposite centric axial forces of magnitude P. (a) Knowing that $E = 70$ GPa and $\sigma_{all} = 200$ MPa, determine the maximum allowable value of P and the corresponding total elongation of the specimen. (b) Solve part a, assuming that the specimen has been replaced by an aluminum bar of the same length and a uniform 60×15-mm rectangular cross section.

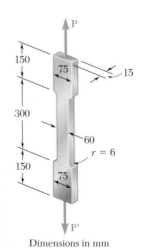

Dimensions in mm

Fig. P2.97

145

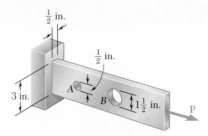

$\frac{1}{2}$ in.

$\frac{1}{2}$ in.

3 in.

A

B

$1\frac{1}{2}$ in.

P

Fig. P2.99 and P2.100

2.98 For the test specimen of Prob. 2.97, determine the maximum value of the normal stress corresponding to a total elongation of 0.75 mm.

2.99 Two holes have been drilled through a long steel bar that is subjected to a centric axial load as shown. For $P = 30$ kN, determine the maximum value of the stress (*a*) at *A*, (*b*) at *B*.

2.100 Knowing that $\sigma_{\text{all}} = 110$ MPa, determine the maximum allowable value of the centric axial load **P**.

2.101 Rod *ABC* consists of two cylindrical portions *AB* and *BC*; it is made of a mild steel that is assumed to be elastoplastic with $E = 200$ GPa and $\sigma_Y = 250$ MPa. A force **P** is applied to the rod and then removed to give it a permanent set $\delta_p = 2$ mm. Determine the maximum value of the force **P** and the maximum amount δ_m by which the rod should be stretched to give it the desired permanent set.

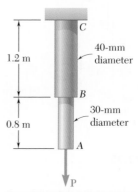

C

40-mm diameter

1.2 m

B

30-mm diameter

0.8 m

A

P

Fig. P2.101 and P2.102

2.102 Rod *ABC* consists of two cylindrical portions *AB* and *BC*; it is made of a mild steel that is assumed to be elastoplastic with $E = 200$ GPa and $\sigma_Y = 250$ MPa. A force **P** is applied to the rod until its end *A* has moved down by an amount $\delta_m = 5$ mm. Determine the maximum value of the force **P** and the permanent set of the rod after the force has been removed.

2.103 The 30-mm-square bar *AB* has a length $L = 2.2$ m; it is made of a mild steel that is assumed to be elastoplastic with $E = 200$ GPa and $\sigma_Y = 345$ MPa. A force **P** is applied to the bar until end *A* has moved down by an amount δ_m. Determine the maximum value of the force **P** and the permanent set of the bar after the force has been removed, knowing that (*a*) $\delta_m = 4.5$ mm, (*b*) $\delta_m = 8$ mm.

B

L

A

P

Fig. P2.103 and P2.104

2.104 The 30-mm-square bar *AB* has a length $L = 2.5$ m; it is made of a mild steel that is assumed to be elastoplastic with $E = 200$ GPa and $\sigma_Y = 345$ MPa. A force **P** is applied to the bar and then removed to give it a permanent set δ_p. Determine the maximum value of the force **P** and the maximum amount δ_m by which the bar should be stretched if the desired value of δ_p is (*a*) 3.5 mm, (*b*) 6.5 mm.

2.105 Rod *AB* is made of a mild steel that is assumed to be elastoplastic with $E = 200$ GPa and $\sigma_Y = 345$ MPa. After the rod has been attached to the rigid lever *CD*, it is found that end *C* is 6 mm too high. A vertical force **Q** is then applied at *C* until this point has moved to position *C'*. Determine the required magnitude of **Q** and the deflection δ_1 if the lever is to *snap* back to a horizontal position after **Q** is removed.

2.106 Solve Prob. 2.105, assuming that the yield point of the mild steel is 250 MPa.

2.107 Each of the three 6-mm-diameter steel cables is made of an elasto-plastic material for which $\sigma_Y = 345$ MPa and $E = 200$ GPa. A force **P** is applied to the rigid bar *ABC* until the bar has moved downward a distance $\delta = 2$ mm. Knowing that the cables were initially taut, determine (*a*) the maximum value of **P**, (*b*) the maximum stress that occurs in cable *AD*, (*c*) the final displacement of the bar after the load is removed. (*Hint:* In part *c*, cable *BE* is not taut.)

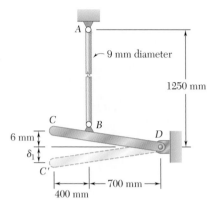

Fig. P2.105

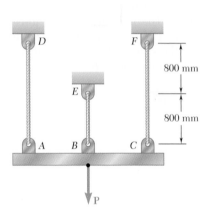

Fig. P2.107

2.108 Solve Prob. 2.107, assuming that the cables are replaced by rods of the same cross-sectional area and material. Further assume that the rods are braced so that they can carry compressive forces.

2.109 Rod *AB* consists of two cylindrical portions *AC* and *BC*, each with a cross-sectional area of 1750 mm². Portion *AC* is made of a mild steel with $E = 200$ GPa and $\sigma_Y = 250$ MPa, and portion *CB* is made of a high-strength steel with $E = 200$ GPa and $\sigma_Y = 345$ MPa. A load **P** is applied at *C* as shown. Assuming both steels to be elastoplastic, determine (*a*) the maximum deflection of *C* if *P* is gradually increased from zero to 975 kN and then reduced back to zero, (*b*) the maximum stress in each portion of the rod, (*c*) the permanent deflection of *C*.

2.110 For the composite rod of Prob. 2.109, if *P* is gradually increased from zero until the deflection of point *C* reaches a maximum value of $\delta_m = 0.3$ mm and then decreased back to zero, determine (*a*) the maximum value of *P*, (*b*) the maximum stress in each portion of the rod, (*c*) the permanent deflection of *C* after the load is removed.

Fig. P2.109

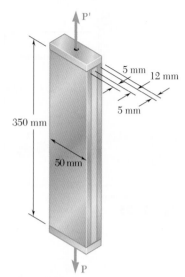

Fig. P2.111

2.111 Two tempered-steel bars, each 5 mm thick, are bonded to a 12-mm mild-steel bar. This composite bar is subjected as shown to a centric axial load of magnitude P. Both steels are elastoplastic with $E = 200$ GPa and with yield strengths equal to 690 MPa and 345 MPa, respectively, for the tempered and mild steel. The load P is gradually increased from zero until the deformation of the bar reaches a maximum value $\delta_m = 1$ mm and then decreased back to zero. Determine (*a*) the maximum value of P, (*b*) the maximum stress in the tempered-steel bars, (*c*) the permanent set after the load is removed.

2.112 For the composite bar of Prob. 2.111, if P is gradually increased from zero to 440 kN and then decreased back to zero, determine (*a*) the maximum deformation of the bar, (*b*) the maximum stress in the tempered-steel bars, (*c*) the permanent set after the load is removed.

2.113 The rigid bar ABC is supported by two links, AD and BE, of uniform 37.5×6-mm rectangular cross section and made of a mild steel that is assumed to be elastoplastic with $E = 200$ GPa and $\sigma_Y = 250$ MPa. The magnitude of the force **Q** applied at B is gradually increased from zero to 260 kN. Knowing that $a = 0.640$ m, determine (*a*) the value of the normal stress in each link, (*b*) the maximum deflection of point B.

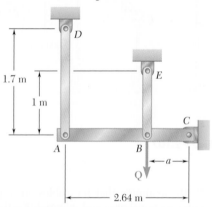

Fig. P2.113

2.114 Solve Prob. 2.113, knowing that $a = 1.76$ m and that the magnitude of the force **Q** applied at B is gradually increased from zero to 135 kN.

*2.115 Solve Prob. 2.113, assuming that the magnitude of the force **Q** applied at B is gradually increased from zero to 260 kN and then decreased back to zero. Knowing that $a = 0.640$ m, determine (*a*) the residual stress in each link, (*b*) the final deflection of point B. Assume that the links are braced so that they can carry compressive forces without buckling.

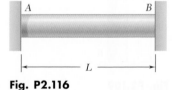

Fig. P2.116

2.116 A uniform steel rod of cross-sectional area A is attached to rigid supports and is unstressed at a temperature of 7°C. The steel is assumed to be elastoplastic with $\sigma_Y = 250$ MPa and $E = 200$ GPa. Knowing that $\alpha = 11.7 \times 10^{-6}$/°C, determine the stress in the bar (*a*) when the temperature is raised to 160°C, (*b*) after the temperature has returned to 7°C.

2.117 The steel rod ABC is attached to rigid supports and is unstressed at a temperature of 3°C. The steel is assumed elastoplastic, with $\sigma_Y = 248$ MPa and $E = 200$ GPa. The temperature of both portions of the rod is then raised to 120°C. Knowing that $\alpha = 11.7 \times 10^{-6}$/°C, determine (a) the stress in portion AC, (b) the deflection of point C.

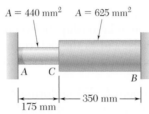

$A = 440$ mm² $\quad A = 625$ mm²

175 mm

350 mm

Fig. P2.117

*2.118 Solve Prob. 2.117, assuming that the temperature of the rod is raised to 120°C and then returned to 3°C.

*2.119 For the composite bar of Prob. 2.111, determine the residual stresses in the tempered-steel bars if P is gradually increased from zero to 440 kN and then decreased back to zero.

*2.120 For the composite bar in Prob. 2.111, determine the residual stresses in the tempered-steel bars if P is gradually increased from zero until the deformation of the bar reaches a maximum value $\delta_m = 1.0$ mm and is then decreased back to zero.

*2.121 Narrow bars of aluminum are bonded to the two sides of a thick steel plate as shown. Initially, at $T_1 = 21$°C, all stresses are zero. Knowing that the temperature will be slowly raised to T_2 and then reduced to T_1, determine (a) the highest temperature T_2 that does *not* result in residual stresses, (b) the temperature T_2 that will result in a residual stress in the aluminum equal to 400 MPa. Assume $\alpha_a = 23 \times 10^{-6}$/°C for the aluminum and $\alpha_s = 11.7 \times 10^{-6}$/°C for the steel. Further assume that the aluminum is elastoplastic with $E = 75$ GPa and $\sigma_Y = 400$ MPa. (*Hint:* Neglect the small stresses in the plate.)

Fig. P2.121

*2.122 Bar AB has a cross-sectional area of 1200 mm² and is made of a steel that is assumed to be elastoplastic with $E = 200$ GPa and $\sigma_Y = 250$ MPa. Knowing that the force $\mathbf{F}$ increases from 0 to 520 kN and then decreases to zero, determine (a) the permanent deflection of point C, (b) the residual stress in the bar.

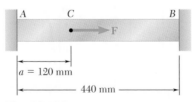

$a = 120$ mm

440 mm

Fig. P2.122

*2.123 Solve Prob. 2.122, assuming that $a = 180$ mm.

This chapter was devoted to the introduction of the concept of *strain*, to the discussion of the relationship between stress and strain in various types of materials, and to the determination of the deformations of structural components under axial loading.

Normal strain

Considering a rod of length L and uniform cross section and denoting by δ its deformation under an axial load **P** (Fig. 2.68), we defined the *normal strain* ϵ in the rod as the *deformation per unit length* [Sec. 2.2]:

$$\epsilon = \frac{\delta}{L} \tag{2.1}$$

In the case of a rod of variable cross section, the normal strain was defined at any given point Q by considering a small element of rod at Q. Denoting by Δx the length of the element and by $\Delta\delta$ its deformation under the given load, we wrote

$$\epsilon = \lim_{\Delta x \to 0} \frac{\Delta\delta}{\Delta x} = \frac{d\delta}{dx} \tag{2.2}$$

Plotting the stress σ versus the strain ϵ as the load increased, we obtained a *stress-strain diagram* for the material used [Sec. 2.3]. From such a diagram, we were able to distinguish between *brittle* and *ductile* materials: A specimen made of a brittle material ruptures without any noticeable prior change in the rate of elongation (Fig. 2.69), while a specimen made of a ductile material *yields* after a critical stress σ_Y, called the *yield strength*, has been reached, i.e., the specimen undergoes a large deformation before rupturing, with a relatively small increase in the applied load (Fig. 2.70). An example of brittle material with different properties in tension and in compression was provided by *concrete*.

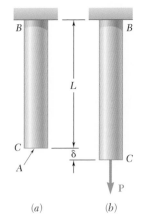

(a) (b)

Fig. 2.68

Stress-strain diagram

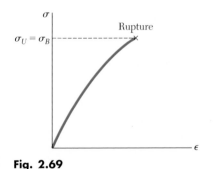

Fig. 2.69

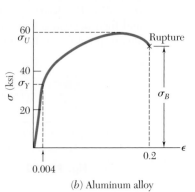

(a) Low-carbon steel (b) Aluminum alloy

Fig. 2.70

We noted in Sec. 2.5 that the initial portion of the stress-strain diagram is a straight line. This means that for small deformations, the stress is directly proportional to the strain:

$$\sigma = E\epsilon \qquad (2.4)$$

This relation is known as *Hooke's law* and the coefficient E as the *modulus of elasticity* of the material. The largest stress for which Eq. (2.4) applies is the *proportional limit* of the material.

Materials considered up to this point were *isotropic*, i.e., their properties were independent of direction. In Sec. 2.5 we also considered a class of *anisotropic* materials, i.e., materials whose properties depend upon direction. They were *fiber-reinforced composite materials*, made of fibers of a strong, stiff material embedded in layers of a weaker, softer material (Fig. 2.71). We saw that different moduli of elasticity had to be used, depending upon the direction of loading.

Hooke's law
Modulus of elasticity

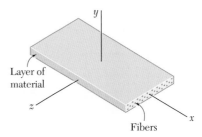

Fig. 2.71

If the strains caused in a test specimen by the application of a given load disappear when the load is removed, the material is said to behave *elastically*, and the largest stress for which this occurs is called the *elastic limit* of the material [Sec. 2.6]. If the elastic limit is exceeded, the stress and strain decrease in a linear fashion when the load is removed and the strain does not return to zero (Fig. 2.72), indicating that a *permanent set* or *plastic deformation* of the material has taken place.

Elastic limit. Plastic deformation

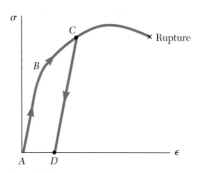

Fig. 2.72

In Sec. 2.7, we discussed the phenomenon of *fatigue*, which causes the failure of structural or machine components after a very large number of repeated loadings, even though the stresses remain in the elastic range. A standard fatigue test consists in determining the number n of successive loading-and-unloading cycles required to cause the failure of a specimen for any given maximum stress level σ, and plotting the resulting σ-n curve. The value of σ for which failure does not occur, even for an indefinitely large number of cycles, is known as the *endurance limit* of the material used in the test.

Fatigue. Endurance limit

Section 2.8 was devoted to the determination of the elastic deformations of various types of machine and structural components under various conditions of axial loading. We saw that if a rod of length L and uniform cross section of area A is subjected at its end to a centric axial load **P** (Fig. 2.73), the corresponding deformation is

$$\delta = \frac{PL}{AE} \qquad (2.7)$$

Elastic deformation under axial loading

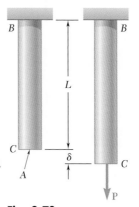

If the rod is loaded at several points or consists of several parts of various cross sections and possibly of different materials, the deformation δ of the rod must be expressed as the sum of the deformations of its component parts [Example 2.01]:

$$\delta = \sum_i \frac{P_i L_i}{A_i E_i} \qquad (2.8)$$

← when you have multible maverial

Fig. 2.73

Statically indeterminate problems

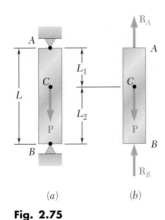

(a)   *(b)*

Fig. 2.75

Problems with temperature changes

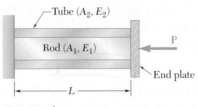

Fig. 2.74

Section 2.9 was devoted to the solution of *statically indeterminate problems,* i.e., problems in which the reactions and the internal forces *cannot* be determined from statics alone. The equilibrium equations derived from the free-body diagram of the member under consideration were complemented by relations involving deformations and obtained from the geometry of the problem. The forces in the rod and in the tube of Fig. 2.74, for instance, were determined by observing, on one hand, that their sum is equal to P, and on the other, that they cause equal deformations in the rod and in the tube [Example 2.02]. Similarly, the reactions at the supports of the bar of Fig. 2.75 could not be obtained from the free-body diagram of the bar alone [Example 2.03]; but they could be determined by expressing that the total elongation of the bar must be equal to zero.

In Sec. 2.10, we considered problems involving *temperature changes.* We first observed that if the temperature of an *unrestrained rod AB* of length L is increased by ΔT, its elongation is

$$\delta_T = \alpha(\Delta T)L \qquad (2.21)$$

where α is the *coefficient of thermal expansion* of the material. We noted that the corresponding strain, called *thermal strain,* is

$$\epsilon_T = \alpha\Delta T \qquad (2.22)$$

and that *no stress* is associated with this strain. However, if the rod AB is *restrained* by fixed supports (Fig. 2.76), stresses develop in the

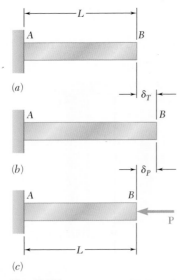

(a)

(b)

(c)

Fig. 2.77

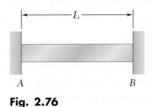

Fig. 2.76

rod as the temperature increases, because of the reactions at the supports. To determine the magnitude P of the reactions, we detached the rod from its support at B (Fig. 2.77) and considered separately the deformation δ_T of the rod as it expands freely because of the temperature change, and the deformation δ_P caused by the force **P** required to bring it back to its original length, so that it may be reattached to the support at B. Writing that the total deformation $\delta = \delta_T + \delta_P$ is equal to zero, we obtained an equation that could be solved for P. While the final strain in rod AB is clearly zero, this will generally not be the case for rods and bars consisting of elements of different cross sections or materials, since the deformations of the various elements will usually *not* be zero [Example 2.06].

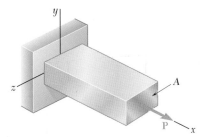

Fig. 2.78

When an axial load **P** is applied to a homogeneous, slender bar (Fig. 2.78), it causes a strain, not only along the axis of the bar but in any transverse direction as well [Sec. 2.11]. This strain is referred to as the *lateral strain,* and the ratio of the lateral strain over the axial strain is called *Poisson's ratio* and is denoted by ν (Greek letter nu). We wrote

Lateral strain. Poisson's ratio

$$\nu = -\frac{\text{lateral strain}}{\text{axial strain}} \qquad (2.25)$$

Recalling that the axial strain in the bar is $\epsilon_x = \sigma_x/E$, we expressed as follows the condition of strain under an axial loading in the x direction:

$$\epsilon_x = \frac{\sigma_x}{E} \quad \epsilon_y = \epsilon_z = -\frac{\nu\sigma_x}{E} \qquad (2.27)$$

This result was extended in Sec. 2.12 to the case of a *multiaxial loading* causing the state of stress shown in Fig. 2.79. The resulting strain condition was described by the following relations, referred to as the *generalized Hooke's law* for a multiaxial loading.

Multiaxial loading

$$
\begin{aligned}
\epsilon_x &= +\frac{\sigma_x}{E} - \frac{\nu\sigma_y}{E} - \frac{\nu\sigma_z}{E} \\[2mm]
\epsilon_y &= -\frac{\nu\sigma_x}{E} + \frac{\sigma_y}{E} - \frac{\nu\sigma_z}{E} \\[2mm]
\epsilon_z &= -\frac{\nu\sigma_x}{E} - \frac{\nu\sigma_y}{E} + \frac{\sigma_z}{E}
\end{aligned}
\qquad (2.28)
$$

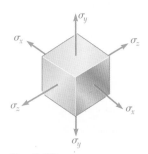

Fig. 2.79

If an element of material is subjected to the stresses σ_x, σ_y, σ_z, it will deform and a certain change of volume will result [Sec. 2.13]. The *change in volume per unit volume* is referred to as the *dilatation* of the material and is denoted by e. We showed that

Dilatation

$$e = \frac{1 - 2\nu}{E}(\sigma_x + \sigma_y + \sigma_z) \qquad (2.31)$$

When a material is subjected to a hydrostatic pressure p, we have

$$e = -\frac{p}{k} \qquad (2.34)$$

where k is known as the *bulk modulus* of the material:

Bulk modulus

$$k = \frac{E}{3(1 - 2\nu)} \qquad (2.33)$$

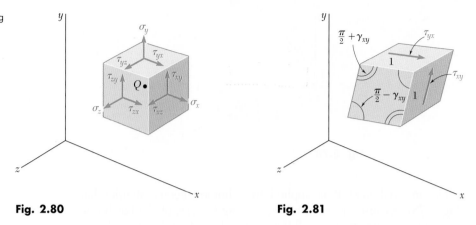

Fig. 2.80   **Fig. 2.81**

Shearing strain. Modulus of rigidity

As we saw in Chap. 1, the state of stress in a material under the most general loading condition involves shearing stresses, as well as normal stresses (Fig. 2.80). The shearing stresses tend to deform a cubic element of material into an oblique parallelepiped [Sec. 2.14]. Considering, for instance, the stresses τ_{xy} and τ_{yx} shown in Fig. 2.81 (which, we recall, are equal in magnitude), we noted that they cause the angles formed by the faces on which they act to either increase or decrease by a small angle γ_{xy}; this angle, expressed in radians, defines the *shearing strain* corresponding to the x and y directions. Defining in a similar way the shearing strains γ_{yz} and γ_{zx}, we wrote the relations

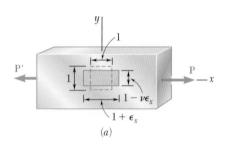

(a)

$$\tau_{xy} = G\gamma_{xy} \qquad \tau_{yz} = G\gamma_{yz} \qquad \tau_{zx} = G\gamma_{zx} \qquad (2.36, 37)$$

which are valid for any homogeneous isotropic material within its proportional limit in shear. The constant G is called the *modulus of rigidity* of the material and the relations obtained express *Hooke's law for shearing stress and strain*. Together with Eqs. (2.28), they form a group of equations representing the generalized Hooke's law for a homogeneous isotropic material under the most general stress condition.

We observed in Sec. 2.15 that while an axial load exerted on a slender bar produces only normal strains—both axial and transverse—on an element of material oriented along the axis of the bar, it will produce both normal and shearing strains on an element rotated through 45° (Fig. 2.82). We also noted that the three constants E, ν, and G are not independent; they satisfy the relation.

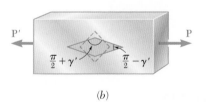

(b)

Fig. 2.82

$$\frac{E}{2G} = 1 + \nu \qquad (2.43)$$

which may be used to determine any of the three constants in terms of the other two.

Fiber-reinforced composite materials

Stress-strain relationships for fiber-reinforced composite materials were discussed in an optional section (Sec. 2.16). Equations similar to Eqs. (2.28) and (2.36, 2.37) were derived for these materials, but we noted that direction-dependent moduli of elasticity, Poisson's ratios, and moduli of rigidity had to be used.

In Sec. 2.17, we discussed *Saint-Venant's principle*, which states that except in the immediate vicinity of the points of application of the loads, the distribution of stresses in a given member is independent of the actual mode of application of the loads. This principle makes it possible to assume a uniform distribution of stresses in a member subjected to concentrated axial loads, except close to the points of application of the loads, where stress concentrations will occur.

Saint-Venant's principle

Stress concentrations will also occur in structural members near a discontinuity, such as a hole or a sudden change in cross section [Sec. 2.18]. The ratio of the maximum value of the stress occurring near the discontinuity over the average stress computed in the critical section is referred to as the *stress-concentration factor* of the discontinuity and is denoted by K:

Stress concentrations

$$K = \frac{\sigma_{max}}{\sigma_{ave}} \tag{2.48}$$

Values of K for circular holes and fillets in flat bars were given in Fig. 2.64 on p. 140.

In Sec. 2.19, we discussed the *plastic deformations* which occur in structural members made of a ductile material when the stresses in some part of the member exceed the yield strength of the material. Our analysis was carried out for an idealized *elastoplastic material* characterized by the stress-strain diagram shown in Fig. 2.83 [Examples 2.13, 2.14, and 2.15]. Finally, in Sec. 2.20, we observed that when an indeterminate structure undergoes plastic deformations, the stresses do not, in general, return to zero after the load has been removed. The stresses remaining in the various parts of the structure are called *residual stresses* and may be determined by adding the maximum stresses reached during the loading phase and the reverse stresses corresponding to the unloading phase [Example 2.16].

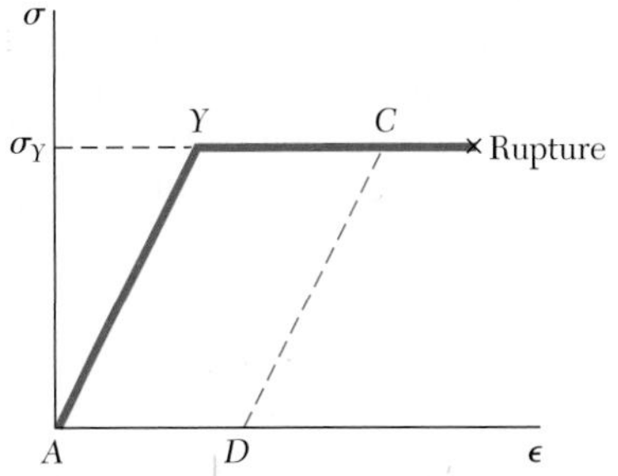

Plastic deformations

Fig. 2.83

REVIEW PROBLEMS

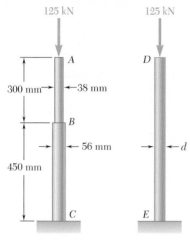

Fig. P2.124

2.124 The aluminum rod *ABC* (*E* = 70 GPa), which consists of two cylindrical portions *AB* and *BC*, is to be replaced with a cylindrical steel rod *DE* (*E* = 200 GPa) of the same overall length. Determine the minimum required diameter of *d* of the steel rod if its vertical deformation is not to exceed the deformation of the aluminum rod under the same load and if the allowable stress in the steel rod is not to exceed 165 MPa.

2.125 Two solid cylindrical rods are joined at *B* and loaded as shown. Rod *AB* is made of steel (*E* = 200 GPa) and rod *BC* of brass (*E* = 105 GPa). Determine (*a*) the total deformation of the composite rod *ABC*, (*b*) the deflection of point *B*.

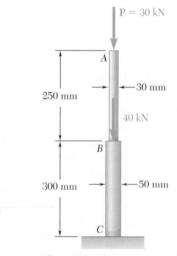

Fig. P2.125

2.126 The specimen shown is made from a 24-mm-diameter cylindrical steel rod with two 36-mm-outer-diameter sleeves bonded to the rod as shown. Knowing that *E* = 200 GPa, determine (*a*) the load **P** so that the total deformation is 0.04 mm, (*b*) the corresponding deformation of the central portion *BC*.

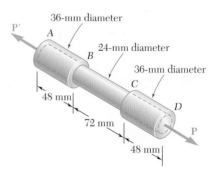

Fig. P2.126

2.127 The uniform wire ABC, of unstretched length $2l$, is attached to the supports shown and a vertical load **P** is applied at the midpoint B. Denoting by A the cross-sectional area of the wire and by E the modulus of elasticity, show that, for $\delta \ll l$, the deflection at the midpoint B is

$$\delta = l \sqrt[3]{\frac{P}{AE}}$$

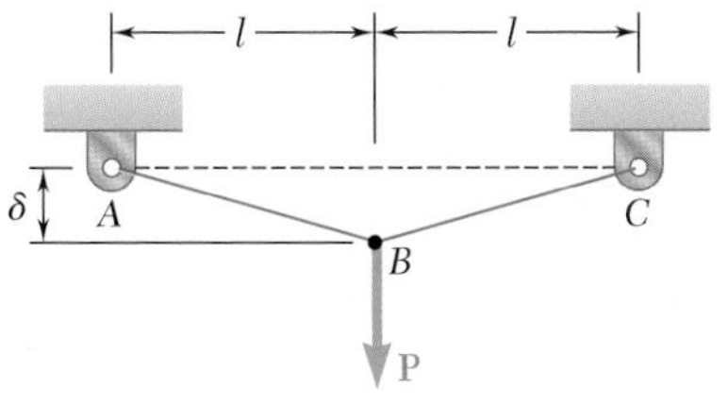

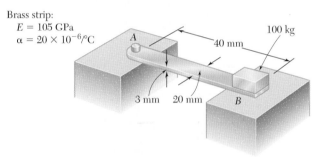

Fig. P2.127

2.128 The brass strip AB has been attached to a fixed support at A and rests on a rough support at B. Knowing that the coefficient of friction is 0.60 between the strip and the support at B, determine the decrease in temperature for which slipping will impend.

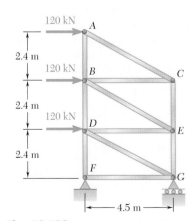

Fig. P2.128

2.129 For the steel truss $(E = 200 \text{ GPa})$ and loading shown, determine the deformations of the members BD and DE, knowing that their cross-sectional areas are 1250 mm^2 and 1875 mm^2, respectively.

$$\gamma_{xy} = \frac{\partial u}{\partial y} + \frac{\partial v}{\partial x}$$

$$\gamma_{xz} = \frac{\partial u}{\partial z} + \frac{\partial w}{\partial z}$$

$$\gamma_{yz} = \frac{\partial v}{\partial z} + \frac{\partial w}{\partial z}$$

Fig. P2.129

2.130 A 250-mm bar of 150×30-mm rectangular cross section consists of two aluminum layers, 5 mm thick, brazed to a center brass layer of the same thickness. If it is subjected to centric forces of magnitude $P = 30$ kN, and knowing that $E_a = 70$ GPa and $E_b = 105$ GPa, determine the normal stress (a) in the aluminum layers, (b) in the brass layer.

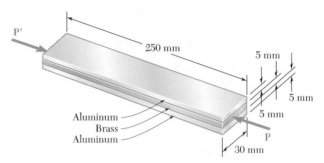

Fig. P2.130

2.131 The brass shell ($\alpha_b = 20.9 \times 10^{-6}/°C$) is fully bonded to the steel core ($\alpha_s = 11.7 \times 10^{-6}/°C$). Determine the largest allowable increase in temperature if the stress in the steel core is not to exceed 55 MPa.

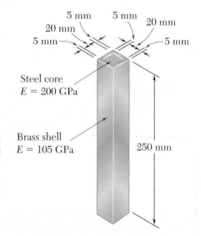

Fig. P2.131

2.132 A fabric used in air-inflated structures is subjected to a biaxial loading that results in normal stresses $\sigma_x = 120$ MPa and $\sigma_z = 160$ MPa. Knowing that the properties of the fabric can be approximated as $E = 87$ GPa and $\nu = 0.34$, determine the change in length of (a) side AB, (b) side BC, (c) diagonal AC.

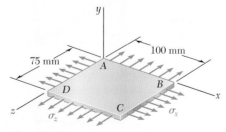

Fig. P2.132

2.133 A vibration isolation unit consists of two blocks of hard rubber bonded to a plate AB and to rigid supports as shown. Knowing that a force of magnitude $P = 25$ kN causes a deflection $\delta = 1.5$ mm of plate AB, determine the modulus of rigidity of the rubber used.

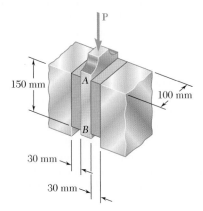

Fig. P2.133 and P2.134

2.134 A vibration isolation unit consists of two blocks of hard rubber with a modulus of rigidity $G = 19$ MPa bonded to a plate AB and to rigid supports as shown. Denoting by P the magnitude of the force applied to the plate and by δ the corresponding deflection, determine the effective spring constant, $k = P/\delta$, of the system.

2.135 The uniform rod BC has cross-sectional area A and is made of a mild steel that can be assumed to be elastoplastic with a modulus of elasticity E and a yield strength σ_Y. Using the block-and-spring system shown, it is desired to simulate the deflection of end C of the rod as the axial force $\mathbf{P}$ is gradually applied and removed, that is, the deflection of points C and C' should be the same for all values of P. Denoting by μ the coefficient of friction between the block and the horizontal surface, derive an expression for (a) the required mass m of the block, (b) the required constant k of the spring.

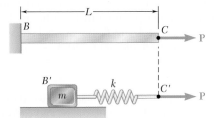

Fig. P2.135

COMPUTER PROBLEMS

The following problems are designed to be solved with a computer. Write each program so that it can be used with either SI or U.S. customary units and in such a way that solid cylindrical elements may be defined by either their diameter or their cross-sectional area.

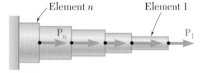

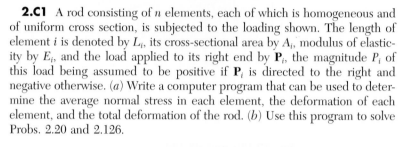

Fig. P2.C1

2.C1 A rod consisting of n elements, each of which is homogeneous and of uniform cross section, is subjected to the loading shown. The length of element i is denoted by L_i, its cross-sectional area by A_i, modulus of elasticity by E_i, and the load applied to its right end by $\mathbf{P}_i$, the magnitude P_i of this load being assumed to be positive if $\mathbf{P}_i$ is directed to the right and negative otherwise. (*a*) Write a computer program that can be used to determine the average normal stress in each element, the deformation of each element, and the total deformation of the rod. (*b*) Use this program to solve Probs. 2.20 and 2.126.

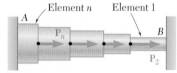

Fig. P2.C2

2.C2 Rod AB is horizontal with both ends fixed; it consists of n elements, each of which is homogeneous and of uniform cross section, and is subjected to the loading shown. The length of element i is denoted by L_i, its cross-sectional area by A_i, its modulus of elasticity by E_i, and the load applied to its right end by $\mathbf{P}_i$, the magnitude P_i of this load being assumed to be positive if $\mathbf{P}_i$ is directed to the right and negative otherwise. (Note that $P_1 = 0$.) (*a*) Write a computer program that can be used to determine the reactions at A and B, the average normal stress in each element, and the deformation of each element. (*b*) Use this program to solve Probs. 2.41 and 2.42.

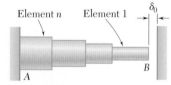

Fig. P2.C3

2.C3 Rod AB consists of n elements, each of which is homogeneous and of uniform cross section. End A is fixed, while initially there is a gap δ_0 between end B and the fixed vertical surface on the right. The length of element i is denoted by L_i, its cross-sectional area by A_i, its modulus of elasticity by E_i, and its coefficient of thermal expansion by α_i. After the temperature of the rod has been increased by ΔT, the gap at B is closed and the vertical surfaces exert equal and opposite forces on the rod. (*a*) Write a computer program that can be used to determine the magnitude of the reactions at A and B, the normal stress in each element, and the deformation of each element. (*b*) Use this program to solve Probs. 2.51, 2.59, and 2.60.

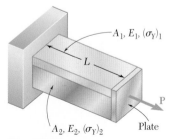

Fig. P2.C4

2.C4 Bar AB has a length L and is made of two different materials of given cross-sectional area, modulus of elasticity, and yield strength. The bar is subjected as shown to a load $\mathbf{P}$ that is gradually increased from zero until the deformation of the bar has reached a maximum value δ_m and then decreased back to zero. (*a*) Write a computer program that, for each of 25 values of δ_m equally spaced over a range extending from 0 to a value equal to 120 percent of the deformation causing both materials to yield, can be used to determine the maximum value P_m of the load, the maximum normal stress in each material, the permanent deformation δ_p of the bar, and the residual stress in each material. (*b*) Use this program to solve Probs. 2.111 and 2.112.

2.C5 The plate has a hole centered across the width. The stress concentration factor for a flat bar under axial loading with a centric hole is:

$$K = 3.00 - 3.13\left(\frac{2r}{D}\right) + 3.66\left(\frac{2r}{D}\right)^2 - 1.53\left(\frac{2r}{D}\right)^3$$

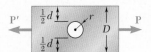

Fig. P2.C5

where r is the radius of the hole and D is the width of the bar. Write a computer program to determine the allowable load **P** for the given values of r, D, the thickness t of the bar, and the allowable stress σ_{all} of the material. Knowing that $t = 6$ mm, $D = 75$ mm, and $\sigma_{all} = 110$ MPa, determine the allowable load **P** for values of r from 3 mm to 18 mm, using 3-mm increments.

2.C6 A solid truncated cone is subjected to an axial force **P** as shown. The exact elongation is $(PL)/(2\pi c^2 E)$. By replacing the cone by n circular cylinders of equal thickness, write a computer program that can be used to calculate the elongation of the truncated cone. What is the percentage error in the answer obtained from the program using (a) $n = 6$, (b) $n = 12$, (c) $n = 60$?

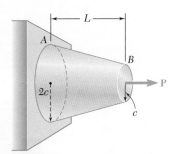

Fig. P2.C6

This chapter is devoted to the study of torsion and of the stresses and deformations it causes. In the jet engine shown here, the central shaft links the components of the engine to develop the thrust that propels the plane.

3.1 INTRODUCTION

In the two preceding chapters you studied how to calculate the stresses and strains in structural members subjected to axial loads, that is, to forces directed along the axis of the member. In this chapter structural members and machine parts that are in *torsion* will be considered. More specifically, you will analyze the stresses and strains in members of circular cross section subjected to twisting couples, or *torques*, **T** and **T′** (Fig. 3.1). These couples have a common magnitude T, and opposite senses. They are vector quantities and can be represented either by curved arrows as in Fig. 3.1*a*, or by couple vectors as in Fig. 3.1*b*.

Members in torsion are encountered in many engineering applications. The most common application is provided by *transmission shafts*, which are used to transmit power from one point to another. For example, the shaft shown in Photo 3.1 is used to transmit power from the engine to the rear wheels of an automobile. These shafts can be either solid, as shown in Fig. 3.1, or hollow.

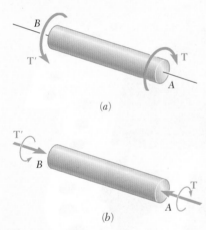

Fig. 3.1 Shaft subject to torsion.

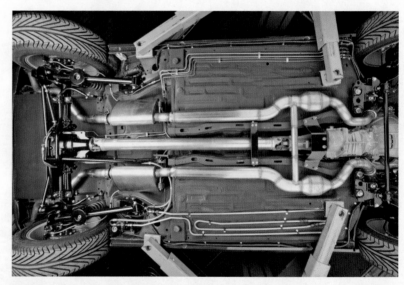

Photo 3.1 In the automotive power train shown, the shaft transmits power from the engine to the rear wheels.

Consider the system shown in Fig. 3.2*a*, which consists of a steam turbine *A* and an electric generator *B* connected by a transmission shaft *AB*. By breaking the system into its three component parts (Fig. 3.2*b*), you can see that the turbine exerts a twisting couple or torque **T** on the shaft and that the shaft exerts an equal torque on the generator. The generator reacts by exerting the equal and opposite torque **T′** on the shaft, and the shaft by exerting the torque **T′** on the turbine.

You will first analyze the stresses and deformations that take place in circular shafts. In Sec. 3.3, an important property of circular shafts is demonstrated: *When a circular shaft is subjected to torsion,*

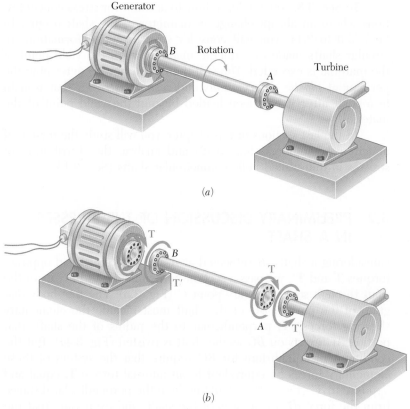

Generator

B Rotation

Turbine

(a)

T

B

T′

T

A T′

(b)

Fig. 3.2 Transmission shaft.

every cross section remains plane and undistorted. In other words, while the various cross sections along the shaft rotate through different angles, each cross section rotates as a solid rigid slab. This property will enable you to determine the *distribution of shearing strains in a circular shaft and to conclude that the shearing strain varies linearly with the distance from the axis of the shaft.*

Considering deformations in the *elastic range* and using Hooke's law for shearing stress and strain, you will determine the *distribution of shearing stresses* in a circular shaft and derive the *elastic torsion formulas* (Sec. 3.4).

In Sec. 3.5, you will learn how to find the *angle of twist* of a circular shaft subjected to a given torque, assuming again elastic deformations. The solution of problems involving *statically indeterminate shafts* is considered in Sec. 3.6.

In Sec. 3.7, you will study the *design of transmission shafts.* In order to accomplish the design, you will learn to determine the required physical characteristics of a shaft in terms of its speed of rotation and the power to be transmitted.

The torsion formulas cannot be used to determine stresses near sections where the loading couples are applied or near a section where an abrupt change in the diameter of the shaft occurs. Moreover, these formulas apply only within the elastic range of the material.

In Sec. 3.8, you will learn how to account for stress concentrations where an abrupt change in diameter of the shaft occurs. In Secs. 3.9 to 3.11, you will consider stresses and deformations in circular shafts made of a ductile material when the yield point of the material is exceeded. You will then learn how to determine the permanent *plastic deformations* and *residual stresses* that remain in a shaft after it has been loaded beyond the yield point of the material.

In the last sections of this chapter, you will study the torsion of noncircular members (Sec. 3.12) and analyze the distribution of stresses in thin-walled hollow noncircular shafts (Sec. 3.13).

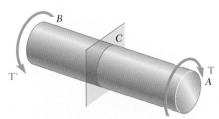

Fig. 3.3 Shaft subject to torques.

3.2 PRELIMINARY DISCUSSION OF THE STRESSES IN A SHAFT

Considering a shaft AB subjected at A and B to equal and opposite torques **T** and **T′**, we pass a section perpendicular to the axis of the shaft through some arbitrary point C (Fig. 3.3). The free-body diagram of the portion BC of the shaft must include the elementary shearing forces $d\mathbf{F}$, perpendicular to the radius of the shaft, that portion AC exerts on BC as the shaft is twisted (Fig. 3.4a). But the conditions of equilibrium for BC require that the system of these elementary forces be equivalent to an internal torque **T,** equal and opposite to **T′** (Fig. 3.4b). Denoting by ρ the perpendicular distance from the force $d\mathbf{F}$ to the axis of the shaft, and expressing that the sum of the moments of the shearing forces $d\mathbf{F}$ about the axis of the shaft is equal in magnitude to the torque **T,** we write

$$\int \rho \, dF = T$$

or, since $dF = \tau \, dA$, where τ is the shearing stress on the element of area dA,

$$\int \rho (\tau \, dA) = T \tag{3.1}$$

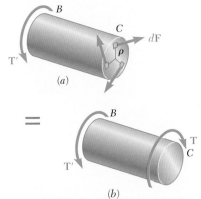

(a)

$=$

(b)

Fig. 3.4

While the relation obtained expresses an important condition that must be satisfied by the shearing stresses in any given cross section of the shaft, it does *not* tell us how these stresses are distributed in the cross section. We thus observe, as we already did in Sec. 1.5, that the actual distribution of stresses under a given load is *statically indeterminate*, i.e., this distribution *cannot be determined by the methods of statics*. However, having assumed in Sec. 1.5 that the normal stresses produced by an axial centric load were uniformly distributed, we found later (Sec. 2.17) that this assumption was justified, except in the neighborhood of concentrated loads. A similar assumption with respect to the distribution of shearing stresses in an elastic shaft *would be wrong*. We must withhold any judgment regarding the distribution of stresses in a shaft until we have analyzed the *deformations* that are produced in the shaft. This will be done in the next section.

One more observation should be made at this point. As was indicated in Sec. 1.12, shear cannot take place in one plane only. Consider the very small element of shaft shown in Fig. 3.5. We know that the torque applied to the shaft produces shearing stresses τ on

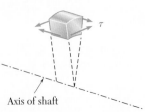

Axis of shaft

Fig. 3.5 Element in shaft.

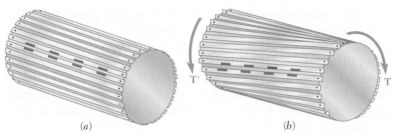

Fig. 3.6 Model of shaft.

the faces perpendicular to the axis of the shaft. But the conditions of equilibrium discussed in Sec. 1.12 require the existence of equal stresses on the faces formed by the two planes containing the axis of the shaft. That such shearing stresses actually occur in torsion can be demonstrated by considering a "shaft" made of separate slats pinned at both ends to disks as shown in Fig. 3.6*a*. If markings have been painted on two adjoining slats, it is observed that the slats slide with respect to each other when equal and opposite torques are applied to the ends of the "shaft" (Fig. 3.6*b*). While sliding will not actually take place in a shaft made of a homogeneous and cohesive material, the tendency for sliding will exist, showing that stresses occur on longitudinal planes as well as on planes perpendicular to the axis of the shaft.†

3.3 DEFORMATIONS IN A CIRCULAR SHAFT

Consider a circular shaft that is attached to a fixed support at one end (Fig. 3.7*a*). If a torque **T** is applied to the other end, the shaft will twist, with its free end rotating through an angle ϕ called *the angle of twist* (Fig. 3.7*b*). Observation shows that, within a certain range of values of *T*, the angle of twist ϕ is proportional to *T*. It also shows that ϕ is proportional to the length *L* of the shaft. In other words, the angle of twist for a shaft of the same material and same cross section, but twice as long, will be twice as large under the same torque **T.** One purpose of our analysis will be to find the specific relation existing among ϕ, *L*, and *T*; another purpose will be to determine the distribution of shearing stresses in the shaft, which we were unable to obtain in the preceding section on the basis of statics alone.

At this point, an important property of circular shafts should be noted: When a circular shaft is subjected to torsion, *every cross section remains plane and undistorted*. In other words, while the various cross sections along the shaft rotate through different amounts, each cross section rotates as a solid rigid slab. This is illustrated in Fig. 3.8*a*, which shows the deformations in a rubber model subjected to torsion. The property we are discussing is characteristic of circular shafts, whether solid or hollow; it is not enjoyed by members of noncircular cross section. For example, when a bar of square cross section is subjected to torsion, its various cross sections warp and do not remain plane (Fig. 3.8*b*).

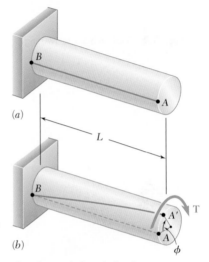

Fig. 3.7 Shaft with fixed support.

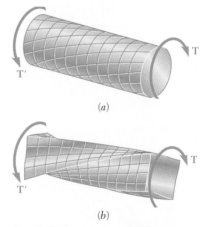

Fig. 3.8 Comparison of deformations in circular and square shafts.

†The twisting of a cardboard tube that has been slit lengthwise provides another demonstration of the existence of shearing stresses on longitudinal planes.

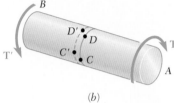

Fig. 3.9 Shaft subject to twisting.

Fig. 3.10 Concentric circles.

The cross sections of a circular shaft remain plane and undistorted because a circular shaft is *axisymmetric,* i.e., its appearance remains the same when it is viewed from a fixed position and rotated about its axis through an arbitrary angle. (Square bars, on the other hand, retain the same appearance only if they are rotated through 90° or 180°.) As we will see presently, the axisymmetry of circular shafts may be used to prove theoretically that their cross sections remain plane and undistorted.

Consider the points C and D located on the circumference of a given cross section of the shaft, and let C' and D' be the positions they will occupy after the shaft has been twisted (Fig. 3.9a). The axisymmetry of the shaft and of the loading requires that the rotation which would have brought D into D' should now bring C into C'. Thus C' and D' must lie on the circumference of a circle, and the arc $C'D'$ must be equal to the arc CD (Fig. 3.9b). We will now examine whether the circle on which C' and D' lie is different from the original circle. Let us assume that C' and D' do lie on a different circle and that the new circle is located to the left of the original circle, as shown in Fig. 3.9b. The same situation will prevail for any other cross section, since all the cross sections of the shaft are subjected to the same internal torque T, and an observer looking at the shaft from its end A will conclude that the loading causes any given circle drawn on the shaft to move *away*. But an observer located at B, to whom the given loading looks the same (a clockwise couple in the foreground and a counterclockwise couple in the background) will reach the opposite conclusion, i.e., that the circle moves *toward* him. This contradiction proves that our assumption is wrong and that C' and D' lie on the same circle as C and D. Thus, as the shaft is twisted, the original circle just rotates in its own plane. Since the same reasoning may be applied to any smaller, concentric circle located in the cross section under consideration, we conclude that the entire cross section remains plane (Fig. 3.10).

The argument does not preclude the possibility for the various concentric circles of Fig. 3.10 to rotate by different amounts when the shaft is twisted. But if that were so, a given diameter of the cross section would be distorted into a curve which might look as shown in Fig. 3.11a. An observer looking at this curve from A would conclude that the outer layers of the shaft get more twisted than the inner ones, while an observer looking from B would reach the opposite conclusion (Fig. 3.11b). This inconsistency leads us to conclude that any diameter of a given cross section remains straight (Fig. 3.11c) and, therefore, that any given cross section of a circular shaft remains plane and undistorted.

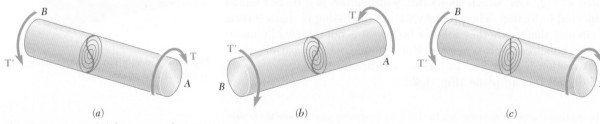

Fig. 3.11 Potential deformations of cross section.

Our discussion so far has ignored the mode of application of the twisting couples **T** and **T′**. If *all* sections of the shaft, from one end to the other, are to remain plane and undistorted, we must make sure that the couples are applied in such a way that the ends of the shaft themselves remain plane and undistorted. This may be accomplished by applying the couples **T** and **T′** to rigid plates, which are solidly attached to the ends of the shaft (Fig. 3.12a). We can then be sure that all sections will remain plane and undistorted when the loading is applied, and that the resulting deformations will occur in a uniform fashion throughout the entire length of the shaft. All of the equally spaced circles shown in Fig. 3.12a will rotate by the same amount relative to their neighbors, and each of the straight lines will be transformed into a curve (helix) intersecting the various circles at the same angle (Fig. 3.12b).

The derivations given in this and the following sections will be based on the assumption of rigid end plates. Loading conditions encountered in practice may differ appreciably from those corresponding to the model of Fig. 3.12. The chief merit of this model is that it helps us define a torsion problem for which we can obtain an exact solution, just as the rigid-end-plates model of Sec. 2.17 made it possible for us to define an axial-load problem which could be easily and accurately solved. By virtue of Saint-Venant's principle, the results obtained for our idealized model may be extended to most engineering applications. However, we should keep these results associated in our mind with the specific model shown in Fig. 3.12.

We will now determine the distribution of *shearing strains* in a circular shaft of length L and radius c that has been twisted through an angle ϕ (Fig. 3.13a). Detaching from the shaft a cylinder of radius ρ, we consider the small square element formed by two adjacent circles and two adjacent straight lines traced on the surface of the cylinder before any load is applied (Fig. 3.13b). As the shaft is subjected to a torsional load, the element deforms into a rhombus (Fig. 3.13c). We now recall from Sec. 2.14 that the shearing strain γ in a given element is measured by the change in the angles formed by the sides of that element. Since the circles defining two of the sides of the element considered here remain unchanged, the shearing strain γ must be equal to the angle between lines AB and $A'B$. (We recall that γ should be expressed in radians.)

We observe from Fig. 3.13c that, for small values of γ, we can express the arc length AA' as $AA' = L\gamma$. But, on the other hand, we have $AA' = \rho\phi$. It follows that $L\gamma = \rho\phi$, or

$$\gamma = \frac{\rho\phi}{L} \tag{3.2}$$

where γ and ϕ are both expressed in radians. The equation obtained shows, as we could have anticipated, that the shearing strain γ at a given point of a shaft in torsion is proportional to the angle of twist ϕ. It also shows that γ is proportional to the distance ρ from the axis of the shaft to the point under consideration. Thus, *the shearing strain in a circular shaft varies linearly with the distance from the axis of the shaft*.

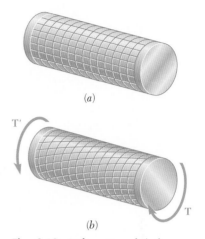

Fig. 3.12 Deformation of shaft subject to twisting couples.

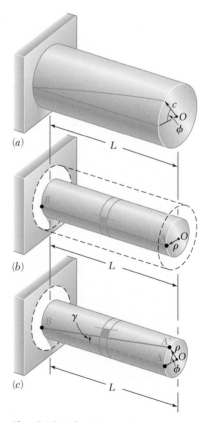

Fig. 3.13 Shearing strain.

It follows from Eq. (3.2) that the shearing strain is maximum on the surface of the shaft, where $\rho = c$. We have

$$\gamma_{max} = \frac{c\phi}{L} \tag{3.3}$$

Eliminating ϕ from Eqs. (3.2) and (3.3), we can express the shearing strain γ at a distance ρ from the axis of the shaft as

$$\gamma = \frac{\rho}{c}\gamma_{max} \tag{3.4}$$

3.4 STRESSES IN THE ELASTIC RANGE

No particular stress-strain relationship has been assumed so far in our discussion of circular shafts in torsion. Let us now consider the case when the torque **T** is such that all shearing stresses in the shaft remain below the yield strength τ_Y. We know from Chap. 2 that, for all practical purposes, this means that the stresses in the shaft will remain below the proportional limit and below the elastic limit as well. Thus, Hooke's law will apply and there will be no permanent deformation.

Recalling Hooke's law for shearing stress and strain from Sec. 2.14, we write

$$\tau = G\gamma \tag{3.5}$$

where G is the modulus of rigidity or shear modulus of the material. Multiplying both members of Eq. (3.4) by G, we write

$$G\gamma = \frac{\rho}{c}G\gamma_{max}$$

or, making use of Eq. (3.5),

$$\tau = \frac{\rho}{c}\tau_{max} \tag{3.6}$$

The equation obtained shows that, as long as the yield strength (or proportional limit) is not exceeded in any part of a circular shaft, *the shearing stress in the shaft varies linearly with the distance ρ from the axis of the shaft.* Figure 3.14a shows the stress distribution in a solid circular shaft of radius c, and Fig. 3.14b in a hollow circular shaft of inner radius c_1 and outer radius c_2. From Eq. (3.6), we find that, in the latter case,

$$\tau_{min} = \frac{c_1}{c_2}\tau_{max} \tag{3.7}$$

We now recall from Sec. 3.2 that the sum of the moments of the elementary forces exerted on any cross section of the shaft must be equal to the magnitude T of the torque exerted on the shaft:

$$\int \rho(\tau \, dA) = T \tag{3.1}$$

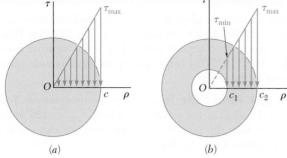

Fig. 3.14 Distribution of shearing stresses.

Substituting for τ from (3.6) into (3.1), we write

$$T = \int \rho \tau \, dA = \frac{\tau_{\max}}{c} \int \rho^2 \, dA$$

But the integral in the last member represents the polar moment of inertia J of the cross section with respect to its center O. We have therefore

$$T = \frac{\tau_{\max} J}{c} \tag{3.8}$$

or, solving for $\tau_{\max}$,

$$\tau_{\max} = \frac{Tc}{J} \tag{3.9}$$

Substituting for $\tau_{\max}$ from (3.9) into (3.6), we express the shearing stress at any distance ρ from the axis of the shaft as

$$\tau = \frac{T\rho}{J} \tag{3.10}$$

Equations (3.9) and (3.10) are known as the *elastic torsion formulas*. We recall from statics that the polar moment of inertia of a circle of radius c is $J = \frac{1}{2}\pi c^4$. In the case of a hollow circular shaft of inner radius c_1 and outer radius c_2, the polar moment of inertia is

$$J = \tfrac{1}{2}\pi c_2^4 - \tfrac{1}{2}\pi c_1^4 = \tfrac{1}{2}\pi(c_2^4 - c_1^4) \tag{3.11}$$

We note that, if SI metric units are used in Eq. (3.9) or (3.10), T will be expressed in N $\cdot$ m, c or ρ in meters, and J in m^4; we check that the resulting shearing stress will be expressed in N/m^2, i.e., pascals (Pa).

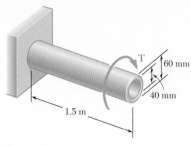

Fig. 3.15

EXAMPLE 3.01

A hollow cylindrical steel shaft is 1.5 m long and has inner and outer diameters respectively equal to 40 and 60 mm (Fig. 3.15). (*a*) What is the largest torque that can be applied to the shaft if the shearing stress is not to exceed 120 MPa? (*b*) What is the corresponding minimum value of the shearing stress in the shaft?

(a) Largest Permissible Torque. The largest torque **T** that can be applied to the shaft is the torque for which $\tau_{max} = 120$ MPa. Since this value is less than the yield strength for steel, we can use Eq. (3.9). Solving this equation for T, we have

$$T = \frac{J\tau_{max}}{c} \tag{3.12}$$

Recalling that the polar moment of inertia J of the cross section is given by Eq. (3.11), where $c_1 = \frac{1}{2}(40 \text{ mm}) = 0.02$ m and $c_2 = \frac{1}{2}(60 \text{ mm}) = 0.03$ m, we write

$$J = \frac{1}{2}\pi \left(c_2^4 - c_1^4\right) = \frac{1}{2}\pi(0.03^4 - 0.02^4) = 1.021 \times 10^{-6} \text{ m}^4$$

Substituting for J and τ_{max} into (3.12), and letting $c = c_2 = 0.03$ m, we have

$$T = \frac{J\tau_{max}}{c} = \frac{(1.021 \times 10^{-6} \text{ m}^4)(120 \times 10^6 \text{ Pa})}{0.03 \text{ m}} = 4.08 \text{ kN} \cdot \text{m}$$

(b) Minimum Shearing Stress. The minimum value of the shearing stress occurs on the inner surface of the shaft. It is obtained from Eq. (3.7), which expresses that τ_{min} and τ_{max} are respectively proportional to c_1 and c_2:

$$\tau_{min} = \frac{c_1}{c_2}\tau_{max} = \frac{0.02 \text{ m}}{0.03 \text{ m}}(120 \text{ MPa}) = 80 \text{ MPa}$$

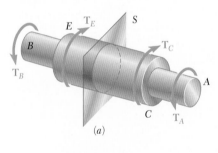

(*a*)

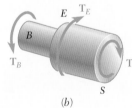

(*b*)

Fig. 3.16 Shaft with variable cross section.

The torsion formulas (3.9) and (3.10) were derived for a shaft of uniform circular cross section subjected to torques at its ends. However, they can also be used for a shaft of variable cross section or for a shaft subjected to torques at locations other than its ends (Fig. 3.16*a*). The distribution of shearing stresses in a given cross section S of the shaft is obtained from Eq. (3.9), where J denotes the polar moment of inertia of that section, and where T represents the *internal torque* in that section. The value of T is obtained by drawing the free-body diagram of the portion of shaft located on one side of the section (Fig. 3.16*b*) and writing that the sum of the torques applied to that portion, including the internal torque **T,** is zero (see Sample Prob. 3.1).

Up to this point, our analysis of stresses in a shaft has been limited to shearing stresses. This is due to the fact that the element we had selected was oriented in such a way that its faces were either parallel or perpendicular to the axis of the shaft (Fig. 3.5). We know from earlier discussions (Secs. 1.11 and 1.12) that normal stresses, shearing stresses, or a combination of both may be found under the same loading condition, depending upon the orientation of the element that has been chosen. Consider the two elements *a* and *b* located on the surface of a circular shaft subjected to torsion

(Fig. 3.17). Since the faces of element a are respectively parallel and perpendicular to the axis of the shaft, the only stresses on the element will be the shearing stresses defined by formula (3.9), namely $\tau_{max} = Tc/J$. On the other hand, the faces of element b, which form arbitrary angles with the axis of the shaft, will be subjected to a combination of normal and shearing stresses.

Let us consider the stresses and resulting forces on faces that are at 45° to the axis of the shaft. In order to determine the stresses on the faces of this element, we consider the two triangular elements shown in Fig. 3.18 and draw their free-body diagrams. In the case of the element of Fig. 3.18a, we know that the stresses exerted on the faces BC and BD are the shearing stresses $\tau_{max} = Tc/J$. The magnitude of the corresponding shearing forces is thus $\tau_{max} A_0$, where A_0 denotes the area of the face. Observing that the components along DC of the two shearing forces are equal and opposite, we conclude that the force **F** exerted on DC must be perpendicular to that face. It is a tensile force, and its magnitude is

$$F = 2(\tau_{max}A_0)\cos 45° = \tau_{max}A_0\sqrt{2} \qquad (3.13)$$

The corresponding stress is obtained by dividing the force F by the area A of face DC. Observing that $A = A_0\sqrt{2}$, we write

$$\sigma = \frac{F}{A} = \frac{\tau_{max}A_0\sqrt{2}}{A_0\sqrt{2}} = \tau_{max} \qquad (3.14)$$

A similar analysis of the element of Fig. 3.18b shows that the stress on the face BE is $\sigma = -\tau_{max}$. We conclude that the stresses exerted on the faces of an element c at 45° to the axis of the shaft (Fig. 3.19) are normal stresses equal to $\pm\tau_{max}$. Thus, while the element a in Fig. 3.19 is in pure shear, the element c in the same figure is subjected to a tensile stress on two of its faces, and to a compressive stress on the other two. We also note that all the stresses involved have the same magnitude, Tc/J.†

As you learned in Sec. 2.3, ductile materials generally fail in shear. Therefore, when subjected to torsion, a specimen J made of a ductile material breaks along a plane perpendicular to its longitudinal axis (Photo 3.2a). On the other hand, brittle materials are weaker in tension than in shear. Thus, when subjected to torsion, a specimen made of a brittle material tends to break along surfaces that are perpendicular to the direction in which tension is maximum, i.e., along surfaces forming a 45° angle with the longitudinal axis of the specimen (Photo 3.2b).

Fig. 3.17 Circular shaft with elements at different orientations.

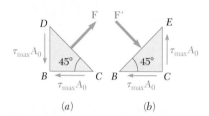

Fig. 3.18 Forces on faces at 45° to shaft axis.

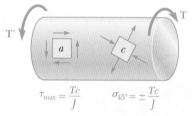

Fig. 3.19 Shaft with elements with only shear stresses or normal stresses.

(a) Ductile failure

(b) Brittle failure

Photo 3.2 Shear failure of shaft subject to torque.

†Stresses on elements of arbitrary orientation, such as element b of Fig. 3.18, will be discussed in Chap. 7.

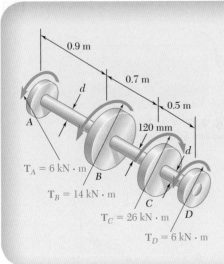

SAMPLE PROBLEM 3.1

Shaft BC is hollow with inner and outer diameters of 90 mm and 120 mm, respectively. Shafts AB and CD are solid and of diameter d. For the loading shown, determine (*a*) the maximum and minimum shearing stress in shaft BC, (*b*) the required diameter d of shafts AB and CD if the allowable shearing stress in these shafts is 65 MPa.

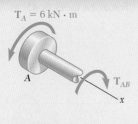

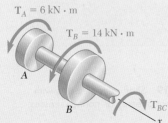

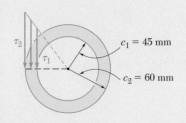

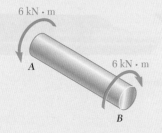

SOLUTION

Equations of Statics.   Denoting by $\mathbf{T}_{AB}$ the torque in shaft AB, we pass a section through shaft AB and, for the free body shown, we write

$$\Sigma M_x = 0: \qquad (6 \text{ kN} \cdot \text{m}) - T_{AB} = 0 \qquad T_{AB} = 6 \text{ kN} \cdot \text{m}$$

We now pass a section through shaft BC and, for the free body shown, we have

$$\Sigma M_x = 0: \quad (6 \text{ kN} \cdot \text{m}) + (14 \text{ kN} \cdot \text{m}) - T_{BC} = 0 \quad T_{BC} = 20 \text{ kN} \cdot \text{m}$$

a. Shaft BC.   For this hollow shaft we have

$$J = \frac{\pi}{2}(c_2^4 - c_1^4) = \frac{\pi}{2}[(0.060)^4 - (0.045)^4] = 13.92 \times 10^{-6} \text{ m}^4$$

Maximum Shearing Stress.   On the outer surface, we have

$$\tau_{\max} = \tau_2 = \frac{T_{BC} c_2}{J} = \frac{(20 \text{ kN} \cdot \text{m})(0.060 \text{ m})}{13.92 \times 10^{-6} \text{ m}^4} \qquad \tau_{\max} = 86.2 \text{ MPa} \blacktriangleleft$$

Minimum Shearing Stress.   We write that the stresses are proportional to the distance from the axis of the shaft.

$$\frac{\tau_{\min}}{\tau_{\max}} = \frac{c_1}{c_2} \qquad \frac{\tau_{\min}}{86.2 \text{ MPa}} = \frac{45 \text{ mm}}{60 \text{ mm}} \qquad \tau_{\min} = 64.7 \text{ MPa} \blacktriangleleft$$

b. Shafts AB and CD.   We note that in both of these shafts the magnitude of the torque is $T = 6 \text{ kN} \cdot \text{m}$ and $\tau_{\text{all}} = 65 \text{ MPa}$. Denoting by c the radius of the shafts, we write

$$\tau = \frac{Tc}{J} \qquad 65 \text{ MPa} = \frac{(6 \text{ kN} \cdot \text{m})c}{\frac{\pi}{2}c^4}$$

$$c^3 = 58.8 \times 10^{-6} \text{ m}^3 \qquad c = 38.9 \times 10^{-3} \text{ m}$$

$$d = 2c = 2(38.9 \text{ mm}) \qquad\qquad d = 77.8 \text{ mm} \blacktriangleleft$$

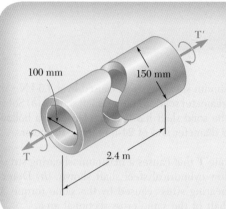

SAMPLE PROBLEM 3.2

The preliminary design of a large shaft connecting a motor to a generator calls for the use of a hollow shaft with inner and outer diameters of 100 mm and 150 mm, respectively. Knowing that the allowable shearing stress is 84 MPa, determine the maximum torque that can be transmitted (a) by the shaft as designed, (b) by a solid shaft of the same weight, (c) by a hollow shaft of the same weight and of 200-mm outer diameter.

SOLUTION

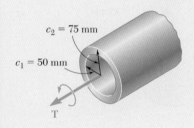

a. Hollow Shaft as Designed. For the hollow shaft we have

$$J = \frac{\pi}{2}(c_2^4 - c_1^4) = \frac{\pi}{2}[(75 \text{ mm})^4 - (50 \text{ mm})^4] = 39.88 \times 10^6 \text{ mm}^4$$

Using Eq. (3.9), we write

$$\tau_{max} = \frac{Tc_2}{J} \qquad 84 \text{ MPa} = \frac{T(75 \text{ mm})}{39.88 \times 10^6 \text{ mm}^4} \qquad T = 44.7 \text{ kN} \cdot \text{m} \blacktriangleleft$$

b. Solid Shaft of Equal Weight. For the shaft as designed and this solid shaft to have the same weight and length, their cross-sectional areas must be equal.

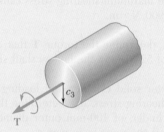

$$A_{(a)} = A_{(b)}$$
$$\pi[(75 \text{ mm})^2 - (50 \text{ mm})^2] = \pi c_3^2 \qquad c_3 = 55.9 \text{ mm}$$

Since $\tau_{all} = 84$ MPa, we write

$$\tau_{max} = \frac{Tc_3}{J} \qquad 84 \text{ MPa} = \frac{T(55.9 \text{ mm})}{\frac{\pi}{2}(55.9 \text{ mm})^4} \qquad T = 23.1 \text{ kN} \cdot \text{m} \blacktriangleleft$$

c. Hollow Shaft of 200-mm Diameter. For equal weight, the cross-sectional areas again must be equal. We determine the inside diameter of the shaft by writing

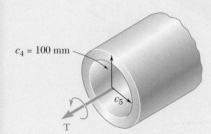

$$A_{(a)} = A_{(c)}$$
$$\pi[(75 \text{ mm})^2 - (50 \text{ mm})^2] = \pi[(100 \text{ mm})^2 - c_5^2] \qquad c_5 = 82.92 \text{ mm}$$

For $c_5 = 82.9$ mm and $c_4 = 100$ mm,

$$J = \frac{\pi}{2}[(100 \text{ mm})^4 - (82.92 \text{ mm})^4] = 82.82 \times 10^6 \text{ mm}^4$$

With $\tau_{all} = 84$ MPa and $c_4 = 100$ mm,

$$\tau_{max} = \frac{Tc_4}{J} \qquad 84 \text{ MPa} = \frac{T(100 \text{ mm})}{82.82 \times 10^6 \text{ mm}^4} \qquad T = 69.6 \text{ kN} \cdot \text{m} \blacktriangleleft$$

PROBLEMS

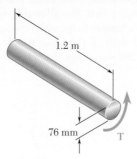

Fig. P3.1

3.1 (*a*) Determine the maximum shearing stress caused by a 4.6-kN · m torque **T** in the 76-mm-diameter solid aluminum shaft shown. (*b*) Solve part *a*, assuming that the solid shaft has been replaced by a hollow shaft of the same outer diameter and of 24-mm inner diameter.

3.2 (*a*) Determine the torque **T** that causes a maximum shearing stress of 45 MPa in the hollow cylindrical steel shaft shown. (*b*) Determine the maximum shearing stress caused by the same torque **T** in a solid cylindrical shaft of the same cross-sectional area.

Fig. P3.2

3.3 Knowing that the internal diameter of the hollow shaft shown is $d = 22$ mm, determine the maximum shearing stress caused by a torque of magnitude $T = 900$ N · m.

3.4 Knowing that $d = 30$ mm, determine the torque **T** that causes a maximum shearing stress of 52 MPa in the hollow shaft shown.

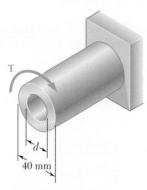

Fig. P3.3 and P3.4

3.5 (*a*) For the 75-mm-diameter solid cylinder and loading shown, determine the maximum shearing stress. (*b*) Determine the inner diameter of the hollow cylinder, of 100-mm outer diameter, for which the maximum stress is the same as in part *a*.

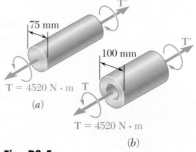

Fig. P3.5

3.6 (*a*) Determine the torque that can be applied to a solid shaft of 20-mm diameter without exceeding an allowable shearing stress of 80 MPa. (*b*) Solve part *a*, assuming that the solid shaft has been replaced by a hollow shaft of the same cross-sectional area and with an inner diameter equal to half of its outer diameter.

3.7 The solid spindle *AB* is made of a steel with an allowable shearing stress of 84 MPa, and sleeve *CD* is made of a brass with an allowable shearing stress of 50 MPa. Determine (*a*) the largest torque **T** that can be applied at *A* if the allowable shearing stress is not to be exceeded in sleeve *CD*, (*b*) the corresponding required value of the diameter d_s of spindle *AB*.

3.8 The solid spindle *AB* has a diameter d_s = 38 mm and is made of a steel with an allowable shearing stress of 84 MPa, while sleeve *CD* is made of a brass with an allowable shearing stress of 50 MPa. Determine the largest torque **T** that can be applied at *A*.

3.9 The torques shown are exerted on pulleys *A* and *B*. Knowing that both shafts are solid, determine the maximum shearing stress in (*a*) in shaft *AB*, (*b*) in shaft *BC*.

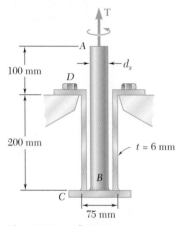

Fig. P3.7 and P3.8

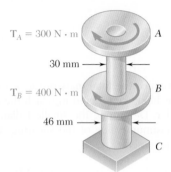

Fig. P3.9

3.10 In order to reduce the total mass of the assembly of Prob. 3.9, a new design is being considered in which the diameter of shaft *BC* will be smaller. Determine the smallest diameter of shaft *BC* for which the maximum value of the shearing stress in the assembly will not increase.

3.11 Knowing that each of the shafts *AB*, *BC*, and *CD* consists of a solid circular rod, determine (*a*) the shaft in which the maximum shearing stress occurs, (*b*) the magnitude of that stress.

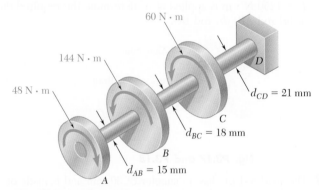

Fig. P3.11 and P3.12

3.12 Knowing that an 8-mm-diameter hole has been drilled through each of the shafts *AB*, *BC*, and *CD*, determine (*a*) the shaft in which the maximum shearing stress occurs, (*b*) the magnitude of that stress.

3.13 Under normal operating conditions, the electric motor exerts a torque of 2.8 kN · m on shaft AB. Knowing that each shaft is solid, determine the maximum shearing stress in (*a*) shaft AB, (*b*) shaft BC, (*c*) shaft CD.

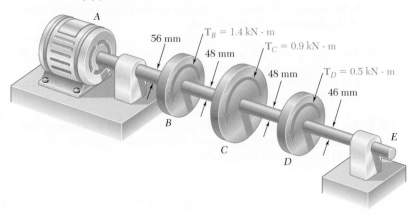

Fig. P3.13

3.14 In order to reduce the total mass of the assembly of Prob. 3.13, a new design is being considered in which the diameter of shaft BC will be smaller. Determine the smallest diameter of shaft BC for which the maximum value of the shearing stress in the assembly will not be increased.

3.15 The allowable shearing stress is 100 MPa in the steel rod AB and 60 MPa in the brass rod BC. Knowing that a torque of magnitude $T = 900$ N · m is applied at A, determine the required diameter of (*a*) rod AB, (*b*) rod BC.

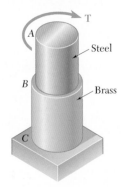

Fig. P3.15 and P3.16

3.16 The allowable shearing stress is 100 MPa in the 36-mm-diameter steel rod AB and 60 MPa in the 40-mm-diameter rod BC. Neglecting the effect of stress concentrations, determine the largest torque that can be applied at A.

3.17 The allowable stress is 50 MPa in the brass rod AB and 25 MPa in the aluminum rod BC. Knowing that a torque of magnitude $T = 1250$ N · m is applied at A, determine the required diameter of (*a*) rod AB, (*b*) rod BC.

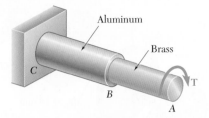

Fig. *P3.17* and *P3.18*

3.18 The *solid* rod BC has a diameter of 30 mm and is made of an aluminum for which the allowable shearing stress is 25 MPa. Rod AB is *hollow* and has an outer diameter of 25 mm; it is made of a brass for which the allowable shearing stress is 50 MPa. Determine (*a*) the largest inner diameter of rod AB for which the factor of safety is the same for each rod, (*b*) the largest torque that can be applied at A.

3.19 The solid rod AB has a diameter d_{AB} = 60 mm. The pipe CD has an outer diameter of 90 mm and a wall thickness of 6 mm. Knowing that both the rod and the pipe are made of steel for which the allowable shearing stress is 75 MPa, determine the largest torque **T** that can be applied at A.

3.20 The solid rod AB has a diameter d_{AB} = 60 mm and is made of a steel for which the allowable shearing stress is 85 MPa. The pipe CD, which has an outer diameter of 90 mm and a wall thickness of 6 mm, is made of an aluminum for which the allowable shearing stress is 54 MPa. Determine the largest torque **T** that can be applied at A.

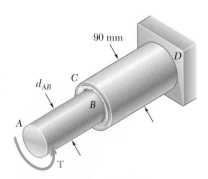

Fig. P3.19 and P3.20

3.21 The two solid shafts are connected by gears as shown and are made of a steel for which the allowable shearing stress is 50 MPa. Knowing the diameters of the two shafts are, respectively, d_{BC} = 40 mm and d_{EF} = 30 mm, determine the largest torque $\mathbf{T}_C$ that can be applied at C.

3.22 The two solid shafts are connected by gears as shown and are made of a steel for which the allowable shearing stress is 60 MPa. Knowing that a torque of magnitude T_C = 600 N · m is applied at C and that the assembly is in equilibrium, determine the required diameter of (a) shaft BC, (b) shaft EF.

3.23 Two solid steel shafts are connected by the gears shown. A torque of magnitude T = 900 N · m is applied to shaft AB. Knowing that the allowable shearing stress is 50 MPa and considering only stresses due to twisting, determine the required diameter of (a) shaft AB, (b) shaft CD.

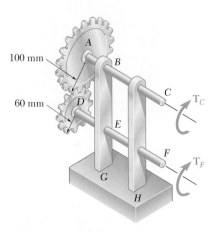

Fig. P3.21 and P3.22

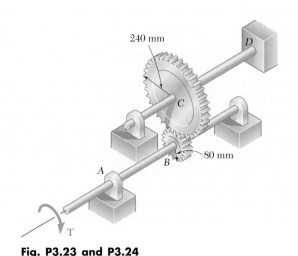

Fig. P3.23 and P3.24

3.24 Shaft CD is made from a 66-mm-diameter rod and is connected to the 48-mm-diameter shaft AB as shown. Considering only stresses due to twisting and knowing that the allowable shearing stress is 60 MPa for each shaft, determine the largest torque **T** that can be applied.

3.25 A torque of magnitude $T = 1000$ N $\cdot$ m is applied at D as shown. Knowing that the diameter of shaft AB is 56 mm and that the diameter of shaft CD is 42 mm, determine the maximum shearing stress in (a) shaft AB, (b) shaft CD.

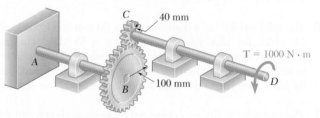

Fig. P3.25 and P3.26

3.26 A torque of magnitude $T = 1000$ N $\cdot$ m is applied at D as shown. Knowing that the allowable shearing stress is 60 MPa in each shaft, determine the required diameter of (a) shaft AB, (b) shaft CD.

3.27 A torque of magnitude $T = 100$ N $\cdot$ m is applied to shaft AB of the gear train shown. Knowing that the diameters of the three solid shafts are, respectively, $d_{AB} = 21$ mm, $d_{CD} = 30$ mm, and $d_{EF} = 40$ mm, determine the maximum shearing stress in (a) shaft AB, (b) shaft CD, (c) shaft EF.

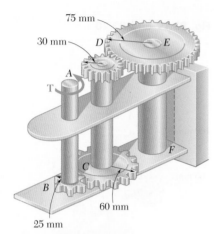

3.28 A torque of magnitude $T = 120$ N $\cdot$ m is applied to shaft AB of the gear train shown. Knowing that the allowable shearing stress is 75 MPa in each of the three solid shafts, determine the required diameter of (a) shaft AB, (b) shaft CD, (c) shaft EF.

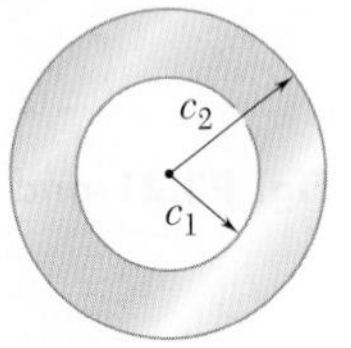

Fig. P3.27 and P3.28

3.29 (a) For a given allowable shearing stress, determine the ratio T/w of the maximum allowable torque T and the weight per unit length w for the hollow shaft shown. (b) Denoting by $(T/w)_0$ the value of this ratio for a solid shaft of the same radius c_2, express the ratio T/w for the hollow shaft in terms of $(T/w)_0$ and c_1/c_2.

Fig. P3.29

3.30 While the exact distribution of the shearing stresses in a hollow cylindrical shaft is as shown in Fig. P3.30a, an approximate value can be obtained for τ_{max} by assuming that the stresses are uniformly distributed over the area A of the cross section, as shown in Fig. P3.30b, and then further assuming that all of the elementary shearing forces act at a distance from O equal to the mean radius $\frac{1}{2}(c_1 + c_2)$ of the cross section. This approximate value $\tau_0 = T/Ar_m$, where T is the applied torque. Determine the ratio τ_{max}/τ_0 of the true value of the maximum shearing stress and its approximate value τ_0 for values of c_1/c_2 respectively equal to 1.00, 0.95, 0.75, 0.50, and 0.

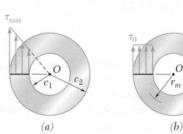

Fig. P3.30

3.5 ANGLE OF TWIST IN THE ELASTIC RANGE

In this section, a relation will be derived between the angle of twist ϕ of a circular shaft and the torque $\mathbf{T}$ exerted on the shaft. The entire shaft will be assumed to remain elastic. Considering first the case of a shaft of length L and of uniform cross section of radius c subjected to a torque $\mathbf{T}$ at its free end (Fig. 3.20), we recall from Sec. 3.3 that the angle of twist ϕ and the maximum shearing strain γ_{max} are related as follows:

$$\gamma_{max} = \frac{c\phi}{L} \tag{3.3}$$

But, in the elastic range, the yield stress is not exceeded anywhere in the shaft, Hooke's law applies, and we have $\gamma_{max} = \tau_{max}/G$ or, recalling Eq. (3.9),

$$\gamma_{max} = \frac{\tau_{max}}{G} = \frac{Tc}{JG} \tag{3.15}$$

Equating the right-hand members of Eqs. (3.3) and (3.15), and solving for ϕ, we write

$$\phi = \frac{TL}{JG} \tag{3.16}$$

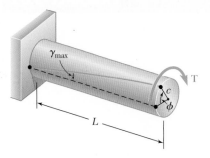

Fig. 3.20 Angle of twist ϕ.

where ϕ is expressed in radians. The relation obtained shows that, within the elastic range, *the angle of twist ϕ is proportional to the torque T applied to the shaft*. This is in accordance with the experimental evidence cited at the beginning of Sec. 3.3.

Equation (3.16) provides us with a convenient method for determining the modulus of rigidity of a given material. A specimen of the material, in the form of a cylindrical rod of known diameter and length, is placed in a *torsion testing machine* (Photo 3.3). Torques of increasing magnitude T are applied to the specimen, and the corresponding values of the angle of twist ϕ in a length L of the specimen are recorded. As long as the yield stress of the material is not exceeded, the points obtained by plotting ϕ against T will fall on a straight line. The slope of this line represents the quantity JG/L, from which the modulus of rigidity G may be computed.

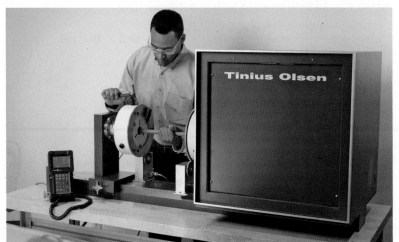

Photo 3.3 Torsion testing machine.

EXAMPLE 3.02

What torque should be applied to the end of the shaft of Example 3.01 to produce a twist of 2°? Use the value $G = 77$ GPa for the modulus of rigidity of steel.

Solving Eq. (3.16) for T, we write

$$T = \frac{JG}{L}\phi$$

Substituting the given values

$$G = 77 \times 10^9 \text{ Pa} \qquad L = 1.5 \text{ m}$$

$$\phi = 2°\left(\frac{2\pi \text{ rad}}{360°}\right) = 34.9 \times 10^{-3} \text{ rad}$$

and recalling from Example 3.01 that, for the given cross section,

$$J = 1.021 \times 10^{-6} \text{ m}^4$$

we have

$$T = \frac{JG}{L}\phi = \frac{(1.021 \times 10^{-6} \text{ m}^4)(77 \times 10^9 \text{ Pa})}{1.5 \text{ m}}(34.9 \times 10^{-3} \text{ rad})$$

$$T = 1.829 \times 10^3 \text{ N} \cdot \text{m} = 1.829 \text{ kN} \cdot \text{m}$$

EXAMPLE 3.03

What angle of twist will create a shearing stress of 70 MPa on the inner surface of the hollow steel shaft of Examples 3.01 and 3.02?

The method of attack for solving this problem that first comes to mind is to use Eq. (3.10) to find the torque T corresponding to the given value of τ, and Eq. (3.16) to determine the angle of twist ϕ corresponding to the value of T just found.

A more direct solution, however, may be used. From Hooke's law, we first compute the shearing strain on the inner surface of the shaft:

$$\gamma_{\min} = \frac{\tau_{\min}}{G} = \frac{70 \times 10^6 \text{ Pa}}{77 \times 10^9 \text{ Pa}} = 909 \times 10^{-6}$$

Recalling Eq. (3.2), which was obtained by expressing the length of arc AA' in Fig. 3.13c in terms of both γ and ϕ, we have

$$\phi = \frac{L\gamma_{\min}}{c_1} = \frac{1500 \text{ mm}}{20 \text{ mm}}(909 \times 10^{-6}) = 68.2 \times 10^{-3} \text{ rad}$$

To obtain the angle of twist in degrees, we write

$$\phi = (68.2 \times 10^{-3} \text{ rad})\left(\frac{360°}{2\pi \text{ rad}}\right) = 3.91°$$

Formula (3.16) for the angle of twist can be used only if the shaft is homogeneous (constant G), has a uniform cross section, and is loaded only at its ends. If the shaft is subjected to torques at locations other than its ends, or if it consists of several portions with various cross sections and possibly of different materials, we must divide it into component parts that satisfy individually the required

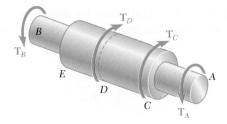

Fig. 3.21 Multiple sections and multiple torques.

conditions for the application of formula (3.16). In the case of the shaft AB shown in Fig. 3.21, for example, four different parts should be considered: AC, CD, DE, and EB. The total angle of twist of the shaft, i.e., the angle through which end A rotates with respect to end B, is obtained by adding *algebraically* the angles of twist of each component part. Denoting, respectively, by T_i, L_i, J_i, and G_i the internal torque, length, cross-sectional polar moment of inertia, and modulus of rigidity corresponding to part i, the total angle of twist of the shaft is expressed as

$$\phi = \sum_i \frac{T_i L_i}{J_i G_i} \tag{3.17}$$

The internal torque T_i in any given part of the shaft is obtained by passing a section through that part and drawing the free-body diagram of the portion of shaft located on one side of the section. This procedure, which has already been explained in Sec. 3.4 and illustrated in Fig. 3.16, is applied in Sample Prob. 3.3.

In the case of a shaft with a variable circular cross section, as shown in Fig. 3.22, formula (3.16) may be applied to a disk of thickness dx. The angle by which one face of the disk rotates with respect to the other is thus

$$d\phi = \frac{T\,dx}{JG}$$

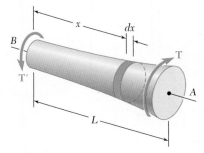

Fig. 3.22 Shaft with variable cross section.

where J is a function of x, which may be determined. Integrating in x from 0 to L, we obtain the total angle of twist of the shaft:

$$\phi = \int_0^L \frac{T \, dx}{JG} \qquad (3.18)$$

The shaft shown in Fig. 3.20, which was used to derive formula (3.16), and the shaft of Fig. 3.15, which was discussed in Examples 3.02 and 3.03, both had one end attached to a fixed support. In each case, therefore, the angle of twist ϕ of the shaft was equal to the angle of rotation of its free end. When both ends of a shaft rotate, however, the angle of twist of the shaft is equal to the angle through which one end of the shaft rotates *with respect to the other*. Consider, for instance, the assembly shown in Fig. 3.23a, consisting of two elastic shafts AD and BE, each of length L, radius c, and modulus of rigidity G, which are attached to gears meshed at C. If a torque $\mathbf{T}$ is applied at E (Fig. 3.23b), both shafts will be twisted. Since the end D of shaft AD is fixed, the angle of twist of AD is measured by the angle of rotation ϕ_A of end A. On the other hand, since both ends of shaft BE rotate, the angle of twist of BE is equal to the difference between the angles of rotation ϕ_B and ϕ_E, i.e., the angle of twist is equal to the angle through which end E rotates with respect to end B. Denoting this relative angle of rotation by $\phi_{E/B}$, we write

$$\phi_{E/B} = \phi_E - \phi_B = \frac{TL}{JG}$$

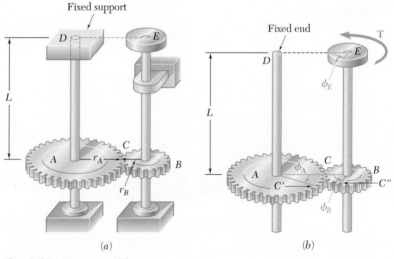

Fig. 3.23 Gear assembly.

EXAMPLE 3.04

For the assembly of Fig. 3.23, knowing that $r_A = 2r_B$, determine the angle of rotation of end E of shaft BE when the torque $\mathbf{T}$ is applied at E.

We first determine the torque $\mathbf{T}_{AD}$ exerted on shaft AD. Observing that equal and opposite forces $\mathbf{F}$ and $\mathbf{F}'$ are applied on the two gears at C (Fig. 3.24), and recalling that $r_A = 2r_B$, we conclude that the torque exerted on shaft AD is twice as large as the torque exerted on shaft BE; thus, $T_{AD} = 2T$.

Since the end D of shaft AD is fixed, the angle of rotation ϕ_A of gear A is equal to the angle of twist of the shaft and is obtained by writing

$$\phi_A = \frac{T_{AD}L}{JG} = \frac{2TL}{JG}$$

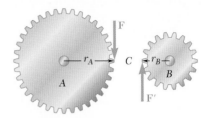

Fig. 3.24

Observing that the arcs CC' and CC'' in Fig. 3.23b must be equal, we write $r_A\phi_A = r_B\phi_B$ and obtain

$$\phi_B = (r_A/r_B)\phi_A = 2\phi_A$$

We have, therefore,

$$\phi_B = 2\phi_A = \frac{4TL}{JG}$$

Considering now shaft BE, we recall that the angle of twist of the shaft is equal to the angle $\phi_{E/B}$ through which end E rotates with respect to end B. We have

$$\phi_{E/B} = \frac{T_{BE}L}{JG} = \frac{TL}{JG}$$

The angle of rotation of end E is obtained by writing

$$\phi_E = \phi_B + \phi_{E/B}$$

$$= \frac{4TL}{JG} + \frac{TL}{JG} = \frac{5TL}{JG}$$

3.6 STATICALLY INDETERMINATE SHAFTS

You saw in Sec. 3.4 that, in order to determine the stresses in a shaft, it was necessary to first calculate the internal torques in the various parts of the shaft. These torques were obtained from statics by drawing the free-body diagram of the portion of shaft located on one side of a given section and writing that the sum of the torques exerted on that portion was zero.

There are situations, however, where the internal torques cannot be determined from statics alone. In fact, in such cases the external torques themselves, i.e., the torques exerted on the shaft by the supports and connections, cannot be determined from the free-body diagram of the entire shaft. The equilibrium equations must be complemented by relations involving the deformations of the shaft and obtained by considering the geometry of the problem. Because statics is not sufficient to determine the external and internal torques, the shafts are said to be *statically indeterminate*. The following example, as well as Sample Prob. 3.5, will show how to analyze statically indeterminate shafts.

EXAMPLE 3.05

A circular shaft AB consists of a 250-mm-long, 22-mm-diameter steel cylinder, in which a 125-mm-long, 16-mm-diameter cavity has been drilled from end B. The shaft is attached to fixed supports at both ends, and a 120-N · m torque is applied at its midsection (Fig. 3.25). Determine the torque exerted on the shaft by each of the supports.

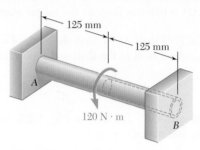

Fig. 3.25

Drawing the free-body diagram of the shaft and denoting by $\mathbf{T}_A$ and $\mathbf{T}_B$ the torques exerted by the supports (Fig. 3.26a), we obtain the equilibrium equation

$$T_A + T_B = 120 \text{ N} \cdot \text{m}$$

Since this equation is not sufficient to determine the two unknown torques $\mathbf{T}_A$ and $\mathbf{T}_B$, the shaft is statically indeterminate.

However, $\mathbf{T}_A$ and $\mathbf{T}_B$ can be determined if we observe that the total angle of twist of shaft AB must be zero, since both of its ends are restrained. Denoting by ϕ_1 and ϕ_2, respectively, the angles of twist of portions AC and CB, we write

$$\phi = \phi_1 + \phi_2 = 0$$

From the free-body diagram of a small portion of shaft including end A (Fig. 3.26b), we note that the internal torque T_1 in AC is equal to T_A; from the free-body diagram of a small portion of shaft including end B (Fig. 3.26c), we note that the internal torque T_2 in CB is equal to T_B. Recalling Eq. (3.16) and observing that portions AC and CB of the shaft are twisted in opposite senses, we write

$$\phi = \phi_1 + \phi_2 = \frac{T_A L_1}{J_1 G} - \frac{T_B L_2}{J_2 G} = 0$$

Solving for T_B, we have

$$T_B = \frac{L_1 J_2}{L_2 J_1} T_A$$

Substituting the numerical data

$$L_1 = L_2 = 125 \text{ mm}$$
$$J_1 = \tfrac{1}{2}\pi(0.011 \text{ m})^4 = 230 \times 10^{-6} \text{ m}^4$$
$$J_2 = \tfrac{1}{2}\pi[(0.011 \text{ m})^4 - (0.008 \text{ m})^4] = 165.6 \times 10^{-6} \text{ m}^4$$

we obtain

$$T_B = 0.72 \ T_A$$

Substituting this expression into the original equilibrium equation, we write

$$1.72 \ T_A = 120 \text{ N} \cdot \text{m}$$

$$T_A = 69.8 \text{ N} \cdot \text{m} \qquad T_B = 50.2 \text{ N} \cdot \text{m}$$

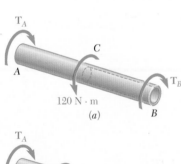

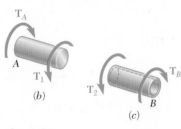

Fig. 3.26

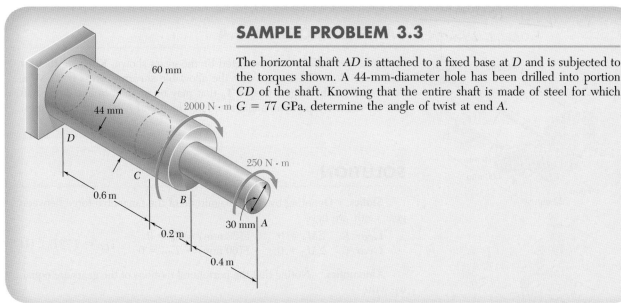

SAMPLE PROBLEM 3.3

The horizontal shaft AD is attached to a fixed base at D and is subjected to the torques shown. A 44-mm-diameter hole has been drilled into portion CD of the shaft. Knowing that the entire shaft is made of steel for which $G = 77$ GPa, determine the angle of twist at end A.

SOLUTION

Since the shaft consists of three portions AB, BC, and CD, each of uniform cross section and each with a constant internal torque, Eq. (3.17) may be used.

Statics. Passing a section through the shaft between A and B and using the free body shown, we find

$$\Sigma M_x = 0: \qquad (250 \text{ N} \cdot \text{m}) - T_{AB} = 0 \qquad T_{AB} = 250 \text{ N} \cdot \text{m}$$

Passing now a section between B and C, we have

$$\Sigma M_x = 0: (250 \text{ N} \cdot \text{m}) + (2000 \text{ N} \cdot \text{m}) - T_{BC} = 0 \qquad T_{BC} = 2250 \text{ N} \cdot \text{m}$$

Since no torque is applied at C,

$$T_{CD} = T_{BC} = 2250 \text{ N} \cdot \text{m}$$

Polar Moments of Inertia

$$J_{AB} = \frac{\pi}{2} c^4 = \frac{\pi}{2} (0.015 \text{ m})^4 = 0.0795 \times 10^{-6} \text{ m}^4$$

$$J_{BC} = \frac{\pi}{2} c^4 = \frac{\pi}{2} (0.030 \text{ m})^4 = 1.272 \times 10^{-6} \text{ m}^4$$

$$J_{CD} = \frac{\pi}{2} (c_2^4 - c_1^4) = \frac{\pi}{2} [(0.030 \text{ m})^4 - (0.022 \text{ m})^4] = 0.904 \times 10^{-6} \text{ m}^4$$

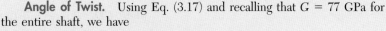

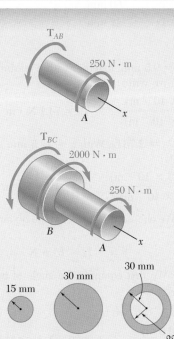

Angle of Twist. Using Eq. (3.17) and recalling that $G = 77$ GPa for the entire shaft, we have

$$\phi_A = \sum_i \frac{T_i L_i}{J_i G} = \frac{1}{G}\left(\frac{T_{AB}L_{AB}}{J_{AB}} + \frac{T_{BC}L_{BC}}{J_{BC}} + \frac{T_{CD}L_{CD}}{J_{CD}}\right)$$

$$\phi_A = \frac{1}{77 \text{ GPa}}\left[\frac{(250 \text{ N} \cdot \text{m})(0.4 \text{ m})}{0.0795 \times 10^{-6} \text{ m}^4} + \frac{(2250)(0.2)}{1.272 \times 10^{-6}} + \frac{(2250)(0.6)}{0.904 \times 10^{-6}}\right]$$

$$= 0.01634 + 0.00459 + 0.01939 = 0.0403 \text{ rad}$$

$$\phi_A = (0.0403 \text{ rad}) \frac{360°}{2\pi \text{ rad}} \qquad\qquad \phi_A = 2.31° \blacktriangleleft$$

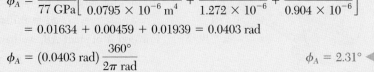

SAMPLE PROBLEM 3.4

Two solid steel shafts are connected by the gears shown. Knowing that for each shaft $G = 77$ GPa and that the allowable shearing stress is 55 MPa, determine (*a*) the largest torque $\mathbf{T}_0$ that may be applied to end A of shaft AB, (*b*) the corresponding angle through which end A of shaft AB rotates.

SOLUTION

Statics. Denoting by F the magnitude of the tangential force between gear teeth, we have

Gear B. $\Sigma M_B = 0$: $F(22 \text{ mm}) - T_0 = 0$

Gear C. $\Sigma M_C = 0$: $F(60 \text{ mm}) - T_{CD} = 0$ $T_{CD} = 2.73T_0$ (1)

Kinematics. Noting that the peripheral motions of the gears are equal, we write

$$r_B\phi_B = r_C\phi_C \qquad \phi_B = \phi_C\frac{r_C}{r_B} = \phi_C\frac{60 \text{ mm}}{22 \text{ mm}} = 2.73\phi_C \qquad (2)$$

a. Torque T_0

Shaft AB. With $T_{AB} = T_0$ and $c = 9.5$ mm, together with a maximum permissible shearing stress of 55 MPa, we write

$$\tau = \frac{T_{AB}c}{J} \qquad 55 \text{ MPa} = \frac{T_0(9.5 \times 10^{-3} \text{ m})}{\frac{1}{2}\pi(9.5 \times 10^{-3} \text{ m})^4} \qquad T_0 = 74.1 \text{ N} \cdot \text{m} \quad \blacktriangleleft$$

Shaft CD. From (1) we have $T_{CD} = 2.73T_0$. With $c = 12.5$ mm and $\tau_{\text{all}} = 55$ MPa, we write

$$\tau = \frac{T_{CD}c}{J} \qquad 55 \text{ MPa} = \frac{2.73T_0(12.5 \times 10^{-3} \text{ m})}{\frac{1}{2}\pi(12.5 \times 10^{-3} \text{ m})^4} \qquad T_0 = 61.8 \text{ N} \cdot \text{m} \quad \blacktriangleleft$$

Maximum Permissible Torque. We choose the smaller value obtained for T_0

$$T_0 = 61.8 \text{ N} \cdot \text{m} \quad \blacktriangleleft$$

b. Angle of Rotation at End A.

We first compute the angle of twist for each shaft.

Shaft AB. For $T_{AB} = T_0 = 61.8$ N $\cdot$ m, we have

$$\phi_{A/B} = \frac{T_{AB}L}{JG} = \frac{(61.8 \text{ N} \cdot \text{m})(0.6 \text{ m})}{\frac{1}{2}\pi(0.0095 \text{ m})^4(77 \times 10^9 \text{ Pa})} = 0.0376 \text{ rad} = 2.16°$$

Shaft CD. $T_{CD} = 2.73T_0 = 2.73(61.8 \text{ N} \cdot \text{m}) = 168.7 \text{ N} \cdot \text{m}$

$$\phi_{C/D} = \frac{T_{CD}L}{JG} = \frac{(168.7 \text{ N} \cdot \text{m})(0.9 \text{ m})}{\frac{1}{2}\pi(0.0125 \text{ m})^4(77 \times 10^9 \text{ Pa})} = 0.0514 \text{ rad} = 2.95°$$

Since end D of shaft CD is fixed, we have $\phi_C = \phi_{C/D} = 2.95°$. Using (2), we find the angle of rotation of gear B to be

$$\phi_B = 2.73\phi_C = 2.73(2.95°) = 8.04°$$

For end A of shaft AB, we have

$$\phi_A = \phi_B + \phi_{A/B} = 8.04° + 2.16° \qquad \phi_A = 10.2° \quad \blacktriangleleft$$

188

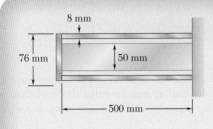

8 mm

76 mm

50 mm

500 mm

SAMPLE PROBLEM 3.5

A steel shaft and an aluminum tube are connected to a fixed support and to a rigid disk as shown in the cross section. Knowing that the initial stresses are zero, determine the maximum torque $\mathbf{T}_0$ that can be applied to the disk if the allowable stresses are 120 MPa in the steel shaft and 70 MPa in the aluminum tube. Use $G = 77$ GPa for steel and $G = 27$ GPa for aluminum.

T_1

T_0

T_2

SOLUTION

Statics. *Free Body of Disk.* Denoting by $\mathbf{T}_1$ the torque exerted by the tube on the disk and by $\mathbf{T}_2$ the torque exerted by the shaft, we find

$$T_0 = T_1 + T_2 \tag{1}$$

Deformations. Since both the tube and the shaft are connected to the rigid disk, we have

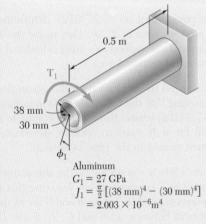

0.5 m

T_1

38 mm
30 mm

ϕ_1

Aluminum
$G_1 = 27$ GPa
$J_1 = \frac{\pi}{2}[(38\text{ mm})^4 - (30\text{ mm})^4]$
$\quad = 2.003 \times 10^{-6}\text{m}^4$

$$\phi_1 = \phi_2: \qquad \frac{T_1 L_1}{J_1 G_1} = \frac{T_2 L_2}{J_2 G_2}$$

$$\frac{T_1(0.5\text{ m})}{(2.003 \times 10^{-6}\text{ m}^4)(27\text{ GPa})} = \frac{T_2(0.5\text{ m})}{(0.614 \times 10^{-6}\text{ m}^4)(77\text{ GPa})}$$

$$T_2 = 0.874 T_1 \tag{2}$$

Shearing Stresses. We *assume* that the requirement $\tau_{\text{alum}} \leq 70$ MPa is critical. For the aluminum tube, we have

$$T_1 = \frac{\tau_{\text{alum}} J_1}{c_1} = \frac{(70\text{ MPa})(2.003 \times 10^{-6}\text{ m}^4)}{0.038\text{ m}} = 3690\text{ N} \cdot \text{m}$$

Using Eq. (2), we compute the corresponding value T_2 and then find the maximum shearing stress in the steel shaft.

$$T_2 = 0.874 T_1 = 0.874(3690) = 3225\text{ N} \cdot \text{m}$$

$$\tau_{\text{steel}} = \frac{T_2 c_2}{J_2} = \frac{(3225\text{ N} \cdot \text{m})(0.025\text{ m})}{0.614 \times 10^{-6}\text{ m}^4} = 131.3\text{ MPa}$$

We note that the allowable steel stress of 120 MPa is exceeded; our assumption was *wrong*. Thus the maximum torque $\mathbf{T}_0$ will be obtained by making $\tau_{\text{steel}} = 120$ MPa. We first determine the torque $\mathbf{T}_2$.

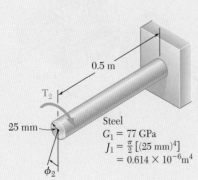

0.5 m

T_2

25 mm

ϕ_2

Steel
$G_1 = 77$ GPa
$J_1 = \frac{\pi}{2}[(25\text{ mm})^4]$
$\quad = 0.614 \times 10^{-6}\text{m}^4$

$$T_2 = \frac{\tau_{\text{steel}} J_2}{c_2} = \frac{(120\text{ MPa})(0.614 \times 10^{-6}\text{ m}^4)}{0.025\text{ m}} = 2950\text{ N} \cdot \text{m}$$

From Eq. (2), we have

$$2950\text{ N} \cdot \text{m} = 0.874 T_1 \qquad T_1 = 3375\text{ N} \cdot \text{m}$$

Using Eq. (1), we obtain the maximum permissible torque

$$T_0 = T_1 + T_2 = 3375\text{ N} \cdot \text{m} + 2950\text{ N} \cdot \text{m}$$

$$\boxed{T_0 = 6.325\text{ kN} \cdot \text{m}} \blacktriangleleft$$

3.33

a) $\phi = \dfrac{TL}{JG}$

$T_0 = \dfrac{\phi JG}{L} = \dfrac{(0.0349 \, rad)(\frac{\pi L}{\ })(50^4 - 40^4)(27E3 \, MPa)}{(2500 \, mm)}$

$mm^4 \cdot \dfrac{N}{mm^2}$

$\boxed{T_0 = 2184.72 \, Nm}$

b) $\phi = \dfrac{TL}{JG} = $ (Hollow circ shaft)
$A = \pi(c_2^2 - c_1^2)$ solid $A = \pi c^2$

$\pi(c_2^2 - c_1^2) = c^2 \pi$
$c = 30 \, mm$

$\therefore J = \dfrac{\pi}{2}(30)^4$

$\phi = 9.13°$

PROBLEMS

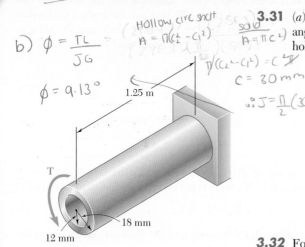

Fig. *P3.32*

3.31 (*a*) For the solid steel shaft shown ($G = 77$ GPa), determine the angle of twist at *A*. (*b*) Solve part *a*, assuming that the steel shaft is hollow with a 30-mm-outer diameter and a 20-mm-inner diameter.

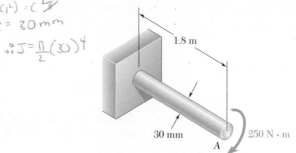

Fig. **P3.31**

3.32 For the aluminum shaft shown ($G = 27$ GPa), determine (*a*) the torque **T** that causes an angle of twist of 4°, (*b*) the angle of twist caused by the same torque **T** in a solid cylindrical shaft of the same length and cross-sectional area.

3.33 (*a*) For the aluminum pipe shown ($G = 27$ GPa), determine the torque **T**$_0$ causing an angle of twist of 2°. (*b*) Determine the angle of twist if the same torque **T**$_0$ is applied to a solid cylindrical shaft of the same length and cross-sectional area.

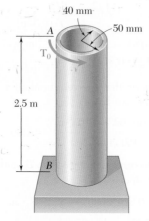

Fig. **P3.33**

3.34 The ship at *A* has just started to drill for oil on the ocean floor at a depth of 1500 m. Knowing that the top of the 200-mm-diameter steel drill pipe ($G = 77.2$ GPa) rotates through two complete revolutions before the drill bit at *B* starts to operate, determine the maximum shearing stress caused in the pipe by torsion.

3.35 The electric motor exerts a 500-N · m torque on the aluminum shaft *ABCD* when it is rotating at a constant speed. Knowing that $G = 27$ GPa and that the torques exerted on pulleys *B* and *C* are as shown, determine the angle of twist between (*a*) *B* and *C*, (*b*) *B* and *D*.

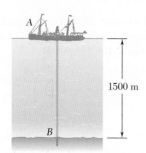

Fig. **P3.34**

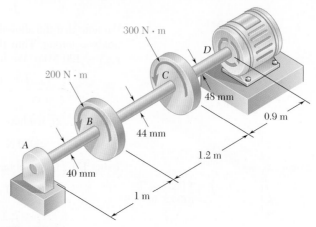

Fig. **P3.35**

3.49 The design specifications for the gear-and-shaft system shown require that the same diameter be used for both shafts and that the angle through which pulley A will rotate when subjected to a 200-N · m torque **T**$_A$ while pulley D is held fixed will not exceed 7.5°. Determine the required diameter of the shafts if both shafts are made of a steel with G = 77 GPa and τ_{all} = 84 MPa.

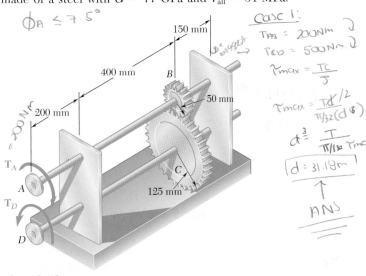

(handwritten annotations)

FBD of B

200 Nm

Bx

$\circlearrowleft \Sigma M_O = 0$

$-200 \text{Nm} + F_{BC}(0.05\text{m}) = 0$

$F_{BC} = 4000 \text{N}$

FBD of C

F_{BC}

$\circlearrowleft \Sigma M_O = 0$

$F_{BC}(0.125) + T_D = 0$

$T_D = -500 \text{Nm}$

$\phi_A \leq 7.5°$

Case 1:
$T_{AB} = 200 \text{Nm}$
$T_{BD} = 500 \text{Nm}$

$\tau_{max} = \dfrac{TC}{J}$

$\tau_{max} = \dfrac{Td/2}{\pi/32(d^8)}$

$d^3 = \dfrac{T}{\pi/16\,\tau_{max}}$

$\boxed{d = 31.19\text{m}}$

↑

ANS

Case 2:
$r_B \phi_B = r_C \phi_C$ ①
$\phi_B/\phi_C = 0.4\,\phi_B$

$\phi_{C/D} = \phi_C - \phi_D$ fixed.

$\phi_C = \dfrac{T_C L_{CD}}{JG}$ = solve ②

$\phi_{B/A} = \phi_B - \phi_A$
ϕ_A = given
$\phi_B = \phi_{B/A} + \phi_A$ ③
solve.
plug ① → ②
④
equate ③ & ④
$d = 28.24\text{m}$

3.50 The design specifications of a 1.2-m-long solid transmission shaft require that the angle of twist of the shaft not exceed 4° when a torque of 680 N · m is applied. Determine the required diameter of the shaft, knowing that the shaft is made of steel with an allowable shearing stress of 83 MPa and a modulus of rigidity of 77 GPa.

3.51 A torque of magnitude T = 4 kN · m is applied at end A of the composite shaft shown. Knowing that the modulus of rigidity is 77 GPa for the steel and 27 GPa for the aluminum, determine (a) the maximum shearing stress in the steel core, (b) the maximum shearing stress in the aluminum jacket, (c) the angle of twist at A.

(handwritten)

$J_A = \pi/2(36^4 - 27^4)$
$= 1.809 \times 10^6 \text{mm}^4$

$J_S = \pi/2(27^4)$
$= 8.348 \times 10^5 \text{mm}^4$

3.52.
$\phi = \Sigma \dfrac{T_i L_i}{J_i G_i}$

$\tau_{max} = \dfrac{TC}{J}$

$T_i = \dfrac{\tau_i J_i}{C_i}$

$\phi = \Sigma \dfrac{\tau_i J_i L_i}{J_i G_i C_i} \Rightarrow \phi = \Sigma \dfrac{\tau_i L_i}{G_i C}$

$\phi = 0.72 \text{ rad} + 1.157 \text{ rad}.$

$\phi = 0.72 \text{ rad} + 1.157 \text{ rad}.$

↑

ANS.

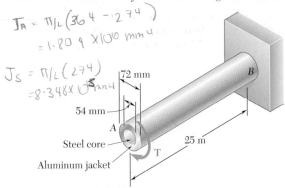

Fig. P3.51 and P3.52

3.52 The composite shaft shown is to be twisted by applying a torque **T** at end A. Knowing that the modulus of rigidity is 77 GPa for the steel and 27 GPa for the aluminum, determine the largest angle through which end A can be rotated if the following allowable stresses are not to be exceeded: τ_{steel} = 60 MPa and $\tau_{aluminum}$ = 45 MPa.

$1 \text{ MPa} = 1 \text{N/mm}^2$

for hollow rod
$$J = \frac{\pi}{2}(r_{outer}^4 - r_{inner}^4)$$

$$\phi = \sum \frac{T_i L_i}{J_i G_i}$$

3.53

a) $\phi = \frac{TL}{JG}$ T_c $(565N)($

$T_B = (\phi/L) J_B G_B$ $T_S = (\phi/L) J_S G_S$

$J_B = \pi/2 (20^4 - 15^4)$

$J_S = \pi/2 (15^4)$

$T_T = T_B + T_S$

$565 N = (\dot\phi/L) J_B (39E3 MPa) + (\phi/L) J_S (77 \times 10^3 MPa)$

$\phi/L = 4.4059 \times 10^{-5}$ rad/min

$\tau = G\delta$; $\delta_{max} = \frac{C\phi}{L}$

$\tau_{max} = \frac{G C\phi}{L} = $

a) $\tau_{max \, brass} = (39E3)(20)(4.4059 \times 10^{-5})$

b) $\tau max \, steel =$

c) $\phi = L(\phi/L)$

3.55

$\phi_{B/A} = \phi_B - \phi_A$ ⟵ 0 b/c fixed

$\phi_{B/A} = \phi_B$

constant rotated $1.5° = \frac{T_{AB} L_{AB}}{JG}$

$T_{AB} = 267.17$ Nm ↞ this torque will close the 1.5° gap then both shafts will rotate.

$570 - 267.17 = 302.816$

$302.8 = T_{AB} + T_{CD}$

$302.8 = \phi \left(\frac{\pi/32 (30^4)(7E3)}{600}\right) + $

$\phi \left(\frac{\pi/32 (36^4)(77E3)}{900}\right)$

$\phi = 0.012455$ rad

$\frac{AB}{\phi_B = \phi_{B/A} + \phi_A}$ ⟋ fixed

$\phi_B = 1.5 \times \frac{\pi}{180} + 0.01245$ rad

$\frac{CD}{\phi_C = \phi_{C/D} + \phi_D}$

$\phi_C = 0.01245$ rad

$\boxed{\tau_{max} = \frac{GC\phi}{L}}$

3.53 The composite shaft shown consists of a 5-mm-thick brass jacket ($G = 39$ GPa) bonded to a 30-mm-diameter steel core ($G = 77$ GPa). Knowing that the shaft is subjected to 565-N · m torques, determine (a) the maximum shearing stress in the brass jacket, (b) the maximum shearing stress in the steel core, (c) the angle of twist of end B relative to end A.

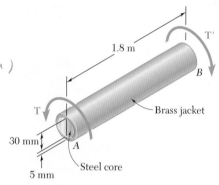

Fig. P3.53 and P3.54

3.54 The composite shaft shown is to be twisted by applying the torques shown. Knowing that the modulus of rigidity is 77 GPa for the steel and 39 GPa for the brass, determine the largest angle of twist of end B relative to end A if the following allowable stresses are not to be exceeded: $\tau_{steel} = 100$ MPa and $\tau_{brass} = 55$ MPa.

3.55 and 3.56 Two solid steel shafts are fitted with flanges that are then connected by bolts as shown. The bolts are slightly undersized and permit a 1.5° rotation of one flange with respect to the other before the flanges begin to rotate as a single unit. Knowing that $G = 77$ GPa, determine the maximum shearing stress in each shaft when a torque **T** of magnitude 570 N · m is applied to the flange indicated.

3.55 The torque **T** is applied to flange B.
3.56 The torque **T** is applied to flange C.

want to find what value of torque closes the 1.5° angle

$\phi = \frac{TL}{JG}$ → $T = \frac{JG\phi}{L}$

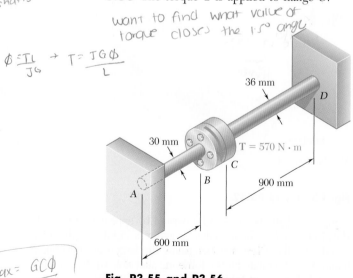

Fig. P3.55 and P3.56

3.57 Ends A and D of the two solid steel shafts AB and CD are fixed, while ends B and C are connected to gears as shown. Knowing that a 4-kN · m torque **T** is applied to gear B, determine the maximum shearing stress (a) in shaft AB, (b) in shaft CD.

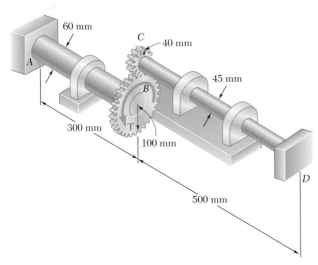

60 mm

C — 40 mm

A

45 mm

B

300 mm

T

100 mm

D

500 mm

Fig. P3.57 and P3.58

3.58 Ends A and D of the two solid steel shafts AB and CD are fixed, while ends B and C are connected to gears as shown. Knowing that the allowable shearing stress is 50 MPa in each shaft, determine the largest torque **T** that can be applied to gear B.

3.59 The steel jacket CD has been attached to the 40-mm-diameter steel shaft AE by means of *rigid* flanges welded to the jacket and to the rod. The outer diameter of the jacket is 80 mm and its wall thickness is 4 mm. If 500-N · m torques are applied as shown, determine the maximum shearing stress in the jacket.

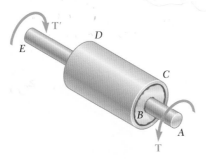

T'

D

E

C

B

A

T

Fig. P3.59

3.60 A solid shaft and a hollow shaft are made of the same material and are of the same weight and length. Denoting by n the ratio c_1/c_2, show that the ratio T_s/T_h of the torque T_s in the solid shaft to the torque T_h in the hollow shaft is (a) $\sqrt{(1 - n^2)}/(1 + n^2)$ if the maximum shearing stress is the same in each shaft, (b) $(1 - n^2)/(1 + n^2)$ if the angle of twist is the same for each shaft.

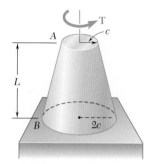

Fig. P3.61

3.61 A torque **T** is applied as shown to a solid tapered shaft AB. Show by integration that the angle of twist at A is

$$\phi = \frac{7TL}{12\pi Gc^4}$$

3.62 The mass moment of inertia of a gear is to be determined experimentally by using a torsional pendulum consisting of a 1.8-m steel wire. Knowing that $G = 77$ GPa, determine the diameter of the wire for which the torsional spring constant will be 6 N · m/rad.

Fig. P3.62

3.63 An annular plate of thickness t and modulus G is used to connect shaft AB of radius r_1 to tube CD of radius r_2. Knowing that a torque **T** is applied to end A of shaft AB and that end D of tube CD is fixed, (a) determine the magnitude and location of the maximum shearing stress in the annular plate, (b) show that the angle through which end B of the shaft rotates with respect to end C of the tube is

$$\phi_{BC} = \frac{T}{4\pi Gt}\left(\frac{1}{r_1^2} - \frac{1}{r_2^2}\right)$$

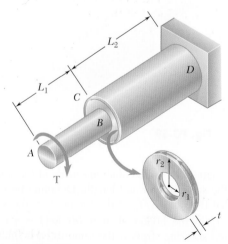

Fig. P3.63

3.7 DESIGN OF TRANSMISSION SHAFTS

The principal specifications to be met in the design of a transmission shaft are the *power* to be transmitted and the *speed of rotation* of the shaft. The role of the designer is to select the material and the dimensions of the cross section of the shaft, so that the maximum shearing stress allowable in the material will not be exceeded when the shaft is transmitting the required power at the specified speed.

To determine the torque exerted on the shaft, we recall from elementary dynamics that the power P associated with the rotation of a rigid body subjected to a torque $\mathbf{T}$ is

$$P = T\omega \tag{3.19}$$

where ω is the angular velocity of the body expressed in radians per second. But $\omega = 2\pi f$, where f is the frequency of the rotation, i.e., the number of revolutions per second. The unit of frequency is thus $1\ \text{s}^{-1}$ and is called a *hertz* (Hz). Substituting for ω into Eq. (3.19), we write

$$P = 2\pi f T \tag{3.20}$$

If SI units are used we verify that, with f expressed in Hz and T in N $\cdot$ m, the power will be expressed in N $\cdot$ m/s, that is, in *watts* (W). Solving Eq. (3.20) for T, we obtain the torque exerted on a shaft transmitting the power P at a frequency of rotation f,

$$T = \frac{P}{2\pi f} \tag{3.21}$$

where P, f, and T are expressed in the units indicated above.

After having determined the torque $\mathbf{T}$ that will be applied to the shaft and having selected the material to be used, the designer will carry the values of T and of the maximum allowable stress into the elastic torsion formula (3.9). Solving for J/c, we have

$$\frac{J}{c} = \frac{T}{\tau_{\max}} \tag{3.22}$$

and obtain in this way the minimum value allowable for the parameter J/c. We check that, if SI units are used, T will be expressed in N $\cdot$ m, $\tau_{\max}$ in Pa (or N/m^2), and J/c will be obtained in m^3. In the case of a solid circular shaft, $J = \frac{1}{2}\pi c^4$, and $J/c = \frac{1}{2}\pi c^3$; substituting this value for J/c into Eq. (3.22) and solving for c yields the minimum allowable value for the radius of the shaft. In the case of a hollow circular shaft, the critical parameter is J/c_2, where c_2 is the outer radius of the shaft; the value of this parameter may be computed from Eq. (3.11) of Sec. 3.4 to determine whether a given cross section will be acceptable.

EXAMPLE 3.06

What size of shaft should be used for the rotor of a 3.7-kW motor operating at 3600 rpm if the shearing stress is not to exceed 60 MPa in the shaft?

We first express the power of the motor in N · m/s and its frequency in cycles per second (or hertzes).

$$P = 3.7 \text{ kW} = 3700 \text{ N} \cdot \text{m/s}$$

$$f = (3600 \text{ rpm})\frac{1 \text{ Hz}}{60 \text{ rpm}} = 60 \text{ Hz} = 60 \text{ s}^{-1}$$

The torque exerted on the shaft is given by Eq. (3.21):

$$T = \frac{P}{2\pi f} = \frac{3700 \text{ N} \cdot \text{m/s}}{2\pi (60 \text{ s}^{-1})} = 9.815 \text{ N} \cdot \text{m}$$

Substituting for T and $\tau_{\max}$ into Eq. (3.22), we write

$$\frac{J}{c} = \frac{T}{\tau_{\max}} = \frac{9.815 \text{ N} \cdot \text{m}}{60 \text{ MPa}} = 163.58 \text{ mm}^3$$

But $J/c = \frac{1}{2}\pi c^3$ for a solid shaft. We have, therefore,

$$\frac{1}{2}\pi c^3 = 163.58 \text{ mm}^3$$

$$c = 4.705 \text{ mm}$$

$$d = 2c = 9.41 \text{ mm}$$

A 10-mm shaft should be used.

EXAMPLE 3.07

A shaft consisting of a steel tube of 50-mm outer diameter is to transmit 100 kW of power while rotating at a frequency of 20 Hz. Determine the tube thickness that should be used if the shearing stress is not to exceed 60 MPa.

The torque exerted on the shaft is given by Eq. (3.21):

$$T = \frac{P}{2\pi f} = \frac{100 \times 10^3 \text{ W}}{2\pi (20 \text{ Hz})} = 795.8 \text{ N} \cdot \text{m}$$

From Eq. (3.22) we conclude that the parameter J/c_2 must be at least equal to

$$\frac{J}{c_2} = \frac{T}{\tau_{\max}} = \frac{795.8 \text{ N} \cdot \text{m}}{60 \times 10^6 \text{ N/m}^2} = 13.26 \times 10^{-6} \text{ m}^3 \qquad (3.23)$$

But, from Eq. (3.10) we have

$$\frac{J}{c_2} = \frac{\pi}{2c_2}(c_2^4 - c_1^4) = \frac{\pi}{0.050}[(0.025)^4 - c_1^4] \qquad (3.24)$$

Equating the right-hand members of Eqs. (3.23) and (3.24), we obtain:

$$(0.025)^4 - c_1^4 = \frac{0.050}{\pi}(13.26 \times 10^{-6})$$

$$c_1^4 = 390.6 \times 10^{-9} - 211.0 \times 10^{-9} = 179.6 \times 10^{-9} \text{ m}^4$$

$$c_1 = 20.6 \times 10^{-3} \text{ m} = 20.6 \text{ mm}$$

The corresponding tube thickness is

$$c_2 - c_1 = 25 \text{ mm} - 20.6 \text{ mm} = 4.4 \text{ mm}$$

A tube thickness of 5 mm should be used.

3.8 STRESS CONCENTRATIONS IN CIRCULAR SHAFTS

The torsion formula $\tau_{max} = Tc/J$ was derived in Sec. 3.4 for a circular shaft of uniform cross section. Moreover, we had assumed earlier in Sec. 3.3 that the shaft was loaded at its ends through rigid end plates solidly attached to it. In practice, however, the torques are usually applied to the shaft through flange couplings (Fig. 3.27a) or through gears connected to the shaft by keys fitted into keyways (Fig. 3.27b). In both cases one should expect the distribution of stresses, in and near the section where the torques are applied, to be different from that given by the torsion formula. High concentrations of stresses, for example, will occur in the neighborhood of the keyway shown in Fig. 3.27b. The determination of these localized stresses may be carried out by experimental stress analysis methods or, in some cases, through the use of the mathematical theory of elasticity.

As we indicated in Sec. 3.4, the torsion formula can also be used for a shaft of variable circular cross section. In the case of a shaft with an abrupt change in the diameter of its cross section, however, stress concentrations will occur near the discontinuity, with the highest stresses occurring at A (Fig. 3.28). These stresses may

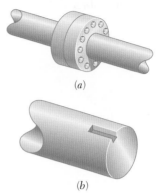

(a)

(b)

Fig. 3.27 Shaft examples.

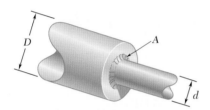

Fig. 3.28 Shaft with change in diameter.

be reduced through the use of a fillet, and the maximum value of the shearing stress at the fillet can be expressed as

$$\tau_{max} = K\frac{Tc}{J} \tag{3.25}$$

where the stress Tc/J is the stress computed for the smaller-diameter shaft, and where K is a stress-concentration factor. Since the factor K depends only upon the ratio of the two diameters and the ratio of the radius of the fillet to the diameter of the smaller shaft, it may be computed once and for all and recorded in the form of a table or a graph, as shown in Fig. 3.29. We should note, however, that this procedure for determining localized shearing stresses is valid only as long as the value of τ_{max} given by Eq. (3.25) does not exceed the proportional limit of the material, since the values of K plotted in Fig. 3.29 were obtained under the assumption of a linear relation between shearing stress and shearing strain. If plastic deformations occur, they will result in values of the maximum stress lower than those indicated by Eq. (3.25).

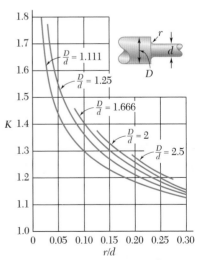

Fig. 3.29 Stress-concentration factors for fillets in circular shafts.†

†W. D. Pilkey, *Peterson's Stress Concentration Factors*, 2nd ed., John Wiley & Sons, New York, 1997.

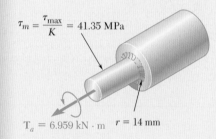

190 mm

95 mm

$r = 14$ mm

SAMPLE PROBLEM 3.6

The stepped shaft shown is to rotate at 900 rpm as it transmits power from a turbine to a generator. The grade of steel specified in the design has an allowable shearing stress of 55 MPa. (*a*) For the preliminary design shown, determine the maximum power that can be transmitted. (*b*) If in the final design the radius of the fillet is increased so that $r = 24$ mm, what will be the percent change, relative to the preliminary design, in the power that can be transmitted?

SOLUTION

a. Preliminary Design. Using the notation of Fig. 3.29, we have: $D = 190$ mm, $d = 95$ mm, $r = 14$ mm.

$$\frac{D}{d} = \frac{190 \text{ mm}}{95 \text{ mm}} = 2 \qquad \frac{r}{d} = \frac{14 \text{ mm}}{95 \text{ mm}} = 0.15$$

A stress concentration factor $K = 1.33$ is found from Fig. 3.29.

Torque. Recalling Eq. (3.25), we write

$$\tau_{max} = K\frac{Tc}{J} \qquad T = \frac{J}{c}\frac{\tau_{max}}{K} \qquad (1)$$

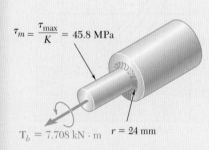

$\tau_m = \dfrac{\tau_{max}}{K} = 41.35$ MPa

$T_a = 6.959$ kN · m $r = 14$ mm

where J/c refers to the smaller-diameter shaft:

$$J/c = \tfrac{1}{2}\pi c^3 = \tfrac{1}{2}\pi(47.5 \text{ mm})^3 = 168.3 \times 10^3 \text{ mm}^3$$

and where

$$\frac{\tau_{max}}{K} = \frac{55 \text{ MPa}}{1.33} = 41.35 \text{ MPa}$$

Substituting into Eq. (1), we find $T = (168.3 \times 10^3 \text{ mm}^3)(41.35 \text{ MPa}) = 6.959$ kN · m.

Power. Since $f = (900 \text{ rpm})\dfrac{1 \text{ Hz}}{60 \text{ rpm}} = 15 \text{ Hz} = 15 \text{ s}^{-1}$, we write

$$P_a = 2\pi fT = 2\pi(15 \text{ s}^{-1})(6959 \text{ N} \cdot \text{m}) = 656 \text{ kN} \cdot \text{m/s} = 656 \text{ kW}$$

$$P_a = 656 \text{ kW} \quad \blacktriangleleft$$

b. Final Design. For $r = 24$ mm,

$$\frac{D}{d} = 2 \qquad \frac{r}{d} = \frac{24 \text{ mm}}{95 \text{ mm}} = 0.253 \qquad K = 1.20$$

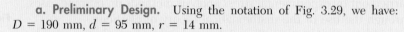

$\tau_m = \dfrac{\tau_{max}}{K} = 45.8$ MPa

$T_b = 7.708$ kN · m $r = 24$ mm

Following the procedure used above, we write

$$\frac{\tau_{max}}{K} = \frac{55 \text{ MPa}}{1.20} = 45.8 \text{ MPa}$$

$$T = \frac{J}{c}\frac{\tau_{max}}{K} = (168.3 \times 10^3 \text{ mm}^3)(45.8 \text{ MPa}) = 7.708 \text{ kN} \cdot \text{m}$$

$$P_b = 2\pi fT = 2\pi(15 \text{ s}^{-1})(7708 \text{ N} \cdot \text{m}) = 726.5 \text{ kW}$$

Percent Change in Power

$$\text{Percent change} = 100\frac{P_b - P_a}{P_a} = 100\frac{726.5 - 656}{656} = +10.75\% \quad \blacktriangleleft$$

PROBLEMS

3.64 Determine the maximum shearing stress in a solid shaft of 12-mm diameter as it transmits 2.5 kW at a frequency of (a) 25 Hz, (b) 50 Hz.

3.65 Determine the maximum shearing stress in a solid shaft of 38-mm diameter as it transmits 55 kW at a speed of (a) 750 rpm, (b) 1500 rpm.

3.66 Design a solid steel shaft to transmit 0.375 kW at a frequency of 29 Hz, if the shearing stress in the shaft is not to exceed 35 MPa.

3.67 A hollow shaft is to transmit 250 kW at a frequency of 30 Hz. Knowing that the shearing stress must not exceed 50 MPa, design a shaft for which the ratio of the inner diameter to the outer diameter is 0.75.

3.68 Determine the required thickness of the 50-mm tubular shaft of Example 3.07, if it is to transmit the same power while rotating at a frequency of 30 Hz.

3.69 While a steel shaft of the cross section shown rotates at 120 rpm, a stroboscopic measurement indicates that the angle of twist is 2° in a 4-m length. Using $G = 77$ GPa, determine the power being transmitted.

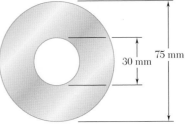

Fig. P3.69

3.70 One of two hollow drive shafts of an ocean liner is 38 m long, and its outer and inner diameters are 400 mm and 200 mm, respectively. The shaft is made of a steel for which $\tau_{all} = 60$ MPa and $G = 77$ GPa. Knowing that the maximum speed of rotation of the shaft is 2.75 Hz, determine (a) the maximum power that can be transmitted by the one shaft to its propeller, (b) the corresponding angle of twist of the shaft.

3.71 The hollow steel shaft shown ($G = 77.2$ GPa, $\tau_{all} = 50$ MPa) rotates at 240 rpm. Determine (a) the maximum power that can be transmitted, (b) the corresponding angle of twist of the shaft.

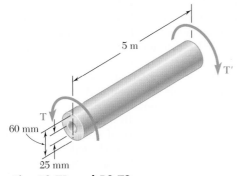

Fig. P3.71 and P3.72

3.72 As the hollow steel shaft shown rotates at 180 rpm, a stroboscopic measurement indicates that the angle of twist of the shaft is 3°. Knowing that $G = 77.2$ GPa, determine (a) the power being transmitted, (b) the maximum shearing stress in the shaft.

203

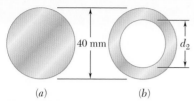

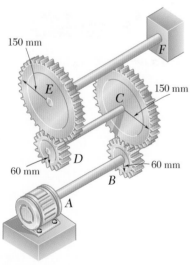

40 mm

d_2

(a)

(b)

Fig. P3.73

Fig. P3.77

3.73 The design of a machine element calls for a 40-mm-outer-diameter shaft to transmit 45 kW. (a) If the speed of rotation is 720 rpm, determine the maximum shearing stress in shaft a. (b) If the speed of rotation can be increased 50 percent to 1080 rpm, determine the largest inner diameter of shaft b for which the maximum shearing stress will be the same in each shaft.

3.74 The two solid shafts and gears shown are used to transmit 12 kW from the motor at A operating at a speed of 1260 rpm to a machine tool at D. Knowing that the maximum allowable shearing stress is 55 MPa, determine the required diameter (a) of shaft AB, (b) of shaft CD.

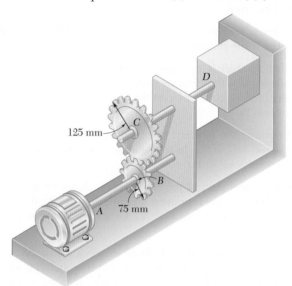

125 mm

D

C

B

A   75 mm

Fig. P3.74 and P3.75

3.75 The two solid shafts and gears shown are used to transmit 12 kW from the motor at A operating at a speed of 1260 rpm to a machine tool at D. Knowing that each shaft has a diameter of 25 mm, determine the maximum shearing stress (a) in shaft AB, (b) in shaft CD.

3.76 Three shafts and four gears are used to form a gear train that will transmit 7.5 kW from the motor at A to a machine tool at F. (Bearings for the shafts are omitted in the sketch.) Knowing that the frequency of the motor is 30 Hz and that the allowable stress for each shaft is 60 MPa, determine the required diameter of each shaft.

3.77 Three shafts and four gears are used to form a gear train that will transmit power from the motor at A to a machine tool at F. (Bearings for the shafts are omitted in the sketch.) The diameter of each shaft is as follows: $d_{AB} = 16$ mm, $d_{CD} = 20$ mm, $d_{EF} = 28$ mm. Knowing that the frequency of the motor is 24 Hz and that the allowable shearing stress for each shaft is 75 MPa, determine the maximum power that can be transmitted.

3.78 A 1.5-m-long solid steel shaft of 48-mm diameter is to transmit 36 kW between a motor and a machine tool. Determine the lowest speed at which the shaft can rotate, knowing that $G = 77.2$ GPa, that the maximum shearing stress must not exceed 60 MPa, and the angle of twist must not exceed 2.5°.

150 mm

E

C

150 mm

D

60 mm

60 mm

B

A

3.79 The shaft-disk-belt arrangement shown is used to transmit 2 kW from point A to point D. (*a*) Using an allowable shearing stress of 66 MPa, determine the required speed of shaft AB. (*b*) Solve part *a*, assuming that the diameters of shafts AB and CD are, respectively, 18 mm and 15 mm.

3.80 A 2.5-m-long steel shaft of 30-mm diameter rotates at a frequency of 30 Hz. Determine the maximum power that the shaft can transmit, knowing that $G = 77.2$ GPa, that the allowable shearing stress is 50 MPa, and that the angle of twist must not exceed $7.5°$.

3.81 A steel shaft must transmit 150 kW at a speed of 360 rpm. Knowing that $G = 77.2$ GPa, design a solid shaft so that the maximum shearing stress will not exceed 50 MPa and the angle of twist in a 2.5-m length will not exceed $3°$.

3.82 A 1.6-m-long tubular steel shaft of 42-mm outer diameter d_1 is to be made of a steel for which $\tau_{all} = 75$ MPa and $G = 77.2$ GPa. Knowing that the angle of twist must not exceed $4°$ when the shaft is subjected to a torque of 900 N · m, determine the largest inner diameter d_2 that can be specified in the design.

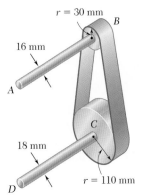
r = 30 mm
16 mm
18 mm
r = 110 mm

Fig. P3.79

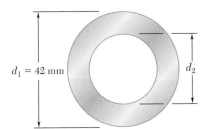

$d_1 = 42$ mm d_2

Fig. *P3.82* and *P3.83*

3.83 A 1.6-m-long tubular steel shaft ($G = 77.2$ GPa) of 42-mm outer diameter d_1 and 30-mm inner diameter d_2 is to transmit 120 kW between a turbine and a generator. Knowing that the allowable shearing stress is 65 MPa and that the angle of twist must not exceed $3°$, determine the minimum frequency at which the shaft can rotate.

3.84 The stepped shaft shown rotates at 450 rpm. Knowing that $r = 12$ mm, determine the largest torque **T** that can be transmitted without exceeding an allowable shearing stress of 50 MPa.

3.85 The stepped shaft shown rotates at 450 rpm. Knowing that $r = 5$ mm, determine the maximum power that can be transmitted without exceeding an allowable shearing stress of 50 MPa.

125 mm   150 mm   *r*

Fig. *P3.84* and *P3.85*

3.86 The stepped shaft shown must transmit 40 kW at a speed of 720 rpm. Determine the minimum radius r of the fillet if an allowable stress of 36 MPa is not to be exceeded.

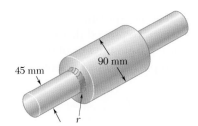
45 mm   90 mm   *r*

Fig. P3.86

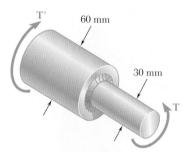

Fig. P3.87 and P3.88

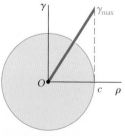

Full quarter-circular fillet
extends to edge of larger shaft.
Fig. *P3.89, P3.90,* and *P3.91*

3.87 The stepped shaft shown must transmit 45 kW. Knowing that the allowable shearing stress in the shaft is 40 MPa and that the radius of the fillet is $r = 6$ mm, determine the smallest permissible speed of the shaft.

3.88 The stepped shaft shown must rotate at a frequency of 50 Hz. Knowing that the radius of the fillet is $r = 8$ mm and the allowable shearing stress is 45 MPa, determine the maximum power that can be transmitted.

3.89 In the stepped shaft shown, which has a full quarter-circular fillet, $D = 30$ mm and $d = 25$ mm. Knowing that the speed of the shaft is 2400 rpm and that the allowable shearing stress is 50 MPa, determine the maximum power that can be transmitted by the shaft.

3.90 A torque of magnitude $T = 22$ N · m is applied to the stepped shaft shown, which has a full quarter-circular fillet. Knowing that $D = 25$ mm, determine the maximum shearing stress in the shaft when (*a*) $d = 20$ mm, (*b*) $d = 23$ mm.

3.91 In the stepped shaft shown, which has a full quarter-circular fillet, the allowable shearing stress is 80 MPa. Knowing that $D = 30$ mm, determine the largest allowable torque that can be applied to the shaft if (*a*) $d = 26$ mm, (*b*) $d = 24$ mm.

*3.9 PLASTIC DEFORMATIONS IN CIRCULAR SHAFTS

When we derived Eqs. (3.10) and (3.16), which define, respectively, the stress distribution and the angle of twist for a circular shaft subjected to a torque **T,** we assumed that Hooke's law applied throughout the shaft. If the yield strength is exceeded in some portion of the shaft, or if the material involved is a brittle material with a nonlinear shearing-stress-strain diagram, these relations cease to be valid. The purpose of this section is to develop a more general method—which may be used when Hooke's law does not apply—for determining the distribution of stresses in a solid circular shaft, and for computing the torque required to produce a given angle of twist.

We first recall that no specific stress-strain relationship was assumed in Sec. 3.3, when we proved that the shearing strain γ varies linearly with the distance ρ from the axis of the shaft (Fig. 3.30). Thus, we may still use this property in our present analysis and write

$$\gamma = \frac{\rho}{c}\gamma_{max} \tag{3.4}$$

where c is the radius of the shaft.

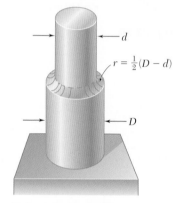

Fig. 3.30 Shearing strain variation.

Assuming that the maximum value τ_{max} of the shearing stress τ has been specified, the plot of τ versus ρ may be obtained as follows. We first determine from the shearing-stress-strain diagram the value of γ_{max} corresponding to τ_{max} (Fig. 3.31), and carry this value into Eq. (3.4). Then, for each value of ρ, we determine the corresponding value of γ from Eq. (3.4) or Fig. 3.30 and obtain from the stress-strain diagram of Fig. 3.31 the shearing stress τ corresponding to this value of γ. Plotting τ against ρ yields the desired distribution of stresses (Fig. 3.32).

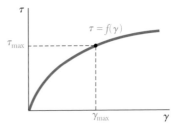

Fig. 3.31 Nonlinear, shear stress-strain diagram.

We now recall that, when we derived Eq. (3.1) in Sec. 3.2, we assumed no particular relation between shearing stress and strain. We may therefore use Eq. (3.1) to determine the torque **T** corresponding to the shearing-stress distribution obtained in Fig. 3.32. Considering an annular element of radius ρ and thickness $d\rho$, we express the element of area in Eq. (3.1) as $dA = 2\pi\rho \, d\rho$ and write

$$T = \int_0^c \rho\tau(2\pi\rho \, d\rho)$$

or

$$T = 2\pi \int_0^c \rho^2\tau \, d\rho \qquad (3.26)$$

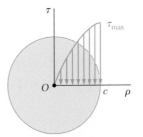

Fig. 3.32 Shearing strain variation for shaft with nonlinear stress-strain diagram.

where τ is the function of ρ plotted in Fig. 3.32.

If τ is a known analytical function of γ, Eq. (3.4) may be used to express τ as a function of ρ, and the integral in (3.26) may be determined analytically. Otherwise, the torque **T** may be obtained through a numerical integration. This computation becomes more meaningful if we note that the integral in Eq. (3.26) represents the second moment, or moment of inertia, with respect to the vertical axis of the area in Fig. 3.32 located above the horizontal axis and bounded by the stress-distribution curve.

An important value of the torque is the ultimate torque T_U which causes failure of the shaft. This value may be determined from the ultimate shearing stress τ_U of the material by choosing $\tau_{max} = \tau_U$ and carrying out the computations indicated earlier. However, it is found more convenient in practice to determine T_U experimentally by twisting a specimen of a given material until it breaks. Assuming a fictitious linear distribution of stresses, Eq. (3.9) is then used to determine the corresponding maximum shearing stress R_T:

$$R_T = \frac{T_U c}{J} \qquad (3.27)$$

The fictitious stress R_T is called the *modulus of rupture in torsion* of the given material. It may be used to determine the ultimate torque T_U of a shaft made of the same material, but of different dimensions, by solving Eq. (3.27) for T_U. Since the actual and the fictitious linear stress distributions shown in Fig. 3.33 must yield the same value T_U for the ultimate torque, the areas they define must have the same moment of inertia with respect to the vertical axis. It is thus clear that the modulus of rupture R_T will always be larger than the actual ultimate shearing stress τ_U.

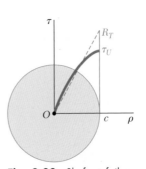

Fig. 3.33 Shaft at failure.

In some cases, we may wish to determine the stress distribution and the torque **T** corresponding to a given angle of twist ϕ. This may be done by recalling the expression obtained in Sec. 3.3 for the shearing strain γ in terms of ϕ, ρ, and the length L of the shaft:

$$\gamma = \frac{\rho\phi}{L} \tag{3.2}$$

With ϕ and L given, we may determine from Eq. (3.2) the value of γ corresponding to any given value of ρ. Using the stress-strain diagram of the material, we may then obtain the corresponding value of the shearing stress τ and plot τ against ρ. Once the shearing-stress distribution has been obtained, the torque **T** may be determined analytically or numerically as explained earlier.

*3.10 CIRCULAR SHAFTS MADE OF AN ELASTOPLASTIC MATERIAL

Further insight into the plastic behavior of a shaft in torsion is obtained by considering the idealized case of a *solid circular shaft made of an elastoplastic material*. The shearing-stress-strain diagram of such a material is shown in Fig. 3.34. Using this diagram, we can proceed as indicated earlier and find the stress distribution across a section of the shaft for any value of the torque **T**.

As long as the shearing stress τ does not exceed the yield strength τ_Y, Hooke's law applies, and the stress distribution across the section is linear (Fig. 3.35a), with $\tau_{\max}$ given by Eq. (3.9):

$$\tau_{\max} = \frac{Tc}{J} \tag{3.9}$$

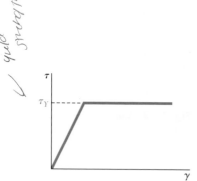

Fig. 3.34 Elastoplastic stress-strain diagram.

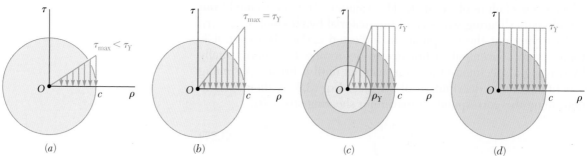

Fig. 3.35 Stress-strain diagrams for shaft made of elastoplastic material.

As the torque increases, $\tau_{\max}$ eventually reaches the value τ_Y (Fig. 3.35b). Substituting this value into Eq. (3.9), and solving for the corresponding value of T, we obtain the value T_Y of the torque at the onset of yield:

$$T_Y = \frac{J}{c}\tau_Y \tag{3.28}$$

The value obtained is referred to as the *maximum elastic torque*, since it is the largest torque for which the deformation remains fully elastic. Recalling that for a solid circular shaft $J/c = \frac{1}{2}\pi c^3$, we have

$$T_Y = \frac{1}{2}\pi c^3 \tau_Y \tag{3.29}$$

As the torque is further increased, a plastic region develops in the shaft, around an elastic core of radius ρ_Y (Fig. 3.35c). In the plastic region the stress is uniformly equal to τ_Y, while in the elastic core the stress varies linearly with ρ and may be expressed as

$$\tau = \frac{\tau_Y}{\rho_Y}\rho \tag{3.30}$$

As T is increased, the plastic region expands until, at the limit, the deformation is fully plastic (Fig. 3.35d).

Equation (3.26) will be used to determine the value of the torque T corresponding to a given radius ρ_Y of the elastic core. Recalling that τ is given by Eq. (3.30) for $0 \le \rho \le \rho_Y$, and is equal to τ_Y for $\rho_Y \le \rho \le c$, we write

$$T = 2\pi \int_0^{\rho_Y} \rho^2 \left(\frac{\tau_Y}{\rho_Y}\rho\right) d\rho + 2\pi \int_{\rho_Y}^c \rho^2 \tau_Y \, d\rho$$

$$= \frac{1}{2}\pi \rho_Y^3 \tau_Y + \frac{2}{3}\pi c^3 \tau_Y - \frac{2}{3}\pi \rho_Y^3 \tau_Y$$

$$T = \frac{2}{3}\pi c^3 \tau_Y \left(1 - \frac{1}{4}\frac{\rho_Y^3}{c^3}\right) \tag{3.31}$$

or, in view of Eq. (3.29),

$$T = \frac{4}{3}T_Y \left(1 - \frac{1}{4}\frac{\rho_Y^3}{c^3}\right) \tag{3.32}$$

where T_Y is the maximum elastic torque. We note that, as ρ_Y approaches zero, the torque approaches the limiting value

$$T_p = \frac{4}{3}T_Y \tag{3.33}$$

This value of the torque, which corresponds to a fully plastic deformation (Fig. 3.35d), is called the *plastic torque* of the shaft considered. We note that Eq. (3.33) is valid only for a *solid circular shaft made of an elastoplastic material*.

Since the distribution of *strain* across the section remains linear after the onset of yield, Eq. (3.2) remains valid and can be used to express the radius ρ_Y of the elastic core in terms of the angle of twist ϕ. If ϕ is large enough to cause a plastic deformation, the radius ρ_Y of the elastic core is obtained by making γ equal to the yield strain γ_Y in Eq. (3.2) and solving for the corresponding value ρ_Y of the distance ρ. We have

$$\rho_Y = \frac{L\gamma_Y}{\phi} \tag{3.34}$$

Let us denote by ϕ_Y the angle of twist at the onset of yield, i.e., when $\rho_Y = c$. Making $\phi = \phi_Y$ and $\rho_Y = c$ in Eq. (3.34), we have

$$c = \frac{L\gamma_Y}{\phi_Y} \tag{3.35}$$

Dividing (3.34) by (3.35), member by member, we obtain the following relation:†

$$\frac{\rho_Y}{c} = \frac{\phi_Y}{\phi} \tag{3.36}$$

If we carry into Eq. (3.32) the expression obtained for ρ_Y/c, we express the torque T as a function of the angle of twist ϕ,

$$T = \frac{4}{3}T_Y\left(1 - \frac{1}{4}\frac{\phi_Y^3}{\phi^3}\right) \tag{3.37}$$

where T_Y and ϕ_Y represent, respectively, the torque and the angle of twist at the onset of yield. Note that Eq. (3.37) may be used only for values of ϕ larger than ϕ_Y. For $\phi < \phi_Y$, the relation between T and ϕ is linear and given by Eq. (3.16). Combining both equations, we obtain the plot of T against ϕ represented in Fig. 3.39. We check that, as ϕ increases indefinitely, T approaches the limiting value $T_p = \frac{4}{3}T_Y$ corresponding to the case of a fully developed plastic zone (Fig. 3.35d). While the value T_p cannot actually be reached, we note from Eq. (3.37) that it is rapidly approached as ϕ increases. For $\phi = 2\phi_Y$, T is within about 3 percent of T_p, and for $\phi = 3\phi_Y$ within about 1 percent.

Since the plot of T against ϕ that we have obtained for an idealized elastoplastic material (Fig. 3.36) differs greatly from the shearing-stress-strain diagram of that material (Fig. 3.34), it is clear that the shearing-stress-strain diagram of an actual material cannot be obtained directly from a torsion test carried out on a solid circular rod made of that material. However, a fairly accurate diagram may be obtained from a torsion test if the specimen used incorporates a portion consisting of a thin circular tube.‡ Indeed, we may assume that the shearing stress will have a constant value τ in that portion. Equation (3.1) thus reduces to

$$T = \rho A\tau$$

where ρ denotes the average radius of the tube and A its cross-sectional area. The shearing stress is thus proportional to the torque, and successive values of τ can be easily computed from the corresponding values of T. On the other hand, the values of the shearing strain γ may be obtained from Eq. (3.2) and from the values of ϕ and L measured on the tubular portion of the specimen.

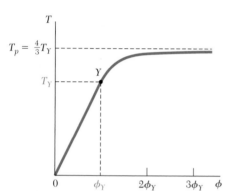

Fig. 3.36 Load displacement relation for elastoplastic material.

---

†Equation (3.36) applies to any ductile material with a well-defined yield point, since its derivation is independent of the shape of the stress-strain diagram beyond the yield point.

‡In order to minimize the possibility of failure by buckling, the specimen should be made so that the length of the tubular portion is no longer than its diameter.

EXAMPLE 3.08

A solid circular shaft, 1.2 m long and 50 mm in diameter, is subjected to a 4.60-kN · m torque at each end (Fig. 3.37). Assuming the shaft to be made of an elastoplastic material with a yield strength in shear of 150 MPa and a modulus of rigidity of 77 GPa, determine (a) the radius of the elastic core, (b) the angle of twist of the shaft.

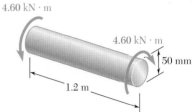

(a) Radius of Elastic Core. We first determine the torque T_Y at the onset of yield. Using Eq. (3.28) with $\tau_Y = 150$ MPa, $c = 25$ mm, and

$$J = \tfrac{1}{2}\pi c^4 = \tfrac{1}{2}\pi(25 \times 10^{-3}\text{ m})^4 = 614 \times 10^{-9}\text{ m}^4$$

we write

$$T_Y = \frac{J\tau_Y}{c} = \frac{(614 \times 10^{-9}\text{ m}^4)(150 \times 10^6\text{ Pa})}{25 \times 10^{-3}\text{ m}} = 3.68\text{ kN} \cdot \text{m}$$

Fig. 3.37

Solving Eq. (3.32) for $(\rho_Y/c)^3$ and substituting the values of T and T_Y, we have

$$\left(\frac{\rho_Y}{c}\right)^3 = 4 - \frac{3T}{T_Y} = 4 - \frac{3(4.60\text{ kN} \cdot \text{m})}{3.68\text{ kN} \cdot \text{m}} = 0.250$$

$$\frac{\rho_Y}{c} = 0.630 \qquad \rho_Y = 0.630(25\text{ mm}) = 15.8\text{ mm}$$

(b) Angle of Twist. We first determine the angle of twist ϕ_Y at the onset of yield from Eq. (3.16):

$$\phi_Y = \frac{T_Y L}{JG} = \frac{(3.68 \times 10^3\text{ N} \cdot \text{m})(1.2\text{ m})}{(614 \times 10^{-9}\text{ m}^4)(77 \times 10^9\text{ Pa})} = 93.4 \times 10^{-3}\text{ rad}$$

Solving Eq. (3.36) for ϕ and substituting the values obtained for ϕ_Y and ρ_Y/c, we write

$$\phi = \frac{\phi_Y}{\rho_Y/c} = \frac{93.4 \times 10^{-3}\text{ rad}}{0.630} = 148.3 \times 10^{-3}\text{ rad}$$

or

$$\phi = (148.3 \times 10^{-3}\text{ rad})\left(\frac{360°}{2\pi\text{ rad}}\right) = 8.50°$$

*3.11 RESIDUAL STRESSES IN CIRCULAR SHAFTS

In the two preceding sections, we saw that a plastic region will develop in a shaft subjected to a large enough torque, and that the shearing stress τ at any given point in the plastic region may be obtained from the shearing-stress-strain diagram of Fig. 3.31. If the torque is removed, the resulting reduction of stress and strain at the point considered will take place along a straight line (Fig. 3.38). As you will see further in this section, the final value of the stress will not, in general, be zero. There will be a residual stress at most points, and that stress may be either positive or negative. We note that, as was the case for the normal stress, the shearing stress will keep decreasing until it has reached a value equal to its maximum value at C minus twice the yield strength of the material.

Consider again the idealized case of the elastoplastic material characterized by the shearing-stress-strain diagram of Fig. 3.34.

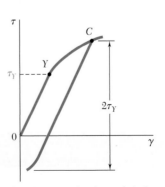

Fig. 3.38 Unloading of shaft with nonlinear stress-strain diagram.

211

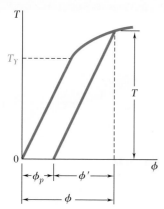

Fig. 3.39 Unloading of shaft with elastoplastic material.

Assuming that the relation between τ and γ at any point of the shaft remains linear as long as the stress does not decrease by more than $2\tau_Y$, we can use Eq. (3.16) to obtain the angle through which the shaft untwists as the torque decreases back to zero. As a result, the unloading of the shaft will be represented by a straight line on the T-ϕ diagram (Fig. 3.39). We note that the angle of twist does not return to zero after the torque has been removed. Indeed, the loading and unloading of the shaft result in a permanent deformation characterized by the angle

$$\phi_p = \phi - \phi' \qquad (3.38)$$

where ϕ corresponds to the loading phase and may be obtained from T by solving Eq. (3.38), and where ϕ' corresponds to the unloading phase and may be obtained from Eq. (3.16).

The residual stresses in an elastoplastic material are obtained by applying the principle of superposition in a manner similar to that described in Sec. 2.20 for an axial loading. We consider, on one hand, the stresses due to the application of the given torque $\mathbf{T}$ and, on the other, the stresses due to the equal and opposite torque which is applied to unload the shaft. The first group of stresses reflects the elastoplastic behavior of the material during the loading phase (Fig. 3.40a), and the second group the linear behavior of the same material during the unloading phase (Fig. 3.40b). Adding the two groups of stresses, we obtain the distribution of the residual stresses in the shaft (Fig. 3.40c).

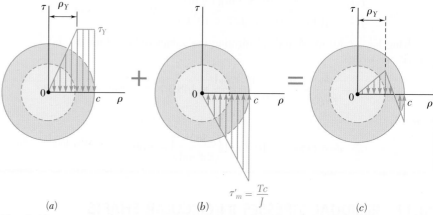

(a) $\tau'_m = \dfrac{Tc}{J}$ (b) (c)

Fig. 3.40 Stress distributions for unloading of shaft with elastoplastic material.

We note from Fig. 3.40c that some residual stresses have the same sense as the original stresses, while others have the opposite sense. This was to be expected since, according to Eq. (3.1), the relation

$$\int \rho(\tau \, dA) = 0 \qquad (3.39)$$

must be verified after the torque has been removed.

EXAMPLE 3.09 For the shaft of Example 3.08 determine (a) the permanent twist, (b) the distribution of residual stresses, after the 4.60-kN · m torque has been removed.

(a) Permanent Twist. We recall from Example 3.08 that the angle of twist corresponding to the given torque is $\phi = 8.50°$. The angle ϕ'

through which the shaft untwists as the torque is removed is obtained from Eq. (3.16). Substituting the given data,

$$T = 4.60 \times 10^3 \text{ N} \cdot \text{m}$$
$$L = 1.2 \text{ m}$$
$$G = 77 \times 10^9 \text{ Pa}$$

and the value $J = 614 \times 10^{-9} \text{ m}^4$ obtained in the solution of Example 3.08, we have

$$\phi' = \frac{TL}{JG} = \frac{(4.60 \times 10^3 \text{ N} \cdot \text{m})(1.2 \text{ m})}{(614 \times 10^{-9} \text{ m}^4)(77 \times 10^9 \text{ Pa})}$$
$$= 116.8 \times 10^{-3} \text{ rad}$$

or

$$\phi' = (116.8 \times 10^{-3} \text{ rad}) \frac{360°}{2\pi \text{ rad}} = 6.69°$$

The permanent twist is therefore

$$\phi_p = \phi - \phi' = 8.50° - 6.69° = 1.81°$$

(b) Residual Stresses. We recall from Example 3.08 that the yield strength is $\tau_Y = 150$ MPa and that the radius of the elastic core corresponding to the given torque is $\rho_Y = 15.8$ mm. The distribution of the stresses in the loaded shaft is thus as shown in Fig. 3.41a.

The distribution of stresses due to the opposite 4.60-kN · m torque required to unload the shaft is linear and as shown in Fig. 3.41b. The maximum stress in the distribution of the reverse stresses is obtained from Eq. (3.9):

$$\tau'_{max} = \frac{Tc}{J} = \frac{(4.60 \times 10^3 \text{ N} \cdot \text{m})(25 \times 10^{-3} \text{ m})}{614 \times 10^{-9} \text{ m}^4}$$
$$= 187.3 \text{ MPa}$$

Superposing the two distributions of stresses, we obtain the residual stresses shown in Fig. 3.41c. We check that, even though the reverse stresses exceed the yield strength τ_Y, the assumption of a linear distribution of these stresses is valid, since they do not exceed $2\tau_Y$.

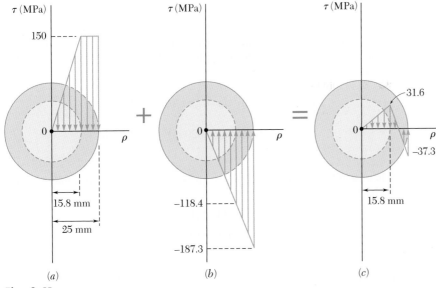

Fig. 3.41

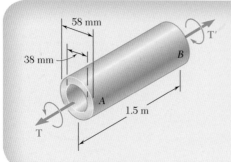

SAMPLE PROBLEM 3.7

Shaft AB is made of a mild steel which is assumed to be elastoplastic with $G = 77$ GPa and $\tau_Y = 145$ MPa. A torque $\mathbf{T}$ is applied and gradually increased in magnitude. Determine the magnitude of $\mathbf{T}$ and the corresponding angle of twist (a) when yield first occurs, (b) when the deformation has become fully plastic.

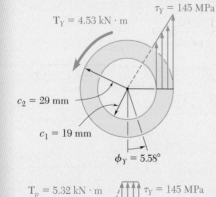

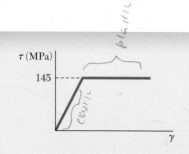

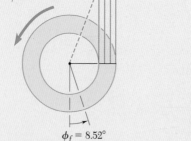

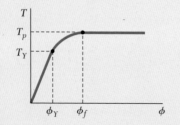

SOLUTION

Geometric Properties
The geometric properties of the cross section are

$$c_1 = \tfrac{1}{2}(38 \text{ mm}) = 19 \text{ mm} \qquad c_2 = \tfrac{1}{2}(58 \text{ mm}) = 29 \text{ mm}$$

$$J = \tfrac{1}{2}\pi(c_2^4 - c_1^4) = \tfrac{1}{2}\pi[(29 \text{ mm})^4 - (19 \text{ mm})^4] = 906.3 \times 10^3 \text{ mm}^3$$

a. Onset of Yield. For $\tau_{\max} = \tau_Y = 145$ MPa, we find

$$T_Y = \frac{\tau_Y J}{c_2} = \frac{(145 \text{ MPa})(906.3 \times 10^3 \text{ mm}^3)}{29 \text{ mm}}$$

$$T_Y = 4.53 \text{ kN} \cdot \text{m} \quad \blacktriangleleft$$

Making $\rho = c_2$ and $\gamma = \gamma_Y$ in Eq. (3.2) and solving for ϕ, we obtain the value of ϕ_Y:

$$\phi_Y = \frac{\gamma_Y L}{c_2} = \frac{\tau_Y L}{c_2 G} = \frac{(145 \text{ MPa})(1500 \text{ mm})}{(29 \text{ mm})(77 \text{ GPa})} = 0.0974 \text{ rad}$$

$$\phi_Y = 5.58° \quad \blacktriangleleft$$

b. Fully Plastic Deformation. When the plastic zone reaches the inner surface, the stresses are uniformly distributed as shown. Using Eq. (3.26), we write

$$T_p = 2\pi\tau_Y \int_{c_1}^{c_2} \rho^2 \, d\rho = \tfrac{2}{3}\pi\tau_Y(c_2^3 - c_1^3)$$

$$= \tfrac{2}{3}\pi(145 \text{ MPa})[(0.029 \text{ m})^3 - (0.019 \text{ m})^3]$$

$$T_p = 5.32 \text{ kN} \cdot \text{m} \quad \blacktriangleleft$$

When yield first occurs on the inner surface, the deformation is fully plastic; we have from Eq. (3.2):

$$\phi_f = \frac{\gamma_Y L}{c_1} = \frac{\tau_Y L}{c_1 G} = \frac{(145 \text{ MPa})(1.5 \text{ m})}{(0.019 \text{ m})(77 \text{ GPa})} = 0.1487 \text{ rad}$$

$$\phi_f = 8.52° \quad \blacktriangleleft$$

For larger angles of twist, the torque remains constant; the T-ϕ diagram of the shaft is as shown.

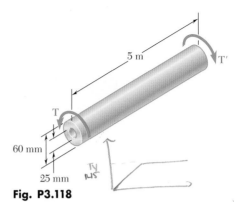

3.117 After the solid shaft of Prob. 3.116 has been loaded and unloaded as described in that problem, a torque $\mathbf{T}_1$ of sense opposite to the original torque $\mathbf{T}$ is applied to the shaft. Assuming no change in the value of ϕ_Y, determine the angle of twist ϕ_1 for which yield is initiated in this second loading and compare it with the angle ϕ_Y for which the shaft started to yield in the original loading.

3.118 The hollow shaft shown is made of a steel that is assumed to be elastoplastic with $\tau_Y = 145$ MPa and $G = 77.2$ GPa. The magnitude T of the torques is slowly increased until the plastic zone first reaches the inner surface of the shaft; the torques are then removed. Determine the magnitude and location of the maximum residual shearing stress in the rod.

Fig. P3.118

3.119 In Prob. 3.118, determine the permanent angle of twist of the rod.

$\phi_P = $ permanent angle of twist.

3.120 A torque $\mathbf{T}$ applied to a solid rod made of an elastoplastic material is increased until the rod is fully plastic and then removed. (*a*) Show that the distribution of residual shearing stresses is as represented in the figure. (*b*) Determine the magnitude of the torque due to the stresses acting on the portion of the rod located within a circle of radius c_0.

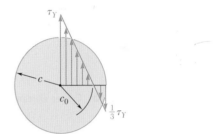

Fig. P3.120

$\phi_P = \phi_f + \phi_{unload}$
$= 43.157° - 22.86° = 20.295°$

*3.12   TORSION OF NONCIRCULAR MEMBERS

The formulas obtained in Secs. 3.3 and 3.4 for the distributions of strain and stress under a torsional loading apply only to members with a circular cross section. Indeed, their derivation was based on the assumption that the cross section of the member remained plane and undistorted, and we saw in Sec. 3.3 that the validity of this assumption depends upon the *axisymmetry* of the member, i.e., upon the fact that its appearance remains the same when it is viewed from a fixed position and rotated about its axis through an arbitrary angle.

A square bar, on the other hand, retains the same appearance only when it is rotated through 90° or 180°. Following a line of reasoning similar to that used in Sec. 3.3, one could show that the diagonals of the square cross section of the bar and the lines joining the midpoints of the sides of that section remain straight (Fig. 3.42). However, because of the lack of axisymmetry of the bar, any other line drawn in its cross section will deform when the bar is twisted, and the cross section itself will be warped out of its original plane.

It follows that Eqs. (3.4) and (3.6), which define, respectively, the distributions of strain and stress in an elastic circular shaft, cannot be used for noncircular members. For example, it would be wrong to assume that the shearing stress in the cross section of a square bar varies linearly with the distance from the axis of the bar and is, therefore, largest at the corners of the cross section. As you will see presently, the shearing stress is actually zero at these points.

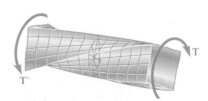

Fig. 3.42 Twisting of shaft with square cross section.

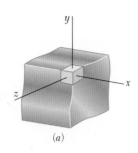

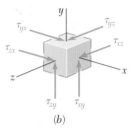

Fig. 3.43 Corner element.

Consider a small cubic element located at a corner of the cross section of a square bar in torsion and select coordinate axes parallel to the edges of the element (Fig. 3.43*a*). Since the face of the element perpendicular to the *y* axis is part of the free surface of the bar, all stresses on this face must be zero. Referring to Fig. 3.43*b*, we write

$$\tau_{yx} = 0 \qquad \tau_{yz} = 0 \tag{3.40}$$

For the same reason, all stresses on the face of the element perpendicular to the *z* axis must be zero, and we write

$$\tau_{zx} = 0 \qquad \tau_{zy} = 0 \tag{3.41}$$

It follows from the first of Eqs. (3.40) and the first of Eqs. (3.41) that

$$\tau_{xy} = 0 \qquad \tau_{xz} = 0 \tag{3.42}$$

Thus, both components of the shearing stress on the face of the element perpendicular to the axis of the bar are zero. We conclude that there is no shearing stress at the corners of the cross section of the bar.

By twisting a rubber model of a square bar, one easily verifies that no deformations—and, thus, no stresses—occur along the edges of the bar, while the largest deformations—and, thus, the largest stresses—occur along the center line of each of the faces of the bar (Fig. 3.44).

Fig. 3.44 Deformation of square bar.

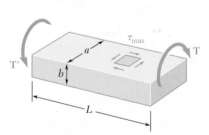

Fig. 3.45 Shaft with rectangular cross section.

The determination of the stresses in noncircular members subjected to a torsional loading is beyond the scope of this text. However, results obtained from the mathematical theory of elasticity for straight bars with a *uniform rectangular cross section* will be indicated here for convenience.† Denoting by *L* the length of the bar, by *a* and *b*, respectively, the wider and narrower side of its cross section, and by *T* the magnitude of the torques applied to the bar (Fig. 3.45), we find that the maximum shearing stress occurs along the center line of the *wider* face of the bar and is equal to

$$\tau_{\max} = \frac{T}{c_1 a b^2} \tag{3.43}$$

The angle of twist, on the other hand, may be expressed as

$$\phi = \frac{TL}{c_2 a b^3 G} \tag{3.44}$$

†See S. P. Timoshenko and J. N. Goodier, *Theory of Elasticity*, 3d ed., McGraw-Hill, New York, 1969, sec. 109.

The coefficients c_1 and c_2 depend only upon the ratio a/b and are given in Table 3.1 for a number of values of that ratio. Note that Eqs. (3.43) and (3.44) are valid only within the elastic range.

We note from Table 3.1 that for $a/b \geq 5$, the coefficients c_1 and c_2 are equal. It may be shown that for such values of a/b, we have

$$c_1 = c_2 = \tfrac{1}{3}(1 - 0.630b/a) \qquad \textbf{(for } a/b \geq 5 \textbf{ only)} \qquad (3.45)$$

The distribution of shearing stresses in a noncircular member may be visualized more easily by using the *membrane analogy*. A homogeneous elastic membrane attached to a fixed frame and subjected to a uniform pressure on one of its sides happens to constitute an *analog* of the bar in torsion, i.e., the determination of the deformation of the membrane depends upon the solution of the same partial differential equation as the determination of the shearing stresses in the bar.† More specifically, if Q is a point of the cross section of the bar and Q' the corresponding point of the membrane (Fig. 3.46), the shearing stress τ at Q will have the same direction as the horizontal tangent to the membrane at Q', and its magnitude will be proportional to the maximum slope of the membrane at Q'.‡ Furthermore, the applied torque will be proportional to the volume between the membrane and the plane of the fixed frame. In the case of the membrane of Fig. 3.46, which is attached to a rectangular frame, the steepest slope occurs at the midpoint N' of the larger side of the frame. Thus, we verify that the maximum shearing stress in a bar of rectangular cross section will occur at the midpoint N of the larger side of that section.

The membrane analogy may be used just as effectively to visualize the shearing stresses in any straight bar of uniform, noncircular cross section. In particular, let us consider several thin-walled members with the cross sections shown in Fig. 3.47, which are subjected

TABLE 3.1. Coefficients for Rectangular Bars in Torsion

a/b	c_1	c_2
1.0	0.208	0.1406
1.2	0.219	0.1661
1.5	0.231	0.1958
2.0	0.246	0.229
2.5	0.258	0.249
3.0	0.267	0.263
4.0	0.282	0.281
5.0	0.291	0.291
10.0	0.312	0.312
∞	0.333	0.333

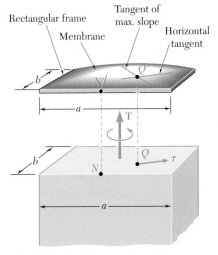

Fig. 3.46 Application of *membrane analogy* to shaft with rectangular cross section.

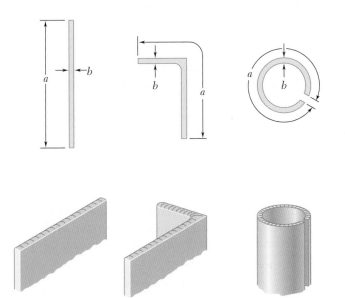

Fig. 3.47 Various thin-walled members.

†See ibid. Sec. 107.

‡This is the slope measured in a direction perpendicular to the horizontal tangent at Q'.

to the same torque. Using the membrane analogy to help us visualize the shearing stresses, we note that, since the same torque is applied to each member, the same volume will be located under each membrane, and the maximum slope will be about the same in each case. Thus, for a thin-walled member of uniform thickness and arbitrary shape, the maximum shearing stress is the same as for a rectangular bar with a very large value of a/b and may be determined from Eq. (3.43) with $c_1 = 0.333$.[†]

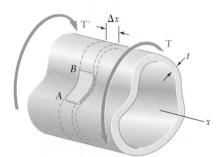

Fig. 3.48 Thin-walled hollow shaft.

*3.13 THIN-WALLED HOLLOW SHAFTS

In the preceding section we saw that the determination of stresses in noncircular members generally requires the use of advanced mathematical methods. In the case of thin-walled hollow noncircular shafts, however, a good approximation of the distribution of stresses in the shaft can be obtained by a simple computation.

Consider a hollow cylindrical member of *noncircular* section subjected to a torsional loading (Fig. 3.48).[‡] While the thickness t of the wall may vary within a transverse section, it will be assumed that it remains small compared to the other dimensions of the member. We now detach from the member the colored portion of wall AB bounded by two transverse planes at a distance Δx from each other, and by two longitudinal planes perpendicular to the wall. Since the portion AB is in equilibrium, the sum of the forces exerted on it in the longitudinal x direction must be zero (Fig. 3.49). But the only forces involved are the shearing forces $\mathbf{F}_A$ and $\mathbf{F}_B$ exerted on the ends of portion AB. We have therefore

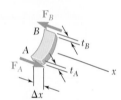

Fig. 3.49 Segment of thin-walled hollow shaft.

$$\Sigma F_x = 0: \qquad F_A - F_B = 0 \qquad (3.46)$$

We now express F_A as the product of the longitudinal shearing stress τ_A on the small face at A and of the area $t_A \Delta x$ of that face:

$$F_A = \tau_A(t_A \, \Delta x)$$

We note that, while the shearing stress is independent of the x coordinate of the point considered, it may vary across the wall; thus, τ_A represents the average value of the stress computed across the wall. Expressing F_B in a similar way and substituting for F_A and F_B into (3.46), we write

$$\tau_A(t_A \, \Delta x) - \tau_B(t_B \, \Delta x) = 0$$

or $$\tau_A t_A = \tau_B t_B \qquad (3.47)$$

Since A and B were chosen arbitrarily, Eq. (3.47) expresses that the product τt of the longitudinal shearing stress τ and of the wall thickness t is constant throughout the member. Denoting this product by q, we have

$$q = \tau t = \text{constant} \qquad (3.48)$$

[†]It could also be shown that the angle of twist may be determined from Eq. (3.44) with $c_2 = 0.333$.

[‡]The wall of the member must enclose a single cavity and must not be slit open. In other words, the member should be topologically equivalent to a hollow circular shaft.

We now detach a small element from the wall portion AB (Fig. 3.50). Since the upper and lower faces of this element are part of the free surface of the hollow member, the stresses on these faces are equal to zero. Recalling relations (1.21) and (1.22) of Sec. 1.12, it follows that the stress components indicated on the other faces by dashed arrows are also zero, while those represented by solid arrows are equal. Thus, the shearing stress at any point of a transverse section of the hollow member is parallel to the wall surface (Fig. 3.51) and its average value computed across the wall satisfies Eq. (3.48).

Fig. 3.50 Small element from segment.

At this point we can note an analogy between the distribution of the shearing stresses τ in the transverse section of a thin-walled hollow shaft and the distribution of the velocities v in water flowing through a closed channel of unit depth and variable width. While the velocity v of the water varies from point to point on account of the variation in the width t of the channel, the rate of flow, $q = vt$, remains constant throughout the channel, just as τt in Eq. (3.48). Because of this analogy, the product $q = \tau t$ is referred to as the *shear flow* in the wall of the hollow shaft.

Fig. 3.51 Direction of shearing stress on cross section.

We will now derive a relation between the torque T applied to a hollow member and the shear flow q in its wall. We consider a small element of the wall section, of length ds (Fig. 3.52). The area of the element is $dA = t\,ds$, and the magnitude of the shearing force $d\mathbf{F}$ exerted on the element is

$$dF = \tau\,dA = \tau(t\,ds) = (\tau t)\,ds = q\,ds \tag{3.49}$$

The moment dM_O of this force about an arbitrary point O within the cavity of the member may be obtained by multiplying dF by the perpendicular distance p from O to the line of action of $d\mathbf{F}$. We have

$$dM_O = p\,dF = p(q\,ds) = q(p\,ds) \tag{3.50}$$

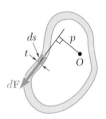

Fig. 3.52

But the product $p\,ds$ is equal to twice the area $d\mathcal{Q}$ of the colored triangle in Fig. 3.53. We thus have

$$dM_O = q(2d\mathcal{Q}) \tag{3.51}$$

Since the integral around the wall section of the left-hand member of Eq. (3.51) represents the sum of the moments of all the elementary shearing forces exerted on the wall section, and since this sum is equal to the torque T applied to the hollow member, we have

$$T = \oint dM_O = \oint q(2d\mathcal{Q})$$

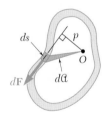

Fig. 3.53

The shear flow q being a constant, we write

$$T = 2q\mathcal{Q} \tag{3.52}$$

where $\mathcal{Q}$ is the area bounded by the center line of the wall cross section (Fig. 3.54).

The shearing stress τ at any given point of the wall may be expressed in terms of the torque T if we substitute for q from (3.48) into (3.52) and solve for τ the equation obtained. We have

$$\tau = \frac{T}{2t\mathcal{Q}} \tag{3.53}$$

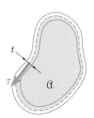

Fig. 3.54 Area for shear flow.

where t is the wall thickness at the point considered and $\mathcal{A}$ the area bounded by the center line. We recall that τ represents the average value of the shearing stress across the wall. However, for elastic deformations the distribution of stresses across the wall may be assumed uniform, and Eq. (3.53) will yield the actual value of the shearing stress at a given point of the wall.

The angle of twist of a thin-walled hollow shaft may be obtained by using the method of energy (Chap. 11). Assuming an elastic deformation, it may be shown† that the angle of twist of a thin-walled shaft of length L and modulus of rigidity G is

$$\phi = \frac{TL}{4\mathcal{A}^2 G} \oint \frac{ds}{t} \tag{3.54}$$

where the integral is computed along the center line of the wall section.

EXAMPLE 3.10

Structural aluminum tubing of 60×100-mm rectangular cross section was fabricated by extrusion. Determine the shearing stress in each of the four walls of a portion of such tubing when it is subjected to a torque of 2.7 kN · m, assuming (a) a uniform 4-mm wall thickness (Fig. 3.55a), (b) that, as a result of defective fabrication, walls AB and AC are 3-mm thick, and walls BD and CD are 5-mm thick (Fig. 3.55b).

(a) Tubing of Uniform Wall Thickness. The area bounded by the center line (Fig. 3.56) is

$$\mathcal{A} = (96 \text{ mm})(56 \text{ mm}) = 5376 \text{ mm}^2$$

Since the thickness of each of the four walls is $t = 4$ mm, we find from Eq. (3.53) that the shearing stress in each wall is

$$\tau = \frac{T}{2t\mathcal{A}} = \frac{2700 \text{ N} \cdot \text{m}}{2(0.004 \text{ m})(5376 \times 10^{-6} \text{ m}^2)} = 62.8 \text{ MPa}$$

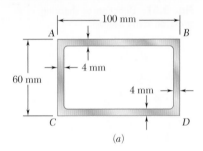

(a)

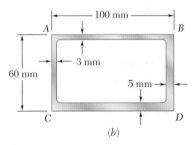

(b)

Fig. 3.55

Fig. 3.56

(b) Tubing with Variable Wall Thickness. Observing that the area $\mathcal{A}$ bounded by the center line is the same as in part a, and substituting successively $t = 3$ mm and $t = 5$ mm into Eq. (3.53), we have

$$\tau_{AB} = \tau_{AC} = \frac{2700 \text{ N} \cdot \text{m}}{2(0.003 \text{ m})(5376 \times 10^{-6} \text{ m}^2)} = 83.7 \text{ MPa}$$

and

$$\tau_{BD} = \tau_{CD} = \frac{2700 \text{ N} \cdot \text{m}}{2(0.005 \text{ m})(5376 \times 10^{-6} \text{ m}^2)} = 50.2 \text{ MPa}$$

We note that the stress in a given wall depends only upon its thickness.

†See Prob. 11.70.

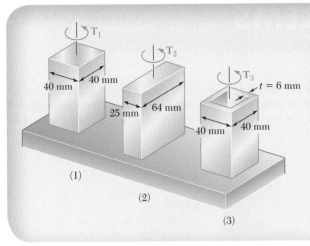

(1)

(2)

(3)

Using $\tau_{all} = 40$ MPa, determine the largest torque that may be applied to each of the brass bars and to the brass tube shown. Note that the two solid bars have the same cross-sectional area, and that the square bar and square tube have the same outside dimensions.

SOLUTION

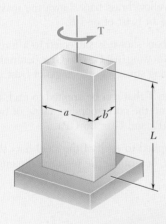

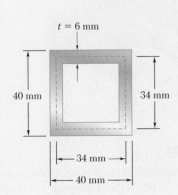

1. Bar with Square Cross Section. For a solid bar of rectangular cross section the maximum shearing stress is given by Eq. (3.43)

$$\tau_{max} = \frac{T}{c_1 a b^2}$$

where the coefficient c_1 is obtained from Table 3.1 in Sec. 3.12. We have

$$a = b = 0.040 \text{ m} \qquad \frac{a}{b} = 1.00 \qquad c_1 = 0.208$$

For $\tau_{max} = \tau_{all} = 40$ MPa, we have

$$\tau_{max} = \frac{T_1}{c_1 a b^2} \qquad 40 \text{ MPa} = \frac{T_1}{0.208(0.040 \text{ m})^3} \qquad T_1 = 532 \text{ N} \cdot \text{m} \quad \blacktriangleleft$$

2. Bar with Rectangular Cross Section. We now have

$$a = 0.064 \text{ m} \qquad b = 0.025 \text{ m} \qquad \frac{a}{b} = 2.56$$

Interpolating in Table 3.1: $c_1 = 0.259$

$$\tau_{max} = \frac{T_2}{c_1 a b^2} \qquad 40 \text{ MPa} = \frac{T_2}{0.259(0.064 \text{ m})(0.025 \text{ m})^2} \qquad T_2 = 414 \text{ N} \cdot \text{m} \quad \blacktriangleleft$$

3. Square Tube. For a tube of thickness t, the shearing stress is given by Eq. (3.53)

$$\tau = \frac{T}{2t\mathcal{A}}$$

where $\mathcal{A}$ is the area bounded by the center line of the cross section. We have

$$\mathcal{A} = (0.034 \text{ m})(0.034 \text{ m}) = 1.156 \times 10^{-3} \text{ m}^2$$

We substitute $\tau = \tau_{all} = 40$ MPa and $t = 0.006$ m and solve for the allowable torque:

$$\tau = \frac{T}{2t\mathcal{A}} \qquad 40 \text{ MPa} = \frac{T_3}{2(0.006 \text{ m})(1.156 \times 10^{-3} \text{ m}^2)} \qquad T_3 = 555 \text{ N} \cdot \text{m} \quad \blacktriangleleft$$

PROBLEMS

3.121 Using $\tau_{all} = 50$ MPa and knowing that $G = 39$ GPa, determine for each of the cold-rolled yellow brass bars shown the largest torque **T** that can be applied and the corresponding angle of twist at end B.

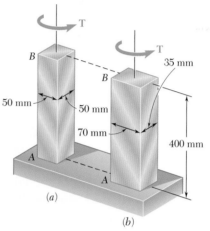

Fig. P3.121 and P3.122

3.122 Knowing that $T = 800$ N · m and that $G = 39$ GPa, determine for each of the cold-rolled yellow brass bars shown the maximum shearing stress and the angle of twist of end B.

3.123 Each of the two aluminum bars shown is subjected to a torque of magnitude $T = 1800$ N · m. Knowing that $G = 26$ GPa, determine for each bar the maximum shearing stress and the angle of twist at B.

3.124 Determine the largest torque **T** that can be applied to each of the two aluminum bars shown and the corresponding angle of twist at B, knowing that $\tau_{all} = 50$ MPa and $G = 26$ GPa.

3.125 A 226-N · m torque **T** is applied to each of the steel bars shown. Knowing that $\tau_{all} = 40$ MPa, determine the required dimension b for each bar.

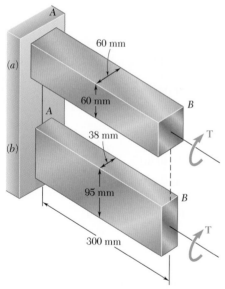

Fig. P3.123 and P3.124

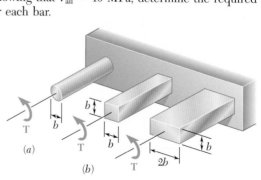

Fig. P3.125 and P3.126

3.126 A 300-N · m torque **T** is applied to each of the aluminum bars shown. Knowing that $\tau_{all} = 60$ MPa, determine the required dimension b for each bar.

3.127 The torque **T** causes a rotation of 2° at end B of the stainless steel bar shown. Knowing that $b = 20$ mm and $G = 75$ GPa, determine the maximum shearing stress in the bar.

3.128 The torque **T** causes a rotation of 0.6° at end B of the aluminum bar shown. Knowing that $b = 15$ mm and $G = 26$ GPa, determine the maximum shearing stress in the bar.

3.129 Two shafts are made of the same material. The cross section of shaft A is a square of side b and that of shaft B is a circle of diameter b. Knowing that the shafts are subjected to the same torque, determine the ratio τ_A/τ_B of maximum shearing stresses occurring in the shafts.

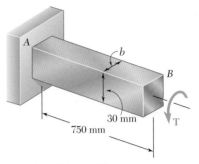

Fig. P3.127 and P3.128

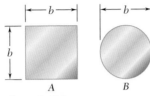

Fig. P3.129

3.130 Shafts A and B are made of the same material and have the same cross-sectional area, but A has a circular cross section and B has a square cross section. Determine the ratio of the maximum shearing stresses occurring in A and B, respectively, when the two shafts are subjected to the same torque $(T_A = T_B)$. Assume both deformations to be elastic.

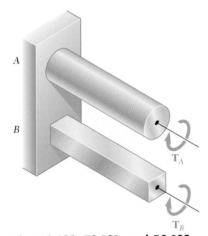

Fig. *P3.130*, P3.131, and P3.132

3.131 Shafts A and B are made of the same material and have the same cross-sectional area, but A has a circular cross section and B has a square cross section. Determine the ratio of the maximum torques T_A and T_B that can be safely applied to A and B, respectively.

3.132 Shafts A and B are made of the same material and have the same length and cross-sectional area, but A has a circular cross section and B has a square cross section. Determine the ratio of the maximum values of the angles ϕ_A and ϕ_B through which shafts A and B, respectively, can be twisted.

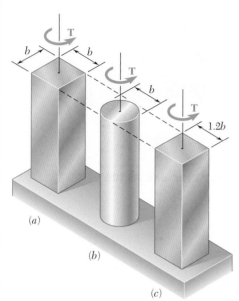

(a)

(b)

(c)

Fig. P3.133 and P3.134

3.133 Each of the three aluminum bars shown is to be twisted through an angle of 2°. Knowing that $b = 30$ mm, $\tau_{all} = 50$ MPa, and $G = 27$ GPa, determine the shortest allowable length of each bar.

3.134 Each of the three steel bars is subjected to a torque as shown. Knowing that the allowable shearing stress is 56 MPa and that $b = 38$ mm, determine the maximum torque **T** that can be applied to each bar.

3.135 A 340-N · m torque is applied to a 1.8-m-long steel angle with an L102 × 102 × 9.5 cross section. From Appendix C we find that the thickness of the section is 9.5 mm and that its area is 1850 mm². Knowing that $G = 77$ GPa, determine (a) the maximum shearing stress along line a-a, (b) the angle of twist.

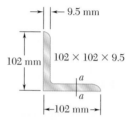

Fig. P3.135

3.136 A 3-m-long steel angle has an L203 × 152 × 12.7 cross section. From Appendix C we find that the thickness of the section is 12.7 mm and that its area is 4350 mm². Knowing that $\tau_{all} = 50$ MPa and that $G = 77.2$ GPa, and ignoring the effect of stress concentrations, determine (a) the largest torque **T** that can be applied, (b) the corresponding angle of twist.

L203 × 152 × 12.7

Fig. P3.136

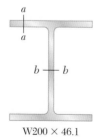

W200 × 46.1

Fig. P3.137

3.137 An 2.4-m-long steel member with a W200 × 46.1 cross section is subjected to a 560-N · m torque. The properties of the rolled-steel section are given in Appendix C. Knowing that $G = 77$ GPa, determine (a) the maximum shearing stress along line a-a, (b) the maximum shearing stress along line b-b, (c) the angle of twist. (*Hint*: consider the web and flanges separately and obtain a relation between the torques exerted on the web and a flange, respectively, by expressing that the resulting angles of twist are equal.)

3.138 A 4-m-long steel member has a W310 × 60 cross section. Knowing that G = 77.2 GPa and that the allowable shearing stress is 40 MPa, determine (a) the largest torque **T** that can be applied, (b) the corresponding angle of twist. Refer to Appendix C for the dimensions of the cross section and neglect the effect of stress concentrations. (See hint of Prob. 3.137.)

3.139 A torque T = 750 kN · m is applied to the hollow shaft shown that has a uniform 8-mm wall thickness. Neglecting the effect of stress concentrations, determine the shearing stress at points a and b.

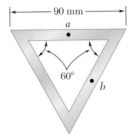

Fig. P3.139

W310 × 60

Fig. P3.138

3.140 A 750-N · m torque is applied to a hollow shaft having the cross section shown and a uniform 6-mm wall thickness. Neglecting the effect of stress concentrations, determine the shearing stress at points a and b.

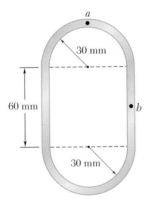

Fig. P3.140

3.141 A 90-N · m torque is applied to a hollow shaft having the cross section shown. Neglecting the effect of stress concentrations, determine the shearing stress at points a and b.

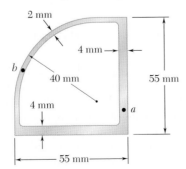

Fig. P3.141

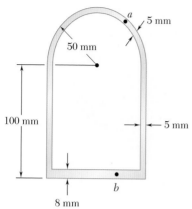

5 mm

50 mm

100 mm

5 mm

8 mm

b

a

Fig. P3.142

3.142 A 5.6-kN · m torque is applied to a hollow shaft having the cross section shown. Neglecting the effect of stress concentrations, determine the shearing stress at points *a* and *b*.

3.143 A hollow member having the cross section shown is formed from sheet metal of 2-mm thickness. Knowing that the shearing stress must not exceed 3 MPa, determine the largest torque that can be applied to the member.

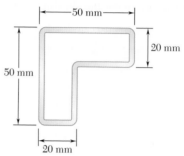

Fig. P3.143

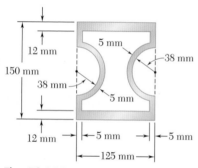

12 mm

150 mm

5 mm

38 mm

38 mm

5 mm

12 mm

5 mm

5 mm

125 mm

Fig. P3.144

3.144 A hollow brass shaft has the cross section shown. Knowing that the shearing stress must not exceed 84 MPa and neglecting the effect of stress concentrations, determine the largest torque that can be applied to the shaft.

3.145 and 3.146 A hollow member having the cross section shown is to be formed from sheet metal of 1.5-mm thickness. Knowing that a 140-N · m torque will be applied to the member, determine the smallest dimension *d* that can be used if the shearing stress is not to exceed 5 MPa.

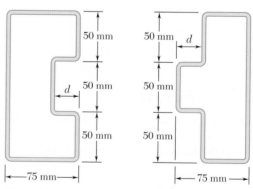

Fig. *P3.145*   Fig. *P3.146*

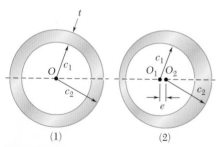

Fig. P3.147

3.147 A hollow cylindrical shaft was designed with the cross section shown in Fig. (*1*) to withstand a maximum torque T_0. Defective fabrication, however, resulted in a slight eccentricity *e* between the inner and outer cylindrical surfaces of the shaft as shown in Fig. (*2*). (*a*) Express the maximum torque *T* that can be safely applied to the defective shaft in terms of T_0, *e*, and *t*. (*b*) Calculate the percent decrease in the allowable torque for values of the ratio e/t equal to 0.1, 0.5, and 0.9.

3.148 A cooling tube having the cross section shown is formed from a sheet of stainless steel of 3-mm thickness. The radii $c_1 = 150$ mm and $c_2 = 100$ mm are measured to the center line of the sheet metal. Knowing that a torque of magnitude $T = 3$ kN $\cdot$ m is applied to the tube, determine (a) the maximum shearing stress in the tube, (b) the magnitude of the torque carried by the outer circular shell. Neglect the dimension of the small opening where the outer and inner shells are connected.

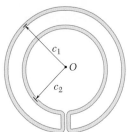

Fig. P3.148

3.149 A hollow cylindrical shaft of length L, mean radius c_m, and uniform thickness t is subjected to a torque of magnitude T. Consider, on the one hand, the values of the average shearing stress τ_{ave} and the angle of twist ϕ obtained from the elastic torsion formulas developed in Secs. 3.4 and 3.5 and, on the other hand, the corresponding values obtained from the formulas developed in Sec. 3.13 for thin-walled shafts. (a) Show that the relative error introduced by using the thin-walled-shaft formulas rather than the elastic torsion formulas is the same for τ_{ave} and ϕ and that the relative error is positive and proportional to the ratio t/c_m. (b) Compare the percent error corresponding to values of the ratio t/c_m of 0.1, 0.2, and 0.4.

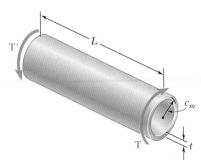

Fig. P3.149

3.150 Equal torques are applied to thin-walled tubes of the same length L, same thickness t, and same radius c. One of the tubes has been slit lengthwise as shown. Determine (a) the ratio τ_b/τ_a of the maximum shearing stresses in the tubes, (b) the ratio ϕ_b/ϕ_a of the angles of twist of the tubes.

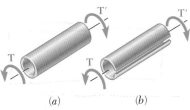

(a)   (b)

Fig. P3.150

REVIEW AND SUMMARY

Both elastic & plastic

$T_T = T_e + T_p$ ↓

J_{core}

Deformations in circular shafts

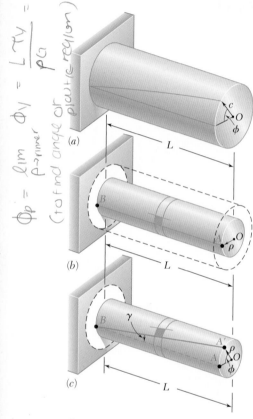

Fig. 3.57

Shearing stresses in elastic range

$\phi_p = \lim_{\rho \to inner} \phi_Y = \dfrac{L \tau_Y}{\rho G}$

(to find angle of plastic region)

This chapter was devoted to the analysis and design of *shafts* subjected to twisting couples, or *torques*. Except for the last two sections of the chapter, our discussion was limited to *circular shafts*.

In a preliminary discussion [Sec. 3.2], it was pointed out that the distribution of stresses in the cross section of a circular shaft is *statically indeterminate*. The determination of these stresses, therefore, requires a prior analysis of the *deformations* occurring in the shaft [Sec. 3.3]. Having demonstrated that in a circular shaft subjected to torsion, *every cross section remains plane and undistorted*, we derived the following expression for the *shearing strain* in a small element with sides parallel and perpendicular to the axis of the shaft and at a distance ρ from that axis:

$$\gamma = \frac{\rho\phi}{L} \tag{3.2}$$

where ϕ is the angle of twist for a length L of the shaft (Fig. 3.57). Equation (3.2) shows that the *shearing strain in a circular shaft varies linearly with the distance from the axis of the shaft*. It follows that the strain is maximum at the surface of the shaft, where ρ is equal to the radius c of the shaft. We wrote

$\rho_Y = \dfrac{L\tau_Y}{G\phi}$ $\phi_Y = \dfrac{L\tau_Y}{c G}$

$$\gamma_{max} = \frac{c\phi}{L} \qquad \gamma = \frac{\rho}{c}\gamma_{max} \tag{3.3, 4}$$

Considering *shearing stresses* in a circular shaft within the <u>elastic range</u> [Sec. 3.4] and recalling Hooke's law for shearing stress and strain, $\tau = G\gamma$, we derived the relation

$$\tau = \frac{\rho}{c}\tau_{max} \tag{3.6}$$

which shows that within the elastic range, the *shearing stress τ in a circular shaft also varies linearly with the distance from the axis of the shaft*. Equating the sum of the moments of the elementary forces exerted on any section of the shaft to the magnitude T of the torque applied to the shaft, we derived the *elastic torsion formulas*

elastic material!

$$\tau_{max} = \frac{Tc}{J} \qquad \tau = \frac{T\rho}{J} \tag{3.9, 10}$$

where c is the radius of the cross section and J its centroidal polar moment of inertia. We noted that $J = \frac{1}{2}\pi c^4$ for a solid shaft and $J = \frac{1}{2}\pi(c_2^4 - c_1^4)$ for a hollow shaft of inner radius c_1 and outer radius c_2.

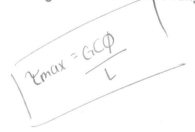

$\tau_{max} = \dfrac{Gc\phi}{L}$

We noted that while the element a in Fig. 3.58 is in pure shear, the element c in the same figure is subjected to normal stresses of the same magnitude, Tc/J, two of the normal stresses being tensile and two compressive. This explains why in a torsion test ductile materials, which generally fail in shear, will break along a plane perpendicular to the axis of the specimen, while brittle materials, which are weaker in tension than in shear, will break along surfaces forming a 45° angle with that axis.

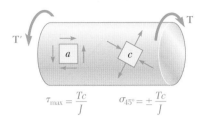

$$\tau_{max} = \frac{Tc}{J} \qquad \sigma_{45°} = \pm \frac{Tc}{J}$$

Fig. 3.58

Angle of twist

In Sec. 3.5, we found that within the elastic range, the angle of twist ϕ of a circular shaft is proportional to the torque T applied to it (Fig. 3.59). Expressing ϕ in *radians*, we wrote

$$\phi = \frac{TL}{JG} \tag{3.16}$$

where L = length of shaft
J = polar moment of inertia of cross section
G = modulus of rigidity of material

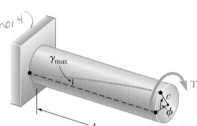

Fig. 3.59

If the shaft is subjected to torques at locations other than its ends or consists of several parts of various cross sections and possibly of different materials, the angle of twist of the shaft must be expressed as the *algebraic sum* of the angles of twist of its component parts [Sample Prob. 3.3]:

$$\phi = \sum_i \frac{T_i L_i}{J_i G_i} \tag{3.17}$$

We observed that when both ends of a shaft BE rotate (Fig. 3.60), the angle of twist of the shaft is equal to the *difference* between the angles of rotation ϕ_B and ϕ_E of its ends. We also noted that when two shafts AD and BE are connected by gears A and B, the torques applied, respectively, by gear A on shaft AD and by gear B on shaft BE are *directly proportional* to the radii r_A and r_B of the two gears— since the forces applied on each other by the gear teeth at C are equal and opposite. On the other hand, the angles ϕ_A and ϕ_B through which the two gears rotate are *inversely proportional* to r_A and r_B— since the arcs CC' and CC'' described by the gear teeth are equal [Example 3.04 and Sample Prob. 3.4].

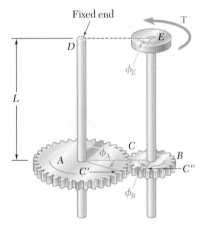

Fig. 3.60

Statically indeterminate shafts

If the reactions at the supports of a shaft or the internal torques cannot be determined from statics alone, the shaft is said to be *statically indeterminate* [Sec. 3.6]. The equilibrium equations obtained from free-body diagrams must then be complemented by relations involving the deformations of the shaft and obtained from the geometry of the problem [Example 3.05, Sample Prob. 3.5].

Transmission shafts

In Sec. 3.7, we discussed the *design of transmission shafts*. We first observed that the power P transmitted by a shaft is

$$P = 2\pi f T \tag{3.20}$$

where T is the torque exerted at each end of the shaft and f the *frequency* or speed of rotation of the shaft. The unit of frequency is the revolution per second (s^{-1}) or *hertz* (Hz). If SI units are used, T is expressed in newton-meters (N · m) and P in *watts* (W).

To design a shaft to transmit a given power P at a frequency f, you should first solve Eq. (3.20) for T. Carrying this value and the maximum allowable value of τ for the material used into the elastic formula (3.9), you will obtain the corresponding value of the parameter J/c, from which the required diameter of the shaft may be calculated [Examples 3.06 and 3.07].

Stress concentrations

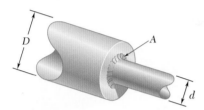

Fig. 3.61

In Sec. 3.8, we discussed *stress concentrations* in circular shafts. We saw that the stress concentrations resulting from an abrupt change in the diameter of a shaft can be reduced through the use of a *fillet* (Fig. 3.61). The maximum value of the shearing stress at the fillet is

$$\tau_{\max} = K\frac{Tc}{J} \tag{3.25}$$

where the stress Tc/J is computed for the smaller-diameter shaft, and where K is a stress-concentration factor. Values of K were plotted in Fig. 3.29 on p. 201 against the ratio r/d, where r is the radius of the fillet, for various values of D/d.

Plastic deformations

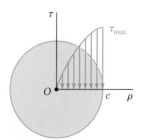

Fig. 3.62

Sections 3.9 through 3.11 were devoted to the discussion of *plastic deformations* and *residual stresses* in circular shafts. We first recalled that even when Hooke's law does not apply, the distribution of *strains* in a circular shaft is always linear [Sec. 3.9]. If the shearing-stress-strain diagram for the material is known, it is then possible to plot the shearing stress τ against the distance ρ from the axis of the shaft for any given value of $\tau_{\max}$ (Fig. 3.62). Summing the contributions to the torque of annular elements of radius ρ and thickness $d\rho$, we expressed the torque T as

$$T = \int_0^c \rho\tau(2\pi\rho\,d\rho) = 2\pi\int_0^c \rho^2\tau\,d\rho \tag{3.26}$$

where τ is the function of ρ plotted in Fig. 3.62.

Modulus of rupture

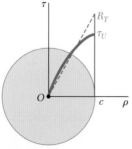

Fig. 3.63

An important value of the torque is the ultimate torque T_U which causes failure of the shaft. This value can be determined, either experimentally, or by carrying out the computations indicated above with $\tau_{\max}$ chosen equal to the ultimate shearing stress τ_U of the material. From T_U, and assuming a linear stress distribution (Fig 3.63), we determined the corresponding fictitious stress $R_T = T_Uc/J$, known as the *modulus of rupture in torsion* of the given material.

Considering the idealized case of a *solid circular shaft* made of an *elastoplastic material* [Sec. 3.10], we first noted that, as long as

τ_{max} does not exceed the yield strength τ_Y of the material, the stress distribution across a section of the shaft is linear (Fig. 3.64a). The torque T_Y corresponding to $\tau_{max} = \tau_Y$ (Fig. 3.64b) is known as the *maximum elastic torque;* for a solid circular shaft of radius c, we have

↳ only for a solid circular shaft

$$T_Y = \tfrac{1}{2}\pi c^3 \tau_Y \tag{3.29}$$

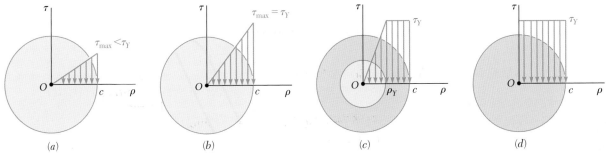

| (a) | (b) | (c) | (d) |

Fig. 3.64

As the torque increases, a plastic region develops in the shaft around an elastic core of radius ρ_Y. The torque T corresponding to a given value of ρ_Y was found to be

$$T = \frac{4}{3}T_Y\left(1 - \frac{1}{4}\frac{\rho_Y^3}{c^3}\right) \tag{3.32}$$

We noted that as ρ_Y approaches zero, the torque approaches a limiting value T_p, called the *plastic torque* of the shaft considered:

$$T_p = \frac{4}{3}T_Y \tag{3.33}$$

Plotting the torque T against the angle of twist ϕ of a solid circular shaft (Fig. 3.65), we obtained the segment of straight line 0Y defined by Eq. (3.16), followed by a curve approaching the straight line $T = T_p$ and defined by the equation

$$T = \frac{4}{3}T_Y\left(1 - \frac{1}{4}\frac{\phi_Y^3}{\phi^3}\right) \tag{3.37}$$

Solid shaft of elastoplastic material

For non-solid shaft

$$T = 2\pi \int_{c_1}^{c_2} \rho^2 \tau \, d\rho$$

Plastic Torque.

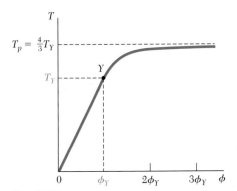

Fig. 3.65

Permanent deformation

Loading a circular shaft beyond the onset of yield and unloading it [Sec. 3.11] results in a *permanent deformation* characterized by the angle of twist $\phi_p = \phi - \phi'$, where ϕ corresponds to the loading phase described in the previous paragraph, and ϕ' to the unloading phase represented by a straight line in Fig. 3.66. There will also be *residual stresses* in the shaft, which can be determined by adding the maximum stresses reached during the loading phase and the reverse stresses corresponding to the unloading phase [Example 3.09].

Residual stresses

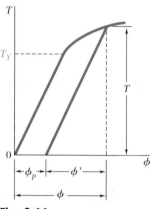

Fig. 3.66

Torsion of noncircular members

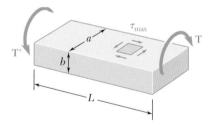

Fig. 3.67

The last two sections of the chapter dealt with the torsion of *noncircular members*. We first recalled that the derivation of the formulas for the distribution of strain and stress in circular shafts was based on the fact that due to the axisymmetry of these members, cross sections remain plane and undistorted. Since this property does not hold for noncircular members, such as the square bar of Fig. 3.67, none of the formulas derived earlier can be used in their analysis [Sec. 3.12].

Bars of rectangular cross section

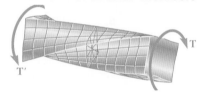

Fig. 3.68

It was indicated in Sec. 3.12 that in the case of straight bars with a *uniform rectangular cross section* (Fig. 3.68), the maximum shearing stress occurs along the center line of the *wider* face of the bar. Formulas for the maximum shearing stress and the angle of twist were given without proof. The *membrane analogy* for visualizing the distribution of stresses in a noncircular member was also discussed.

Thin-walled hollow shafts

We next analyzed the distribution of stresses in *noncircular thin-walled hollow shafts* [Sec. 3.13]. We saw that the shearing stress is parallel to the wall surface and varies both across the wall and along the wall cross section. Denoting by τ the average value of the shearing stress computed across the wall at a given point of the cross section, and by t the thickness of the wall at that point (Fig. 3.69), we showed that the product $q = \tau t$, called the *shear flow*, is constant along the cross section.

Furthermore, denoting by T the torque applied to the hollow shaft and by $\mathfrak{a}$ the area bounded by the center line of the wall cross section, we expressed as follows the average shearing stress τ at any given point of the cross section:

$$\tau = \frac{T}{2t\mathfrak{a}} \tag{3.53}$$

Fig. 3.69

REVIEW PROBLEMS

3.151 For the cylindrical shaft shown, determine the maximum shearing stress caused by a torque of magnitude $T = 1.5$ kN $\cdot$ m.

3.152 Determine the largest allowable diameter of a 3-m-long steel rod ($G = 77$ GPa) if the rod is to be twisted through 30° without exceeding a shearing stress of 84 MPa.

3.153 A steel pipe of 300-mm outer diameter is fabricated from 6-mm-thick plate by welding along a helix that forms an angle of 45° with a plane perpendicular to the axis of the pipe. Knowing that the maximum allowable tensile stress in the weld is 84 MPa, determine the largest torque that can be applied to the pipe.

Fig. P3.151

Fig. P3.153

3.154 Two solid shafts are connected by gears as shown. Knowing that $G = 77.2$ GPa for each shaft, determine the angle through which end A rotates when $T_A = 1200$ N $\cdot$ m.

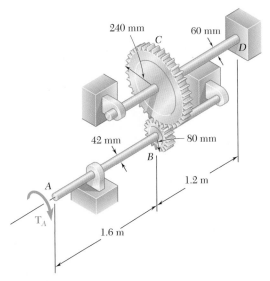

Fig. P3.154

237

3.155 Two solid steel shafts (G = 77.2 GPa) are connected to a coupling disk B and to fixed supports at A and C. For the loading shown, determine (a) the reaction at each support, (b) the maximum shearing stress in shaft AB, (c) the maximum shearing stress in shaft BC.

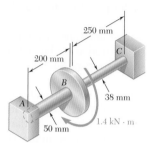

250 mm

200 mm

B

C

38 mm

A

50 mm

1.4 kN · m

Fig. P3.155

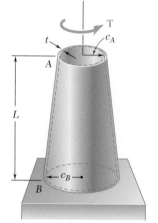

Fig. P3.156

3.156 The long, hollow, tapered shaft AB has a uniform thickness t. Denoting by G the modulus of rigidity, show that the angle of twist at end A is

$$\phi_A = \frac{TL}{4\pi Gt} \frac{c_A + c_B}{c_A^2 c_B^2}$$

3.157 The solid cylindrical rod BC is attached to the rigid lever AB and to the fixed support at C. The vertical force **P** applied at A causes a small displacement at point A. Show that the corresponding maximum shearing stress in the rod is

$$\tau = \frac{Gd}{2La}\Delta$$

where d is the diameter of the rod and G its modulus of rigidity.

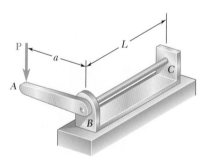

P

a

L

A

C

B

Fig. P3.157

3.158 The design specifications of a 1.2-m-long solid transmission shaft require that the angle of twist of the shaft not exceed 4° when a torque of 750 N · m is applied. Determine the required diameter of the shaft, knowing that the shaft is made of a steel with an allowable shearing stress of 90 MPa and a modulus of rigidity of 77.2 GPa.

3.159 Two solid brass rods AB and CD are brazed to a brass sleeve EF. Determine the ratio d_2/d_1 for which the same maximum shearing stress occurs in the rods and in the sleeve.

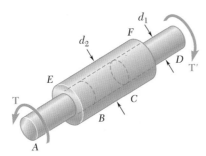

Fig. P3.159

3.160 A torque $T = 5$ kN $\cdot$ m is applied to a hollow shaft having the cross section shown. Neglecting the effect of stress concentrations, determine the shearing stress at points a and b.

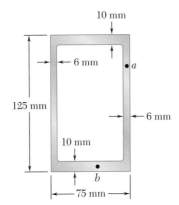

Fig. P3.160

3.161 The composite shaft shown is twisted by applying a torque **T** at end A. Knowing that the maximum shearing stress in the steel shell is 150 MPa, determine the corresponding maximum shearing stress in the aluminum core. Use $G = 77.2$ GPa for steel and $G = 27$ GPa for aluminum.

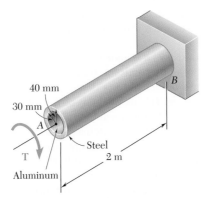

Fig. P3.161

COMPUTER PROBLEMS

The following problems are designed to be solved with a computer.

3.C1 Shaft AB consists of n homogeneous cylindrical elements, which can be solid or hollow. Its end A is fixed, while its end B is free, and it is subjected to the loading shown. The length of element i is denoted by L_i, its outer diameter by OD_i, its inner diameter by ID_i, its modulus of rigidity by G_i, and the torque applied to its right end by $\mathbf{T}_i$, the magnitude T_i of this torque being assumed to be positive if $\mathbf{T}_i$ is observed as counterclockwise from end B and negative otherwise. (Note that $ID_i = 0$ if the element is solid.) (*a*) Write a computer program that can be used to determine the maximum shearing stress in each element, the angle of twist of each element, and the angle of twist of the entire shaft. (*b*) Use this program to solve Probs. 3.35 and 3.38.

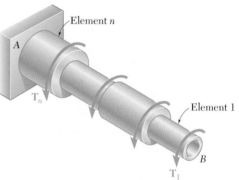

Fig. P3.C1

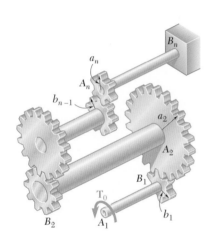

Fig. P3.C2

3.C2 The assembly shown consists of n cylindrical shafts, which can be solid or hollow, connected by gears and supported by brackets (not shown). End A_1 of the first shaft is free and is subjected to a torque $\mathbf{T}_0$, while end B_n of the last shaft is fixed. The length of shaft A_iB_i is denoted by L_i, its outer diameter by OD_i, its inner diameter by ID_i, and its modulus of rigidity by G_i. (Note that $ID_i = 0$ if the element is solid.) The radius of gear A_i is denoted by a_i, and the radius of gear B_i by b_i. (*a*) Write a computer program that can be used to determine the maximum shearing stress in each shaft, the angle of twist of each shaft, and the angle through which end A_i rotates. (*b*) Use this program to solve Probs. 3.41 and 3.44.

3.C3 Shaft AB consists of n homogeneous cylindrical elements, which can be solid or hollow. Both of its ends are fixed, and it is subjected to the loading shown. The length of element i is denoted by L_i, its outer diameter by OD_i, its inner diameter by ID_i, its modulus of rigidity by G_i, and the torque applied to its right end by T_i, the magnitude T_i of this torque being assumed to be positive if $\mathbf{T}_i$ is observed as counterclockwise from end B and negative otherwise. Note that $ID_i = 0$ if the element is solid and also that $T_1 = 0$. Write a computer program that can be used to determine the reactions at A and B, the maximum shearing stress in each element, and the angle of twist of each element. Use this program (*a*) to solve Prob. 3.155, (*b*) to determine the maximum shearing stress in the shaft of Example 3.05.

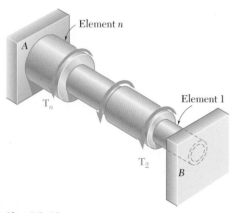

Fig. P3.C3

3.C4 The homogeneous, solid cylindrical shaft AB has a length L, a diameter d, a modulus of rigidity G, and a yield strength τ_Y. It is subjected to a torque $\mathbf{T}$ that is gradually increased from zero until the angle of twist of the shaft has reached a maximum value ϕ_m and then decreased back to zero. (*a*) Write a computer program that, for each of 16 values of ϕ_m equally spaced over a range extending from 0 to a value 3 times as large as the angle of twist at the onset of yield, can be used to determine the maximum value T_m of the torque, the radius of the elastic core, the maximum shearing stress, the permanent twist, and the residual shearing stress both at the surface of the shaft and at the interface of the elastic core and the plastic region. (*b*) Use this program to obtain approximate answers to Probs. 3.114, 3.115, 3.116.

Fig. P3.C4

3.C5 The exact expression is given in Prob. 3.61 for the angle of twist of the solid tapered shaft AB when a torque $\mathbf{T}$ is applied as shown. Derive an approximate expression for the angle of twist by replacing the tapered shaft by n cylindrical shafts of equal length and of radius $r_i = (n + i - \frac{1}{2})(c/n)$, where $i = 1, 2, \ldots, n$. Using for T, L, G, and c values of your choice, determine the percentage error in the approximate expression when (*a*) $n = 4$, (*b*) $n = 8$, (*c*) $n = 20$, (*d*) $n = 100$.

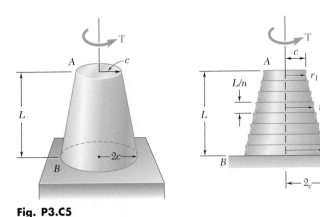
Fig. P3.C5

3.C6 A torque $\mathbf{T}$ is applied as shown to the long, hollow, tapered shaft AB of uniform thickness t. Derive an approximate expression for the angle of twist by replacing the tapered shaft by n cylindrical rings of equal length and of radius $r_i = (n + i - \frac{1}{2})(c/n)$, where $i = 1, 2, \ldots, n$. Using for T, L, G, c, and t values of your choice, determine the percentage error in the approximate expression when (*a*) $n = 4$, (*b*) $n = 8$, (*c*) $n = 20$, (*d*) $n = 100$.

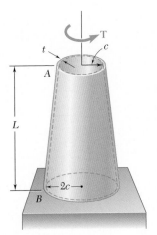
Fig. P3.C6

The athlete shown holds the barbell with his hands placed at equal distances from the weights. This results in pure bending in the center portion of the bar. The normal stresses and the curvature resulting from pure bending will be determined in this chapter.

Pure Bending

4.1 INTRODUCTION

In the preceding chapters you studied how to determine the stresses in prismatic members subjected to axial loads or to twisting couples. In this chapter and in the following two you will analyze the stresses and strains in prismatic members subjected to *bending*. Bending is a major concept used in the design of many machine and structural components, such as beams and girders.

This chapter will be devoted to the analysis of prismatic members subjected to equal and opposite couples $\mathbf{M}$ and $\mathbf{M}'$ acting in the same longitudinal plane. Such members are said to be in *pure bending*. In most of the chapter, the members will be assumed to possess a plane of symmetry and the couples $\mathbf{M}$ and $\mathbf{M}'$ to be acting in that plane (Fig. 4.1).

Fig. 4.1 Member in pure bending.

An example of pure bending is provided by the bar of a typical barbell as it is held overhead by a weight lifter as shown in the opening photo for this chapter. The bar carries equal weights at equal distances from the hands of the weight lifter. Because of the symmetry of the free-body diagram of the bar (Fig. 4.2*a*), the reactions at the hands must be equal and opposite to the weights. Therefore, as far as the middle portion CD of the bar is concerned, the weights and the reactions can be replaced by two equal and opposite $120\text{-N} \cdot \text{m}$ couples (Fig. 4.2*b*), showing that the middle portion of the bar is in pure bending. A similar analysis of the axle of a small sport buggy (Photo 4.1) would show that, between the two points where it is attached to the frame, the axle is in pure bending.

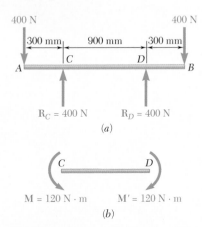

Fig. 4.2 Beam in which portion *CD* is in pure bending.

Photo 4.1 For the sport buggy shown, the center portion of the rear axle is in pure bending.

As interesting as the direct applications of pure bending may be, devoting an entire chapter to its study would not be justified if it were not for the fact that the results obtained will be used in the analysis of other types of loadings as well, such as *eccentric axial loadings* and *transverse loadings.*

Photo 4.2 shows a 300 mm steel bar clamp used to exert 600-N forces on two pieces of lumber as they are being glued together. Figure 4.3*a* shows the equal and opposite forces exerted by the lumber on the clamp. These forces result in an *eccentric loading* of the straight portion of the clamp. In Fig. 4.3*b* a section *CC′* has been passed through the clamp and a free-body diagram has been drawn of the upper half of the clamp, from which we conclude that the internal forces in the section are equivalent to a 600-N axial tensile force **P** and a 72-N · m couple **M.** We can thus combine our knowledge of the stresses under a *centric* load and the results of our forthcoming analysis of stresses in pure bending to obtain the distribution of stresses under an *eccentric* load. This will be further discussed in Sec. 4.12.

Photo 4.2 Clamp used to glue lumber pieces together.

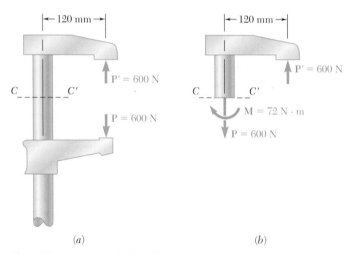

(a) (b)

Fig. 4.3 Forces exerted on clamp.

The study of pure bending will also play an essential role in the study of beams, i.e., the study of prismatic members subjected to various types of *transverse loads*. Consider, for instance, a cantilever beam *AB* supporting a concentrated load **P** at its free end (Fig. 4.4*a*). If we pass a section through *C* at a distance *x* from *A*, we observe from the free-body diagram of *AC* (Fig. 4.4*b*) that the internal forces in the section consist of a force **P′** equal and opposite to **P** and a couple **M** of magnitude $M = Px$. The distribution of normal stresses in the section can be obtained from the couple **M** as if the beam were in pure bending. On the other hand, the shearing stresses in the section depend on the force **P′**, and you will learn in Chap. 6 how to determine their distribution over a given section.

The first part of the chapter is devoted to the analysis of the stresses and deformations caused by pure bending in a homogeneous member possessing a plane of symmetry and made of a material following Hooke's law. In a preliminary discussion of the stresses due to bending (Sec. 4.2), the methods of statics will be used to derive

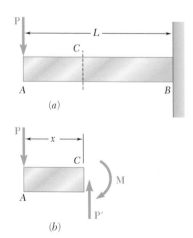

Fig. 4.4 Cantilever beam, not in pure bending.

three fundamental equations which must be satisfied by the normal stresses in any given cross section of the member. In Sec. 4.3, it will be proved that *transverse sections remain plane* in a member subjected to pure bending, while in Sec. 4.4 formulas will be developed that can be used to determine the *normal stresses,* as well as the *radius of curvature* for that member within the elastic range.

In Sec. 4.6, you will study the stresses and deformations in *composite members* made of more than one material, such as reinforced-concrete beams, which utilize the best features of steel and concrete and are extensively used in the construction of buildings and bridges. You will learn to draw a *transformed section* representing the section of a member made of a homogeneous material that undergoes the same deformations as the composite member under the same loading. The transformed section will be used to find the stresses and deformations in the original composite member. Section 4.7 is devoted to the determination of the *stress concentrations* occurring at locations where the cross section of a member undergoes a sudden change.

In the next part of the chapter you will study *plastic deformations* in bending, i.e., the deformations of members which are made of a material which does not follow Hooke's law and are subjected to bending. After a general discussion of the deformations of such members (Sec. 4.8), you will investigate the stresses and deformations in members made of an *elastoplastic material* (Sec. 4.9). Starting with the *maximum elastic moment* M_Y, which corresponds to the onset of yield, you will consider the effects of increasingly larger moments until the *plastic moment* M_p is reached, at which time the member has yielded fully. You will also learn to determine the *permanent deformations* and *residual stresses* that result from such loadings (Sec. 4.11). It should be noted that during the past half-century the elastoplastic property of steel has been widely used to produce designs resulting in both improved safety and economy.

In Sec. 4.12, you will learn to analyze an *eccentric axial loading* in a plane of symmetry, such as the one shown in Fig. 4.4, by superposing the stresses due to pure bending and the stresses due to a centric axial loading.

Your study of the bending of prismatic members will conclude with the analysis of *unsymmetric bending* (Sec. 4.13), and the study of the general case of *eccentric axial loading* (Sec. 4.14). The final section of the chapter will be devoted to the determination of the stresses in *curved members* (Sec. 4.15).

4.2 SYMMETRIC MEMBER IN PURE BENDING

Consider a prismatic member AB possessing a plane of symmetry and subjected to equal and opposite couples $\mathbf{M}$ and $\mathbf{M}'$ acting in that plane (Fig. 4.5a). We observe that if a section is passed through the member AB at some arbitrary point C, the conditions of equilibrium of the portion AC of the member require that the internal forces in the section be equivalent to the couple $\mathbf{M}$ (Fig. 4.5b). Thus, the internal forces in any cross section of a symmetric member in pure bending are equivalent to a couple. The moment M of that couple

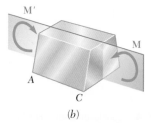

(a)

(b)

Fig. 4.5 Member in pure bending.

is referred to as the *bending moment* in the section. Following the usual convention, a positive sign will be assigned to M when the member is bent as shown in Fig. 4.5a, i.e., when the concavity of the beam faces upward, and a negative sign otherwise.

Denoting by σ_x the normal stress at a given point of the cross section and by τ_{xy} and τ_{xz} the components of the shearing stress, we express that the system of the elementary internal forces exerted on the section is equivalent to the couple **M** (Fig. 4.6).

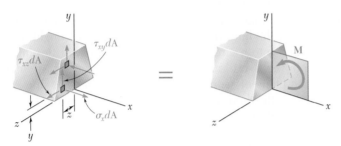

Fig. 4.6

We recall from statics that a couple **M** actually consists of two equal and opposite forces. The sum of the components of these forces in any direction is therefore equal to zero. Moreover, the moment of the couple is the same about *any* axis perpendicular to its plane, and is zero about any axis contained in that plane. Selecting arbitrarily the z axis as shown in Fig. 4.6, we express the equivalence of the elementary internal forces and of the couple **M** by writing that the sums of the components and of the moments of the elementary forces are equal to the corresponding components and moments of the couple **M**:

x components: $\qquad\qquad \int \sigma_x \, dA = 0 \qquad\qquad$ (4.1)

moments about y axis: $\qquad \int z\sigma_x \, dA = 0 \qquad\qquad$ (4.2)

moments about z axis: $\qquad \int (-y\sigma_x \, dA) = M \qquad$ (4.3)

Three additional equations could be obtained by setting equal to zero the sums of the y components, z components, and moments about the x axis, but these equations would involve only the components of the shearing stress and, as you will see in the next section, the components of the shearing stress are both equal to zero.

Two remarks should be made at this point: (1) The minus sign in Eq. (4.3) is due to the fact that a tensile stress ($\sigma_x > 0$) leads to a negative moment (clockwise) of the normal force $\sigma_x \, dA$ about the z axis. (2) Equation (4.2) could have been anticipated, since the application of couples in the plane of symmetry of member AB will result in a distribution of normal stresses that is symmetric about the y axis.

Once more, we note that the actual distribution of stresses in a given cross section cannot be determined from statics alone. It is *statically indeterminate* and may be obtained only by analyzing the *deformations* produced in the member.

4.3 DEFORMATIONS IN A SYMMETRIC MEMBER IN PURE BENDING

Let us now analyze the deformations of a prismatic member possessing a plane of symmetry and subjected at its ends to equal and opposite couples **M** and **M'** acting in the plane of symmetry. The member will bend under the action of the couples, but will remain symmetric with respect to that plane (Fig. 4.7). Moreover, since the bending moment M is the same in any cross section, the member will bend uniformly. Thus, the line AB along which the upper face of the member intersects the plane of the couples will have a constant curvature. In other words, the line AB, which was originally a straight line, will be transformed into a circle of center C, and so will the line $A'B'$ (not shown in the figure) along which the lower face of the member intersects the plane of symmetry. We also note that the line AB will decrease in length when the member is bent as shown in the figure, i.e., when $M > 0$, while $A'B'$ will become longer.

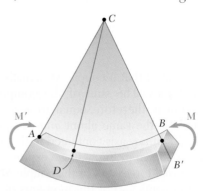

Fig. 4.7 Deformation of member in pure bending.

Next we will prove that any cross section perpendicular to the axis of the member remains plane, and that the plane of the section passes through C. If this were not the case, we could find a point E of the original section through D (Fig. 4.8a) which, after the member has been bent, would *not* lie in the plane perpendicular to the plane of symmetry that contains line CD (Fig. 4.8b). But, because of the symmetry of the member, there would be another point E' that would be transformed exactly in the same way. Let us assume that, after the beam has been bent, both points would be located to the left of the plane defined by CD, as shown in Fig. 4.8b. Since the bending moment M is the same throughout the member, a similar situation would prevail in any other cross section, and the points corresponding to E and E' would also move to the left. Thus, an observer at A would conclude that the loading causes the points E and E' in the various cross sections to move forward (toward the observer). But an observer at B, to whom the loading looks the same, and who observes the points E and E' in the same positions (except that they are now inverted) would reach the opposite conclusion. This inconsistency leads us to conclude that E and E' will lie in the plane defined by CD and, therefore, that the section remains plane and passes through C. We should note,

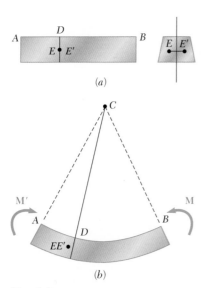

Fig. 4.8

however, that this discussion does not rule out the possibility of deformations *within* the plane of the section (see Sec. 4.5).

Suppose that the member is divided into a large number of small cubic elements with faces respectively parallel to the three coordinate planes. The property we have established requires that these elements be transformed as shown in Fig. 4.9 when the member is subjected to the couples **M** and **M'**. Since all the faces represented in the two projections of Fig. 4.9 are at 90° to each other, we conclude that $\gamma_{xy} = \gamma_{zx} = 0$ and, thus, that $\tau_{xy} = \tau_{xz} = 0$. Regarding the three stress components that we have not yet discussed, namely, σ_y, σ_z, and τ_{yz}, we note that they must be zero on the surface of the member. Since, on the other hand, the deformations involved do not require any interaction between the elements of a given transverse cross section, we can assume that these three stress components are equal to zero throughout the member. This assumption is verified, both from experimental evidence and from the theory of elasticity, for slender members undergoing small deformations.† We conclude that the only nonzero stress component exerted on any of the small cubic elements considered here is the normal component σ_x. Thus, at any point of a slender member in pure bending, we have a state of *uniaxial stress*. Recalling that, for $M > 0$, lines AB and $A'B'$ are observed, respectively, to decrease and increase in length, we note that the strain ϵ_x and the stress σ_x are negative in the upper portion of the member (*compression*) and positive in the lower portion (*tension*).

It follows from the above that there must exist a surface parallel to the upper and lower faces of the member, where ϵ_x and σ_x are zero. This surface is called the *neutral surface*. The neutral surface intersects the plane of symmetry along an arc of circle DE (Fig. 4.10a), and it intersects a transverse section along a straight line called the *neutral axis* of the section (Fig. 4.10b). The origin of coordinates will now be selected on the neutral surface, rather than on the lower face of the member as done earlier, so that the distance from any point to the neutral surface will be measured by its coordinate y.

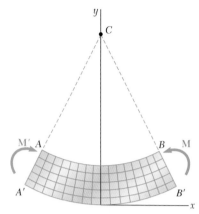

(a) Longitudinal, vertical section
(plane of symmetry)

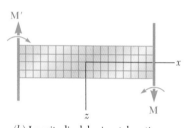

(b) Longitudinal, horizontal section

Fig. 4.9 Member subject to pure bending.

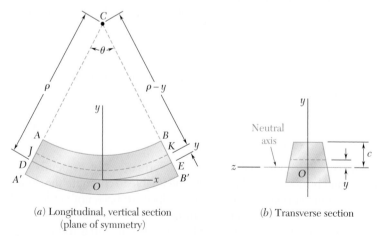

(a) Longitudinal, vertical section
(plane of symmetry)

(b) Transverse section

Fig. 4.10 Deformation with respect to neutral axis.

†Also see Prob. 4.32.

Denoting by ρ the radius of arc DE (Fig. 4.10a), by θ the central angle corresponding to DE, and observing that the length of DE is equal to the length L of the undeformed member, we write

$$L = \rho\theta \tag{4.4}$$

Considering now the arc JK located at a distance y above the neutral surface, we note that its length L' is

$$L' = (\rho - y)\theta \tag{4.5}$$

Since the original length of arc JK was equal to L, the deformation of JK is

$$\delta = L' - L \tag{4.6}$$

or, if we substitute from (4.4) and (4.5) into (4.6),

$$\delta = (\rho - y)\theta - \rho\theta = -y\theta \tag{4.7}$$

The longitudinal strain ϵ_x in the elements of JK is obtained by dividing δ by the original length L of JK. We write

$$\epsilon_x = \frac{\delta}{L} = \frac{-y\theta}{\rho\theta}$$

or

$$\epsilon_x = -\frac{y}{\rho} \tag{4.8}$$

The minus sign is due to the fact that we have assumed the bending moment to be positive and, thus, the beam to be concave upward.

Because of the requirement that transverse sections remain plane, identical deformations will occur in all planes parallel to the plane of symmetry. Thus the value of the strain given by Eq. (4.8) is valid anywhere, and we conclude that the *longitudinal normal strain ϵ_x varies linearly with the distance y from the neutral surface.*

The strain ϵ_x reaches its maximum absolute value when y itself is largest. Denoting by c the largest distance from the neutral surface (which corresponds to either the upper or the lower surface of the member), and by ϵ_m the *maximum absolute value* of the strain, we have

$$\epsilon_m = \frac{c}{\rho} \tag{4.9}$$

Solving (4.9) for ρ and substituting the value obtained into (4.8), we can also write

$$\epsilon_x = -\frac{y}{c}\epsilon_m \tag{4.10}$$

We conclude our analysis of the deformations of a member in pure bending by observing that we are still unable to compute the strain or stress at a given point of the member, since we have not yet located the neutral surface in the member. In order to locate this surface, we must first specify the stress-strain relation of the material used.†

---

†Let us note, however, that if the member possesses both a vertical and a horizontal plane of symmetry (e.g., a member with a rectangular cross section), and if the stress-strain curve is the same in tension and compression, the neutral surface will coincide with the plane of symmetry (cf. Sec. 4.8).

4.4 STRESSES AND DEFORMATIONS IN THE ELASTIC RANGE

We now consider the case when the bending moment M is such that the normal stresses in the member remain below the yield strength σ_Y. This means that, for all practical purposes, the stresses in the member will remain below the proportional limit and the elastic limit as well. There will be no permanent deformation, and Hooke's law for uniaxial stress applies. Assuming the material to be homogeneous, and denoting by E its modulus of elasticity, we have in the longitudinal x direction

$$\sigma_x = E\epsilon_x \qquad (4.11)$$

Recalling Eq. (4.10), and multiplying both members of that equation by E, we write

$$E\epsilon_x = -\frac{y}{c}(E\epsilon_m)$$

or, using (4.11),

$$\sigma_x = -\frac{y}{c}\sigma_m \qquad (4.12)$$

where σ_m denotes the *maximum absolute value* of the stress. This result shows that, *in the elastic range, the normal stress varies linearly with the distance from the neutral surface* (Fig. 4.11).

It should be noted that, at this point, we do not know the location of the neutral surface, nor the maximum value σ_m of the stress. Both can be found if we recall the relations (4.1) and (4.3) which were obtained earlier from statics. Substituting first for σ_x from (4.12) into (4.1), we write

$$\int \sigma_x \, dA = \int \left(-\frac{y}{c}\sigma_m\right) dA = -\frac{\sigma_m}{c}\int y \, dA = 0$$

from which it follows that

$$\int y \, dA = 0 \qquad (4.13)$$

This equation shows that the first moment of the cross section about its neutral axis must be zero.† In other words, for a member subjected to pure bending, and *as long as the stresses remain in the elastic range, the neutral axis passes through the centroid of the section.*

We now recall Eq. (4.3), which was derived in Sec. 4.2 with respect to an *arbitrary* horizontal z axis,

$$\int (-y\sigma_x \, dA) = M \qquad (4.3)$$

Specifying that the z axis should coincide with the neutral axis of the cross section, we substitute for σ_x from (4.12) into (4.3) and write

$$\int (-y)\left(-\frac{y}{c}\sigma_m\right) dA = M$$

†See Appendix A for a discussion of the moments of areas.

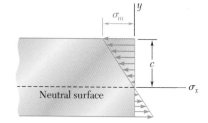

Fig. 4.11 Bending stresses.

or

$$\frac{\sigma_m}{c} \int y^2 \, dA = M \qquad (4.14)$$

Recalling that in the case of pure bending the neutral axis passes through the centroid of the cross section, we note that I is the moment of inertia, or second moment, of the cross section with respect to a centroidal axis perpendicular to the plane of the couple **M.** Solving (4.14) for σ_m, we write therefore†

$$\sigma_m = \frac{Mc}{I} \qquad (4.15)$$

Substituting for σ_m from (4.15) into (4.12), we obtain the normal stress σ_x at any distance y from the neutral axis:

$$\sigma_x = -\frac{My}{I} \qquad (4.16)$$

Equations (4.15) and (4.16) are called the *elastic flexure formulas,* and the normal stress σ_x caused by the bending or "flexing" of the member is often referred to as the *flexural stress.* We verify that the stress is compressive ($\sigma_x < 0$) above the neutral axis ($y > 0$) when the bending moment M is positive, and tensile ($\sigma_x > 0$) when M is negative.

Returning to Eq. (4.15), we note that the ratio I/c depends only upon the geometry of the cross section. This ratio is called the *elastic section modulus* and is denoted by S. We have

$$\text{Elastic section modulus} = S = \frac{I}{c} \qquad (4.17)$$

Substituting S for I/c into Eq. (4.15), we write this equation in the alternative form

$$\sigma_m = \frac{M}{S} \qquad (4.18)$$

Since the maximum stress σ_m is inversely proportional to the elastic section modulus S, it is clear that beams should be designed with as large a value of S as practicable. For example, in the case of a wooden beam with a rectangular cross section of width b and depth h, we have

$$S = \frac{I}{c} = \frac{\frac{1}{12}bh^3}{h/2} = \frac{1}{6}bh^2 = \frac{1}{6}Ah \qquad (4.19)$$

---

†We recall that the bending moment was assumed to be positive. If the bending moment is negative, M should be replaced in Eq. (4.15) by its absolute value $|M|$.

where A is the cross-sectional area of the beam. This shows that, of two beams with the same cross-sectional area A (Fig. 4.12), the beam with the larger depth h will have the larger section modulus and, thus, will be the more effective in resisting bending.†

In the case of structural steel, American standard beams (S-beams) and wide-flange beams (W-beams), Photo 4.3, are preferred

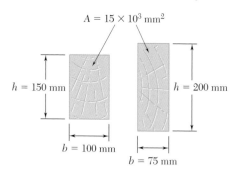

Fig. 4.12 Wood beam cross sections.

Photo 4.3 Wide-flange steel beams form the frame of many buildings.

to other shapes because a large portion of their cross section is located far from the neutral axis (Fig. 4.13). Thus, for a given cross-sectional area and a given depth, their design provides large values

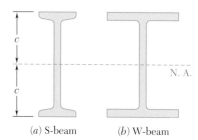

(a) S-beam   (b) W-beam

Fig. 4.13 Steel beam cross sections.

†However, large values of the ratio h/b could result in lateral instability of the beam.

of I and, consequently, of S. Values of the elastic section modulus of commonly manufactured beams can be obtained from tables listing the various geometric properties of such beams. To determine the maximum stress σ_m in a given section of a standard beam, the engineer needs only to read the value of the elastic section modulus S in a table, and divide the bending moment M in the section by S.

The deformation of the member caused by the bending moment M is measured by the *curvature* of the neutral surface. The curvature is defined as the reciprocal of the radius of curvature ρ, and can be obtained by solving Eq. (4.9) for $1/\rho$:

$$\frac{1}{\rho} = \frac{\epsilon_m}{c} \tag{4.20}$$

But, in the elastic range, we have $\epsilon_m = \sigma_m/E$. Substituting for ϵ_m into (4.20), and recalling (4.15), we write

$$\frac{1}{\rho} = \frac{\sigma_m}{Ec} = \frac{1}{Ec}\frac{Mc}{I}$$

or

$$\frac{1}{\rho} = \frac{M}{EI} \tag{4.21}$$

EXAMPLE 4.01

A steel bar of 20 × 60-mm rectangular cross section is subjected to two equal and opposite couples acting in the vertical plane of symmetry of the bar (Fig. 4.14). Determine the value of the bending moment M that causes the bar to yield. Assume $\sigma_Y = 250$ MPa.

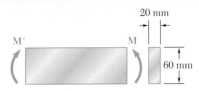

Fig. 4.14

Since the neutral axis must pass through the centroid C of the cross section, we have $c = 30$ mm (Fig. 4.15). On the other hand, the centroidal moment of inertia of the rectangular cross section is

$$I = \tfrac{1}{12}bh^3 = \tfrac{1}{12}(20\text{ mm})(60\text{ mm})^3 = 360 \times 10^3 \text{ mm}^4$$

Solving Eq. (4.15) for M, and substituting the above data, we have

$$M = \frac{I}{c}\sigma_m = \frac{360 \times 10^{-9}\text{ m}^4}{0.03\text{ m}}(250\text{ MPa})$$

$$M = 3 \text{ kN} \cdot \text{m}$$

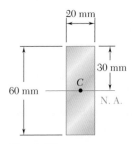

Fig. 4.15

EXAMPLE 4.02

An aluminum rod with a semicircular cross section of radius $r = 12$ mm (Fig. 4.16) is bent into the shape of a circular arc of mean radius $\rho = 2.5$ m. Knowing that the flat face of the rod is turned toward the center of curvature of the arc, determine the maximum tensile and compressive stress in the rod. Use $E = 70$ GPa.

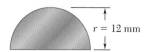

$r = 12$ mm

Fig. 4.16

We could use Eq. (4.21) to determine the bending moment M corresponding to the given radius of curvature ρ, and then Eq. (4.15) to determine σ_m. However, it is simpler to use Eq. (4.9) to determine ϵ_m, and Hooke's law to obtain σ_m.

The ordinate $\bar{y}$ of the centroid C of the semicircular cross section is

$$\bar{y} = \frac{4r}{3\pi} = \frac{4(12 \text{ mm})}{3\pi} = 5.093 \text{ mm}$$

The neutral axis passes through C (Fig. 4.17) and the distance c to the point of the cross section farthest away from the neutral axis is

Fig. 4.17

$$c = r - \bar{y} = 12 \text{ mm} - 5.093 \text{ mm} = 6.907 \text{ mm}$$

Using Eq. (4.9), we write

$$\epsilon_m = \frac{c}{\rho} = \frac{6.907 \times 10^{-3} \text{ m}}{2.5 \text{ m}} = 2.763 \times 10^{-3}$$

and, applying Hooke's law,

$$\sigma_m = E\epsilon_m = (70 \times 10^9 \text{ Pa})(2.763 \times 10^{-3}) = 193.4 \text{ MPa}$$

Since this side of the rod faces away from the center of curvature, the stress obtained is a tensile stress. The maximum compressive stress occurs on the flat side of the rod. Using the fact that the stress is proportional to the distance from the neutral axis, we write

$$\sigma_{\text{comp}} = -\frac{\bar{y}}{c}\sigma_m = -\frac{5.093 \text{ mm}}{6.907 \text{ mm}}(193.4 \text{ MPa})$$

$$= -142.6 \text{ MPa}$$

4.5 DEFORMATIONS IN A TRANSVERSE CROSS SECTION

When we proved in Sec. 4.3 that the transverse cross section of a member in pure bending remains plane, we did not rule out the possibility of deformations within the plane of the section. That such deformations will exist is evident, if we recall from Sec. 2.11 that elements in a state of uniaxial stress, $\sigma_x \neq 0$, $\sigma_y = \sigma_z = 0$, are deformed in the transverse y and z directions, as well as in the axial x direction. The normal strains ϵ_y and ϵ_z depend upon Poisson's ratio ν for the material used and are expressed as

$$\epsilon_y = -\nu\epsilon_x \qquad \epsilon_z = -\nu\epsilon_x$$

or, recalling Eq. (4.8),

$$\epsilon_y = \frac{\nu y}{\rho} \qquad \epsilon_z = \frac{\nu y}{\rho} \qquad (4.22)$$

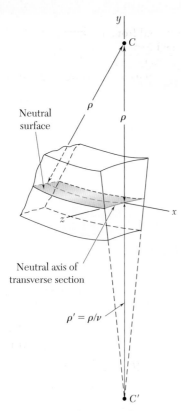

Fig. 4.18 Deformation of transverse cross section.

The relations we have obtained show that the elements located above the neutral surface ($y > 0$) will expand in both the y and z directions, while the elements located below the neutral surface ($y < 0$) will contract. In the case of a member of rectangular cross section, the expansion and contraction of the various elements in the vertical direction will compensate, and no change in the vertical dimension of the cross section will be observed. As far as the deformations in the horizontal transverse z direction are concerned, however, the expansion of the elements located above the neutral surface and the corresponding contraction of the elements located below that surface will result in the various horizontal lines in the section being bent into arcs of circle (Fig. 4.18). The situation observed here is similar to that observed earlier in a longitudinal cross section. Comparing the second of Eqs. (4.22) with Eq. (4.8), we conclude that the neutral axis of the transverse section will be bent into a circle of radius $\rho' = \rho/\nu$. The center C' of this circle is located below the neutral surface (assuming $M > 0$), i.e., on the side opposite to the center of curvature C of the member. The reciprocal of the radius of curvature ρ' represents the curvature of the transverse cross section and is called the *anticlastic curvature*. We have

$$\text{Anticlastic curvature} = \frac{1}{\rho'} = \frac{\nu}{\rho} \tag{4.23}$$

In our discussion of the deformations of a symmetric member in pure bending, in this section and in the preceding ones, we have ignored the manner in which the couples $\mathbf{M}$ and $\mathbf{M}'$ were actually applied to the member. If *all* transverse sections of the member, from one end to the other, are to remain plane and free of shearing stresses, we must make sure that the couples are applied in such a way that the ends of the member themselves remain plane and free of shearing stresses. This can be accomplished by applying the couples $\mathbf{M}$ and $\mathbf{M}'$ to the member through the use of rigid and smooth plates (Fig. 4.19). The elementary forces exerted by the plates on the member will be normal to the end sections, and these sections, while remaining plane, will be free to deform as described earlier in this section.

We should note that these loading conditions cannot be actually realized, since they require each plate to exert tensile forces on the corresponding end section below its neutral axis, while allowing the section to freely deform in its own plane. The fact that the rigid-end-plates model of Fig. 4.19 cannot be physically realized, however, does not detract from its importance, which is to allow us to *visualize* the loading conditions corresponding to the relations derived in the preceding sections. Actual loading conditions may differ appreciably from this idealized model. By virtue of Saint-Venant's principle, however, the relations obtained can be used to compute stresses in engineering situations, as long as the section considered is not too close to the points where the couples are applied.

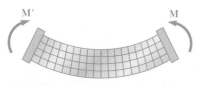

Fig. 4.19 Deformation of longitudinal segment.

while the force dF_2 exerted on an element of the same area dA of the lower portion is

$$dF_2 = \sigma_2 dA = -\frac{E_2 y}{\rho} dA \qquad (4.26)$$

But, denoting by n the ratio E_2/E_1 of the two moduli of elasticity, we can express dF_2 as

$$dF_2 = -\frac{(nE_1)y}{\rho} dA = -\frac{E_1 y}{\rho}(n\, dA) \qquad (4.27)$$

Comparing Eqs. (4.25) and (4.27), we note that the same force dF_2 would be exerted on an element of area $n\, dA$ of the first material. In other words, the resistance to bending of the bar would remain the same if both portions were made of the first material, provided that the width of each element of the lower portion were multiplied by the factor n. Note that this widening (if $n > 1$), or narrowing (if $n < 1$), must be effected *in a direction parallel to the neutral axis of the section*, since it is essential that the distance y of each element from the neutral axis remain the same. The new cross section obtained in this way is called the *transformed section* of the member (Fig. 4.22).

Since the transformed section represents the cross section of a member made of a *homogeneous material* with a modulus of elasticity E_1, the method described in Sec. 4.4 can be used to determine the neutral axis of the section and the normal stress at various points of the section. The neutral axis will be drawn *through the centroid of the transformed section* (Fig. 4.23), and the stress σ_x at any point of the corresponding fictitious homogeneous member will be obtained from Eq. (4.16)

$$\sigma_x = -\frac{My}{I} \qquad (4.16)$$

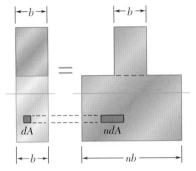

Fig. 4.22 Transformed section for composite bar.

where y is the distance from the neutral surface, and I *the moment of inertia of the transformed section* with respect to its centroidal axis.

To obtain the stress σ_1 at a point located in the upper portion of the cross section of the original composite bar, we simply compute the stress σ_x at the corresponding point of the transformed section. However, to obtain the stress σ_2 at a point in the lower portion of the cross section, we must *multiply by n* the stress σ_x computed at the corresponding point of the transformed section. Indeed, as we saw earlier, the same elementary force dF_2 is applied to an element of area $n\, dA$ of the transformed section and to an element of area dA of the original section. Thus, the stress σ_2 at a point of the original section must be n times larger than the stress at the corresponding point of the transformed section.

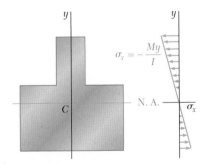

Fig. 4.23 Distribution of stresses in transformed section.

The deformations of a composite member can also be determined by using the transformed section. We recall that the transformed section represents the cross section of a member, made of a homogeneous material of modulus E_1, which deforms in the same manner as the composite member. Therefore, using Eq. (4.21), we write that the curvature of the composite member is

$$\frac{1}{\rho} = \frac{M}{E_1 I}$$

where I is the moment of inertia of the transformed section with respect to its neutral axis.

EXAMPLE 4.03

A bar obtained by bonding together pieces of steel (E_s = 200 GPa) and brass (E_b = 100 GPa) has the cross section shown (Fig. 4.24). Determine the maximum stress in the steel and in the brass when the bar is in pure bending with a bending moment M = 4.5 kN · m.

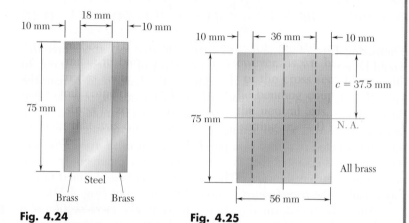

Fig. 4.24   Fig. 4.25

The transformed section corresponding to an equivalent bar made entirely of brass is shown in Fig. 4.25. Since

$$n = \frac{E_s}{E_b} = \frac{200 \text{ GPa}}{100 \text{ GPa}} = 2.0$$

the width of the central portion of brass, which replaces the original steel portion, is obtained by multiplying the original width by 2.0, we have

$$(18 \text{ mm})(2) = 36 \text{ mm}$$

Note that this change in dimension occurs in a direction parallel to the neutral axis. The moment of inertia of the transformed section about its centroidal axis is

$$I = \tfrac{1}{12}bh^3 = \tfrac{1}{12}(56 \text{ mm})(75 \text{ mm})^3 = 1.9688 \times 10^6 \text{ mm}^4$$

and the maximum distance from the neutral axis is c = 37.5 mm. Using Eq. (4.15), we find the maximum stress in the transformed section:

$$\sigma_m = \frac{Mc}{I} = \frac{(4.5 \times 10^3 \text{ N} \cdot \text{m})(37.5 \times 10^{-3} \text{ m})}{1.9688 \times 10^{-6} \text{ m}^4} = 85.7 \text{ MPa}$$

The value obtained also represents the maximum stress in the brass portion of the original composite bar. The maximum stress in the steel portion, however, will be larger than the value obtained for the transformed section, since the area of the central portion must be reduced by the factor n = 2 when we return from the transformed section to the original one. We thus conclude that

$$(\sigma_{\text{brass}})_{\text{max}} = 85.7 \text{ MPa}$$
$$(\sigma_{\text{steel}})_{\text{max}} = (2)(85.7 \text{ MPa}) = 171.4 \text{ MPa}$$

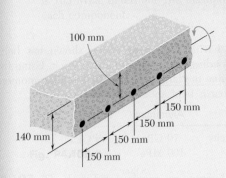

100 mm

150 mm
150 mm
150 mm
150 mm
140 mm

SAMPLE PROBLEM 4.4

A concrete floor slab is reinforced by 16-mm-diameter steel rods placed 38 mm above the lower face of the slab and spaced 150 mm on centers. The modulus of elasticity is 25 GPa for the concrete used and 200 GPa for the steel. Knowing that a bending moment of 4.5 kN · m is applied to each 0.3-m width of the slab, determine (a) the maximum stress in the concrete, (b) the stress in the steel.

SOLUTION

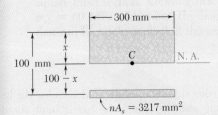

300 mm
100 mm
x
C N. A.
$100 - x$
$nA_s = 3217$ mm^2

Transformed Section. We consider a portion of the slab 300 mm wide, in which there are two 16-mm-diameter rods having a total cross-sectional area

$$A_s = 2\left[\frac{\pi}{4}(16 \text{ mm})^2\right] = 402.1 \text{ mm}^2$$

Since concrete acts only in compression, all the tensile forces are carried by the steel rods, and the transformed section consists of the two areas shown. One is the portion of concrete in compression (located above the neutral axis), and the other is the transformed steel area nA_s. We have

$$n = \frac{E_s}{E_c} = \frac{200 \text{ GPa}}{25 \text{ GPa}} = 8$$

$$nA_s = 8(402.1 \text{ mm})^2 = 3217 \text{ mm}^2$$

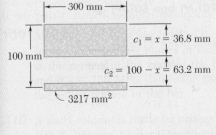

300 mm
100 mm
$c_1 = x = 36.8$ mm
$c_2 = 100 - x = 63.2$ mm
3217 mm^2

Neutral Axis. The neutral axis of the slab passes through the centroid of the transformed section. Summing moments of the transformed area about the neutral axis, we write

$$300x\left(\frac{x}{2}\right) - 3217(100 - x) = 0 \qquad x = 36.8 \text{ mm}$$

Moment of Inertia. The centroidal moment of inertia of the transformed area is

$$I = \tfrac{1}{3}(300)(36.8)^3 + 3217(100 - 36.8)^2 = 17.83 \times 10^6 \text{ mm}^4$$

a. Maximum Stress in Concrete. At the top of the slab, we have $c_1 = 36.8$ mm and

$$\sigma_c = \frac{Mc_1}{I} = \frac{(4.5 \times 10^3 \text{ N} \cdot \text{m})(36.8 \times 10^{-3} \text{ m})}{17.83 \times 10^{-6} \text{ m}^4} \qquad \sigma_c = 9.29 \text{ MPa} \blacktriangleleft$$

b. Stress in Steel. For the steel, we have $c_2 = 63.2$ mm, $n = 8$ and

$$\sigma_s = n\frac{Mc_2}{I} = \frac{8(4.5 \times 10^3 \text{ N} \cdot \text{m})(63.2 \times 10^{-3} \text{ m})}{17.83 \times 10^{-6} \text{ m}^4} \qquad \sigma_s = 127.6 \text{ MPa} \blacktriangleleft$$

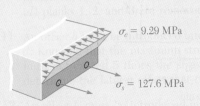

$\sigma_c = 9.29$ MPa

$\sigma_s = 127.6$ MPa

PROBLEMS

4.33 and 4.34 A bar having the cross section shown has been formed by securely bonding brass and aluminum stock. Using the data provided, determine the largest permissible bending moment when the composite bar is bent about a horizontal axis.

	Aluminum	Brass
Modulus of elasticity	70 GPa	105 GPa
Allowable stress	100 MPa	160 MPa

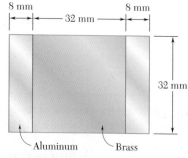

8 mm — 32 mm — 8 mm

32 mm

Aluminum   Brass

Fig. P4.33

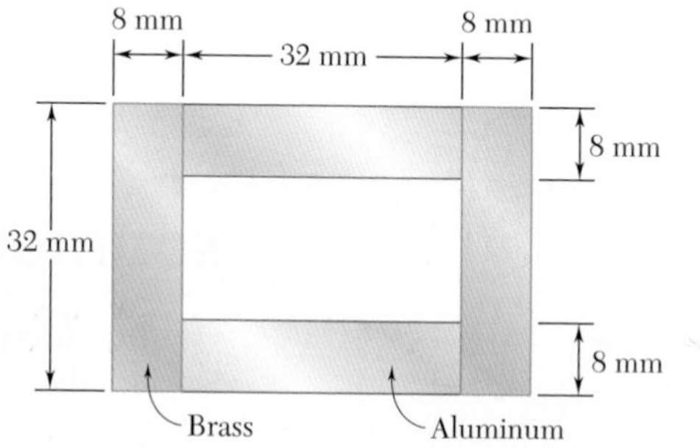

8 mm — 32 mm — 8 mm

8 mm

32 mm

8 mm

Brass   Aluminum

Fig. P4.34

4.35 and 4.36 For the composite bar indicated, determine the largest permissible bending moment when the bar is bent about a vertical axis.

    4.35 Bar of Prob. 4.33.
    4.36 Bar of Prob. 4.34.

4.37 Three wooden beams and two steel plates are securely bolted together to form the composite member shown. Using the data provided, determine the largest permissible bending moment when the member is bent about a <u>horizontal axis</u>.

	Wood	Steel
Modulus of elasticity	14 GPa	200 GPa
Allowable stress	14 MPa	150 MPa

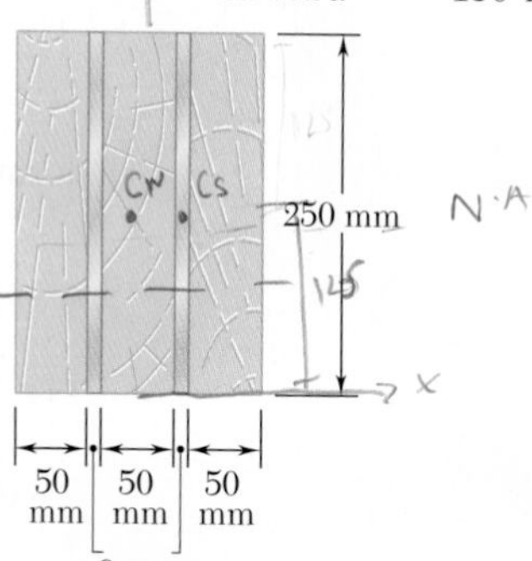

250 mm   N·A

125

50 mm | 50 mm | 50 mm

6 mm

Fig. P4.37

4.38 For the composite member of Prob. 4.37, determine the largest permissible bending moment when the member is bent about a vertical axis.

272

(handwritten notes in left margin and below Fig. P4.33:)

4.37

$3E_w \bar{y}_w A_w + 2E_s \bar{y}_s A_s = 0$

$3(14\text{E}3)(125 - h_1)(250)(50)$

$+ 2(200\text{E}3)(250)(6)(125 - h_1) = 0$

$(125 - h_1) + 9.5238(125 - h_1) = 0$

$1315.475 = 10.5238 h_1$

$h_1 = 125$

I. $\sigma_i = \dfrac{-M y E_i}{\Sigma E_i I_i}$ ← b/c centroid is on N·A

$I_s = \left(\frac{1}{12}\right)(6)(250)^3 + 0$

$I_w = \left(\frac{1}{12}\right)(50)(250)^3$ → 125 (max dist from N·A to surface of bending)

$|\sigma_s| = \left| \dfrac{-M y \Sigma E_s}{2E_s I_s + 3E_w I_w} \right| \ne \boxed{35 \text{ kNm}}$

ANS (smallest value!)

$\sigma_w = 46875 \text{ kNm}$

*4.8 PLASTIC DEFORMATIONS

When we derived the fundamental relation $\sigma_x = -My/I$ in Sec. 4.4, we assumed that Hooke's law applied throughout the member. If the yield strength is exceeded in some portion of the member, or if the material involved is a brittle material with a nonlinear stress-strain diagram, this relation ceases to be valid. The purpose of this section is to develop a more general method for the determination of the distribution of stresses in a member in pure bending, which can be used when Hooke's law does not apply.

We first recall that no specific stress-strain relationship was assumed in Sec. 4.3, when we proved that the normal strain ϵ_x varies linearly with the distance y from the neutral surface. Thus, we can still use this property in our present analysis and write

$$\epsilon_x = -\frac{y}{c}\epsilon_m \tag{4.10}$$

where y represents the distance of the point considered from the neutral surface, and c the maximum value of y.

However, we cannot assume anymore that, in a given section, the neutral axis passes through the centroid of that section, since this property was derived in Sec. 4.4 under the assumption of elastic deformations. In general, the neutral axis must be located by trial and error, until a distribution of stresses has been found, that satisfies Eqs. (4.1) and (4.3) of Sec. 4.2. However, in the particular case of a member possessing both a vertical and a horizontal plane of symmetry, and made of a material characterized by the same stress-strain relation in tension and in compression, the neutral axis will coincide with the horizontal axis of symmetry of the section. Indeed, the properties of the material require that the stresses be symmetric with respect to the neutral axis, i.e., with respect to *some* horizontal axis, and it is clear that this condition will be met, and Eq. (4.1) satisfied at the same time, only if that axis is the horizontal axis of symmetry itself.

Our analysis will first be limited to the special case we have just described. The distance y in Eq. (4.10) is thus measured from the horizontal axis of symmetry z of the cross section, and the distribution of strain ϵ_x is linear and symmetric with respect to that axis (Fig. 4.30). On the other hand, the stress-strain curve is symmetric with respect to the origin of coordinates (Fig. 4.31).

The distribution of stresses in the cross section of the member, i.e., the plot of σ_x versus y, is obtained as follows. Assuming that σ_{max} has been specified, we first determine the corresponding value of ϵ_m from the stress-strain diagram and carry this value into Eq. (4.10). Then, for each value of y, we determine the corresponding value of ϵ_x from Eq. (4.10) or Fig. 4.30, and obtain from the stress-strain diagram of Fig. 4.31 the stress σ_x corresponding to this value of ϵ_x. Plotting σ_x against y yields the desired distribution of stresses (Fig. 4.32).

We now recall that, when we derived Eq. (4.3) in Sec. 4.2, we assumed no particular relation between stress and strain. We can therefore use Eq. (4.3) to determine the bending moment M corresponding to the stress distribution obtained in Fig. 4.32. Considering the particular

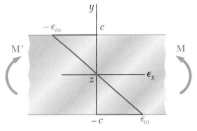

Fig. 4.30 Linear strain distribution in beam.

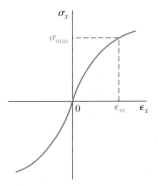

Fig. 4.31 Nonlinear stress-strain material diagram.

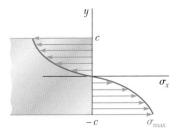

Fig. 4.32 Nonlinear stress distribution in beam.

case of a member with a rectangular cross section of width b, we express the element of area in Eq. (4.3) as $dA = b\,dy$ and write

$$M = -b \int_{-c}^{c} y\sigma_x\,dy \tag{4.30}$$

where σ_x is the function of y plotted in Fig. 4.32. Since σ_x is an odd function of y, we can write Eq. (4.30) in the alternative form

$$M = -2b \int_{0}^{c} y\sigma_x\,dy \tag{4.31}$$

If σ_x is a known analytical function of ϵ_x, Eq. (4.10) can be used to express σ_x as a function of y, and the integral in (4.31) can be determined analytically. Otherwise, the bending moment M can be obtained through a numerical integration. This computation becomes more meaningful if we note that the integral in Eq. (4.31) represents the first moment with respect to the horizontal axis of the area in Fig. 4.32 that is located above the horizontal axis and is bounded by the stress-distribution curve and the vertical axis.

An important value of the bending moment is the ultimate bending moment M_U that causes failure of the member. This value can be determined from the ultimate strength σ_U of the material by choosing $\sigma_{max} = \sigma_U$ and carrying out the computations indicated earlier. However, it is found more convenient in practice to determine M_U experimentally for a specimen of a given material. Assuming a fictitious linear distribution of stresses, Eq. (4.15) is then used to determine the corresponding maximum stress R_B:

$$R_B = \frac{M_U c}{I} \tag{4.32}$$

The fictitious stress R_B is called the *modulus of rupture in bending* of the given material. It can be used to determine the ultimate bending moment M_U of a member made of the same material and having a cross section of the same shape, but of different dimensions, by solving Eq. (4.32) for M_U. Since, in the case of a member with a rectangular cross section, the actual and the fictitious linear stress distributions shown in Fig. 4.33 must yield the same value M_U for the ultimate bending moment, the areas they define must have the same first moment with respect to the horizontal axis. It is thus clear that the modulus of rupture R_B will always be larger than the actual ultimate strength σ_U.

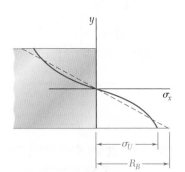

Fig. 4.33 Beam stress distribution at ultimate moment M_U.

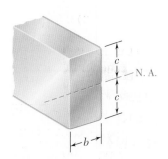

Fig. 4.34 Beam with rectangular cross section.

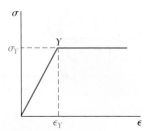

Fig. 4.35 Idealized steel stress-strain diagram.

*4.9 MEMBERS MADE OF AN ELASTOPLASTIC MATERIAL

In order to gain a better insight into the plastic behavior of a member in bending, let us consider the case of a member made of an *elastoplastic material* and first assume the member to have a *rectangular cross section* of width b and depth $2c$ (Fig. 4.34). We recall from Sec. 2.17 that the stress-strain diagram for an idealized elastoplastic material is as shown in Fig. 4.35.

As long as the normal stress σ_x does not exceed the yield strength σ_Y, Hooke's law applies, and the stress distribution across the section is linear (Fig. 4.36a). The maximum value of the stress is

$$\sigma_m = \frac{Mc}{I} \tag{4.15}$$

As the bending moment increases, σ_m eventually reaches the value σ_Y (Fig. 4.36b). Substituting this value into Eq. (4.15), and solving for the corresponding value of M, we obtain the value M_Y of the bending moment at the onset of yield:

$$M_Y = \frac{I}{c}\sigma_Y \tag{4.33}$$

The moment M_Y is referred to as the *maximum elastic moment*, since it is the largest moment for which the deformation remains fully elastic. Recalling that, for the rectangular cross section considered here,

$$\frac{I}{c} = \frac{b(2c)^3}{12c} = \frac{2}{3}bc^2 \tag{4.34}$$

we write

$$M_Y = \frac{2}{3}bc^2\sigma_Y \tag{4.35}$$

As the bending moment further increases, plastic zones develop in the member, with the stress uniformly equal to $-\sigma_Y$ in the upper zone, and to $+\sigma_Y$ in the lower zone (Fig. 4.36c). Between the plastic zones, an elastic core subsists, in which the stress σ_x varies linearly with y,

$$\sigma_x = -\frac{\sigma_Y}{y_Y}y \tag{4.36}$$

where y_Y represents half the thickness of the elastic core. As M increases, the plastic zones expand until, at the limit, the deformation is fully plastic (Fig. 4.36d).

Equation (4.31) will be used to determine the value of the bending moment M corresponding to a given thickness $2y_Y$ of the elastic core. Recalling that σ_x is given by Eq. (4.36) for $0 \le y \le y_Y$, and is equal to $-\sigma_Y$ for $y_Y \le y \le c$, we write

$$M = -2b\int_0^{y_Y} y\left(-\frac{\sigma_Y}{y_Y}y\right)dy - 2b\int_{y_Y}^c y(-\sigma_Y)\,dy$$

$$= \frac{2}{3}by_Y^2\sigma_Y + bc^2\sigma_Y - by_Y^2\sigma_Y$$

$$M = bc^2\sigma_Y\left(1 - \frac{1}{3}\frac{y_Y^2}{c^2}\right) \tag{4.37}$$

or, in view of Eq. (4.35),

$$M = \frac{3}{2}M_Y\left(1 - \frac{1}{3}\frac{y_Y^2}{c^2}\right) \tag{4.38}$$

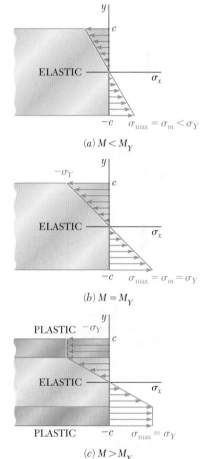

(a) $M < M_Y$

(b) $M = M_Y$

(c) $M > M_Y$

(d) $M = M_p$

Fig. 4.36 Bending stress distribution in beam for different moments.

where M_Y is the maximum elastic moment. Note that as y_Y approaches zero, the bending moment approaches the limiting value

$$M_p = \frac{3}{2}M_Y \tag{4.39}$$

This value of the bending moment, which corresponds to a fully plastic deformation (Fig. 4.36d), is called the *plastic moment* of the member considered. Note that Eq. (4.39) is valid only for a *rectangular member made of an elastoplastic material*.

You should keep in mind that the distribution of *strain* across the section remains linear after the onset of yield. Therefore, Eq. (4.8) of Sec. 4.3 remains valid and can be used to determine the half-thickness y_Y of the elastic core. We have

$$y_Y = \epsilon_Y \rho \tag{4.40}$$

where ϵ_Y is the yield strain and ρ the radius of curvature corresponding to a bending moment $M \geq M_Y$. When the bending moment is equal to M_Y, we have $y_Y = c$ and Eq. (4.40) yields

$$c = \epsilon_Y \rho_Y \tag{4.41}$$

where ρ_Y is the radius of curvature corresponding to the maximum elastic moment M_Y. Dividing (4.40) by (4.41) member by member, we obtain the relation[†]

$$\frac{y_Y}{c} = \frac{\rho}{\rho_Y} \tag{4.42}$$

Substituting for y_Y/c from (4.42) into Eq. (4.38), we express the bending moment M as a function of the radius of curvature ρ of the neutral surface:

$$M = \frac{3}{2}M_Y\left(1 - \frac{1}{3}\frac{\rho^2}{\rho_Y^2}\right) \tag{4.43}$$

Note that Eq. (4.43) is valid only after the onset of yield, i.e., for values of M larger than M_Y. For $M < M_Y$, Eq. (4.21) of Sec. 4.4 should be used.

We observe from Eq. (4.43) that the bending moment reaches the value $M_p = \frac{3}{2}M_Y$ only when $\rho = 0$. Since we clearly cannot have a zero radius of curvature at every point of the neutral surface, we conclude that a fully plastic deformation cannot develop in pure bending. As you will see in Chap. 5, however, such a situation may occur at one point in the case of a beam under a transverse loading.

The stress distributions in a rectangular member corresponding respectively to the maximum elastic moment M_Y and to the limiting case of the plastic moment M_p have been represented in three dimensions in Fig. 4.37. Since, in both cases, the resultants of the elementary tensile and compressive forces must pass through the centroids of the volumes representing the stress distributions and be equal in magnitude to these volumes, we check that

$$R_Y = \tfrac{1}{2}bc\sigma_Y$$

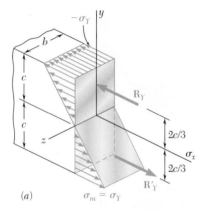

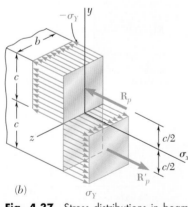

Fig. 4.37 Stress distributions in beam at maximum elastic moment and at plastic moment.

[†]Equation (4.42) applies to any member made of any ductile material with a well-defined yield point, since its derivation is independent of the shape of the cross section and of the shape of the stress-strain diagram beyond the yield point.

4.94 A solid bar of rectangular cross section is made of a material that is assumed to be elastoplastic. Denoting by M_Y and ρ_Y, respectively, the bending moment and radius of curvature at the onset of yield, determine (a) the radius of curvature when a couple of moment $M = 1.25\,M_Y$ is applied to the bar, (b) the radius of curvature after the couple is removed. Check the results obtained by using the relation derived in Prob. 4.93.

4.95 The prismatic bar shown is made of a steel that is assumed to be elastoplastic and for which $E = 200$ GPa. Knowing that the radius of curvature of the bar is 2.4 m when a couple of moment $M = 420$ N $\cdot$ m is applied as shown, determine (a) the yield strength σ_Y of the steel, (b) the thickness of the elastic core of the bar.

Fig. P4.95

4.96 The prismatic bar AB is made of an aluminum alloy for which the tensile stress-strain diagram is as shown. Assuming that the σ-ϵ diagram is the same in compression as in tension, determine (a) the radius of curvature of the bar when the maximum stress is 250 MPa, (b) the corresponding value of the bending moment. (*Hint:* For part b, plot σ versus y and use an approximate method of integration.)

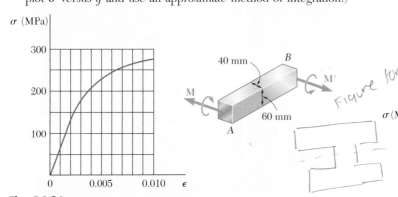

Figure for 4.90

Fig. P4.96

4.97 The prismatic bar AB is made of a bronze alloy for which the tensile stress-strain diagram is as shown. Assuming that the σ-ϵ diagram is the same in compression as in tension, determine (a) the maximum stress in the bar when the radius of curvature of the bar is 2.5 m, (b) the corresponding value of the bending moment. (See hint given in Prob. 4.96.)

Fig. P4.97

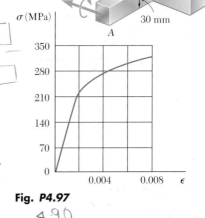

4.98 A prismatic bar of rectangular cross section is made of an alloy for which the stress-strain diagram can be represented by the relation $\epsilon = k\sigma^n$ for $\sigma > 0$ and $\epsilon = -|k\sigma^n|$ for $\sigma < 0$. If a couple $\mathbf{M}$ is applied to the bar, show that the maximum stress is

$$\sigma_m = \frac{1 + 2n}{3n}\frac{Mc}{I}$$

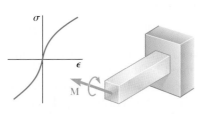

Fig. P4.98

4.90

$\sigma_Y = 290$ MPa

$M_P = \sigma_Y\, \Sigma_i y_i^{} A_i^{}$

$M_P = (290)\,[\,2(25)(75)(37.5) + 2(25)(25)(12.5)\,]$

4.53×10^7 Nmm

* would have to recalculate N.A for elastic unloading to find I

then $\sigma'\, = \dfrac{M\rho_Y}{I}$

a) $y = 25$ b) $y = 50$

$\boxed{\sigma_{res} = \sigma' - \sigma_Y}$

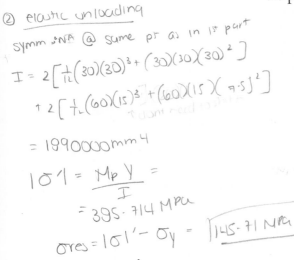

4.98

$M_P = \sigma_Y \Sigma y_i A_i$

① Plastic loading

$M_P = [240][2(30)(30)(30) + (2)(60)(15)(7.5)]$

$\boxed{1.62 \times 10^7 \text{ N·mm}}$

② elastic unloading

symm $\therefore$ NA @ same pt as in 1st part

$I = 2\left[\frac{1}{12}(30)(30)^3 + (30)(30)(30)^2\right]$
$+ 2\left[\frac{1}{12}(60)(15)^3 + (60)(15)(7.5)^2\right]$

I dont need to...

$= 1990000 \text{ mm}^4$

$|\sigma'| = \frac{M_P y}{I} =$
$= 395.714 \text{ MPa}$

$\sigma_{res} = |\sigma'| - \sigma_Y = \boxed{145.71 \text{ MPa}}$

(a)

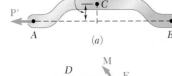

(b)

Fig. 4.42 Member with eccentric loading.

(a)

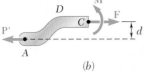

(b)

Fig. 4.43 Internal forces in member with eccentric loading.

4.12 ECCENTRIC AXIAL LOADING IN A PLANE OF SYMMETRY

We saw in Sec. 1.5 that the distribution of stresses in the cross section of a member under axial loading can be assumed uniform only if the line of action of the loads **P** and **P′** passes through the centroid of the cross section. Such a loading is said to be *centric*. Let us now analyze the distribution of stresses when the line of action of the loads does *not* pass through the centroid of the cross section, i.e., when the loading is *eccentric*.

Two examples of an eccentric loading are shown in Photos 4.5 and 4.6. In the case of the walkway light, the weight of the lamp causes an eccentric loading on the post. Likewise, the vertical forces exerted on the press cause an eccentric loading on the back column of the press.

Photo 4.5

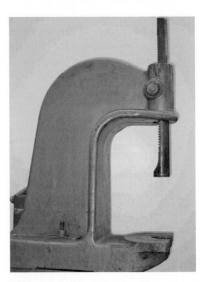

Photo 4.6

In this section, our analysis will be limited to members that possess a plane of symmetry, and it will be assumed that the loads are applied in the plane of symmetry of the member (Fig. 4.42*a*). The internal forces acting on a given cross section may then be represented by a force **F** applied at the centroid *C* of the section and a couple **M** acting in the plane of symmetry of the member (Fig. 4.42*b*). The conditions of equilibrium of the free body *AC* require that the force **F** be equal and opposite to **P′** and that the moment of the couple **M** be equal and opposite to the moment of **P′** about *C*. Denoting by *d* the distance from the centroid *C* to the line of action *AB* of the forces **P** and **P′**, we have

$$F = P \qquad \text{and} \qquad M = Pd \qquad (4.49)$$

We now observe that the internal forces in the section would have been represented by the same force and couple if the straight portion *DE* of member *AB* had been detached from *AB* and subjected simultaneously to the centric loads **P** and **P′** and to the bending couples **M** and **M′** (Fig. 4.43). Thus, the stress distribution due to the original eccentric loading can be obtained by superposing the uniform stress distribution corresponding to the centric loads **P** and **P′** and the linear distribution corresponding to the bending couples **M** and **M′** (Fig. 4.44). We write

$$\sigma_x = (\sigma_x)_{\text{centric}} + (\sigma_x)_{\text{bending}}$$

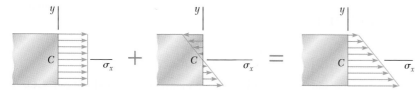

Fig. 4.44 Stress distribution—eccentric loading.

or, recalling Eqs. (1.5) and (4.16):

$$\sigma_x = \frac{P}{A} - \frac{My}{I} \tag{4.50}$$

where A is the area of the cross section and I its centroidal moment of inertia, and where y is measured from the centroidal axis of the cross section. The relation obtained shows that the distribution of stresses across the section is *linear but not uniform*. Depending upon the geometry of the cross section and the eccentricity of the load, the combined stresses may all have the same sign, as shown in Fig. 4.44, or some may be positive and others negative, as shown in Fig. 4.45. In the latter case, there will be a line in the section, along which $\sigma_x = 0$. This line represents the *neutral axis* of the section. We note that the neutral axis does *not* coincide with the centroidal axis of the section, since $\sigma_x \neq 0$ for $y = 0$.

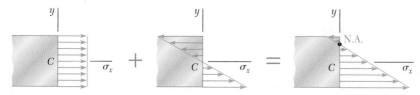

Fig. 4.45 Alternative stress distribution—eccentric loading.

The results obtained are valid only to the extent that the conditions of applicability of the superposition principle (Sec. 2.12) and of Saint-Venant's principle (Sec. 2.17) are met. This means that the stresses involved must not exceed the proportional limit of the material, that the deformations due to bending must not appreciably affect the distance d in Fig. 4.42a, and that the cross section where the stresses are computed must not be too close to points D or E in the same figure. The first of these requirements clearly shows that the superposition method cannot be applied to plastic deformations.

EXAMPLE 4.07

An open-link chain is obtained by bending low-carbon steel rods of 12-mm diameter into the shape shown (Fig. 4.46). Knowing that the chain carries a load of 700 N, determine (a) the largest tensile and compressive stresses in the straight portion of a link, (b) the distance between the centroidal and the neutral axis of a cross section.

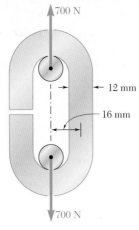

Fig. 4.46

(a) Largest Tensile and Compressive Stresses. The internal forces in the cross section are equivalent to a centric force **P** and a bending couple **M** (Fig. 4.47) of magnitudes

$$P = 700 \text{ N}$$
$$M = Pd = (700 \text{ N})(0.016 \text{ m}) = 11.2 \text{ N} \cdot \text{m}$$

The corresponding stress distributions are shown in parts a and b of Fig. 4.48. The distribution due to the centric force **P** is uniform and equal to $\sigma_0 = P/A$. We have

$$A = \pi c^2 = \pi (6 \text{ mm})^2 = 113.1 \text{ mm}^2$$
$$\sigma_0 = \frac{P}{A} = \frac{700 \text{ N}}{113.1 \text{ mm}^2} = 6.189 \text{ MPa}$$

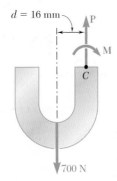

Fig. 4.47

The distribution due to the bending couple **M** is linear with a maximum stress $\sigma_m = Mc/I$. We write

$$I = \tfrac{1}{4}\pi c^4 = \tfrac{1}{4}\pi (6 \text{ mm})^4 = 1017.9 \text{ mm}^4$$
$$\sigma_m = \frac{Mc}{I} = \frac{(11.2 \text{ N} \cdot \text{m})(6 \times 10^{-3} \text{ m})}{1017.9 \times 10^{-12} \text{ m}^4} = 66.02 \text{ MPa}$$

Fig. 4.48

Superposing the two distributions, we obtain the stress distribution corresponding to the given eccentric loading (Fig. 4.48c). The largest tensile and compressive stresses in the section are found to be, respectively,

$$\sigma_t = \sigma_0 + \sigma_m = 6.189 + 66.02 = 72.2 \text{ MPa}$$
$$\sigma_c = \sigma_0 - \sigma_m = 6.189 - 66.02 = -59.8 \text{ MPa}$$

(b) Distance Between Centroidal and Neutral Axes. The distance y_0 from the centroidal to the neutral axis of the section is obtained by setting $\sigma_x = 0$ in Eq. (4.50) and solving for y_0:

$$0 = \frac{P}{A} - \frac{My_0}{I}$$
$$y_0 = \left(\frac{P}{A}\right)\left(\frac{I}{M}\right) = (6.189 \times 10^6 \text{ Pa})\frac{(1017.9 \times 10^{-12} \text{ m}^4)}{11.2 \text{ N} \cdot \text{m}}$$
$$y_0 = 5.62 \times 10^{-4} \text{ m} = 0.562 \text{ mm}$$

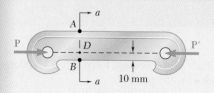

Knowing that for the cast iron link shown the allowable stresses are 30 MPa in tension and 120 MPa in compression, determine the largest force **P** which can be applied to the link. (*Note*: The T-shaped cross section of the link has previously been considered in Sample Prob. 4.2.)

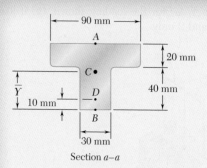

Section a–a

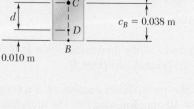

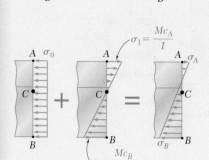

(1)   (2)   (3)

SOLUTION

Properties of Cross Section. From Sample Prob. 4.2, we have

$$A = 3000 \text{ mm}^2 = 3 \times 10^{-3} \text{ m}^2 \qquad \overline{Y} = 38 \text{ mm} = 0.038 \text{ m}$$
$$I = 868 \times 10^{-9} \text{ m}^4$$

We now write: $d = (0.038 \text{ m}) - (0.010 \text{ m}) = 0.028 \text{ m}$

Force and Couple at C. We replace **P** by an equivalent force-couple system at the centroid C.

$$P = P \qquad M = P(d) = P(0.028 \text{ m}) = 0.028P$$

The force **P** acting at the centroid causes a uniform stress distribution (Fig. 1). The bending couple **M** causes a linear stress distribution (Fig. 2).

$$\sigma_0 = \frac{P}{A} = \frac{P}{3 \times 10^{-3}} = 333P \qquad \text{(Compression)}$$

$$\sigma_1 = \frac{Mc_A}{I} = \frac{(0.028P)(0.022)}{868 \times 10^{-9}} = 710P \qquad \text{(Tension)}$$

$$\sigma_2 = \frac{Mc_B}{I} = \frac{(0.028P)(0.038)}{868 \times 10^{-9}} = 1226P \qquad \text{(Compression)}$$

Superposition. The total stress distribution (Fig. 3) is found by superposing the stress distributions caused by the centric force **P** and by the couple **M.** Since tension is positive, and compression negative, we have

$$\sigma_A = -\frac{P}{A} + \frac{Mc_A}{I} = -333P + 710P = +377P \qquad \text{(Tension)}$$

$$\sigma_B = -\frac{P}{A} - \frac{Mc_B}{I} = -333P - 1226P = -1559P \qquad \text{(Compression)}$$

Largest Allowable Force. The magnitude of **P** for which the tensile stress at point A is equal to the allowable tensile stress of 30 MPa is found by writing

$$\sigma_A = 377P = 30 \text{ MPa} \qquad P = 79.6 \text{ kN} \quad \blacktriangleleft$$

We also determine the magnitude of **P** for which the stress at B is equal to the allowable compressive stress of 120 MPa.

$$\sigma_B = -1559P = -120 \text{ MPa} \qquad P = 77.0 \text{ kN} \quad \blacktriangleleft$$

The magnitude of the largest force **P** that can be applied without exceeding either of the allowable stresses is the smaller of the two values we have found.

$$P = 77.0 \text{ kN} \quad \blacktriangleleft$$

PROBLEMS

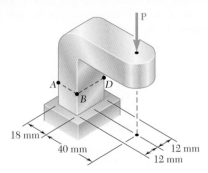

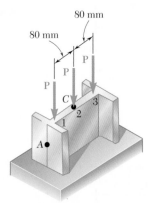

Fig. P4.99

4.99 Knowing that the magnitude of the vertical force **P** is 2 kN, determine the stress at (*a*) point *A*, (*b*) point *B*.

4.100 As many as three axial loads, each of magnitude $P = 50$ kN, can be applied to the end of a W200 × 31.1 rolled-steel shape. Determine the stress at point *A*, (*a*) for the loading shown, (*b*) if loads are applied at points 1 and 2 only.

Fig. P4.100

4.101 Knowing that the magnitude of the horizontal force **P** is 8 kN, determine the stress at (*a*) point *A*, (*b*) point *B*.

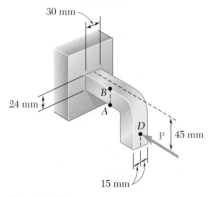

Fig. P4.101

4.102 The vertical portion of the press shown consists of a rectangular tube of wall thickness $t = 10$ mm. Knowing that the press has been tightened on wooden planks being glued together until $P = 20$ kN, determine the stress at (*a*) point *A*, (*b*) point *B*.

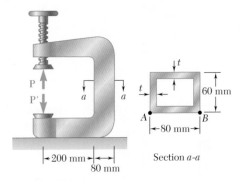

Fig. P4.102

4.103 Solve Prob. 4.102, assuming that $t = 8$ mm.

4.104 Determine the stress at points *A* and *B*, (*a*) for the loading shown, (*b*) if the 60-kN loads are applied at points 1 and 2 only.

4.105 Determine the stress at points *A* and *B*, (*a*) for the loading shown, (*b*) if the 60-kN loads applied at points 2 and 3 are removed.

Fig. P4.104 and P4.105

4.106 Portions of a 12 × 12-mm square bar have been bent to form the two machine components shown. Knowing that the allowable stress is 105 MPa, determine the maximum load that can be applied to each component.

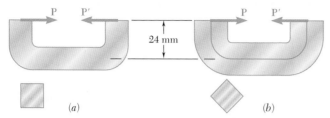

24 mm

(a)   (b)

Fig. P4.106

4.107 The four forces shown are applied to a rigid plate supported by a solid steel post of radius a. Knowing that $P = 100$ kN and $a = 40$ mm, determine the maximum stress in the post when (a) the force at D is removed, (b) the forces at C and D are removed.

Fig. P4.107

4.108 A milling operation was used to remove a portion of a solid bar of square cross section. Knowing that $a = 30$ mm, $d = 20$ mm, and $\sigma_{all} = 60$ MPa, determine the magnitude P of the largest forces that can be safely applied at the centers of the ends of the bar.

Fig. P4.108 and P4.109

4.109 A milling operation was used to remove a portion of a solid bar of square cross section. Forces of magnitude $P = 18$ kN are applied at the centers of the ends of the bar. Knowing that $a = 30$ mm and $\sigma_{all} = 135$ MPa, determine the smallest allowable depth d of the milled portion of the bar.

4.110 A short column is made by nailing four 25 × 100-mm planks to a 100 × 100-mm timber. Determine the largest compressive stress created in the column by a 64-kN load applied as shown in the center of the top section of the timber if (a) the column is as described, (b) plank 1 is removed, (c) planks 1 and 2 are removed, (d) planks 1, 2, and 3 are removed, (e) all planks are removed.

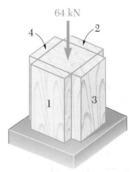

64 kN

Fig. P4.110

4.111 An offset h must be introduced into a solid circular rod of diameter d. Knowing that the maximum stress after the offset is introduced must not exceed 5 times the stress in the rod when it is straight, determine the largest offset that can be used.

4.112 An offset h must be introduced into a metal tube of 18-mm outer diameter and 2-mm wall thickness. Knowing that the maximum stress after the offset is introduced must not exceed 4 times the stress in the tube when it is straight, determine the largest offset that can be used.

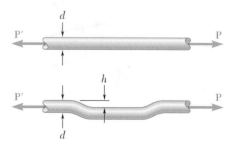

Fig. P4.111 and P4.112

4.113 A steel rod is welded to a steel plate to form the machine element shown. Knowing that the allowable stress is 135 MPa, determine (a) the largest force **P** that can be applied to the element, (b) the corresponding location of the neutral axis. (*Given*: The centroid of the cross section is at C and $I_z = 4195$ mm^4.)

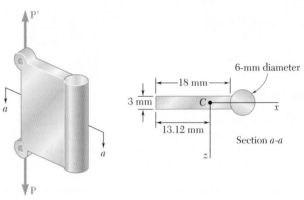

Fig. P4.113

4.114 Knowing that the allowable stress is 150 MPa in section *a-a* of the hanger shown, determine (a) the largest vertical force **P** that can be applied at point A, (b) the corresponding location of the neutral axis of section *a-a*.

4.115 Solve Prob. 4.114, assuming that the vertical force **P** is applied at point B.

4.116 Three steel plates, each of 25 × 150-mm cross section, are welded together to form a short H-shaped column. Later, for architectural reasons, a 25-mm strip is removed from each side of one of the flanges. Knowing that the load remains centric with respect to the original cross section and that the allowable stress is 100 MPa, determine the largest force **P** (a) that could be applied to the original column, (b) that can be applied to the modified column.

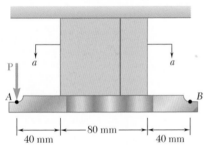

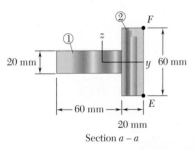

Section *a – a*

Fig. P4.114

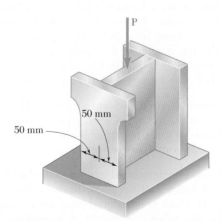

Fig. P4.116

derived in Secs. 4.3 and 4.4 for symmetric members can be used to determine the stresses in this case as well.

As you will see presently, the principle of superposition can be used to determine stresses in the most general case of unsymmetric bending. Consider first a member with a vertical plane of symmetry, which is subjected to bending couples $\mathbf{M}$ and $\mathbf{M}'$ acting in a plane forming an angle θ with the vertical plane (Fig. 4.54). The couple

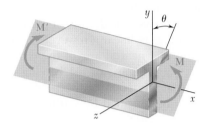

Fig. 4.54 Unsymmetric bending.

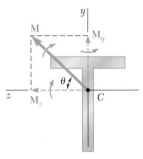

Fig. 4.55

vector $\mathbf{M}$ representing the forces acting on a given cross section will form the same angle θ with the horizontal z axis (Fig. 4.55). Resolving the vector $\mathbf{M}$ into component vectors $\mathbf{M}_z$ and $\mathbf{M}_y$ along the z and y axes, respectively, we write

$$M_z = M \cos \theta \qquad M_y = M \sin \theta \qquad (4.52)$$

Since the y and z axes are the principal centroidal axes of the cross section, we can use Eq. (4.16) to determine the stresses resulting from the application of either of the couples represented by $\mathbf{M}_z$ and $\mathbf{M}_y$. The couple $\mathbf{M}_z$ acts in a vertical plane and bends the member in that plane (Fig. 4.56). The resulting stresses are

$$\sigma_x = -\frac{M_z y}{I_z} \qquad (4.53)$$

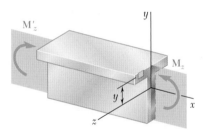

Fig. 4.56

where I_z is the moment of inertia of the section about the principal centroidal z axis. The negative sign is due to the fact that we have compression above the xz plane ($y > 0$) and tension below ($y < 0$). On the other hand, the couple $\mathbf{M}_y$ acts in a horizontal plane and bends the member in that plane (Fig. 4.57). The resulting stresses are found to be

$$\sigma_x = +\frac{M_y z}{I_y} \qquad (4.54)$$

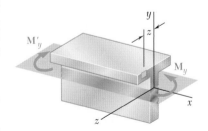

Fig. 4.57

where I_y is the moment of inertia of the section about the principal centroidal y axis, and where the positive sign is due to the fact that we have tension to the left of the vertical xy plane ($z > 0$) and compression to its right ($z < 0$). The distribution of the stresses caused by the original couple $\mathbf{M}$ is obtained by superposing the stress distributions defined by Eqs. (4.53) and (4.54), respectively. We have

$$\sigma_x = -\frac{M_z y}{I_z} + \frac{M_y z}{I_y} \qquad (4.55)$$

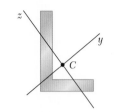

Fig. 4.58 Unsymmetric cross section.

We note that the expression obtained can also be used to compute the stresses in an unsymmetric section, such as the one shown in Fig. 4.58, once the principal centroidal y and z axes have been determined. On the other hand, Eq. (4.55) is valid only if the conditions of applicability of the principle of superposition are met. In other words, it should not be used if the combined stresses exceed the proportional limit of the material, or if the deformations caused by one of the component couples appreciably affect the distribution of the stresses due to the other.

Equation (4.55) shows that the distribution of stresses caused by unsymmetric bending is linear. However, as we have indicated earlier in this section, the neutral axis of the cross section will not, in general, coincide with the axis of the bending couple. Since the normal stress is zero at any point of the neutral axis, the equation defining that axis can be obtained by setting $\sigma_x = 0$ in Eq. (4.55). We write

$$-\frac{M_z y}{I_z} + \frac{M_y z}{I_y} = 0$$

or, solving for y and substituting for M_z and M_y from Eqs. (4.52),

$$y = \left(\frac{I_z}{I_y} \tan \theta\right) z \tag{4.56}$$

The equation obtained is that of a straight line of slope $m = (I_z/I_y) \tan \theta$. Thus, the angle ϕ that the neutral axis forms with the z axis (Fig. 4.59) is defined by the relation

$$\tan \phi = \frac{I_z}{I_y} \tan \theta \tag{4.57}$$

where θ is the angle that the couple vector **M** forms with the same axis. Since I_z and I_y are both positive, ϕ and θ have the same sign. Furthermore, we note that $\phi > \theta$ when $I_z > I_y$, and $\phi < \theta$ when $I_z < I_y$. Thus, the neutral axis is always located between the couple vector **M** and the principal axis corresponding to the minimum moment of inertia.

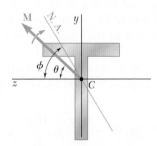

Fig. 4.59

EXAMPLE 4.08

A 180-N · m couple is applied to a wooden beam, of rectangular cross section 40 by 90 mm, in a plane forming an angle of 30° with the vertical (Fig. 4.60). Determine (a) the maximum stress in the beam, (b) the angle that the neutral surface forms with the horizontal plane.

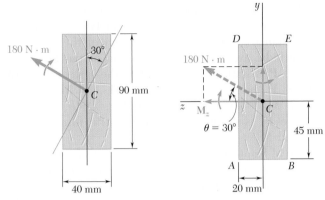

Fig. 4.60   **Fig. 4.61**

(a) **Maximum Stress.** The components $\mathbf{M}_z$ and $\mathbf{M}_y$ of the couple vector are first determined (Fig. 4.61):

$$M_z = (180 \text{ N} \cdot \text{m}) \cos 30° = 155.9 \text{ N} \cdot \text{m}$$
$$M_y = (180 \text{ N} \cdot \text{m}) \sin 30° = 90 \text{ N} \cdot \text{m}$$

We also compute the moments of inertia of the cross section with respect to the z and y axes:

$$I_z = \tfrac{1}{12}(0.04 \text{ m})(0.09 \text{ m})^3 = 2.43 \times 10^{-6} \text{ m}^4$$
$$I_y = \tfrac{1}{12}(0.09 \text{ m})(0.04 \text{ m})^3 = 0.48 \times 10^{-6} \text{ m}^4$$

The largest tensile stress due to $\mathbf{M}_z$ occurs along AB and is

$$\sigma_1 = \frac{M_z y}{I_z} = \frac{(155.9 \text{ N} \cdot \text{m})(0.045 \text{ m})}{2.43 \times 10^{-6} \text{ m}^4} = 2.887 \text{ MPa}$$

The largest tensile stress due to $\mathbf{M}_y$ occurs along AD and is

$$\sigma_2 = \frac{M_y z}{I_y} = \frac{(90 \text{ N} \cdot \text{m})(0.02 \text{ m})}{0.48 \times 10^{-6} \text{ m}^4} = 3.75 \text{ MPa}$$

The largest tensile stress due to the combined loading, therefore, occurs at A and is

$$\sigma_{\max} = \sigma_1 + \sigma_2 = 2.887 + 3.75 = 6.637 \text{ MPa}$$
$$\sigma_{\max} = 6.64 \text{ MPa}$$

The largest compressive stress has the same magnitude and occurs at E.

(b) **Angle of Neutral Surface with Horizontal Plane.** The angle ϕ that the neutral surface forms with the horizontal plane (Fig. 4.62) is obtained from Eq. (4.57):

$$\tan \phi = \frac{I_z}{I_y} \tan \theta = \frac{2.43 \times 10^{-6} \text{ m}^4}{0.48 \times 10^{-6} \text{ m}^4} \tan 30° = 2.923$$
$$\phi = 71.1°$$

The distribution of the stresses across the section is shown in Fig. 4.63.

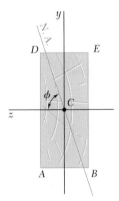

Fig. 4.62

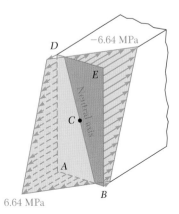

Fig. 4.63

305

4.14 GENERAL CASE OF ECCENTRIC AXIAL LOADING

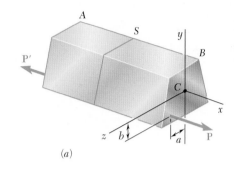

(a)

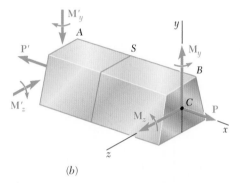

(b)

Fig. 4.64 Eccentric axial loading.

In Sec. 4.12 you analyzed the stresses produced in a member by an eccentric axial load applied in a plane of symmetry of the member. You will now study the more general case when the axial load is not applied in a plane of symmetry.

Consider a straight member AB subjected to equal and opposite eccentric axial forces **P** and **P′** (Fig. 4.64a), and let a and b denote the distances from the line of action of the forces to the principal centroidal axes of the cross section of the member. The eccentric force **P** is statically equivalent to the system consisting of a centric force **P** and of the two couples **M**$_y$ and **M**$_z$ of moments $M_y = Pa$ and $M_z = Pb$ represented in Fig. 4.64b. Similarly, the eccentric force **P′** is equivalent to the centric force **P′** and the couples **M′**$_y$ and **M′**$_z$.

By virtue of Saint-Venant's principle (Sec. 2.17), we can replace the original loading of Fig. 4.64a by the statically equivalent loading of Fig. 4.64b in order to determine the distribution of stresses in a section S of the member, as long as that section is not too close to either end of the member. Furthermore, the stresses due to the loading of Fig. 4.64b can be obtained by superposing the stresses corresponding to the centric axial load **P** and to the bending couples **M**$_y$ and **M**$_z$, as long as the conditions of applicability of the principle of superposition are satisfied (Sec. 2.12). The stresses due to the centric load **P** are given by Eq. (1.5), and the stresses due to the bending couples by Eq. (4.55), since the corresponding couple vectors are directed along the principal centroidal axes of the section. We write, therefore,

$$\sigma_x = \frac{P}{A} - \frac{M_z y}{I_z} + \frac{M_y z}{I_y} \tag{4.58}$$

where y and z are measured from the principal centroidal axes of the section. The relation obtained shows that the distribution of stresses across the section is *linear*.

In computing the combined stress σ_x from Eq. (4.58), care should be taken to correctly determine the sign of each of the three terms in the right-hand member, since each of these terms can be positive or negative, depending upon the sense of the loads **P** and **P′** and the location of their line of action with respect to the principal centroidal axes of the cross section. Depending upon the geometry of the cross section and the location of the line of action of **P** and **P′**, the combined stresses σ_x obtained from Eq. (4.58) at various points of the section may all have the same sign, or some may be positive and others negative. In the latter case, there will be a line in the section, along which the stresses are zero. Setting $\sigma_x = 0$ in Eq. (4.58), we obtain the equation of a straight line, which represents the *neutral axis* of the section:

$$\frac{M_z}{I_z} y - \frac{M_y}{I_y} z = \frac{P}{A}$$

*4.141 through *4.143   The couple **M** acts in a vertical plane and is applied to a beam oriented as shown. Determine the stress at point A.

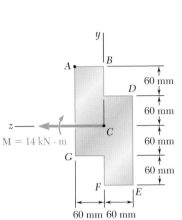

Fig. P4.141

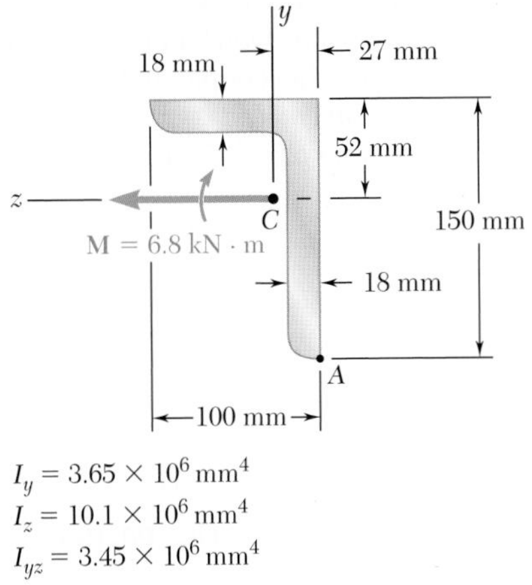

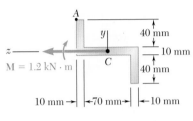

$$I_y = 3.65 \times 10^6 \text{ mm}^4$$
$$I_z = 10.1 \times 10^6 \text{ mm}^4$$
$$I_{yz} = 3.45 \times 10^6 \text{ mm}^4$$

Fig. P4.142

$$I_y = 1.894 \times 10^6 \text{ mm}^4$$
$$I_z = 0.614 \times 10^6 \text{ mm}^4$$
$$I_{yz} = +0.800 \times 10^6 \text{ mm}^4$$

Fig. P4.143

4.144   The tube shown has a uniform wall thickness of 12 mm. For the loading given, determine (a) the stress at points A and B, (b) the point where the neutral axis intersects line ABD.

4.145   Solve Prob. 4.144, assuming that the 28-kN force at point E is removed.

4.146   A rigid circular plate of 125-mm radius is attached to a solid 150 × 200-mm rectangular post, with the center of the plate directly above the center of the post. If a 4-kN force **P** is applied at E with θ = 30°, determine (a) the stress at point A, (b) the stress at point B, (c) the point where the neutral axis intersects line ABD.

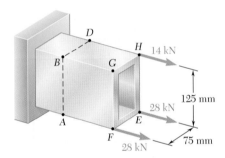

Fig. P4.144

4.147   In Prob. 4.146, determine (a) the value of θ for which the stress at D reaches its largest value, (b) the corresponding values of the stress at A, B, C, and D.

4.148   An axial load **P** of magnitude 50 kN is applied as shown to a short section of a W150 × 24 rolled-steel member. Determine the largest distance a for which the maximum compressive stress does not exceed 90 MPa.

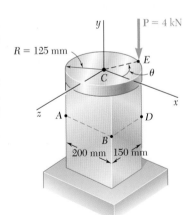

Fig. P4.146

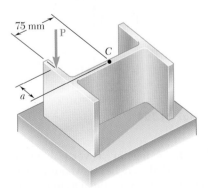

Fig. P4.148

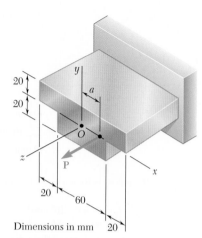

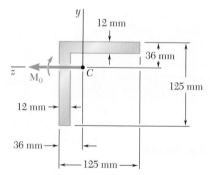

Dimensions in mm

Fig. P4.149

Fig. P4.152

4.149 A horizontal load **P** of magnitude 100 kN is applied to the beam shown. Determine the largest distance a for which the maximum tensile stress in the beam does not exceed 75 MPa.

4.150 The Z section shown is subjected to a couple $\mathbf{M}_0$ acting in a vertical plane. Determine the largest permissible value of the moment M_0 of the couple if the maximum stress is not to exceed 80 MPa. *Given:* $I_{max} = 2.28 \times 10^{-6}$ m^4, $I_{min} = 0.23 \times 10^{-6}$ m^4, principal axes 25.7° ⊿ and 64.3° ⊿.

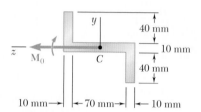

Fig. P4.150

4.151 Solve Prob. 4.150, assuming that the couple $\mathbf{M}_0$ acts in a horizontal plane.

4.152 A beam having the cross section shown is subjected to a couple $\mathbf{M}_0$ that acts in a vertical plane. Determine the largest permissible value of the moment M_0 of the couple if the maximum stress in the beam is not to exceed 84 MPa. *Given:* $I_y = I_z = 4.7 \times 10^6$ mm^4, $A = 3064$ mm^2, $k_{min} = 25$ mm. (*Hint:* By reason of symmetry, the principal axes form an angle of 45° with the coordinate axes. Use the relations $I_{min} = Ak_{min}^2$ and $I_{min} + I_{max} = I_y + I_z$.)

4.153 Solve Prob. 4.152, assuming that the couple $\mathbf{M}_0$ acts in a horizontal plane.

4.154 A beam having the cross section shown is subjected to a couple $\mathbf{M}_0$ acting in a vertical plane. Determine the largest permissible value of the moment M_0 of the couple if the maximum stress is not to exceed 100 MPa. (*Given:* $I_y = I_z = b^4/36$ and $I_{yz} = b^4/72$.)

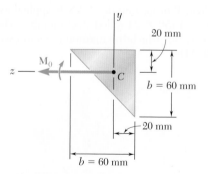

Fig. P4.154

Fig. P4.155

4.155 A couple $\mathbf{M}_0$ acting in a vertical plane is applied to a W310 × 23.8 rolled-steel beam, whose web forms an angle θ with the vertical. Denoting by σ_0 the maximum stress in the beam when $\theta = 0$, determine the angle of inclination θ of the beam for which the maximum stress is $2\sigma_0$.

4.156 Show that, if a solid rectangular beam is bent by a couple applied in a plane containing one diagonal of a rectangular cross section, the neutral axis will lie along the other diagonal.

4.157 A beam of unsymmetric cross section is subjected to a couple $\mathbf{M}_0$ acting in the horizontal plane xz. Show that the stress at point A, of coordinates y and z, is

$$\sigma_A = \frac{zI_z - yI_{yz}}{I_yI_z - I_{yz}^2}M_y$$

where I_y, I_z, and I_{yz} denote the moments and product of inertia of the cross section with respect to the coordinate axes, and M_y the moment of the couple.

4.158 A beam of unsymmetric cross section is subjected to a couple $\mathbf{M}_0$ acting in the vertical plane xy. Show that the stress at point A, of coordinates y and z, is

$$\sigma_A = -\frac{yI_y - zI_{yz}}{I_yI_z - I_{yz}^2}M_z$$

where I_y, I_z, and I_{yz} denote the moments and product of inertia of the cross section with respect to the coordinate axes, and M_z the moment of the couple.

4.159 (*a*) Show that, if a vertical force $\mathbf{P}$ is applied at point A of the section shown, the equation of the neutral axis BD is

$$\left(\frac{x_A}{r_z^2}\right)x + \left(\frac{z_A}{r_x^2}\right)z = -1$$

where r_z and r_x denote the radius of gyration of the cross section with respect to the z axis and the x axis, respectively. (*b*) Further show that, if a vertical force $\mathbf{Q}$ is applied at any point located on line BD, the stress at point A will be zero.

4.160 (*a*) Show that the stress at corner A of the prismatic member shown in Fig. P4.160a will be zero if the vertical force $\mathbf{P}$ is applied at a point located on the line

$$\frac{x}{b/6} + \frac{z}{h/6} = 1$$

(*b*) Further show that, if no tensile stress is to occur in the member, the force $\mathbf{P}$ must be applied at a point located within the area bounded by the line found in part a and three similar lines corresponding to the condition of zero stress at B, C, and D, respectively. This area, shown in Fig. P4.160b, is known as the *kern* of the cross section.

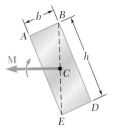

Fig. P4.156

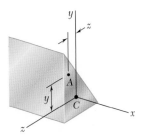

Fig. P4.157 and P4.158

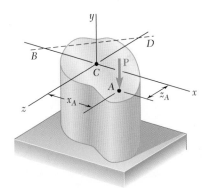

Fig. P4.159

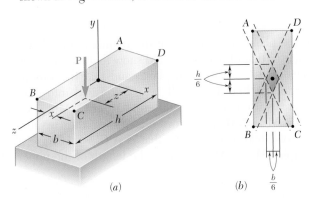

(*a*)

(*b*)

Fig. P4.160

*4.15 BENDING OF CURVED MEMBERS

Our analysis of stresses due to bending has been restricted so far to straight members. In this section we will consider the stresses caused by the application of equal and opposite couples to members that are initially curved. Our discussion will be limited to curved members of uniform cross section possessing a plane of symmetry in which the bending couples are applied, and it will be assumed that all stresses remain below the proportional limit.

If the initial curvature of the member is small, i.e., if its radius of curvature is large compared to the depth of its cross section, a good approximation can be obtained for the distribution of stresses by assuming the member to be straight and using the formulas derived in Secs. 4.3 and 4.4.† However, when the radius of curvature and the dimensions of the cross section of the member are of the same order of magnitude, we must use a different method of analysis, which was first introduced by the German engineer E. Winkler (1835–1888).

Consider the curved member of uniform cross section shown in Fig. 4.70. Its transverse section is symmetric with respect to the y axis (Fig. 4.70b) and, in its unstressed state, its upper and lower surfaces intersect the vertical xy plane along arcs of circle AB and FG centered at C (Fig. 4.70a). We now apply two equal and opposite couples $\mathbf{M}$

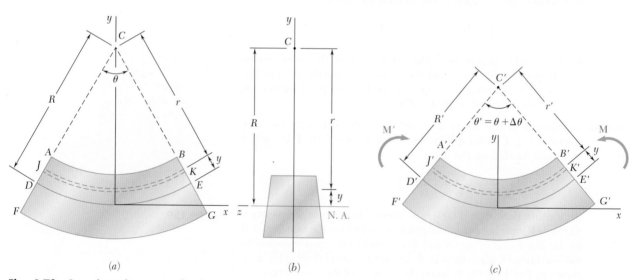

(a)   (b)   (c)

Fig. 4.70 Curved member in pure bending.

and $\mathbf{M}'$ in the plane of symmetry of the member (Fig. 4.70c). A reasoning similar to that of Sec. 4.3 would show that any transverse plane section containing C will remain plane, and that the various arcs of circle indicated in Fig. 4.70a will be transformed into circular and concentric arcs with a center C' different from C. More specifically, if the couples $\mathbf{M}$ and $\mathbf{M}'$ are directed as shown, the curvature of the various arcs of circle will increase; that is $A'C' < AC$. We also note that the couples $\mathbf{M}$ and $\mathbf{M}'$ will cause the length of the upper surface

†See Prob. 4.166.

of the member to decrease ($A'B' < AB$) and the length of the lower surface to increase ($F'G' > FG$). We conclude that a *neutral surface* must exist in the member, the length of which remains constant. The intersection of the neutral surface with the xy plane has been represented in Fig. 4.70a by the arc DE of radius R, and in Fig. 4.70c by the arc $D'E'$ of radius R'. Denoting by θ and θ' the central angles corresponding respectively to DE and $D'E'$, we express the fact that the length of the neutral surface remains constant by writing

$$R\theta = R'\theta' \tag{4.59}$$

Considering now the arc of circle JK located at a distance y above the neutral surface, and denoting respectively by r and r' the radius of this arc before and after the bending couples have been applied, we express the deformation of JK as

$$\delta = r'\theta' - r\theta \tag{4.60}$$

Observing from Fig. 4.70 that

$$r = R - y \qquad r' = R' - y \tag{4.61}$$

and substituting these expressions into Eq. (4.60), we write

$$\delta = (R' - y)\theta' - (R - y)\theta$$

or, recalling Eq. (4.59) and setting $\theta' - \theta = \Delta\theta$,

$$\delta = -y\,\Delta\theta \tag{4.62}$$

The normal strain ϵ_x in the elements of JK is obtained by dividing the deformation δ by the original length $r\theta$ of arc JK. We write

$$\epsilon_x = \frac{\delta}{r\theta} = -\frac{y\,\Delta\theta}{r\theta}$$

or, recalling the first of the relations (4.61),

$$\epsilon_x = -\frac{\Delta\theta}{\theta}\frac{y}{R - y} \tag{4.63}$$

The relation obtained shows that, while each transverse section remains plane, the normal strain ϵ_x *does not vary linearly* with the distance y from the neutral surface.

The normal stress σ_x can now be obtained from Hooke's law, $\sigma_x = E\epsilon_x$, by substituting for ϵ_x from Eq. (4.63). We have

$$\sigma_x = -\frac{E\,\Delta\theta}{\theta}\frac{y}{R - y} \tag{4.64}$$

or, alternatively, recalling the first of Eqs. (4.61),

$$\sigma_x = -\frac{E\,\Delta\theta}{\theta}\frac{R - r}{r} \tag{4.65}$$

Equation (4.64) shows that, like ϵ_x, the normal stress σ_x *does not vary linearly* with the distance y from the neutral surface. Plotting σ_x versus y, we obtain an arc of hyperbola (Fig. 4.71).

In order to determine the location of the neutral surface in the member and the value of the coefficient $E\,\Delta\theta/\theta$ used in Eqs. (4.64)

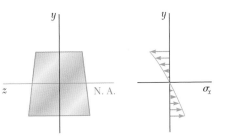

Fig. 4.71

and (4.65), we now recall that the elementary forces acting on any transverse section must be statically equivalent to the bending couple **M.** Expressing, as we did in Sec. 4.2 for a straight member, that the sum of the elementary forces acting on the section must be zero, and that the sum of their moments about the transverse z axis must be equal to the bending moment M, we write the equations

$$\int \sigma_x \, dA = 0 \qquad (4.1)$$

and

$$\int (-y\sigma_x \, dA) = M \qquad (4.3)$$

Substituting for σ_x from (4.65) into Eq. (4.1), we write

$$-\int \frac{E \, \Delta\theta}{\theta} \frac{R-r}{r} dA = 0$$

$$\int \frac{R-r}{r} dA = 0$$

$$R \int \frac{dA}{r} - \int dA = 0$$

from which it follows that the distance R from the center of curvature C to the neutral surface is defined by the relation

$$R = \frac{A}{\displaystyle\int \frac{dA}{r}} \qquad (4.66)$$

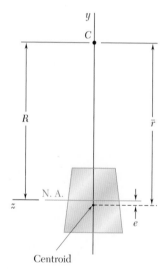

Fig. 4.72

We note that the value obtained for R is not equal to the distance $\bar{r}$ from C to the centroid of the cross section, since $\bar{r}$ is defined by a different relation, namely,

$$\bar{r} = \frac{1}{A} \int r \, dA \qquad (4.67)$$

We thus conclude that, in a curved member, *the neutral axis of a transverse section does not pass through the centroid of that section* (Fig. 4.72).† Expressions for the radius R of the neutral surface will be derived for some specific cross-sectional shapes in Example 4.10 and in Probs. 4.188 through 4.190. For convenience, these expressions are shown in Fig. 4.73.

Substituting now for σ_x from (4.65) into Eq. (4.3), we write

$$\int \frac{E \, \Delta\theta}{\theta} \frac{R-r}{r} y \, dA = M$$

†However, an interesting property of the neutral surface can be noted if we write Eq. (4.66) in the alternative form

$$\frac{1}{R} = \frac{1}{A} \int \frac{1}{r} dA \qquad (4.66')$$

Equation (4.66′) shows that, if the member is divided into a large number of fibers of cross-sectional area dA, the curvature $1/R$ of the neutral surface will be equal to the average value of the curvature $1/r$ of the various fibers.

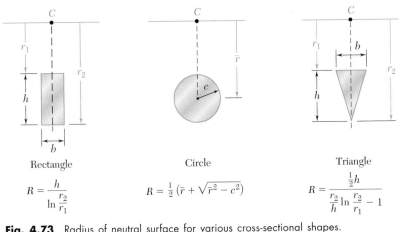
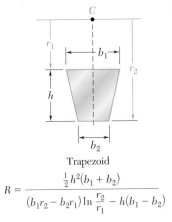

Rectangle	Circle	Triangle	Trapezoid
$R = \dfrac{h}{\ln \dfrac{r_2}{r_1}}$	$R = \dfrac{1}{2}\left(\bar{r} + \sqrt{\bar{r}^2 - c^2}\right)$	$R = \dfrac{\frac{1}{2}h}{\dfrac{r_2}{h}\ln \dfrac{r_2}{r_1} - 1}$	$R = \dfrac{\frac{1}{2}h^2(b_1 + b_2)}{(b_1 r_2 - b_2 r_1)\ln \dfrac{r_2}{r_1} - h(b_1 - b_2)}$

Fig. 4.73 Radius of neutral surface for various cross-sectional shapes.

or, since $y = R - r$,

$$\frac{E\,\Delta\theta}{\theta}\int \frac{(R-r)^2}{r}\,dA = M$$

Expanding the square in the integrand, we obtain after reductions

$$\frac{E\,\Delta\theta}{\theta}\left[R^2 \int \frac{dA}{r} - 2RA + \int r\,dA\right] = M$$

Recalling Eqs. (4.66) and (4.67), we note that the first term in the brackets is equal to RA, while the last term is equal to $\bar{r}A$. We have, therefore,

$$\frac{E\,\Delta\theta}{\theta}(RA - 2RA + \bar{r}A) = M$$

and, solving for $E\,\Delta\theta/\theta$,

$$\frac{E\,\Delta\theta}{\theta} = \frac{M}{A(\bar{r} - R)} \tag{4.68}$$

Referring to Fig. 4.70, we note that $\Delta\theta > 0$ for $M > 0$. It follows that $\bar{r} - R > 0$, or $R < \bar{r}$, regardless of the shape of the section. Thus, the neutral axis of a transverse section is always located between the centroid of the section and the center of curvature of the member (Fig. 4.72). Setting $\bar{r} - R = e$, we write Eq. (4.68) in the form

$$\frac{E\,\Delta\theta}{\theta} = \frac{M}{Ae} \tag{4.69}$$

Substituting now for $E\,\Delta\theta/\theta$ from (4.69) into Eqs. (4.64) and (4.65), we obtain the following alternative expressions for the normal stress σ_x in a curved beam:

$$\sigma_x = -\frac{My}{Ae(R-y)} \tag{4.70}$$

and

$$\sigma_x = \frac{M(r-R)}{Aer} \tag{4.71}$$

319

We should note that the parameter e in the previous equations is a small quantity obtained by subtracting two lengths of comparable size, R and $\bar{r}$. In order to determine σ_x with a reasonable degree of accuracy, it is therefore necessary to compute R and $\bar{r}$ very accurately, particularly when both of these quantities are large, i.e., when the curvature of the member is small. However, as we indicated earlier, it is possible in such a case to obtain a good approximation for σ_x by using the formula $\sigma_x = -My/I$ developed for straight members.

Let us now determine the change in curvature of the neutral surface caused by the bending moment M. Solving Eq. (4.59) for the curvature $1/R'$ of the neutral surface in the deformed member, we write

$$\frac{1}{R'} = \frac{1}{R}\frac{\theta'}{\theta}$$

or, setting $\theta' = \theta + \Delta\theta$ and recalling Eq. (4.69),

$$\frac{1}{R'} = \frac{1}{R}\left(1 + \frac{\Delta\theta}{\theta}\right) = \frac{1}{R}\left(1 + \frac{M}{EAe}\right)$$

from which it follows that the change in curvature of the neutral surface is

$$\frac{1}{R'} - \frac{1}{R} = \frac{M}{EAeR} \qquad (4.72)$$

EXAMPLE 4.10

A curved rectangular bar has a mean radius $\bar{r} = 150$ mm and a cross section of width $b = 60$ mm and depth $h = 36$ mm (Fig. 4.74). Determine the distance e between the centroid and the neutral axis of the cross section.

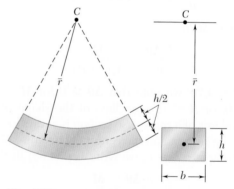

Fig. 4.74

We first derive the expression for the radius R of the neutral surface. Denoting by r_1 and r_2, respectively, the inner and outer radius of the bar (Fig. 4.75), we use Eq. (4.66) and write

$$R = \frac{A}{\displaystyle\int_{r_1}^{r_2}\frac{dA}{r}} = \frac{bh}{\displaystyle\int_{r_1}^{r_2}\frac{b\,dr}{r}} = \frac{h}{\displaystyle\int_{r_1}^{r_2}\frac{dr}{r}}$$

$$R = \frac{h}{\ln\dfrac{r_2}{r_1}} \qquad (4.73)$$

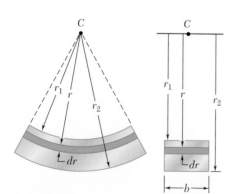

Fig. 4.75

For the given data, we have

$$r_1 = \bar{r} - \tfrac{1}{2}h = 150 - 18 = 132 \text{ mm}$$
$$r_2 = \bar{r} + \tfrac{1}{2}h = 150 + 18 = 168 \text{ mm}$$

Substituting for h, r_1, and r_2 into Eq. (4.73), we have

$$R = \frac{h}{\ln\dfrac{r_2}{r_1}} = \frac{36 \text{ mm}}{\ln\dfrac{168}{132}} = 149.277 \text{ mm}$$

The distance between the centroid and the neutral axis of the cross section (Fig. 4.76) is thus

$$e = \bar{r} - R = 150 - 149.277 = 0.723 \text{ mm}$$

We note that it was necessary to calculate R with six significant figures in order to obtain e with the usual degree of accuracy.

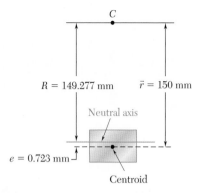

Fig. 4.76

EXAMPLE 4.11

For the bar of Example 4.10, determine the largest tensile and compressive stresses, knowing that the bending moment in the bar is $M = 900 \text{ N} \cdot \text{m}$.

We use Eq. (4.71) with the given data,

$$M = 900 \text{ N} \cdot \text{m} \qquad A = bh = (60 \text{ mm})(36 \text{ mm}) = 2160 \text{ mm}^2$$

and the values obtained in Example 4.10 for R and e,

$$R = 149.28 \text{ mm} \qquad e = 0.723 \text{ mm}$$

Making first $r = r_2 = 168 \text{ mm}$ in Eq. (4.71), we write

$$\sigma_{max} = \frac{M(r_2 - R)}{Aer_2}$$
$$= \frac{(900 \text{ N} \cdot \text{m})(0.168 \text{ m} - 0.14928 \text{ m})}{(2160 \times 10^{-6} \text{ m}^2)(0.723 \times 10^{-3} \text{ m})(0.168 \text{ m})}$$
$$\sigma_{max} = 64.2 \text{ MPa}$$

Making now $r = r_1 = 132 \text{ mm}$ in Eq. (4.71), we have

$$\sigma_{min} = \frac{M(r_1 - R)}{Aer_1}$$
$$= \frac{(900 \text{ N} \cdot \text{m})(0.132 \text{ m} - 0.14928 \text{ m})}{(2160 \times 10^{-6} \text{ m}^2)(0.723 \times 10^{-3} \text{ m})(0.132 \text{ m})}$$
$$\sigma_{min} = -75.4 \text{ MPa}$$

Remark. Let us compare the values obtained for σ_{max} and σ_{min} with the result we would get for a straight bar. Using Eq. (4.15) of Sec. 4.4, we write

$$\sigma_{max, min} = \pm\frac{Mc}{I}$$
$$= \pm\frac{(900 \text{ N} \cdot \text{m})(0.018 \text{ m})}{\tfrac{1}{12}(0.06 \text{ m})(0.036 \text{ m})^3} = \pm 69.4 \text{ MPa}$$

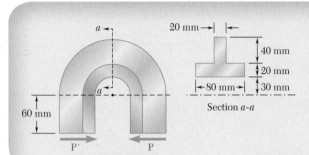

Section a-a

SAMPLE PROBLEM 4.11

A machine component has a T-shaped cross section and is loaded as shown. Knowing that the allowable compressive stress is 50 MPa, determine the largest force **P** that can be applied to the component.

SOLUTION

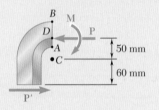

Centroid of the Cross Section. We locate the centroid D of the cross section

	A_i, mm²	$\bar{r}_i$, mm	$\bar{r}_i A_i$, mm³	$\bar{r}\Sigma A_i = \Sigma \bar{r}_i A_i$
1	$(20)(80) = 1600$	40	64×10^3	$\bar{r}(2400) = 120 \times 10^3$
2	$\underline{(40)(20) = \ \ 800}$	70	$\underline{56 \times 10^3}$	$\bar{r} = 50$ mm $= 0.050$ m
	$\Sigma A_i = 2400$		$\Sigma \bar{r}_i A_i = 120 \times 10^3$	

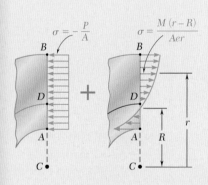

Force and Couple at D. The internal forces in section a-a are equivalent to a force **P** acting at D and a couple **M** of moment

$$M = P(50 \text{ mm} + 60 \text{ mm}) = (0.110 \text{ m})P$$

Superposition. The centric force **P** causes a uniform compressive stress on section a-a. The bending couple **M** causes a varying stress distribution [Eq. (4.71)]. We note that the couple **M** tends to increase the curvature of the member and is therefore positive (cf. Fig. 4.70). The total stress at a point of section a-a located at distance r from the center of curvature C is

$$\sigma = -\frac{P}{A} + \frac{M(r - R)}{Aer} \qquad (1)$$

$$\sigma = -\frac{P}{A} \qquad \sigma = \frac{M(r-R)}{Aer}$$

Radius of Neutral Surface. We now determine the radius R of the neutral surface by using Eq. (4.66).

$$R = \frac{A}{\displaystyle\int \frac{dA}{r}} = \frac{2400 \text{ mm}^2}{\displaystyle\int_{r_1}^{r_2} \frac{(80 \text{ mm}) \, dr}{r} + \int_{r_2}^{r_3} \frac{(20 \text{ mm}) \, dr}{r}}$$

$$= \frac{2400}{80 \ln \dfrac{50}{30} + 20 \ln \dfrac{90}{50}} = \frac{2400}{40.866 + 11.756} = 45.61 \text{ mm}$$

$$= 0.04561 \text{ m}$$

We also compute: $e = \bar{r} - R = 0.05000 \text{ m} - 0.04561 \text{ m} = 0.00439 \text{ m}$

Allowable Load. We observe that the largest compressive stress will occur at point A where $r = 0.030$ m. Recalling that $\sigma_{\text{all}} = 50$ MPa and using Eq. (1), we write

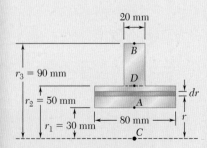

$$-50 \times 10^6 \text{ Pa} = -\frac{P}{2.4 \times 10^{-3} \text{ m}^2} + \frac{(0.110 \ P)(0.030 \text{ m} - 0.04561 \text{ m})}{(2.4 \times 10^{-3} \text{ m}^2)(0.00439 \text{ m})(0.030 \text{ m})}$$

$$-50 \times 10^6 = -417P - 5432P \qquad\qquad P = 8.55 \text{ kN} \blacktriangleleft$$

PROBLEMS

4.161 For the curved bar and loading shown, determine the stress at point A when (a) $r_1 = 30$ mm, (b) $r_1 = 50$ mm.

4.162 For the curved bar and loading shown, determine the stress at points A and B when $r_1 = 40$ mm.

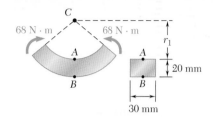

Fig. P4.161 and P4.162

4.163 The curved portion of the bar shown has an inner radius of 20 mm. Knowing that the allowable stress in the bar is 150 MPa, determine the largest permissible distance a from the line of action of the 3-kN force to the vertical plane containing the center of curvature of the bar.

4.164 The curved portion of the bar shown has an inner radius of 20 mm. Knowing that the line of action of the 3-kN force is located at a distance $a = 60$ mm from the vertical plane containing the center of curvature of the bar, determine the largest compressive stress in the bar.

4.165 The curved bar shown has a cross section of 40×60 mm and an inner radius $r_1 = 15$ mm. For the loading shown determine the largest tensile and compressive stresses.

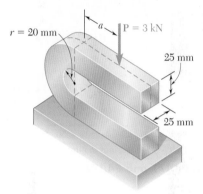

Fig. P4.163 and P4.164

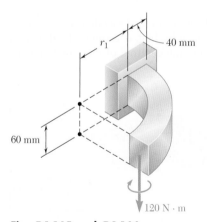

Fig. P4.165 and P4.166

4.166 For the curved bar and loading shown, determine the percent error introduced in the computation of the maximum stress by assuming that the bar is straight. Consider the case when (a) $r_1 = 20$ mm, (b) $r_1 = 200$ mm, (c) $r_1 = 2$ m.

4.167 The curved bar shown has a cross section of 30×30 mm. Knowing that $a = 60$ mm, determine the stress at (a) point A, (b) point B.

4.168 The curved bar shown has a cross section of 30×30 mm. Knowing that the allowable compressive stress is 175 MPa, determine the largest allowable distance a.

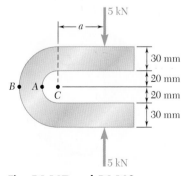

Fig. P4.167 and P4.168

4.169 Steel links having the cross section shown are available with different central angles β. Knowing that the allowable stress is 100 MPa, determine the largest force **P** that can be applied to a link for which $\beta = 90°$.

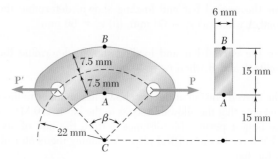

Fig. P4.169

4.170 Solve Prob. 4.169, assuming that $\beta = 60°$.

4.171 A machine component has a T-shaped cross section that is orientated as shown. Knowing that $M = 2.5$ kN · m, determine the stress at (a) point A, (b) point B.

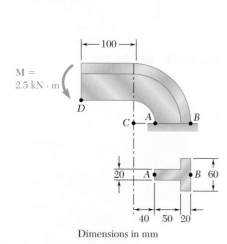

Dimensions in mm

Fig. P4.171 and P4.172

4.172 Assuming that the couple shown is replaced by a vertical 10-kN force attached at point D and acting downward, determine the stress at (a) point A, (b) point B.

4.173 Three plates are welded together to form the curved beam shown. For the given loading, determine the distance e between the neutral axis and the centroid of the cross section.

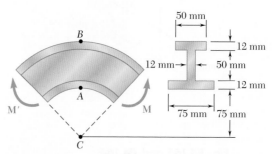

Fig. P4.173 and P4.174

4.174 Three plates are welded together to form the curved beam shown. For $M = 900$ N · m, determine the stress at (a) point A, (b) point B, (c) the centroid of the cross section.

4.175 The split ring shown has an inner radius $r_1 = 20$ mm and a circular cross section of diameter $d = 32$ mm. For the loading shown, determine the stress at (a) point A, (b) point B.

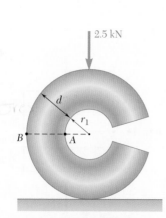

Fig. P4.175 and P4.176

4.176 The split ring shown has an inner radius $r_1 = 16$ mm and a circular cross section of diameter $d = 32$ mm. For the loading shown, determine the stress at (a) point A, (b) point B.

4.177 The curved bar shown has a circular cross section of 32-mm diameter. Determine the largest couple **M** that can be applied to the bar about a horizontal axis if the maximum stress is not to exceed 60 MPa.

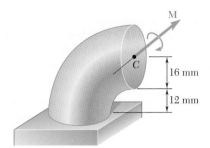

Fig. P4.177

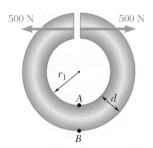

Fig. P4.178

4.178 The split ring shown has an inner radius $r_1 = 20$ mm and a *circular* cross section of diameter $d = 15$ mm. Knowing that each of the 500-N forces is applied at the centroid of the cross section, determine the stress at (*a*) point *A*, (*b*) point *B*.

4.179 Solve Prob. 4.178, assuming that the ring has an inner radius $r_1 = 15$ mm and a cross-sectional diameter $d = 20$ mm.

4.180 Knowing that $P = 10$ kN, determine the stress at (*a*) point *A*, (*b*) point *B*.

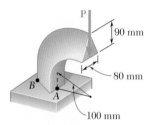

Fig. P4.180

4.181 and 4.182 Knowing that $M = 560$ N · m, determine the stress at (*a*) point *A*, (*b*) point *B*.

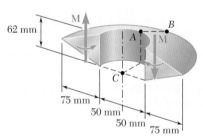

Fig. *P4.181*

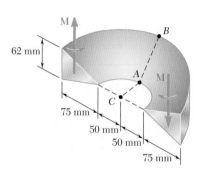

Fig. *P4.182*

4.183 For the curved beam and loading shown, determine the stress at (*a*) point *A*, (*b*) point *B*.

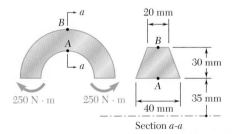

Fig. P4.183

4.184 For the crane hook shown, determine the largest tensile stress in section *a-a*.

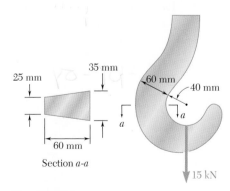

Fig. *P4.184*

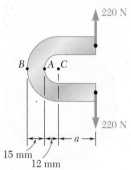

Fig. P4.185 and P4.186

4.185 The bar shown has a circular cross section of 15-mm diameter. Knowing that $a = 30$ mm, determine the stress at (a) point A, (b) point B.

4.186 The bar shown has a circular cross section of 15-mm diameter. Knowing that the allowable tensile stress is 55 MPa, determine the largest permissible distance a from the line of action of the 220-N forces to the plane containing the center of curvature of the bar.

4.187 Show that if the cross section of a curved beam consists of two or more rectangles, the radius R of the neutral surface can be expressed as

$$R = \frac{A}{\ln\left[\left(\dfrac{r_2}{r_1}\right)^{b_1}\left(\dfrac{r_3}{r_2}\right)^{b_2}\left(\dfrac{r_4}{r_3}\right)^{b_3}\right]}$$

where A is the total area of the cross section.

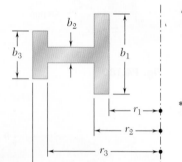

Fig. P4.187

4.188 through 4.190 Using Eq. (4.66), derive the expression for R given in Fig. 4.73 for
 *4.188 A circular cross section.
 4.189 A trapezoidal cross section.
 4.190 A triangular cross section.

*4.191 For a curved bar of rectangular cross section subjected to a bending couple **M,** show that the radial stress at the neutral surface is

$$\sigma_r = \frac{M}{Ae}\left(1 - \frac{r_1}{R} - \ln\frac{R}{r_1}\right)$$

and compute the value of σ_r for the curved bar of Examples 4.10 and 4.11.

(*Hint:* Consider the free-body diagram of the portion of the beam located above the neutral surface.)

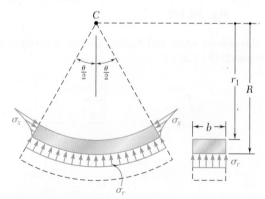

Fig. P4.191

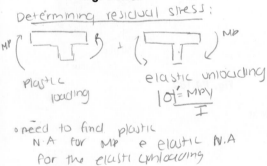

Handwritten top annotations:

$$\frac{dM}{dx} = +V \qquad \frac{dV}{dx} = -q(x) \qquad \frac{dM^2}{dx^2} = -q(x)$$

REVIEW AND SUMMARY

This chapter was devoted to the analysis of members in *pure bending*. That is, we considered the stresses and deformation in members subjected to equal and opposite couples **M** and **M'** acting in the same longitudinal plane (Fig. 4.77).

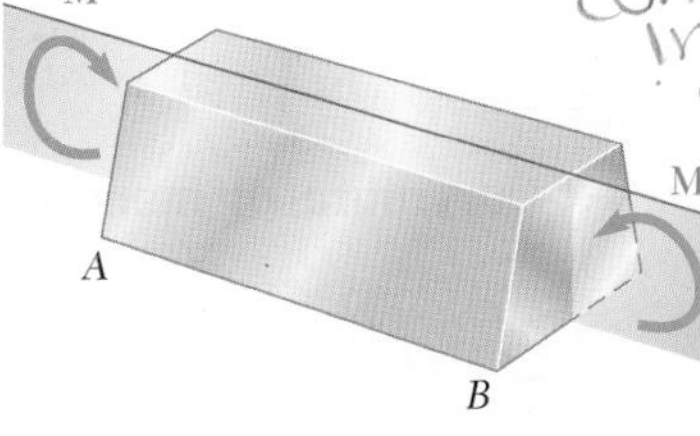

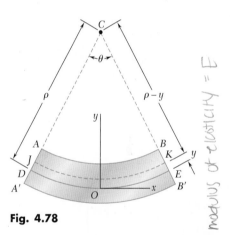

*Handwritten: * Neglect concrete in tension if reinforced w/ steel +*
$I_c = \frac{1}{3} bh^3$

Fig. 4.77

Normal strain in bending

We first studied members possessing a plane of symmetry and subjected to couples acting in that plane. Considering possible deformations of the member, we proved that *transverse sections remain plane* as a member is deformed [Sec. 4.3]. We then noted that a member in pure bending has a *neutral surface* along which normal strains and stresses are zero and that the longitudinal *normal strain* ϵ_x *varies linearly* with the distance y from the neutral surface:

$$\epsilon_x = -\frac{y}{\rho} \qquad (4.8)$$

where ρ is the *radius of curvature* of the neutral surface (Fig. 4.78). The intersection of the neutral surface with a transverse section is known as the *neutral axis* of the section.

Handwritten (right margin, vertical): modulus of elasticity = E

Fig. 4.78

Normal stress in elastic range

For members made of a material that follows Hooke's law [Sec. 4.4], we found that the *normal stress* σ_x *varies linearly* with the distance from the neutral axis (Fig. 4.79). Denoting by σ_m the maximum stress we wrote

$$\sigma_x = -\frac{y}{c}\sigma_m \qquad (4.12)$$

where c is the largest distance from the neutral axis to a point in the section.

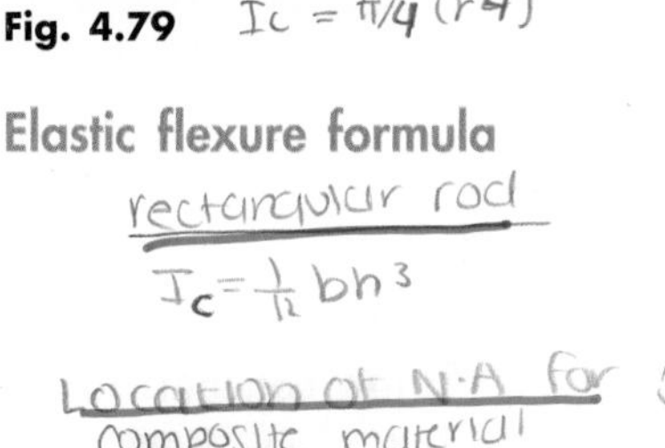

By setting the sum of the elementary forces, $\sigma_x\, dA$, equal to zero, we proved that the *neutral axis passes through the centroid* of the cross section of a member in pure bending. Then by setting the sum of the moments of the elementary forces equal to the bending moment, we derived the *elastic flexure formula* for the maximum normal stress

Handwritten: circle $I_c = \pi/4 (r^4)$

Fig. 4.79

Elastic flexure formula

Handwritten: rectangular rod $I_c = \frac{1}{12} bh^3$

$$\sigma_m = \frac{Mc}{I} \qquad (4.15)$$

where I is the moment of inertia of the cross section with respect to the neutral axis. We also obtained the normal stress at any distance y from the neutral axis:

$$\sigma_x = -\frac{My}{I} \qquad (4.16)$$

Handwritten left margin (vertical): y = dist from centroid to original cardinal/incline

Handwritten: location of N.A for one material:
$$\bar{y} = \frac{\Sigma \bar{y}A}{\Sigma A}$$

Handwritten: y = dist from N.A to where your trying to find (include sign)

Handwritten: Composite material
$$\sigma_{xx}^{(1)} = -\frac{M_z y E_1}{E_1 I_1 + E_2 I_2}$$
$$\sigma_{xx}^{(2)} = -\frac{M_z y E_2}{E_1 I_1 + E_2 I_2}$$

Handwritten: Parallel axis theorem $I = I_c + Ad^2$ d = dist from N.A to centroid of material

Handwritten: Location of N.A for composite material
$$E_1 \bar{y}_1 A_1 + E_2 \bar{y}_2 A_2 = \emptyset$$
must include sign

Handwritten right margin (vertical): $\bar{y}_1, \bar{y}_2 = $ dist to centroid of each material from N.A

Handwritten bottom: (only time you don't shift is if centroid lies on N.A)

327

Handwritten margin notes:
@ yield still in elastic region

$\sigma_x = \left| \sigma_{max} \right| = \left| \frac{-My}{I} \right|$

✗ Plastic region starts @ end and goes to center

Elastic section modulus

Noting that I and c depend only on the geometry of the cross section, we introduced the *elastic section modulus*

$$S = \frac{I}{c} \qquad (4.17)$$

and then used the section modulus to write an alternative expression for the maximum normal stress:

$$\sigma_m = \frac{M}{S} \qquad (4.18)$$

Curvature of member

Recalling that the curvature of a member is the reciprocal of its radius of curvature, we expressed the *curvature* of the member as

$$\frac{1}{\rho} = \frac{M}{EI} \qquad (4.21)$$

In Sec. 4.5, we completed our study of the bending of homogeneous members possessing a plane of symmetry by noting that deformations occur in the plane of a transverse cross section and result in *anticlastic curvature* of the members.

Anticlastic curvature

Members made of several materials

Next we considered the bending of members made of several materials with *different moduli of elasticity* [Sec. 4.6]. While transverse sections remain plane, we found that, in general, the *neutral axis does not pass through the centroid* of the composite cross section (Fig. 4.80). Using the ratio of the moduli of elasticity of the materials,

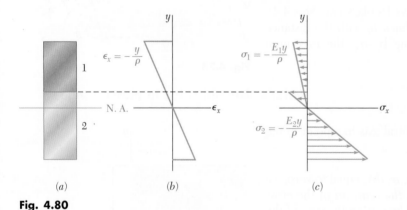

Fig. 4.80

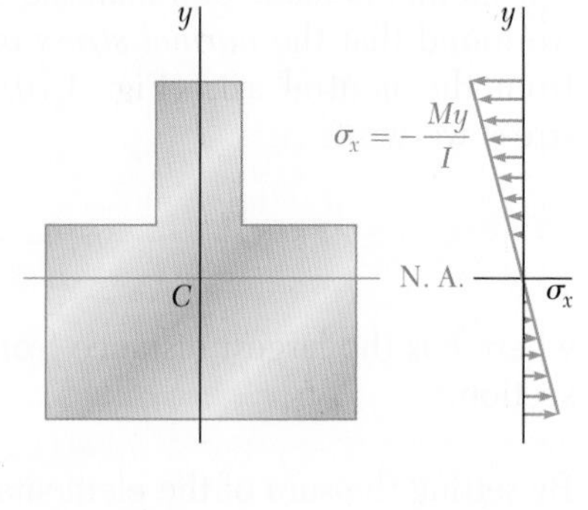

Fig. 4.81

we obtained a *transformed section* corresponding to an equivalent member made entirely of one material. We then used the methods previously developed to determine the stresses in this equivalent homogeneous member (Fig. 4.81) and then again used the ratio of the moduli of elasticity to determine the stresses in the composite beam [Sample Probs. 4.3 and 4.4].

Stress concentrations

In Sec. 4.7, *stress concentrations* that occur in members in pure bending were discussed and charts giving stress-concentration factors for flat bars with fillets and grooves were presented in Figs. 4.27 and 4.28.

[Handwritten top notes:]

If need to calculate total moment of an elastoplastic material

$$M = -2b \int_0^{yy} y \left(\frac{-\sigma_Y (y)}{yy} \right) dy + -2b \int_{yy} y (-\sigma_Y) dy$$

- first term → elastic core
- second term → plastic region
- b = horizontal (or thickness) of region

We next investigated members made of materials that do not follow Hooke's law [Sec. 4.8]. A rectangular beam made of an *elastoplastic material* (Fig. 4.82) was analyzed as the magnitude of the bending moment was increased. The *maximum elastic moment M_Y* occurred when yielding was initiated in the beam (Fig. 4.83). As the bending moment was further increased, plastic zones developed and the size of the elastic core of the member decreased [Sec. 4.9]. Finally the beam became fully plastic and we obtained the maximum or *plastic moment M_p*. In Sec. 4.11, we found that *permanent deformations* and *residual stresses* remain in a member after the loads that caused yielding have been removed.

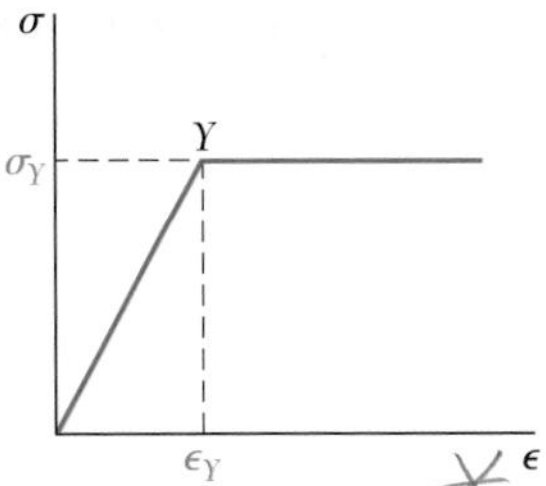

Fig. 4.82

[Handwritten note near Fig 4.82:] d = dist from centroid to N.A (DON'T INCLUDE SIGN)

[Handwritten boxed:] OR $M_p = \sigma_Y \Sigma A_i \bar{y}_i$ (fully plastic moment)

[Handwritten vertical:] FULLY PLASTIC

Plastic deformations

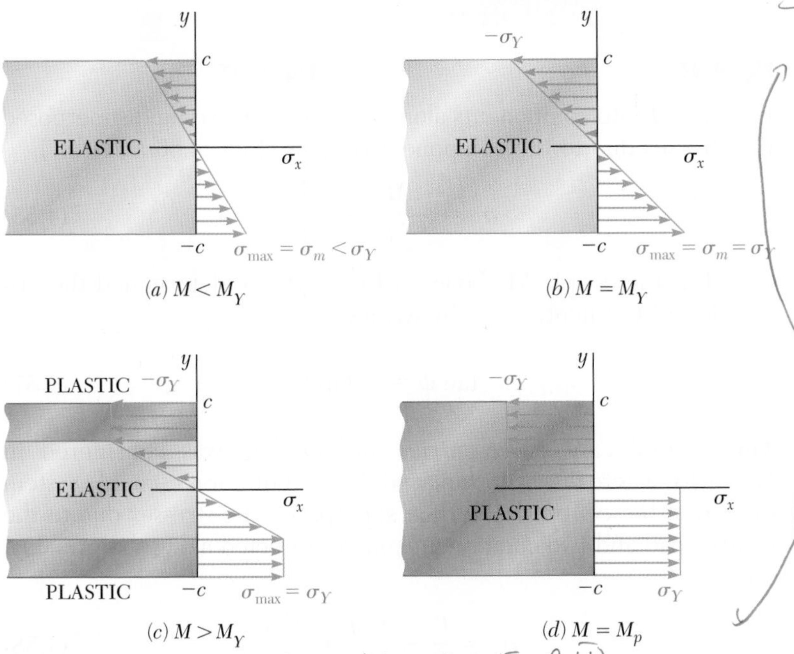

[Handwritten:]
* under fully plastic conditions the N.A divides the cross section into two equal parts *

$$M_p = \frac{\sigma_Y A (\bar{q}_1 + \bar{q}_2)}{2}$$

DO NOT USE $E\bar{y}_1 A_1 + E\bar{y}_2 A_2 = \emptyset$

the sign is already taken into account in this

[Fig 4.83 panels:]

(a) $M < M_Y$ — ELASTIC — $\sigma_{max} = \sigma_m < \sigma_Y$

(b) $M = M_Y$ — ELASTIC — $\sigma_{max} = \sigma_m = \sigma_Y$

(c) $M > M_Y$ — PLASTIC / ELASTIC / PLASTIC — $\sigma_{max} = \sigma_Y$

(d) $M = M_p$ — PLASTIC

Fig. 4.83

[Handwritten:] to determine $\bar{y}_1$ & $\bar{y}_2$
$$\bar{y}_i = \frac{\Sigma \sqrt{\sigma}}{2A}$$

In Sec. 4.12, we studied the stresses in members loaded *eccentrically in a plane of symmetry*. Our analysis made use of methods developed earlier. We replaced the *eccentric load* by a force-couple system located at the centroid of the cross section (Fig. 4.84) and then superposed stresses due to the centric load and the bending couple (Fig. 4.85):

$$\sigma_x = \frac{P}{A} - \frac{My}{I} \qquad (4.50)$$

Eccentric axial loading

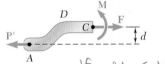

Fig. 4.84

[Handwritten:]
If $M_Y \le M \le M_p$
$$M = \frac{bh^2 \sigma_Y}{6} \left(\frac{3}{2} - \frac{2e^2}{h^2} \right)$$

e = radius of elastic core

size of elastic core:
$$e = h \sqrt{\frac{1}{2} \left(\frac{3}{2} - \frac{M}{M_Y} \right)}$$

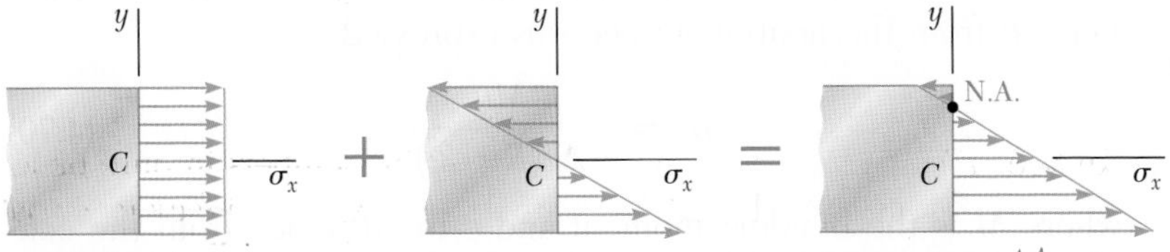

Fig. 4.85

[Handwritten:] @ plastic collapse $M_p \neq M_{max}$

Unsymmetric bending

The bending of members of *unsymmetric cross section* was considered next [Sec. 4.13]. We found that the flexure formula may be used, provided that the couple vector **M** is directed along one of the principal centroidal axes of the cross section. When necessary we

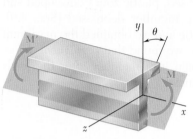

Fig. 4.86 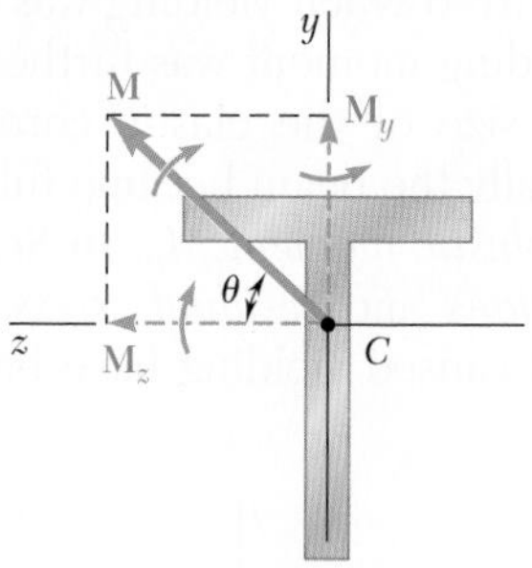   **Fig. 4.87**

resolved **M** into components along the principal axes and superposed the stresses due to the component couples (Figs. 4.86 and 4.87).

$$\sigma_x = -\frac{M_z y}{I_z} + \frac{M_y z}{I_y} \qquad (4.55)$$

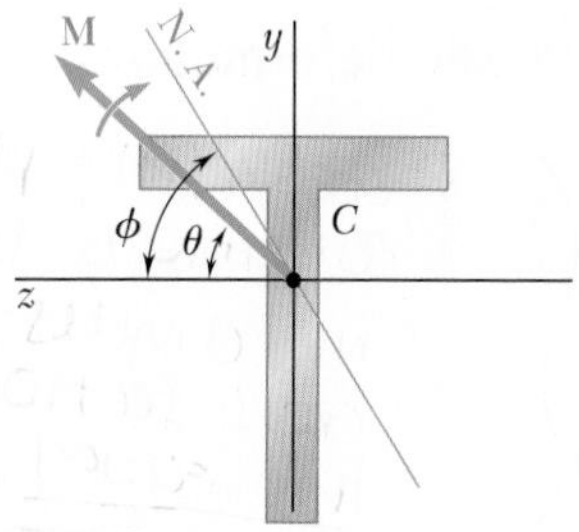

For the couple **M** shown in Fig. 4.88, we determined the orientation of the neutral axis by writing

$$\tan \phi = \frac{I_z}{I_y}\tan \theta \qquad (4.57)$$

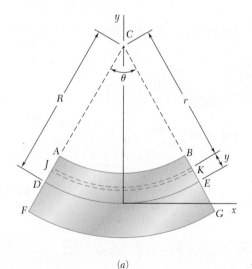

Fig. 4.88

General eccentric axial loading

The general case of *eccentric axial loading* was considered in Sec. 4.14, where we again replaced the load by a force-couple system located at the centroid. We then superposed the stresses due to the centric load and two component couples directed along the principal axes:

$$\sigma_x = \frac{P}{A} - \frac{M_z y}{I_z} + \frac{M_y z}{I_y} \qquad (4.58)$$

Curved members

The chapter concluded with the analysis of stresses in *curved members* (Fig. 4.89). While transverse sections remain plane when the member is subjected to bending, we found that the *stresses do not vary linearly* and the neutral surface does not pass through the centroid of the section. The distance R from the center of curvature of the member to the neutral surface was found to be

$$R = \frac{A}{\displaystyle\int \frac{dA}{r}} \qquad (4.66)$$

where A is the area of the cross section. The normal stress at a distance y from the neutral surface was expressed as

$$\sigma_x = -\frac{My}{Ae(R - y)} \qquad (4.70)$$

(a)

Fig. 4.89

where M is the bending moment and e the distance from the centroid of the section to the neutral surface.

REVIEW PROBLEMS

4.192 A beam of the cross section shown is extruded from an aluminum alloy for which $\sigma_Y = 250$ MPa and $\sigma_U = 450$ MPa. Using a factor of safety of 3.00, determine the largest couple that can be applied to the beam when it is bent about the z axis.

4.193 Straight rods of 8-mm diameter and 60-m length are sometimes used to clear underground conduits of obstructions or to thread wires through a new conduit. The rods are made of high-strength steel and, for storage and transportation, are wrapped on spools of 1.5-m diameter. Assuming that the yield strength is not exceeded, determine (a) the maximum stress in a rod, when the rod, which was initially straight, is wrapped on a spool, (b) the corresponding bending moment in the rod. Use $E = 200$ GPa.

4.194 Knowing that for the beam shown the allowable stress is 84 MPa in tension and 110 MPa in compression, determine the largest couple **M** that can be applied.

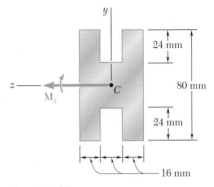

Fig. P4.192

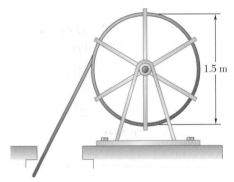

Fig. P4.193

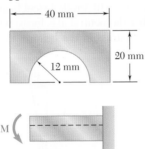

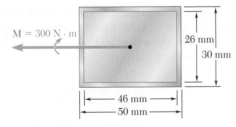

Fig. P4.194

4.195 In order to increase corrosion resistance, a 2-mm-thick cladding of aluminum has been added to a steel bar as shown. The modulus of elasticity is 200 GPa for steel and 70 GPa for aluminum. For a bending moment of 300 N · m, determine (a) the maximum stress in the steel, (b) the maximum stress in the aluminum, (c) the radius of curvature of the bar.

Fig. P4.195

4.196 A single vertical force **P** is applied to a short steel post as shown. Gages located at A, B, and C indicate the following strains:

$$\epsilon_A = -500\ \mu \qquad \epsilon_B = -1000\ \mu \qquad \epsilon_C = -200\ \mu$$

Knowing that $E = 200$ GPa, determine (a) the magnitude of **P**, (b) the line of action of **P**, (c) the corresponding strain at the hidden edge of the post, where $x = -62$ mm and $z = -38$ mm.

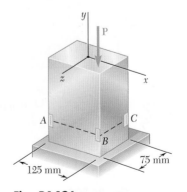

Fig. P4.196

4.197 For the split ring shown, determine the stress at (*a*) point *A*, (*b*) point *B*.

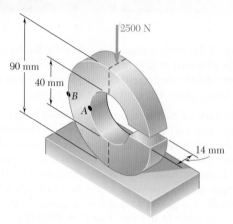

Fig. P4.197

4.198 A couple **M** of moment 8 kN · m acting in a vertical plane is applied to a W200 × 19.3 rolled-steel beam as shown. Determine (*a*) the angle that the neutral axis forms with the horizontal plane, (*b*) the maximum stress in the beam.

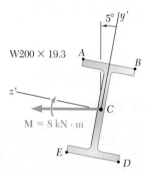

Fig. P4.198

4.199 A beam of the cross section shown is extruded from an aluminum alloy for which *E* = 72 GPa. Knowing that the couple shown acts in a vertical plane, determine (*a*) the maximum stress in the beam, (*b*) the corresponding radius of curvature.

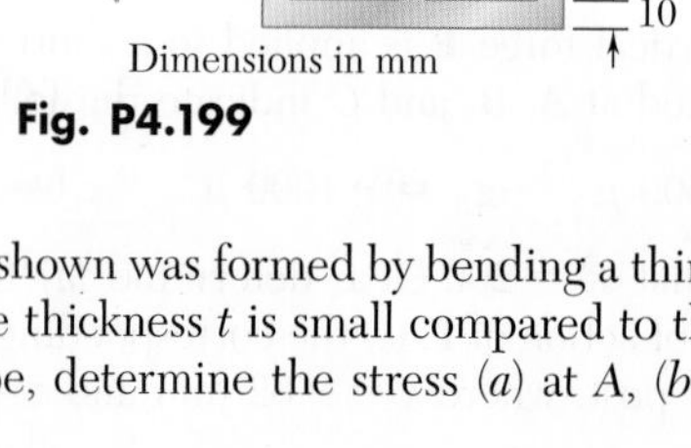

M = 3 kN · m

10 40 10 40 10

10
20
10

Dimensions in mm

Fig. P4.199

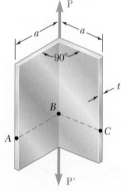

Fig. P4.200

4.200 The shape shown was formed by bending a thin steel plate. Assuming that the thickness *t* is small compared to the length *a* of a side of the shape, determine the stress (*a*) at *A*, (*b*) at *B*, (*c*) at *C*.

4.201 Three 120 × 10-mm steel plates have been welded together to form the beam shown. Assuming that the steel is elastoplastic with $E = 200$ GPa and $\sigma_Y = 300$ MPa, determine (a) the bending moment for which the plastic zones at the top and bottom of the beam are 40 mm thick, (b) the corresponding radius of curvature of the beam.

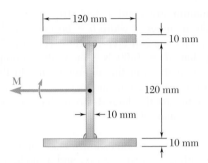

Fig. P4.201

4.202 The couple **M** is applied to a beam of the cross section shown in a plane forming an angle β with the vertical. Determine the stress at (a) point A, (b) point B, (c) point D.

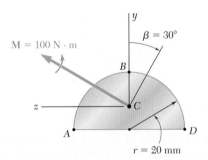

Fig. P4.202

4.203 Two thin strips of the same material and same cross section are bent by couples of the same magnitude and glued together. After the two surfaces of contact have been securely bonded, the couples are removed. Denoting by σ_1 the maximum stress and by ρ_1 the radius of curvature of each strip while the couples were applied, determine (a) the final stresses at points A, B, C, and D, (b) the final radius of curvature.

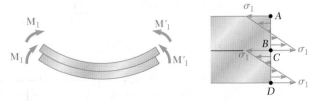

Fig. P4.203

COMPUTER PROBLEMS

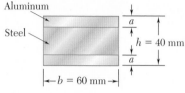

Aluminum

Steel

a

$h = 40$ mm

a

$b = 60$ mm

Fig. P4.C1

The following problems are designed to be solved with a computer.

4.C1 Two aluminum strips and a steel strip are to be bonded together to form a composite member of width $b = 60$ mm and depth $h = 40$ mm. The modulus of elasticity is 200 GPa for the steel and 75 GPa for the aluminum. Knowing that $M = 1500$ N · m, write a computer program to calculate the maximum stress in the aluminum and in the steel for values of a from 0 to 20 mm using 2-mm increments. Using appropriate smaller increments, determine (a) the largest stress that can occur in the steel, (b) the corresponding value of a.

4.C2 A beam of the cross section shown, made of a steel that is assumed to be elastoplastic with a yield strength σ_Y and a modulus of elasticity E, is bent about the x axis. (a) Denoting by y_Y the half thickness of the elastic core, write a computer program to calculate the bending moment M and the radius of curvature ρ for values of y_Y from $\frac{1}{2}d$ to $\frac{1}{6}d$ using decrements equal to $\frac{1}{2}t_f$. Neglect the effect of fillets. (b) Use this program to solve Prob. 4.201.

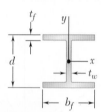

t_f

y

d

x

t_w

b_f

Fig. P4.C2

4.C3 A 900-N · m couple **M** is applied to a beam of the cross section shown in a plane forming an angle β with the vertical. Noting that the centroid of the cross section is located at C and that the y and z axes are principal axes, write a computer program to calculate the stress at A, B, C, and D for values of β from 0 to 180° using 10° increments. (*Given: $I_y = 2.59 \times 10^6$ mm^4 and $I_z = 0.62 \times 10^6$ mm^4.*)

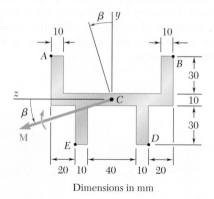

Fig. P4.C3

Dimensions in mm

4.C4 Couples of moment $M = 2$ kN $\cdot$ m are applied as shown to a curved bar having a rectangular cross section with $h = 100$ mm and $b = 25$ mm. Write a computer program and use it to calculate the stresses at points A and B for values of the ratio r_1/h from 10 to 1 using decrements of 1, and from 1 to 0.1 using decrements of 0.1. Using appropriate smaller increments, determine the ratio r_1/h for which the maximum stress in the curved bar is 50 percent larger than the maximum stress in a straight bar of the same cross section.

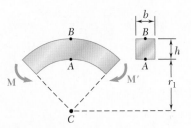

Fig. P4.C4

4.C5 The couple **M** is applied to a beam of the cross section shown. (a) Write a computer program that can be used to calculate the maximum tensile and compressive stresses in the beam. (b) Use this program to solve Prob. 4.11.

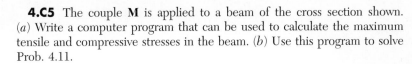

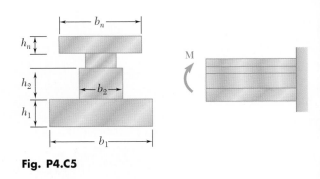

Fig. P4.C5

4.C6 A solid rod of radius $c = 30$ mm is made of a steel that is assumed to be elastoplastic with $E = 200$ GPa and $\sigma_Y = 290$ MPa. The rod is subjected to a couple of moment M that increases from zero to the maximum elastic moment M_Y and then to the plastic moment M_p. Denoting by y_Y the half thickness of the elastic core, write a computer program and use it to calculate the bending moment M and the radius of curvature ρ for values of y_Y from 30 mm to 0 using 5-mm decrements. (*Hint:* Divide the cross section into 60 horizontal elements of 1-mm height.)

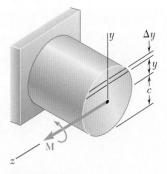

Fig. P4.C6

4.C7 The machine element of Prob. 4.182 is to be redesigned by removing part of the triangular cross section. It is believed that the removal of a small triangular area of width a will lower the maximum stress in the element. In order to verify this design concept, write a computer program to calculate the maximum stress in the element for values of a from 0 to 25 mm using 2.5-mm increments. Using appropriate smaller increments, determine the distance a for which the maximum stress is as small as possible and the corresponding value of the maximum stress.

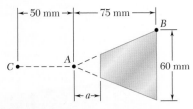

Fig. P4.C7

The beams supporting the multiple overhead cranes system shown in this picture are subjected to transverse loads causing the beams to bend. The normal stresses resulting from such loadings will be determined in this chapter.

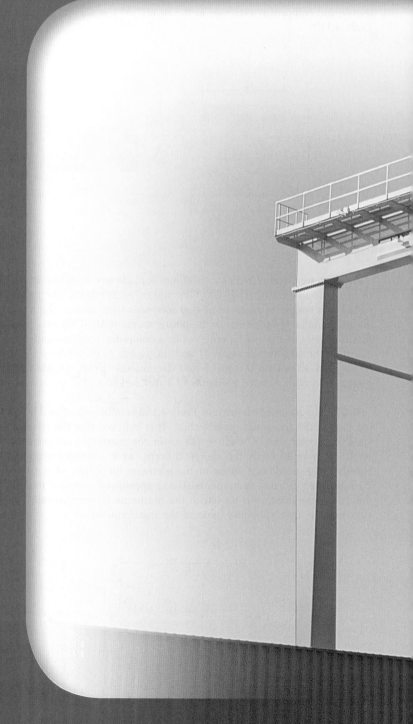

Analysis and Design of Beams for Bending

5.1 INTRODUCTION

This chapter and most of the next one will be devoted to the analysis and the design of *beams*, i.e., structural members supporting loads applied at various points along the member. Beams are usually long, straight prismatic members, as shown in the chapter opening photo. Steel and aluminum beams play an important part in both structural and mechanical engineering. Timber beams are widely used in home construction (Photo 5.1). In most cases, the loads are perpendicular to the axis of the beam. Such a *transverse loading* causes only bending and shear in the beam. When the loads are not at a right angle to the beam, they also produce axial forces in the beam.

Photo 5.1 Timber beams used in residential dwelling.

The transverse loading of a beam may consist of *concentrated loads* P_1, P_2, . . . , expressed in newtons, or its multiple, kilonewtons (Fig. 5.1a), of a *distributed load* w, expressed in N/m or kN/m (Fig. 5.1b), or of a combination of both. When the load w per unit length has a constant value over part of the beam (as between A and B in Fig. 5.1b), the load is said to be *uniformly distributed* over that part of the beam.

Beams are classified according to the way in which they are supported. Several types of beams frequently used are shown in Fig. 5.2. The distance L shown in the various parts of the figure is

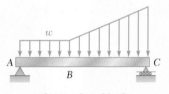

(a) Concentrated loads

(b) Distributed load

Fig. 5.1 Transversely loaded beams.

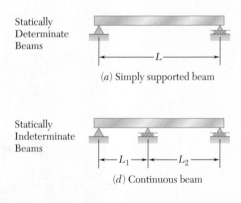

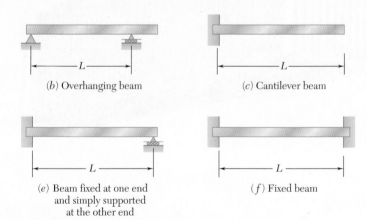

Statically Determinate Beams

(a) Simply supported beam
(b) Overhanging beam
(c) Cantilever beam

Statically Indeterminate Beams

(d) Continuous beam
(e) Beam fixed at one end and simply supported at the other end
(f) Fixed beam

Fig. 5.2 Common beam support configurations.

called the *span*. Note that the reactions at the supports of the beams in parts *a*, *b*, and *c* of the figure involve a total of only three unknowns and, therefore, can be determined by the methods of statics. Such beams are said to be *statically determinate* and will be discussed in this chapter and the next. On the other hand, the reactions at the supports of the beams in parts *d*, *e*, and *f* of Fig. 5.2 involve more than three unknowns and cannot be determined by the methods of statics alone. The properties of the beams with regard to their resistance to deformations must be taken into consideration. Such beams are said to be *statically indeterminate* and their analysis will be postponed until Chap. 9, where deformations of beams will be discussed.

Sometimes two or more beams are connected by hinges to form a single continuous structure. Two examples of beams hinged at a point *H* are shown in Fig. 5.3. It will be noted that the reactions at the supports involve four unknowns and cannot be determined from the free-body diagram of the two-beam system. They can be determined, however, by recognizing that the internal moment at the hinge is zero. Then, after considering the free-body diagram of each beam separately, six unknowns are involved (including two force components at the hinge), and six equations are available.

When a beam is subjected to transverse loads, the internal forces in any section of the beam will generally consist of a shear force **V** and a bending couple **M**. Consider, for example, a simply supported beam *AB* carrying two concentrated loads and a uniformly distributed load (Fig. 5.4*a*). To determine the internal forces in a section through point *C* we first draw the free-body diagram of the entire beam to obtain the reactions at the supports (Fig. 5.4*b*). Passing a section through *C*, we then draw the free-body diagram of *AC* (Fig. 5.4*c*), from which we determine the shear force **V** and the bending couple **M**.

The bending couple **M** creates *normal stresses* in the cross section, while the shear force **V** creates *shearing stresses* in that section. In most cases the dominant criterion in the design of a beam for strength is the maximum value of the normal stress in the beam. The determination of the normal stresses in a beam will be the subject of this chapter, while shearing stresses will be discussed in Chap. 6.

Since the distribution of the normal stresses in a given section depends only upon the value of the bending moment M in that section and the geometry of the section,[†] the elastic flexure formulas derived in Sec. 4.4 can be used to determine the maximum stress, as well as the stress at any given point, in the section. We write[‡]

$$\sigma_m = \frac{|M|c}{I} \qquad \sigma_x = -\frac{My}{I} \qquad (5.1, 5.2)$$

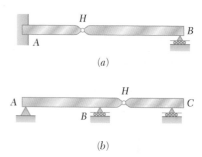

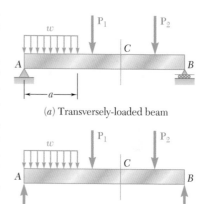

(*a*) Transversely-loaded beam

(*b*) Free-body diagram to find support reactions

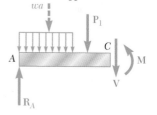

(*c*) Free-body diagram to find internal forces at *C*

Fig. 5.3 Beams connected by hinges.

Fig. 5.4 Analysis of a simply supported beam.

[†] It is assumed that the distribution of the normal stresses in a given cross section is not affected by the deformations caused by the shearing stresses. This assumption will be verified in Sec. 6.5.

[‡] We recall from Sec. 4.2 that M can be positive or negative, depending upon whether the concavity of the beam at the point considered faces upward or downward. Thus, in the case considered here of a transverse loading, the sign of M can vary along the beam. On the other hand, since σ_m is a positive quantity, the absolute value of M is used in Eq. (5.1).

where I is the moment of inertia of the cross section with respect to a centroidal axis perpendicular to the plane of the couple, y is the distance from the neutral surface, and c is the maximum value of that distance (Fig. 4.11). We also recall from Sec. 4.4 that, introducing the elastic section modulus $S = I/c$ of the beam, the maximum value σ_m of the normal stress in the section can be expressed as

$$\sigma_m = \frac{|M|}{S} \tag{5.3}$$

The fact that σ_m is inversely proportional to S underlines the importance of selecting beams with a large section modulus. Section moduli of various rolled-steel shapes are given in Appendix C, while the section modulus of a rectangular shape can be expressed, as shown in Sec. 4.4, as

$$S = \tfrac{1}{6}bh^2 \tag{5.4}$$

where b and h are, respectively, the width and the depth of the cross section.

Equation (5.3) also shows that, for a beam of uniform cross section, σ_m is proportional to $|M|$: Thus, the maximum value of the normal stress in the beam occurs in the section where $|M|$ is largest. It follows that one of the most important parts of the design of a beam for a given loading condition is the determination of the location and magnitude of the largest bending moment.

This task is made easier if a *bending-moment diagram* is drawn, i.e., if the value of the bending moment M is determined at various points of the beam and plotted against the distance x measured from one end of the beam. It is further facilitated if a *shear diagram* is drawn at the same time by plotting the shear V against x.

The sign convention to be used to record the values of the shear and bending moment will be discussed in Sec. 5.2. The values of V and M will then be obtained at various points of the beam by drawing free-body diagrams of successive portions of the beam. In Sec. 5.3 relations among load, shear, and bending moment will be derived and used to obtain the shear and bending-moment diagrams. This approach facilitates the determination of the largest absolute value of the bending moment and, thus, the determination of the maximum normal stress in the beam.

In Sec. 5.4 you will learn to design a beam for bending, i.e., so that the maximum normal stress in the beam will not exceed its allowable value. As indicated earlier, this is the dominant criterion in the design of a beam.

Another method for the determination of the maximum values of the shear and bending moment, based on expressing V and M in terms of *singularity functions,* will be discussed in Sec. 5.5. This approach lends itself well to the use of computers and will be expanded in Chap. 9 to facilitate the determination of the slope and deflection of beams.

Finally, the design of *nonprismatic beams,* i.e., beams with a variable cross section, will be discussed in Sec. 5.6. By selecting the shape and size of the variable cross section so that its elastic section modulus $S = I/c$ varies along the length of the beam in the same way as $|M|$, it is possible to design beams for which the maximum normal stress in each section is equal to the allowable stress of the material. Such beams are said to be of *constant strength*.

5.2 SHEAR AND BENDING-MOMENT DIAGRAMS

As indicated in Sec. 5.1, the determination of the maximum absolute values of the shear and of the bending moment in a beam are greatly facilitated if V and M are plotted against the distance x measured from one end of the beam. Besides, as you will see in Chap. 9, the knowledge of M as a function of x is essential to the determination of the deflection of a beam.

   In the examples and sample problems of this section, the shear and bending-moment diagrams will be obtained by determining the values of V and M at selected points of the beam. These values will be found in the usual way, i.e., by passing a section through the point where they are to be determined (Fig. 5.5a) and considering the equilibrium of the portion of beam located on either side of the section (Fig. 5.5b). Since the shear forces **V** and **V′** have opposite senses, recording the shear at point C with an up or down arrow would be meaningless, unless we indicated at the same time which of the free bodies AC and CB we are considering. For this reason, the shear V will be recorded with a sign: a *plus sign* if the shearing forces are directed as shown in Fig. 5.5b, and a *minus sign* otherwise. A similar convention will apply for the bending moment M. It will be considered as positive if the bending couples are directed as shown in that figure, and negative otherwise.† Summarizing the sign conventions we have presented, we state:

   The shear V and the bending moment M at a given point of a beam are said to be positive when the internal forces and couples acting on each portion of the beam are directed as shown in Fig. 5.6a.

   These conventions can be more easily remembered if we note that

1. *The shear at any given point of a beam is positive when the* **external** *forces (loads and reactions) acting on the beam tend to shear off the beam at that point as indicated in Fig. 5.6b.*

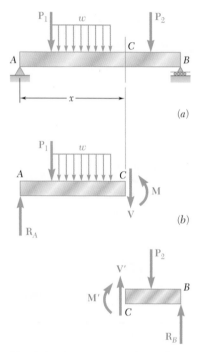

Fig. 5.5   Determination of V and M.

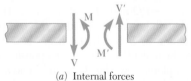

(a) Internal forces
(positive shear and positive bending moment)

(b) Effect of external forces
(positive shear)

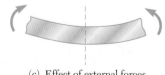

(c) Effect of external forces
(positive bending moment)

Fig. 5.6   Sign convention for shear and bending moment.

†Note that this convention is the same that we used earlier in Sec. 4.2

2. *The bending moment at any given point of a beam is positive when the **external** forces acting on the beam tend to bend the beam at that point as indicated in Fig. 5.6c.*

It is also of help to note that the situation described in Fig. 5.6, in which the values of the shear and of the bending moment are positive, is precisely the situation that occurs in the left half of a simply supported beam carrying a single concentrated load at its midpoint. This particular case is fully discussed in the next example.

EXAMPLE 5.01

Draw the shear and bending-moment diagrams for a simply supported beam AB of span L subjected to a single concentrated load $\mathbf{P}$ at its midpoint C (Fig. 5.7).

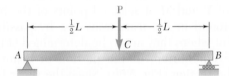

Fig. 5.7

We first determine the reactions at the supports from the free-body diagram of the entire beam (Fig. 5.8a); we find that the magnitude of each reaction is equal to $P/2$.

Next we cut the beam at a point D between A and C and draw the free-body diagrams of AD and DB (Fig. 5.8b). *Assuming that shear and bending moment are positive,* we direct the internal forces $\mathbf{V}$ and $\mathbf{V}'$ and the internal couples $\mathbf{M}$ and $\mathbf{M}'$ as indicated in Fig. 5.6a. Considering the free body AD and writing that the sum of the vertical components and the sum of the moments about D of the forces acting on the free body are zero, we find $V = +P/2$ and $M = +Px/2$. Both the shear and the bending moment are therefore positive; this may be checked by observing that the reaction at A tends to shear off and to bend the beam at D as indicated in Figs. 5.6b and c. We now plot V and M between A and C (Figs. 5.8d and e); the shear has a constant value $V = P/2$, while the bending moment increases linearly from $M = 0$ at $x = 0$ to $M = PL/4$ at $x = L/2$.

Cutting, now, the beam at a point E between C and B and considering the free body EB (Fig. 5.8c), we write that the sum of the vertical components and the sum of the moments about E of the forces acting on the free body are zero. We obtain $V = -P/2$ and $M = P(L - x)/2$. The shear is therefore negative and the bending moment positive; this can be checked by observing that the reaction at B bends the beam at E as indicated in Fig. 5.6c but tends to shear it off in a manner opposite to that shown in Fig. 5.6b. We can complete, now, the shear and bending-moment diagrams of Figs. 5.8d and e; the shear has a constant value $V = -P/2$ between C and B, while the bending moment decreases linearly from $M = PL/4$ at $x = L/2$ to $M = 0$ at $x = L$.

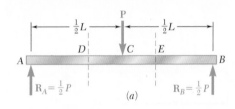

(a)

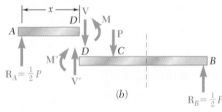

(b)

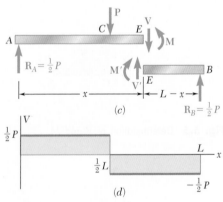

(c)

(d)

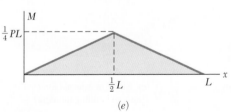

(e)

Fig. 5.8

5.9 and 5.10 Draw the shear and bending-moment diagrams for the beam and loading shown, and determine the maximum absolute value (*a*) of the shear, (*b*) of the bending moment.

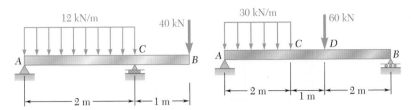

Fig. P5.9

Fig. *P5.10*

5.11 and 5.12 Draw the shear and bending-moment diagrams for the beam and loading shown, and determine the maximum absolute value (*a*) of the shear, (*b*) of the bending moment.

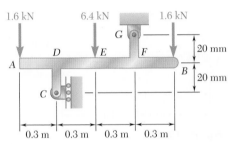

Fig. P5.11

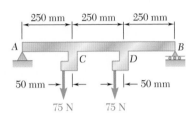

Fig. P5.12

5.13 and 5.14 Assuming that the reaction of the ground is uniformly distributed, draw the shear and bending-moment diagrams for the beam *AB* and determine the maximum absolute value (*a*) of the shear, (*b*) of the bending moment.

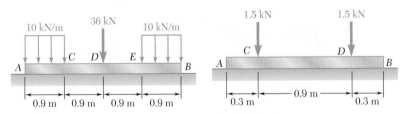

Fig. *P5.13*

Fig. P5.14

5.15 and 5.16 For the beam and loading shown, determine the maximum normal stress due to bending on a transverse section at *C*.

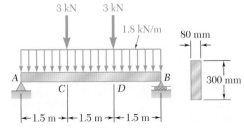

Fig. P5.15

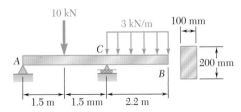

Fig. P5.16

5.17 For the beam and loading shown, determine the maximum normal stress due to bending on a transverse section at C.

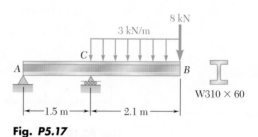

Fig. P5.17

5.18 For the beam and loading shown, determine the maximum normal stress due to bending on section a-a.

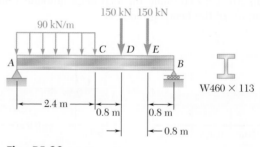

Fig. P5.18

5.19 and 5.20 For the beam and loading shown, determine the maximum normal stress due to bending on a transverse section at C.

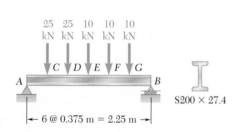

Fig. P5.19

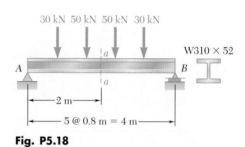

Fig. P5.20

5.21 Draw the shear and bending-moment diagrams for the beam and loading shown and determine the maximum normal stress due to bending.

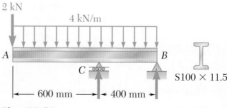

Fig. P5.21

5.22 and 5.23 Draw the shear and bending-moment diagrams for the beam and loading shown and determine the maximum normal stress due to bending.

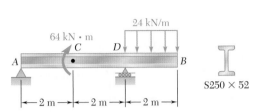

Fig. P5.22

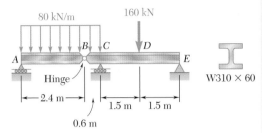

Fig. *P5.23*

5.24 and 5.25 Draw the shear and bending-moment diagrams for the beam and loading shown and determine the maximum normal stress due to bending.

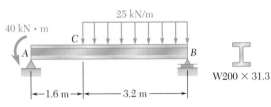

Fig. *P5.24*

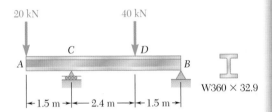

Fig. *P5.25*

5.26 Knowing that $W = 12$ kN, draw the shear and bending-moment diagrams for beam AB and determine the maximum normal stress due to bending.

5.27 Determine (*a*) the magnitude of the counterweight W for which the maximum absolute value of the bending moment in the beam is as small as possible, (*b*) the corresponding maximum normal stress due to bending. (*Hint:* Draw the bending-moment diagram and equate the absolute values of the largest positive and negative bending moments obtained.)

5.28 Determine (*a*) the distance a for which the absolute value of the bending moment in the beam is as small as possible, (*b*) the corresponding maximum normal stress due to bending. (See hint of Prob. 5.27.)

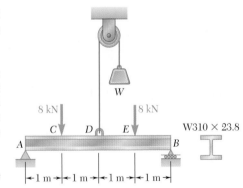

Figs. P5.26 and P5.27

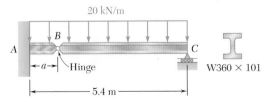

Fig. **P5.28**

5.29 Determine (*a*) the distance *a* for which the absolute value of the bending moment in the beam is as small as possible, (*b*) the corresponding maximum normal stress due to bending. (See hint of Prob. 5.27.)

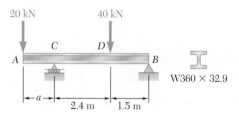

Fig. P5.29

5.30 Knowing that $P = Q = 480$ N, determine (*a*) the distance *a* for which the absolute value of the bending moment in the beam is as small as possible, (*b*) the corresponding maximum normal stress due to bending. (See hint of Prob. 5.27.)

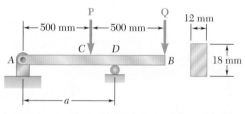

Fig. P5.30

5.31 Solve Prob. 5.30, assuming that $P = 480$ N and $Q = 320$ N.

5.32 A solid steel bar has a square cross section of side *b* and is supported as shown. Knowing that for steel $\rho = 7860$ kg/m³, determine the dimension *b* for which the maximum normal stress due to bending is (*a*) 10 MPa, (*b*) 50 MPa.

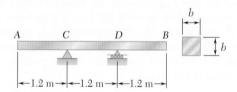

Fig. P5.32

5.33 A solid steel rod of diameter *d* is supported as shown. Knowing that for steel $\rho = 7860$ kg/m³, determine the smallest diameter *d* that can be used if the normal stress due to bending is not to exceed 28 MPa.

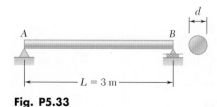

Fig. P5.33

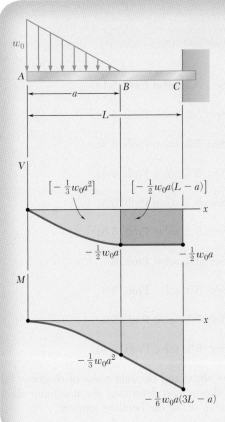

SAMPLE PROBLEM 5.5

Sketch the shear and bending-moment diagrams for the cantilever beam shown.

SOLUTION

Shear Diagram. At the free end of the beam, we find $V_A = 0$. Between A and B, the area under the load curve is $\frac{1}{2}w_0 a$; we find V_B by writing

$$V_B - V_A = -\tfrac{1}{2}w_0 a \qquad V_B = -\tfrac{1}{2}w_0 a$$

Between B and C, the beam is not loaded; thus $V_C = V_B$. At A, we have $w = w_0$ and, according to Eq. (5.5), the slope of the shear curve is $dV/dx = -w_0$, while at B the slope is $dV/dx = 0$. Between A and B, the loading decreases linearly, and the shear diagram is parabolic. Between B and C, $w = 0$, and the shear diagram is a horizontal line.

Bending-Moment Diagram. The bending moment M_A at the free end of the beam is zero. We compute the area under the shear curve and write

$$M_B - M_A = -\tfrac{1}{3}w_0 a^2 \qquad M_B = -\tfrac{1}{3}w_0 a^2$$
$$M_C - M_B = -\tfrac{1}{2}w_0 a(L - a)$$
$$M_C = -\tfrac{1}{6}w_0 a(3L - a)$$

The sketch of the bending-moment diagram is completed by recalling that $dM/dx = V$. We find that between A and B the diagram is represented by a cubic curve with zero slope at A, and between B and C by a straight line.

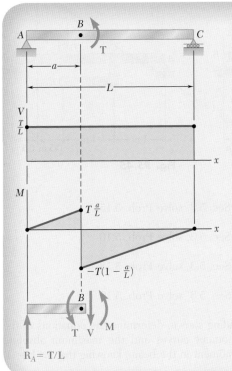

SAMPLE PROBLEM 5.6

The simple beam AC is loaded by a couple of moment T applied at point B. Draw the shear and bending-moment diagrams of the beam.

SOLUTION

The entire beam is taken as a free body, and we obtain

$$\mathbf{R}_A = \frac{T}{L} \uparrow \qquad \mathbf{R}_C = \frac{T}{L} \downarrow$$

The shear at any section is constant and equal to T/L. Since a couple is applied at B, the bending-moment diagram is discontinuous at B; it is represented by two oblique straight lines and decreases suddenly at B by an amount equal to T. The character of this discontinuity can also be verified by equilibrium analysis. For example, considering the free body of the portion of the beam from A to just beyond the right of B as shown, we find the value of M by

$$+\uparrow\Sigma M_B = 0: \quad -\frac{T}{L}a + T + M = 0 \quad M = -T\left(1 - \frac{a}{L}\right)$$

357

PROBLEMS

5.34 Using the method of Sec. 5.3, solve Prob. 5.1*a*.

5.35 Using the method of Sec. 5.3, solve Prob. 5.2*a*.

5.36 Using the method of Sec. 5.3, solve Prob. 5.3*a*.

5.37 Using the method of Sec. 5.3, solve Prob. 5.4*a*.

5.38 Using the method of Sec. 5.3, solve Prob. 5.5*a*.

5.39 Using the method of Sec. 5.3, solve Prob. 5.6*a*.

5.40 Using the method of Sec. 5.3, solve Prob. 5.7.

5.41 Using the method of Sec. 5.3, solve Prob. 5.8.

5.42 Using the method of Sec. 5.3, solve Prob. 5.9.

5.43 Using the method of Sec. 5.3, solve Prob. 5.10.

5.44 and 5.45 Draw the shear and bending-moment diagrams for the beam and loading shown, and determine the maximum absolute value (*a*) of the shear, (*b*) of the bending moment.

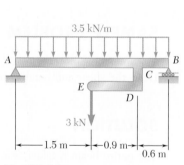

Fig. P5.44

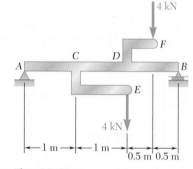

Fig. P5.45

5.46 Using the method of Sec. 5.3, solve Prob. 5.15.

5.47 Using the method of Sec. 5.3, solve Prob. 5.16.

5.48 Using the method of Sec. 5.3, solve Prob. 5.18.

5.49 Using the method of Sec. 5.3, solve Prob. 5.19.

5.50 For the beam and loading shown, determine the equations of the shear and bending-moment curves and the maximum absolute value of the bending moment in the beam, knowing that (*a*) $k = 1$, (*b*) $k = 0.5$.

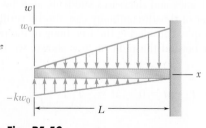

Fig. P5.50

5.51 and 5.52 Determine (a) the equations of the shear and bending-moment curves for the beam and loading shown, (b) the maximum absolute value of the bending moment in the beam.

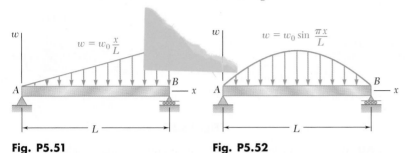

Fig. P5.51

Fig. P5.52

5.53 Determine (a) the equations of the shear and bending-moment curves for the beam and loading shown, (b) the maximum absolute value of the bending moment in the beam.

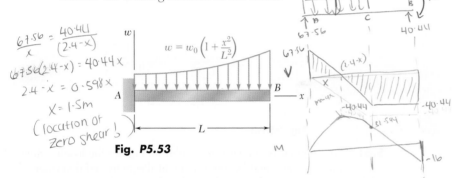

Fig. P5.53

5.54 and 5.55 Draw the shear and bending-moment diagrams for the beam and loading shown and determine the maximum normal stress due to bending.

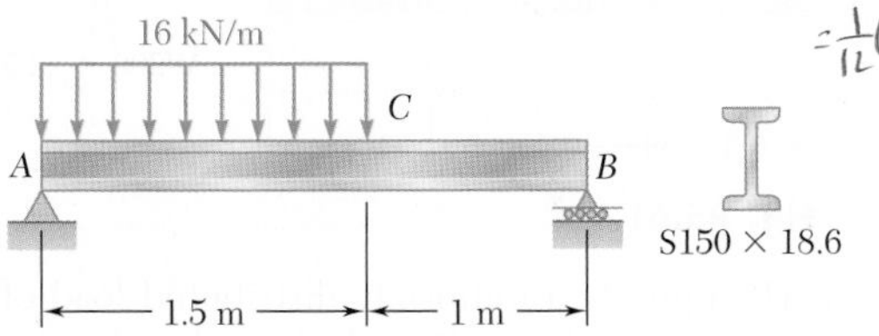

Fig. P5.54

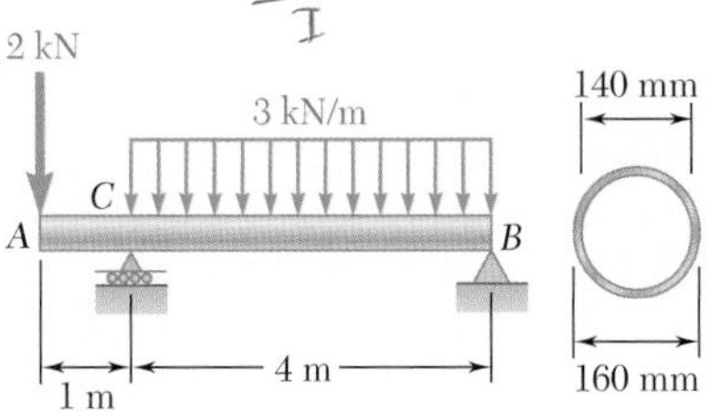

Fig. P5.55

5.56 and 5.57 Draw the shear and bending-moment diagrams for the beam and loading shown and determine the maximum normal stress due to bending.

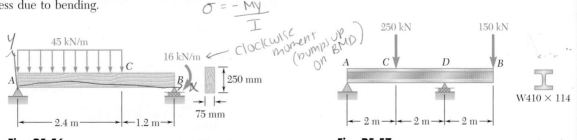

Fig. P5.56

Fig. P5.57

Handwritten notes:

5.56

FBD of entire system $1 \cdot 2 + \frac{2.4}{L}$

108

16

R_A R_B

$\Sigma M_B = 0$.

$-16 - R_A (3.6) + 108(2 \cdot 4) = 0$

$R_A = 67.56 \text{ kN}$

$\Sigma F_Y = 0$ $R_B = 40.44 \text{ kN}$

Section btw A & C $(0 \le x < 2.4)$

45x M_1

V   67.56

x

$\rightarrow \Sigma F_Y = 0$

$\boxed{V_1 = 67.56 - 45x}$

$\Sigma M_0 = 0$

$M_1 + 45x \left(\frac{x}{2}\right) - 67.56(x)$

$\boxed{M_1 = 67.56x - 22.5x^2}$

Section btw C & B $\left(2.4 < x \le 3.6\right)$

1.2   108

67.56   2.4   x-2.4 V_2 M_2

$\boxed{V_2 = -40.44}$

$M_2 = -40.44x + 129.6$

$\dfrac{M_{max}}{\dfrac{dM}{dx} = V = 0}$ occurs when $x = 1.50m$

$M_{max} = 67.56(1.5) - 22.5(1.5)^2$

$= 50.715 \text{ kNm}$

$\sigma = -\dfrac{My}{I}$ where $y = \dfrac{250}{2}$

$I = \dfrac{1}{12}bh^3$

$= \dfrac{1}{12}(75mm)(250mm)^3$

$\sigma = -\dfrac{My}{I}$

clockwise moment (bumps up on BMD)

5.58 and 5.59 Draw the shear and bending-moment diagrams for the beam and loading shown and determine the maximum normal stress due to bending.

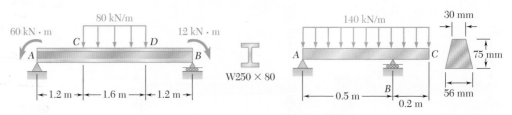

Fig. P5.58   **Fig. P5.59**

5.60 Beam AB, of length L and square cross section of side a, is supported by a pivot at C and loaded as shown. (a) Check that the beam is in equilibrium. (b) Show that the maximum stress due to bending occurs at C and is equal to $w_0L^2/(1.5a)^3$.

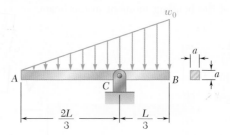

Fig. P5.60

5.61 Knowing that beam AB is in equilibrium under the loading shown, draw the shear and bending-moment diagrams and determine the maximum normal stress due to bending.

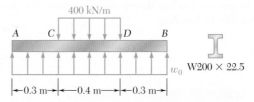

Fig. P5.61

*5.62 Beam AB supports a uniformly distributed load of 15 kN/m and two concentrated loads $\mathbf{P}$ and $\mathbf{Q}$. It has been experimentally determined that the normal stress due to bending on the bottom edge of the lower flange of the W250 × 32.7 rolled-steel beam is −14 MPa at D and −5.35 MPa at E. (a) Draw the shear and bending-moment diagrams for the beam. (b) Determine the maximum normal stress due to bending that occurs in the beam.

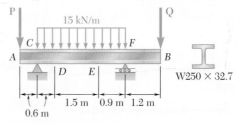

Fig. P5.62

*5.63 Beam AB supports a uniformly distributed load of 2 kN/m and two concentrated loads **P** and **Q**. It has been experimentally determined that the normal stress due to bending in the bottom edge of the beam is -56.9 MPa at A and -29.9 MPa at C. Draw the shear and bending-moment diagrams for the beam and determine the magnitudes of the loads **P** and **Q**.

*5.64 The beam AB supports two concentrated loads **P** and **Q**. The normal stress due to bending on the bottom edge of the beam is $+55$ MPa at D and $+37.5$ MPa at F. (a) Draw the shear and bending-moment diagrams for the beam. (b) Determine the maximum normal stress due to bending that occurs in the beam.

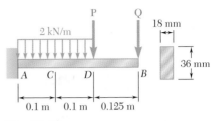

Fig. P5.63

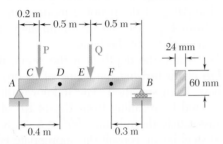

Fig. P5.64

5.4 DESIGN OF PRISMATIC BEAMS FOR BENDING

As indicated in Sec. 5.1, the design of a beam is usually controlled by the maximum absolute value $|M|_{\max}$ of the bending moment that will occur in the beam. The largest normal stress σ_m in the beam is found at the surface of the beam in the critical section where $|M|_{\max}$ occurs and can be obtained by substituting $|M|_{\max}$ for $|M|$ in Eq. (5.1) or Eq. (5.3).† We write

$$\sigma_m = \frac{|M|_{\max} c}{I} \qquad \sigma_m = \frac{|M|_{\max}}{S} \qquad (5.1', \ 5.3')$$

A safe design requires that $\sigma_m \leq \sigma_{\text{all}}$, where σ_{all} is the allowable stress for the material used. Substituting σ_{all} for σ_m in (5.3') and solving for S yields the minimum allowable value of the section modulus for the beam being designed:

$$S_{\min} = \frac{|M|_{\max}}{\sigma_{\text{all}}} \qquad (5.9)$$

The design of common types of beams, such as timber beams of rectangular cross section and rolled-steel beams of various cross-sectional shapes, will be considered in this section. A proper procedure should lead to the most economical design. This means that, among beams of the same type and the same material, and other

---

†For beams that are not symmetrical with respect to their neutral surface, the largest of the distances from the neutral surface to the surfaces of the beam should be used for c in Eq. (5.1) and in the computation of the section modulus $S = I/c$.

things being equal, the beam with the smallest weight per unit length—and, thus, the smallest cross-sectional area—should be selected, since this beam will be the least expensive.

The design procedure will include the following steps†:

1. First determine the value of σ_{all} for the material selected from a table of properties of materials or from design specifications. You can also compute this value by dividing the ultimate strength σ_U of the material by an appropriate factor of safety (Sec. 1.13). Assuming for the time being that the value of σ_{all} is the same in tension and in compression, proceed as follows.

2. Draw the shear and bending-moment diagrams corresponding to the specified loading conditions, and determine the maximum absolute value $|M|_{max}$ of the bending moment in the beam.

3. Determine from Eq. (5.9) the minimum allowable value S_{min} of the section modulus of the beam.

4. For a timber beam, the depth h of the beam, its width b, or the ratio h/b characterizing the shape of its cross section will probably have been specified. The unknown dimensions may then be selected by recalling from Eq. (4.19) of Sec. 4.4 that b and h must satisfy the relation $\frac{1}{6}bh^2 = S \geq S_{min}$.

5. For a rolled-steel beam, consult the appropriate table in Appendix C. Of the available beam sections, consider only those with a section modulus $S \geq S_{min}$ and select from this group the section with the smallest weight per unit length. This is the most economical of the sections for which $S \geq S_{min}$. Note that this is not necessarily the section with the smallest value of S (see Example 5.04). In some cases, the selection of a section may be limited by other considerations, such as the allowable depth of the cross section, or the allowable deflection of the beam (cf. Chap. 9).

The foregoing discussion was limited to materials for which σ_{all} is the same in tension and in compression. If σ_{all} is different in tension and in compression, you should make sure to select the beam section in such a way that $\sigma_m \leq \sigma_{all}$ for both tensile and compressive stresses. If the cross section is not symmetric about its neutral axis, the largest tensile and the largest compressive stresses will not necessarily occur in the section where $|M|$ is maximum. One may occur where M is maximum and the other where M is minimum. Thus, step 2 should include the determination of both M_{max} and M_{min}, and step 3 should be modified to take into account both tensile and compressive stresses.

Finally, keep in mind that the design procedure described in this section takes into account only the normal stresses occurring on the surface of the beam. Short beams, especially those made of timber, may fail in shear under a transverse loading. The determination of shearing stresses in beams will be discussed in Chap. 6. Also, in the case of rolled-steel beams, normal stresses larger than those considered here may occur at the junction of the web with the flanges. This will be discussed in Chap. 8.

†We assume that all beams considered in this chapter are adequately braced to prevent lateral buckling, and that bearing plates are provided under concentrated loads applied to rolled-steel beams to prevent local buckling (crippling) of the web.

EXAMPLE 5.04

Select a wide-flange beam to support the 60-kN load as shown in Fig. 5.14. The allowable normal stress for the steel used is 165 MPa.

1. The allowable normal stress is given: σ_{all} = 165 MPa.
2. The shear is constant and equal to 60 kN. The bending moment is maximum at B. We have

$$|M|_{max} = (60 \text{ kN})(2.4 \text{ m}) = 144 \text{ kN} \cdot \text{m}$$

3. The minimum allowable section modulus is

$$S_{min} = \frac{|M|_{max}}{\sigma_{all}} = \frac{144 \text{ kN} \cdot \text{m}}{165 \text{ MPa}} = 872.7 \times 10^3 \text{ mm}^3$$

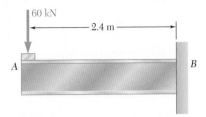

Fig. 5.14

4. Referring to the table of *Properties of Rolled-Steel Shapes* in Appendix C, we note that the shapes are arranged in groups of the same depth and that in each group they are listed in order of decreasing weight. We choose in each group the lightest beam having a section modulus $S = I/c$ at least as large as S_{min} and record the results in the following table.

Shape	$S \times 10^{-3}$, mm^3
W530 × 66	1340
W460 × 52	942
W410 × 60	1060
W360 × 57.8	899
W310 × 74	1060
W250 × 80	984

The most economical is the W460 × 52 shape since it weighs only 52 kg/m, even though it has a larger section modulus than one of the other shapes. We also note that the total weight of the beam will be (2.4 m) × (52 kg/m) × (9.81 m/s^2) = 1224 N. This weight is small compared to the 60,000 N load and can be neglected in our analysis.

Load and Resistance Factor Design. This alternative method of design was briefly described in Sec. 1.13 and applied to members under axial loading. It can readily be applied to the design of beams in bending. Replacing in Eq. (1.26) the loads P_D, P_L, and P_U, respectively, by the bending moments M_D, M_L, and M_U, we write

$$\gamma_D M_D + \gamma_L M_L \leq \phi M_U \tag{5.10}$$

The coefficients γ_D and γ_L are referred to as the *load factors* and the coefficient ϕ as the *resistance factor*. The moments M_D and M_L are the bending moments due, respectively, to the dead and the live loads, while M_U is equal to the product of the ultimate strength σ_U of the material and the section modulus S of the beam: $M_U = S\sigma_U$.

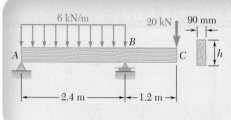

SAMPLE PROBLEM 5.7

A 3.6-m-long overhanging timber beam AC with an 2.4-m span AB is to be designed to support the distributed and concentrated loads shown. Knowing that timber of 100-mm nominal width (90-mm actual width) with a 12-MPa allowable stress is to be used, determine the minimum required depth h of the beam.

SOLUTION

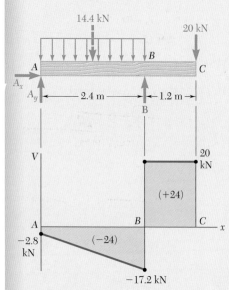

Reactions. Considering the entire beam as a free body, we write

$+\uparrow\Sigma M_A = 0$: $B(2.4\text{ m}) - (14.4\text{ kN})(1.2\text{ m}) - (20\text{ kN})(3.6\text{ m}) = 0$

$\qquad\qquad B = 37.2\text{ kN} \qquad \mathbf{B} = 37.2\text{ kN} \uparrow$

$\xrightarrow{+}\Sigma F_x = 0$: $\qquad\qquad\qquad A_x = 0$

$+\uparrow\Sigma F_y = 0$: $A_y + 37.2\text{ kN} - 14.4\text{ kN} - 20\text{ kN} = 0$

$\qquad\qquad A_y = -2.8\text{ kN} \qquad \mathbf{A} = 2.8\text{ kN} \downarrow$

Shear Diagram. The shear just to the right of A is $V_A = A_y = -2.8$ kN. Since the change in shear between A and B is equal to *minus* the area under the load curve between these two points, we obtain V_B by writing

$$V_B - V_A = -(6\text{ kN/m})(2.4\text{ m}) = -14.4\text{ kN}$$
$$V_B = V_A - 14.4\text{ kN} = -2.8\text{ kN} - 14.4\text{ kN} = -17.2\text{ kN}$$

The reaction at B produces a sudden increase of 37.2 kN in V, resulting in a value of the shear equal to 20 kN to the right of B. Since no load is applied between B and C, the shear remains constant between these two points.

Determination of $|M|_{max}$. We first observe that the bending moment is equal to zero at both ends of the beam: $M_A = M_C = 0$. Between A and B the bending moment decreases by an amount equal to the area under the shear curve, and between B and C it increases by a corresponding amount. Thus, the maximum absolute value of the bending moment is $|M|_{max} = 24$ kN · m.

Minimum Allowable Section Modulus. Substituting into Eq. (5.9) the given value of σ_{all} and the value of $|M|_{max}$ that we have found, we write

$$S_{min} = \frac{|M|_{max}}{\sigma_{all}} = \frac{24\text{ kN} \cdot \text{m}}{12\text{ MPa}} = 2 \times 10^6 \text{ mm}^3$$

Minimum Required Depth of Beam. Recalling the formula developed in part 4 of the design procedure described in Sec. 5.4 and substituting the values of b and S_{min}, we have

$$\tfrac{1}{6}bh^2 \geq S_{min} \quad \tfrac{1}{6}(90\text{ mm})h^2 \geq 2 \times 10^6 \text{ mm}^3 \quad h \geq 365.2\text{ mm}$$

The minimum required depth of the beam is $\qquad\qquad h = 366\text{ mm}$ ◄

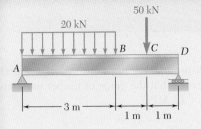

SAMPLE PROBLEM 5.8

A 5-m-long, simply supported steel beam AD is to carry the distributed and concentrated loads shown. Knowing that the allowable normal stress for the grade of steel to be used is 160 MPa, select the wide-flange shape that should be used.

SOLUTION

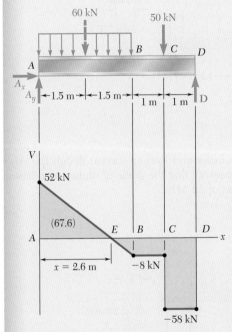

Reactions. Considering the entire beam as a free body, we write

$$+\circlearrowleft\Sigma M_A = 0: \quad D(5\text{ m}) - (60\text{ kN})(1.5\text{ m}) - (50\text{ kN})(4\text{ m}) = 0$$
$$D = 58.0\text{ kN} \qquad \mathbf{D} = 58.0\text{ kN} \uparrow$$

$$\xrightarrow{+}\Sigma F_x = 0: \qquad A_x = 0$$

$$+\uparrow\Sigma F_y = 0: \quad A_y + 58.0\text{ kN} - 60\text{ kN} - 50\text{ kN} = 0$$
$$A_y = 52.0\text{ kN} \qquad \mathbf{A} = 52.0\text{ kN} \uparrow$$

Shear Diagram. The shear just to the right of A is $V_A = A_y = +52.0$ kN. Since the change in shear between A and B is equal to *minus* the area under the load curve between these two points, we have

$$V_B = 52.0\text{ kN} - 60\text{ kN} = -8\text{ kN}$$

The shear remains constant between B and C, where it drops to -58 kN, and keeps this value between C and D. We locate the section E of the beam where $V = 0$ by writing

$$V_E - V_A = -wx$$
$$0 - 52.0\text{ kN} = -(20\text{ kN/m})x$$

Solving for x we find $x = 2.60$ m.

Determination of $|M|_{\max}$. The bending moment is maximum at E, where $V = 0$. Since M is zero at the support A, its maximum value at E is equal to the area under the shear curve between A and E. We have, therefore, $|M|_{\max} = M_E = 67.6$ kN · m.

Minimum Allowable Section Modulus. Substituting into Eq. (5.9) the given value of σ_{all} and the value of $|M|_{\max}$ that we have found, we write

$$S_{\min} = \frac{|M|_{\max}}{\sigma_{\text{all}}} = \frac{67.6\text{ kN · m}}{160\text{ MPa}} = 422.5 \times 10^{-6}\text{ m}^3 = 422.5 \times 10^3\text{ mm}^3$$

Selection of Wide-Flange Shape. From Appendix C we compile a list of shapes that have a section modulus larger than $S_{\min}$ and are also the lightest shape in a given depth group.

Shape	$S \times 10^{-3}$, mm^3
W410 × 38.8	629
W360 × 32.9	475
W310 × 38.7	547
W250 × 44.8	531
W200 × 46.1	451

We select the lightest shape available, namely   W360 × 32.9 ◄

365

PROBLEMS

5.65 and 5.66 For the beam and loading shown, design the cross section of the beam, knowing that the grade of timber used has an allowable normal stress of 12 MPa.

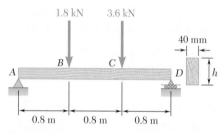

Fig. P5.65

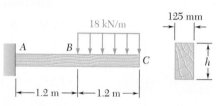

Fig. P5.66

5.67 and 5.68 For the beam and loading shown, design the cross section of the beam, knowing that the grade of timber used has an allowable normal stress of 12 MPa.

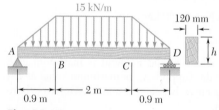

Fig. P5.67

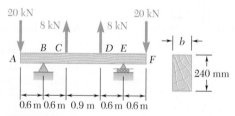

Fig. P5.68

5.69 and 5.70 For the beam and loading shown, design the cross section of the beam, knowing that the grade of timber used has an allowable normal stress of 12 MPa.

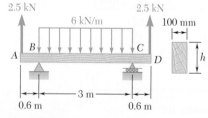

Fig. P5.69

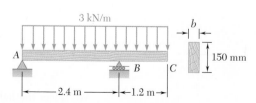

Fig. P5.70

5.71 and 5.72 Knowing that the allowable stress for the steel used is 160 MPa, select the most economical wide-flange beam to support the loading shown.

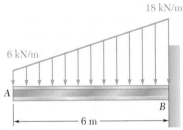

Fig. P5.71

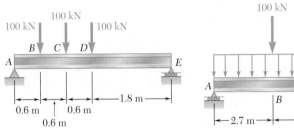

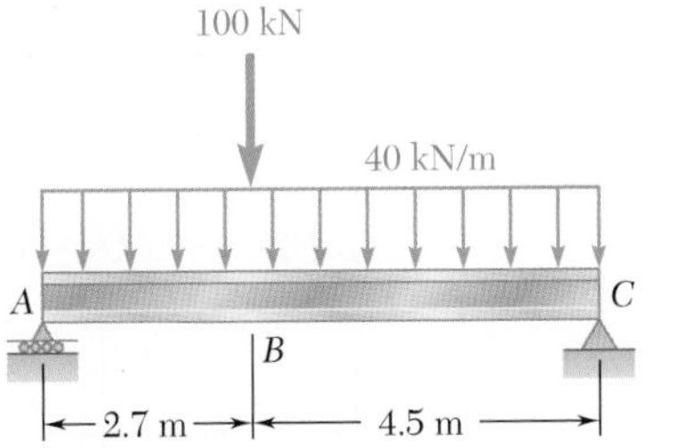

Fig. P5.72

5.73 and 5.74 Knowing that the allowable stress for the steel used is 160 MPa, select the most economical wide-flange beam to support the loading shown.

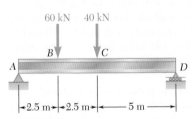

Fig. P5.73

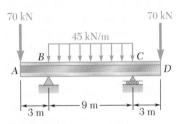

Fig. P5.74

5.75 and 5.76 Knowing that the allowable stress for the steel used is 160 MPa, select the most economical S-shape beam to support the loading shown.

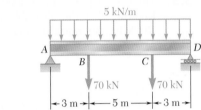

Fig. P5.75

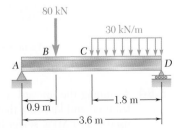

Fig. P5.76

5.77 and 5.78 Knowing that the allowable stress for the steel used is 160 MPa, select the most economical S-shape beam to support the loading shown.

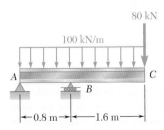

Fig. P5.77

Fig. P5.78

5.79 Two L102 × 76 rolled-steel angles are bolted together and used to support the loading shown. Knowing that the allowable normal stress for the steel used is 140 MPa, determine the minimum angle thickness that can be used.

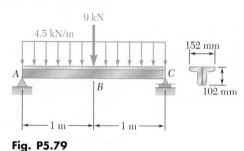

Fig. P5.79

5.80 Two rolled-steel channels are to be welded along their edges and used to support the loading shown. Knowing that the allowable normal stress for the steel used is 150 MPa, determine the most economical channels that can be used.

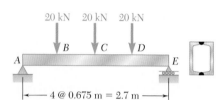

Fig. P5.80

5.81 Three steel plates are welded together to form the beam shown. Knowing that the allowable normal stress for the steel used is 154 MPa, determine the minimum flange width b that can be used.

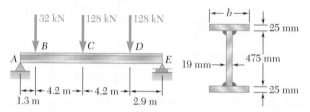

Fig. P5.81

5.82 A steel pipe of 100-mm diameter is to support the loading shown. Knowing that the stock of pipes available has thicknesses varying from 6 mm to 24 mm in 3-mm increments, and that the allowable normal stress for the steel used is 150 MPa, determine the minimum wall thickness t that can be used.

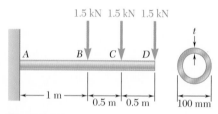

Fig. P5.82

5.83 Assuming the upward reaction of the ground to be uniformly distributed and knowing that the allowable normal stress for the steel used is 165 MPa, select the most economical wide-flange beam to support the loading shown.

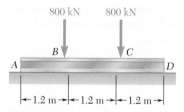

Fig. P5.83

5.84 Assuming the upward reaction of the ground to be uniformly distributed and knowing that the allowable normal stress for the steel used is 170 MPa, select the most economical wide-flange beam to support the loading shown.

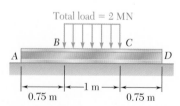

Fig. P5.84

5.85 Determine the allowable value of **P** for the loading shown, knowing that the allowable normal stress is +55 MPa in tension and −125 MPa in compression.

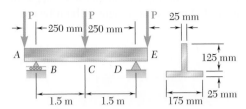

Fig. P5.85

5.86 Solve Prob. 5.85, assuming that the T-shaped beam is inverted.

5.87 Determine the largest permissible distributed load w for the beam shown, knowing that the allowable normal stress is +80 MPa in tension and −130 MPa in compression.

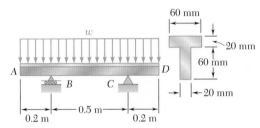

Fig. P5.87

5.88 Solve Prob. 5.87, assuming that the cross section of the beam is reversed, with the flange of the beam resting on the supports at B and C.

5.89 A 240-kN load is to be supported at the center of the 5-m span shown. Knowing that the allowable normal stress for the steel used is 165 MPa, determine (*a*) the smallest allowable length l of beam CD if the W310 × 74 beam AB is not to be overstressed, (*b*) the most economical W shape that can be used for beam CD. Neglect the weight of both beams.

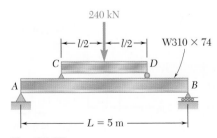

Fig. P5.89

5.90 A uniformly distributed load of 66 kN/m is to be supported over the 6-m span shown. Knowing that the allowable normal stress for the steel used is 140 MPa, determine (*a*) the smallest allowable length *l* of beam *CD* if the W460 × 74 beam *AB* is not to be overstressed, (*b*) the most economical W shape that can be used for beam *CD*. Neglect the weight of both beams.

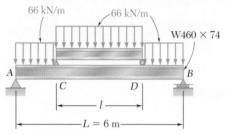

Fig. P5.90

5.91 Beam *ABC* is bolted to beams *DBE* and *FCG*. Knowing that the allowable normal stress is 165 MPa, select the most economical wide-flange shape that can be used (*a*) for beam *ABC*, (*b*) for beam *DBE*, (*c*) for beam *FCG*.

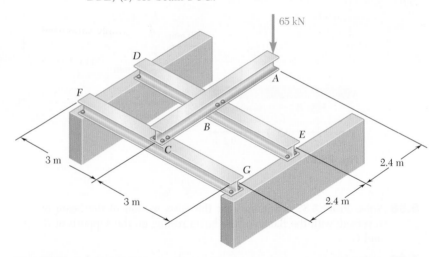

Fig. P5.91

5.92 Beams *AB*, *BC*, and *CD* have the cross section shown and are pin-connected at *B* and *C*. Knowing that the allowable normal stress is +110 MPa in tension and −150 MPa in compression, determine (*a*) the largest permissible value of *w* if beam *BC* is not to be overstressed, (*b*) the corresponding maximum distance *a* for which the cantilever beams *AB* and *CD* are not overstressed.

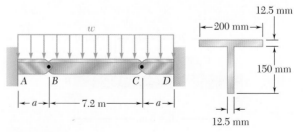

Fig. P5.92

5.93 Beams *AB*, *BC*, and *CD* have the cross section shown and are pin-connected at *B* and *C*. Knowing that the allowable normal stress is +110 MPa in tension and −150 MPa in compression, determine (*a*) the largest permissible value of **P** if beam *BC* is not to be overstressed, (*b*) the corresponding maximum distance *a* for which the cantilever beams *AB* and *CD* are not overstressed.

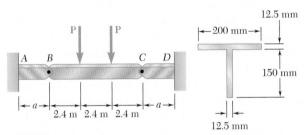

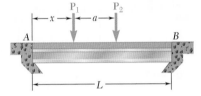

Fig. P5.93

*5.94 A bridge of length *L* = 15 m is to be built on a secondary road whose access to trucks is limited to two-axle vehicles of medium weight. It will consist of a concrete slab and of simply supported steel beams with an ultimate strength σ_U = 420 MPa. The combined weight of the slab and beams can be approximated by a uniformly distributed load *w* = 11 kN/m on each beam. For the purpose of the design, it is assumed that a truck with axles located at a distance *a* = 4 m from each other will be driven across the bridge and that the resulting concentrated loads P_1 and P_2 exerted on each beam could be as large as 95 kN and 25 kN, respectively. Determine the most economical wide-flange shape for the beams, using LRFD with the load factors γ_D = 1.25, γ_L = 1.75 and the resistance factor ϕ = 0.9. [*Hint:* It can be shown that the maximum value of $|M_L|$ occurs under the larger load when that load is located to the left of the center of the beam at a distance equal to $aP_2/2(P_1 + P_2)$.]

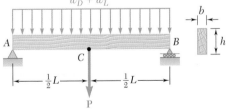

Fig. P5.94

*5.95 Assuming that the front and rear axle loads remain in the same ratio as for the truck of Prob. 5.96, determine how much heavier a truck could safely cross the bridge designed in that problem.

*5.96 A roof structure consists of plywood and roofing material supported by several timber beams of length *L* = 16 m. The dead load carried by each beam, including the estimated weight of the beam, can be represented by a uniformly distributed load w_D = 350 N/m. The live load consists of a snow load, represented by a uniformly distributed load w_L = 600 N/m, and a 6-kN concentrated load **P** applied at the midpoint *C* of each beam. Knowing that the ultimate strength for the timber used is σ_U = 50 MPa and that the width of each beam is *b* = 75 mm, determine the minimum allowable depth *h* of the beams, using LRFD with the load factors γ_D = 1.2, γ_L = 1.6 and the resistance factor ϕ = 0.9.

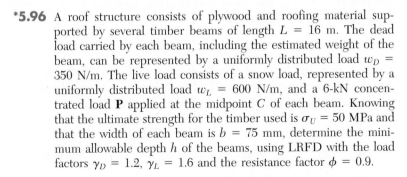

Fig. P5.96

*5.97 Solve Prob. 5.96, assuming that the 6-kN concentrated load **P** applied to each beam is replaced by 3-kN concentrated loads P_1 and P_2 applied at a distance of 4 m from each end of the beams.

*5.5 USING SINGULARITY FUNCTIONS TO DETERMINE SHEAR AND BENDING MOMENT IN A BEAM

Reviewing the work done in the preceding sections, we note that the shear and bending moment could only rarely be described by single analytical functions. In the case of the cantilever beam of Example 5.02 (Fig. 5.9), which supported a uniformly distributed load w, the shear and bending moment *could* be represented by single analytical functions, namely, $V = -wx$ and $M = -\frac{1}{2}wx^2$; this was due to the fact that *no discontinuity* existed in the loading of the beam. On the other hand, in the case of the simply supported beam of Example 5.01, which was loaded only at its midpoint C, the load **P** applied at C represented a *singularity* in the beam loading. This singularity resulted in discontinuities in the shear and bending moment and required the use of different analytical functions to represent V and M in the portions of beam located, respectively, to the left and to the right of point C. In Sample Prob. 5.2, the beam had to be divided into three portions, in each of which different functions were used to represent the shear and the bending moment. This situation led us to rely on the graphical representation of the functions V and M provided by the shear and bending-moment diagrams and, later in Sec. 5.3, on a graphical method of integration to determine V and M from the distributed load w.

The purpose of this section is to show how the use of *singularity functions* makes it possible to represent the shear V and the bending moment M by single mathematical expressions.

Consider the simply supported beam AB, of length $2a$, which carries a uniformly distributed load w_0 extending from its midpoint C to its right-hand support B (Fig. 5.15). We first draw the free-body diagram of the entire beam (Fig. 5.16a); replacing the distributed load by an equivalent concentrated load and, summing moments about B, we write

$$+\curvearrowleft \Sigma M_B = 0: \qquad (w_0 a)(\tfrac{1}{2}a) - R_A(2a) = 0 \qquad R_A = \tfrac{1}{4}w_0 a$$

Next we cut the beam at a point D between A and C. From the free-body diagram of AD (Fig. 5.16b) we conclude that, over the interval $0 < x < a$, the shear and bending moment are expressed, respectively, by the functions

$$V_1(x) = \tfrac{1}{4}w_0 a \qquad \text{and} \qquad M_1(x) = \tfrac{1}{4}w_0 ax$$

Cutting, now, the beam at a point E between C and B, we draw the free-body diagram of portion AE (Fig. 5.16c). Replacing the distributed load by an equivalent concentrated load, we write

$$+\uparrow \Sigma F_y = 0: \qquad \tfrac{1}{4}w_0 a - w_0(x - a) - V_2 = 0$$

$$+\curvearrowleft \Sigma M_E = 0: \qquad -\tfrac{1}{4}w_0 ax + w_0(x - a)[\tfrac{1}{2}(x - a)] + M_2 = 0$$

and conclude that, over the interval $a < x < 2a$, the shear and bending moment are expressed, respectively, by the functions

$$V_2(x) = \tfrac{1}{4}w_0 a - w_0(x - a) \qquad \text{and} \qquad M_2(x) = \tfrac{1}{4}w_0 ax - \tfrac{1}{2}w_0(x - a)^2$$

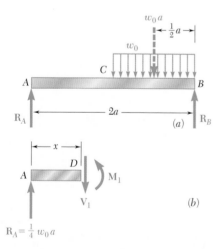

Fig. 5.15

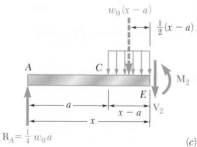

Fig. 5.16

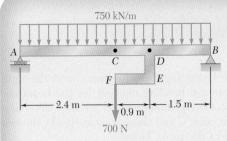

750 kN/m

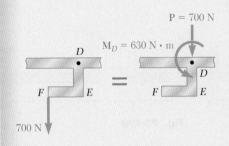

2.4 m

0.9 m

1.5 m

700 N

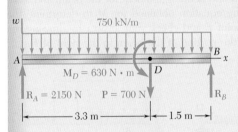

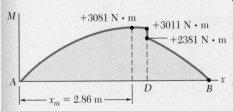

SAMPLE PROBLEM 5.10

The rigid bar *DEF* is welded at point *D* to the steel beam *AB*. For the loading shown, determine (*a*) the equations defining the shear and bending moment at any point of the beam, (*b*) the location and magnitude of the largest bending moment.

SOLUTION

Reactions. We consider the beam and bar as a free body and observe that the total load is 4300 N. Because of symmetry, each reaction is equal to 2150 N.

Modified Loading Diagram. We replace the 700-N load applied at *F* by an equivalent force-couple system at *D*. We thus obtain a loading diagram consisting of a concentrated couple, three concentrated loads (including the two reactions), and a uniformly distributed load

$$w(x) = 750 \text{ kN/m} \qquad (1)$$

a. Equations for Shear and Bending Moment. We obtain $V(x)$ by integrating (1), changing the sign, and adding constants representing the respective contributions of $\mathbf{R}_A$ and $\mathbf{P}$ to the shear. Since $\mathbf{P}$ affects $V(x)$ only for values of *x* larger than 3.3 m, we use a step function to express its contribution.

$$V(x) = -750x + 2150 - 700\langle x - 3.3\rangle^0 \qquad (2) \quad \blacktriangleleft$$

We obtain $M(x)$ by integrating (2) and using a step function to represent the contribution of the concentrated couple $\mathbf{M}_D$:

$$M(x) = -375x^2 + 2150x - 700\langle x - 3.3\rangle^1 - 630\langle x - 3.3\rangle^0 \quad (3) \quad \blacktriangleleft$$

b. Largest Bending Moment. Since *M* is maximum or minimum when $V = 0$, we set $V = 0$ in (2) and solve that equation for *x* to find the location of the largest bending moment. Considering first values of *x* less than 3.3 m and noting that for such values the bracket is equal to zero, we write

$$-750x + 2150 = 0 \qquad x = 2.86 \text{ m}$$

Considering now values of *x* larger than 3.3 m, for which the bracket is equal to 1, we have

$$-750x + 2150 - 700 = 0 \qquad x = 1.93 \text{ m}$$

Since this value is *not* larger than 3.3 m, it must be rejected. Thus, the value of *x* corresponding to the largest bending moment is

$$X_m = 2.86 \text{ m} \quad \blacktriangleleft$$

Substituting this value for *x* into Eq. (3), we obtain

$$M_{max} = -375(2.86)^2 + 2150(2.86) - 700\langle -0.44\rangle^1 - 630\langle -0.44\rangle^0$$

and, recalling that brackets containing a negative quantity are equal to zero,

$$M_{max} = -375(2.86)^2 + 2150(2.86) \quad M_{max} = 3080 \text{ N} \cdot \text{m} \quad \blacktriangleleft$$

The bending-moment diagram has been plotted. Note the discontinuity at point *D* due to the concentrated couple applied at that point. The values of *M* just to the left and just to the right of *D* were obtained by making $x = 3.3$ in Eq. (3) and replacing the step function $\langle x - 3.3\rangle^0$ by 0 and 1, respectively.

PROBLEMS

5.98 through 5.100 (*a*) Using singularity functions, write the equations defining the shear and bending moment for the beam and loading shown. (*b*) Use the equation obtained for *M* to determine the bending moment at point *C* and check your answer by drawing the free-body diagram of the entire beam.

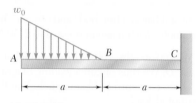

Fig. P5.98

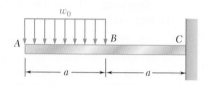

Fig. P5.99

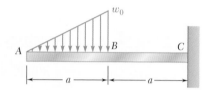

Fig. P5.100

5.101 through 5.103 (*a*) Using singularity functions, write the equations defining the shear and bending moment for the beam and loading shown. (*b*) Use the equation obtained for *M* to determine the bending moment at point *E* and check your answer by drawing the free-body diagram of the portion of the beam to the right of *E*.

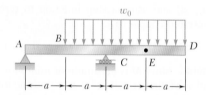

Fig. P5.101

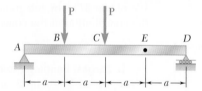

Fig. P5.102

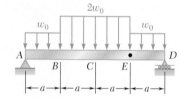

Fig. P5.103

5.104 (*a*) Using singularity functions, write the equations for the shear and bending moment for beam *ABC* under the loading shown. (*b*) Use the equation obtained for *M* to determine the bending moment just to the right of point *B*.

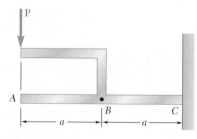

Fig. P5.104

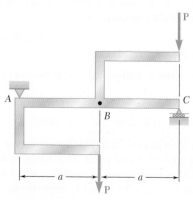

Fig. P5.105

5.105 (*a*) Using singularity functions, write the equations for the shear and bending moment for beam *ABC* under the loading shown. (*b*) Use the equation obtained for *M* to determine the bending moment just to the right of point *D*.

5.106 through 5.109 (*a*) Using singularity functions, write the equations for the shear and bending moment for the beam and loading shown. (*b*) Determine the maximum value of the bending moment in the beam.

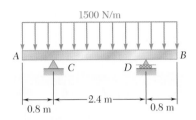

Fig. P5.106

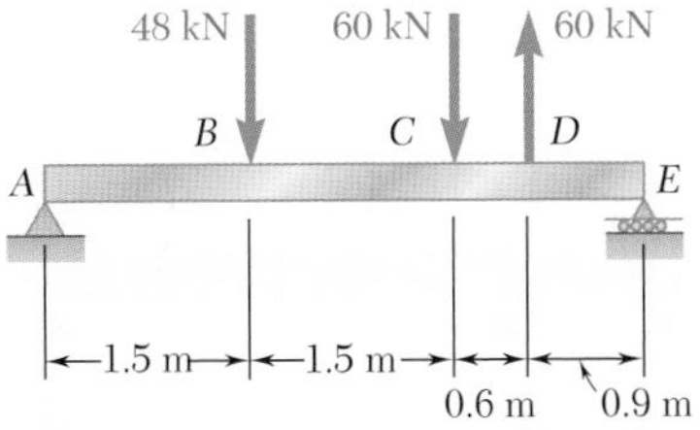

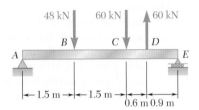

Fig. P5.107

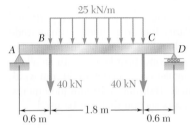

Fig. P5.108

Fig. P5.109

5.110 and 5.111 (*a*) Using singularity functions, write the equations for the shear and bending moment for the beam and loading shown. (*b*) Determine the maximum normal stress due to bending.

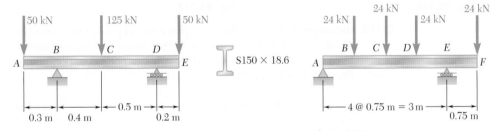

Fig. P5.110

Fig. P5.111

5.112 and 5.113 (*a*) Using singularity functions, find the magnitude and location of the maximum bending moment for the beam and loading shown. (*b*) Determine the maximum normal stress due to bending.

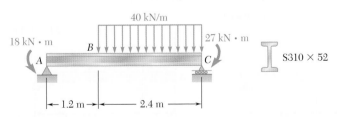

Fig. P5.112

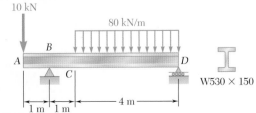

Fig. P5.113

5.114 and 5.115 A beam is being designed to be supported and loaded as shown. (*a*) Using singularity functions, find the magnitude and location of the maximum bending moment in the beam. (*b*) Knowing that the allowable normal stress for the steel to be used is 165 MPa, find the most economical wide-flange shape that can be used.

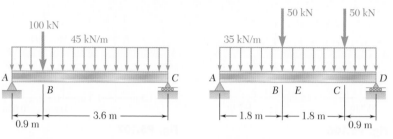

Fig. P5.114   **Fig. P5.115**

5.116 and 5.117 A timber beam is being designed with supports and loads as shown. (*a*) Using singularity functions, find the magnitude and location of the maximum bending moment in the beam. (*b*) Knowing that the available stock consists of beams with an allowable stress of 12 MPa and a rectangular cross section of 30-mm width and depth h varying from 80 mm to 160 mm in 10-mm increments, determine the most economical cross section that can be used.

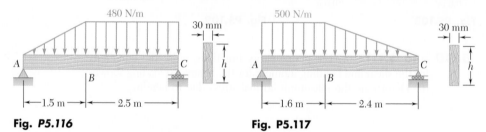

Fig. P5.116   **Fig. P5.117**

5.118 through 5.121 Using a computer and step functions, calculate the shear and bending moment for the beam and loading shown. Use the specified increment ΔL, starting at point A and ending at the right-hand support.

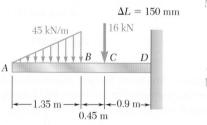

Fig. P5.118

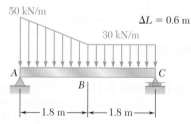

Fig. P5.119

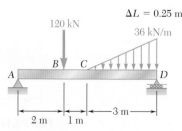

Fig. P5.120

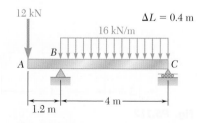

Fig. P5.121

5.122 and 5.123 For the beam and loading shown, and using a computer and step functions, (a) tabulate the shear, bending moment, and maximum normal stress in sections of the beam from $x = 0$ to $x = L$, using the increments ΔL indicated, (b) using smaller increments if necessary, determine with a 2 percent accuracy the maximum normal stress in the beam. Place the origin of the x axis at end A of the beam.

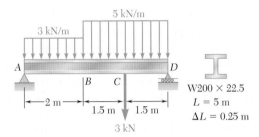

W200 × 22.5
$L = 5$ m
$\Delta L = 0.25$ m

Fig. P5.122

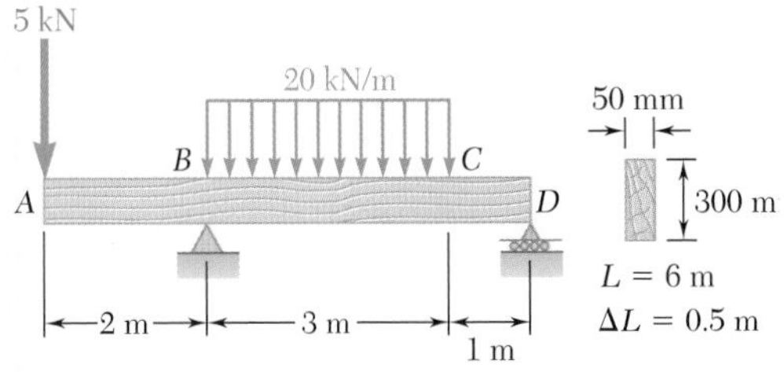

5 kN
20 kN/m
50 mm
300 m
$L = 6$ m
$\Delta L = 0.5$ m

Fig. P5.123

5.124 and 5.125 For the beam and loading shown, and using a computer and step functions, (a) tabulate the shear, bending moment, and maximum normal stress in sections of the beam from $x = 0$ to $x = L$, using the increments ΔL indicated, (b) using smaller increments if necessary, determine with a 2 percent accuracy the maximum normal stress in the beam. Place the origin of the x axis at end A of the beam.

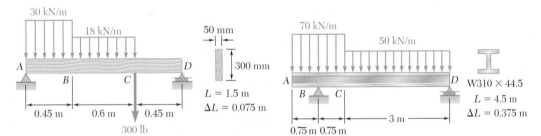

30 kN/m
18 kN/m
50 mm
300 mm
$L = 1.5$ m
$\Delta L = 0.075$ m
0.45 m   0.6 m   0.45 m
300 lb

Fig. P5.124

70 kN/m
50 kN/m
W310 × 44.5
$L = 4.5$ m
$\Delta L = 0.375$ m
3 m
0.75 m 0.75 m

Fig. P5.125

*5.6 NONPRISMATIC BEAMS

Our analysis has been limited so far to prismatic beams, i.e., to beams of uniform cross section. As we saw in Sec. 5.4, prismatic beams are designed so that the normal stresses in their critical sections are at most equal to the allowable value of the normal stress for the material being used. It follows that, in all other sections, the normal stresses will be smaller, possibly much smaller, than their allowable value. A prismatic beam, therefore, is almost always overdesigned, and considerable savings of material can be realized by using nonprismatic beams, i.e., beams of variable cross section. The cantilever beams shown in the bridge during construction in Photo 5.2 are examples of nonprismatic beams.

Since the maximum normal stresses σ_m usually control the design of a beam, the design of a nonprismatic beam will be optimum if the

Photo 5.2 Nonprismatic cantilever beams of bridge during construction.

section modulus $S = I/c$ of every cross section satisfies Eq. (5.3) of Sec. 5.1. Solving that equation for S, we write

$$S = \frac{|M|}{\sigma_{all}} \qquad (5.18)$$

A beam designed in this manner is referred to as a *beam of constant strength*.

For a forged or cast structural or machine component, it is possible to vary the cross section of the component along its length and to eliminate most of the unnecessary material (see Example 5.07). For a timber beam or a rolled-steel beam, however, it is not possible to vary the cross section of the beam. But considerable savings of material can be achieved by gluing wooden planks of appropriate lengths to a timber beam (see Sample Prob. 5.11) and using cover plates in portions of a rolled-steel beam where the bending moment is large (see Sample Prob. 5.12).

EXAMPLE 5.07

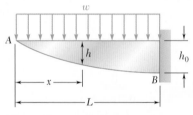

Fig. 5.21

A cast-aluminum plate of uniform thickness b is to support a uniformly distributed load w as shown in Fig. 5.21. (a) Determine the shape of the plate that will yield the most economical design. (b) Knowing that the allowable normal stress for the aluminum used is 72 MPa and that $b = 40$ mm, $L = 800$ mm, and $w = 135$ kN/m, determine the maximum depth h_0 of the plate.

Bending Moment. Measuring the distance x from A and observing that $V_A = M_A = 0$, we use Eqs. (5.6) and (5.8) of Sec. 5.3 and write

$$V(x) = -\int_0^x w\,dx = -wx$$

$$M(x) = \int_0^x V(x)\,dx = -\int_0^x wx\,dx = -\tfrac{1}{2}wx^2$$

(a) Shape of Plate. We recall from Sec. 5.4 that the modulus S of a rectangular cross section of width b and depth h is $S = \frac{1}{6}bh^2$. Carrying this value into Eq. (5.18) and solving for h^2, we have

$$h^2 = \frac{6|M|}{b\sigma_{all}} \qquad (5.19)$$

and, after substituting $|M| = \tfrac{1}{2}wx^2$,

$$h^2 = \frac{3wx^2}{b\sigma_{all}} \quad \text{or} \quad h = \left(\frac{3w}{b\sigma_{all}}\right)^{1/2} x \qquad (5.20)$$

Since the relation between h and x is linear, the lower edge of the plate is a straight line. Thus, the plate providing the most economical design is of *triangular shape*.

(b) Maximum Depth h_0. Making $x = L$ in Eq. (5.20) and substituting the given data, we obtain

$$h_0 = \left[\frac{3(135 \text{ kN/m})}{(0.040 \text{ m})(72 \text{ MPa})}\right]^{1/2} (800 \text{ mm}) = 300 \text{ mm}$$

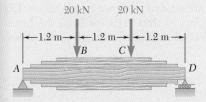

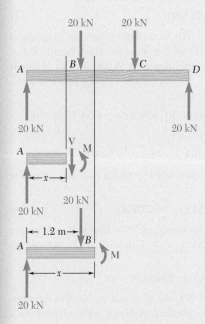

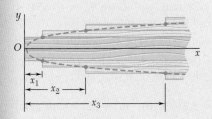

SAMPLE PROBLEM 5.11

A 3.6-m-long beam made of a timber with an allowable normal stress of 16 MPa and an allowable shearing stress of 3 MPa is to carry two 20-kN loads located at its third points. As shown in Chap. 6, a beam of uniform rectangular cross section, 100 mm wide and 115 mm deep, would satisfy the allowable shearing stress requirement. Since such a beam would not satisfy the allowable normal stress requirement, it will be reinforced by gluing planks of the same timber, 100 mm wide and 32 mm thick, to the top and bottom of the beam in a symmetric manner. Determine (*a*) the required number of pairs of planks, (*b*) the length of the planks in each pair that will yield the most economical design.

SOLUTION

Bending Moment. We draw the free-body diagram of the beam and find the following expressions for the bending moment:

From A to B ($0 \leq x \leq 1.2$ m): $M = (20 \text{ kN})x$

From B to C (1.2 m $\leq x \leq$ 2.4 m):
$$M = (20 \text{ kN}) x - (20 \text{ kN})(x - 1.2 \text{ m}) = 24 \text{ kN} \cdot \text{m}$$

a. Number of Pairs of Planks. We first determine the required total depth of the reinforced beam between B and C. We recall from Sec. 5.4 that $S = \frac{1}{6} bh^2$ for a beam with a rectangular cross section of width b and depth h. Substituting this value into Eq. (5.17) and solving for h^2, we have

$$h^2 = \frac{6|M|}{b\sigma_{\text{all}}} \tag{1}$$

Substituting the value obtained for M from B to C and the given values of b and σ_{all}, we write

$$h^2 = \frac{6(24 \text{ kN} \cdot \text{m})}{(0.1 \text{ m})(16 \text{ MPa})} = 0.09 \text{ m}^2 \qquad h = 300 \text{ mm}$$

Since the original beam has a depth of 115 mm, the planks must provide an additional depth of 185 mm. Recalling that each pair of planks is 64 mm thick:

Required number of pairs of planks = 3 ◄

b. Length of Planks. The bending moment was found to be $M = (20 \text{ kN})x$ in the portion AB of the beam. Substituting this expression and the given values of b and σ_{all}, into Eq. (1) and solving for x, we have

$$x = \frac{(0.1 \text{ m})(16 \text{ MPa})}{6(20 \text{ kN})}h^2 \qquad x = \frac{h^2}{0.075 \text{ m}} \tag{2}$$

Equation (2) defines the maximum distance x from end A at which a given depth h of the cross section is acceptable. Making $h = 115$ mm, we find the distance x_1 from A at which the original prismatic beam is safe: $x_1 = 0.1763$ m. From that point on, the original beam should be reinforced by the first pair of planks. Making $h = 115$ mm + 64 mm = 179 mm yields the distance $x_2 = 0.427$ m from which the second pair of planks should be used, and making $h = 243$ mm yields the distance $x_3 = 0.787$ m from which the third pair of planks should be used. The length l_i of the planks of the pair i, where $i = 1, 2, 3$, is obtained by subtracting $2x_i$ from the 3.6-m length of the beam. We find

$$l_1 = 3.25 \text{ m}, l_2 = 2.75 \text{ m}, l_3 = 2.03 \text{ m} ◄$$

The corners of the various planks lie on the parabola defined by Eq. (2).

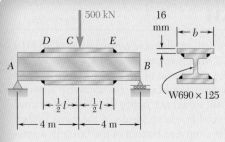

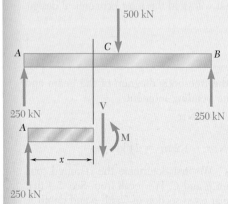

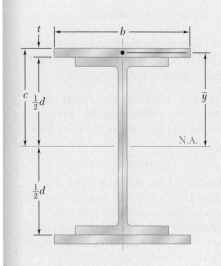

SAMPLE PROBLEM 5.12

Two steel plates, each 16 mm thick, are welded as shown to a W690 × 125 beam to reinforce it. Knowing that σ_{all} = 160 MPa for both the beam and the plates, determine the required value of (a) the length of the plates, (b) the width of the plates.

SOLUTION

Bending Moment. We first find the reactions. From the free body of a portion of beam of length $x \le 4$ m, we obtain M between A and C:

$$M = (250 \text{ kN})x \tag{1}$$

a. Required Length of Plates. We first determine the maximum allowable length x_m of the portion AD of the unreinforced beam. From Appendix C we find that the section modulus of a W690 × 125 beam is $S = 3490 \times 10^6 \text{ mm}^3$, or $S = 3.49 \times 10^{-3} \text{ m}^3$. Substituting for S and σ_{all} into Eq. (5.17) and solving for M, we write

$$M = S\sigma_{all} = (3.49 \times 10^{-3} \text{ m}^3)(160 \times 10^3 \text{ kN/m}^2) = 558.4 \text{ kN} \cdot \text{m}$$

Substituting for M in Eq. (1), we have

$$558.4 \text{ kN} \cdot \text{m} = (250 \text{ kN})x_m \qquad x_m = 2.234 \text{ m}$$

The required length l of the plates is obtained by subtracting $2x_m$ from the length of the beam:

$$l = 8 \text{ m} - 2(2.234 \text{ m}) = 3.532 \text{ m} \qquad l = 3.53 \text{ m} \blacktriangleleft$$

b. Required Width of Plates. The maximum bending moment occurs in the midsection C of the beam. Making $x = 4$ m in Eq. (1), we obtain the bending moment in that section:

$$M = (250 \text{ kN})(4 \text{ m}) = 1000 \text{ kN} \cdot \text{m}$$

In order to use Eq. (5.1) of Sec. 5.1, we now determine the moment of inertia of the cross section of the reinforced beam with respect to a centroidal axis and the distance c from that axis to the outer surfaces of the plates. From Appendix C we find that the moment of inertia of a W690 × 125 beam is $I_b = 1190 \times 10^6 \text{ mm}^4$ and its depth is $d = 678$ mm. On the other hand, denoting by t the thickness of one plate, by b its width, and by $\overline{y}$ the distance of its centroid from the neutral axis, we express the moment of inertia I_p of the two plates with respect to the neutral axis:

$$I_p = 2(\tfrac{1}{12}bt^3 + A\overline{y}^2) = (\tfrac{1}{6}t^3)b + 2bt(\tfrac{1}{2}d + \tfrac{1}{2}t)^2$$

Substituting $t = 16$ mm and $d = 678$ mm, we obtain $I_p = (3.854 \times 10^6 \text{ mm}^3)b$. The moment of inertia I of the beam and plates is

$$I = I_b + I_p = 1190 \times 10^6 \text{ mm}^4 + (3.854 \times 10^6 \text{ mm}^3)b \tag{2}$$

and the distance from the neutral axis to the surface is $c = \tfrac{1}{2}d + t = 355$ mm. Solving Eq. (5.1) for I and substituting the values of M, σ_{all}, and c, we write

$$I = \frac{|M|c}{\sigma_{all}} = \frac{(1000 \text{ kN} \cdot \text{m})(355 \text{ mm})}{160 \text{ MPa}} = 2.219 \times 10^{-3} \text{ m}^4 = 2219 \times 10^6 \text{ mm}^4$$

Replacing I by this value in Eq. (2) and solving for b, we have

$$2219 \times 10^6 \text{ mm}^4 = 1190 \times 10^6 \text{ mm}^4 + (3.854 \times 10^6 \text{ mm}^3)b$$

$$b = 267 \text{ mm} \blacktriangleleft$$

PROBLEMS

5.126 and 5.127 The beam AB, consisting of an aluminum plate of uniform thickness b and length L, is to support the load shown. (*a*) Knowing that the beam is to be of constant strength, express h in terms of x, L, and h_0 for portion AC of the beam. (*b*) Determine the maximum allowable load if $L = 800$ mm, $h_0 = 200$ mm, $b = 25$ mm, and $\sigma_{all} = 72$ MPa.

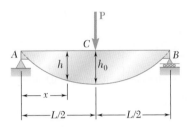

Fig. P5.126

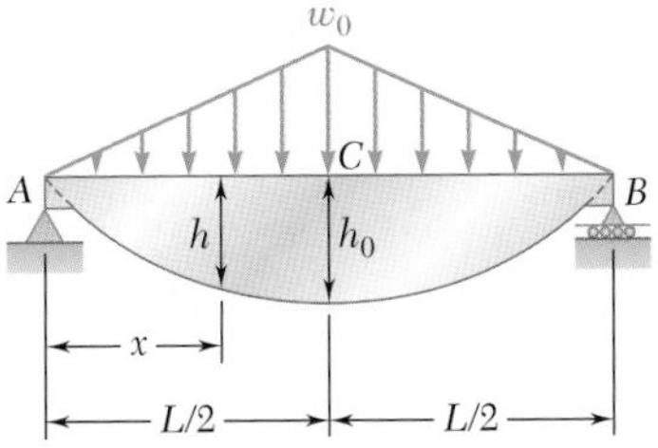

Fig. P5.127

5.128 and 5.129 The beam AB, consisting of a cast-iron plate of uniform thickness b and length L, is to support the load shown. (*a*) Knowing that the beam is to be of constant strength, express h in terms of x, L, and h_0. (*b*) Determine the maximum allowable load if $L = 0.9$ m, $h_0 = 300$ mm, $b = 30$ mm, and $\sigma_{all} = 165$ MPa.

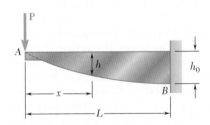

Fig. P5.128

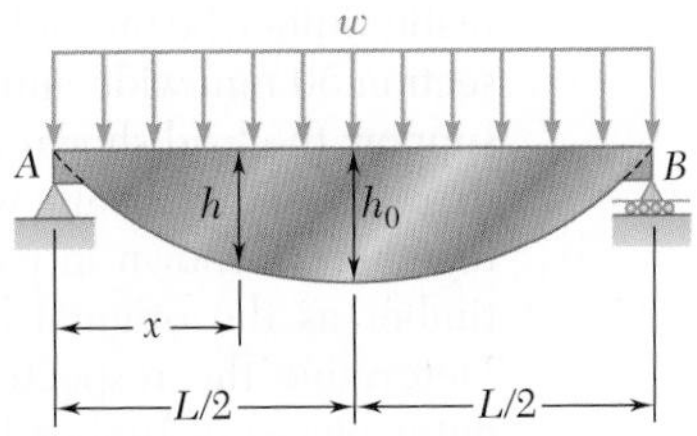

Fig. P5.129

5.130 and 5.131 The beam AB, consisting of a cast-iron plate of uniform thickness b and length L, is to support the distributed load $w(x)$ shown. (*a*) Knowing that the beam is to be of constant strength, express h in terms of x, L, and h_0. (*b*) Determine the smallest value of h_0 if $L = 750$ mm, $b = 30$ mm, $w_0 = 300$ kN/m, and $\sigma_{all} = 200$ MPa.

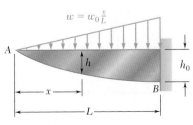

$$w = w_0 \frac{x}{L}$$

Fig. P5.130

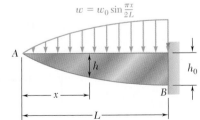

$$w = w_0 \sin \frac{\pi x}{2L}$$

Fig. P5.131

5.132 and 5.133 A preliminary design on the use of a simply supported prismatic timber beam indicated that a beam with a rectangular cross section 50 mm wide and 200 mm deep would be required to safely support the load shown in part *a* of the figure. It was then decided to replace that beam with a built-up beam obtained by gluing together, as shown in part *b* of the figure, four pieces of the same timber as the original beam and of 50×50-mm cross section. Determine the length *l* of the two outer pieces of timber that will yield the same factor of safety as the original design.

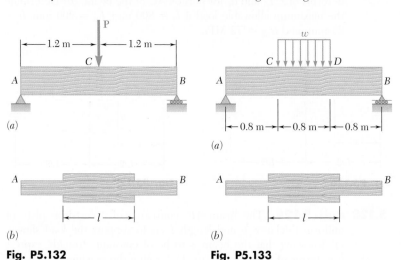

(a)   (a)

(b)   (b)

Fig. P5.132   **Fig. P5.133**

5.134 and *5.135* A preliminary design on the use of a cantilever prismatic timber beam indicated that a beam with a rectangular cross section 50 mm wide and 250 mm deep would be required to safely support the load shown in part *a* of the figure. It was then decided to replace that beam with a built-up beam obtained by gluing together, as shown in part *b* of the figure, five pieces of the same timber as the original beam and of 50×50-mm cross section. Determine the respective lengths l_1 and l_2 of the two inner and outer pieces of timber that will yield the same factor of safety as the original design.

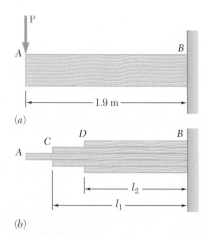

(a)

(b)

Fig. P5.134

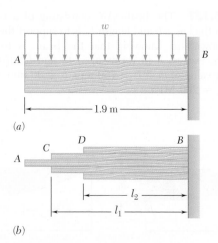

(a)

(b)

Fig. P5.135

5.155 Determine (a) the equations of the shear and bending-moment curves for the beam and loading shown, (b) the maximum absolute value of the bending moment in the beam.

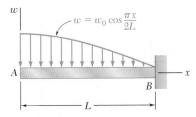

Fig. P5.155

5.156 Draw the shear and bending-moment diagrams for the beam and loading shown and determine the maximum normal stress due to bending.

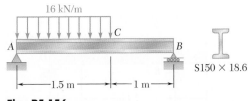

Fig. P5.156

5.157 Knowing that beam AB is in equilibrium under the loading shown, draw the shear and bending-moment diagrams and determine the maximum normal stress due to bending.

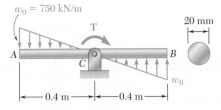

Fig. P5.157

5.158 For the beam and loading shown, design the cross section of the beam, knowing that the grade of timber used has an allowable normal stress of 12 MPa.

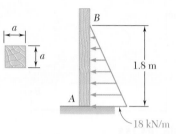

Fig. P5.158

5.159 Knowing that the allowable stress for the steel used is 160 MPa, select the most economical wide-flange beam to support the loading shown.

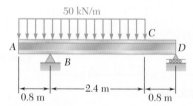

Fig. P5.159

5.160 Determine the largest permissible value of **P** for the beam and loading shown, knowing that the allowable normal stress is +80 MPa in tension and −140 MPa in compression.

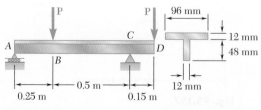

Fig. P5.160

5.161 (*a*) Using singularity functions, find the magnitude and location of the maximum bending moment for the beam and loading shown. (*b*) Determine the maximum normal stress due to bending.

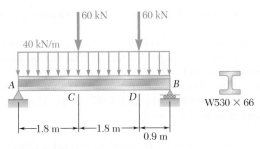

W530 × 66

Fig. P5.161

5.162 The beam *AB*, consisting of an aluminum plate of uniform thickness *b* and length *L*, is to support the load shown. (*a*) Knowing that the beam is to be of constant strength, express *h* in terms of *x*, *L*, and h_0 for portion *AC* of the beam. (*b*) Determine the maximum allowable load if $L = 800$ mm, $h_0 = 200$ mm, $b = 25$ mm, and $\sigma_{\text{all}} = 72$ MPa.

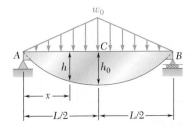

Fig. P5.162

5.163 A transverse force **P** is applied as shown at end *A* of the conical taper *AB*. Denoting by d_0 the diameter of the taper at *A*, show that the maximum normal stress occurs at point *H*, which is contained in a transverse section of diameter $d = 1.5\,d_0$.

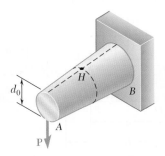

Fig. P5.163

COMPUTER PROBLEMS

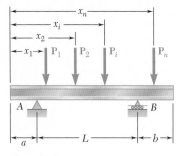

Fig. P5.C1

The following problems are designed to be solved with a computer.

5.C1 Several concentrated loads P_i ($i = 1, 2, \ldots , n$) can be applied to a beam as shown. Write a computer program that can be used to calculate the shear, bending moment, and normal stress at any point of the beam for a given loading of the beam and a given value of its section modulus. Use this program to solve Probs. 5.18 and 5.25. (*Hint:* Maximum values will occur at a support or under a load.)

5.C2 A timber beam is to be designed to support a distributed load and up to two concentrated loads as shown. One of the dimensions of its uniform rectangular cross section has been specified and the other is to be determined so that the maximum normal stress in the beam will not exceed a given allowable value σ_{all}. Write a computer program that can be used to calculate at given intervals ΔL the shear, the bending moment, and the smallest acceptable value of the unknown dimension. Apply this program to solve the following problems, using the intervals ΔL indicated: (*a*) Prob. 5.65 ($\Delta L = 0.1$ m), (*b*) Prob. 5.69 ($\Delta L = 0.3$ m), (*c*) Prob. 5.70 ($\Delta L = 0.2$ m).

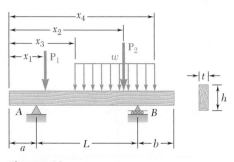

Fig. P5.C2

5.C3 Two cover plates, each of thickness t, are to be welded to a wide-flange beam of length L that is to support a uniformly distributed load w. Denoting by σ_{all} the allowable normal stress in the beam and in the plates, by d the depth of the beam, and by I_b and S_b, respectively, the moment of inertia and the section modulus of the cross section of the unreinforced beam about a horizontal centroidal axis, write a computer program that can be used to calculate the required value of (*a*) the length a of the plates, (*b*) the width b of the plates. Use this program to solve Prob. 5.145.

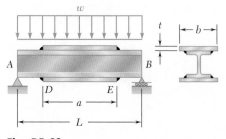

Fig. P5.C3

5.C4 Two 100-kN loads are maintained 1.8 m apart as they are moved slowly across the 6-m beam *AB*. Write a computer program and use it to calculate the bending moment under each load and at the midpoint *C* of the beam for values of *x* from 0 to 7 m at intervals $\Delta x = 0.45$ m.

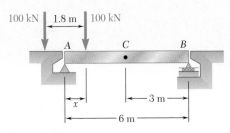

Fig. P5.C4

5.C5 Write a computer program that can be used to plot the shear and bending-moment diagrams for the beam and loading shown. Apply this program with a plotting interval $\Delta L = 0.05$ m to the beam and loading of Prob. 5.72.

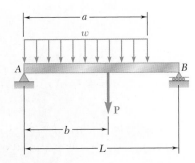

Fig. P5.C5

5.C6 Write a computer program that can be used to plot the shear and bending-moment diagrams for the beam and loading shown. Apply this program with a plotting interval $\Delta L = 0.025$ m to the beam and loading of Prob. 5.112.

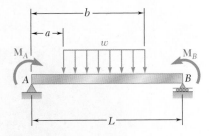

Fig. P5.C6

A reinforced concrete deck will be attached to each of the steel sections shown to form a composite box girder bridge. In this chapter the shearing stresses will be determined in various types of beams and girders.

Shearing Stresses in Beams and Thin-Walled Members

6.1 INTRODUCTION

You saw in Sec. 5.1 that a transverse loading applied to a beam will result in normal and shearing stresses in any given transverse section of the beam. The normal stresses are created by the bending couple **M** in that section and the shearing stresses by the shear **V**. Since the dominant criterion in the design of a beam for strength is the maximum value of the normal stress in the beam, our analysis was limited in Chap. 5 to the determination of the normal stresses. Shearing stresses, however, can be important, particularly in the design of short, stubby beams, and their analysis will be the subject of the first part of this chapter.

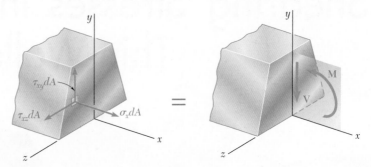

Fig. 6.1 Beam cross section.

Figure 6.1 expresses graphically that the elementary normal and shearing forces exerted on a given transverse section of a prismatic beam with a vertical plane of symmetry are equivalent to the bending couple **M** and the shearing force **V**. Six equations can be written to express that fact. Three of these equations involve only the normal forces $\sigma_x \, dA$ and have already been discussed in Sec. 4.2; they are Eqs. (4.1), (4.2), and (4.3), which express that the sum of the normal forces is zero and that the sums of their moments about the y and z axes are equal to zero and M, respectively. Three more equations involving the shearing forces $\tau_{xy} \, dA$ and $\tau_{xz} \, dA$ can now be written. One of them expresses that the sum of the moments of the shearing forces about the x axis is zero and can be dismissed as trivial in view of the symmetry of the beam with respect to the xy plane. The other two involve the y and z components of the elementary forces and are

$$y \text{ components:} \quad \int \tau_{xy} \, dA = -V \tag{6.1}$$

$$z \text{ components:} \quad \int \tau_{xz} \, dA = 0 \tag{6.2}$$

The first of these equations shows that vertical shearing stresses must exist in a transverse section of a beam under transverse loading. The second equation indicates that the average horizontal shearing stress in any section is zero. However, this does not mean that the shearing stress τ_{xz} is zero everywhere.

Let us now consider a small cubic element located in the vertical plane of symmetry of the beam (where we know that τ_{xz} must be zero) and examine the stresses exerted on its faces (Fig. 6.2). As we

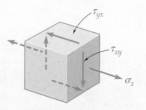

Fig. 6.2 Element from beam.

have just seen, a normal stress σ_x and a shearing stress τ_{xy} are exerted on each of the two faces perpendicular to the x axis. But we know from Chap. 1 that, when shearing stresses τ_{xy} are exerted on the vertical faces of an element, equal stresses must be exerted on the horizontal faces of the same element. We thus conclude that longitudinal shearing stresses must exist in any member subjected to a transverse loading. This can be verified by considering a cantilever beam made of separate planks clamped together at one end (Fig. 6.3a). When a transverse load **P** is applied to the free end of this composite beam, the planks are observed to slide with respect to each other (Fig. 6.3b). In contrast, if a couple **M** is applied to the free end of the same composite beam (Fig. 6.3c), the various planks will bend into concentric arcs of circle and will not slide with respect to each other, thus verifying the fact that shear does not occur in a beam subjected to pure bending (cf. Sec. 4.3).

While sliding does not actually take place when a transverse load **P** is applied to a beam made of a homogeneous and cohesive material such as steel, the tendency to slide does exist, showing that stresses occur on horizontal longitudinal planes as well as on vertical transverse planes. In the case of timber beams, whose resistance to shear is weaker between fibers, failure due to shear will occur along a longitudinal plane rather than a transverse plane (Photo 6.1).

In Sec. 6.2, a beam element of length Δx bounded by two transverse planes and a horizontal one will be considered and the shearing force $\Delta \mathbf{H}$ exerted on its horizontal face will be determined, as well as the shear per unit length, q, also known as *shear flow*. A formula for the shearing stress in a beam with a vertical plane of symmetry will be derived in Sec. 6.3 and used in Sec. 6.4 to determine the shearing stresses in common types of beams. The distribution of stresses in a narrow rectangular beam will be further discussed in Sec. 6.5.

The derivation given in Sec. 6.2 will be extended in Sec. 6.6 to cover the case of a beam element bounded by two transverse planes and a curved surface. This will allow us in Sec. 6.7 to determine the shearing stresses at any point of a symmetric thin-walled member, such as the flanges of wide-flange beams and box beams. The effect of plastic deformations on the magnitude and distribution of shearing stresses will be discussed in Sec. 6.8.

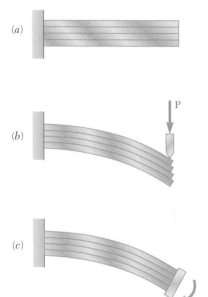

(a)

(b)

(c)

Fig. 6.3 Beam made from planks.

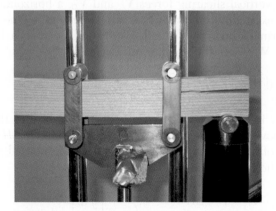

Photo 6.1 Longitudinal shear failure in timber beam.

In the last section of the chapter (Sec. 6.9), the unsymmetric
loading of thin-walled members will be considered and the concept
of *shear center* will be introduced. You will then learn to determine
the distribution of shearing stresses in such members.

6.2 SHEAR ON THE HORIZONTAL FACE OF A BEAM ELEMENT

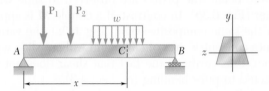

Fig. 6.4 Beam example.

Consider a prismatic beam *AB* with a vertical plane of symmetry that
supports various concentrated and distributed loads (Fig. 6.4). At a
distance x from end *A* we detach from the beam an element $CDD'C'$
of length Δx extending across the width of the beam from the upper
surface of the beam to a horizontal plane located at a distance y_1
from the neutral axis (Fig. 6.5). The forces exerted on this element

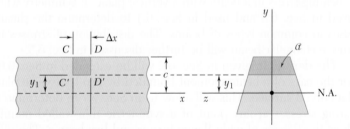

Fig. 6.5 Short segment of beam example.

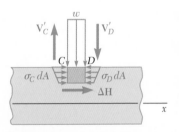

Fig. 6.6 Forces exerted on
element.

consist of vertical shearing forces $\mathbf{V}'_C$ and $\mathbf{V}'_D$, a horizontal shearing
force $\Delta\mathbf{H}$ exerted on the lower face of the element, elementary horizontal normal forces $\sigma_C\,dA$ and $\sigma_D\,dA$, and possibly a load $w\,\Delta x$
(Fig. 6.6). We write the equilibrium equation

$$\xrightarrow{+}\Sigma F_x = 0: \qquad \Delta H + \int_a (\sigma_C - \sigma_D)\,dA = 0$$

where the integral extends over the shaded area a of the section
located above the line $y = y_1$. Solving this equation for ΔH and using
Eq. (5.2) of Sec. 5.1, $\sigma = My/I$, to express the normal stresses in
terms of the bending moments at C and D, we have

$$\Delta H = \frac{M_D - M_C}{I} \int_a y\,dA \qquad (6.3)$$

The integral in (6.3) represents the *first moment* with respect to the neutral axis of the portion $\mathfrak{a}$ of the cross section of the beam that is located above the line $y = y_1$ and will be denoted by Q. On the other hand, recalling Eq. (5.7) of Sec. 5.3, we can express the increment $M_D - M_C$ of the bending moment as

$$M_D - M_C = \Delta M = (dM/dx)\,\Delta x = V\,\Delta x$$

Substituting into (6.3), we obtain the following expression for the horizontal shear exerted on the beam element

$$\Delta H = \frac{VQ}{I}\,\Delta x \tag{6.4}$$

The same result would have been obtained if we had used as a free body the lower element $C'D'D''C''$, rather than the upper element $CDD'C'$ (Fig. 6.7), since the shearing forces $\Delta\mathbf{H}$ and $\Delta\mathbf{H}'$

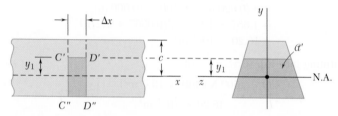

Fig. 6.7 Short segment of beam example.

exerted by the two elements on each other are equal and opposite. This leads us to observe that the first moment Q of the portion $\mathfrak{a}'$ of the cross section located below the line $y = y_1$ (Fig. 6.7) is equal in magnitude and opposite in sign to the first moment of the portion $\mathfrak{a}$ located above that line (Fig. 6.5). Indeed, the sum of these two moments is equal to the moment of the area of the entire cross section with respect to its centroidal axis and, thus, must be zero. This property can sometimes be used to simplify the computation of Q. We also note that Q is maximum for $y_1 = 0$, since the elements of the cross section located above the neutral axis contribute positively to the integral in (6.3) that defines Q, while the elements located below that axis contribute negatively.

The *horizontal shear per unit length*, which will be denoted by the letter q, is obtained by dividing both members of Eq. (6.4) by Δx:

$$q = \frac{\Delta H}{\Delta x} = \frac{VQ}{I} \tag{6.5}$$

We recall that Q is the first moment with respect to the neutral axis of the portion of the cross section located either above or below the point at which q is being computed, and that I is the centroidal moment of inertia of the *entire* cross-sectional area. For a reason that will become apparent later (Sec. 6.7), the horizontal shear per unit length q is also referred to as the *shear flow*.

EXAMPLE 6.01

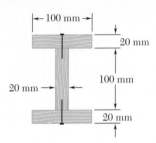

Fig. 6.8

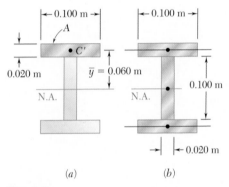

(a)   (b)

Fig. 6.9

A beam is made of three planks, 20 by 100 mm in cross section, nailed together (Fig. 6.8). Knowing that the spacing between nails is 25 mm and that the vertical shear in the beam is $V = 500$ N, determine the shearing force in each nail.

We first determine the horizontal force per unit length, q, exerted on the lower face of the upper plank. We use Eq. (6.5), where Q represents the first moment with respect to the neutral axis of the shaded area A shown in Fig. 6.9a, and where I is the moment of inertia about the same axis of the entire cross-sectional area (Fig. 6.9b). Recalling that the first moment of an area with respect to a given axis is equal to the product of the area and of the distance from its centroid to the axis,† we have

$$Q = A\bar{y} = (0.020 \text{ m} \times 0.100 \text{ m})(0.060 \text{ m})$$
$$= 120 \times 10^{-6} \text{ m}^3$$
$$I = \tfrac{1}{12}(0.020 \text{ m})(0.100 \text{ m})^3$$
$$+2[\tfrac{1}{12}(0.100 \text{ m})(0.020 \text{ m})^3$$
$$+(0.020 \text{ m} \times 0.100 \text{ m})(0.060 \text{ m})^2]$$
$$= 1.667 \times 10^{-6} + 2(0.0667 + 7.2)10^{-6}$$
$$= 16.20 \times 10^{-6} \text{ m}^4$$

Substituting into Eq. (6.5), we write

$$q = \frac{VQ}{I} = \frac{(500 \text{ N})(120 \times 10^{-6} \text{ m}^3)}{16.20 \times 10^{-6} \text{ m}^4} = 3704 \text{ N/m}$$

Since the spacing between the nails is 25 mm, the shearing force in each nail is

$$F = (0.025 \text{ m})q = (0.025 \text{ m})(3704 \text{ N/m}) = 92.6 \text{ N}$$

6.3 DETERMINATION OF THE SHEARING STRESSES IN A BEAM

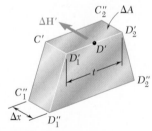

Fig. 6.10 Beam element.

Consider again a beam with a vertical plane of symmetry, subjected to various concentrated or distributed loads applied in that plane. We saw in the preceding section that if, through two vertical cuts and one horizontal cut, we detach from the beam an element of length Δx (Fig. 6.10), the magnitude ΔH of the shearing force exerted on the horizontal face of the element can be obtained from Eq. (6.4). The *average shearing stress* τ_{ave} on that face of the element is obtained by dividing ΔH by the area ΔA of the face. Observing that $\Delta A = t \, \Delta x$, where t is the width of the element at the cut, we write

$$\tau_{\text{ave}} = \frac{\Delta H}{\Delta A} = \frac{VQ}{I} \frac{\Delta x}{t \, \Delta x}$$

or

$$\tau_{\text{ave}} = \frac{VQ}{It} \qquad (6.6)$$

†See Appendix A.

We note that, since the shearing stresses τ_{xy} and τ_{yx} exerted respectively on a transverse and a horizontal plane through D' are equal, the expression obtained also represents the average value of τ_{xy} along the line $D_1'D_2'$ (Fig. 6.11).

We observe that $\tau_{yx} = 0$ on the upper and lower faces of the beam, since no forces are exerted on these faces. It follows that $\tau_{xy} = 0$ along the upper and lower edges of the transverse section (Fig. 6.12). We also note that, while Q is maximum for $y = 0$ (see Sec. 6.2), we cannot conclude that τ_{ave} will be maximum along the neutral axis, since τ_{ave} depends upon the width t of the section as well as upon Q.

As long as the width of the beam cross section remains small compared to its depth, the shearing stress varies only slightly along the line $D_1'D_2'$ (Fig. 6.11) and Eq. (6.6) can be used to compute τ_{xy} at any point along $D_1'D_2'$. Actually, τ_{xy} is larger at points D_1' and D_2' than at D', but the theory of elasticity shows† that, for a beam of rectangular section of width b and depth h, and as long as $b \le h/4$, the value of the shearing stress at points C_1 and C_2 (Fig. 6.13) does not exceed by more than 0.8 percent the average value of the stress computed along the neutral axis.‡

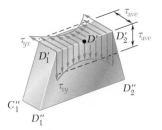

Fig. 6.11 Beam segment.

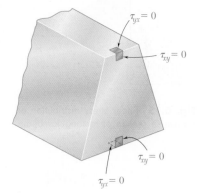

Fig. 6.12 Beam cross section.

6.4 SHEARING STRESSES τ_{xy} IN COMMON TYPES OF BEAMS

We saw in the preceding section that, for a *narrow rectangular beam*, i.e., for a beam of rectangular section of width b and depth h with $b \le \frac{1}{4}h$, the variation of the shearing stress τ_{xy} across the width of the beam is less than 0.8 percent of τ_{ave}. We can, therefore, use Eq. (6.6) in practical applications to determine the shearing stress at any point of the cross section of a narrow rectangular beam and write

$$\tau_{xy} = \frac{VQ}{It} \tag{6.7}$$

where t is equal to the width b of the beam, and where Q is the first moment with respect to the neutral axis of the shaded area A (Fig. 6.14).

Observing that the distance from the neutral axis to the centroid C' of A is $\bar{y} = \frac{1}{2}(c + y)$, and recalling that $Q = A\bar{y}$, we write

$$Q = A\bar{y} = b(c - y)\tfrac{1}{2}(c + y) = \tfrac{1}{2}b(c^2 - y^2) \tag{6.8}$$

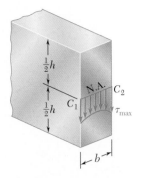

Fig. 6.13 Rectangular beam cross section.

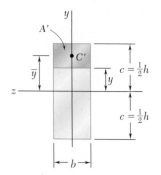

Fig. 6.14 Beam cross section.

†See S. P. Timoshenko and J. N. Goodier, *Theory of Elasticity*, McGraw-Hill, New York, 3d ed., 1970, sec. 124.

‡On the other hand, for large values of b/h, the value τ_{max} of the stress at C_1 and C_2 may be many times larger then the average value τ_{ave} computed along the neutral axis, as we may see from the following table:

b/h	0.25	0.5	1	2	4	6	10	20	50
τ_{max}/τ_{ave}	1.008	1.033	1.126	1.396	1.988	2.582	3.770	6.740	15.65
τ_{min}/τ_{ave}	0.996	0.983	0.940	0.856	0.805	0.800	0.800	0.800	0.800

Recalling, on the other hand, that $I = bh^3/12 = \frac{2}{3}bc^3$, we have

$$\tau_{xy} = \frac{VQ}{Ib} = \frac{3}{4}\frac{c^2 - y^2}{bc^3}V$$

or, noting that the cross-sectional area of the beam is $A = 2bc$,

$$\tau_{xy} = \frac{3}{2}\frac{V}{A}\left(1 - \frac{y^2}{c^2}\right) \tag{6.9}$$

Equation (6.9) shows that the distribution of shearing stresses in a transverse section of a rectangular beam is *parabolic* (Fig. 6.15). As we have already observed in the preceding section, the shearing stresses are zero at the top and bottom of the cross section ($y = \pm c$). Making $y = 0$ in Eq. (6.9), we obtain the value of the maximum shearing stress in a given section of a *narrow rectangular beam:*

$$\tau_{max} = \frac{3}{2}\frac{V}{A} \tag{6.10}$$

The relation obtained shows that the maximum value of the shearing stress in a beam of rectangular cross section is 50 percent larger than the value V/A that would be obtained by wrongly assuming a uniform stress distribution across the entire cross section.

In the case of an *American standard beam* (S-beam) or a *wide-flange beam* (W-beam), Eq. (6.6) can be used to determine the average value of the shearing stress τ_{xy} over a section aa' or bb' of the transverse cross section of the beam (Figs. 6.16a and b). We write

$$\tau_{ave} = \frac{VQ}{It} \tag{6.6}$$

where V is the vertical shear, t the width of the section at the elevation considered, Q the first moment of the shaded area with respect to the neutral axis cc', and I the moment of inertia of the entire cross-sectional area about cc'. Plotting τ_{ave} against the vertical distance y, we obtain the curve shown in Fig. 6.16c. We note the discontinuities existing in this curve, which reflect the difference between the values of t corresponding respectively to the flanges $ABGD$ and $A'B'G'D'$ and to the web $EFF'E'$.

In the case of the web, the shearing stress τ_{xy} varies only very slightly across the section bb', and can be assumed equal to its average

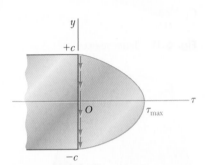

Fig. 6.15 Shear stress distribution on transverse section of rectangular beam.

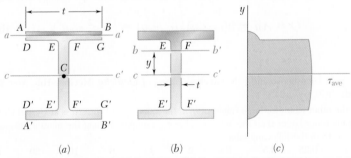

Fig. 6.16 Shear stress distribution on transverse section of wide-flange beam.

value τ_{ave}. This is not true, however, for the flanges. For example, considering the horizontal line $DEFG$, we note that τ_{xy} is zero between D and E and between F and G, since these two segments are part of the free surface of the beam. On the other hand the value of τ_{xy} between E and F can be obtained by making $t = EF$ in Eq. (6.6). In practice, one usually assumes that the entire shear load is carried by the web, and that a good approximation of the maximum value of the shearing stress in the cross section can be obtained by dividing V by the cross-sectional area of the web.

$$\tau_{\text{max}} = \frac{V}{A_{\text{web}}} \tag{6.11}$$

We should note, however, that while the vertical component τ_{xy} of the shearing stress in the flanges can be neglected, its horizontal component τ_{xz} has a significant value that will be determined in Sec. 6.7.

EXAMPLE 6.02

Knowing that the allowable shearing stress for the timber beam of Sample Prob. 5.7 is $\tau_{\text{all}} = 1.75$ MPa, check that the design obtained in that sample problem is acceptable from the point of view of the shearing stresses.

We recall from the shear diagram of Sample Prob. 5.7 that $V_{\text{max}} = 20$ kN. The actual width of the beam was given as $b = 90$ mm, and the value obtained for its depth was $h = 366$ mm. Using Eq. (6.10) for the maximum shearing stress in a narrow rectangular beam, we write

$$\tau_{\text{max}} = \frac{3}{2}\frac{V}{A} = \frac{3}{2}\frac{V}{bh} = \frac{3(20 \text{ kN})}{2(0.09 \text{ m})(0.366 \text{ m})} = 0.911 \text{ MPa}$$

Since $\tau_{\text{max}} < \tau_{\text{all}}$, the design obtained in Sample Prob. 5.7 is acceptable.

EXAMPLE 6.03

Knowing that the allowable shearing stress for the steel beam of Sample Prob. 5.8 is $\tau_{\text{all}} = 90$ MPa, check that the W360 × 32.9 shape obtained in that sample problem is acceptable from the point of view of the shearing stresses.

We recall from the shear diagram of Sample Prob. 5.8 that the maximum absolute value of the shear in the beam is $|V|_{\text{max}} = 58$ kN. As we saw in Sec. 6.4, it may be assumed in practice that the entire shear load is carried by the web and that the maximum value of the shearing stress in the beam can be obtained from Eq. (6.11). From Appendix C we find that for a W360 × 32.9 shape the depth of the beam and the thickness of its web are, respectively, $d = 349$ mm and $t_w = 5.8$ mm. We thus have

$$A_{\text{web}} = d\, t_w = (349 \text{ mm})(5.8 \text{ mm}) = 2024 \text{ mm}^2$$

Substituting the values of $|V|_{\text{max}}$ and A_{web} into Eq. (6.11), we obtain

$$\tau_{\text{max}} = \frac{|V|_{\text{max}}}{A_{\text{web}}} = \frac{58 \text{ kN}}{2024 \text{ mm}^2} = 28.7 \text{ MPa}$$

Since $\tau_{\text{max}} < \tau_{\text{all}}$, the design obtained in Sample Prob. 5.8 is acceptable.

*6.5 FURTHER DISCUSSION OF THE DISTRIBUTION OF STRESSES IN A NARROW RECTANGULAR BEAM

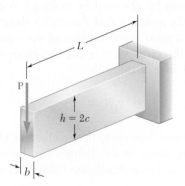

Fig. 6.17 Cantilever beam.

Consider a narrow cantilever beam of rectangular cross section of width b and depth h subjected to a load $\mathbf{P}$ at its free end (Fig. 6.17). Since the shear V in the beam is constant and equal in magnitude to the load $\mathbf{P}$, Eq. (6.9) yields

$$\tau_{xy} = \frac{3}{2}\frac{P}{A}\left(1 - \frac{y^2}{c^2}\right) \tag{6.12}$$

We note from Eq. (6.12) that the shearing stresses depend only upon the distance y from the neutral surface. They are independent, therefore, of the distance from the point of application of the load; it follows that all elements located at the same distance from the neutral surface undergo the same shear deformation (Fig. 6.18). While plane sections do *not* remain plane, the distance between two corresponding points D and D' located in different sections remains the same. This indicates that the normal strains ϵ_x, and thus the normal stresses σ_x, are unaffected by the shearing stresses, and that the assumption made in Sec. 5.1 is justified for the loading condition of Fig. 6.17.

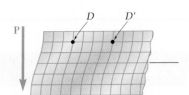

Fig. 6.18 Deformation of segment of cantilever beam.

We conclude that our analysis of the stresses in a cantilever beam of rectangular cross section, subjected to a concentrated load $\mathbf{P}$ at its free end, is valid. The correct values of the shearing stresses in the beam are given by Eq. (6.12), and the normal stresses at a distance x from the free end are obtained by making $M = -Px$ in Eq. (5.2) of Sec. 5.1. We have

$$\sigma_x = +\frac{Pxy}{I} \tag{6.13}$$

The validity of the above statement, however, depends upon the end conditions. If Eq. (6.12) is to apply everywhere, then the load P must be distributed parabolically over the free-end section. Moreover, the fixed-end support must be of such a nature that it will allow the type of shear deformation indicated in Fig. 6.18. The resulting model (Fig. 6.19) is highly unlikely to be encountered in practice. However, it follows from Saint-Venant's principle that, for other modes of application of the load and for other types of fixed-end supports, Eqs. (6.12) and (6.13) still provide us with the correct distribution of stresses, except close to either end of the beam.

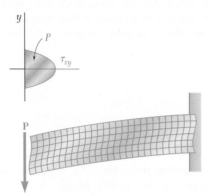

Fig. 6.19 Deformation of cantilever beam with concentrated load.

When a beam of rectangular cross section is subjected to several concentrated loads (Fig. 6.20), the principle of superposition can be used to determine the normal and shearing stresses in sections located between the points of application of the loads. However, since the loads P_2, P_3, etc., are applied on the surface of the beam and cannot be assumed to be distributed parabolically throughout the cross section, the results obtained cease to be valid in the immediate vicinity of the points of application of the loads.

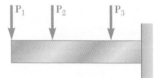

Fig. 6.20 Cantilever beam.

When the beam is subjected to a distributed load (Fig. 6.21), the shear varies with the distance from the end of the beam, and so does the shearing stress at a given elevation y. The resulting shear deformations are such that the distance between two corresponding points of different cross sections, such as D_1 and D_1', or D_2 and D_2', will depend upon their elevation. This indicates that the assumption that plane sections remain plane, under which Eqs. (6.12) and (6.13) were derived, must be rejected for the loading condition of Fig. 6.21. The error involved, however, is small for the values of the span-depth ratio encountered in practice.

We should also note that, in portions of the beam located under a concentrated or distributed load, normal stresses σ_y will be exerted on the horizontal faces of a cubic element of material, in addition to the stresses τ_{xy} shown in Fig. 6.2.

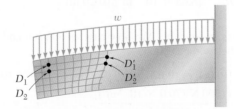

Fig. 6.21 Deformation of cantilever beam with distributed load.

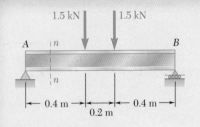

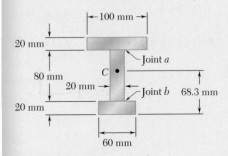

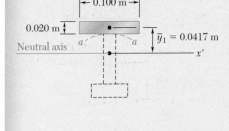

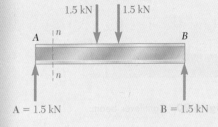

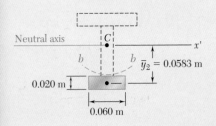

SAMPLE PROBLEM 6.1

Beam AB is made of three planks glued together and is subjected, in its plane of symmetry, to the loading shown. Knowing that the width of each glued joint is 20 mm, determine the average shearing stress in each joint at section n-n of the beam. The location of the centroid of the section is given in the sketch and the centroidal moment of inertia is known to be $I = 8.63 \times 10^{-6} \text{ m}^4$.

SOLUTION

Vertical Shear at Section n-n. Since the beam and loading are both symmetric with respect to the center of the beam, we have $\mathbf{A} = \mathbf{B} = 1.5 \text{ kN} \uparrow$.

Considering the portion of the beam to the left of section n-n as a free body, we write

$$+\uparrow \Sigma F_y = 0: \qquad\qquad 1.5 \text{ kN} - V = 0 \qquad V = 1.5 \text{ kN}$$

Shearing Stress in Joint a. We pass the section a-a through the glued joint and separate the cross-sectional area into two parts. We choose to determine Q by computing the first moment with respect to the neutral axis of the area above section a-a.

$$Q = A\bar{y}_1 = [(0.100 \text{ m})(0.020 \text{ m})](0.0417 \text{ m}) = 83.4 \times 10^{-6} \text{ m}^3$$

Recalling that the width of the glued joint is $t = 0.020$ m, we use Eq. (6.7) to determine the average shearing stress in the joint.

$$\tau_{\text{ave}} = \frac{VQ}{It} = \frac{(1500 \text{ N})(83.4 \times 10^{-6} \text{ m}^3)}{(8.63 \times 10^{-6} \text{ m}^4)(0.020 \text{ m})} \qquad \tau_{\text{ave}} = 725 \text{ kPa} \blacktriangleleft$$

Shearing Stress in Joint b. We now pass section b-b and compute Q by using the area below the section.

$$Q = A\bar{y}_2 = [(0.060 \text{ m})(0.020 \text{ m})](0.0583 \text{ m}) = 70.0 \times 10^{-6} \text{ m}^3$$

$$\tau_{\text{ave}} = \frac{VQ}{It} = \frac{(1500 \text{ N})(70.0 \times 10^{-6} \text{ m}^3)}{(8.63 \times 10^{-6} \text{ m}^4)(0.020 \text{ m})} \qquad \tau_{\text{ave}} = 608 \text{ kPa} \blacktriangleleft$$

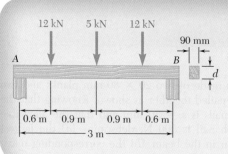

12 kN   5 kN   12 kN

90 mm

A   B

d

0.6 m   0.9 m   0.9 m   0.6 m

3 m

SAMPLE PROBLEM 6.2

A timber beam AB of span 3 m and nominal width 100 mm (actual width = 90 mm) is to support the three concentrated loads shown. Knowing that for the grade of timber used $\sigma_{all} = 12$ MPa and $\tau_{all} = 0.8$ MPa, determine the minimum required depth d of the beam.

SOLUTION

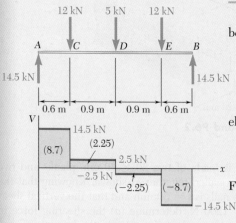

12 kN   5 kN   12 kN

A   C   D   E   B

14.5 kN   14.5 kN

0.6 m   0.9 m   0.9 m   0.6 m

V

14.5 kN

(2.25)

(8.7)

2.5 kN

−2.5 kN

(−2.25)   (−8.7)

−14.5 kN

x

M

8.7 kN · m   10.95 kN · m

8.7 kN · m

x

Maximum Shear and Bending Moment. After drawing the shear and bending-moment diagrams, we note that

$$M_{max} = 10.95 \text{ kN} \cdot \text{m}$$
$$V_{max} = 14.5 \text{ kN}$$

Design Based on Allowable Normal Stress. We first express the elastic section modulus S in terms of the depth d. We have

$$I = \frac{1}{12}bd^3 \qquad S = \frac{I}{c} = \frac{1}{6}bd^2 = \frac{1}{6}(90)d^2 = 15d^2$$

For $M_{max} = 10.95 \text{ kN} \cdot \text{m}$ and $\sigma_{all} = 12$ MPa, we write

$$S = \frac{M_{max}}{\sigma_{all}} \qquad 15d^2 = \frac{10.95 \text{ kN} \cdot \text{m}}{12 \text{ MPa}}$$
$$d^2 = 60833 \qquad d = 246.6 \text{ mm}$$

We have satisfied the requirement that $\sigma_m \leq 12$ MPa.

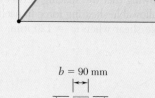

b = 90 mm

c = d/2

d

Check Shearing Stress. For $V_{max} = 14.5$ kN and $d = 246.6$ mm, we find

$$\tau_m = \frac{3}{2}\frac{V_{max}}{A} = \frac{3}{2}\frac{14.5 \text{ kN}}{(90 \text{ mm})(246 \text{ mm})} \qquad \tau_m = 0.980 \text{ MPa}$$

Since $\tau_{all} = 0.8$ MPa, the depth $d = 246.6$ mm is *not* acceptable and we must redesign the beam on the basis of the requirement that $\tau_m \leq 0.8$ MPa.

Design Based on Allowable Shearing Stress. Since we now know that the allowable shearing stress controls the design, we write

$$\tau_m = \tau_{all} = \frac{3}{2}\frac{V_{max}}{A} \qquad 0.8 \text{ MPa} = \frac{3}{2}\frac{14.5 \text{ kN}}{(90 \text{ mm})d}$$

$$d = 302 \text{ mm} \quad \blacktriangleleft$$

90 mm

302 mm

100 mm × 350 mm
Nominal size

The normal stress is, of course, less than $\sigma_{all} = 12$ MPa, and the depth of 302 mm is fully acceptable.

Comment. Since timber is normally available in depth increments of 50 mm, a 100 mm × 350-mm nominal size timber should be used. The actual cross section would then be 90 mm × 302 mm.

PROBLEMS

6.1 A square box beam is made of two 20 × 80-mm planks and two 20 × 120-mm planks nailed together as shown. Knowing that the spacing between the nails is $s = 50$ mm and that the allowable shearing force in each nail is 300 N, determine (*a*) the largest allowable vertical shear in the beam, (*b*) the corresponding maximum shearing stress in the beam.

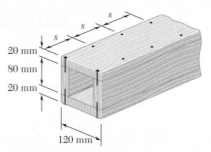

20 mm
80 mm
20 mm

120 mm

Fig. P6.1 and P6.2

6.2 A square box beam is made of two 20 × 80-mm planks and two 20 × 120-mm planks nailed together as shown. Knowing that the spacing between the nails is $s = 30$ mm and that the vertical shear in the beam is $V = 1200$ N, determine (*a*) the shearing force in each nail, (*b*) the maximum shearing stress in the beam.

6.3 Three boards are nailed together to form a beam shown, which is subjected to a vertical shear. Knowing that the spacing between the nails is $s = 75$ mm and that the allowable shearing force in each nail is 400 N, determine the allowable shear when $w = 120$ mm.

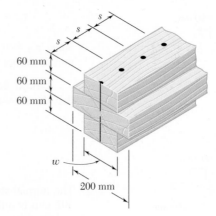

60 mm
60 mm
60 mm

w

200 mm

Fig. P6.3

6.4 Solve Prob. 6.3, assuming that the width of the top and bottom boards is changed to $w = 100$ mm.

EXAMPLE 6.04

A square box beam is made of two 18 × 76-mm planks and two 18 × 112-mm planks, nailed together as shown (Fig. 6.25). Knowing that the spacing between nails is 44 mm and that the beam is subjected to a vertical shear of magnitude $V = 2.5$ kN, determine the shearing force in each nail.

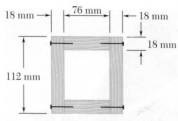

Fig. 6.25

We isolate the upper plank and consider the total force per unit length, q, exerted on its two edges. We use Eq. (6.5), where Q represents the first moment with respect to the neutral axis of the shaded area A' shown in Fig. 6.26a, and where I is the moment of inertia about the same axis of the entire cross-sectional area of the box beam (Fig. 6.26b). We have

$$Q = A'\bar{y} = (18 \text{ mm})(76 \text{ mm})(47 \text{ mm}) = 64296 \text{ mm}^3$$

Recalling that the moment of inertia of a square of side a about a centroidal axis is $I = \frac{1}{12}a^4$, we write

$$I = \tfrac{1}{12}(112 \text{ mm})^4 - \tfrac{1}{12}(76 \text{ mm})^4 = 10.332 \times 10^6 \text{ mm}^4$$

Substituting into Eq. (6.5), we obtain

$$q = \frac{VQ}{I} = \frac{(2500 \text{ N})(64296 \text{ mm}^3)}{10.332 \times 10^6 \text{ mm}^4} = 15.56 \text{ N/mm}$$

Because both the beam and the upper plank are symmetric with respect to the vertical plane of loading, equal forces are exerted on both edges of the plank. The force per unit length on each of these edges is thus $\frac{1}{2}q = \frac{1}{2}(15.56) = 7.78$ N/mm. Since the spacing between nails is 44 mm, the shearing force in each nail is

$$F = (44 \text{ mm})(7.78 \text{ N/mm}) = 342 \text{ N}$$

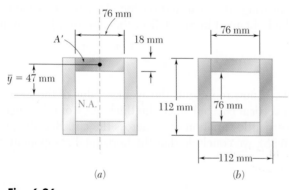

Fig. 6.26

6.7 SHEARING STRESSES IN THIN-WALLED MEMBERS

We saw in the preceding section that Eq. (6.4) may be used to determine the longitudinal shear $\mathbf{\Delta H}$ exerted on the walls of a beam element of arbitrary shape and Eq. (6.5) to determine the corresponding shear flow q. These equations will be used in this section to calculate both the shear flow and the average shearing stress in thin-walled members such as the flanges of wide-flange beams (Photo 6.2) and box beams, or the walls of structural tubes (Photo 6.3).

Photo 6.2 Wide-flange beams.

Photo 6.3 Box beams and tubes.

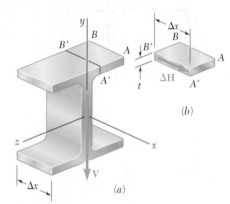

Fig. 6.27 Wide-flange beam segment.

Consider, for instance, a segment of length Δx of a wide-flange beam (Fig. 6.27a) and let **V** be the vertical shear in the transverse section shown. Let us detach an element $ABB'A'$ of the upper flange (Fig. 6.27b). The longitudinal shear ΔH exerted on that element can be obtained from Eq. (6.4):

$$\Delta H = \frac{VQ}{I} \Delta x \tag{6.4}$$

Dividing ΔH by the area $\Delta A = t \, \Delta x$ of the cut, we obtain for the average shearing stress exerted on the element the same expression that we had obtained in Sec. 6.3 in the case of a horizontal cut:

$$\tau_{\text{ave}} = \frac{VQ}{It} \tag{6.6}$$

Note that τ_{ave} now represents the average value of the shearing stress τ_{zx} over a vertical cut, but since the thickness t of the flange is small, there is very little variation of τ_{zx} across the cut. Recalling that $\tau_{xz} = \tau_{zx}$ (Fig. 6.28), we conclude that the horizontal component τ_{xz} of the

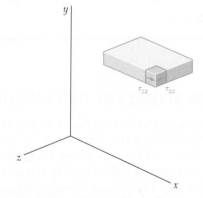

Fig. 6.28 Segment of beam flange.

shearing stress at any point of a transverse section of the flange can be obtained from Eq. (6.6), where Q is the first moment of the shaded area about the neutral axis (Fig. 6.29a). We recall that a similar result was obtained in Sec. 6.4 for the vertical component τ_{xy} of the shearing stress in the web (Fig. 6.29b). Equation (6.6) can be used to determine shearing stresses in box beams (Fig. 6.30), half pipes (Fig. 6.31), and other thin-walled members, as long as the loads are applied in a plane of symmetry of the member. In each case, the cut must be perpendicular to the surface of the member, and Eq. (6.6) will yield the component of the shearing stress in the direction of the tangent to that surface. (The other component may be assumed equal to zero, in view of the proximity of the two free surfaces.)

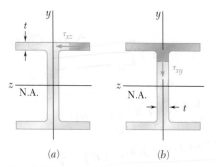

Fig. 6.29 Wide-flange beam.

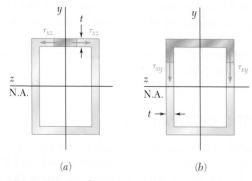

Fig. 6.30 Box beam.

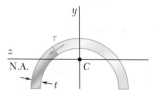

Fig. 6.31 Half pipe beam.

Comparing Eqs. (6.5) and (6.6), we note that the product of the shearing stress τ at a given point of the section and of the thickness t of the section at that point is equal to q. Since V and I are constant in any given section, q depends only upon the first moment Q and, thus, can easily be sketched on the section. In the case of a box beam, for example (Fig. 6.32), we note that q grows smoothly from zero at A to a maximum value at C and C' on the neutral axis, and then decreases back to zero as E is reached. We also note that there is no sudden variation in the magnitude of q as we pass a corner at B, D, B', or D', and that the sense of q in the horizontal portions of the section may be easily obtained from its sense in the vertical portions (which is the same as the sense of the shear $\mathbf{V}$). In the case of a wide-flange section (Fig. 6.33), the values of q in portions AB and $A'B$ of the upper flange are distributed symmetrically. As we turn at B into the web, the values of q corresponding to the two halves of the flange must be combined to obtain the value of q at the top of the web. After reaching a maximum value at C on the neutral axis, q decreases, and at D splits into two equal parts corresponding to the two halves of the lower flange. The name of *shear flow* commonly used to refer to the shear per unit length, q, reflects the similarity between the properties of q that we have just described and some of the characteristics of a fluid flow through an open channel or pipe.[†]

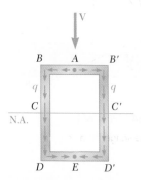

Fig. 6.32 Shear flow q in box beam section.

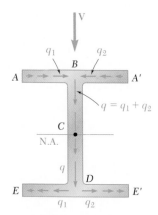

Fig. 6.33 Shear flow q in wide-flange beam section.

[†]We recall that the concept of shear flow was used to analyze the distribution of shearing stresses in thin-walled hollow shafts (Sec. 3.13). However, while the shear flow in a hollow shaft is constant, the shear flow in a member under a transverse loading is not.

So far we have assumed that all the loads were applied in a plane of symmetry of the member. In the case of members possessing two planes of symmetry, such as the wide-flange beam of Fig. 6.29 or the box beam of Fig. 6.30, any load applied through the centroid of a given cross section can be resolved into components along the two axes of symmetry of the section. Each component will cause the member to bend in a plane of symmetry, and the corresponding shearing stresses can be obtained from Eq. (6.6). The principle of superposition can then be used to determine the resulting stresses.

However, if the member considered possesses no plane of symmetry, or if it possesses a single plane of symmetry and is subjected to a load that is not contained in that plane, the member is observed to *bend and twist* at the same time, except when the load is applied at a specific point, called the *shear center*. Note that the shear center generally does *not* coincide with the centroid of the cross section. The determination of the shear center of various thin-walled shapes is discussed in Sec. 6.9.

*6.8 PLASTIC DEFORMATIONS

Consider a cantilever beam AB of length L and rectangular cross section, subjected at its free end A to a concentrated load $\mathbf{P}$ (Fig. 6.34). The largest value of the bending moment occurs at the fixed end B and is equal to $M = PL$. As long as this value does not exceed the maximum elastic moment M_Y, i.e., as long as $PL \leq M_Y$, the normal stress σ_x will not exceed the yield strength σ_Y anywhere in the beam. However, as P is increased beyond the value M_Y/L, yield is initiated at points B and B' and spreads toward the free end of the beam. Assuming the material to be elastoplastic, and considering a cross section CC' located at a distance x from the free end A of the beam (Fig. 6.35), we obtain the half-thickness y_Y of the elastic core in that section by making $M = Px$ in Eq. (4.38) of Sec. 4.9. We have

$$Px = \frac{3}{2}M_Y\left(1 - \frac{1}{3}\frac{y_Y^2}{c^2}\right) \tag{6.14}$$

where c is the half-depth of the beam. Plotting y_Y against x, we obtain the boundary between the elastic and plastic zones.

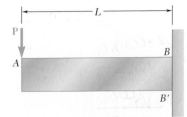

Fig. 6.34 $(PL \leq M_Y)$

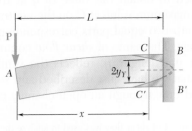

Fig. 6.35 $(PL > M_Y)$

As long as $PL < \frac{3}{2}M_Y$, the parabola defined by Eq. (6.14) intersects the line BB', as shown in Fig. 6.36. However, when PL reaches the value $\frac{3}{2}M_Y$, i.e., when $PL = M_p$, where M_p is the plastic moment defined in Sec. 4.9, Eq. (6.14) yields $y_Y = 0$ for $x = L$, which shows that the vertex of the parabola is now located in section BB', and that this section has become fully plastic (Fig. 6.36). Recalling Eq. (4.40) of Sec. 4.9, we also note that the radius of curvature ρ of the neutral surface at that point is equal to zero, indicating the presence of a sharp bend in the beam at its fixed end. We say that a *plastic hinge* has developed at that point. The load $P = M_p/L$ is the largest load that can be supported by the beam.

The above discussion was based only on the analysis of the normal stresses in the beam. Let us now examine the distribution of the shearing stresses in a section that has become partly plastic. Consider the portion of beam $CC''D''D$ located between the transverse sections CC' and DD', and above the horizontal plane $D''C''$ (Fig. 6.37a). If this portion is located entirely in the plastic zone, the normal stresses exerted on the faces CC'' and DD'' will be uniformly distributed and equal to the yield strength σ_Y (Fig. 6.37b). The equilibrium of the free body $CC''D''D$ thus requires that the horizontal shearing force $\Delta\mathbf{H}$ exerted on its lower face be equal to zero. It follows that the average value of the horizontal shearing stress τ_{yx} across the beam at C'' is zero, as well as the average value of the vertical shearing stress τ_{xy}. We thus conclude that the vertical shear $V = P$ in section CC' must be distributed entirely over the portion EE' of that section that is located within the elastic zone (Fig. 6.38). It can be shown† that the distribution of the shearing stresses over EE' is the same as in an elastic rectangular beam of the same width b as beam AB, and of depth equal to the thickness $2y_Y$ of the elastic zone. Denoting by A' the area $2by_Y$ of the elastic portion of the cross section, we have

$$\tau_{xy} = \frac{3}{2}\frac{P}{A'}\left(1 - \frac{y^2}{y_Y^2}\right) \qquad (6.15)$$

The maximum value of the shearing stress occurs for $y = 0$ and is

$$\tau_{\max} = \frac{3}{2}\frac{P}{A'} \qquad (6.16)$$

As the area A' of the elastic portion of the section decreases, $\tau_{\max}$ increases and eventually reaches the yield strength in shear τ_Y. Thus, shear contributes to the ultimate failure of the beam. A more exact analysis of this mode of failure should take into account the combined effect of the normal and shearing stresses.

†See Prob. 6.60.

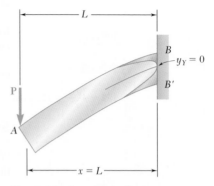

Fig. 6.36 $(PL = M_P = \frac{3}{2}M_Y)$

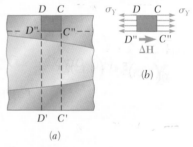

Fig. 6.37 Beam segment.

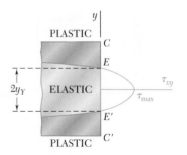

Fig. 6.38

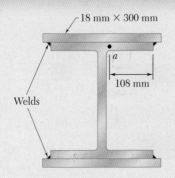

$t_f = 19.6$ mm

108 mm

a

132 mm

$132 - \dfrac{19.6}{2} = 122.2$ mm

264 mm

C

$I_x = 164 \times 10^6\,\text{mm}^4$

SAMPLE PROBLEM 6.3

Knowing that the vertical shear is 200 kN in a W250 × 101 rolled-steel beam, determine the horizontal shearing stress in the top flange at a point a located 108 mm from the edge of the beam. The dimensions and other geometric data of the rolled-steel section are given in Appendix C.

SOLUTION

We isolate the shaded portion of the flange by cutting along the dashed line that passes through point a.

$$Q = (108\text{ mm})(19.6\text{ mm})(122.2\text{ mm}) = 258.7 \times 10^3 \text{ mm}^3$$

$$\tau = \frac{VQ}{It} = \frac{(200\text{ kN})(258.7 \times 10^3 \text{ mm}^3)}{(164 \times 10^6 \text{ mm}^4)(19.6\text{ mm})} = 16.096 \text{ MPa} \qquad \tau = 16.1 \text{ MPa} \quad \blacktriangleleft$$

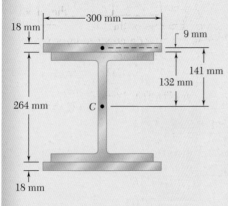

18 mm × 300 mm

a

108 mm

Welds

300 mm

18 mm

9 mm

141 mm

132 mm

264 mm

C

18 mm

SAMPLE PROBLEM 6.4

Solve Sample Prob. 6.3, assuming that 18 × 300-mm plates have been attached to the flanges of the W250 × 101 beam by continuous fillet welds as shown.

SOLUTION

For the composite beam the centroidal moment of inertia is

$$I = 164 \times 10^6 \text{ mm}^4 + 2\big[\tfrac{1}{12}(300\text{ mm})(18\text{ mm})^3 + (300\text{ mm})(18\text{ mm})(141\text{ mm})^2\big]$$
$$I = 379 \times 10^6 \text{ mm}^4$$

Since the top plate and the flange are connected only at the welds, we find the shearing stress at a by passing a section through the flange at a, *between* the plate and the flange, and again through the flange at the symmetric point a'.

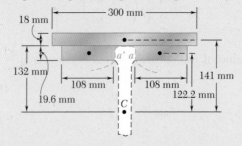

18 mm

300 mm

a' a

132 mm

108 mm   108 mm

19.6 mm

141 mm

122.2 mm

C

For the shaded area that we have isolated, we have

$$t = 2t_f = 2(19.6\text{ mm}) = 39.2 \text{ mm}$$
$$Q = 2[(108\text{ mm})(19.6\text{ mm})(122.2\text{ mm})] + (300\text{ mm})(18\text{ mm})(141\text{ mm})$$
$$Q = 1.2787 \times 10^6 \text{ mm}^3$$
$$\tau = \frac{VQ}{It} = \frac{(200\text{ kN})(1.2787 \times 10^6 \text{ mm}^3)}{(379 \times 10^6 \text{ mm}^4)(39.2\text{ mm})} \qquad \tau = 17.21 \text{ MPa} \quad \blacktriangleleft$$

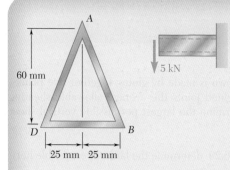

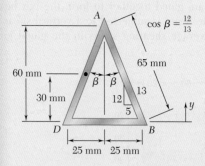

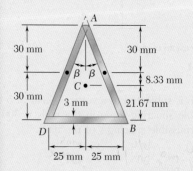

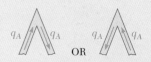

OR

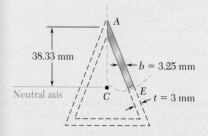

SAMPLE PROBLEM 6.5

The thin-walled extruded beam shown is made of aluminum and has a uniform 3-mm wall thickness. Knowing that the shear in the beam is 5 kN, determine (a) the shearing stress at point A, (b) the maximum shearing stress in the beam. (*Note:* The dimensions given are to lines midway between the outer and inner surfaces of the beam.)

SOLUTION

Centroid. We note that $AB = AD = 65$ mm.

$$\overline{Y} = \frac{\Sigma \overline{y} A}{\Sigma A} = \frac{2[(65 \text{ mm})(3 \text{ mm})(30 \text{ mm})]}{2[(65 \text{ mm})(3 \text{ mm})] + (50 \text{ mm})(3 \text{ mm})}$$

$$\overline{Y} = 21.67 \text{ mm}$$

Centroidal Moment of Inertia. Each side of the thin-walled beam can be considered as a parallelogram, and we recall that for the case shown $I_{nn} = bh^3/12$ where b is measured parallel to the axis nn.

$$b = (3 \text{ mm})/\cos \beta = (3 \text{ mm})/(12/13) = 3.25 \text{ mm}$$

$$I = \Sigma(\overline{I} + Ad^2) = 2[\tfrac{1}{12}(3.25 \text{ mm})(60 \text{ mm})^3$$
$$+ (3.25 \text{ mm})(60 \text{ mm})(8.33 \text{ mm})^2] + [\tfrac{1}{12}(50 \text{ mm})(3 \text{ mm})^3$$
$$+ (50 \text{ mm})(3 \text{ mm})(21.67 \text{ mm})^2]$$

$$I = 214.6 \times 10^3 \text{ mm}^4 \qquad I = 0.2146 \times 10^{-6} \text{ m}^4$$

a. Shearing Stress at A. If a shearing stress τ_A occurs at A, the shear flow will be $q_A = \tau_A t$ and must be directed in one of the two ways shown. But the cross section and the loading are symmetric about a vertical line through A, and thus the shear flow must also be symmetric. Since neither of the possible shear flows is symmetric, we conclude that $\tau_A = 0$ ◄

b. Maximum Shearing Stress. Since the wall thickness is constant, the maximum shearing stress occurs at the neutral axis, where Q is maximum. Since we know that the shearing stress at A is zero, we cut the section along the dashed line shown and isolate the shaded portion of the beam. In order to obtain the largest shearing stress, the cut at the neutral axis is made perpendicular to the sides, and is of length $t = 3$ mm.

$$Q = [(3.25 \text{ mm})(38.33 \text{ mm})]\left(\frac{38.33 \text{ mm}}{2}\right) = 2387 \text{ mm}^3$$

$$Q = 2.387 \times 10^{-6} \text{ m}^3$$

$$\tau_E = \frac{VQ}{It} = \frac{(5 \text{ kN})(2.387 \times 10^{-6} \text{ m}^3)}{(0.2146 \times 10^{-6} \text{ m}^4)(0.003 \text{ m})} \qquad \tau_{max} = \tau_E = 18.54 \text{ MPa} \ ◄$$

PROBLEMS

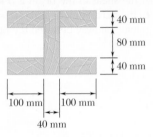

Fig. P6.29

6.29 The built-up beam shown is made by gluing together five planks. Knowing that in the glued joints the average allowable shearing stress is 350 kPa, determine the largest permissible vertical shear in the beam.

6.30 For the beam of Prob. 6.29, determine the largest permissible horizontal shear.

6.31 Several wooden planks are glued together to form the box beam shown. Knowing that the beam is subjected to a vertical shear of 3 kN, determine the average shearing stress in the glued joint (a) at A, (b) at B.

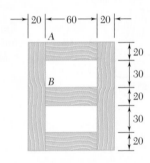

Dimensions in mm

Fig. P6.31

6.32 The built-up timber beam is subjected to a vertical shear of 5 kN. Knowing that the allowable shearing force in the nails is 300 N, determine the largest permissible spacing s of the nails.

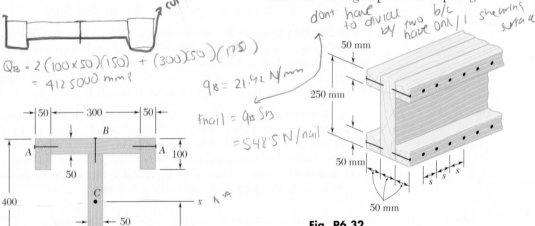

Fig. P6.32

6.33 The built-up wooden beam shown is subjected to a vertical shear of 8 kN. Knowing that the nails are spaced longitudinally every 60 mm at A and every 25 mm at B, determine the shearing force in the nails (a) at A, (b) at B. (Given: $I_x = 1.504 \times 10^9$ mm⁴.)

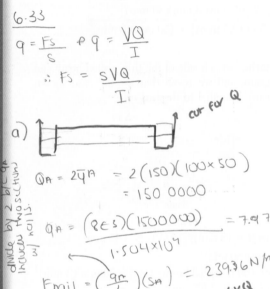

Fig. P6.33

— handwritten notes —

6.33

$q = \dfrac{F_s}{S}$ & $q = \dfrac{VQ}{I}$

$\therefore F_s = \dfrac{SVQ}{I}$

cut for Q

a)

$Q_A = 2\bar{y}A = 2(150)(100 \times 50)$
$= 150\,0000$

$q_A = \dfrac{(8E3)(1500000)}{1.504 \times 10^9} = 7.978$ N/mm

divide by 2 b/c of 2 shear surfaces. Includes two nails.

$F_{nail} = \left(\dfrac{q_A}{2}\right)(S_A) = 239.36$ N/nail

cut for Q

b)

$Q_B = 2(100 \times 50)(150) + (300)(50)(175)$
$= 4125000$ mm³

$q_B = 21.92$ N/mm

$F_{nail} = q_B \, S_B$
$= 548.5$ N/nail

dont have to divide by two b/c have only 1 shearing surface.

$V = 8$kN

$S = 25$mm

430

6.34 Knowing that a vertical shear **V** of 250 kN is exerted on a W360 × 122 rolled-steel beam, determine the shearing stress (a) at point a, (b) at the centroid C.

6.35 An extruded aluminum beam has the cross section shown. Knowing that the vertical shear in the beam is 40 kN, determine the shearing stress at (a) point a, (b) point b.

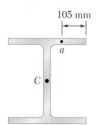

Fig. P6.34

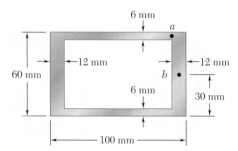

Fig. P6.35

6.36 Knowing that a given vertical shear **V** causes a maximum shearing stress of 75 MPa in the hat-shaped extrusion shown, determine the corresponding shearing stress at (a) point a, (b) point b.

6.37 Knowing that a given vertical shear **V** causes a maximum shearing stress of 75 MPa in an extruded beam having the cross section shown, determine the shearing stress at the three points indicated.

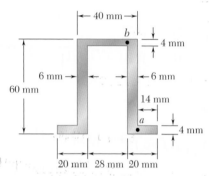

Fig. P6.36

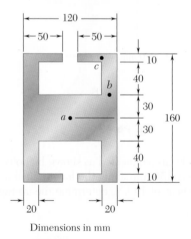

Dimensions in mm

Fig. P6.37

6.38 An extruded beam has the cross section shown and a uniform wall thickness of 5 mm. Knowing that a given vertical shear **V** causes a maximum shearing stress $\tau = 60$ MPa, determine the shearing stress at the four points indicated.

6.39 Solve Prob. 6.38 assuming that the beam is subjected to a horizontal shear **V**.

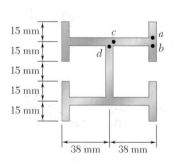

Fig. P6.38

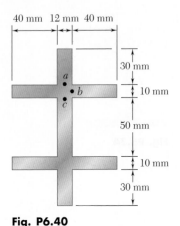

Fig. P6.40

6.40 Knowing that a given vertical shear **V** causes a maximum shearing stress of 50 MPa in a thin-walled member having the cross section shown, determine the corresponding shearing stress at (a) point a, (b) point b, (c) point c.

6.41 and 6.42 The extruded aluminum beam has a uniform wall thickness of 3 mm. Knowing that the vertical shear in the beam is 9 kN, determine the corresponding shearing stress at each of the five points indicated.

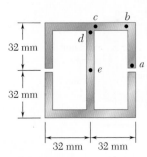

Fig. P6.41

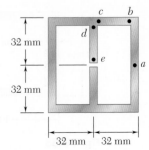

Fig. P6.42

6.43 Four L102 × 102 × 9.5 steel angle shapes and a 12 × 400-mm steel plate are bolted together to form a beam with the cross section shown. The bolts are of 22-mm diameter and are spaced longitudinally every 120 mm. Knowing that the beam is subjected to a vertical shear of 240 kN, determine the average shearing stress in each bolt.

Fig. P6.43

6.44 Three planks are connected as shown by bolts of 14-mm diameter spaced every 150 mm along the longitudinal axis of the beam. For a vertical shear of 10 kN, determine the average shearing stress in the bolts.

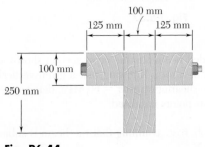

Fig. P6.44

6.45

N·A

Area	$\bar{y}$	$A\bar{y}$
1) (1500)	75	1125000
2) 7500	175	1312500
3) 7500	175	"

$\bar{y} = 125\ mm$

$$I = \left[\tfrac{1}{12}(100)(150)^2 + (100)(150)(50)^2\right]$$

$$+ 2\left[\tfrac{1}{12}(50)(150)^3 + (50)(150)(50^2)\right]$$

$$= 431\,250\,000\,mm^4\ (100)$$

$$Q = 2\bar{y}A = 2(50)(150)(50)$$

$$= 750\,000\,mm^3$$

$$= (50\,000)(200)(6)$$

$$q = \frac{VQ}{I} \qquad V = 6\,kN$$

$$= 34.2857\ N/mm$$

$$FS = \left(\frac{q}{2}\right)(s) = 3857.14\ N$$

$$\Rightarrow FS = 3857.14\ N$$

$$\pi/4\,d^2 \quad \tau_{ave} = \frac{V}{A}$$

$$\hookleftarrow A = \frac{V}{\tau_{av}}$$

6.45 A beam consists of three planks connected as shown by steel bolts with a longitudinal spacing of 225 mm. Knowing that the shear in the beam is vertical and equal to 6 kN and that the allowable average shearing stress in each bolt is 60 MPa, determine the smallest permissible bolt diameter that can be used.

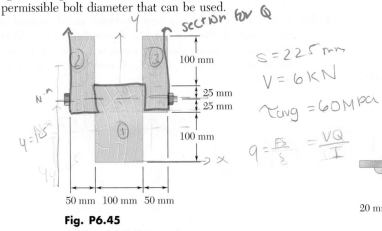

(handwritten annotations)

section for Q

$S = 225\,mm$

$V = 6\,KN$

$\tau_{avg} = 60\,MPa$

$\tau_{avg} = \dfrac{V}{A}$

$q = \dfrac{F}{S} = \dfrac{VQ}{I}$

100 mm
25 mm
25 mm
100 mm

50 mm   100 mm   50 mm

Fig. P6.45

6.46 Three 20 × 450-mm steel plates are bolted to four L152 × 152 × 19 angles to form a beam with the cross section shown. The bolts have a 22-mm diameter and are spaced longitudinally every 125 mm. Knowing that the allowable average shearing stress in the bolts is 90 MPa, determine the largest permissible vertical shear in the beam. (*Given:* $I_x = 1896 \times 10^6$ mm^4.)

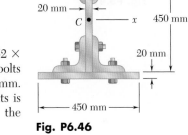

20 mm

20 mm →

C •——— x 450 mm

20 mm

450 mm

Fig. P6.46

6.47 A plate of 6-mm thickness is corrugated as shown and then used as a beam. For a vertical shear of 5 kN, determine (*a*) the maximum shearing stress in the section, (*b*) the shearing stress at point *B*. Also sketch the shear flow in the cross section.

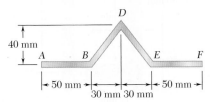

40 mm
A B D E F
← 50 mm →|← 50 mm →
30 mm 30 mm

Fig. P6.47

6.48 A plate of 4-mm thickness is bent as shown and then used as a beam. For a vertical shear of 12 kN, determine (*a*) the shearing stress at point *A*, (*b*) the maximum shearing stress in the beam. Also sketch the shear flow in the cross section.

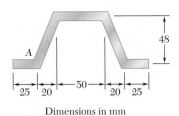

48

A

25   20   —50—   20   25

Dimensions in mm

Fig. P6.48

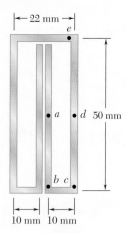

Fig. P6.49

6.49 A plate of 2-mm thickness is bent as shown and then used as a beam. For a vertical shear of 5 kN, determine the shearing stress at the five points indicated and sketch the shear flow in the cross section.

6.50 A plate of thickness t is bent as shown and then used as a beam. For a vertical shear of 2.4 kN, determine (*a*) the thickness t for which the maximum shearing stress is 2 MPa, (*b*) the corresponding shearing stress at point E. Also sketch the shear flow in the cross section.

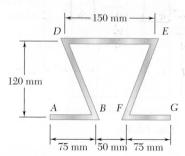

Fig. P6.50

6.51 The design of a beam calls for connecting two vertical rectangular 10 × 100-mm plates by welding them to two horizontal 12 × 50-mm plates as shown. For a vertical shear **V**, determine the dimension a for which the shear flow through the welded surfaces is maximum.

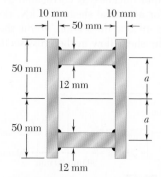

Fig. P6.51

6.52 and 6.53 An extruded beam has a uniform wall thickness t. Denoting by **V** the vertical shear and by A the cross-sectional area of the beam, express the maximum shearing stress as $\tau_{\max} = k(V/A)$ and determine the constant k for each of the two orientations shown.

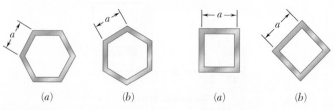

(*a*)   (*b*)   (*a*)   (*b*)

Fig. P6.52   **Fig. P6.53**

6.54 (*a*) Determine the shearing stress at point P of a thin-walled pipe of the cross section shown caused by a vertical shear **V**. (*b*) Show that the maximum shearing stress occurs for $\theta = 90°$ and is equal to $2V/A$, where A is the cross-sectional area of the pipe.

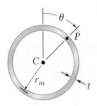

Fig. P6.54

6.55 For a beam made of two or more materials with different moduli of elasticity, show that Eq. (6.6)

$$\tau_{ave} = \frac{VQ}{It}$$

remains valid provided that both Q and I are computed by using the transformed section of the beam (see Sec. 4.6) and provided further that t is the actual width of the beam where τ_{ave} is computed.

6.56 and 6.57 A steel bar and an aluminum bar are bonded together as shown to form a composite beam. Knowing that the vertical shear in the beam is 18 kN and that the modulus of elasticity is 200 GPa for the steel and 73 GPa for the aluminum, determine (*a*) the average stress at the bonded surface, (*b*) the maximum shearing stress in the beam. (*Hint:* Use the method indicated in Prob. 6.55.)

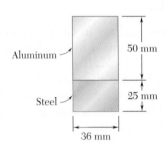

Fig. P6.56

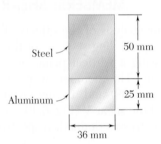

Fig. P6.57

6.58 and 6.59 A composite beam is made by attaching the timber and steel portions shown with bolts of 12-mm diameter spaced longitudinally every 200 mm. The modulus of elasticity is 10 GPa for the wood and 200 GPa for the steel. For a vertical shear of 4 kN, determine (*a*) the average shearing stress in the bolts, (*b*) the shearing stress at the center of the cross section. (*Hint:* Use the method indicated in Prob. 6.55.)

Fig. P6.58

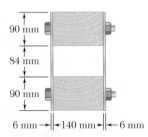

Fig. P6.59

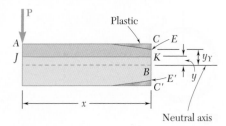

Fig. P6.60

6.60 Consider the cantilever beam AB discussed in Sec. 6.8 and the portion $ACKJ$ of the beam that is located to the left of the transverse section CC' and above the horizontal plane JK, where K is a point at a distance $y < y_Y$ above the neutral axis (Fig. P6.60). (a) Recalling that $\sigma_x = \sigma_Y$ between C and E and $\sigma_x = (\sigma_Y/y_Y)y$ between E and K, show that the magnitude of the horizontal shearing force $\mathbf{H}$ exerted on the lower face of the portion of beam $ACKJ$ is

$$H = \frac{1}{2}b\sigma_Y\left(2c - y_Y - \frac{y^2}{y_Y}\right)$$

(b) Observing that the shearing stress at K is

$$\tau_{xy} = \lim_{\Delta A \to 0} \frac{\Delta H}{\Delta A} = \lim_{\Delta x \to 0} \frac{1}{b}\frac{\Delta H}{\Delta x} = \frac{1}{b}\frac{\partial H}{\partial x}$$

and recalling that y_Y is a function of x defined by Eq. (6.14), derive Eq. (6.15).

Fig. 6.39 Channel beam.

*6.9 UNSYMMETRIC LOADING OF THIN-WALLED MEMBERS; SHEAR CENTER

Our analysis of the effects of transverse loadings in Chap. 5 and in the preceding sections of this chapter was limited to members possessing a vertical plane of symmetry and to loads applied in that plane. The members were observed to bend in the plane of loading (Fig. 6.39) and, in any given cross section, the bending couple $\mathbf{M}$ and the shear $\mathbf{V}$ (Fig. 6.40) were found to result in normal and shearing stresses defined, respectively, by the formulas

$$\sigma_x = -\frac{My}{I} \tag{4.16}$$

and

$$\tau_{ave} = \frac{VQ}{It} \tag{6.6}$$

In this section, the effects of transverse loadings on *thin-walled members that do not possess a vertical plane of symmetry* will be

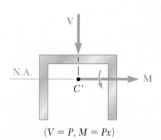

$(V = P, M = Px)$

Fig. 6.40 Loaded in vertical plane of symmetry.

examined. Let us assume, for example, that the channel member of Fig. 6.39 has been rotated through 90° and that the line of action of **P** still passes through the centroid of the end section. The couple vector **M** representing the bending moment in a given cross section is still directed along a principal axis of the section (Fig. 6.41), and the neutral axis will coincide with that axis (cf. Sec. 4.13). Equation (4.16), therefore, is applicable and can be used to compute the normal stresses in the section. However, Eq. (6.6) cannot be used to determine the shearing stresses in the section, since this equation was derived for a member possessing a vertical plane of symmetry (cf. Sec. 6.7). Actually, the member will be observed to *bend and twist* under the applied load (Fig. 6.42), and the resulting distribution of shearing stresses will be quite different from that defined by Eq. (6.6).

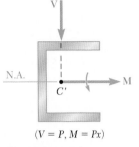

$(V = P, M = Px)$

Fig. 6.41 Load perpendicular to vertical plane of symmetry.

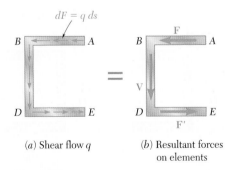

Fig. 6.42 Deformation of channel beam when not loaded in vertical plane of symmetry.

The following question now arises: Is it possible to apply the vertical load **P** in such a way that the channel member of Fig. 6.42 will *bend without twisting* and, if so, where should the load **P** be applied? If the member bends without twisting, then the shearing stress at any point of a given cross section can be obtained from Eq. (6.6), where Q is the first moment of the shaded area with respect to the neutral axis (Fig. 6.43a), and the distribution of stresses will look as shown in Fig. 6.43b, with $\tau = 0$ at both A and E. We note that the shearing force exerted on a small element of cross-sectional area $dA = t\,ds$ is $dF = \tau\,dA = \tau t\,ds$, or $dF = q\,ds$ (Fig. 6.44a),

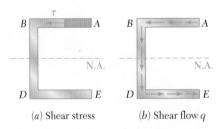

(a) Shear stress   (b) Shear flow q

Fig. 6.43 Stresses applied to cross section as a result of load shown in Fig. 6.42.

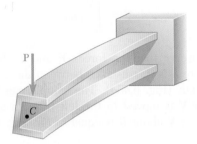

(a) Shear flow q (b) Resultant forces on elements

Fig. 6.44

where q is the shear flow $q = \tau t = VQ/I$ at the point considered. The resultant of the shearing forces exerted on the elements of the upper flange AB of the channel is found to be a horizontal force $\mathbf{F}$ (Fig. 6.44b) of magnitude

$$F = \int_A^B q\, ds \tag{6.17}$$

Because of the symmetry of the channel section about its neutral axis, the resultant of the shearing forces exerted on the lower flange DE is a force $\mathbf{F}'$ of the same magnitude as $\mathbf{F}$ but of opposite sense. We conclude that the resultant of the shearing forces exerted on the web BD must be equal to the vertical shear $\mathbf{V}$ in the section:

$$V = \int_B^D q\, ds \tag{6.18}$$

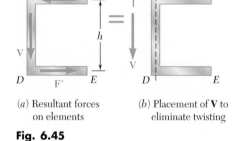

(a) Resultant forces on elements

(b) Placement of **V** to eliminate twisting

Fig. 6.45

We now observe that the forces $\mathbf{F}$ and $\mathbf{F}'$ form a couple of moment Fh, where h is the distance between the center lines of the flanges AB and DE (Fig. 6.45a). This couple can be eliminated if the vertical shear $\mathbf{V}$ is moved to the left through a distance e such that the moment of $\mathbf{V}$ about B is equal to Fh (Fig. 6.45b). We write $Ve = Fh$ or

$$e = \frac{Fh}{V} \tag{6.19}$$

and conclude that, when the force $\mathbf{P}$ is applied at a distance e to the left of the center line of the web BD, the member bends in a vertical plane without twisting (Fig. 6.46).

The point O where the line of action of $\mathbf{P}$ intersects the axis of symmetry of the end section is called the *shear center* of that section. We note that, in the case of an oblique load $\mathbf{P}$ (Fig. 6.47a), the member will also be free of any twist if the load $\mathbf{P}$ is applied at the shear center of the section. Indeed, the load $\mathbf{P}$ can then be resolved into two components $\mathbf{P}_z$ and $\mathbf{P}_y$ (Fig. 6.47b) corresponding respectively to the loading conditions of Figs. 6.39 and 6.46, neither of which causes the member to twist.

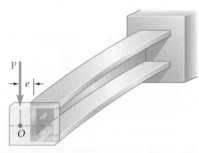

Fig. 6.46 Placement of load to eliminate twisting.

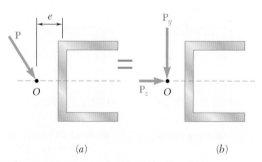

(a)

(b)

Fig. 6.47 Beam with oblique load.

Determine the shear center O of a channel section of uniform thickness (Fig. 6.48), knowing that $b = 100$ mm, $h = 150$ mm, and $t = 4$ mm.

EXAMPLE 6.05

Assuming that the member does not twist, we first determine the shear flow q in flange AB at a distance s from A (Fig. 6.49). Recalling Eq. (6.5) and observing that the first moment Q of the shaded area with respect to the neutral axis is $Q = (st)(h/2)$, we write

$$q = \frac{VQ}{I} = \frac{Vsth}{2I} \tag{6.20}$$

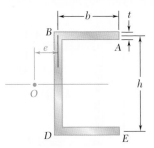

Fig. 6.48

where V is the vertical shear and I the moment of inertia of the section with respect to the neutral axis.

Recalling Eq. (6.17), we determine the magnitude of the shearing force $\mathbf{F}$ exerted on flange AB by integrating the shear flow q from A to B:

$$F = \int_0^b q \, ds = \int_0^b \frac{Vsth}{2I} \, ds = \frac{Vth}{2I} \int_0^b s \, ds$$

$$F = \frac{Vthb^2}{4I} \tag{6.21}$$

The distance e from the center line of the web BD to the shear center O can now be obtained from Eq. (6.19):

$$e = \frac{Fh}{V} = \frac{Vthb^2}{4I} \frac{h}{V} = \frac{th^2b^2}{4I} \tag{6.22}$$

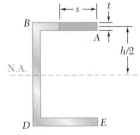

Fig. 6.49

The moment of inertia I of the channel section can be expressed as follows:

$$I = I_{\text{web}} + 2I_{\text{flange}}$$

$$= \frac{1}{12}th^3 + 2\left[\frac{1}{12}bt^3 + bt\left(\frac{h}{2}\right)^2\right]$$

Neglecting the term containing t^3, which is very small, we have

$$I = \tfrac{1}{12}th^3 + \tfrac{1}{2}tbh^2 = \tfrac{1}{12}th^2(6b + h) \tag{6.23}$$

Substituting this expression into (6.22), we write

$$e = \frac{3b^2}{6b + h} = \frac{b}{2 + \dfrac{h}{3b}} \tag{6.24}$$

We note that the distance e does not depend upon t and can vary from 0 to $b/2$, depending upon the value of the ratio $h/3b$. For the given channel section, we have

$$\frac{h}{3b} = \frac{150 \text{ mm}}{3(100 \text{ mm})} = 0.5$$

and

$$e = \frac{100 \text{ mm}}{2 + 0.5} = 40 \text{ mm}$$

EXAMPLE 6.06

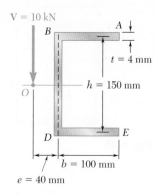

V = 10 kN

Fig. 6.50

For the channel section of Example 6.05 determine the distribution of the shearing stresses caused by a 10-kN vertical shear **V** applied at the shear center O (Fig. 6.50).

Shearing stresses in flanges. Since **V** is applied at the shear center, there is no torsion, and the stresses in flange AB are obtained from Eq. (6.20) of Example 6.05. We have

$$\tau = \frac{q}{t} = \frac{VQ}{It} = \frac{Vh}{2I} s \tag{6.25}$$

which shows that the stress distribution in flange AB is linear. Letting $s = b$ and substituting for I from Eq. (6.23), we obtain the value of the shearing stress at B:

$$\tau_B = \frac{Vhb}{2(\frac{1}{12}th^2)(6b + h)} = \frac{6Vb}{th(6b + h)} \tag{6.26}$$

Letting $V = 10$ kN, and using the given dimensions, we have

$$\tau_B = \frac{6(10 \text{ kN})(100 \text{ mm})}{(4 \text{ mm})(150 \text{ mm})(6 \times 100 \text{ mm} + 150 \text{ mm})}$$
$$= 13.33 \text{ MPa}$$

Shearing stresses in web. The distribution of the shearing stresses in the web BD is parabolic, as in the case of a W-beam, and the maximum stress occurs at the neutral axis. Computing the first moment of the upper half of the cross section with respect to the neutral axis (Fig. 6.51), we write

$$Q = bt(\tfrac{1}{2}h) + \tfrac{1}{2}ht(\tfrac{1}{4}h) = \tfrac{1}{8}ht(4b + h) \tag{6.27}$$

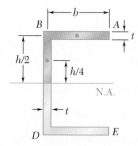

Fig. 6.51

Substituting for I and Q from (6.23) and (6.27), respectively, into the expression for the shearing stress, we have

$$\tau_{\max} = \frac{VQ}{It} = \frac{V(\tfrac{1}{8}ht)(4b + h)}{\tfrac{1}{12}th^2(6b + h)t} = \frac{3V(4b + h)}{2th(6b + h)}$$

or, with the given data,

$$\tau_{\max} = \frac{3(10 \text{ kN})(4 \times 100 \text{ mm} + 150 \text{ mm})}{2(4 \text{ mm})(150 \text{ mm})(6 \times 100 \text{ mm} + 150 \text{ mm})}$$
$$= 18.33 \text{ MPa}$$

Distribution of stresses over the section. The distribution of the shearing stresses over the entire channel section has been plotted in Fig. 6.52.

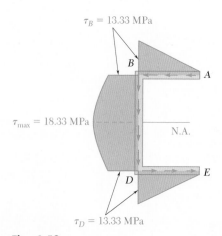

Fig. 6.52

EXAMPLE 6.07

For the channel section of Example 6.05, and neglecting stress concentrations, determine the maximum shearing stress caused by a 10-kN vertical shear **V** applied at the centroid C of the section, which is located 28.5 mm to the right of the center line of the web BD (Fig. 6.53).

Equivalent force-couple system at shear center. The shear center O of the cross section was determined in Example 6.05 and found to be at a distance $e = 40$ mm to the left of the center line of the web BD. We replace the shear **V** (Fig. 6.54a) by an equivalent force-couple system at the shear center O (Fig. 6.54b). This system consists of a 10-kN force **V** and of a torque **T** of magnitude

$$T = V(OC) = (10 \text{ kN})(40 \text{ mm} + 28.5 \text{ mm})$$
$$= 685 \text{ N} \cdot \text{m}$$

Stresses due to bending. The 10-kN force **V** causes the member to bend, and the corresponding distribution of shearing stresses in the section (Fig. 6.54c) was determined in Example 6.06. We recall that the maximum value of the stress due to this force was found to be

$$(\tau_{max})_{bending} = 18.33 \text{ MPa}$$

Stresses due to twisting. The torque **T** causes the member to twist, and the corresponding distribution of stresses is shown in Fig. 6.54d. We recall from Sec. 3.12 that the membrane analogy shows that, in a thin-walled member of uniform thickness, the stress caused by a torque **T** is maximum along the edge of the section. Using Eqs. (3.45) and (3.43) with

$$a = 100 \text{ mm} + 150 \text{ mm} + 100 \text{ mm} = 350 \text{ mm}$$
$$b = t = 4 \text{ mm} \qquad b/a = 0.01143$$

we have

$$c_1 = \tfrac{1}{3}(1 - 0.630 b/a) = \tfrac{1}{3}(1 - 0.630 \times 0.01143) = 0.331$$

$$(\tau_{max})_{twisting} = \frac{T}{c_1 a b^2} = \frac{685 \text{ N} \cdot \text{m}}{(0.331)(350 \text{ mm})(4 \text{ mm})^2} = 370 \text{ MPa}$$

Combined stresses. The maximum stress due to the combined bending and twisting occurs at the neutral axis, on the inside surface of the web, and is

$$\tau_{max} = 18.33 \text{ MPa} + 370 \text{ MPa} = 388 \text{ MPa}$$

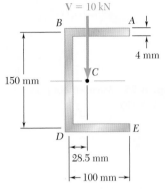

Fig. 6.53

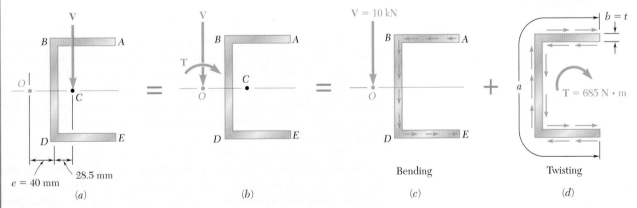

Fig. 6.54

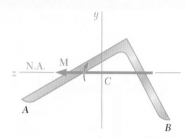

Fig. 6.55 Beam without plane of symmetry.

Turning our attention to thin-walled members possessing no plane of symmetry, we now consider the case of an angle shape subjected to a vertical load **P.** If the member is oriented in such a way that the load **P** is perpendicular to one of the principal centroidal axes Cz of the cross section, the couple vector **M** representing the bending moment in a given section will be directed along Cz (Fig. 6.55), and the neutral axis will coincide with that axis (cf. Sec. 4.13). Equation (4.16), therefore, is applicable and can be used to compute the normal stresses in the section. We now propose to determine where the load **P** should be applied if Eq. (6.6) is to define the shearing stresses in the section, i.e., if the member is to *bend without twisting.*

Let us *assume* that the shearing stresses in the section are defined by Eq. (6.6). As in the case of the channel member considered earlier, the elementary shearing forces exerted on the section can be expressed as $dF = q\ ds$, with $q = VQ/I$, where Q represents a first moment with respect to the neutral axis (Fig. 6.56a). We note that the resultant of the shearing forces exerted on portion OA of the cross section is a force $\mathbf{F}_1$ directed along OA, and that the resultant of the shearing forces exerted on portion OB is a force $\mathbf{F}_2$ along OB (Fig. 6.56b). Since both $\mathbf{F}_1$ and $\mathbf{F}_2$ pass through point O at the corner of the angle, it follows that their own resultant, which is the shear **V** in the section, must also pass through O (Fig. 6.56c). We conclude that the member will not be twisted if the line of action of the load **P** passes through the corner O of the section in which it is applied.

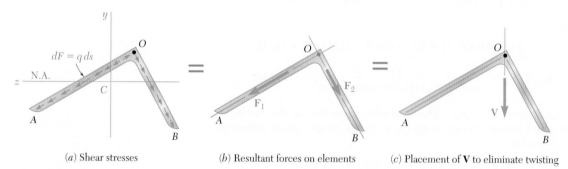

| (a) Shear stresses | (b) Resultant forces on elements | (c) Placement of **V** to eliminate twisting |

Fig. 6.56

The same reasoning can be applied when the load **P** is perpendicular to the other principal centroidal axis Cy of the angle section. And, since any load **P** applied at the corner O of a cross section can be resolved into components perpendicular to the principal axes, it follows that the member will not be twisted if each load is applied at the corner O of a cross section. We thus conclude that O is the shear center of the section.

Angle shapes with one vertical and one horizontal leg are encountered in many structures. It follows from the preceding discussion that such members will not be twisted if vertical loads are applied along the center line of their vertical leg. We note from Fig. 6.57 that the resultant of the elementary shearing forces exerted on the vertical portion OA of a given section will be equal to the

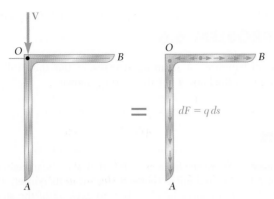

Fig. 6.57 Angle section.

shear **V,** while the resultant of the shearing forces on the horizontal portion *OB* will be zero:

$$\int_O^A q \, ds = V \qquad \int_O^B q \, ds = 0$$

This does *not* mean, however, that there will be no shearing stress in the horizontal leg of the member. By resolving the shear **V** into components perpendicular to the principal centroidal axes of the section and computing the shearing stress at every point, we would verify that τ is zero at only one point between *O* and *B* (see Sample Prob. 6.6).

Another type of thin-walled member frequently encountered in practice is the Z shape. While the cross section of a Z shape does not possess any axis of symmetry, it does possess a *center of symmetry O* (Fig. 6.58). This means that, to any point *H* of the cross section corresponds another point *H'* such that the segment of straight line *HH'* is bisected by *O*. Clearly, the center of symmetry *O* coincides with the centroid of the cross section. As you will see presently, point *O* is also the shear center of the cross section.

As we did earlier in the case of an angle shape, we assume that the loads are applied in a plane perpendicular to one of the principal axes of the section, so that this axis is also the neutral axis of the section (Fig. 6.59). We further assume that the shearing stresses in the section are defined by Eq. (6.6), i.e., that the member is bent without being twisted. Denoting by *Q* the first moment about the neutral axis of portion *AH* of the cross section, and by *Q'* the first moment of portion *EH'*, we note that $Q' = -Q$. Thus the shearing stresses at *H* and *H'* have the same magnitude and the same direction, and the shearing forces exerted on small elements of area *dA* located respectively at *H* and *H'* are equal forces that have equal and opposite moments about *O* (Fig. 6.60). Since this is true for any pair of symmetric elements, it follows that the resultant of the shearing forces exerted on the section has a zero moment about *O*. This means that the shear **V** in the section is directed along a line that passes through *O*. Since this analysis can be repeated when the loads are applied in a plane perpendicular to the other principal axis, we conclude that point *O* is the shear center of the section.

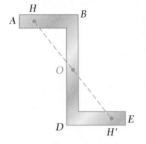

Fig. 6.58 Z section.

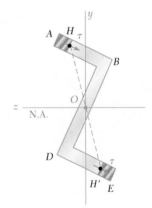

Fig. 6.59

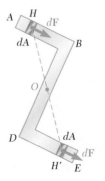

Fig. 6.60

SAMPLE PROBLEM 6.6

Determine the distribution of shearing stresses in the thin-walled angle shape DE of uniform thickness t for the loading shown.

SOLUTION

Shear Center. We recall from Sec. 6.9 that the shear center of the cross section of a thin-walled angle shape is located at its corner. Since the load **P** is applied at D, it causes bending but no twisting of the shape.

Principal Axes. We locate the centroid C of a given cross section AOB. Since the y' axis is an axis of symmetry, the y' and z' axes are the principal centroidal axes of the section. We recall that for the parallelogram shown $I_{nn} = \frac{1}{12}bh^3$ and $I_{mm} = \frac{1}{3}bh^3$. Considering each leg of the section as a parallelogram, we now determine the centroidal moments of inertia $I_{y'}$ and $I_{z'}$:

$$I_{y'} = 2\left[\frac{1}{3}\left(\frac{t}{\cos 45°}\right)(a\cos 45°)^3\right] = \frac{1}{3}ta^3$$

$$I_{z'} = 2\left[\frac{1}{12}\left(\frac{t}{\cos 45°}\right)(a\cos 45°)^3\right] = \frac{1}{12}ta^3$$

Superposition. The shear **V** in the section is equal to the load **P**. We resolve it into components parallel to the principal axes.

Shearing Stresses Due to $V_{y'}$. We determine the shearing stress at point e of coordinate y:

$$\bar{y}' = \frac{1}{2}(a + y)\cos 45° - \frac{1}{2}a\cos 45° = \frac{1}{2}y\cos 45°$$

$$Q = t(a - y)\bar{y}' = \frac{1}{2}t(a - y)y\cos 45°$$

$$\tau_1 = \frac{V_{y'}Q}{I_{z'}t} = \frac{(P\cos 45°)[\frac{1}{2}t(a-y)y\cos 45°]}{(\frac{1}{12}ta^3)t} = \frac{3P(a-y)y}{ta^3}$$

The shearing stress at point f is represented by a similar function of z.

Shearing Stresses Due to $V_{z'}$. We again consider point e:

$$\bar{z}' = \frac{1}{2}(a + y)\cos 45°$$

$$Q = (a - y)t\bar{z}' = \frac{1}{2}(a^2 - y^2)t\cos 45°$$

$$\tau_2 = \frac{V_{z'}Q}{I_{y'}t} = \frac{(P\cos 45°)[\frac{1}{2}(a^2 - y^2)t\cos 45°]}{(\frac{1}{3}ta^3)t} = \frac{3P(a^2 - y^2)}{4ta^3}$$

The shearing stress at point f is represented by a similar function of z.

Combined Stresses. *Along the Vertical Leg.* The shearing stress at point e is

$$\tau_e = \tau_2 + \tau_1 = \frac{3P(a^2 - y^2)}{4ta^3} + \frac{3P(a-y)y}{ta^3} = \frac{3P(a-y)}{4ta^3}[(a+y) + 4y]$$

$$\tau_e = \frac{3P(a-y)(a+5y)}{4ta^3} \quad \blacktriangleleft$$

Along the Horizontal Leg. The shearing stress at point f is

$$\tau_f = \tau_2 - \tau_1 = \frac{3P(a^2 - z^2)}{4ta^3} - \frac{3P(a-z)z}{ta^3} = \frac{3P(a-z)}{4ta^3}[(a+z) - 4z]$$

$$\tau_f = \frac{3P(a-z)(a-3z)}{4ta^3} \quad \blacktriangleleft$$

6.61 through 6.64 Determine the location of the shear center O of a thin-walled beam of uniform thickness having the cross section shown.

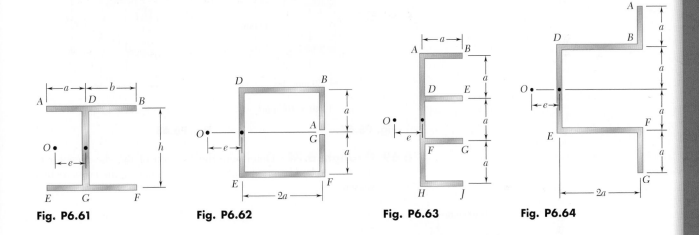

Fig. P6.61   **Fig. P6.62**   **Fig. P6.63**   **Fig. P6.64**

6.65 and 6.66 An extruded beam has the cross section shown. Determine (*a*) the location of the shear center O, (*b*) the distribution of the shearing stresses caused by the 12-kN vertical shearing force applied at O.

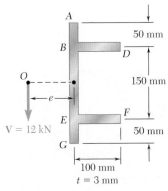

Fig. P6.65

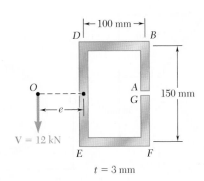

Fig. P6.66

6.67 and 6.68 An extruded beam has the cross section shown. Determine (*a*) the location of the shear center *O*, (*b*) the distribution of the shearing stresses caused by the vertical shearing force **V** shown applied at *O*.

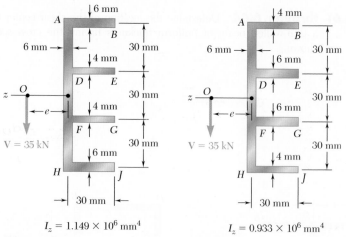

$I_z = 1.149 \times 10^6 \text{ mm}^4$

Fig. P6.67

$I_z = 0.933 \times 10^6 \text{ mm}^4$

Fig. P6.68

6.69 through 6.74 Determine the location of the shear center *O* of a thin-walled beam of uniform thickness having the cross section shown.

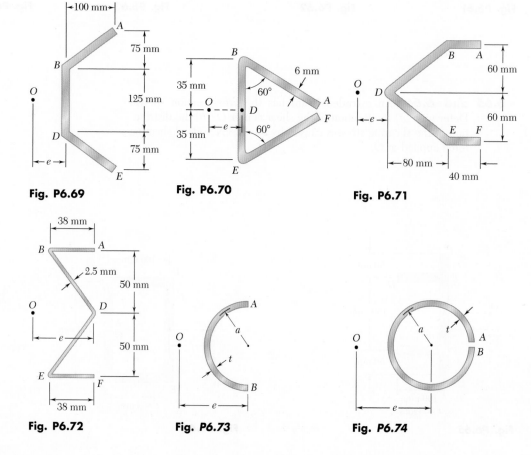

Fig. P6.69

Fig. P6.70

Fig. P6.71

Fig. P6.72

Fig. P6.73

Fig. P6.74

6.75 and 6.76 A thin-walled beam has the cross section shown. Determine the location of the shear center O of the cross section.

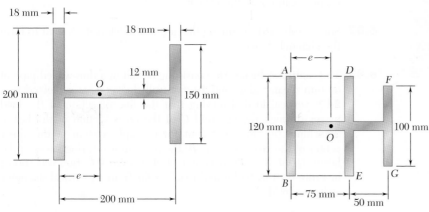

Fig. P6.75   **Fig. P6.76**

6.77 and 6.78 A thin-walled beam of uniform thickness has the cross section shown. Determine the dimension *b* for which the shear center *O* of the cross section is located at the point indicated.

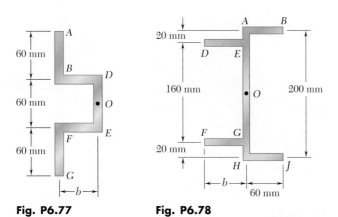

Fig. P6.77   **Fig. P6.78**

6.79 For the angle shape and loading of Sample Prob. 6.6, check that $\int q\, dz = 0$ along the horizontal leg of the angle and $\int q\, dy = P$ along its vertical leg.

6.80 For the angle shape and loading of Sample Prob. 6.6, (*a*) determine the points where the shearing stress is maximum and the corresponding values of the stress, (*b*) verify that the points obtained are located on the neutral axis corresponding to the given loading.

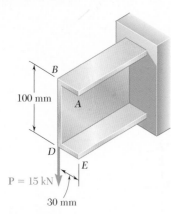

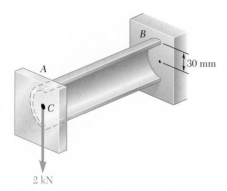

Fig. P6.81

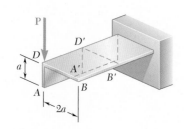

2 kN

Fig. P6.83 and P6.84

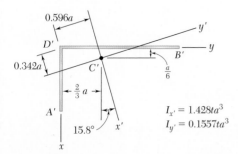

Fig. P6.87

*6.81 A steel plate, 160 mm wide and 8 mm thick, is bent to form the channel shown. Knowing that the vertical load **P** acts at a point in the midplane of the web of the channel, determine (*a*) the torque **T** that would cause the channel to twist in the same way that it does under the load **P**, (*b*) the maximum shearing stress in the channel caused by the load **P**.

*6.82 Solve Prob. 6.81, assuming that a 6-mm-thick plate is bent to form the channel shown.

*6.83 The cantilever beam *AB*, consisting of half of a thin-walled pipe of 30-mm mean radius and 10-mm wall thickness, is subjected to a 2-kN vertical load. Knowing that the line of action of the load passes through the centroid *C* of the cross section of the beam, determine (*a*) the equivalent force-couple system at the shear center of the cross section, (*b*) the maximum shearing stress in the beam. (*Hint:* The shear center *O* of this cross section was shown in Prob. 6.73 to be located twice as far from its vertical diameter as its centroid *C*.)

*6.84 Solve Prob. 6.83, assuming that the thickness of the beam is reduced to 6 mm.

*6.85 The cantilever beam shown consists of a Z shape of 6-mm thickness. For the given loading, determine the distribution of the shearing stresses along line *A'B'* in the upper horizontal leg of the Z shape. The *x'* and *y'* axes are the principal centroidal axes of the cross section and the corresponding moments of inertia are $I_{x'} = 69.2 \times 10^6$ mm^4 and $I_{y'} = 5.7 \times 10^6$ mm^4.

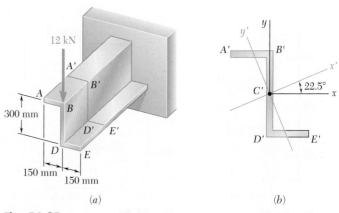

(*a*)   (*b*)

Fig. P6.85

*6.86 For the cantilever beam and loading of Prob. 6.85, determine the distribution of the shearing stress along line *B'D'* in the vertical web of the Z shape.

*6.87 Determine the distribution of the shearing stresses along line *D'B'* in the horizontal leg of the angle shape for the loading shown. The *x'* and *y'* axes are the principal centroidal axes of the cross section.

*6.88 For the angle shape and loading of Prob. 6.87, determine the distribution of the shearing stresses along line *D'A'* in the vertical leg.

REVIEW AND SUMMARY

This chapter was devoted to the analysis of beams and thin-walled members under transverse loadings.

Stresses on a beam element

In Sec. 6.1 we considered a small element located in the vertical plane of symmetry of a beam under a transverse loading (Fig. 6.61) and found that normal stresses σ_x and shearing stresses τ_{xy} were exerted on the transverse faces of that element, while shearing stresses τ_{yx}, equal in magnitude to τ_{xy}, were exerted on its horizontal faces.

In Sec. 6.2 we considered a prismatic beam AB with a vertical plane of symmetry supporting various concentrated and distributed loads (Fig. 6.62). At a distance x from end A we detached from the

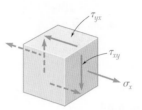

Fig. 6.61

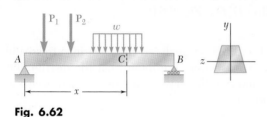

Fig. 6.62

beam an element $CDD'C'$ of length Δx extending across the width of the beam from the upper surface of the beam to a horizontal plane located at a distance y_1 from the neutral axis (Fig. 6.63). We found

Horizontal shear in a beam

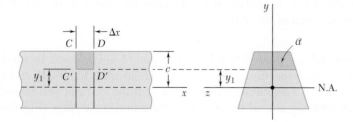

Fig. 6.63

that the magnitude of the shearing force $\Delta\mathbf{H}$ exerted on the lower face of the beam element was

$$\Delta H = \frac{VQ}{I}\Delta x \qquad (6.4)$$

where V = vertical shear in the given transverse section

Q = first moment with respect to the neutral axis of the shaded portion $\mathfrak{a}$ of the section

I = centroidal moment of inertia of the entire cross-sectional area

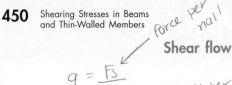

force per nail

Shear flow

$q = \dfrac{Fs}{s}$ ← *spacing per nail*

Shearing stresses in a beam

$\bar{y} = $ *dist to centroid from N.A*

$Q = \bar{y}A$

$Q = \Sigma \bar{y_i}A_i$

$t = $ *thickness of cut.*

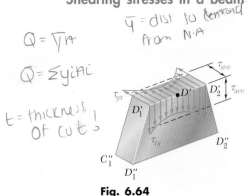

Fig. 6.64

The *horizontal shear per unit length,* or *shear flow,* which was denoted by the letter q, was obtained by dividing both members of Eq. (6.4) by Δx:

$$q = \frac{\Delta H}{\Delta x} = \frac{VQ}{I} \qquad (6.5)$$

Dividing both members of Eq. (6.4) by the area ΔA of the horizontal face of the element and observing that $\Delta A = t\,\Delta x$, where t is the width of the element at the cut, we obtained in Sec. 6.3 the following expression for the *average shearing stress* on the horizontal face of the element

$Q = $ *above the plane where shear occurs*

$$\tau_{ave} = \frac{VQ}{It} \qquad (6.6)$$

$\tau_{av} = \dfrac{V \cdot Fs}{A}$

avg shear flow bars

We further noted that, since the shearing stresses τ_{xy} and τ_{yx} exerted, respectively, on a transverse and a horizontal plane through D' are equal, the expression in (6.6) also represents the average value of τ_{xy} along the line $D_1' D_2'$ (Fig. 6.64).

Shearing stresses in a beam of rectangular cross section

In Secs. 6.4 and 6.5 we analyzed the shearing stresses in a beam of rectangular cross section. We found that the distribution of stresses is parabolic and that the maximum stress, which occurs at the center of the section, is

$$\tau_{max} = \frac{3}{2}\frac{V}{A} \qquad (6.10)$$

where A is the area of the rectangular section. For wide-flange beams, we found that a good approximation of the maximum shearing stress can be obtained by dividing the shear V by the cross-sectional area of the web.

Longitudinal shear on curved surface

In Sec. 6.6 we showed that Eqs. (6.4) and (6.5) could still be used to determine, respectively, the longitudinal shearing force $\Delta \mathbf{H}$ and the shear flow q exerted on a beam element if the element was bounded by an arbitrary curved surface instead of a horizontal plane (Fig. 6.65).

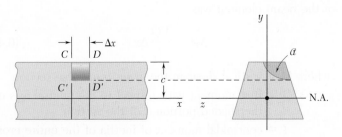

Fig. 6.65

This made it possible for us in Sec. 6.7 to extend the use of Eq. (6.6) to the determination of the average shearing stress in thin-walled members such as wide-flange beams and box beams, in the flanges of such members, and in their webs (Fig. 6.66).

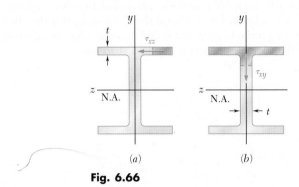

Fig. 6.66

In Sec. 6.8 we considered the effect of plastic deformations on the magnitude and distribution of shearing stresses. From Chap. 4 we recalled that once plastic deformation has been initiated, additional loading causes plastic zones to penetrate into the elastic core of a beam. After demonstrating that shearing stresses can occur only in the elastic core of a beam, we noted that both an increase in loading and the resulting decrease in the size of the elastic core contribute to an increase in shearing stresses.

In Sec. 6.9 we considered prismatic members that are *not* loaded in their plane of symmetry and observed that, in general, both bending and twisting will occur. You learned to locate the point O of the cross section, known as the *shear center*, where the loads should be applied if the member is to bend without twisting (Fig. 6.67) and found that if the loads are applied at that point, the following equations remain valid:

$$\sigma_x = -\frac{My}{I} \qquad \tau_{ave} = \frac{VQ}{It} \qquad (4.16, 6.6)$$

Using the principle of superposition, you also learned to determine the stresses in unsymmetric thin-walled members such as channels, angles, and extruded beams [Example 6.07 and Sample Prob. 6.6].

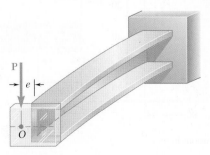

Fig. 6.67

REVIEW PROBLEMS

6.89 Three boards, each 50 mm thick, are nailed together to form a beam that is subjected to a vertical shear. Knowing that the allowable shearing force in each nail is 600 N, determine the allowable shear if the spacing s between the nails is 75 mm.

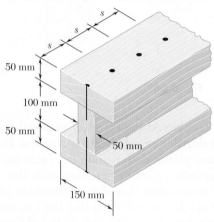

Fig. P6.89

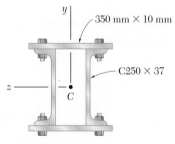

Fig. P6.90

6.90 The beam shown is fabricated by connecting two channel shapes and two plates, using bolts of 18-mm diameter spaced longitudinally every 125 mm. Determine the average shearing stress in the bolts caused by a shearing force of 120 kN parallel to the y axis.

6.91 For the beam and loading shown, consider section n-n and determine (a) the largest shearing stress in that section, (b) the shearing stress at point a.

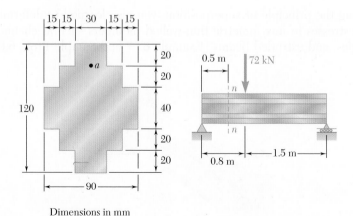

Dimensions in mm

Fig. P6.91

6.92 For the beam and loading shown, determine the minimum required width b, knowing that for the grade of timber used, $\sigma_{all} = 12$ MPa and $\tau_{all} = 825$ kPa.

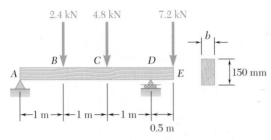

Fig. P6.92

6.93 For the beam and loading shown, consider section n-n and determine the shearing stress at (a) point a, (b) point b.

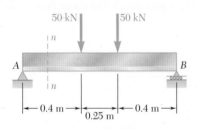 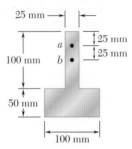

Fig. P6.93

6.94 For the beam and loading shown in Prob. 6.93, determine the largest shearing stress in section n-n.

6.95 The composite beam shown is made by welding C200 × 17.1 rolled-steel channels to the flanges of a W250 × 80 wide-flange rolled-steel shape. Knowing that the beam is subjected to a vertical shear of 200 kN, determine (a) the horizontal shearing force per meter at each weld, (b) the shearing stress at point a of the flange of the wide-flange shape.

Fig. P6.95

6.96 An extruded beam has the cross section shown and a uniform wall thickness of 3 mm. For a vertical shear of 10 kN, determine (a) the shearing stress at point A, (b) the maximum shearing stress in the beam. Also sketch the shear flow in the cross section.

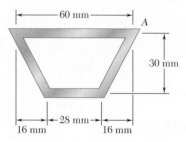

Fig. P6.96

6.97 Three plates, each 12 mm thick, are welded together to form the section shown. For a vertical shear of 100 kN, determine the shear flow through the welded surfaces and sketch the shear flow in the cross section.

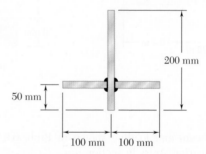

Fig. P6.97

6.98 Determine the location of the shear center O of a thin-walled beam of uniform thickness having the cross section shown.

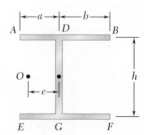

Fig. P6.98

6.99 Determine the location of the shear center O of a thin-walled beam of uniform thickness having the cross section shown.

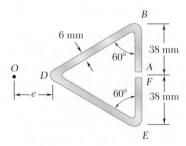

Fig. P6.99

6.100 A thin-walled beam of uniform thickness has the cross section shown. Determine the dimension b for which the shear center O of the cross section is located at the point indicated.

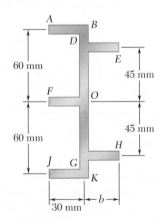

Fig. P6.100

COMPUTER PROBLEMS

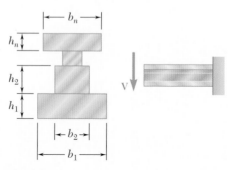

The following problems are designed to be solved with a computer.

6.C1 A timber beam is to be designed to support a distributed load and up to two concentrated loads as shown. One of the dimensions of its uniform rectangular cross section has been specified and the other is to be determined so that the maximum normal stress and the maximum shearing stress in the beam will not exceed given allowable values σ_{all} and τ_{all}. Measuring x from end A, write a computer program to calculate for successive cross sections, from $x = 0$ to $x = L$ and using given increments Δx, the shear, the bending moment, and the smallest value of the unknown dimension that satisfies in that section (1) the allowable normal stress requirement, (2) the allowable shearing stress requirement. Use this program to solve Prob. 5.65 assuming $\sigma_{all} = 12$ MPa and $\tau_{all} = 825$ kPa, using $\Delta x = 0.1$ m.

Fig. P6.C1

6.C2 A cantilever timber beam AB of length L and of uniform rectangular section shown supports a concentrated load $\mathbf{P}$ at its free end and a uniformly distributed load w along its entire length. Write a computer program to determine the length L and the width b of the beam for which both the maximum normal stress and the maximum shearing stress in the beam reach their largest allowable values. Assuming $\sigma_{all} = 12.4$ MPa and $\tau_{all} = 0.8$ MPa, use this program to determine the dimensions L and b when (a) $P = 4.5$ kN and $w = 0$, (b) $P = 0$ and $w = 2.2$ N/mm, (c) $P = 2.25$ kN and $w = 2.2$ N/mm.

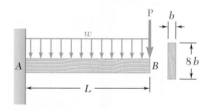

Fig. P6.C2

6.C3 A beam having the cross section shown is subjected to a vertical shear $\mathbf{V}$. Write a computer program that can be used to calculate the shearing stress along the line between any two adjacent rectangular areas forming the cross section. Use this program to solve (a) Prob. 6.10, (b) Prob. 6.12, (c) Prob. 6.21.

Fig. P6.C3

6.C4 A plate of uniform thickness t is bent as shown into a shape with a vertical plane of symmetry and is then used as a beam. Write a computer program that can be used to determine the distribution of shearing stresses caused by a vertical shear **V**. Use this program (*a*) to solve Prob. 6.47, (*b*) to find the shearing stress at a point E for the shape and load of Prob. 6.50, assuming a thickness $t = 6$ mm.

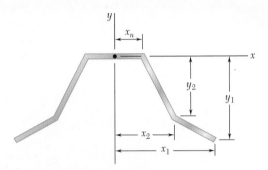

Fig. P6.C4

6.C5 The cross section of an extruded beam is symmetric with respect to the x axis and consists of several straight segments as shown. Write a computer program that can be used to determine (*a*) the location of the shear center O, (*b*) the distribution of shearing stresses caused by a vertical force applied at O. Use this program to solve Probs. 6.66 and 6.70.

6.C6 A thin-walled beam has the cross section shown. Write a computer program that can be used to determine the location of the shear center O of the cross section. Use the program to solve Prob. 6.75.

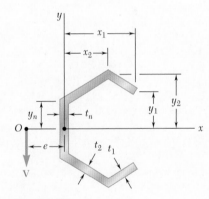

Fig. P6.C5

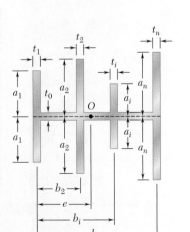

Fig. P6.C6

The aircraft shown is being tested to determine how the forces due to lift would be distributed over the wing. This chapter deals with stresses and strains in structures and machine components.

CHAPTER

7

Transformations of Stress and Strain

7.1   INTRODUCTION

We saw in Sec. 1.12 that the most general state of stress at a given point Q may be represented by six components. Three of these components, σ_x, σ_y, and σ_z, define the normal stresses exerted on the faces of a small cubic element centered at Q and of the same orientation as the coordinate axes (Fig. 7.1a), and the other three, τ_{xy}, τ_{yz}, and τ_{zx},† the components of the shearing stresses on the same element. As we remarked at the time, the same state of stress will be represented by a different set of components if the coordinate axes are rotated (Fig. 7.1b). We propose in the first part of this chapter to determine how the components of stress are transformed under a rotation of the coordinate axes. The second part of the chapter will be devoted to a similar analysis of the transformation of the components of strain.

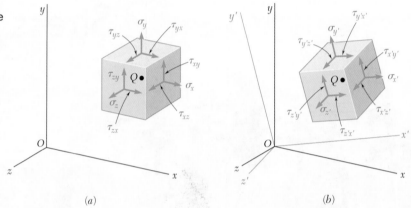

(a)   (b)

Fig. 7.1   General state of stress at a point.

Our discussion of the transformation of stress will deal mainly with *plane stress*, i.e., with a situation in which two of the faces of the cubic element are free of any stress. If the z axis is chosen perpendicular to these faces, we have $\sigma_z = \tau_{zx} = \tau_{zy} = 0$, and the only remaining stress components are σ_x, σ_y, and τ_{xy} (Fig. 7.2). Such a situation occurs in a thin plate subjected to forces acting in the midplane of the plate (Fig. 7.3). It also occurs on the free surface of a structural element or machine component, i.e., at any point of the surface of that element or component that is not subjected to an external force (Fig. 7.4).

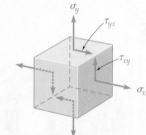

Fig. 7.2   Plane stress.

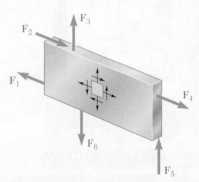

Fig. 7.3   Example of plane stress.

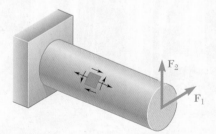

Fig. 7.4   Example of plane stress.

†We recall that $\tau_{yx} = \tau_{xy}$, $\tau_{zy} = \tau_{yz}$, and $\tau_{xz} = \tau_{zx}$.

Considering in Sec. 7.2 a state of plane stress at a given point Q characterized by the stress components σ_x, σ_y, and τ_{xy} associated with the element shown in Fig. 7.5a, you will learn to determine the components $\sigma_{x'}$, $\sigma_{y'}$, and $\tau_{x'y'}$ associated with that element after it has been rotated through an angle θ about the z axis (Fig. 7.5b). In Sec. 7.3, you will determine the value θ_p of θ for which the stresses $\sigma_{x'}$ and $\sigma_{y'}$ are, respectively, maximum and minimum; these values of the normal stress are the *principal stresses* at point Q, and the faces of the corresponding element define the *principal planes of stress* at that point. You will also determine the value θ_s of the angle of rotation for which the shearing stress is maximum, as well as the value of that stress.

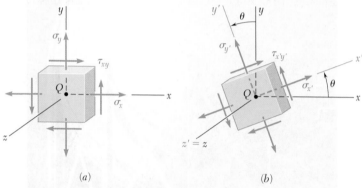

Fig. 7.5 Transformation of stress.

In Sec. 7.4, an alternative method for the solution of problems involving the transformation of plane stress, based on the use of *Mohr's circle*, will be presented.

In Sec. 7.5, the *three-dimensional state of stress* at a given point will be considered and a formula for the determination of the normal stress on a plane of arbitrary orientation at that point will be developed. In Sec. 7.6, you will consider the rotations of a cubic element about each of the principal axes of stress and note that the corresponding transformations of stress can be described by three different Mohr's circles. You will also observe that, in the case of a state of *plane stress* at a given point, the maximum value of the shearing stress obtained earlier by considering rotations in the plane of stress does not necessarily represent the maximum shearing stress at that point. This will bring you to distinguish between *in-plane* and *out-of-plane* maximum shearing stresses.

Yield criteria for ductile materials under plane stress will be developed in Sec. 7.7. To predict whether a material will yield at some critical point under given loading conditions, you will determine the principal stresses σ_a and σ_b at that point and check whether σ_a, σ_b, and the yield strength σ_Y of the material satisfy some criterion. Two criteria in common use are: the *maximum-shearing-strength criterion* and the *maximum-distortion-energy criterion*. In Sec. 7.8, *fracture criteria* for brittle materials under plane stress will be developed in a similar fashion; they will involve the principal stresses σ_a and σ_b at some critical point and the ultimate strength σ_U of the material. Two criteria will be discussed: the *maximum-normal-stress* criterion and *Mohr's criterion*.

Thin-walled pressure vessels provide an important application of the analysis of plane stress. In Sec. 7.9, we will discuss stresses in both cylindrical and spherical pressure vessels (Photos 7.1 and 7.2).

Photo 7.1
Cylindrical
pressure vessel.

Photo 7.2 Spherical pressure vessel.

Sections 7.10 and 7.11 will be devoted to a discussion of the *transformation of plane strain* and to *Mohr's circle for plane strain*. In Sec. 7.12, we will consider the three-dimensional analysis of strain and see how Mohr's circles can be used to determine the maximum shearing strain at a given point. Two particular cases are of special interest and should not be confused: the case of *plane strain* and the case of *plane stress*.

Finally, in Sec. 7.13, we discuss the use of *strain gages* to measure the normal strain on the surface of a structural element or machine component. You will see how the components ϵ_x, ϵ_y, and γ_{xy} characterizing the state of strain at a given point can be computed from the measurements made with three strain gages forming a *strain rosette*.

7.2 TRANSFORMATION OF PLANE STRESS

Let us assume that a state of plane stress exists at point Q (with $\sigma_z = \tau_{zx} = \tau_{zy} = 0$), and that it is defined by the stress components σ_x, σ_y, and τ_{xy} associated with the element shown in Fig. 7.5*a*. We propose to determine the stress components $\sigma_{x'}$, $\sigma_{y'}$, and $\tau_{x'y'}$ associated with the element after it has been rotated through an angle θ about

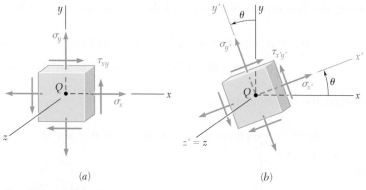

Fig. 7.5 (repeated)

the z axis (Fig. 7.5b), and to express these components in terms of σ_x, σ_y, τ_{xy}, and θ.

In order to determine the normal stress $\sigma_{x'}$ and the shearing stress $\tau_{x'y'}$ exerted on the face perpendicular to the x' axis, we consider a prismatic element with faces respectively perpendicular to the x, y, and x' axes (Fig. 7.6a). We observe that, if the area of the

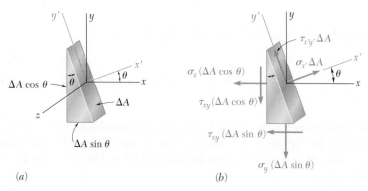

Fig. 7.6

oblique face is denoted by ΔA, the areas of the vertical and horizontal faces are respectively equal to $\Delta A \cos \theta$ and $\Delta A \sin \theta$. It follows that the *forces* exerted on the three faces are as shown in Fig. 7.6b. (No forces are exerted on the triangular faces of the element, since the corresponding normal and shearing stresses have all been assumed equal to zero.) Using components along the x' and y' axes, we write the following equilibrium equations:

$$\Sigma F_{x'} = 0: \quad \sigma_{x'} \Delta A - \sigma_x (\Delta A \cos \theta) \cos \theta - \tau_{xy} (\Delta A \cos \theta) \sin \theta$$
$$- \sigma_y (\Delta A \sin \theta) \sin \theta - \tau_{xy} (\Delta A \sin \theta) \cos \theta = 0$$

$$\Sigma F_{y'} = 0: \quad \tau_{x'y'} \Delta A + \sigma_x (\Delta A \cos \theta) \sin \theta - \tau_{xy} (\Delta A \cos \theta) \cos \theta$$
$$- \sigma_y (\Delta A \sin \theta) \cos \theta + \tau_{xy} (\Delta A \sin \theta) \sin \theta = 0$$

Solving the first equation for $\sigma_{x'}$ and the second for $\tau_{x'y'}$, we have

$$\sigma_{x'} = \sigma_x \cos^2 \theta + \sigma_y \sin^2 \theta + 2\tau_{xy} \sin \theta \cos \theta \qquad (7.1)$$

$$\tau_{x'y'} = -(\sigma_x - \sigma_y) \sin \theta \cos \theta + \tau_{xy}(\cos^2 \theta - \sin^2 \theta) \qquad (7.2)$$

Recalling the trigonometric relations

$$\sin 2\theta = 2 \sin \theta \cos \theta \qquad \cos 2\theta = \cos^2 \theta - \sin^2 \theta \qquad (7.3)$$

and

$$\cos^2 \theta = \frac{1 + \cos 2\theta}{2} \qquad \sin^2 \theta = \frac{1 - \cos 2\theta}{2} \qquad (7.4)$$

we write Eq. (7.1) as follows:

$$\sigma_{x'} = \sigma_x \frac{1 + \cos 2\theta}{2} + \sigma_y \frac{1 - \cos 2\theta}{2} + \tau_{xy} \sin 2\theta$$

or

$$\sigma_{x'} = \frac{\sigma_x + \sigma_y}{2} + \frac{\sigma_x - \sigma_y}{2} \cos 2\theta + \tau_{xy} \sin 2\theta \qquad (7.5)$$

Using the relations (7.3), we write Eq. (7.2) as

$$\tau_{x'y'} = -\frac{\sigma_x - \sigma_y}{2} \sin 2\theta + \tau_{xy} \cos 2\theta \qquad (7.6)$$

The expression for the normal stress $\sigma_{y'}$ is obtained by replacing θ in Eq. (7.5) by the angle $\theta + 90°$ that the y' axis forms with the x axis. Since $\cos (2\theta + 180°) = -\cos 2\theta$ and $\sin (2\theta + 180°) = -\sin 2\theta$, we have

$$\sigma_{y'} = \frac{\sigma_x + \sigma_y}{2} - \frac{\sigma_x - \sigma_y}{2} \cos 2\theta - \tau_{xy} \sin 2\theta \qquad (7.7)$$

Adding Eqs. (7.5) and (7.7) member to member, we obtain

$$\sigma_{x'} + \sigma_{y'} = \sigma_x + \sigma_y \qquad (7.8)$$

Since $\sigma_z = \sigma_{z'} = 0$, we thus verify in the case of plane stress that the sum of the normal stresses exerted on a cubic element of material is independent of the orientation of that element.†

---

†Cf. first footnote on page 460.

7.3 PRINCIPAL STRESSES; MAXIMUM SHEARING STRESS

The equations (7.5) and (7.6) obtained in the preceding section are the parametric equations of a circle. This means that, if we choose a set of rectangular axes and plot a point M of abscissa $\sigma_{x'}$ and ordinate $\tau_{x'y'}$ for any given value of the parameter θ, all the points thus obtained will lie on a circle. To establish this property we eliminate θ from Eqs. (7.5) and (7.6); this is done by first transposing $(\sigma_x + \sigma_y)/2$ in Eq. (7.5) and squaring both members of the equation, then squaring both members of Eq. (7.6), and finally adding member to member the two equations obtained in this fashion. We have

$$\left(\sigma_{x'} - \frac{\sigma_x + \sigma_y}{2}\right)^2 + \tau_{x'y'}^2 = \left(\frac{\sigma_x - \sigma_y}{2}\right)^2 + \tau_{xy}^2 \qquad (7.9)$$

Setting

$$\sigma_{\text{ave}} = \frac{\sigma_x + \sigma_y}{2} \qquad \text{and} \qquad R = \sqrt{\left(\frac{\sigma_x - \sigma_y}{2}\right)^2 + \tau_{xy}^2} \qquad (7.10)$$

we write the identity (7.9) in the form

$$(\sigma_{x'} - \sigma_{\text{ave}})^2 + \tau_{x'y'}^2 = R^2 \qquad (7.11)$$

which is the equation of a circle of radius R centered at the point C of abscissa σ_{ave} and ordinate 0 (Fig. 7.7). It can be observed that, due to the symmetry of the circle about the horizontal axis, the same result would have been obtained if, instead of plotting M, we had plotted a point N of abscissa $\sigma_{x'}$ and ordinate $-\tau_{x'y'}$ (Fig. 7.8). This property will be used in Sec. 7.4.

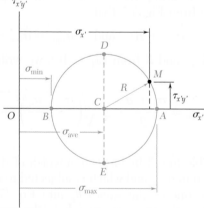
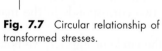

Fig. 7.7 Circular relationship of transformed stresses.

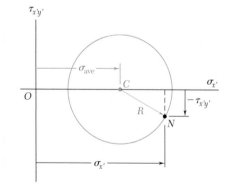

Fig. 7.8 Equivalent formation of stress transformation circle.

The two points A and B where the circle of Fig. 7.7 intersects the horizontal axis are of special interest: Point A corresponds to the maximum value of the normal stress $\sigma_{x'}$, while point B corresponds to its minimum value. Besides, both points correspond to a zero value of the shearing stress $\tau_{x'y'}$. Thus, the values θ_p of the parameter θ which correspond to points A and B can be obtained by setting $\tau_{x'y'} = 0$ in Eq. (7.6). We write†

$$\tan 2\theta_p = \frac{2\tau_{xy}}{\sigma_x - \sigma_y} \qquad (7.12)$$

This equation defines two values $2\theta_p$ that are $180°$ apart, and thus two values θ_p that are $90°$ apart. Either of these values can be used to determine the orientation of the corresponding element (Fig. 7.9).

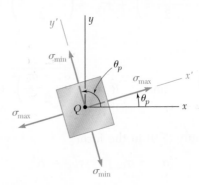

Fig. 7.9 Principal stresses.

The planes containing the faces of the element obtained in this way are called the *principal planes of stress* at point Q, and the corresponding values σ_{max} and σ_{min} of the normal stress exerted on these planes are called the *principal stresses* at Q. Since the two values θ_p defined by Eq. (7.12) were obtained by setting $\tau_{x'y'} = 0$ in Eq. (7.6), it is clear that no shearing stress is exerted on the principal planes.

We observe from Fig. 7.7 that

$$\sigma_{max} = \sigma_{ave} + R \quad \text{and} \quad \sigma_{min} = \sigma_{ave} - R \qquad (7.13)$$

Substituting for σ_{ave} and R from Eq. (7.10), we write

$$\sigma_{max, min} = \frac{\sigma_x + \sigma_y}{2} \pm \sqrt{\left(\frac{\sigma_x - \sigma_y}{2}\right)^2 + \tau_{xy}^2} \qquad (7.14)$$

Unless it is possible to tell by inspection which of the two principal planes is subjected to σ_{max} and which is subjected to σ_{min}, it is necessary to substitute one of the values θ_p into Eq. (7.5) in order to determine which of the two corresponds to the maximum value of the normal stress.

Referring again to the circle of Fig. 7.7, we note that the points D and E located on the vertical diameter of the circle correspond to

†This relation can also be obtained by differentiating $\sigma_{x'}$ in Eq. (7.5) and setting the derivative equal to zero: $d\sigma_{x'}/d\theta = 0$.

the largest numerical value of the shearing stress $\tau_{x'y'}$. Since the abscissa of points D and E is $\sigma_{ave} = (\sigma_x + \sigma_y)/2$, the values θ_s of the parameter θ corresponding to these points are obtained by setting $\sigma_{x'} = (\sigma_x + \sigma_y)/2$ in Eq. (7.5). It follows that the sum of the last two terms in that equation must be zero. Thus, for $\theta = \theta_s$, we write†

$$\frac{\sigma_x - \sigma_y}{2} \cos 2\theta_s + \tau_{xy} \sin 2\theta_s = 0$$

or

$$\tan 2\theta_s = -\frac{\sigma_x - \sigma_y}{2\tau_{xy}} \tag{7.15}$$

This equation defines two values $2\theta_s$ that are 180° apart, and thus two values θ_s that are 90° apart. Either of these values can be used to determine the orientation of the element corresponding to the maximum shearing stress (Fig. 7.10). Observing from Fig. 7.7 that the maximum value of the shearing stress is equal to the radius R of the circle, and recalling the second of Eqs. (7.10), we write

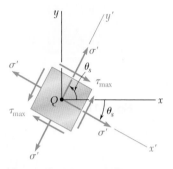

Fig. 7.10 Maximum shearing stress.

$$\tau_{max} = \sqrt{\left(\frac{\sigma_x - \sigma_y}{2}\right)^2 + \tau_{xy}^2} \tag{7.16}$$

As observed earlier, the normal stress corresponding to the condition of maximum shearing stress is

$$\sigma' = \sigma_{ave} = \frac{\sigma_x + \sigma_y}{2} \tag{7.17}$$

Comparing Eqs. (7.12) and (7.15), we note that $\tan 2\theta_s$ is the negative reciprocal of $\tan 2\theta_p$. This means that the angles $2\theta_s$ and $2\theta_p$ are 90° apart and, therefore, that the angles θ_s and θ_p are 45° apart. We thus conclude that *the planes of maximum shearing stress are at 45° to the principal planes*. This confirms the results obtained earlier in Sec. 1.12 in the case of a centric axial loading (Fig. 1.38) and in Sec. 3.4 in the case of a torsional loading (Fig. 3.19.)

We should be aware that our analysis of the transformation of plane stress has been limited to rotations *in the plane of stress*. If the cubic element of Fig. 7.5 is rotated about an axis other than the z axis, its faces may be subjected to shearing stresses larger than the stress defined by Eq. (7.16). As you will see in Sec. 7.5, this occurs when the principal stresses defined by Eq. (7.14) have the same sign, i.e., when they are either both tensile or both compressive. In such cases, the value given by Eq. (7.16) is referred to as the maximum *in-plane* shearing stress.

---

†This relation may also be obtained by differentiating $\tau_{x'y'}$ in Eq. (7.6) and setting the derivative equal to zero: $d\tau_{x'y'}/d\theta = 0$.

EXAMPLE 7.01

For the state of plane stress shown in Fig. 7.11, determine (a) the principal planes, (b) the principal stresses, (c) the maximum shearing stress and the corresponding normal stress.

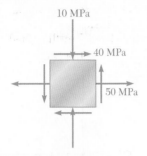

Fig. 7.11

(a) Principal Planes. Following the usual sign convention, we write the stress components as

$$\sigma_x = +50 \text{ MPa} \qquad \sigma_y = -10 \text{ MPa} \qquad \tau_{xy} = +40 \text{ MPa}$$

Substituting into Eq. (7.12), we have

$$\tan 2\theta_p = \frac{2\tau_{xy}}{\sigma_x - \sigma_y} = \frac{2(+40)}{50 - (-10)} = \frac{80}{60}$$

$$2\theta_p = 53.1° \qquad \text{and} \qquad 180° + 53.1° = 233.1°$$

$$\theta_p = 26.6° \qquad \text{and} \qquad 116.6°$$

(b) Principal Stresses. Formula (7.14) yields

$$\sigma_{\text{max, min}} = \frac{\sigma_x + \sigma_y}{2} \pm \sqrt{\left(\frac{\sigma_x - \sigma_y}{2}\right)^2 + \tau_{xy}^2}$$

$$= 20 \pm \sqrt{(30)^2 + (40)^2}$$

$$\sigma_{\text{max}} = 20 + 50 = 70 \text{ MPa}$$

$$\sigma_{\text{min}} = 20 - 50 = -30 \text{ MPa}$$

The principal planes and principal stresses are sketched in Fig. 7.12. Making $\theta = 26.6°$ in Eq. (7.5), we check that the normal stress exerted on face BC of the element is the maximum stress:

$$\sigma_{x'} = \frac{50 - 10}{2} + \frac{50 + 10}{2} \cos 53.1° + 40 \sin 53.1°$$

$$= 20 + 30 \cos 53.1° + 40 \sin 53.1° = 70 \text{ MPa} = \sigma_{\text{max}}$$

(c) Maximum Shearing Stress. Formula (7.16) yields

$$\tau_{\text{max}} = \sqrt{\left(\frac{\sigma_x - \sigma_y}{2}\right)^2 + \tau_{xy}^2} = \sqrt{(30)^2 + (40)^2} = 50 \text{ MPa}$$

Since σ_{max} and σ_{min} have opposite signs, the value obtained for τ_{max} actually represents the maximum value of the shearing stress at the point considered. The orientation of the planes of maximum shearing stress and the sense of the shearing stresses are best determined by passing a section along the diagonal plane AC of the element of Fig. 7.12. Since the faces AB and BC of the element are contained in the principal planes, the diagonal plane AC must be one of the planes of maximum shearing stress (Fig. 7.13). Furthermore, the equilibrium conditions for the prismatic element ABC require that the shearing stress exerted on AC be directed as shown. The cubic element corresponding to the maximum shearing stress is shown in Fig. 7.14. The normal stress on each of the four faces of the element is given by Eq. (7.17):

$$\sigma' = \sigma_{\text{ave}} = \frac{\sigma_x + \sigma_y}{2} = \frac{50 - 10}{2} = 20 \text{ MPa}$$

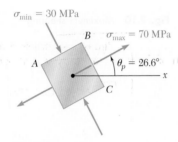

Fig. 7.12

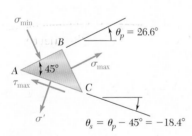

Fig. 7.13

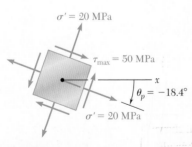

Fig. 7.14

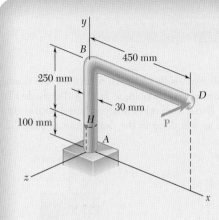

SAMPLE PROBLEM 7.1

A single horizontal force **P** of magnitude 600 N is applied to end D of lever ABD. Knowing that portion AB of the lever has a diameter of 30 mm, determine (a) the normal and shearing stresses on an element located at point H and having sides parallel to the x and y axes, (b) the principal planes and the principal stresses at point H.

SOLUTION

Force-Couple System. We replace the force **P** by an equivalent force-couple system at the center C of the transverse section containing point H:

$$P = 600 \text{ N} \qquad T = (600 \text{ N})(0.45 \text{ m}) = 270 \text{ N} \cdot \text{m}$$
$$M_x = (600 \text{ N})(0.25 \text{ m}) = 150 \text{ N} \cdot \text{m}$$

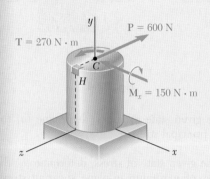

a. Stresses σ_x, σ_y, τ_{xy} at Point H. Using the sign convention shown in Fig. 7.2, we determine the sense and the sign of each stress component by carefully examining the sketch of the force-couple system at point C:

$$\sigma_x = 0 \quad \sigma_y = +\frac{Mc}{I} = +\frac{(150 \text{ N} \cdot \text{m})(0.015 \text{ m})}{\frac{1}{4}\pi (0.015)^4} \qquad \sigma_y = +56.6 \text{ MPa} \blacktriangleleft$$

$$\tau_{xy} = +\frac{Tc}{J} = +\frac{(270 \text{ N} \cdot \text{m})(0.015)}{\frac{1}{2}\pi (0.015)^4} \qquad \tau_{xy} = +50.9 \text{ MPa} \blacktriangleleft$$

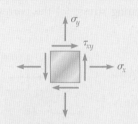

We note that the shearing force **P** does not cause any shearing stress at point H.

b. Principal Planes and Principal Stresses. Substituting the values of the stress components into Eq. (7.12), we determine the orientation of the principal planes:

$$\tan 2\theta_p = \frac{2\tau_{xy}}{\sigma_x - \sigma_y} = \frac{2(50.9)}{0 - 56.6} = -1.799$$

$$2\theta_p = -60.93° \quad \text{and} \quad 180° - 60.93° = +119.1°$$

$$\theta_p = -30.5° \qquad \text{and} \qquad +59.5° \blacktriangleleft$$

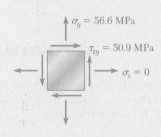

Substituting into Eq. (7.14), we determine the magnitudes of the principal stresses:

$$\sigma_{\text{max, min}} = \frac{\sigma_x + \sigma_y}{2} \pm \sqrt{\left(\frac{\sigma_x - \sigma_y}{2}\right)^2 + \tau_{xy}^2}$$

$$= \frac{0 + 56.6}{2} \pm \sqrt{\left(\frac{0 - 56.6}{2}\right)^2 + (50.9)^2} = +28.3 \pm 58.2$$

$$\sigma_{\text{max}} = +86.5 \text{ MPa} \blacktriangleleft$$
$$\sigma_{\text{min}} = -29.9 \text{ MPa} \blacktriangleleft$$

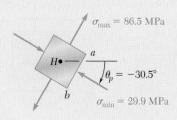

Considering face ab of the element shown, we make $\theta_p = -30.5°$ in Eq. (7.5) and find $\sigma_{x'} = 29.9$ MPa. We conclude that the principal stresses are as shown.

PROBLEMS

7.6   a) orientation of planes of max shear.

$\sigma x = \sigma xx = 28 \text{ MPa}$

$\sigma y = \sigma yy = 140 \text{ MPa}$

$\tau xy = -42 \text{ MPa}$

$\tan 2\theta p = \dfrac{2\tau xy}{\sigma x - \sigma y}$ $\boxed{\theta = 18.43°}$

b) $\tau max = \sqrt{\left(\dfrac{\sigma x - \sigma y}{2}\right)^2 + \tau xy^2}$

$= 70 \text{ MPa}$

c)

plane of max shear

7.1 through 7.4 For the given state of stress, determine the normal and shearing stresses exerted on the oblique face of the shaded triangular element shown. Use a method of analysis based on the equilibrium of that element, as was done in the derivations of Sec. 7.2.

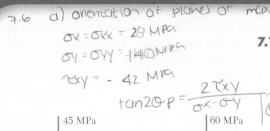

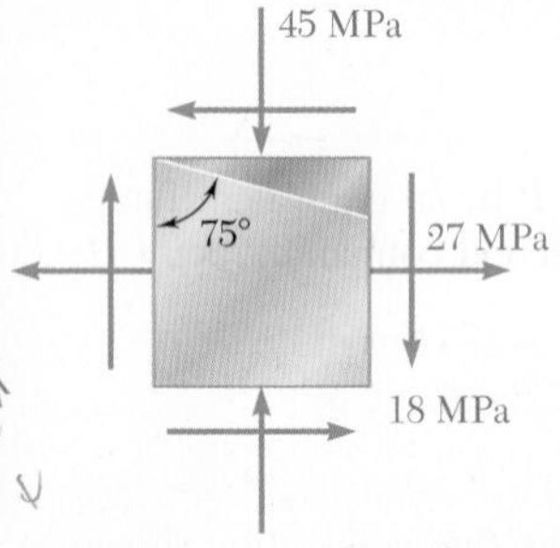

Fig. P7.1

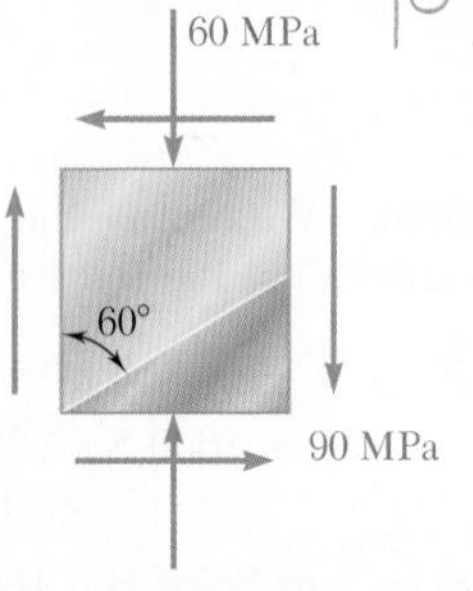

Fig. P7.2

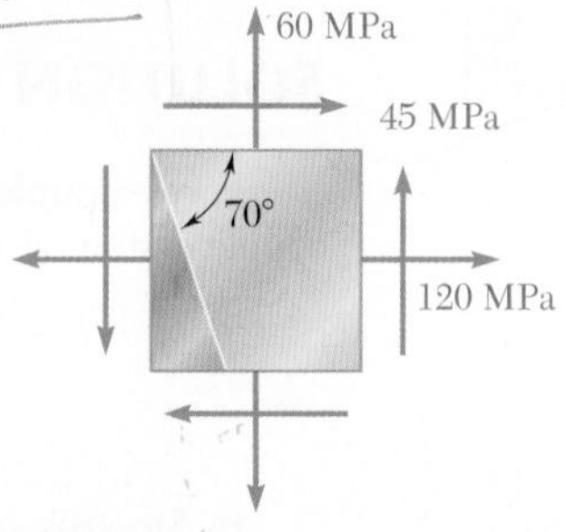

Fig. P7.3

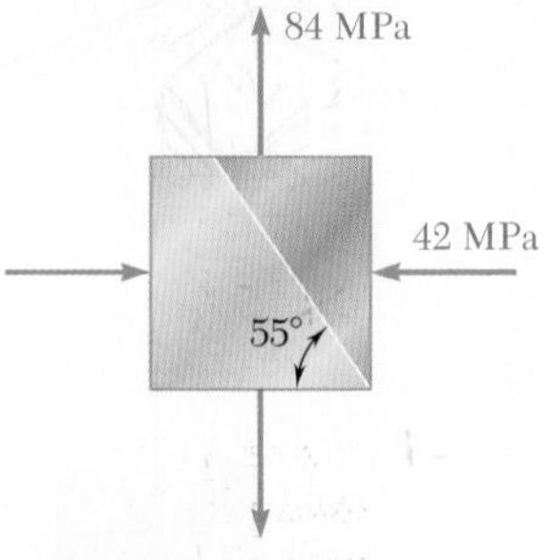

Fig. P7.4

7.5 through 7.8 For the given state of stress, determine (a) the principal planes, (b) the principal stresses.

7.9 through 7.12 For the given state of stress, determine (a) the orientation of the planes of maximum in-plane shearing stress, (b) the maximum in-plane shearing stress, (c) the corresponding normal stress.

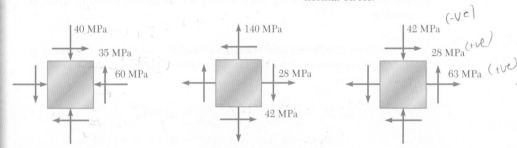

Fig. P7.5 and P7.9

Fig. P7.6 and P7.10

Fig. P7.7 and P7.11

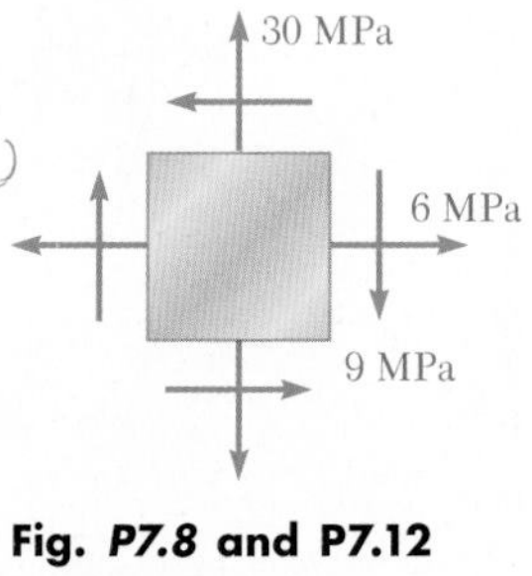

Fig. P7.8 and P7.12

7.13 through 7.16 For the given state of stress, determine the normal and shearing stresses after the element shown has been rotated through (a) 25° clockwise, (b) 10° counterclockwise.

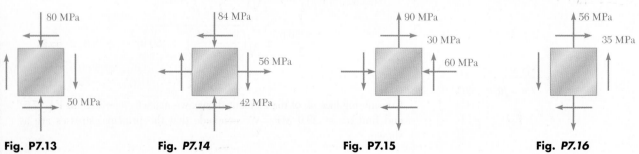

Fig. P7.13

Fig. P7.14

Fig. P7.15

Fig. P7.16

7.17 **and** **7.18** The grain of a wooden member forms an angle of 15°
with the vertical. For the state of stress shown, determine (*a*) the
in-plane shearing stress parallel to the grain, (*b*) the normal stress
perpendicular to the grain.

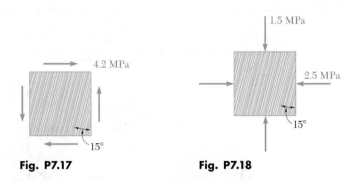

Fig. P7.17   **Fig. P7.18**

7.19 A steel pipe of 300-mm outer diameter is fabricated from 6-mm-
thick plate by welding along a helix which forms an angle of 22.5°
with a plane perpendicular to the axis of the pipe. Knowing that a
160-kN axial force **P** and an 800-N · m torque **T**, each directed as
shown, are applied to the pipe, determine σ and τ in directions,
respectively, normal and tangential to the weld.

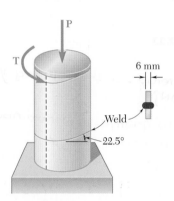

Fig. P7.19

7.20 Two members of uniform cross section 50 × 80 mm are glued
together along plane *a-a* that forms an angle of 25° with the horizontal. Knowing that the allowable stresses for the glued joint are
$\sigma = 800$ kPa and $\tau = 600$ kPa, determine the largest centric load
P that can be applied.

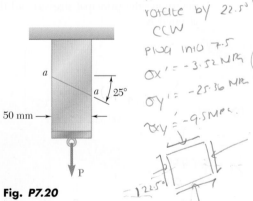

Fig. *P7.20*

7.21 Two steel plates of uniform cross section 10 × 80 mm are welded
together as shown. Knowing that centric 100-kN forces are applied
to the welded plates and that $\beta = 25°$, determine (*a*) the in-plane
shearing stress parallel to the weld, (*b*) the normal stress perpendicular to the weld.

7.22 Two steel plates of uniform cross section 10 × 80 mm are welded
together as shown. Knowing that centric 100-kN forces are applied
to the welded plates and that the in-plane shearing stress parallel
to the weld is 30 MPa, determine (*a*) the angle β, (*b*) the corresponding normal stress perpendicular to the weld.

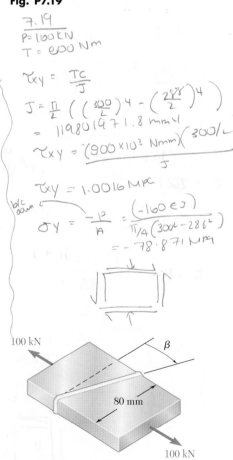

Fig. P7.21 and *P7.22*

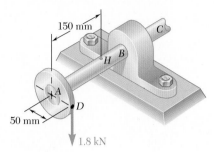

Fig. P7.23

7.23 A 1.8-kN vertical force is applied at D to a gear attached to the solid 25-mm-diameter shaft AB. Determine the principal stresses and the maximum shearing stress at point H located as shown on top of the shaft.

7.24 A mechanic uses a crowfoot wrench to loosen a bolt at E. Knowing that the mechanic applies a vertical 100-N force at A, determine the principal stresses and the maximum shearing stress at point H located as shown on top of the 18-mm-diameter shaft.

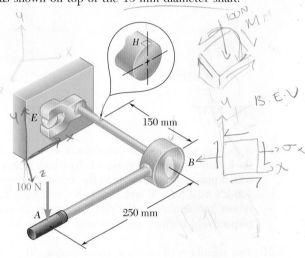

Fig. P7.24

7.25 The steel pipe AB has a 102-mm outer diameter and a 6-mm wall thickness. Knowing that arm CD is rigidly attached to the pipe, determine the principal stresses and the maximum shearing stress at point K.

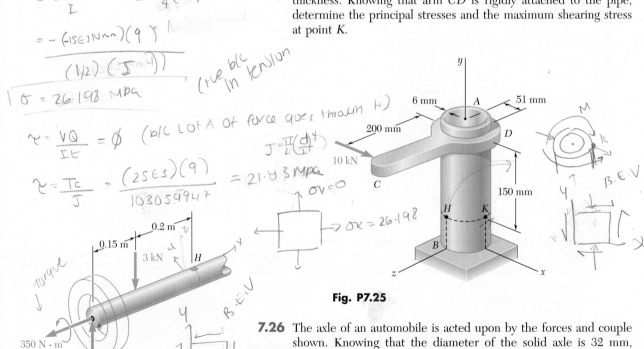

Fig. P7.25

7.26 The axle of an automobile is acted upon by the forces and couple shown. Knowing that the diameter of the solid axle is 32 mm, determine (a) the principal planes and principal stresses at point H located on top of the axle, (b) the maximum shearing stress at the same point.

Handwritten notes:

7.24

Force generates a shear in neg. y
$V = -100 N\hat{j}$

Force also generates a moment about X and a torque about z

$M = r \times w$

$= \begin{vmatrix} \hat{i} & \hat{j} & \hat{k} \\ -250 & 0 & 150 \\ 0 & -100 & 0 \end{vmatrix} (15)\hat{i} + (25)\hat{k}$ Nm

$\sigma_x = -\frac{My}{I}$ $I = \frac{\pi}{4}(r)^4$

$= -\frac{(15E3)(Nmm)(9)}{(1/2)(J \cdot 4)}$ (true b/c in tension)

$\sigma = 26.198$ MPa

$\tau = \frac{VQ}{It} = 0$ (b/c LOFA of force goes through H)

$\tau = \frac{Tc}{J} = \frac{(25E3)(9)}{103059947} = 21.83$ MPa

$\sigma_x = 26.198$

Fig. P7.26

350 N·m

3 kN

0.2 m

0.15 m

3 kN

H

(b) Principal Planes and Principal Stresses. The principal stresses are

$$\sigma_{max} = OA = OC + CA = 20 + 50 = 70 \text{ MPa}$$

$$\sigma_{min} = OB = OC - BC = 20 - 50 = -30 \text{ MPa}$$

Recalling that the angle ACX represents $2\theta_p$ (Fig. 7.19b), we write

$$\tan 2\theta_p = \frac{FX}{CF} = \frac{40}{30}$$

$$2\theta_p = 53.1° \qquad \theta_p = 26.6°$$

Since the rotation which brings CX into CA in Fig. 7.20b is counterclockwise, the rotation that brings Ox into the axis Oa corresponding to σ_{max} in Fig. 7.20a is also counterclockwise.

(c) Maximum Shearing Stress. Since a further rotation of 90° counterclockwise brings CA into CD in Fig. 7.20b, a further rotation of 45° counterclockwise will bring the axis Oa into the axis Od corresponding to the maximum shearing stress in Fig. 7.20a. We note from Fig. 7.20b that $\tau_{max} = R = 50$ MPa and that the corresponding normal stress is $\sigma' = \sigma_{ave} = 20$ MPa. Since point D is located above the σ axis in Fig. 7.20b, the shearing stresses exerted on the faces perpendicular to Od in Fig. 7.20a must be directed so that they will tend to rotate the element clockwise.

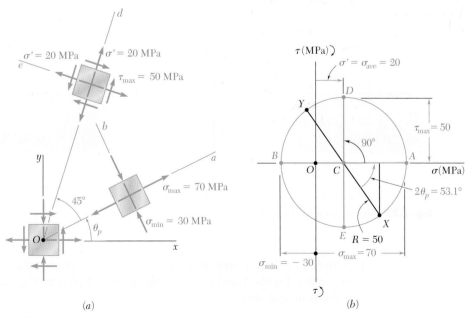

Fig. 7.20

Mohr's circle provides a convenient way of checking the results obtained earlier for stresses under a centric axial loading (Sec. 1.12) and under a torsional loading (Sec. 3.4). In the first case (Fig. 7.21*a*), we have $\sigma_x = P/A$, $\sigma_y = 0$, and $\tau_{xy} = 0$. The corresponding points X and Y define a circle of radius $R = P/2A$ that passes through the

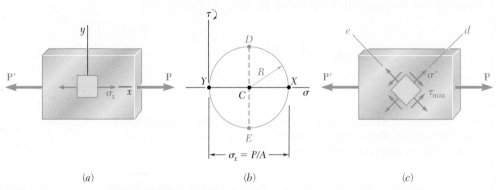

(*a*)	(*b*)	(*c*)

Fig. 7.21 Mohr's circle for centric axial loading.

origin of coordinates (Fig. 7.21*b*). Points D and E yield the orientation of the planes of maximum shearing stress (Fig. 7.21*c*), as well as the values of τ_{max} and of the corresponding normal stresses σ':

$$\tau_{max} = \sigma' = R = \frac{P}{2A} \tag{7.18}$$

In the case of torsion (Fig. 7.22*a*), we have $\sigma_x = \sigma_y = 0$ and $\tau_{xy} = \tau_{max} = Tc/J$. Points X and Y, therefore, are located on the τ axis,

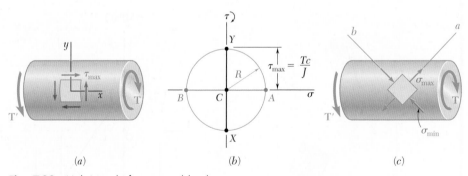

(*a*)	(*b*)	(*c*)

Fig. 7.22 Mohr's circle for torsional loading.

and Mohr's circle is a circle of radius $R = Tc/J$ centered at the origin (Fig. 7.22*b*). Points A and B define the principal planes (Fig. 7.22*c*) and the principal stresses:

$$\sigma_{max, min} = \pm R = \pm \frac{Tc}{J} \tag{7.19}$$

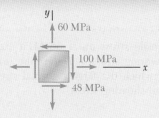

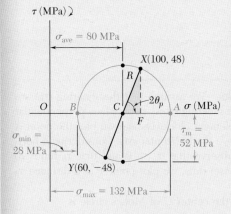

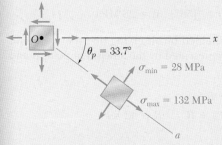

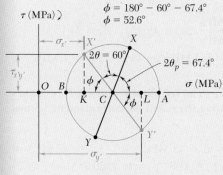

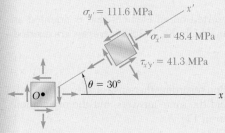

SAMPLE PROBLEM 7.2

For the state of plane stress shown, determine (*a*) the principal planes and the principal stresses, (*b*) the stress components exerted on the element obtained by rotating the given element counterclockwise through 30°.

SOLUTION

Construction of Mohr's Circle. We note that on a face perpendicular to the *x* axis, the normal stress is tensile and the shearing stress tends to rotate the element clockwise; thus, we plot *X* at a point 100 units to the right of the vertical axis and 48 units above the horizontal axis. In a similar fashion, we examine the stress components on the upper face and plot point *Y*(60, −48). Joining points *X* and *Y* by a straight line, we define the center *C* of Mohr's circle. The abscissa of *C*, which represents σ_{ave}, and the radius *R* of the circle can be measured directly or calculated as follows:

$$\sigma_{\text{ave}} = OC = \tfrac{1}{2}(\sigma_x + \sigma_y) = \tfrac{1}{2}(100 + 60) = 80 \text{ MPa}$$
$$R = \sqrt{(CF)^2 + (FX)^2} = \sqrt{(20)^2 + (48)^2} = 52 \text{ MPa}$$

a. Principal Planes and Principal Stresses. We rotate the diameter *XY* clockwise through $2\theta_p$ until it coincides with the diameter *AB*. We have

$$\tan 2\theta_p = \frac{XF}{CF} = \frac{48}{20} = 2.4 \quad 2\theta_p = 67.4° \searrow \quad \theta_p = 33.7° \searrow \quad \blacktriangleleft$$

The principal stresses are represented by the abscissas of points *A* and *B*:

$$\sigma_{\text{max}} = OA = OC + CA = 80 + 52 \qquad \sigma_{\text{max}} = +132 \text{ MPa} \quad \blacktriangleleft$$
$$\sigma_{\text{min}} = OB = OC - BC = 80 - 52 \qquad \sigma_{\text{min}} = +\ 28 \text{ MPa} \quad \blacktriangleleft$$

Since the rotation that brings *XY* into *AB* is clockwise, the rotation that brings *Ox* into the axis *Oa* corresponding to σ_{max} is also clockwise; we obtain the orientation shown for the principal planes.

b. Stress Components on Element Rotated 30° ↺. Points *X'* and *Y'* on Mohr's circle that correspond to the stress components on the rotated element are obtained by rotating *XY* counterclockwise through $2\theta = 60°$. We find

$$\phi = 180° - 60° - 67.4° \qquad\qquad\qquad\qquad \phi = 52.6° \quad \blacktriangleleft$$
$$\sigma_{x'} = OK = OC - KC = 80 - 52 \cos 52.6° \qquad \sigma_{x'} = +\ 48.4 \text{ MPa} \quad \blacktriangleleft$$
$$\sigma_{y'} = OL = OC + CL = 80 + 52 \cos 52.6° \qquad \sigma_{y'} = +111.6 \text{ MPa} \quad \blacktriangleleft$$
$$\tau_{x'y'} = KX' = 52 \sin 52.6° \qquad\qquad\qquad \tau_{x'y'} = \quad 41.3 \text{ MPa} \quad \blacktriangleleft$$

Since *X'* is located above the horizontal axis, the shearing stress on the face perpendicular to *Ox'* tends to rotate the element clockwise.

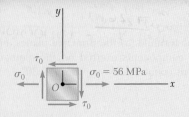

SAMPLE PROBLEM 7.3

A state of plane stress consists of a tensile stress $\sigma_0 = 56$ MPa exerted on vertical surfaces and of unknown shearing stresses. Determine (a) the magnitude of the shearing stress τ_0 for which the largest normal stress is 70 MPa, (b) the corresponding maximum shearing stress.

SOLUTION

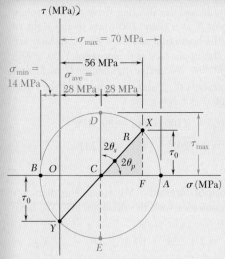

Construction of Mohr's Circle. We assume that the shearing stresses act in the senses shown. Thus, the shearing stress τ_0 on a face perpendicular to the x axis tends to rotate the element clockwise and we plot the point X of coordinates 56 MPa and τ_0 above the horizontal axis. Considering a horizontal face of the element, we observe that $\sigma_y = 0$ and that τ_0 tends to rotate the element counterclockwise; thus, we plot point Y at a distance τ_0 below O.

We note that the abscissa of the center C of Mohr's circle is

$$\sigma_{ave} = \tfrac{1}{2}(\sigma_x + \sigma_y) = \tfrac{1}{2}(56 + 0) = 28 \text{ MPa}$$

The radius R of the circle is determined by observing that the maximum normal stress, $\sigma_{max} = 70$ MPa, is represented by the abscissa of point A and writing

$$\sigma_{max} = \sigma_{ave} + R$$
$$70 \text{ MPa} = 28 \text{ MPa} + R \qquad R = 42 \text{ MPa}$$

a. Shearing Stress τ_0. Considering the right triangle CFX, we find

$$\cos 2\theta_p = \frac{CF}{CX} = \frac{CF}{R} = \frac{28 \text{ MPa}}{42 \text{ MPa}} \qquad 2\theta_p = 48.2° \downarrow \quad \theta_p = 24.1° \downarrow$$

$$\tau_0 = FX = R \sin 2\theta_p = (42 \text{ MPa}) \sin 48.2° \quad \tau_0 = 31.3 \text{ MPa} \blacktriangleleft$$

b. Maximum Shearing Stress. The coordinates of point D of Mohr's circle represent the maximum shearing stress and the corresponding normal stress.

$$\tau_{max} = R = 42 \text{ MPa} \qquad \tau_{max} = 42 \text{ MPa} \blacktriangleleft$$
$$2\theta_s = 90° - 2\theta_p = 90° - 48.2° = 41.8° \nwarrow \qquad \theta_x = 20.9° \nwarrow$$

The maximum shearing stress is exerted on an element that is oriented as shown in Fig. *a*. (The element upon which the principal stresses are exerted is also shown.)

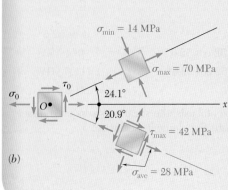

Note. If our original assumption regarding the sense of τ_0 was reversed, we would obtain the same circle and the same answers, but the orientation of the elements would be as shown in Fig. *b*.

PROBLEMS

7.31 Solve Probs. 7.5 and 7.9, using Mohr's circle.

7.32 Solve Probs. 7.6 and 7.10, using Mohr's circle.

7.33 Solve Prob. 7.11, using Mohr's circle.

7.34 Solve Prob. 7.12, using Mohr's circle.

7.35 Solve Prob. 7.13, using Mohr's circle.

7.36 Solve Prob. 7.14, using Mohr's circle.

7.37 Solve Prob. 7.15, using Mohr's circle.

7.38 Solve Prob. 7.16, using Mohr's circle.

7.39 Solve Prob. 7.17, using Mohr's circle.

7.40 Solve Prob. 7.18, using Mohr's circle.

7.41 Solve Prob. 7.19, using Mohr's circle.

7.42 Solve Prob. 7.20, using Mohr's circle.

7.43 Solve Prob. 7.21, using Mohr's circle.

7.44 Solve Prob. 7.22, using Mohr's circle.

7.45 Solve Prob. 7.23, using Mohr's circle.

7.46 Solve Prob. 7.24, using Mohr's circle.

7.47 Solve Prob. 7.25, using Mohr's circle.

7.48 Solve Prob. 7.26, using Mohr's circle.

7.49 Solve Prob. 7.27, using Mohr's circle.

7.50 Solve Prob. 7.28, using Mohr's circle.

7.51 Solve Prob. 7.29, using Mohr's circle.

7.52 Solve Prob. 7.30, using Mohr's circle.

7.53 Solve Prob. 7.30, using Mohr's circle and assuming that the weld forms an angle of 60° with the horizontal.

7.54 and 7.55 Determine the principal planes and the principal stresses for the state of plane stress resulting from the superposition of the two states of stress shown.

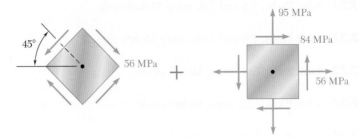

Fig. P7.54

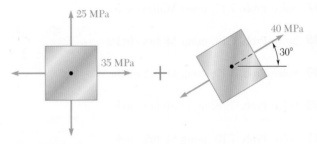

Fig. P7.55

7.56 and 7.57 Determine the principal planes and the principal stresses for the state of plane stress resulting from the superposition of the two states of stress shown.

Fig. P7.56

Fig. P7.57

7.58 For the state of stress shown, determine the range of values of θ for which the magnitude of the shearing stress $\tau_{x'y'}$ is equal to or less than 56 MPa.

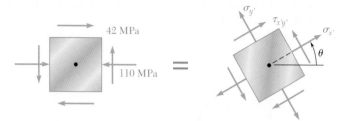

Fig. P7.58

7.59 For the state of stress shown, determine the range of values of θ for which the normal stress $\sigma_{x'}$ is equal to or less than 50 MPa.

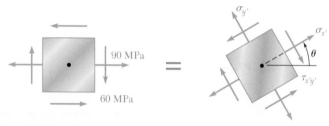

Fig. P7.59 and P7.60

7.60 For the state of stress shown, determine the range of values of θ for which the normal stress $\sigma_{x'}$ is equal to or less than 100 MPa.

7.61 For the element shown, determine the range of values of τ_{xy} for which the maximum tensile stress is equal to or less than 60 MPa.

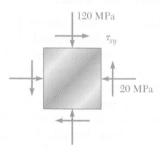

Fig. P7.61 and P7.62

7.62 For the element shown, determine the range of values of τ_{xy} for which the maximum in-plane shearing stress is equal to or less than 150 MPa.

7.63 For the state of stress shown it is known that the normal and shearing stresses are directed as shown and that $\sigma_x = 98$ MPa, $\sigma_y = 63$ MPa, and $\sigma_{\min} = 35$ MPa. Determine (a) the orientation of the principal planes, (b) the principal stress $\sigma_{\max}$, (c) the maximum in-plane shearing stress.

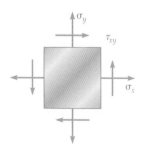

Fig. P7.63

7.64 The Mohr's circle shown corresponds to the state of stress given in Fig. 7.5a and b. Noting that $\sigma_{x'} = OC + (CX') \cos (2\theta_p - 2\theta)$ and that $\tau_{x'y'} = (CX') \sin (2\theta_p - 2\theta)$, derive the expressions for $\sigma_{x'}$ and $\tau_{x'y'}$ given in Eqs. (7.5) and (7.6), respectively. [*Hint:* Use $\sin (A + B) = \sin A \cos B + \cos A \sin B$ and $\cos (A + B) = \cos A \cos B - \sin A \sin B$.]

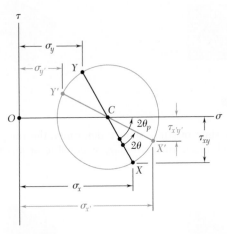

Fig. P7.64

7.65 (*a*) Prove that the expression $\sigma_{x'}\sigma_{y'} - \tau^2_{x'y'}$, where $\sigma_{x'}$, $\sigma_{y'}$, and $\tau_{x'y'}$ are components of the stress along the rectangular axes x' and y', is independent of the orientation of these axes. Also, show that the given expression represents the square of the tangent drawn from the origin of the coordinates to Mohr's circle. (*b*) Using the invariance property established in part *a*, express the shearing stress τ_{xy} in terms of σ_x, σ_y, and the principal stresses $\sigma_{\max}$ and $\sigma_{\min}$.

7.5 GENERAL STATE OF STRESS

In the preceding sections, we have assumed a state of plane stress with $\sigma_z = \tau_{zx} = \tau_{zy} = 0$, and have considered only transformations of stress associated with a rotation about the z axis. We will now consider the general state of stress represented in Fig. 7.1a and the transformation of stress associated with the rotation of axes shown in Fig. 7.1b. However, our analysis will be limited to the determination of the *normal stress* σ_n on a plane of arbitrary orientation.

Consider the tetrahedron shown in Fig. 7.23. Three of its faces are parallel to the coordinate planes, while its fourth face, *ABC*, is perpendicular to the line *QN*. Denoting by ΔA the area of face *ABC*, and by λ_x, λ_y, λ_z the direction cosines of line *QN*, we find that the areas of the faces perpendicular to the x, y, and z axes are, respectively, $(\Delta A)\lambda_x$, $(\Delta A)\lambda_y$, and $(\Delta A)\lambda_z$. If the state of stress at point Q is defined by the stress components σ_x, σ_y, σ_z, τ_{xy}, τ_{yz}, and τ_{zx}, then the *forces* exerted on the faces parallel to the coordinate planes can be obtained by multiplying the appropriate stress components by the area of each face (Fig. 7.24). On the other hand, the forces exerted on face *ABC* consist of a normal force of magnitude $\sigma_n \Delta A$ directed along *QN*, and of a shearing force of magnitude $\tau \Delta A$ perpendicular to *QN* but of otherwise unknown direction. Note that, since *QBC*,

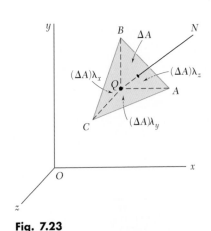

Fig. 7.23

QCA, and QAB, respectively, face the negative x, y, and z axes, the forces exerted on them must be shown with negative senses.

We now express that the sum of the components along QN of all the forces acting on the tetrahedron is zero. Observing that the component along QN of a force parallel to the x axis is obtained by multiplying the magnitude of that force by the direction cosine λ_x, and that the components of forces parallel to the y and z axes are obtained in a similar way, we write

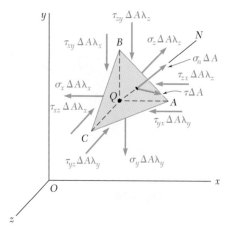

Fig. 7.24

$$\Sigma F_n = 0: \quad \sigma_n \Delta A - (\sigma_x \Delta A \, \lambda_x)\lambda_x - (\tau_{xy} \Delta A \, \lambda_x)\lambda_y - (\tau_{xz} \Delta A \, \lambda_x)\lambda_z$$
$$- (\tau_{yx} \Delta A \, \lambda_y)\lambda_x - (\sigma_y \Delta A \, \lambda_y)\lambda_y - (\tau_{yz} \Delta A \, \lambda_y)\lambda_z$$
$$- (\tau_{zx} \Delta A \, \lambda_z)\lambda_x - (\tau_{zy} \Delta A \, \lambda_z)\lambda_y - (\sigma_z \Delta A \, \lambda_z)\lambda_z = 0$$

Dividing through by ΔA and solving for σ_n, we have

$$\sigma_n = \sigma_x \lambda_x^2 + \sigma_y \lambda_y^2 + \sigma_z \lambda_z^2 + 2\tau_{xy}\lambda_x\lambda_y + 2\tau_{yz}\lambda_y\lambda_z + 2\tau_{zx}\lambda_z\lambda_x \quad (7.20)$$

We note that the expression obtained for the normal stress σ_n is a *quadratic form* in λ_x, λ_y, and λ_z. It follows that we can select the coordinate axes in such a way that the right-hand member of Eq. (7.20) reduces to the three terms containing the squares of the direction cosines.[†] Denoting these axes by a, b, and c, the corresponding normal stresses by σ_a, σ_b, and σ_c, and the direction cosines of QN with respect to these axes by λ_a, λ_b, and λ_c, we write

$$\sigma_n = \sigma_a \lambda_a^2 + \sigma_b \lambda_b^2 + \sigma_c \lambda_c^2 \quad (7.21)$$

The coordinate axes a, b, c are referred to as the *principal axes of stress*. Since their orientation depends upon the state of stress at Q, and thus upon the position of Q, they have been represented in Fig. 7.25 as attached to Q. The corresponding coordinate planes are known as the *principal planes of stress*, and the corresponding normal stresses σ_a, σ_b, and σ_c as the *principal stresses* at Q.[‡]

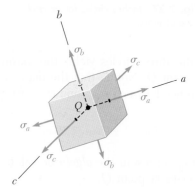

Fig. 7.25 Principal stresses.

[†]In Sec. 9.16 of F. P. Beer and E. R. Johnston, *Vector Mechanics for Engineers*, 9th ed., McGraw-Hill Book Company, 2010, a similar quadratic form is found to represent the moment of inertia of a rigid body with respect to an arbitrary axis. It is shown in Sec. 9.17 that this form is associated with a *quadric surface*, and that reducing the quadratic form to terms containing only the squares of the direction cosines is equivalent to determining the principal axes of that surface.

[‡]For a discussion of the determination of the principal planes of stress and of the principal stresses, see S. P. Timoshenko and J. N. Goodier, *Theory of Elasticity*, 3d ed., McGraw-Hill Book Company, 1970, Sec. 77.

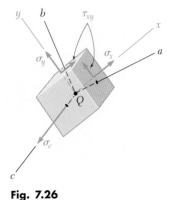

Fig. 7.26

7.6 APPLICATION OF MOHR'S CIRCLE TO THE THREE-DIMENSIONAL ANALYSIS OF STRESS

If the element shown in Fig. 7.25 is rotated about one of the principal axes at Q, say the c axis (Fig. 7.26), the corresponding transformation of stress can be analyzed by means of Mohr's circle as if it were a transformation of plane stress. Indeed, the shearing stresses exerted on the faces perpendicular to the c axis remain equal to zero, and the normal stress σ_c is perpendicular to the plane ab in which the transformation takes place and, thus, does not affect this transformation. We therefore use the circle of diameter AB to determine the normal and shearing stresses exerted on the faces of the element as it is rotated about the c axis (Fig. 7.27). Similarly, circles of diameter BC and CA can be used to determine the stresses on the element as it is rotated about the a and b axes, respectively. While our analysis will be limited to rotations about the principal axes, it could be shown that any other transformation of axes would lead to stresses represented in Fig. 7.27 by a point located within the shaded area. Thus, the radius

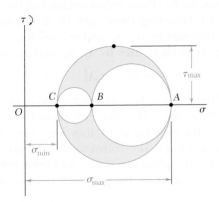

Fig. 7.27 Mohr's circles for general state of stress.

of the largest of the three circles yields the maximum value of the shearing stress at point Q. Noting that the diameter of that circle is equal to the difference between σ_{max} and σ_{min}, we write

$$\tau_{max} = \tfrac{1}{2}|\sigma_{max} - \sigma_{min}| \tag{7.22}$$

where σ_{max} and σ_{min} represent the *algebraic* values of the maximum and minimum stresses at point Q.

Let us now return to the particular case of *plane stress*, which was discussed in Secs. 7.2 through 7.4. We recall that, if the x and y axes are selected in the plane of stress, we have $\sigma_z = \tau_{zx} = \tau_{zy} = 0$. This means that the z axis, i.e., the axis perpendicular to the plane of stress, is one of the three principal axes of stress. In a Mohr-circle diagram, this axis corresponds to the origin O, where $\sigma = \tau = 0$. We also recall that the other two principal axes correspond to points A and B where Mohr's circle for the xy plane intersects the σ axis. If A and B are located on opposite sides of the origin O (Fig. 7.28),

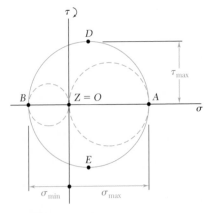

Fig. 7.28

the corresponding principal stresses represent the maximum and minimum normal stresses at point Q, and the maximum shearing stress is equal to the maximum "in-plane" shearing stress. As noted in Sec. 7.3, the planes of maximum shearing stress correspond to points D and E of Mohr's circle and are at 45° to the principal planes corresponding to points A and B. They are, therefore, the shaded diagonal planes shown in Fig. 7.29a and b.

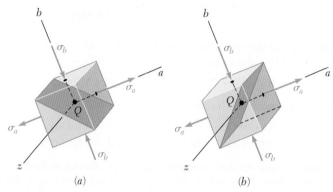

Fig. 7.29

If, on the other hand, A and B are on the same side of O, that is, if σ_a and σ_b have the same sign, then the circle defining $\sigma_{\max}$, $\sigma_{\min}$, and $\tau_{\max}$ is *not* the circle corresponding to a transformation of stress within the xy plane. If $\sigma_a > \sigma_b > 0$, as assumed in Fig. 7.30, we have $\sigma_{\max} = \sigma_a$, $\sigma_{\min} = 0$, and $\tau_{\max}$ is equal to the radius of the circle defined by points O and A, i.e., $\tau_{\max} = \frac{1}{2}\sigma_{\max}$. We also note that the normals Qd' and Qe' to the planes of maximum shearing stress are obtained by rotating the axis Qa through 45° within the za plane. Thus, the planes of maximum shearing stress are the shaded diagonal planes shown in Fig. 7.31a and b.

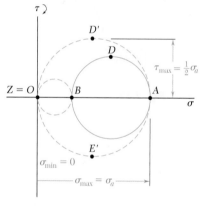

Fig. 7.30

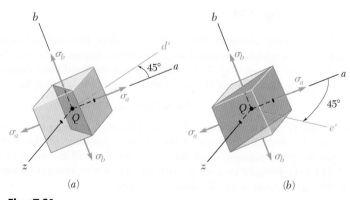

Fig. 7.31

EXAMPLE 7.03

For the state of plane stress shown in Fig. 7.32, determine (*a*) the three principal planes and principal stresses, (*b*) the maximum shearing stress.

(a) Principal Planes and Principal Stresses. We construct Mohr's circle for the transformation of stress in the *xy* plane (Fig. 7.33). Point *X* is plotted 40 units to the right of the τ axis and 20 units above the σ axis (since the corresponding shearing stress tends to rotate the element clockwise). Point *Y* is plotted 25 units to the right of the τ axis and 20 units below the σ axis. Drawing the line *XY*, we obtain the center *C* of Mohr's circle for the *xy* plane; its abscissa is

$$\sigma_{\text{ave}} = \frac{\sigma_x + \sigma_y}{2} = \frac{40 + 25}{2} = 32.5 \text{ MPa}$$

Since the sides of the right triangle *CFX* are $CF = 40 - 32.5 = 7.5$ MPa and $FX = 20$ MPa, the radius of the circle is

$$R = CX = \sqrt{(7.5)^2 + (20)^2} = 21.4 \text{ MPa}$$

The principal stresses in the plane of stress are

$$\sigma_a = OA = OC + CA = 32.5 + 21.4 = 53.9 \text{ MPa}$$
$$\sigma_b = OB = OC - BC = 32.5 - 21.4 = 11.1 \text{ MPa}$$

Since the faces of the element that are perpendicular to the *z* axis are free of stress, these faces define one of the principal planes, and the corresponding principal stress is $\sigma_z = 0$. The other two principal planes are defined by points *A* and *B* on Mohr's circle. The angle θ_p through which the element should be rotated about the *z* axis to bring its faces to coincide with these planes (Fig. 7.34) is half the angle *ACX*. We have

$$\tan 2\theta_p = \frac{FX}{CF} = \frac{20}{7.5}$$
$$2\theta_p = 69.4° \downarrow \qquad \theta_p = 34.7° \downarrow$$

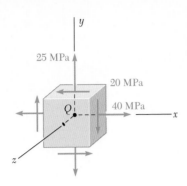

25 MPa
20 MPa
40 MPa

Fig. 7.32

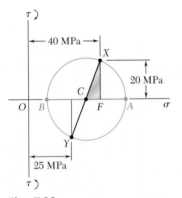

40 MPa
20 MPa
25 MPa

Fig. 7.33

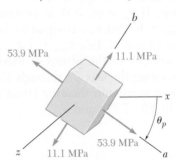

53.9 MPa   11.1 MPa
53.9 MPa   11.1 MPa

Fig. 7.34

(b) Maximum Shearing Stress. We now draw the circles of diameter *OB* and *OA*, which correspond respectively to rotations of the element about the *a* and *b* axes (Fig. 7.35). We note that the maximum shearing stress is equal to the radius of the circle of diameter *OA*. We thus have

$$\tau_{\max} = \tfrac{1}{2}\sigma_a = \tfrac{1}{2}(53.9 \text{ MPa}) = 26.9 \text{ MPa}$$

Since points *D′* and *E′*, which define the planes of maximum shearing stress, are located at the ends of the vertical diameter of the circle corresponding to a rotation about the *b* axis, the faces of the element of Fig. 7.34 can be brought to coincide with the planes of maximum shearing stress through a rotation of 45° about the *b* axis.

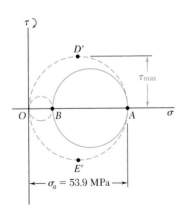

$\tau_{\max}$
$\sigma_a = 53.9$ MPa

Fig. 7.35

*7.7 YIELD CRITERIA FOR DUCTILE MATERIALS UNDER PLANE STRESS

Structural elements and machine components made of a ductile material are usually designed so that the material will not yield under the expected loading conditions. When the element or component is under uniaxial stress (Fig. 7.36), the value of the normal stress σ_x that will cause the material to yield can be obtained readily from a tensile test conducted on a specimen of the same material, since the test specimen and the structural element or machine component are in the same state of stress. Thus, regardless of the actual mechanism that causes the material to yield, we can state that the element or component will be safe as long as $\sigma_x < \sigma_Y$, where σ_Y is the yield strength of the test specimen.

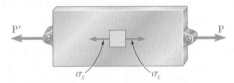

Fig. 7.36 Structural element under uniaxial stress.

On the other hand, when a structural element or machine component is in a state of plane stress (Fig. 7.37a), it is found convenient to use one of the methods developed earlier to determine the principal stresses σ_a and σ_b at any given point (Fig. 7.37b). The material can then be regarded as being in a state of biaxial stress at that point. Since this state is different from the state of uniaxial stress found in a specimen subjected to a tensile test, it is clearly not possible to predict directly from such a test whether or not the structural element or machine component under investigation will fail. Some criterion regarding the actual mechanism of failure of the material must first be established, which will make it possible to compare the effects of both states of stress on the material. The purpose of this section is to present the two yield criteria most frequently used for ductile materials.

Maximum-Shearing-Stress Criterion.

This criterion is based on the observation that yield in ductile materials is caused by slippage of the material along oblique surfaces and is due primarily to shearing stresses (cf. Sec. 2.3). According to this criterion, a given structural component is safe as long as the maximum value τ_{max} of the shearing stress in that component remains smaller than the corresponding value of the shearing stress in a tensile-test specimen of the same material as the specimen starts to yield.

Recalling from Sec. 1.11 that the maximum value of the shearing stress under a centric axial load is equal to half the value of the corresponding normal, axial stress, we conclude that the maximum shearing stress in a tensile-test specimen is $\frac{1}{2}\sigma_Y$ as the specimen starts to yield. On the other hand, we saw in Sec. 7.6 that, for plane stress, the maximum value τ_{max} of the shearing stress is equal to $\frac{1}{2}|\sigma_{max}|$ if the principal stresses are either both positive or both negative, and to $\frac{1}{2}|\sigma_{max} - \sigma_{min}|$ if the maximum stress is positive and the

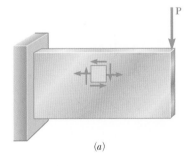

(a)

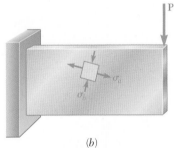

(b)

Fig. 7.37 Structural element in state of plane stress.

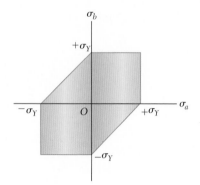

Fig. 7.38 Tresca's hexagon.

minimum stress negative. Thus, if the principal stresses σ_a and σ_b have the same sign, the maximum-shearing-stress criterion gives

$$|\sigma_a| < \sigma_Y \qquad |\sigma_b| < \sigma_Y \qquad (7.23)$$

If the principal stresses σ_a and σ_b have opposite signs, the maximum-shearing-stress criterion yields

$$|\sigma_a - \sigma_b| < \sigma_Y \qquad (7.24)$$

The relations obtained have been represented graphically in Fig. 7.38. Any given state of stress will be represented in that figure by a point of coordinates σ_a and σ_b, where σ_a and σ_b are the two principal stresses. If this point falls within the area shown in the figure, the structural component is safe. If it falls outside this area, the component will fail as a result of yield in the material. The hexagon associated with the initiation of yield in the material is known as *Tresca's hexagon* after the French engineer Henri Edouard Tresca (1814–1885).

Maximum-Distortion-Energy Criterion. This criterion is based on the determination of the distortion energy in a given material, i.e., of the energy associated with changes in shape in that material (as opposed to the energy associated with changes in volume in the same material). According to this criterion, also known as the *von Mises criterion*, after the German-American applied mathematician Richard von Mises (1883–1953), a given structural component is safe as long as the maximum value of the distortion energy per unit volume in that material remains smaller than the distortion energy per unit volume required to cause yield in a tensile-test specimen of the same material. As you will see in Sec. 11.6, the distortion energy per unit volume in an isotropic material under plane stress is

$$u_d = \frac{1}{6G}(\sigma_a^2 - \sigma_a\sigma_b + \sigma_b^2) \qquad (7.25)$$

where σ_a and σ_b are the principal stresses and G the modulus of rigidity. In the particular case of a tensile-test specimen that is starting to yield, we have $\sigma_a = \sigma_Y$, $\sigma_b = 0$, and $(u_d)_Y = \sigma_Y^2/6G$. Thus, the maximum-distortion-energy criterion indicates that the structural component is safe as long as $u_d < (u_d)_Y$, or

$$\sigma_a^2 - \sigma_a\sigma_b + \sigma_b^2 < \sigma_Y^2 \qquad (7.26)$$

i.e., as long as the point of coordinates σ_a and σ_b falls within the area shown in Fig. 7.39. This area is bounded by the ellipse of equation

$$\sigma_a^2 - \sigma_a\sigma_b + \sigma_b^2 = \sigma_Y^2 \qquad (7.27)$$

which intersects the coordinate axes at $\sigma_a = \pm\sigma_Y$ and $\sigma_b = \pm\sigma_Y$. We can verify that the major axis of the ellipse bisects the first and third quadrants and extends from A ($\sigma_a = \sigma_b = \sigma_Y$) to B ($\sigma_a = \sigma_b = -\sigma_Y$), while its minor axis extends from C ($\sigma_a = -\sigma_b = -0.577\sigma_Y$) to D ($\sigma_a = -\sigma_b = 0.577\sigma_Y$).

The maximum-shearing-stress criterion and the maximum-distortion-energy criterion are compared in Fig. 7.40. We note that the ellipse passes through the vertices of the hexagon. Thus, for the states of stress represented by these six points, the two criteria give

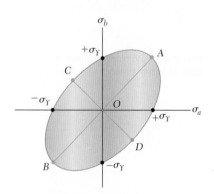

Fig. 7.39 Von Mises criterion.

the same results. For any other state of stress, the maximum-shearing-stress criterion is more conservative than the maximum-distortion-energy criterion, since the hexagon is located within the ellipse.

A state of stress of particular interest is that associated with yield in a torsion test. We recall from Fig. 7.22 of Sec. 7.4 that, for torsion, $\sigma_{min} = -\sigma_{max}$; thus, the corresponding points in Fig. 7.40 are located on the bisector of the second and fourth quadrants. It follows that yield occurs in a torsion test when $\sigma_a = -\sigma_b = \pm 0.5\sigma_Y$ according to the maximum-shearing-stress criterion, and when $\sigma_a = -\sigma_b = \pm 0.577\sigma_Y$ according to the maximum-distortion-energy criterion. But, recalling again Fig. 7.22, we note that σ_a and σ_b must be equal in magnitude to τ_{max}, i.e., to the value obtained from a torsion test for the yield strength τ_Y of the material. Since the values of the yield strength σ_Y in tension and of the yield strength τ_Y in shear are given for various ductile materials in Appendix B, we can compute the ratio τ_Y/σ_Y for these materials and verify that the values obtained range from 0.55 to 0.60. Thus, the maximum-distortion-energy criterion appears somewhat more accurate than the maximum-shearing-stress criterion as far as predicting yield in torsion is concerned.

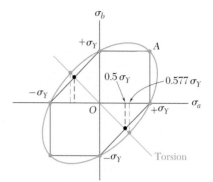

Fig. 7.40

*7.8 FRACTURE CRITERIA FOR BRITTLE MATERIALS UNDER PLANE STRESS

As we saw in Chap. 2, brittle materials are characterized by the fact that, when subjected to a tensile test, they fail suddenly through rupture—or fracture—without any prior yielding. When a structural element or machine component made of a brittle material is under uniaxial tensile stress, the value of the normal stress that causes it to fail is equal to the ultimate strength σ_U of the material as determined from a tensile test, since both the tensile-test specimen and the element or component under investigation are in the same state of stress. However, when a structural element or machine component is in a state of plane stress, it is found convenient to first determine the principal stresses σ_a and σ_b at any given point, and to use one of the criteria indicated in this section to predict whether or not the structural element or machine component will fail.

Maximum-Normal-Stress Criterion.

According to this criterion, a given structural component fails when the maximum normal stress in that component reaches the ultimate strength σ_U obtained from the tensile test of a specimen of the same material. Thus, the structural component will be safe as long as the absolute values of the principal stresses σ_a and σ_b are both less than σ_U:

$$|\sigma_a| < \sigma_U \qquad |\sigma_b| < \sigma_U \qquad (7.28)$$

The maximum-normal-stress criterion can be expressed graphically as shown in Fig. 7.41. If the point obtained by plotting the values σ_a and σ_b of the principal stresses falls within the square area shown in the figure, the structural component is safe. If it falls outside that area, the component will fail.

The maximum-normal-stress criterion, also known as *Coulomb's criterion*, after the French physicist Charles Augustin de Coulomb

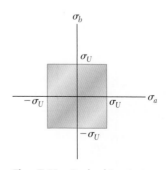

Fig. 7.41 Coulomb's criterion.

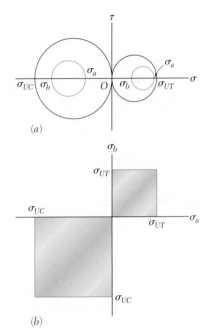

(a)

(b)

Fig. 7.43

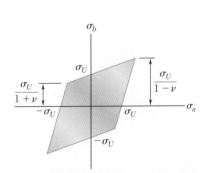

Fig. 7.42 Saint-Venant's criterion.

(1736–1806), suffers from an important shortcoming, since it is based on the assumption that the ultimate strength of the material is the same in tension and in compression. As we noted in Sec. 2.3, this is seldom the case, because of the presence of flaws in the material, such as microscopic cracks or cavities, which tend to weaken the material in tension, while not appreciably affecting its resistance to compressive failure. Besides, this criterion makes no allowance for effects other than those of the normal stresses on the failure mechanism of the material.[†]

Mohr's Criterion. This criterion, suggested by the German engineer Otto Mohr, can be used to predict the effect of a given state of plane stress on a brittle material, when results of various types of tests are available for that material.

Let us first assume that a tensile test and a compressive test have been conducted on a given material, and that the values σ_{UT} and σ_{UC} of the ultimate strength in tension and in compression have been determined for that material. The state of stress corresponding to the rupture of the tensile-test specimen can be represented on a Mohr-circle diagram by the circle intersecting the horizontal axis at O and σ_{UT} (Fig. 7.43a). Similarly, the state of stress corresponding to the failure of the compressive-test specimen can be represented by the circle intersecting the horizontal axis at O and σ_{UC}. Clearly, a state of stress represented by a circle entirely contained in either of these circles will be safe. Thus, if both principal stresses are positive, the state of stress is safe as long as $\sigma_a < \sigma_{UT}$ and $\sigma_b < \sigma_{UT}$; if both principal stresses are negative, the state of stress is safe as long as $|\sigma_a| < |\sigma_{UC}|$ and $|\sigma_b| < |\sigma_{UC}|$. Plotting the point of coordinates σ_a and σ_b (Fig. 7.43b), we verify that the state of stress is safe as long as that point falls within one of the square areas shown in that figure.

In order to analyze the cases when σ_a and σ_b have opposite signs, we now assume that a torsion test has been conducted on the material and that its ultimate strength in shear, τ_U, has been determined. Drawing the circle centered at O representing the state of stress corresponding to the failure of the torsion-test specimen (Fig. 7.44a), we observe that any state of stress represented by a circle entirely contained in that circle is also safe. Mohr's criterion is a logical extension of this observation: According to Mohr's criterion, a state of stress is safe if it is represented by a circle located entirely within the area bounded

---

[†]Another failure criterion known as the *maximum-normal-strain criterion*, or Saint-Venant's criterion, was widely used during the nineteenth century. According to this criterion, a given structural component is safe as long as the maximum value of the normal strain in that component remains smaller than the value ϵ_U of the strain at which a tensile-test specimen of the same material will fail. But, as will be shown in Sec. 7.12, the strain is maximum along one of the principal axes of stress, if the deformation is elastic and the material homogeneous and isotropic. Thus, denoting by ϵ_a and ϵ_b the values of the normal strain along the principal axes in the plane of stress, we write

$$|\epsilon_a| < \epsilon_U \qquad |\epsilon_b| < \epsilon_U \qquad (7.29)$$

Making use of the generalized Hooke's law (Sec. 2.12), we could express these relations in terms of the principal stresses σ_a and σ_b and the ultimate strength σ_U of the material. We would find that, according to the maximum-normal-strain criterion, the structural component is safe as long as the point obtained by plotting σ_a and σ_b falls within the area shown in Fig. 7.42 where ν is Poisson's ratio for the given material.

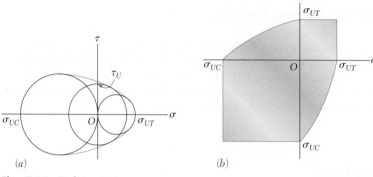

Fig. 7.44 Mohr's criterion.

by the envelope of the circles corresponding to the available data. The remaining portions of the principal-stress diagram can now be obtained by drawing various circles tangent to this envelope, determining the corresponding values of σ_a and σ_b, and plotting the points of coordinates σ_a and σ_b (Fig. 7.44b).

More accurate diagrams can be drawn when additional test results, corresponding to various states of stress, are available. If, on the other hand, the only available data consists of the ultimate strengths σ_{UT} and σ_{UC}, the envelope in Fig. 7.44a is replaced by the tangents AB and $A'B'$ to the circles corresponding respectively to failure in tension and failure in compression (Fig. 7.45a). From the similar triangles drawn in that figure, we note that the abscissa of the center C of a circle tangent to AB and $A'B'$ is linearly related to its radius R. Since $\sigma_a = OC + R$ and $\sigma_b = OC - R$, it follows that σ_a and σ_b are also linearly related. Thus, the shaded area corresponding to this simplified Mohr's criterion is bounded by straight lines in the second and fourth quadrants (Fig. 7.45b).

Note that in order to determine whether a structural component will be safe under a given loading, the state of stress should be calculated at all critical points of the component, i.e., at all points where stress concentrations are likely to occur. This can be done in a number of cases by using the stress-concentration factors given in Figs. 2.60, 3.29, 4.27, and 4.28. There are many instances, however, when the theory of elasticity must be used to determine the state of stress at a critical point.

Special care should be taken when *macroscopic cracks* have been detected in a structural component. While it can be assumed that the test specimen used to determine the ultimate tensile strength of the material contained the same type of flaws (i.e., *microscopic cracks or cavities*) as the structural component under investigation, the specimen was certainly free of any detectable macroscopic cracks. When a crack is detected in a structural component, it is necessary to determine whether that crack will tend to propagate under the expected loading condition and cause the component to fail, or whether it will remain stable. This requires an analysis involving the energy associated with the growth of the crack. Such an analysis is beyond the scope of this text and should be carried out by the methods of fracture mechanics.

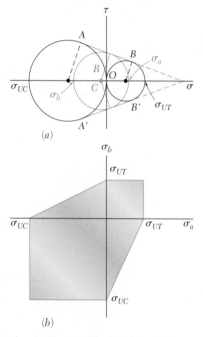

Fig. 7.45 Simplified Mohr's criterion.

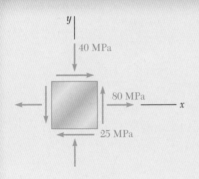

SAMPLE PROBLEM 7.4

The state of plane stress shown occurs at a critical point of a steel machine component. As a result of several tensile tests, it has been found that the tensile yield strength is $\sigma_Y = 250$ MPa for the grade of steel used. Determine the factor of safety with respect to yield, using (a) the maximum-shearing-stress criterion, and (b) the maximum-distortion-energy criterion.

SOLUTION

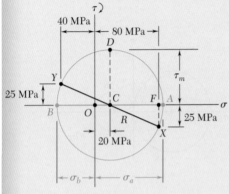

Mohr's Circle. We construct Mohr's circle for the given state of stress and find

$$\sigma_{ave} = OC = \tfrac{1}{2}(\sigma_x + \sigma_y) = \tfrac{1}{2}(80 - 40) = 20 \text{ MPa}$$
$$\tau_m = R = \sqrt{(CF)^2 + (FX)^2} = \sqrt{(60)^2 + (25)^2} = 65 \text{ MPa}$$

Principal Stresses

$$\sigma_a = OC + CA = 20 + 65 = +85 \text{ MPa}$$
$$\sigma_b = OC - BC = 20 - 65 = -45 \text{ MPa}$$

a. Maximum-Shearing-Stress Criterion. Since for the grade of steel used the tensile strength is $\sigma_Y = 250$ MPa, the corresponding shearing stress at yield is

$$\tau_Y = \tfrac{1}{2}\sigma_Y = \tfrac{1}{2}(250 \text{ MPa}) = 125 \text{ MPa}$$

For $\tau_m = 65$ MPa: $\qquad F.S. = \dfrac{\tau_Y}{\tau_m} = \dfrac{125 \text{ MPa}}{65 \text{ MPa}} \qquad F.S. = 1.92 \blacktriangleleft$

b. Maximum-Distortion-Energy Criterion. Introducing a factor of safety into Eq. (7.26), we write

$$\sigma_a^2 - \sigma_a\sigma_b + \sigma_b^2 = \left(\frac{\sigma_Y}{F.S.}\right)^2$$

For $\sigma_a = +85$ MPa, $\sigma_b = -45$ MPa, and $\sigma_Y = 250$ MPa, we have

$$(85)^2 - (85)(-45) + (45)^2 = \left(\frac{250}{F.S.}\right)^2$$
$$114.3 = \frac{250}{F.S.} \qquad F.S. = 2.19 \blacktriangleleft$$

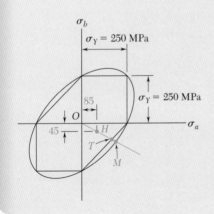

Comment. For a ductile material with $\sigma_Y = 250$ MPa, we have drawn the hexagon associated with the maximum-shearing-stress criterion and the ellipse associated with the maximum-distortion-energy criterion. The given state of plane stress is represented by point H of coordinates $\sigma_a = 85$ MPa and $\sigma_b = -45$ MPa. We note that the straight line drawn through points O and H intersects the hexagon at point T and the ellipse at point M. For each criterion, the value obtained for $F.S.$ can be verified by measuring the line segments indicated and computing their ratios:

$$(a)\ F.S. = \frac{OT}{OH} = 1.92 \qquad (b)\ F.S. = \frac{OM}{OH} = 2.19$$

PROBLEMS

7.66 For the state of plane stress shown, determine the maximum shearing stress when (*a*) $\sigma_x = 0$ and $\sigma_y = 60$ MPa, (*b*) $\sigma_x = 105$ MPa and $\sigma_y = 45$ MPa. (*Hint:* Consider both in-plane and out-of-plane shearing stresses.)

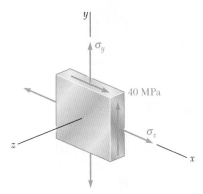

Fig. P7.66 and P7.67

7.67 For the state of plane stress shown, determine the maximum shearing stress when (*a*) $\sigma_x = 30$ MPa and $\sigma_y = 90$ MPa, (*b*) $\sigma_x = 70$ MPa and $\sigma_y = 10$ MPa. (*Hint:* Consider both in-plane and out-of-plane shearing stresses.)

7.68 For the state of stress shown, determine the maximum shearing stress when (*a*) $\sigma_y = 40$ MPa, (*b*) $\sigma_y = 120$ MPa. (*Hint:* Consider both in-plane and out-of-plane shearing stresses.)

7.69 For the state of stress shown, determine the maximum shearing stress when (*a*) $\sigma_y = 20$ MPa, (*b*) $\sigma_y = 140$ MPa. (*Hint:* Consider both in-plane and out-of-plane shearing stresses.)

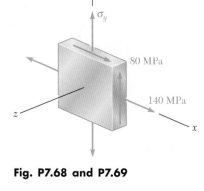

Fig. P7.68 and P7.69

7.70 and 7.71 For the state of stress shown, determine the maximum shearing stress when (*a*) $\sigma_z = +24$ MPa, (*b*) $\sigma_z = -24$ MPa, (*c*) $\sigma_z = 0$.

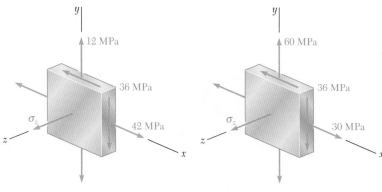

Fig. P7.70

Fig. P7.71

7.72 and 7.73 For the state of stress shown, determine the maximum shearing stress when (a) $\sigma_z = 0$, (b) $\sigma_z = +45$ MPa, (c) $\sigma_z = -45$ MPa.

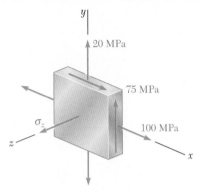

Fig. P7.72

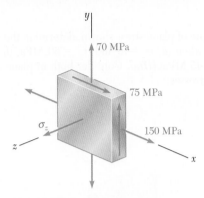

Fig. P7.73

7.74 For the state of stress shown, determine two values of σ_y for which the maximum shearing stress is 75 MPa.

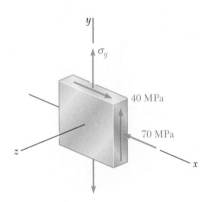

Fig. P7.74

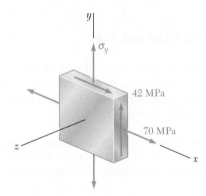

Fig. P7.75

7.75 For the state of stress shown, determine two values of σ_y for which the maximum shearing stress is 52 MPa.

7.76 For the state of stress shown, determine the value of τ_{xy} for which the maximum shearing stress is (a) 63 MPa, (b) 84 MPa.

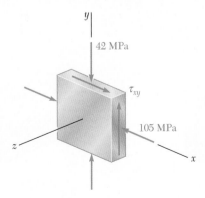

Fig. P7.76

7.77 For the state of stress shown, determine the value of τ_{xy} for which the maximum shearing stress is (a) 60 MPa, (b) 78 MPa.

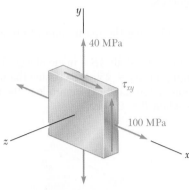

Fig. P7.77

7.78 For the state of stress shown, determine two values of σ_y for which the maximum shearing stress is 80 MPa.

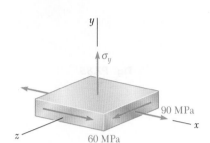

Fig. P7.78

7.79 For the state of stress shown, determine the range of values of τ_{xz} for which the maximum shearing stress is equal to or less than 60 MPa.

*7.80 For the state of stress of Prob. 7.69, determine (a) the value of σ_y for which the maximum shearing stress is as small as possible, (b) the corresponding value of the shearing stress.

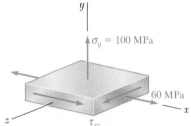

Fig. P7.79

7.81 The state of plane stress shown occurs in a machine component made of a steel with $\sigma_Y = 325$ MPa. Using the maximum-distortion-energy criterion, determine whether yield will occur when (a) $\sigma_0 = 200$ MPa, (b) $\sigma_0 = 240$ MPa, (c) $\sigma_0 = 280$ MPa. If yield does not occur, determine the corresponding factor of safety.

7.82 Solve Prob. 7.81, using the maximum-shearing-stress criterion.

7.83 The state of plane stress shown occurs in a machine component made of a steel with $\sigma_Y = 210$ MPa. Using the maximum-distortion-energy criterion, determine whether yield will occur when (a) $\tau_{xy} = 42$ MPa, (b) $\tau_{xy} = 84$ MPa, (c) $\tau_{xy} = 98$ MPa. If yield does not occur, determine the corresponding factor of safety.

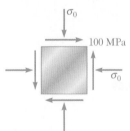

Fig. P7.81

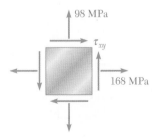

Fig. P7.83

7.84 Solve Prob. 7.83, using the maximum-shearing-stress criterion.

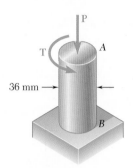

Fig. P7.85

7.85 The 36-mm-diameter shaft is made of a grade of steel with a 250-MPa tensile yield stress. Using the maximum-shearing-stress criterion, determine the magnitude of the torque **T** for which yield occurs when $P = 200$ kN.

7.86 Solve Prob. 7.85, using the maximum-distortion-energy criterion.

7.87 The 38-mm-diameter shaft AB is made of a grade of steel for which the yield strength is $\sigma_Y = 250$ MPa. Using the maximum-shearing-stress criterion, determine the magnitude of the torque **T** for which yield occurs when $P = 240$ kN.

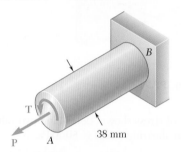

Fig. P7.87

7.88 Solve Prob. 7.87, using the maximum-distortion-energy criterion.

7.89 and 7.90 The state of plane stress shown is expected to occur in an aluminum casting. Knowing that for the aluminum alloy used $\sigma_{UT} = 80$ MPa and $\sigma_{UC} = 200$ MPa and using Mohr's criterion, determine whether rupture of the casting will occur.

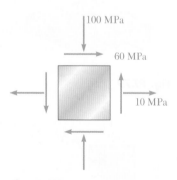

Fig. P7.89

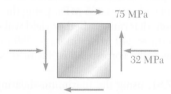

Fig. P7.90

7.91 and 7.92 The state of plane stress shown is expected to occur in an aluminum casting. Knowing that for the aluminum alloy used $\sigma_{UT} = 70$ MPa and $\sigma_{UC} = 210$ MPa and using Mohr's criterion, determine whether rupture of the casting will occur.

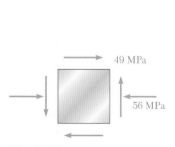

Fig. P7.91

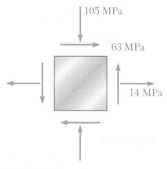

Fig. P7.92

7.93 The state of plane stress shown will occur at a critical point in an aluminum casting that is made of an alloy for which $\sigma_{UT} = 70$ MPa and $\sigma_{UC} = 170$ MPa. Using Mohr's criterion, determine the shearing stress τ_0 for which failure should be expected.

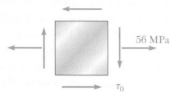

56 MPa

τ_0

Fig. P7.93

7.94 The state of plane stress shown will occur at a critical point in a pipe made of an aluminum alloy for which $\sigma_{UT} = 75$ MPa and $\sigma_{UC} = 150$ MPa. Using Mohr's criterion, determine the shearing stress τ_0 for which failure should be expected.

80 MPa

τ_0

Fig. P7.94

7.95 The cast-aluminum rod shown is made of an alloy for which $\sigma_{UT} = 60$ MPa and $\sigma_{UC} = 120$ MPa. Using Mohr's criterion, determine the magnitude of the torque **T** for which failure should be expected.

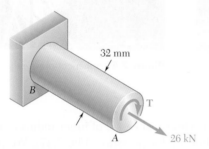

32 mm

B

T

A

26 kN

Fig. P7.95

7.96 The cast-aluminum rod shown is made of an alloy for which $\sigma_{UT} = 70$ MPa and $\sigma_{UC} = 175$ MPa. Knowing that the magnitude T of the applied torques is slowly increased and using Mohr's criterion, determine the shearing stress τ_0 that should be expected at rupture.

T

T'

τ_0

Fig. P7.96

7.97 A machine component is made of a grade of cast iron for which $\sigma_{UT} = 56$ MPa and $\sigma_{UC} = 140$ MPa. For each of the states of stress shown, and using Mohr's criterion, determine the normal stress σ_0 at which rupture of the component should be expected.

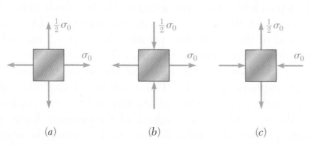

$\frac{1}{2}\sigma_0$ $\frac{1}{2}\sigma_0$ $\frac{1}{2}\sigma_0$

σ_0 σ_0 σ_0

(a)   (b)   (c)

Fig. P7.97

Fig. 7.46 Assumed stress distribution in thin-walled pressure vessels.

7.9 STRESSES IN THIN-WALLED PRESSURE VESSELS

Thin-walled pressure vessels provide an important application of the analysis of plane stress. Since their walls offer little resistance to bending, it can be assumed that the internal forces exerted on a given portion of wall are tangent to the surface of the vessel (Fig. 7.46). The resulting stresses on an element of wall will thus be contained in a plane tangent to the surface of the vessel.

Our analysis of stresses in thin-walled pressure vessels will be limited to the two types of vessels most frequently encountered: cylindrical pressure vessels and spherical pressure vessels (Photos 7.3 and 7.4).

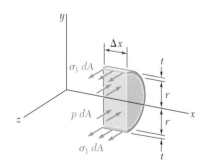

Photo 7.3 Cylindrical pressure vessels.

Photo 7.4 Spherical pressure vessels.

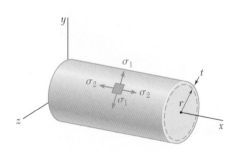

Fig. 7.47 Pressurized cylindrical vessel.

Fig. 7.48 Free body to determine hoop stress.

Consider a cylindrical vessel of inner radius r and wall thickness t containing a fluid under pressure (Fig. 7.47). We propose to determine the stresses exerted on a small element of wall with sides respectively parallel and perpendicular to the axis of the cylinder. Because of the axisymmetry of the vessel and its contents, it is clear that no shearing stress is exerted on the element. The normal stresses σ_1 and σ_2 shown in Fig. 7.47 are therefore principal stresses. The stress σ_1 is known as the *hoop stress*, because it is the type of stress found in hoops used to hold together the various slats of a wooden barrel, and the stress σ_2 is called the *longitudinal stress*.

In order to determine the hoop stress σ_1, we detach a portion of the vessel and its contents bounded by the xy plane and by two planes parallel to the yz plane at a distance Δx from each other (Fig. 7.48). The forces parallel to the z axis acting on the free body defined in this fashion consist of the elementary internal forces $\sigma_1\, dA$ on the wall sections, and of the elementary pressure forces $p\, dA$ exerted on the portion of fluid included in the free body. Note that p denotes the *gage pressure* of the fluid, i.e., the excess of the inside pressure over the outside atmospheric pressure. The resultant of the internal forces $\sigma_1\, dA$ is equal to the product of σ_1 and of the cross-sectional area $2t\, \Delta x$ of the wall, while the resultant of the pressure forces $p\, dA$ is equal to the product of p and of the area $2r\, \Delta x$. Writing the equilibrium equation $\Sigma F_z = 0$, we have

$$\Sigma F_z = 0: \qquad \sigma_1(2t\,\Delta x) - p(2r\,\Delta x) = 0$$

and, solving for the hoop stress σ_1,

$$\sigma_1 = \frac{pr}{t} \qquad (7.30)$$

To determine the longitudinal stress σ_2, we now pass a section perpendicular to the x axis and consider the free body consisting of the portion of the vessel and its contents located to the left of the section

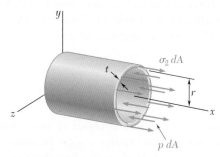

Fig. 7.49 Free body to determine longitudinal stress.

(Fig. 7.49). The forces acting on this free body are the elementary internal forces $\sigma_2\,dA$ on the wall section and the elementary pressure forces $p\,dA$ exerted on the portion of fluid included in the free body. Noting that the area of the fluid section is πr^2 and that the area of the wall section can be obtained by multiplying the circumference $2\pi r$ of the cylinder by its wall thickness t, we write the equilibrium equation:†

$$\Sigma F_x = 0: \qquad \sigma_2(2\pi rt) - p(\pi r^2) = 0$$

and, solving for the longitudinal stress σ_2,

$$\sigma_2 = \frac{pr}{2t} \qquad (7.31)$$

We note from Eqs. (7.30) and (7.31) that the hoop stress σ_1 is twice as large as the longitudinal stress σ_2:

$$\sigma_1 = 2\sigma_2 \qquad (7.32)$$

---

†Using the mean radius of the wall section, $r_m = r + \frac{1}{2}t$, in computing the resultant of the forces on that section, we would obtain a more accurate value of the longitudinal stress, namely,

$$\sigma_2 = \frac{pr}{2t}\,\frac{1}{1 + \dfrac{t}{2r}} \qquad (7.31')$$

However, for a thin-walled pressure vessel, the term $t/2r$ is sufficiently small to allow the use of Eq. (7.31) for engineering design and analysis. If a pressure vessel is not thin-walled (i.e., if $t/2r$ is not small), the stresses σ_1 and σ_2 vary across the wall and must be determined by the methods of the theory of elasticity.

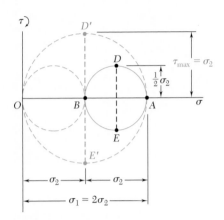

Fig. 7.50 Mohr's circle for element of cylindrical pressure vessel.

Fig. 7.51 Pressurized spherical vessel.

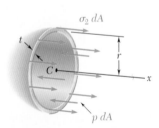

Fig. 7.52 Free body to determine wall stress.

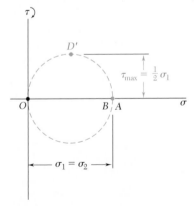

Fig. 7.53 Mohr's circle for element of spherical pressure vessel.

Drawing Mohr's circle through the points A and B that correspond respectively to the principal stresses σ_1 and σ_2 (Fig. 7.50), and recalling that the maximum in-plane shearing stress is equal to the radius of this circle, we have

$$\tau_{\text{max (in plane)}} = \tfrac{1}{2}\sigma_2 = \frac{pr}{4t} \tag{7.33}$$

This stress corresponds to points D and E and is exerted on an element obtained by rotating the original element of Fig. 7.47 through 45° *within the plane* tangent to the surface of the vessel. The maximum shearing stress in the wall of the vessel, however, is larger. It is equal to the radius of the circle of diameter OA and corresponds to a rotation of 45° about a longitudinal axis and *out of the plane* of stress.† We have

$$\tau_{\text{max}} = \sigma_2 = \frac{pr}{2t} \tag{7.34}$$

We now consider a spherical vessel of inner radius r and wall thickness t, containing a fluid under a gage pressure p. For reasons of symmetry, the stresses exerted on the four faces of a small element of wall must be equal (Fig. 7.51). We have

$$\sigma_1 = \sigma_2 \tag{7.35}$$

To determine the value of the stress, we pass a section through the center C of the vessel and consider the free body consisting of the portion of the vessel and its contents located to the left of the section (Fig. 7.52). The equation of equilibrium for this free body is the same as for the free body of Fig. 7.49. We thus conclude that, for a spherical vessel,

$$\sigma_1 = \sigma_2 = \frac{pr}{2t} \tag{7.36}$$

Since the principal stresses σ_1 and σ_2 are equal, Mohr's circle for transformations of stress within the plane tangent to the surface of the vessel reduces to a point (Fig. 7.53); we conclude that the in-plane normal stress is constant and that the in-plane maximum shearing stress is zero. The maximum shearing stress in the wall of the vessel, however, is not zero; it is equal to the radius of the circle of diameter OA and corresponds to a rotation of 45° out of the plane of stress. We have

$$\tau_{\text{max}} = \tfrac{1}{2}\sigma_1 = \frac{pr}{4t} \tag{7.37}$$

†It should be observed that, while the third principal stress is zero on the outer surface of the vessel, it is equal to $-p$ on the inner surface, and is represented by a point $C(-p, 0)$ on a Mohr-circle diagram. Thus, close to the inside surface of the vessel, the maximum shearing stress is equal to the radius of a circle of diameter CA, and we have

$$\tau_{\text{max}} = \frac{1}{2}(\sigma_1 + p) = \frac{pr}{2t}\left(1 + \frac{t}{r}\right)$$

For a thin-walled vessel, however, the term t/r is small, and we can neglect the variation of τ_{max} across the wall section. This remark also applies to spherical pressure vessels.

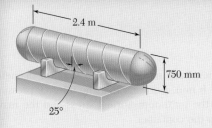

25°

SAMPLE PROBLEM 7.5

A compressed-air tank is supported by two cradles as shown; one of the cradles is designed so that it does not exert any longitudinal force on the tank. The cylindrical body of the tank has a 750-mm outer diameter and is fabricated from a 10-mm steel plate by butt welding along a helix that forms an angle of 25° with a transverse plane. The end caps are spherical and have a uniform wall thickness of 8 mm. For an internal gage pressure of 1.2 MPa, determine (*a*) the normal stress and the maximum shearing stress in the spherical caps, (*b*) the stresses in directions perpendicular and parallel to the helical weld.

SOLUTION

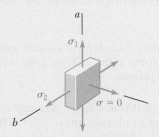

a. Spherical Cap. Using Eq. (7.36), we write

$$p = 1.2 \text{ MPa}, t = 8 \text{ mm}, r = 375 - 8 = 367 \text{ mm}$$

$$\sigma_1 = \sigma_2 = \frac{pr}{2t} = \frac{(1.2 \text{ MPa})(367 \text{ mm})}{2(8 \text{ mm})} \qquad \sigma = 27.5 \text{ MPa} \quad \blacktriangleleft$$

We note that for stresses in a plane tangent to the cap, Mohr's circle reduces to a point (A, B) on the horizontal axis and that all in-plane shearing stresses are zero. On the surface of the cap the third principal stress is zero and corresponds to point O. On a Mohr's circle of diameter AO, point D' represents the maximum shearing stress; it occurs on planes at 45° to the plane tangent to the cap.

$$\tau_{\max} = \tfrac{1}{2}(27.5 \text{ MPa}) \qquad \tau_{\max} = 13.75 \text{ MPa} \quad \blacktriangleleft$$

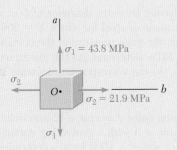

b. Cylindrical Body of the Tank. We first determine the hoop stress σ_1 and the longitudinal stress σ_2. Using Eqs. (7.30) and (7.32), we write

$$p = 1.2 \text{ MPa}, t = 10 \text{ mm}, r = 375 - 10 = 365 \text{ mm}$$

$$\sigma_1 = \frac{pr}{t} = \frac{(1.2 \text{ MPa})(365 \text{ mm})}{10 \text{ mm}} = 43.8 \text{ MPa} \qquad \sigma_2 = \tfrac{1}{2}\sigma_1 = 21.9 \text{ MPa}$$

$$\sigma_{\text{ave}} = \tfrac{1}{2}(\sigma_1 + \sigma_2) = 32.85 \text{ MPa} \qquad R = \tfrac{1}{2}(\sigma_1 - \sigma_2) = 10.95 \text{ MPa}$$

Stresses at the Weld. Noting that both the hoop stress and the longitudinal stress are principal stresses, we draw Mohr's circle as shown.

An element having a face parallel to the weld is obtained by rotating the face perpendicular to the axis Ob counterclockwise through 25°. Therefore, on Mohr's circle we locate the point X' corresponding to the stress components on the weld by rotating radius CB counterclockwise through $2\theta = 50°$.

$$\sigma_w = \sigma_{\text{ave}} - R \cos 50° = 32.85 - 10.95 \cos 50° \qquad \sigma_w = +25.8 \text{ MPa} \quad \blacktriangleleft$$

$$\tau_w = R \sin 50° = 10.95 \sin 50° \qquad \tau_w = 8.39 \text{ MPa} \quad \blacktriangleleft$$

Since X' is below the horizontal axis, τ_w tends to rotate the element counterclockwise.

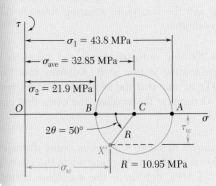

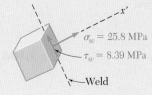

$\sigma_w = 25.8 \text{ MPa}$

$\tau_w = 8.39 \text{ MPa}$

Weld

PROBLEMS

7.98 A spherical gas container made of steel has a 5-m outer diameter and a wall thickness of 6 mm. Knowing that the internal pressure is 350 kPa, determine the maximum normal stress and the maximum shearing stress in the container.

7.99 The maximum gage pressure is known to be 8 MPa in a spherical steel pressure vessel having a 250-mm outer diameter and a 6-mm wall thickness. Knowing that the ultimate stress in the steel used is $\sigma_U = 400$ MPa, determine the factor of safety with respect to tensile failure.

7.100 A basketball has a 300-mm outer diameter and a 3-mm wall thickness. Determine the normal stress in the wall when the basketball is inflated to a 120-kPa gage pressure.

7.101 A spherical pressure vessel of 900-mm outer diameter is to be fabricated from a steel having an ultimate stress $\sigma_U = 400$ MPa. Knowing that a factor of safety of 4.0 is desired and that the gage pressure can reach 3.5 MPa, determine the smallest wall thickness that should be used.

7.102 A spherical pressure vessel has an inner diameter of 3 m and a wall thickness of 12 mm. Knowing that for the steel used $\sigma_{all} = 80$ MPa, $E = 200$ GPa, and $\nu = 0.29$, determine (*a*) the allowable gage pressure, (*b*) the corresponding increase in the diameter of the vessel.

7.103 A spherical gas container having an outer diameter of 5 m and a wall thickness of 22 mm is made of steel for which $E = 200$ GPa and $\nu = 0.29$. Knowing that the gage pressure in the container is increased from zero to 1.7 MPa, determine (*a*) the maximum normal stress in the container, (*b*) the corresponding increase in the diameter of the container.

7.104 A steel penstock has a 750-mm outer diameter, a 12-mm wall thickness, and connects a reservoir at *A* with a generating station at *B*. Knowing that the density of water is 1000 kg/m^3, determine the maximum normal stress and the maximum shearing stress in the penstock under static conditions.

7.105 A steel penstock has a 750-mm outer diameter and connects a reservoir at *A* with a generating station at *B*. Knowing that the density of water is 1000 kg/m^3 and that the allowable normal stress in the steel is 85 MPa, determine the smallest thickness that can be used for the penstock.

7.106 The bulk storage tank shown in Photo 7.3 has an outer diameter of 3.3 m and a wall thickness of 18 mm. At a time when the internal pressure of the tank is 1.5 MPa, determine the maximum normal stress and the maximum shearing stress in the tank.

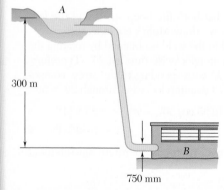

Fig. P7.104 and P7.105

504

7.107 Determine the largest internal pressure that can be applied to a cylindrical tank of 1.75-m outer diameter and 16-mm wall thickness if the ultimate normal stress of the steel used is 450 MPa and a factor of safety of 5.0 is desired.

7.108 A cylindrical storage tank contains liquefied propane under a pressure of 1.5 MPa at a temperature of 38°C. Knowing that the tank has an outer diameter of 320 mm and a wall thickness of 3 mm, determine the maximum normal stress and the maximum shearing stress in the tank.

7.109 The unpressurized cylindrical storage tank shown has a 5-mm wall thickness and is made of steel having a 420-MPa ultimate strength in tension. Determine the maximum height h to which it can be filled with water if a factor of safety of 4.0 is desired. (Specific weight of water = 9810 N/m³.)

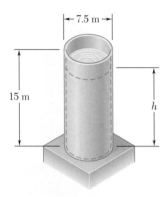

Fig. P7.109

7.110 For the storage tank of Prob. 7.109, determine the maximum normal stress and the maximum shearing stress in the cylindrical wall when the tank is filled to capacity ($h = 14.4$ m).

7.111 A standard-weight steel pipe of 300-mm nominal diameter carries water under a pressure of 2.8 MPa. (*a*) Knowing that the outside diameter is 320 mm and the wall thickness is 10 mm, determine the maximum tensile stress in the pipe. (*b*) Solve part *a*, assuming an extra-strong pipe is used, of 320-mm outside diameter and 12-mm wall thickness.

7.112 The pressure tank shown has a 8-mm wall thickness and butt-welded seams forming an angle $\beta = 20°$ with a transverse plane. For a gage pressure of 600 kPa, determine, (*a*) the normal stress perpendicular to the weld, (*b*) the shearing stress parallel to the weld.

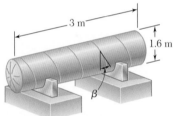

Fig. P7.112

7.113 For the tank of Prob. 7.112, determine the largest allowable gage pressure, knowing that the allowable normal stress perpendicular to the weld is 120 MPa and the allowable shearing stress parallel to the weld is 80 MPa.

7.114 For the tank of Prob. 7.112, determine the range of values of β that can be used if the shearing stress parallel to the weld is not to exceed 12 MPa when the gage pressure is 600 kPa.

7.115 The steel pressure tank shown has a 750-mm inner diameter and a 9-mm wall thickness. Knowing that the butt-welded seams form an angle $\beta = 50°$ with the longitudinal axis of the tank and that the gage pressure in the tank is 1.5 MPa, determine, (*a*) the normal stress perpendicular to the weld, (*b*) the shearing stress parallel to the weld.

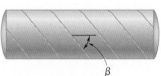

Fig. P7.115 and P7.116

7.116 The pressurized tank shown was fabricated by welding strips of plate along a helix forming an angle β with a transverse plane. Determine the largest value of β that can be used if the normal stress perpendicular to the weld is not to be larger than 85 percent of the maximum stress in the tank.

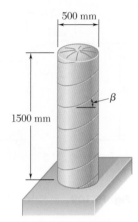

Fig. *P7.117* and *P7.118*

7.117 The cylindrical portion of the compressed-air tank shown is fabricated of 6-mm-thick plate welded along a helix forming an angle $\beta = 30°$ with the horizontal. Knowing that the allowable stress normal to the weld is 75 MPa, determine the largest gage pressure that can be used in the tank.

7.118 The cylindrical portion of the compressed air tank shown is fabricated of 6-mm-thick plate welded along a helix forming an angle $\beta = 30°$ with the horizontal. Determine the gage pressure that will cause a shearing stress parallel to the weld of 30 MPa.

7.119 Square plates, each of 16-mm thickness, can be bent and welded together in either of the two ways shown to form the cylindrical portion of a compressed-air tank. Knowing that the allowable normal stress perpendicular to the weld is 65 MPa, determine the largest allowable gage pressure in each case.

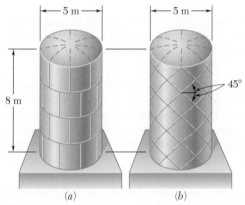

Fig. *P7.119*

7.120 The compressed-air tank *AB* has an inner diameter of 450 mm and a uniform wall thickness of 6 mm. Knowing that the gage pressure inside the tank is 1.2 MPa, determine the maximum normal stress and the maximum in-plane shearing stress at point *a* on the top of the tank.

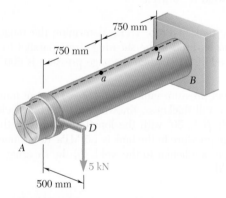

Fig. *P7.120*

7.121 For the compressed-air tank and loading of Prob. 7.120, determine the maximum normal stress and the maximum in-plane shearing stress at point *b* on the top of the tank.

Returning to the particular case of *plane strain,* and selecting the x and y axes in the plane of strain, we have $\epsilon_z = \gamma_{zx} = \gamma_{zy} = 0$. Thus, the z axis is one of the three principal axes at Q, and the corresponding point in the Mohr-circle diagram is the origin O, where $\epsilon = \gamma = 0$. If the points A and B that define the principal axes within the plane of strain fall on opposite sides of O (Fig. 7.70a), the corresponding principal strains represent the maximum and minimum normal strains at point Q, and the maximum shearing strain is equal to the maximum in-plane shearing strain corresponding to points D and E. If, on the other hand, A and B are on the same side of O (Fig. 7.70b), i.e., if ϵ_a and ϵ_b have the same sign, then the maximum shearing strain is defined by points D' and E' on the circle of diameter OA, and we have $\gamma_{\max} = \epsilon_{\max}$.

We now consider the particular case of *plane stress* encountered in a thin plate or on the free surface of a structural element or machine component (Sec. 7.1). Selecting the x and y axes in the plane of stress, we have $\sigma_z = \tau_{zx} = \tau_{zy} = 0$ and verify that the z axis is a principal axis of stress. As we saw earlier, if the deformation is elastic and if the material is homogeneous and isotropic, it follows from Hooke's law that $\gamma_{zx} = \gamma_{zy} = 0$; thus, the z axis is also a principal axis of strain, and Mohr's circle can be used to analyze the transformation of strain in the xy plane. However, as we shall see presently, it does *not* follow from Hooke's law that $\epsilon_z = 0$; indeed, a state of plane stress does not, in general, result in a state of plane strain.†

Denoting by a and b the principal axes within the plane of stress, and by c the principal axis perpendicular to that plane, we let $\sigma_x = \sigma_a$, $\sigma_y = \sigma_b$, and $\sigma_z = 0$ in Eqs. (2.28) for the generalized Hooke's law (Sec. 2.12) and write

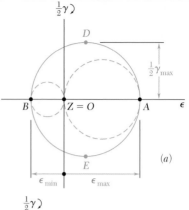
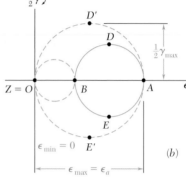

$$\epsilon_a = \frac{\sigma_a}{E} - \frac{\nu\sigma_b}{E} \tag{7.55}$$

$$\epsilon_b = -\frac{\nu\sigma_a}{E} + \frac{\sigma_b}{E} \tag{7.56}$$

$$\epsilon_c = -\frac{\nu}{E}(\sigma_a + \sigma_b) \tag{7.57}$$

Fig. 7.70 Mohr's circle for plane strain.

Adding Eqs. (7.55) and (7.56) member to member, we have

$$\epsilon_a + \epsilon_b = \frac{1 - \nu}{E}(\sigma_a + \sigma_b) \tag{7.58}$$

Solving Eq. (7.58) for $\sigma_a + \sigma_b$ and substituting into Eq. (7.57), we write

$$\epsilon_c = -\frac{\nu}{1 - \nu}(\epsilon_a + \epsilon_b) \tag{7.59}$$

The relation obtained defines the third principal strain in terms of the "in-plane" principal strains. We note that, if B is located between A and C on the Mohr-circle diagram (Fig. 7.71), the maximum shearing strain is equal to the diameter CA of the circle corresponding to a rotation about the b axis, out of the plane of stress.

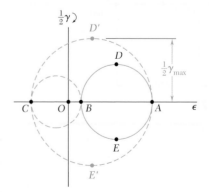

Fig. 7.71 Mohr's circle strain analysis for plane stress.

†See footnote on page 508.

EXAMPLE 7.05

As a result of measurements made on the surface of a machine component with strain gages oriented in various ways, it has been established that the principal strains on the free surface are $\epsilon_a = +400 \times 10^{-6}$ mm/mm and $\epsilon_b = -50 \times 10^{-6}$ mm/mm. Knowing that Poisson's ratio for the given material is $\nu = 0.30$, determine (a) the maximum in-plane shearing strain, (b) the true value of the maximum shearing strain near the surface of the component.

(a) Maximum In-Plane Shearing Strain. We draw Mohr's circle through the points A and B corresponding to the given principal strains (Fig. 7.72). The maximum in-plane shearing strain is defined by points D and E and is equal to the diameter of Mohr's circle:

$$\gamma_{\text{max (in plane)}} = 400 \times 10^{-6} + 50 \times 10^{-6} = 450 \times 10^{-6} \text{ rad}$$

(b) Maximum Shearing Strain. We first determine the third principal strain ϵ_c. Since we have a state of plane *stress* on the surface of the machine component, we use Eq. (7.59) and write

$$\epsilon_c = -\frac{\nu}{1-\nu}(\epsilon_a + \epsilon_b)$$

$$= -\frac{0.30}{0.70}(400 \times 10^{-6} - 50 \times 10^{-6}) = -150 \times 10^{-6} \text{ mm/mm}$$

Drawing Mohr's circles through A and C and through B and C (Fig. 7.73), we find that the maximum shearing strain is equal to the diameter of the circle of diameter CA:

$$\gamma_{\text{max}} = 400 \times 10^{-6} + 150 \times 10^{-6} = 550 \times 10^{-6} \text{ rad}$$

We note that, even though ϵ_a and ϵ_b have opposite signs, the maximum in-plane shearing strain does not represent the true maximum shearing strain.

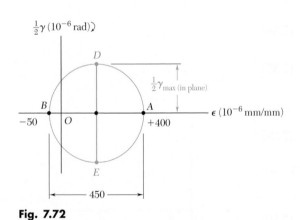

Fig. 7.72

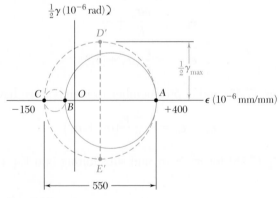

Fig. 7.73

*7.13 MEASUREMENTS OF STRAIN; STRAIN ROSETTE

The normal strain can be determined in any given direction on the surface of a structural element or machine component by scribing two gage marks A and B across a line drawn in the desired direction and measuring the length of the segment AB before and after the load has been applied. If L is the undeformed length of AB and δ its deformation, the normal strain along AB is $\epsilon_{AB} = \delta/L$.

A more convenient and more accurate method for the measurement of normal strains is provided by electrical strain gages. A typical electrical strain gage consists of a length of thin wire arranged as shown in Fig. 7.74 and cemented to two pieces of paper. In order to measure the strain ϵ_{AB} of a given material in the direction AB, the gage is cemented to the surface of the material, with the wire folds running parallel to AB. As the material elongates, the wire increases in length and decreases in diameter, causing the electrical resistance of the gage to increase. By measuring the current passing through a properly calibrated gage, the strain ϵ_{AB} can be determined accurately and continuously as the load is increased.

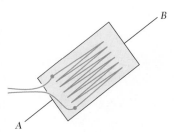

Fig. 7.74 Electrical strain gage.

The strain components ϵ_x and ϵ_y can be determined at a given point of the free surface of a material by simply measuring the normal strain along x and y axes drawn through that point. Recalling Eq. (7.43) of Sec. 7.10, we note that a third measurement of normal strain, made along the bisector OB of the angle formed by the x and y axes, enables us to determine the shearing strain γ_{xy} as well (Fig. 7.75):

$$\gamma_{xy} = 2\epsilon_{OB} - (\epsilon_x + \epsilon_y) \tag{7.43}$$

It should be noted that the strain components ϵ_x, ϵ_y, and γ_{xy} at a given point could be obtained from normal strain measurements made along *any three lines* drawn through that point (Fig. 7.76). Denoting respectively by θ_1, θ_2, and θ_3 the angle each of the three lines forms with the x axis, by ϵ_1, ϵ_2, and ϵ_3 the corresponding strain measurements, and substituting into Eq. (7.41), we write the three equations

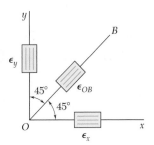

Fig. 7.75

$$\epsilon_1 = \epsilon_x \cos^2 \theta_1 + \epsilon_y \sin^2 \theta_1 + \gamma_{xy} \sin \theta_1 \cos \theta_1$$
$$\epsilon_2 = \epsilon_x \cos^2 \theta_2 + \epsilon_y \sin^2 \theta_2 + \gamma_{xy} \sin \theta_2 \cos \theta_2 \tag{7.60}$$
$$\epsilon_3 = \epsilon_x \cos^2 \theta_3 + \epsilon_y \sin^2 \theta_3 + \gamma_{xy} \sin \theta_3 \cos \theta_3$$

which can be solved simultaneously for ϵ_x, ϵ_y, and γ_{xy}.[†]

The arrangement of strain gages used to measure the three normal strains ϵ_1, ϵ_2, and ϵ_3 is known as a *strain rosette*. The rosette used to measure normal strains along the x and y axes and their bisector is referred to as a 45° rosette (Fig. 7.75). Another rosette frequently used is the 60° rosette (see Sample Prob. 7.7).

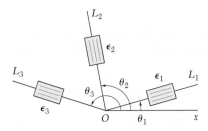

Fig. 7.76 Strain rosette.

[†]It should be noted that the free surface on which the strain measurements are made is in a state of *plane stress*, while Eqs. (7.41) and (7.43) were derived for a state of *plane strain*. However, as observed earlier, the normal to the free surface is a principal axis of strain and the derivations given in Sec. 7.10 remain valid.

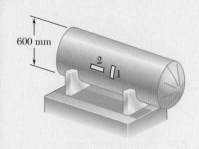

600 mm

SAMPLE PROBLEM 7.6

A cylindrical storage tank used to transport gas under pressure has an inner diameter of 600 mm and a wall thickness of 18 mm. Strain gages attached to the surface of the tank in transverse and longitudinal directions indicate strains of 255×10^{-6} and 60×10^{-6} mm/mm, respectively. Knowing that a torsion test has shown that the modulus of rigidity of the material used in the tank is $G = 77$ GPa, determine (a) the gage pressure inside the tank, (b) the principal stresses and the maximum shearing stress in the wall of the tank.

SOLUTION

a. Gage Pressure Inside Tank. We note that the given strains are the principal strains at the surface of the tank. Plotting the corresponding points A and B, we draw Mohr's circle for strain. The maximum in-plane shearing strain is equal to the diameter of the circle.

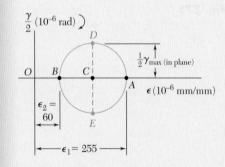

$$\gamma_{max\,(in\,plane)} = \epsilon_1 - \epsilon_2 = 255 \times 10^{-6} - 60 \times 10^{-6} = 195 \times 10^{-6} \text{ rad}$$

From Hooke's law for shearing stress and strain, we have

$$\begin{aligned}\tau_{max\,(in\,plane)} &= G\gamma_{max\,(in\,plane)} \\ &= (77 \text{ GPa})(195 \times 10^{-6} \text{ rad}) \\ &= 15.015 \text{ MPa}\end{aligned}$$

Substituting this value and the given data in Eq. (7.33), we write

$$\tau_{max\,(in\,plane)} = \frac{pr}{4t} \qquad 15.015 \text{ MPa} = \frac{p(300 \text{ mm})}{4(18 \text{ mm})}$$

Solving for the gage pressure p, we have

$$p = 3.60 \text{ MPa} \quad \blacktriangleleft$$

b. Principal Stresses and Maximum Shearing Stress. Recalling that, for a thin-walled cylindrical pressure vessel, $\sigma_1 = 2\sigma_2$, we draw Mohr's circle for stress and obtain

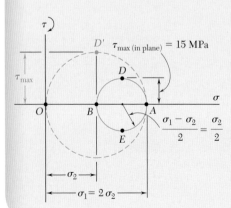

$$\sigma_2 = 2\tau_{max\,(in\,plane)} = 2(15.015 \text{ MPa}) = 30.0 \text{ MPa} \quad \sigma_2 = 30.0 \text{ MPa} \quad \blacktriangleleft$$

$$\sigma_1 = 2\sigma_2 = 2(30.0 \text{ MPa}) \qquad\qquad\qquad \sigma_1 = 60.0 \text{ MPa} \quad \blacktriangleleft$$

The maximum shearing stress is equal to the radius of the circle of diameter OA and corresponds to a rotation of $45°$ about a longitudinal axis.

$$\tau_{max} = \tfrac{1}{2}\sigma_1 = \sigma_2 = 30.0 \text{ MPa} \quad \tau_{max} = 30.0 \text{ MPa} \quad \blacktriangleleft$$

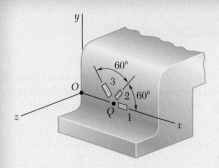

SAMPLE PROBLEM 7.7

Using a 60° rosette, the following strains have been determined at point Q on the surface of a steel machine base:

$$\epsilon_1 = 40\ \mu \qquad \epsilon_2 = 980\ \mu \qquad \epsilon_3 = 330\ \mu$$

Using the coordinate axes shown, determine at point Q, (*a*) the strain components ϵ_x, ϵ_y, and γ_{xy}, (*b*) the principal strains, (*c*) the maximum shearing strain. (Use $\nu = 0.29$.)

SOLUTION

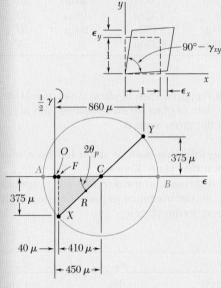

a. Strain Components ϵ_x, ϵ_y, γ_{xy}. For the coordinate axes shown

$$\theta_1 = 0 \qquad \theta_2 = 60° \qquad \theta_3 = 120°$$

Substituting these values into Eqs. (7.60), we have

$$\epsilon_1 = \epsilon_x(1) \qquad\qquad + \epsilon_y(0) \qquad\quad + \gamma_{xy}(0)(1)$$
$$\epsilon_2 = \epsilon_x(0.500)^2 \quad + \epsilon_y(0.866)^2 + \gamma_{xy}(0.866)(0.500)$$
$$\epsilon_3 = \epsilon_x(-0.500)^2 + \epsilon_y(0.866)^2 + \gamma_{xy}(0.866)(-0.500)$$

Solving these equations for ϵ_x, ϵ_y, and γ_{xy}, we obtain

$$\epsilon_x = \epsilon_1 \qquad \epsilon_y = \tfrac{1}{3}(2\epsilon_2 + 2\epsilon_3 - \epsilon_1) \qquad \gamma_{xy} = \frac{\epsilon_2 - \epsilon_3}{0.866}$$

Substituting the given values for ϵ_1, ϵ_2, and ϵ_3, we have

$$\epsilon_x = 40\ \mu \qquad\qquad \epsilon_y = \tfrac{1}{3}[2(980) + 2(330) - 40] \qquad \epsilon_y = +860\ \mu \ \blacktriangleleft$$
$$\gamma_{xy} = (980 - 330)/0.866 \qquad\qquad \gamma_{xy} = 750\ \mu \ \blacktriangleleft$$

These strains are indicated on the element shown.

b. Principal Strains. We note that the side of the element associated with ϵ_x rotates counterclockwise; thus, we plot point X below the horizontal axis, i.e., $X(40, -375)$. We then plot $Y(860, +375)$ and draw Mohr's circle.

$$\epsilon_{ave} = \tfrac{1}{2}(860\ \mu + 40\ \mu) = 450\ \mu$$
$$R = \sqrt{(375\ \mu)^2 + (410\ \mu)^2} = 556\ \mu$$
$$\tan 2\theta_p = \frac{375\ \mu}{410\ \mu} \qquad 2\theta_p = 42.4°\downarrow \qquad \theta_p = 21.2°\downarrow$$

Points A and B correspond to the principal strains. We have

$$\epsilon_a = \epsilon_{ave} - R = 450\ \mu - 556\ \mu \qquad \epsilon_a = -106\ \mu \ \blacktriangleleft$$
$$\epsilon_b = \epsilon_{ave} + R = 450\ \mu + 556\ \mu \qquad \epsilon_b = +1006\ \mu \ \blacktriangleleft$$

Since $\sigma_z = 0$ on the surface, we use Eq. (7.59) to find the principal strain ϵ_c:

$$\epsilon_c = -\frac{\nu}{1-\nu}(\epsilon_a + \epsilon_b) = -\frac{0.29}{1 - 0.29}(-106\ \mu + 1006\ \mu) \quad \epsilon_c = -368\ \mu \ \blacktriangleleft$$

c. Maximum Shearing Strain. Plotting point C and drawing Mohr's circle through points B and C, we obtain point D' and write

$$\tfrac{1}{2}\gamma_{max} = \tfrac{1}{2}(1006\ \mu + 368\ \mu) \qquad \gamma_{max} = 1374\ \mu \ \blacktriangleleft$$

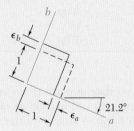

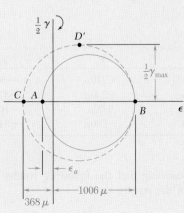

PROBLEMS

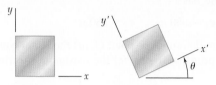

Fig. P7.128 through P7.135

7.128 through 7.131 For the given state of plane strain, use the method of Sec. 7.10 to determine the state of plane strain associated with axes x' and y' rotated through the given angle θ.

	ϵ_x	ϵ_y	γ_{xy}	θ
7.128 and 7.132	-720μ	0	$+300\mu$	$30° \downarrow$
7.129 and 7.133	0	$+320\mu$	-100μ	$30° \nwarrow$
7.130 and 7.134	-800μ	$+450\mu$	$+200\mu$	$25° \downarrow$
7.131 and 7.135	-500μ	$+250\mu$	0	$15° \nwarrow$

7.132 through 7.135 For the given state of plane strain, use Mohr's circle to determine the state of plane strain associated with axes x' and y' rotated through the given angle θ.

7.136 through 7.139 The following state of strain has been measured on the surface of a thin plate. Knowing that the surface of the plate is unstressed, determine (a) the direction and magnitude of the principal strains, (b) the maximum in-plane shearing strain, (c) the maximum shearing strain. (Use $\nu = \frac{1}{3}$.)

	ϵ_x	ϵ_y	γ_{xy}
7.136	-260μ	-60μ	$+480\mu$
7.137	-600μ	-400μ	$+350\mu$
7.138	$+160\mu$	-480μ	-600μ
7.139	$+30\mu$	$+570\mu$	$+720\mu$

7.140 through 7.143 For the given state of plane strain, use Mohr's circle to determine (a) the orientation and magnitude of the principal strains, (b) the maximum in-plane strain, (c) the maximum shearing strain.

	ϵ_x	ϵ_y	γ_{xy}
7.140	$+60\mu$	$+240\mu$	-50μ
7.141	$+400\mu$	$+200\mu$	$+375\mu$
7.142	$+300\mu$	$+60\mu$	$+100\mu$
7.143	-180μ	-260μ	$+315\mu$

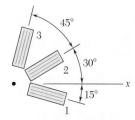

Fig. P7.144

7.144 Determine the strain ϵ_x knowing that the following strains have been determined by use of the rosette shown:

$$\epsilon_1 = +480\mu \qquad \epsilon_2 = -120\mu \qquad \epsilon_3 = +80\mu$$

7.145 Determine the largest in-plane normal strain, knowing that the following strains have been obtained by the use of the rosette shown:

$$\epsilon_1 = -50 \times 10^{-6} \text{ mm/mm} \qquad \epsilon_2 = +360 \times 10^{-6} \text{ mm/mm}$$
$$\epsilon_3 = +315 \times 10^{-6} \text{ mm/mm}$$

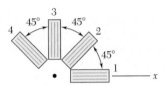

Fig. P7.145

7.146 The rosette shown has been used to determine the following strains at a point on the surface of a crane hook:

$$\epsilon_1 = +420\mu \qquad \epsilon_2 = -45\mu$$
$$\epsilon_4 = +165\mu$$

(a) What should be the reading of gage 3? (b) Determine the principal strains and the maximum in-plane shearing strain.

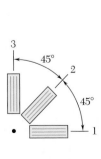

Fig. P7.146

7.147 The strains determined by the use of the rosette attached as shown during the test of a machine element are

$$\epsilon_1 = -93.1\mu \qquad \epsilon_2 = +385\mu$$
$$\epsilon_3 = +210\mu$$

Determine (a) the orientation and magnitude of the principal strains in the plane of the rosette, (b) the maximum in-plane shearing strain.

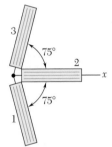

Fig. P7.147

7.148 Using a 45° rosette, the strains $\epsilon_1, \epsilon_2,$ and ϵ_3 have been determined at a given point. Using Mohr's circle, show that the principal strains are:

$$\epsilon_{\text{max, min}} = \frac{1}{2}(\epsilon_1 + \epsilon_3) \pm \frac{1}{\sqrt{2}}\left[(\epsilon_1 - \epsilon_2)^2 + (\epsilon_2 - \epsilon_3)^2\right]^{1/2}$$

(*Hint:* The shaded triangles are congruent.)

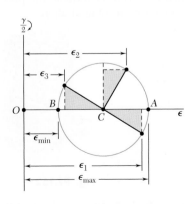

Fig. P7.148

7.149 Show that the sum of the three strain measurements made with a 60° rosette is independent of the orientation of the rosette and equal to

$$\epsilon_1 + \epsilon_2 + \epsilon_3 = 3\epsilon_{avg}$$

where ϵ_{avg} is the abscissa of the center of the corresponding Mohr's circle.

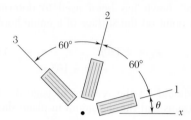

Fig. P7.149

7.150 A single strain gage is cemented to a solid 100-mm-diameter steel shaft at an angle $\beta = 25°$ with a line parallel to the axis of the shaft. Knowing that $G = 79$ GPa, determine the torque **T** indicated by a gage reading of 300μ.

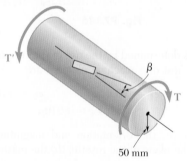

Fig. P7.150

7.151 Solve Prob. 7.150, assuming that the gage forms an angle $\beta = 35°$ with a line parallel to the axis of the shaft.

7.152 A single strain gage forming an angle $\beta = 18°$ with a horizontal plane is used to determine the gage pressure in the cylindrical steel tank shown. The cylindrical wall of the tank is 6-mm thick, has a 600-mm inside diameter, and is made of a steel with $E = 200$ GPa and $\nu = 0.30$. Determine the pressure in the tank indicated by a strain gage reading of 280μ.

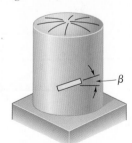

Fig. P7.152

7.153 Solve Prob. 7.152, assuming that the gage forms an angle $\beta = 35°$ with a horizontal plane.

7.165 The compressed-air tank *AB* has a 250-mm outside diameter and an 8-mm wall thickness. It is fitted with a collar by which a 40-kN force **P** is applied at *B* in the horizontal direction. Knowing that the gage pressure inside the tank is 5 MPa, determine the maximum normal stress and the maximum shearing stress at point *K*.

7.166 In Prob. 7.165, determine the maximum normal stress and the maximum shearing stress at point *L*.

7.167 The brass pipe *AD* is fitted with a jacket used to apply a hydrostatic pressure of 3.5 MPa to portion *BC* of the pipe. Knowing that the pressure inside the pipe is 0.7 MPa, determine the maximum normal stress in the pipe.

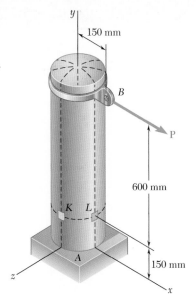

Fig. P7.165

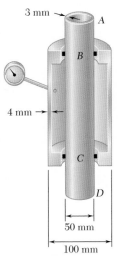

Fig. P7.167

7.168 A square *ABCD* of 60-mm side is scribed on the surface of a thin plate while the plate is unloaded. After the plate is loaded, the lengths of sides *AB* and *AD* are observed to have increased, respectively, by 13.5×10^{-3} mm and 22.5×10^{-3} mm, while the angle *DAB* is observed to have decreased by 360×10^{-6} rad. Knowing that $\nu = \frac{1}{3}$, determine (*a*) the orientation and magnitude of the principal strains, (*b*) the maximum in-plane shearing strain, (*c*) the maximum shearing strain.

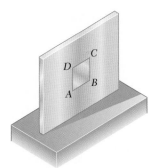

Fig. P7.168

7.169 The strains determined by the use of the rosette shown during the test of a machine element are

$$\epsilon_1 = +600\mu \qquad \epsilon_2 = +450\mu \qquad \epsilon_3 = -75\mu$$

Determine (*a*) the in-plane principal strains, (*b*) the in-plane maximum shearing strain.

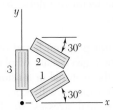

Fig. P7.169

COMPUTER PROBLEMS

The following problems are to be solved with a computer.

7.C1 A state of plane stress is defined by the stress components σ_x, σ_y, and τ_{xy} associated with the element shown in Fig. P7.C1a. (a) Write a computer program that can be used to calculate the stress components $\sigma_{x'}$, $\sigma_{y'}$, and $\tau_{x'y'}$ associated with the element after it has rotated through an angle θ about the z axis (Fig. P7.C1b). (b) Use this program to solve Probs. 7.13 through 7.16.

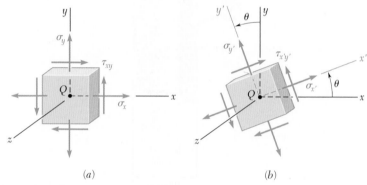

(a)   (b)

Fig. P7.C1

7.C2 A state of plane stress is defined by the stress components σ_x, σ_y, and τ_{xy} associated with the element shown in Fig. P7.C1a. (a) Write a computer program that can be used to calculate the principal axes, the principal stresses, the maximum in-plane shearing stress, and the maximum shearing stress. (b) Use this program to solve Probs. 7.5, 7.9, 7.68, and 7.69.

7.C3 (a) Write a computer program that, for a given state of plane stress and a given yield strength of a ductile material, can be used to determine whether the material will yield. The program should use both the maximum-shearing-strength criterion and the maximum-distortion-energy criterion. It should also print the values of the principal stresses and, if the material does not yield, calculate the factor of safety. (b) Use this program to solve Probs. 7.81, 7.82, and 7.164.

7.C4 (a) Write a computer program based on Mohr's fracture criterion for brittle materials that, for a given state of plane stress and given values of the ultimate strength of the material in tension and compression, can be used to determine whether rupture will occur. The program should also print the values of the principal stresses. (b) Use this program to solve Probs. 7.91 and 7.92 and to check the answers to Probs. 7.93 and 7.94.

7.C5 A state of plane strain is defined by the strain components ϵ_x, ϵ_y, and γ_{xy} associated with the x and y axes. (a) Write a computer program that can be used to calculate the strain components $\epsilon_{x'}$, $\epsilon_{y'}$, and $\gamma_{x'y'}$ associated with the frame of reference $x'y'$ obtained by rotating the x and y axes through an angle θ. (b) Use this program to solve Probs. 7.128 and 7.129.

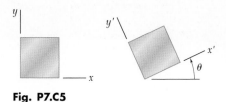

Fig. P7.C5

7.C6 A state of strain is defined by the strain components ϵ_x, ϵ_y, and γ_{xy} associated with the x and y axes. (a) Write a computer program that can be used to determine the orientation and magnitude of the principal strains, the maximum in-plane shearing strain, and the maximum shearing strain. (b) Use this program to solve Probs. 7.136 through 7.139.

7.C7 A state of plane strain is defined by the strain components ϵ_x, ϵ_y, and γ_{xy} measured at a point. (a) Write a computer program that can be used to determine the orientation and magnitude of the principal strains, the maximum in-plane shearing strain, and the magnitude of the shearing strain. (b) Use this program to solve Probs. 7.140 through 7.143.

7.C8 A rosette consisting of three gages forming, respectively, angles of θ_1, θ_2, and θ_3 with the x axis is attached to the free surface of a machine component made of a material with a given Poisson's ratio v. (a) Write a computer program that, for given readings ϵ_1, ϵ_2, and ϵ_3 of the gages, can be used to calculate the strain components associated with the x and y axes and to determine the orientation and magnitude of the three principal strains, the maximum in-plane shearing strain, and the maximum shearing strain. (b) Use this program to solve Probs. 7.144, 7.145, 7.146, and 7.169.

Due to gravity and wind load, the post supporting the sign shown is subjected simultaneously to compression, bending, and torsion. In this chapter you will learn to determine the stresses created by such combined loadings in structures and machine components.

Principal Stresses Under a Given Loading

*8.1 INTRODUCTION

In the first part of this chapter, you will apply to the design of beams and shafts the knowledge that you acquired in Chap. 7 on the transformation of stresses. In the second part of the chapter, you will learn how to determine the principal stresses in structural members and machine elements under given loading conditions.

In Chap. 5 you learned to calculate the maximum normal stress σ_m occurring in a beam under a transverse loading (Fig. 8.1a) and check whether this value exceeded the allowable stress σ_{all} for the given material. If it did, the design of the beam was not acceptable. While the danger for a brittle material is actually to fail in tension, the danger for a ductile material is to fail in shear (Fig. 8.1b). The fact that $\sigma_m > \sigma_{all}$ indicates that $|M|_{max}$ is too large for the cross section selected, but does not provide any information on the actual mechanism of failure. Similarly, the fact that $\tau_m > \tau_{all}$ simply indicates that $|V|_{max}$ is too large for the cross section selected. While the danger for a ductile material is actually to fail in shear (Fig. 8.2a), the danger for a brittle material is to fail in tension under the principal stresses (Fig. 8.2b). The distribution of the principal stresses in a beam will be discussed in Sec. 8.2.

Depending upon the shape of the cross section of the beam and the value of the shear V in the critical section where $|M| = |M|_{max}$, it may happen that the largest value of the normal stress will not occur at the top or bottom of the section, but at some other point within the section. As you will see in Sec. 8.2, a combination of large values of σ_x and τ_{xy} near the junction of the web and the flanges of a W-beam or an S-beam can result in a value of the principal stress σ_{max} (Fig. 8.3) that is larger than the value of σ_m on the surface of the beam.

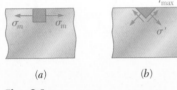

(a)   (b)

Fig. 8.1

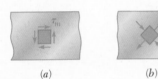

(a)   (b)

Fig. 8.2

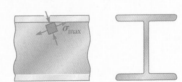

Fig. 8.3 Principal stresses at the junction of a flange and web in an I-shaped beam.

Section 8.3 will be devoted to the design of transmission shafts subjected to transverse loads as well as to torques. The effect of both the normal stresses due to bending and the shearing stresses due to torsion will be taken into account.

In Sec. 8.4 you will learn to determine the stresses at a given point K of a body of arbitrary shape subjected to a combined loading. First, you will reduce the given loading to forces and couples in the section containing K. Next, you will calculate the normal and shearing stresses at K. Finally, using one of the methods for the transformation of stresses that you learned in Chap. 7, you will determine the principal planes, principal stresses, and maximum shearing stress at K.

*8.2 PRINCIPAL STRESSES IN A BEAM

Consider a prismatic beam AB subjected to some arbitrary transverse loading (Fig. 8.4). We denote by V and M, respectively, the shear and bending moment in a section through a given point C. We recall from Chaps. 5 and 6 that, within the elastic limit, the stresses exerted on a small element with faces perpendicular, respectively, to the x and y axes reduce to the normal stresses $\sigma_m = Mc/I$ if the element is at the free surface of the beam, and to the shearing stresses $\tau_m = VQ/It$ if the element is at the neutral surface (Fig. 8.5).

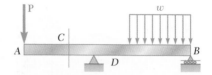

Fig. 8.4 Transversely loaded prismatic beam.

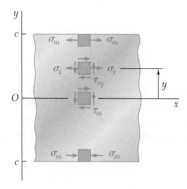

Fig. 8.5 Stress elements at selected points of a beam.

At any other point of the cross section, an element of material is subjected simultaneously to the normal stresses

$$\sigma_x = -\frac{My}{I} \tag{8.1}$$

where y is the distance from the neutral surface and I the centroidal moment of inertia of the section, and to the shearing stresses

$$\tau_{xy} = -\frac{VQ}{It} \tag{8.2}$$

where Q is the first moment about the neutral axis of the portion of the cross-sectional area located above the point where the stresses are computed, and t the width of the cross section at that point. Using either of the methods of analysis presented in Chap. 7, we can obtain the principal stresses at any point of the cross section (Fig. 8.6).

The following question now arises: Can the maximum normal stress σ_{max} at some point within the cross section be larger than the value of $\sigma_m = Mc/I$ computed at the surface of the beam? If it can, then the determination of the largest normal stress in the beam will involve a great deal more than the computation of $|M|_{max}$ and the use of Eq. (8.1). We can obtain an answer to this question by investigating the distribution of the principal stresses in a narrow

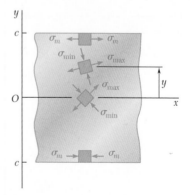

Fig. 8.6 Principal stresses at selected points of a beam.

Fig. 8.7 Narrow rectangular cantilever beam supporting a single concentrated load.

rectangular cantilever beam subjected to a concentrated load **P** at its free end (Fig. 8.7). We recall from Sec. 6.5 that the normal and shearing stresses at a distance x from the load **P** and a distance y above the neutral surface are given, respectively, by Eq. (6.13) and Eq. (6.12). Since the moment of inertia of the cross section is

$$I = \frac{bh^3}{12} = \frac{(bh)(2c)^2}{12} = \frac{Ac^2}{3}$$

where A is the cross-sectional area and c the half-depth of the beam, we write

$$\sigma_x = \frac{Pxy}{I} = \frac{Pxy}{\frac{1}{3}Ac^2} = 3\frac{P}{A}\frac{xy}{c^2} \tag{8.3}$$

and

$$\tau_{xy} = \frac{3}{2}\frac{P}{A}\left(1 - \frac{y^2}{c^2}\right) \tag{8.4}$$

Using the method of Sec. 7.3 or Sec. 7.4, the value of σ_{max} can be determined at any point of the beam. Figure 8.8 shows the results of the computation of the ratios σ_{max}/σ_m and σ_{min}/σ_m in two sections of the beam, corresponding respectively to $x = 2c$ and $x = 8c$. In

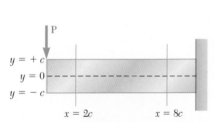

		$x = 2c$		$x = 8c$	
y/c	σ_{min}/σ_m		σ_{max}/σ_m	σ_{min}/σ_m	σ_{max}/σ_m
1.0	0		1.000	0	1.000
0.8	−0.010		0.810	−0.001	0.801
0.6	−0.040		0.640	−0.003	0.603
0.4	−0.090		0.490	−0.007	0.407
0.2	−0.160		0.360	−0.017	0.217
0	−0.250		0.250	−0.063	0.063
−0.2	−0.360		0.160	−0.217	0.017
−0.4	−0.490		0.090	−0.407	0.007
−0.6	−0.640		0.040	−0.603	0.003
−0.8	−0.810		0.010	−0.801	0.001
−1.0	−1.000		0	−1.000	0

Fig. 8.8 Distribution of principal stresses in two transverse sections of a rectangular cantilever beam supporting a single concentrated load.

each section, these ratios have been determined at 11 different points, and the orientation of the principal axes has been indicated at each point.†

It is clear that σ_{max} does not exceed σ_m in either of the two sections considered in Fig. 8.8 and that, if it does exceed σ_m elsewhere, it will be in sections close to the load **P**, where σ_m is small compared to τ_m.‡ But, for sections close to the load **P**, Saint-Venant's principle does not apply, Eqs. (8.3) and (8.4) cease to be valid, except in the very unlikely case of a load distributed parabolically over the end section (cf. Sec. 6.5), and more advanced methods of analysis taking into account the effect of stress concentrations should be used. We thus conclude that, for beams of rectangular cross section, and within the scope of the theory presented in this text, the maximum normal stress can be obtained from Eq. (8.1).

In Fig. 8.8 the directions of the principal axes were determined at 11 points in each of the two sections considered. If this analysis were extended to a larger number of sections and a larger number of points in each section, it would be possible to draw two orthogonal systems of curves on the side of the beam (Fig. 8.9). One system would consist of curves tangent to the principal axes corresponding to σ_{max} and the other of curves tangent to the principal axes corresponding to σ_{min}. The curves obtained in this manner are known as the *stress trajectories*. A trajectory of the first group (solid lines) defines at each of its points the direction of the largest tensile stress, while a trajectory of the second group (dashed lines) defines the direction of the largest compressive stress.§

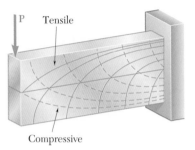

Fig. 8.9 Stress trajectories.

The conclusion we have reached for beams of rectangular cross section, that the maximum normal stress in the beam can be obtained from Eq. (8.1), remains valid for many beams of nonrectangular cross section. However, when the width of the cross section varies in such a way that large shearing stresses τ_{xy} will occur at points close to the surface of the beam, where σ_x is also large, a value of the principal stress σ_{max} larger than σ_m may result at such points. One should be particularly aware of this possibility when selecting W-beams or S-beams, and calculate the principal stress σ_{max} at the junctions b and d of the web with the flanges of the beam (Fig. 8.10). This is done by determining σ_x and τ_{xy} at that point from Eqs. (8.1) and (8.2), respectively, and using either of the methods of analysis of Chap. 7 to obtain σ_{max} (see Sample Prob. 8.1). An alternative procedure, used in design to select an acceptable section, consists of using for τ_{xy} the maximum value of the shearing stress in the section, $\tau_{max} = V/A_{web}$, given by Eq. (6.11) of Sec. 6.4. This leads to a slightly larger, and thus conservative, value of the principal stress σ_{max} at the junction of the web with the flanges of the beam (see Sample Prob. 8.2).

Fig. 8.10 Key stress analysis locations in I-shaped beams.

†See Prob. 8.C2, which refers to a program that can be written to obtain the results shown in Fig. 8.8.

‡As will be verified in Prob. 8.C2, σ_{max} exceeds σ_m if $x \leq 0.544c$.

§A brittle material, such as concrete, will fail in tension along planes that are perpendicular to the tensile-stress trajectories. Thus, to be effective, steel reinforcing bars should be placed so that they intersect these planes. On the other hand, stiffeners attached to the web of a plate girder will be effective in preventing buckling only if they intersect planes perpendicular to the compressive-stress trajectories.

*8.3 DESIGN OF TRANSMISSION SHAFTS

When we discussed the design of transmission shafts in Sec. 3.7, we considered only the stresses due to the torques exerted on the shafts. However, if the power is transferred to and from the shaft by means of gears or sprocket wheels (Fig. 8.11a), the forces exerted on the gear teeth or sprockets are equivalent to force-couple systems applied at the centers of the corresponding cross sections (Fig. 8.11b). This means that the shaft is subjected to a transverse loading, as well as to a torsional loading.

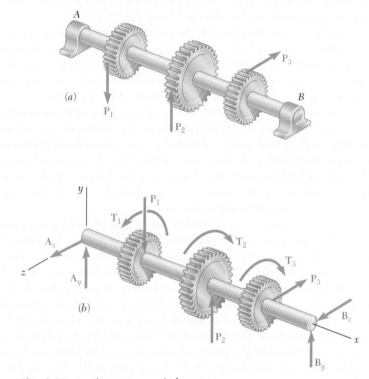

Fig. 8.11 Loadings on gear-shaft systems.

The shearing stresses produced in the shaft by the transverse loads are usually much smaller than those produced by the torques and will be neglected in this analysis.† The normal stresses due to the transverse loads, however, may be quite large and, as you will see presently, their contribution to the maximum shearing stress τ_{max} should be taken into account.

---

†For an application where the shearing stresses produced by the transverse loads must be considered, see Probs. 8.21 and 8.22.

Consider the cross section of the shaft at some point C. We represent the torque $\mathbf{T}$ and the bending couples $\mathbf{M}_y$ and $\mathbf{M}_z$ acting, respectively, in a horizontal and a vertical plane by the couple vectors shown (Fig. 8.12a). Since any diameter of the section is a principal axis of inertia for the section, we can replace $\mathbf{M}_y$ and $\mathbf{M}_z$ by their resultant $\mathbf{M}$ (Fig. 8.12b) in order to compute the normal stresses σ_x exerted on the section. We thus find that σ_x is maximum at the end of the diameter perpendicular to the vector representing $\mathbf{M}$ (Fig. 8.13). Recalling that the values of the normal stresses at that point are, respectively, $\sigma_m = Mc/I$ and zero, while the shearing stress is $\tau_m = Tc/J$, we plot the corresponding points X and Y on a Mohr-circle diagram (Fig. 8.14) and determine the value of the maximum shearing stress:

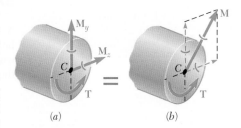

Fig. 8.12 Resultant loading on the cross section of a shaft.

$$\tau_{\max} = R = \sqrt{\left(\frac{\sigma_m}{2}\right)^2 + (\tau_m)^2} = \sqrt{\left(\frac{Mc}{2I}\right)^2 + \left(\frac{Tc}{J}\right)^2}$$

Recalling that, for a circular or annular cross section, $2I = J$, we write

Fig. 8.13 Maximum stress element.

$$\tau_{\max} = \frac{c}{J}\sqrt{M^2 + T^2} \qquad (8.5)$$

It follows that the minimum allowable value of the ratio J/c for the cross section of the shaft is

$$\frac{J}{c} = \frac{(\sqrt{M^2 + T^2})_{\max}}{\tau_{\text{all}}} \qquad (8.6)$$

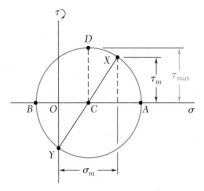

Fig. 8.14 Mohr's circle analysis.

where the numerator in the right-hand member of the expression obtained represents the maximum value of $\sqrt{M^2 + T^2}$ in the shaft, and τ_{all} the allowable shearing stress. Expressing the bending moment M in terms of its components in the two coordinate planes, we can also write

$$\frac{J}{c} = \frac{(\sqrt{M_y^2 + M_z^2 + T^2})_{\max}}{\tau_{\text{all}}} \qquad (8.7)$$

Equations (8.6) and (8.7) can be used to design both solid and hollow circular shafts and should be compared with Eq. (3.22) of Sec. 3.7, which was obtained under the assumption of a torsional loading only.

The determination of the maximum value of $\sqrt{M_y^2 + M_z^2 + T^2}$ will be facilitated if the bending-moment diagrams corresponding to M_y and M_z are drawn, as well as a third diagram representing the values of T along the shaft (see Sample Prob. 8.3).

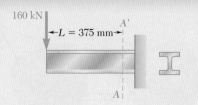

SAMPLE PROBLEM 8.1

A 160-kN force is applied as shown at the end of a W200 × 52 rolled-steel beam. Neglecting the effect of fillets and of stress concentrations, determine whether the normal stresses in the beam satisfy a design specification that they be equal to or less than 150 MPa at section A-A′.

SOLUTION

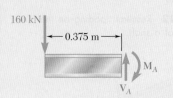

Shear and Bending Moment. At section A-A′, we have

$$M_A = (160 \text{ kN})(0.375 \text{ m}) = 60 \text{ kN} \cdot \text{m}$$
$$V_A = 160 \text{ kN}$$

Normal Stresses on Transverse Plane. Referring to the table of *Properties of Rolled-Steel Shapes* in Appendix C, we obtain the data shown and then determine the stresses σ_a and σ_b.

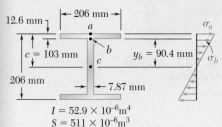

At point *a*:

$$\sigma_a = \frac{M_A}{S} = \frac{60 \text{ kN} \cdot \text{m}}{511 \times 10^{-6} \text{ m}^3} = 117.4 \text{ MPa}$$

At point *b*:

$$\sigma_b = \sigma_a \frac{y_b}{c} = (117.4 \text{ MPa})\frac{90.4 \text{ mm}}{103 \text{ mm}} = 103.0 \text{ MPa}$$

We note that all normal stresses on the transverse plane are less than 150 MPa.

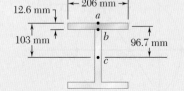

Shearing Stresses on Transverse Plane
At point *a*:

$$Q = 0 \qquad \tau_a = 0$$

At point *b*:

$$Q = (206 \times 12.6)(96.7) = 251.0 \times 10^3 \text{ mm}^3 = 251.0 \times 10^{-6} \text{ m}^3$$
$$\tau_b = \frac{V_A Q}{It} = \frac{(160 \text{ kN})(251.0 \times 10^{-6} \text{ m}^3)}{(52.9 \times 10^{-6} \text{ m}^4)(0.00787 \text{ m})} = 96.5 \text{ MPa}$$

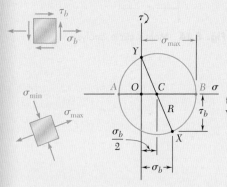

Principal Stress at Point *b*. The state of stress at point *b* consists of the normal stress $\sigma_b = 103.0$ MPa and the shearing stress $\tau_b = 96.5$ MPa. We draw Mohr's circle and find

$$\sigma_{max} = \frac{1}{2}\sigma_b + R = \frac{1}{2}\sigma_b + \sqrt{\left(\frac{1}{2}\sigma_b\right)^2 + \tau_b^2}$$
$$= \frac{103.0}{2} + \sqrt{\left(\frac{103.0}{2}\right)^2 + (96.5)^2}$$
$$\sigma_{max} = 160.9 \text{ MPa}$$

The specification, $\sigma_{max} \le 150$ MPa, is *not* satisfied ◄

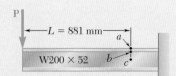

Comment. For this beam and loading, the principal stress at point *b* is 36% larger than the normal stress at point *a*. For $L \ge 881$ mm, the maximum normal stress would occur at point *a*.

8.19 Neglecting the effect of fillets and of stress concentrations, determine the smallest permissible diameters of the solid rods *BC* and *CD*. Use $\tau_{\text{all}} = 60$ MPa.

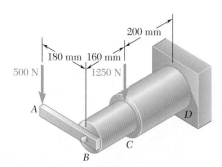

Fig. P8.19 and *P8.20*

8.20 Knowing that rods *BC* and *CD* are of diameter 24 mm and 36 mm, respectively, determine the maximum shearing stress in each rod. Neglect the effect of fillets and of stress concentrations.

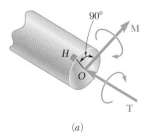

(*a*)

8.21 It was stated in Sec. 8.3 that the shearing stresses produced in a shaft by the transverse loads are usually much smaller than those produced by the torques. In the preceding problems their effect was ignored and it was assumed that the maximum shearing stress in a given section occurred at point *H* (Fig. P8.21*a*) and was equal to the expression obtained in Eq. (8.5), namely,

$$\tau_H = \frac{c}{J}\sqrt{M^2 + T^2}$$

Show that the maximum shearing stress at point *K* (Fig. P8.21*b*), where the effect of the shear *V* is greatest, can be expressed as

$$\tau_K = \frac{c}{J}\sqrt{(M\cos\beta)^2 + \left(\frac{2}{3}cV + T\right)^2}$$

where β is the angle between the vectors **V** and **M**. It is clear that the effect of the shear *V* cannot be ignored when $\tau_K \geq \tau_H$. (*Hint:* Only the component of **M** along **V** contributes to the shearing stress at *K*.)

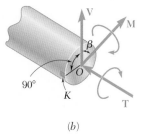

(*b*)

Fig. P8.21

8.22 Assuming that the magnitudes of the forces applied to disks *A* and *C* of Prob. 8.15 are, respectively, $P_1 = 4800$ N and $P_2 = 3600$ N, and using the expressions given in Prob. 8.21, determine the values of τ_H and τ_K in a section (*a*) just to the left of *B*, (*b*) just to the left of *C*.

8.23 The solid shafts *ABC* and *DEF* and the gears shown are used to transmit 15 kW from the motor *M* to a machine tool connected to shaft *DEF*. Knowing that the motor rotates at 240 rpm and that $\tau_{\text{all}} = 50$ MPa, determine the smallest permissible diameter of (*a*) shaft *ABC*, (*b*) shaft *DEF*.

8.24 Solve Prob. 8.23, assuming that the motor rotates at 360 rpm.

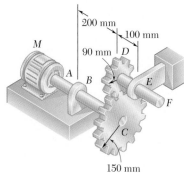

Fig. *P8.23*

8.25 The solid shaft AB rotates at 360 rpm and transmits 20 kW from the motor M to machine tools connected to gears E and F. Knowing that τ_{all} = 45 MPa and assuming that 10 kW is taken off at each gear, determine the smallest permissible diameter of shaft AB.

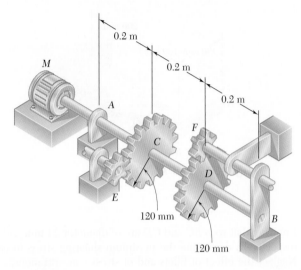

Fig. P8.25

8.26 Solve Prob. 8.25, assuming that the entire 20 kW is taken off at gear E.

8.27 The solid shaft ABC and the gears shown are used to transmit 10 kW from the motor M to a machine tool connected to gear D. Knowing that the motor rotates at 240 rpm and that τ_{all} = 60 MPa, determine the smallest permissible diameter of shaft ABC.

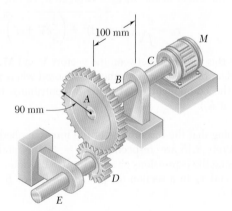

Fig. P8.27

8.28 Assuming that shaft ABC of Prob. 8.27 is hollow and has an outer diameter of 50 mm, determine the largest permissible inner diameter of the shaft.

PROBLEMS

8.31 Two 4-kN forces are applied to an L-shaped machine element AB as shown. Determine the normal and shearing stresses at (a) point a, (b) point b, (c) point c.

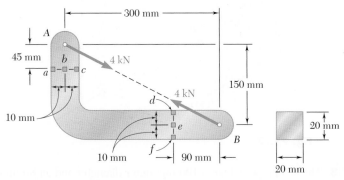
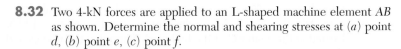
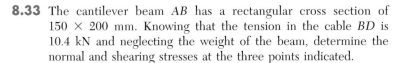
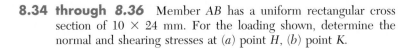

Fig. P8.31 and P8.32

8.32 Two 4-kN forces are applied to an L-shaped machine element AB as shown. Determine the normal and shearing stresses at (a) point d, (b) point e, (c) point f.

8.33 The cantilever beam AB has a rectangular cross section of 150 × 200 mm. Knowing that the tension in the cable BD is 10.4 kN and neglecting the weight of the beam, determine the normal and shearing stresses at the three points indicated.

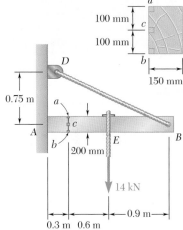

Fig. P8.33

8.34 through 8.36 Member AB has a uniform rectangular cross section of 10 × 24 mm. For the loading shown, determine the normal and shearing stresses at (a) point H, (b) point K.

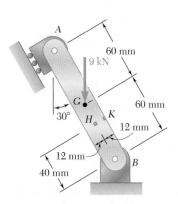

Fig. P8.34

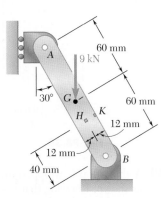

Fig. P8.35

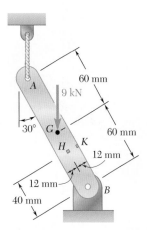

Fig. P8.36

8.37 Several forces are applied to the pipe assembly shown. Knowing that the pipe has inner and outer diameters equal to 40 mm and 48 mm, respectively, determine the normal and shearing stresses at (*a*) point *H*, (*b*) point *K*.

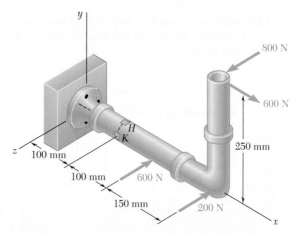

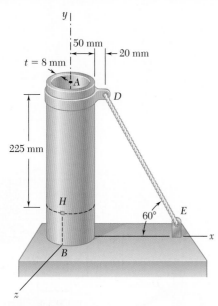

Fig. P8.38

Fig. P8.37

8.38 The steel pipe *AB* has a 100-mm outer diameter and an 8-mm wall thickness. Knowing that the tension in the cable is 40 kN, determine the normal and shearing stresses at point *H*.

8.39 The billboard shown weighs 36 kN and is supported by a structural tube that has a 375-mm outer diameter and a 12-mm wall thickness. At a time when the resultant of the wind pressure is 12 kN located at the center *C* of the billboard, determine the normal and shearing stresses at point *H*.

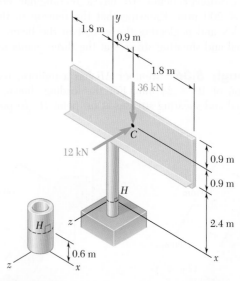

Fig. P8.39

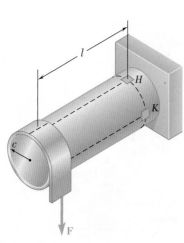

Fig. P8.40

8.40 A thin strap is wrapped around a solid rod of radius $c = 20$ mm as shown. Knowing that $l = 100$ mm and $F = 5$ kN, determine the normal and shearing stresses at (*a*) point *H*, (*b*) point *K*.

8.41 The axle of a small truck is acted upon by the forces and couple shown. Knowing that the diameter of the axle is 36 mm, determine the normal and shearing stresses at point H located on the top of the axle.

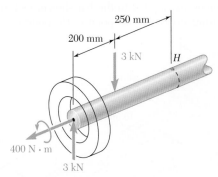

Fig. P8.41

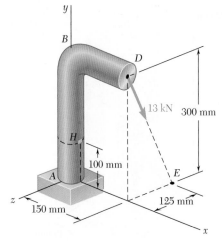

Fig. P8.42

8.42 A 13-kN force is applied as shown to the 60-mm-diameter cast-iron post ABD. At point H, determine (a) the principal stresses and principal planes, (b) the maximum shearing stress.

8.43 A 10-kN force and a 1.4-kN · m couple are applied at the top of the 65-mm-diameter brass post shown. Determine the principal stresses and maximum shearing stress at (a) point H, (b) point K.

8.44 Three forces are applied to a 100-mm-diameter plate that is attached to the solid 45-mm-diameter shaft AB. At point H, determine (a) the principal stresses and principal planes, (b) the maximum shearing stress.

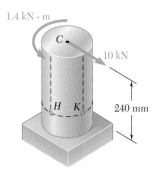

Fig. P8.43

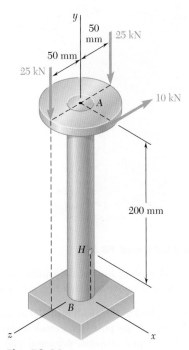

Fig. P8.44

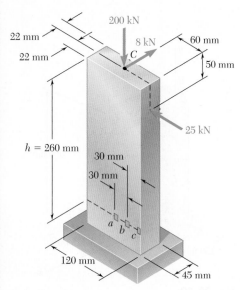

Fig. P8.45

8.45 Three forces are applied to the bar shown. Determine the normal and shearing stresses at (*a*) point *a*, (*b*) point *b*, (*c*) point *c*.

8.46 Solve Prob. 8.45, assuming that *h* = 300 mm.

8.47 Three forces are applied to the bar shown. Determine the normal and shearing stresses at (*a*) point *a*, (*b*) point *b*, (*c*) point *c*.

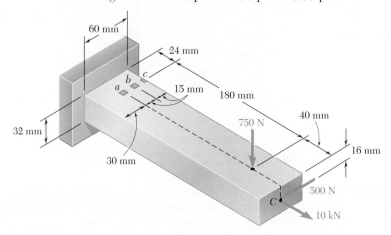

Fig. P8.47

8.48 Solve Prob. 8.47, assuming that the 750-N force is directed vertically upward.

8.49 For the post and loading shown, determine the principal stresses, principal planes, and maximum shearing stress at point *H*.

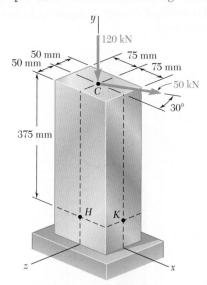

Fig. P8.49 and P8.50

8.50 For the post and loading shown, determine the principal stresses, principal planes, and maximum shearing stress at point *K*.

8.51 Two forces are applied to the small post *BD* as shown. Knowing that the vertical portion of the post has a cross section of 40 × 60 mm, determine the principal stresses, principal planes, and maximum shearing stress at point *H*.

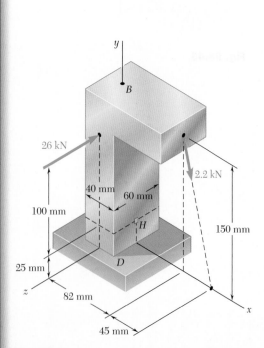

Fig. P8.51

8.52 Solve Prob. 8.51, assuming that the magnitude of the 26-kN force is reduced to 6.5 kN.

8.53 Three steel plates, each 13 mm thick, are welded together to form a cantilever beam. For the loading shown, determine the normal and shearing stresses at points a and b.

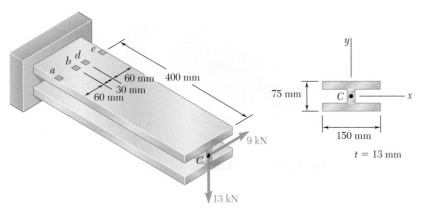

Fig. P8.53 and *P8.54*

8.54 Three steel plates, each 13 mm thick, are welded together to form a cantilever beam. For the loading shown, determine the normal and shearing stresses at points d and e.

8.55 Four forces are applied to a W200 × 41.7 rolled beam as shown. Determine the principal stresses and maximum shearing stress at point a.

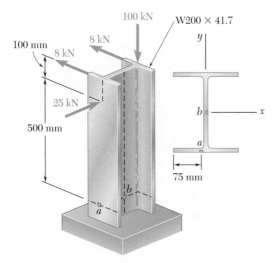

Fig. P8.55 and *P8.56*

8.56 Four forces are applied to a W200 × 41.7 rolled beam as shown. Determine the principal stresses and maximum shearing stress at point b.

8.57 Two forces $\mathbf{P}_1$ and $\mathbf{P}_2$ are applied as shown in directions perpendicular to the longitudinal axis of a W310 × 60 beam. Knowing that $\mathbf{P}_1 = 25$ kN and $\mathbf{P}_2 = 24$ kN, determine the principal stresses and the maximum shearing stress at point a.

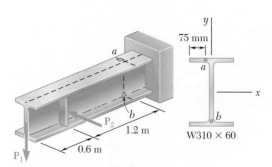

Fig. P8.57 and P8.58

8.58 Two forces $\mathbf{P}_1$ and $\mathbf{P}_2$ are applied as shown in directions perpendicular to the longitudinal axis of a W310 × 60 beam. Knowing that $\mathbf{P}_1 = 25$ kN and $\mathbf{P}_2 = 24$ kN, determine the principal stresses and the maximum shearing stress at point b.

8.59 A vertical force $\mathbf{P}$ is applied at the center of the free end of cantilever beam AB. (*a*) If the beam is installed with the web vertical ($\beta = 0$) and with its longitudinal axis AB horizontal, determine the magnitude of the force $\mathbf{P}$ for which the normal stress at point a is +120 MPa. (*b*) Solve part a, assuming that the beam is installed with $\beta = 3°$.

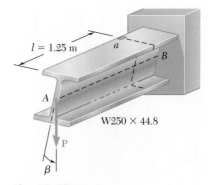

Fig. P8.59

Fig. P8.60

8.60 A force $\mathbf{P}$ is applied to a cantilever beam by means of a cable attached to a bolt located at the center of the free end of the beam. Knowing that $\mathbf{P}$ acts in a direction perpendicular to the longitudinal axis of the beam, determine (*a*) the normal stress at point a in terms of P, b, h, l, and β, (*b*) the values of β for which the normal stress at a is zero.

*8.61 A 5-kN force **P** is applied to a wire that is wrapped around bar AB as shown. Knowing that the cross section of the bar is a square of side $d = 40$ mm, determine the principal stresses and the maximum shearing stress at point a.

Fig. P8.61

*8.62 Knowing that the structural tube shown has a uniform wall thickness of 8 mm, determine the principal stresses, principal planes, and maximum shearing stress at (*a*) point H, (*b*) point K.

*8.63 The structural tube shown has a uniform wall thickness of 8 mm. Knowing that the 60-kN load is applied 4 mm above the base of the tube, determine the shearing stress at (*a*) point a, (*b*) point b.

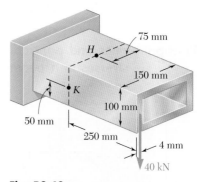

Fig. P8.62

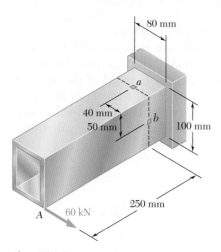

Fig. P8.63

*8.64 For the tube and loading of Prob. 8.63, determine the principal stresses and the maximum shearing stress at point b.

REVIEW AND SUMMARY

$$\sigma_{xx'} = \left(\frac{\sigma_{xx} + \sigma_{yy}}{2}\right)$$

This chapter was devoted to the determination of the principal stresses in beams, transmission shafts, and bodies of arbitrary shape subjected to combined loadings.

We first recalled in Sec. 8.2 the two fundamental relations derived in Chaps. 5 and 6 for the normal stress σ_x and the shearing stress τ_{xy} at any given point of a cross section of a prismatic beam,

$$\sigma_x = -\frac{My}{I} \qquad \tau_{xy} = -\frac{VQ}{It} \qquad (8.1, \ 8.2)$$

where V = shear in the section
$\quad M$ = bending moment in the section
$\quad y$ = distance of the point from the neutral surface
$\quad I$ = centroidal moment of inertia of the cross section
$\quad Q$ = first moment about the neutral axis of the portion of the cross section located above the given point
$\quad t$ = width of the cross section at the given point

Principal planes and principal stresses in a beam

Using one of the methods presented in Chap. 7 for the transformation of stresses, we were able to obtain the principal planes and principal stresses at the given point (Fig. 8.28).

We investigated the distribution of the principal stresses in a narrow rectangular cantilever beam subjected to a concentrated load **P** at its free end and found that in any given transverse section—except close to the point of application of the load—the maximum principal stress $\sigma_{\max}$ did not exceed the maximum normal stress σ_m occurring at the surface of the beam.

While this conclusion remains valid for many beams of non-rectangular cross section, it may not hold for W-beams or S-beams, where $\sigma_{\max}$ at the junctions b and d of the web with the flanges of the beam (Fig. 8.29) may exceed the value of σ_m occurring at points a and e. Therefore, the design of a rolled-steel beam should include the computation of the maximum principal stress at these points. (See Sample Probs. 8.1 and 8.2.)

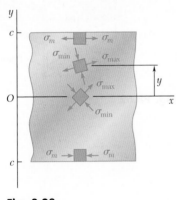

Fig. 8.28

Fig. 8.29

In Sec. 8.3, we considered the design of *transmission shafts* subjected to *transverse loads* as well as to torques. Taking into account the effect of both the normal stresses due to the bending moment M and the shearing stresses due to the torque T in any given transverse section of a cylindrical shaft (either solid or hollow), we found that the minimum allowable value of the ratio J/c for the cross section was

$$\frac{J}{c} = \frac{\left(\sqrt{M^2 + T^2}\right)_{\text{max}}}{\tau_{\text{all}}} \qquad (8.6)$$

In preceding chapters, you learned to determine the stresses in prismatic members caused by axial loadings (Chaps. 1 and 2), torsion (Chap. 3), bending (Chap. 4), and transverse loadings (Chaps. 5 and 6). In the second part of this chapter (Sec. 8.4), we combined this knowledge to determine stresses under more general loading conditions.

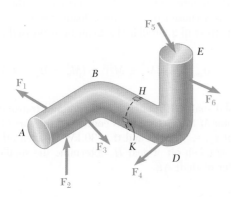

Fig. 8.30

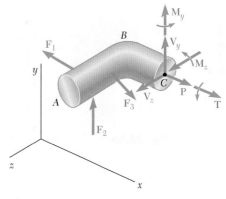

Fig. 8.31

For instance, to determine the stresses at point H or K of the bent member shown in Fig. 8.30, we passed a section through these points and replaced the applied loads by an equivalent force-couple system at the centroid C of the section (Fig. 8.31). The normal and shearing stresses produced at H or K by each of the forces and couples applied at C were determined and then combined to obtain the resulting normal stress σ_x and the resulting shearing stresses τ_{xy} and τ_{xz} at H or K. Finally, the principal stresses, the orientation of the principal planes, and the maximum shearing stress at point H or K were determined by one of the methods presented in Chap. 7 from the values obtained for σ_x, τ_{xy}, and τ_{xz}.

REVIEW PROBLEMS

8.65 (a) Knowing that $\sigma_{all} = 165$ MPa and $\tau_{all} = 100$ MPa, select the most economical wide-flange shape that should be used to support the loading shown. (b) Determine the values to be expected for σ_m, τ_m, and the principal stress σ_{max} at the junction of a flange and the web of the selected beam.

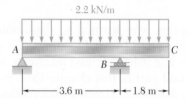

Fig. P8.65

8.66 Determine the smallest allowable diameter of the solid shaft *ABCD*, knowing that $\tau_{all} = 60$ MPa and that the radius of disk *B* is $r = 80$ mm.

8.67 Using the notation of Sec. 8.3 and neglecting the effect of shearing stresses caused by transverse loads, show that the maximum normal stress in a circular shaft can be expressed as follows:

$$\sigma_{max} = \frac{c}{J}[(M_y^2 + M_z^2)^{\frac{1}{2}} + (M_y^2 + M_z^2 + T^2)^{\frac{1}{2}}]_{max}$$

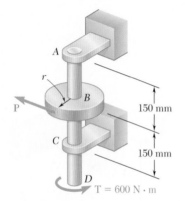

Fig. P8.66

8.68 The solid shaft *AE* rotates at 600 rpm and transmits 45 kW from the motor *M* to machine tools connected to gears *G* and *H*. Knowing that $\tau_{all} = 55$ MPa and that 30 kW is taken off at gear *G* and 15 kW is taken off at gear *H*, determine the smallest permissible diameter of shaft *AE*.

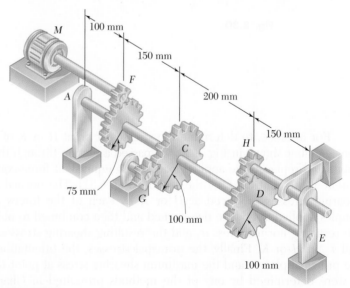

Fig. P8.68

8.69 For the bracket and loading shown, determine the normal and shearing stresses at (*a*) point *a*, (*b*) point *b*.

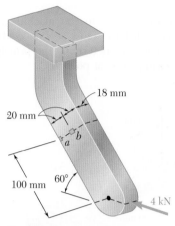

Fig. P8.69

8.70 Two forces are applied to the pipe *AB* as shown. Knowing that the pipe has inner and outer diameters equal to 35 and 42 mm, respectively, determine the normal and shearing stresses at (*a*) point *a*, (*b*) point *b*.

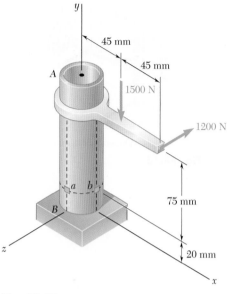

Fig. P8.70

8.71 A close-coiled spring is made of a circular wire of radius *r* that is formed into a helix of radius *R*. Determine the maximum shearing stress produced by the two equal and opposite forces **P** and **P′**. (*Hint:* First determine the shear **V** and the torque **T** in a transverse cross section.)

8.72 The steel pipe *AB* has a 72-mm outer diameter and a 5-mm wall thickness. Knowing that the arm *CDE* is rigidly attached to the pipe, determine the principal stresses, principal planes, and the maximum shearing stress at point *H*.

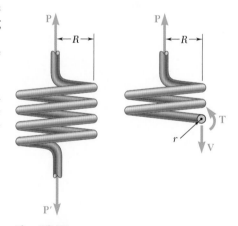

Fig. P8.71

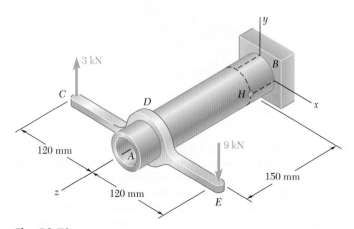

Fig. P8.72

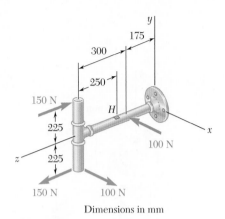

Fig. P8.73

8.73 Several forces are applied to the pipe assembly shown. Knowing that each section of pipe has inner and outer diameters equal to 36 and 42 mm, respectively, determine the normal and shearing stresses at point H located at the top of the outer surface of the pipe.

8.74 Three forces are applied to the machine component ABD as shown. Knowing that the cross section containing point H is a 20×40-mm rectangle, determine the principal stresses and the maximum shearing stress at point H.

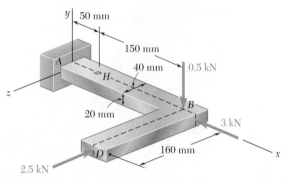

Fig. P8.74

8.75 A vertical force $\mathbf{P}$ of magnitude 250 N is applied to the crank at point A. Knowing that the shaft BDE has a diameter of 18 mm, determine the principal stresses and the maximum shearing stress at point H located at the top of the shaft, 50 mm to the right of support D.

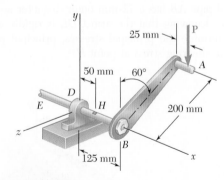

Fig. P8.75

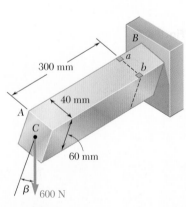

Fig. P8.76

8.76 The cantilever beam AB will be installed so that the 60-mm side forms an angle β between 0 and 90° with the vertical. Knowing that the 600-N vertical force is applied at the center of the free end of the beam, determine the normal stress at point a when (a) $\beta = 0$, (b) $\beta = 90°$. (c) Also, determine the value of β for which the normal stress at point a is a maximum and the corresponding value of that stress.

COMPUTER PROBLEMS

The following problems are designed to be solved with a computer.

8.C1 Let us assume that the shear V and the bending moment M have been determined in a given section of a rolled-steel beam. Write a computer program to calculate in that section, from the data available in Appendix C, (a) the maximum normal stress σ_m, (b) the principal stress σ_{max} at the junction of a flange and the web. Use this program to solve parts a and b of the following problems:
(1) Prob. 8.1 (Use $V = 180$ kN and $M = 45$ kN $\cdot$ m)
(2) Prob. 8.2 (Use $V = 90$ kN and $M = 45$ kN $\cdot$ m)
(3) Prob. 8.3 (Use $V = 700$ kN and $M = 1750$ kN $\cdot$ m)
(4) Prob. 8.4 (Use $V = 850$ kN and $M = 1700$ kN $\cdot$ m)

8.C2 A cantilever beam AB with a rectangular cross section of width b and depth $2c$ supports a single concentrated load $\mathbf{P}$ at its end A. Write a computer program to calculate, for any values of x/c and y/c, (a) the ratios σ_{max}/σ_m and σ_{min}/σ_m, where σ_{max} and σ_{min} are the principal stresses at point $K(x, y)$ and σ_m the maximum normal stress in the same transverse section, (b) the angle θ_p that the principal planes at K form with a transverse and a horizontal plane through K. Use this program to check the values shown in Fig. 8.8 and to verify that σ_{max} exceeds σ_m if $x \leq 0.544c$, as indicated in the second footnote on page 539.

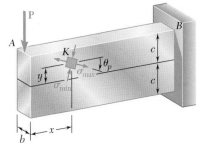

Fig. P8.C2

8.C3 Disks $D_1, D_2, \ldots, D_n$ are attached as shown in Fig. 8.C3 to the solid shaft AB of length L, uniform diameter d, and allowable shearing stress τ_{all}. Forces $\mathbf{P}_1, \mathbf{P}_2, \ldots, \mathbf{P}_n$ of known magnitude (except for one of them) are applied to the disks, either at the top or bottom of their vertical diameter, or at the left or right end of their horizontal diameter. Denoting by r_i the radius of disk D_i and by c_i its distance from the support at A, write a computer program to calculate (a) the magnitude of the unknown force $\mathbf{P}_i$, (b) the smallest permissible value of the diameter d of shaft AB. Use this program to solve Prob. 8.18.

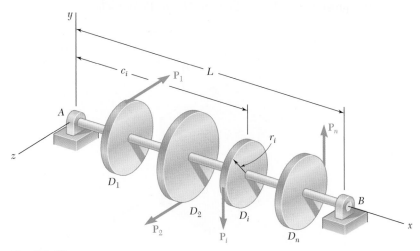

Fig. P8.C3

8.C4 The solid shaft AB of length L, uniform diameter d, and allowable shearing stress τ_{all} rotates at a given speed expressed in rpm (Fig. 8.C4). Gears $G_1, G_2, \ldots, G_n$ are attached to the shaft and each of these gears meshes with another gear (not shown), either at the top or bottom of its vertical diameter, or at the left or right end of its horizontal diameter. One of these gears is connected to a motor and the rest of them to various machine tools. Denoting by r_i the radius of disk G_i, by c_i its distance from the support at A, and by P_i the power transmitted to that gear (+ sign) or taken off that gear (− sign), write a computer program to calculate the smallest permissible value of the diameter d of shaft AB. Use this program to solve Probs. 8.27 and 8.29.

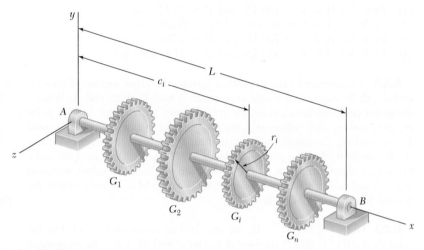

Fig. P8.C4

8.C5 Write a computer program that can be used to calculate the normal and shearing stresses at points with given coordinates y and z located on the surface of a machine part having a rectangular cross section. The internal forces are known to be equivalent to the force-couple system shown. Use this program to solve (a) Prob. 8.45b, (b) Prob. 8.47a.

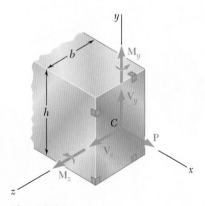

Fig. P8.C5

8.C6 Member *AB* has a rectangular cross section of 10 × 24 mm. For the loading shown, write a computer program that can be used to determine the normal and shearing stresses at points *H* and *K* for values of *d* from 0 to 120 mm, using 15-mm increments. Use this program to solve Prob. 8.35.

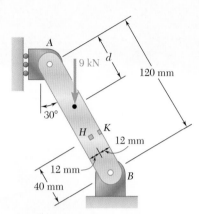

Fig. P8.C6

*8.C7 The structural tube shown has a uniform wall thickness of 8 mm. A 40-kN force is applied at a bar (not shown) that is welded to the end of the tube. Write a computer program that can be used to determine, for any given value of *c*, the principal stresses, principal planes, and maximum shearing stress at point *H* for values of *d* from −75 mm to 75 mm, using 25-mm increments. Use this program to solve Prob. 8.62*a*.

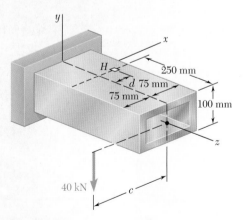

Fig. P8.C7

The photo shows a multiple-girder bridge during construction. The design of the steel girders is based on both strength considerations and deflection evaluations.

CHAPTER

9

Deflection of Beams

9.1 INTRODUCTION

In the preceding chapter we learned to design beams for strength. In this chapter we will be concerned with another aspect in the design of beams, namely, the determination of the *deflection*. Of particular interest is the determination of the *maximum deflection* of a beam under a given loading, since the design specifications of a beam will generally include a maximum allowable value for its deflection. Also of interest is that a knowledge of the deflections is required to analyze *indeterminate beams*. These are beams in which the number of reactions at the supports exceeds the number of equilibrium equations available to determine these unknowns.

We saw in Sec. 4.4 that a prismatic beam subjected to pure bending is bent into an arc of circle and that, within the elastic range, the curvature of the neutral surface can be expressed as

$$\frac{1}{\rho} = \frac{M}{EI} \qquad (4.21)$$

where M is the bending moment, E the modulus of elasticity, and I the moment of inertia of the cross section about its neutral axis.

When a beam is subjected to a transverse loading, Eq. (4.21) remains valid for any given transverse section, provided that Saint-Venant's principle applies. However, both the bending moment and the curvature of the neutral surface will vary from section to section. Denoting by x the distance of the section from the left end of the beam, we write

$$\frac{1}{\rho} = \frac{M(x)}{EI} \qquad (9.1)$$

The knowledge of the curvature at various points of the beam will enable us to draw some general conclusions regarding the deformation of the beam under loading (Sec. 9.2).

To determine the slope and deflection of the beam at any given point, we first derive the following second-order linear differential equation, which governs the *elastic curve* characterizing the shape of the deformed beam (Sec. 9.3):

$$\frac{d^2y}{dx^2} = \frac{M(x)}{EI}$$

If the bending moment can be represented for all values of x by a single function $M(x)$, as in the case of the beams and loadings shown in Fig. 9.1, the slope $\theta = dy/dx$ and the deflection y at any point of the beam may be obtained through two successive integrations. The two constants of integration introduced in the process will be determined from the boundary conditions indicated in the figure.

However, if different analytical functions are required to represent the bending moment in various portions of the beam, different differential equations will also be required, leading to

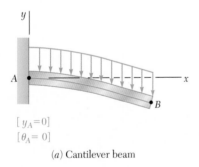

$[y_A=0]$
$[\theta_A=0]$

(a) Cantilever beam

$[y_A=0]$ $[y_B=0]$

(b) Simply supported beam

Fig. 9.1 Situations where bending moment can be given by a single function $M(x)$.

different functions defining the elastic curve in the various portions of the beam. In the case of the beam and loading of Fig. 9.2, for example, two differential equations are required, one for the portion of beam AD and the other for the portion DB. The first equation yields the functions θ_1 and y_1, and the second the functions θ_2 and y_2. Altogether, four constants of integration must be determined; two will be obtained by writing that the deflection is zero at A and B, and the other two by expressing that the portions of beam AD and DB have the same slope and the same deflection at D.

You will observe in Sec. 9.4 that in the case of a beam supporting a distributed load $w(x)$, the elastic curve can be obtained directly from $w(x)$ through four successive integrations. The constants introduced in this process will be determined from the boundary values of V, M, θ, and y.

In Sec. 9.5, we will discuss *statically indeterminate beams* where the reactions at the supports involve four or more unknowns. The three equilibrium equations must be supplemented with equations obtained from the boundary conditions imposed by the supports.

The method described earlier for the determination of the elastic curve when several functions are required to represent the bending moment M can be quite laborious, since it requires matching slopes and deflections at every transition point. You will see in Sec. 9.6 that the use of *singularity functions* (previously discussed in Sec. 5.5) considerably simplifies the determination of θ and y at any point of the beam.

The next part of the chapter (Secs. 9.7 and 9.8) is devoted to the *method of superposition*, which consists of determining separately, and then adding, the slope and deflection caused by the various loads applied to a beam. This procedure can be facilitated by the use of the table in Appendix D, which gives the slopes and deflections of beams for various loadings and types of support.

In Sec. 9.9, certain geometric properties of the elastic curve will be used to determine the deflection and slope of a beam at a given point. Instead of expressing the bending moment as a function $M(x)$ and integrating this function analytically, the diagram representing the variation of M/EI over the length of the beam will be drawn and two moment-area theorems will be derived. The *first moment-area theorem* will enable us to calculate the angle between the tangents to the beam at two points; the *second moment-area theorem* will be used to calculate the vertical distance from a point on the beam to a tangent at a second point.

The moment-area theorems will be used in Sec. 9.10 to determine the slope and deflection at selected points of cantilever beams and beams with symmetric loadings. In Sec. 9.11 you will find that in many cases the areas and moments of areas defined by the M/EI diagram may be more easily determined if you draw the *bending-moment diagram by parts*. As you study the moment-area method, you will observe that this method is particularly effective in the case of *beams of variable cross section*.

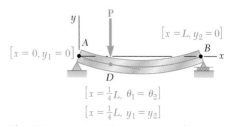

$[x = 0, y_1 = 0]$

$[x = L, y_2 = 0]$

$\left[x = \tfrac{1}{4}L,\ \theta_1 = \theta_2\right]$

$\left[x = \tfrac{1}{4}L,\ y_1 = y_2\right]$

Fig. 9.2 Situation where two sets of equations are required.

Beams with unsymmetric loadings and overhanging beams will be considered in Sec. 9.12. Since for an unsymmetric loading the maximum deflection does not occur at the center of a beam, you will learn in Sec. 9.13 how to locate the point where the tangent is horizontal in order to determine the *maximum deflection*. Section 9.14 will be devoted to the solution of problems involving *statically indeterminate beams*.

9.2 DEFORMATION OF A BEAM UNDER TRANSVERSE LOADING

At the beginning of this chapter, we recalled Eq. (4.21) of Sec. 4.4, which relates the curvature of the neutral surface and the bending moment in a beam in pure bending. We pointed out that this equation remains valid for any given transverse section of a beam subjected to a transverse loading, provided that Saint-Venant's principle applies. However, both the bending moment and the curvature of the neutral surface will vary from section to section. Denoting by x the distance of the section from the left end of the beam, we write

$$\frac{1}{\rho} = \frac{M(x)}{EI} \tag{9.1}$$

Consider, for example, a cantilever beam AB of length L subjected to a concentrated load **P** at its free end A (Fig. 9.3a). We have $M(x) = -Px$ and, substituting into (9.1),

$$\frac{1}{\rho} = -\frac{Px}{EI}$$

which shows that the curvature of the neutral surface varies linearly with x, from zero at A, where ρ_A itself is infinite, to $-PL/EI$ at B, where $|\rho_B| = EI/PL$ (Fig. 9.3b).

Consider now the overhanging beam AD of Fig. 9.4a that supports two concentrated loads as shown. From the free-body diagram of the beam (Fig. 9.4b), we find that the reactions at the supports are $R_A = 1$ kN and $R_C = 5$ kN, respectively, and draw the corresponding bending-moment diagram (Fig. 9.5a). We note from the diagram that M, and thus the curvature of the beam, are both zero at each end of the beam, and also at a point E located at $x = 4$ m. Between A and E the bending moment is positive and the beam is concave upward;

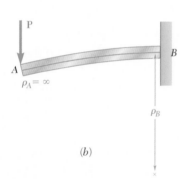

(a)

(b)

Fig. 9.3 Cantilever beam with concentrated load.

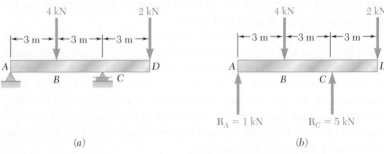

(a)

(b)

Fig. 9.4 Overhanging beam with two concentrated loads.

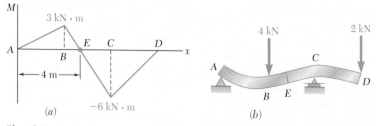

(a)

Fig. 9.5 Moment-curvature relationship for beam of Fig. 9.4.

between E and D the bending moment is negative and the beam is concave downward (Fig. 9.5b). We also note that the largest value of the curvature (i.e., the smallest value of the radius of curvature) occurs at the support C, where $|M|$ is maximum.

From the information obtained on its curvature, we get a fairly good idea of the shape of the deformed beam. However, the analysis and design of a beam usually require more precise information on the *deflection* and the *slope* of the beam at various points. Of particular importance is the knowledge of the *maximum deflection* of the beam. In the next section Eq. (9.1) will be used to obtain a relation between the deflection y measured at a given point Q on the axis of the beam and the distance x of that point from some fixed origin (Fig. 9.6). The relation obtained is the equation of the *elastic curve*, i.e., the equation of the curve into which the axis of the beam is transformed under the given loading (Fig. 9.6b).†

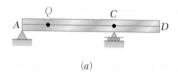

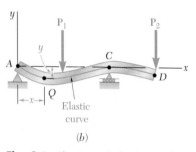

Fig. 9.6 Elastic curve for beam of Fig. 9.4.

9.3 EQUATION OF THE ELASTIC CURVE

We first recall from elementary calculus that the curvature of a plane curve at a point $Q(x,y)$ of the curve can be expressed as

$$\frac{1}{\rho} = \frac{\dfrac{d^2y}{dx^2}}{\left[1 + \left(\dfrac{dy}{dx}\right)^2\right]^{3/2}} \tag{9.2}$$

where dy/dx and d^2y/dx^2 are the first and second derivatives of the function $y(x)$ represented by that curve. But, in the case of the elastic curve of a beam, the slope dy/dx is very small, and its square is negligible compared to unity. We write, therefore,

$$\frac{1}{\rho} = \frac{d^2y}{dx^2} \tag{9.3}$$

Substituting for $1/\rho$ from (9.3) into (9.1), we have

$$\frac{d^2y}{dx^2} = \frac{M(x)}{EI} \tag{9.4}$$

---

†It should be noted that, in this chapter, y represents a vertical displacement, while it was used in previous chapters to represent the distance of a given point in a transverse section from the neutral axis of that section.

The equation obtained is a second-order linear differential equation; it is the governing differential equation for the elastic curve.

The product EI is known as the *flexural rigidity* and, if it varies along the beam, as in the case of a beam of varying depth, we must express it as a function of x before proceeding to integrate Eq. (9.4). However, in the case of a prismatic beam, which is the case considered here, the flexural rigidity is constant. We may thus multiply both members of Eq. (8.4) by EI and integrate in x. We write

$$EI\frac{dy}{dx} = \int_0^x M(x)\, dx + C_1 \tag{9.5}$$

where C_1 is a constant of integration. Denoting by $\theta(x)$ the angle, measured in radians, that the tangent to the elastic curve at Q forms with the horizontal (Fig. 9.7), and recalling that this angle is very small, we have

$$\frac{dy}{dx} = \tan\theta \simeq \theta(x)$$

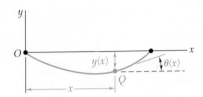

Fig. 9.7 Slope $\theta(x)$ of tangent to the elastic curve.

Thus, we write Eq. (9.5) in the alternative form

$$EI\,\theta(x) = \int_0^x M(x)\, dx + C_1 \tag{9.5'}$$

Integrating both members of Eq. (9.5) in x, we have

$$EI\, y = \int_0^x \left[\int_0^x M(x)\, dx + C_1 \right] dx + C_2$$

$$EI\, y = \int_0^x dx \int_0^x M(x)\, dx + C_1 x + C_2 \tag{9.6}$$

where C_2 is a second constant, and where the first term in the right-hand member represents the function of x obtained by integrating twice in x the bending moment $M(x)$. If it were not for the fact that the constants C_1 and C_2 are as yet undetermined, Eq. (9.6) would define the deflection of the beam at any given point Q, and Eq. (9.5) or (9.5') would similarly define the slope of the beam at Q.

The constants C_1 and C_2 are determined from the *boundary conditions* or, more precisely, from the conditions imposed on the beam by its supports. Limiting our analysis in this section to *statically determinate beams*, i.e., to beams supported in such a way that the reactions at the supports can be obtained by the methods of statics, we note that only three types of beams need to be considered here (Fig. 9.8): (*a*) the *simply supported beam*, (*b*) the *overhanging beam*, and (*c*) the *cantilever beam*.

In the first two cases, the supports consist of a pin and bracket at A and of a roller at B, and require that the deflection be zero at each of these points. Letting first $x = x_A$, $y = y_A = 0$ in Eq. (9.6), and then $x = x_B$, $y = y_B = 0$ in the same equation, we obtain two equations that can be solved for C_1 and C_2. In the case of the cantilever beam (Fig. 9.8*c*), we note that both the deflection and the slope at A must be zero. Letting $x = x_A$, $y = y_A = 0$ in Eq. (9.6), and $x = x_A$, $\theta = \theta_A = 0$ in Eq. (9.5'), we obtain again two equations that can be solved for C_1 and C_2.

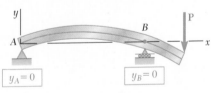

(*a*) Simply supported beam

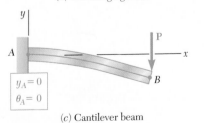

(*b*) Overhanging beam

(*c*) Cantilever beam

Fig. 9.8 Boundary conditions for statically determinate beams.

The cantilever beam AB is of uniform cross section and carries a load $\mathbf{P}$ at its free end A (Fig. 9.9). Determine the equation of the elastic curve and the deflection and slope at A.

EXAMPLE 9.01

Fig. 9.9   **Fig. 9.10**

Using the free-body diagram of the portion AC of the beam (Fig. 9.10), where C is located at a distance x from end A, we find

$$M = -Px \qquad (9.7)$$

Substituting for M into Eq. (9.4) and multiplying both members by the constant EI, we write

$$EI \frac{d^2y}{dx^2} = -Px$$

Integrating in x, we obtain

$$EI \frac{dy}{dx} = -\tfrac{1}{2}Px^2 + C_1 \qquad (9.8)$$

We now observe that at the fixed end B we have $x = L$ and $\theta = dy/dx = 0$ (Fig. 9.11). Substituting these values into (9.8) and solving for C_1, we have

$$C_1 = \tfrac{1}{2}PL^2$$

which we carry back into (9.8):

$$EI \frac{dy}{dx} = -\tfrac{1}{2}Px^2 + \tfrac{1}{2}PL^2 \qquad (9.9)$$

Integrating both members of Eq. (9.9), we write

$$EI\, y = -\tfrac{1}{6}Px^3 + \tfrac{1}{2}PL^2 x + C_2 \qquad (9.10)$$

But, at B we have $x = L$, $y = 0$. Substituting into (9.10), we have

$$0 = -\tfrac{1}{6}PL^3 + \tfrac{1}{2}PL^3 + C_2$$
$$C_2 = -\tfrac{1}{3}PL^3$$

Carrying the value of C_2 back into Eq. (9.10), we obtain the equation of the elastic curve:

$$EI\, y = -\tfrac{1}{6}Px^3 + \tfrac{1}{2}PL^2 x - \tfrac{1}{3}PL^3$$

or

$$y = \frac{P}{6EI}(-x^3 + 3L^2 x - 2L^3) \qquad (9.11)$$

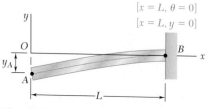

$$[x = L, \theta = 0]$$
$$[x = L, y = 0]$$

Fig. 9.11

The deflection and slope at A are obtained by letting $x = 0$ in Eqs. (9.11) and (9.9). We find

$$y_A = -\frac{PL^3}{3EI} \quad \text{and} \quad \theta_A = \left(\frac{dy}{dx}\right)_A = \frac{PL^2}{2EI}$$

EXAMPLE 9.02

The simply supported prismatic beam AB carries a uniformly distributed load w per unit length (Fig. 9.12). Determine the equation of the elastic curve and the maximum deflection of the beam.

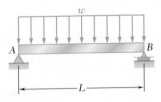

Fig. 9.12

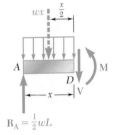

Fig. 9.13

Drawing the free-body diagram of the portion AD of the beam (Fig. 9.13) and taking moments about D, we find that

$$M = \tfrac{1}{2}wLx - \tfrac{1}{2}wx^2 \tag{9.12}$$

Substituting for M into Eq. (9.4) and multiplying both members of this equation by the constant EI, we write

$$EI\frac{d^2y}{dx^2} = -\frac{1}{2}wx^2 + \frac{1}{2}wLx \tag{9.13}$$

Integrating twice in x, we have

$$EI\frac{dy}{dx} = -\frac{1}{6}wx^3 + \frac{1}{4}wLx^2 + C_1 \tag{9.14}$$

$$EI\,y = -\frac{1}{24}wx^4 + \frac{1}{12}wLx^3 + C_1x + C_2 \tag{9.15}$$

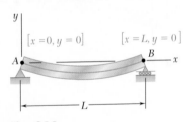

Fig. 9.14

Observing that $y = 0$ at both ends of the beam (Fig. 9.14), we first let $x = 0$ and $y = 0$ in Eq. (9.15) and obtain $C_2 = 0$. We then make $x = L$ and $y = 0$ in the same equation and write

$$0 = -\tfrac{1}{24}wL^4 + \tfrac{1}{12}wL^4 + C_1L$$
$$C_1 = -\tfrac{1}{24}wL^3$$

Carrying the values of C_1 and C_2 back into Eq. (9.15), we obtain the equation of the elastic curve:

$$EI\,y = -\tfrac{1}{24}wx^4 + \tfrac{1}{12}wLx^3 - \tfrac{1}{24}wL^3x$$

or

$$y = \frac{w}{24EI}(-x^4 + 2Lx^3 - L^3x) \tag{9.16}$$

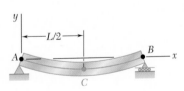

Fig. 9.15

Substituting into Eq. (9.14) the value obtained for C_1, we check that the slope of the beam is zero for $x = L/2$ and that the elastic curve has a minimum at the midpoint C of the beam (Fig. 9.15). Letting $x = L/2$ in Eq. (9.16), we have

$$y_C = \frac{w}{24EI}\left(-\frac{L^4}{16} + 2L\frac{L^3}{8} - L^3\frac{L}{2}\right) = -\frac{5wL^4}{384EI}$$

The maximum deflection or, more precisely, the maximum absolute value of the deflection, is thus

$$|y|_{max} = \frac{5wL^4}{384EI}$$

In each of the two examples considered so far, only one free-body diagram was required to determine the bending moment in the beam. As a result, a single function of x was used to represent M throughout the beam. This, however, is not generally the case. Concentrated loads, reactions at supports, or discontinuities in a distributed load will make it necessary to divide the beam into several portions, and to represent the bending moment by a different function $M(x)$ in each of these portions of beam (Photo 9.1). Each of the functions $M(x)$ will then lead to a different expression for the slope $\theta(x)$ and for the deflection $y(x)$. Since each of the expressions obtained for the deflection must contain two constants of integration, a large number of constants will have to be determined. As you will see in the next example, the required additional boundary conditions can be obtained by observing that, while the shear and bending moment can be discontinuous at several points in a beam, the *deflection* and the *slope* of the beam *cannot be discontinuous* at any point.

Photo 9.1 A different function $M(x)$ is required in each portion of the cantilever arms.

For the prismatic beam and the loading shown (Fig. 9.16), determine the slope and deflection at point D.

EXAMPLE 9.03

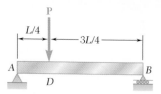

Fig. 9.16

We must divide the beam into two portions, AD and DB, and determine the function $y(x)$ which defines the elastic curve for each of these portions.

1. From A to D $(x < L/4)$. We draw the free-body diagram of a portion of beam AE of length $x < L/4$ (Fig. 9.17). Taking moments about E, we have

$$M_1 = \frac{3P}{4}x \qquad (9.17)$$

or, recalling Eq. (9.4),

$$EI\frac{d^2y_1}{dx^2} = \frac{3}{4}Px \qquad (9.18)$$

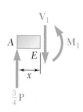

Fig. 9.17

where $y_1(x)$ is the function which defines the elastic curve *for portion AD of the beam.* Integrating in x, we write

$$EI\,\theta_1 = EI\frac{dy_1}{dx} = \frac{3}{8}Px^2 + C_1 \qquad (9.19)$$

$$EI\,y_1 = \frac{1}{8}Px^3 + C_1x + C_2 \qquad (9.20)$$

2. From D to B $(x > L/4)$. We now draw the free-body diagram of a portion of beam AE of length $x > L/4$ (Fig. 9.18) and write

$$M_2 = \frac{3P}{4}x - P\left(x - \frac{L}{4}\right) \qquad (9.21)$$

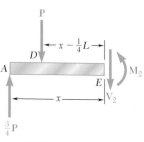

Fig. 9.18

or, recalling Eq. (9.4) and rearranging terms,

$$EI \frac{d^2y_2}{dx^2} = -\frac{1}{4}Px + \frac{1}{4}PL \qquad (9.22)$$

where $y_2(x)$ is the function which defines the elastic curve *for portion DB of the beam.* Integrating in x, we write

$$EI\,\theta_2 = EI\,\frac{dy_2}{dx} = -\frac{1}{8}Px^2 + \frac{1}{4}PLx + C_3 \qquad (9.23)$$

$$EI\,y_2 = -\frac{1}{24}Px^3 + \frac{1}{8}PLx^2 + C_3x + C_4 \qquad (9.24)$$

Determination of the Constants of Integration. The conditions that must be satisfied by the constants of integration have been summarized in Fig. 9.19. At the support A, where the deflection is defined by Eq. (9.20), we must have $x = 0$ and $y_1 = 0$. At the support B, where the deflection is defined by Eq. (9.24), we must have $x = L$ and $y_2 = 0$. Also, the fact that there can be no sudden change in deflection or in slope at point D requires that $y_1 = y_2$ and $\theta_1 = \theta_2$ when $x = L/4$. We have therefore:

$$[x = 0,\ y_1 = 0,]\ \text{Eq. (9.20):} \qquad 0 = C_2 \qquad (9.25)$$

$$[x = L,\ y_2 = 0],\ \text{Eq. (9.24):} \qquad 0 = \frac{1}{12}PL^3 + C_3L + C_4 \qquad (9.26)$$

$[x = L/4,\ \theta_1 = \theta_2]$, Eqs. (9.19) and (9.23):

$$\frac{3}{128}PL^2 + C_1 = \frac{7}{128}PL^2 + C_3 \qquad (9.27)$$

$[x = L/4,\ y_1 = y_2]$, Eqs. (9.20) and (9.24):

$$\frac{PL^3}{512} + C_1\frac{L}{4} = \frac{11PL^3}{1536} + C_3\frac{L}{4} + C_4 \qquad (9.28)$$

Solving these equations simultaneously, we find

$$C_1 = -\frac{7PL^2}{128},\ C_2 = 0,\ C_3 = -\frac{11PL^2}{128},\ C_4 = \frac{PL^3}{384}$$

Substituting for C_1 and C_2 into Eqs. (9.19) and (9.20), we write that for $x \le L/4$,

$$EI\,\theta_1 = \frac{3}{8}Px^2 - \frac{7PL^2}{128} \qquad (9.29)$$

$$EI\,y_1 = \frac{1}{8}Px^3 - \frac{7PL^2}{128}x \qquad (9.30)$$

Letting $x = L/4$ in each of these equations, we find that the slope and deflection at point D are, respectively,

$$\theta_D = -\frac{PL^2}{32EI} \qquad \text{and} \qquad y_D = -\frac{3PL^3}{256EI}$$

We note that, since $\theta_D \ne 0$, the deflection at D is *not* the maximum deflection of the beam.

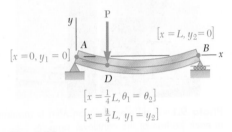

$[x = 0, y_1 = 0]$

$[x = L, y_2 = 0]$

$[x = \tfrac{1}{4}L, \theta_1 = \theta_2]$

$[x = \tfrac{1}{4}L, y_1 = y_2]$

Fig. 9.19

*9.4 DIRECT DETERMINATION OF THE ELASTIC CURVE FROM THE LOAD DISTRIBUTION

We saw in Sec. 9.3 that the equation of the elastic curve can be obtained by integrating twice the differential equation

$$\frac{d^2y}{dx^2} = \frac{M(x)}{EI} \tag{9.4}$$

where $M(x)$ is the bending moment in the beam. We now recall from Sec. 5.3 that, when a beam supports a distributed load $w(x)$, we have $dM/dx = V$ and $dV/dx = -w$ at any point of the beam. Differentiating both members of Eq. (9.4) with respect to x and assuming EI to be constant, we have therefore

$$\frac{d^3y}{dx^3} = \frac{1}{EI}\frac{dM}{dx} = \frac{V(x)}{EI} \tag{9.31}$$

and, differentiating again,

$$\frac{d^4y}{dx^4} = \frac{1}{EI}\frac{dV}{dx} = -\frac{w(x)}{EI}$$

We conclude that, when a prismatic beam supports a distributed load $w(x)$, its elastic curve is governed by the fourth-order linear differential equation

$$\frac{d^4y}{dx^4} = -\frac{w(x)}{EI} \tag{9.32}$$

Multiplying both members of Eq. (9.32) by the constant EI and integrating four times, we write

$$EI\frac{d^4y}{dx^4} = -w(x)$$

$$EI\frac{d^3y}{dx^3} = V(x) = -\int w(x)\,dx + C_1$$

$$EI\frac{d^2y}{dx^2} = M(x) = -\int dx \int w(x)\,dx + C_1 x + C_2 \tag{9.33}$$

$$EI\frac{dy}{dx} = EI\,\theta(x) = -\int dx \int dx \int w(x)\,dx + \frac{1}{2}C_1 x^2 + C_2 x + C_3$$

$$EI\,y(x) = -\int dx \int dx \int dx \int w(x)\,dx + \frac{1}{6}C_1 x^3 + \frac{1}{2}C_2 x^2 + C_3 x + C_4$$

The four constants of integration can be determined from the boundary conditions. These conditions include (a) the conditions imposed on the deflection or slope of the beam by its supports (cf. Sec. 9.3), and (b) the condition that V and M be zero at the free end of a cantilever beam, or that M be zero at both ends of a simply supported beam (cf. Sec. 5.3). This has been illustrated in Fig. 9.20.

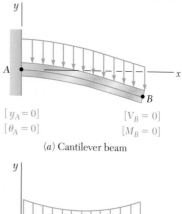

$[y_A = 0]$ $[V_B = 0]$
$[\theta_A = 0]$ $[M_B = 0]$

(a) Cantilever beam

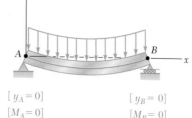

$[y_A = 0]$ $[y_B = 0]$
$[M_A = 0]$ $[M_B = 0]$

(b) Simply supported beam

Fig. 9.20 Boundary conditions.

The method presented here can be used effectively with cantilever or simply supported beams carrying a distributed load. In the case of overhanging beams, however, the reactions at the supports will cause discontinuities in the shear, i.e., in the third derivative of y, and different functions would be required to define the elastic curve over the entire beam.

EXAMPLE 9.04 The simply supported prismatic beam AB carries a uniformly distributed load w per unit length (Fig. 9.21). Determine the equation of the elastic curve and the maximum deflection of the beam. (This is the same beam and loading as in Example 9.02.)

Since $w = $ constant, the first three of Eqs. (9.33) yield

$$EI \frac{d^4y}{dx^4} = -w$$

$$EI \frac{d^3y}{dx^3} = V(x) = -wx + C_1$$

$$EI \frac{d^2y}{dx^2} = M(x) = -\frac{1}{2}wx^2 + C_1x + C_2 \qquad (9.34)$$

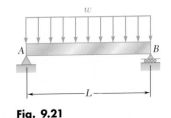

Fig. 9.21

Noting that the boundary conditions require that $M = 0$ at both ends of the beam (Fig. 9.22), we first let $x = 0$ and $M = 0$ in Eq. (9.34) and obtain $C_2 = 0$. We then make $x = L$ and $M = 0$ in the same equation and obtain $C_1 = \frac{1}{2}wL$.

Carrying the values of C_1 and C_2 back into Eq. (9.34), and integrating twice, we write

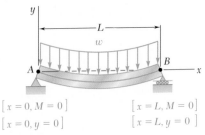

$[x = 0, M = 0]$ $[x = L, M = 0]$
$[x = 0, y = 0]$ $[x = L, y = 0]$

Fig. 9.22

$$EI \frac{d^2y}{dx^2} = -\frac{1}{2}wx^2 + \frac{1}{2}wLx$$

$$EI \frac{dy}{dx} = -\frac{1}{6}wx^3 + \frac{1}{4}wLx^2 + C_3$$

$$EI\, y = -\frac{1}{24}wx^4 + \frac{1}{12}wLx^3 + C_3x + C_4 \qquad (9.35)$$

But the boundary conditions also require that $y = 0$ at both ends of the beam. Letting $x = 0$ and $y = 0$ in Eq. (9.35), we obtain $C_4 = 0$; letting $x = L$ and $y = 0$ in the same equation, we write

$$0 = -\tfrac{1}{24}wL^4 + \tfrac{1}{12}wL^4 + C_3L$$
$$C_3 = -\tfrac{1}{24}wL^3$$

Carrying the values of C_3 and C_4 back into Eq. (9.35) and dividing both members by EI, we obtain the equation of the elastic curve:

$$y = \frac{w}{24EI}(-x^4 + 2Lx^3 - L^3x) \qquad (9.36)$$

The value of the maximum deflection is obtained by making $x = L/2$ in Eq. (9.36). We have

$$|y|_{max} = \frac{5wL^4}{384EI}$$

9.5 STATICALLY INDETERMINATE BEAMS

In the preceding sections, our analysis was limited to statically determinate beams. Consider now the prismatic beam AB (Fig. 9.23a), which has a fixed end at A and is supported by a roller at B. Drawing the free-body diagram of the beam (Fig. 9.23b), we note that the reactions involve four unknowns, while only three equilibrium equations are available, namely

$$\Sigma F_x = 0 \qquad \Sigma F_y = 0 \qquad \Sigma M_A = 0 \qquad (9.37)$$

Since only A_x can be determined from these equations, we conclude that the beam is *statically indeterminate*.

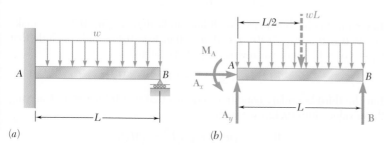

(a)   (b)

Fig. 9.23 Statically indeterminate beam.

However, we recall from Chaps. 2 and 3 that, in a statically indeterminate problem, the reactions can be obtained by considering the *deformations* of the structure involved. We should, therefore, proceed with the computation of the slope and deformation along the beam. Following the method used in Sec. 9.3, we first express the bending moment $M(x)$ at any given point of AB in terms of the distance x from A, the given load, and the unknown reactions. Integrating in x, we obtain expressions for θ and y which contain two additional unknowns, namely the constants of integration C_1 and C_2. But altogether six equations are available to determine the reactions and the constants C_1 and C_2; they are the three equilibrium equations (9.37) and the three equations expressing that the boundary conditions are satisfied, i.e., that the slope and deflection at A are zero, and that the deflection at B is zero (Fig. 9.24). Thus, the reactions at the supports can be determined, and the equation of the elastic curve can be obtained.

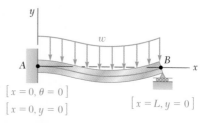

$[x = 0, \theta = 0]$
$[x = 0, y = 0]$ $[x = L, y = 0]$

Fig. 9.24 Boundary conditions for beam of Fig. 9.23.

EXAMPLE 9.05

Determine the reactions at the supports for the prismatic beam of Fig. 9.23a.

Equilibrium Equations. From the free-body diagram of Fig. 9.23b we write

$$\xrightarrow{+} \Sigma F_x = 0: \qquad A_x = 0$$

$$+\uparrow \Sigma F_y = 0: \qquad A_y + B - wL = 0 \qquad (9.38)$$

$$+\curvearrowleft \Sigma M_A = 0: \qquad M_A + BL - \tfrac{1}{2}wL^2 = 0$$

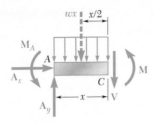

Fig. 9.25

Equation of Elastic Curve. Drawing the free-body diagram of a portion of beam AC (Fig. 9.25), we write

$$+\circlearrowleft \Sigma M_C = 0: \qquad M + \tfrac{1}{2}wx^2 + M_A - A_y x = 0 \tag{9.39}$$

Solving Eq. (9.39) for M and carrying into Eq. (9.4), we write

$$EI \frac{d^2y}{dx^2} = -\frac{1}{2}wx^2 + A_y x - M_A$$

Integrating in x, we have

$$EI\,\theta = EI\frac{dy}{dx} = -\frac{1}{6}wx^3 + \frac{1}{2}A_y x^2 - M_A x + C_1 \tag{9.40}$$

$$EI\,y = -\frac{1}{24}wx^4 + \frac{1}{6}A_y x^3 - \frac{1}{2}M_A x^2 + C_1 x + C_2 \tag{9.41}$$

Referring to the boundary conditions indicated in Fig. 9.24, we make $x = 0$, $\theta = 0$ in Eq. (9.40), $x = 0$, $y = 0$ in Eq. (9.41), and conclude that $C_1 = C_2 = 0$. Thus, we rewrite Eq. (9.41) as follows:

$$EI\,y = -\tfrac{1}{24}wx^4 + \tfrac{1}{6}A_y x^3 - \tfrac{1}{2}M_A x^2 \tag{9.42}$$

But the third boundary condition requires that $y = 0$ for $x = L$. Carrying these values into (9.42), we write

$$0 = -\tfrac{1}{24}wL^4 + \tfrac{1}{6}A_y L^3 - \tfrac{1}{2}M_A L^2$$

or

$$3M_A - A_y L + \tfrac{1}{4}wL^2 = 0 \tag{9.43}$$

Solving this equation simultaneously with the three equilibrium equations (9.38), we obtain the reactions at the supports:

$$A_x = 0 \qquad A_y = \tfrac{5}{8}wL \qquad M_A = \tfrac{1}{8}wL^2 \qquad B = \tfrac{3}{8}wL$$

In the example we have just considered, there was one redundant reaction, i.e., there was one more reaction than could be determined from the equilibrium equations alone. The corresponding beam is said to be *statically indeterminate to the first degree*. Another example of a beam indeterminate to the first degree is provided in Sample Prob. 9.3. If the beam supports are such that two reactions are redundant (Fig. 9.26a), the beam is said to be *indeterminate to the second degree*. While there are now five unknown reactions (Fig. 9.26b), we find that four equations may be obtained from the boundary conditions (Fig. 9.26c). Thus, altogether seven equations are available to determine the five reactions and the two constants of integration.

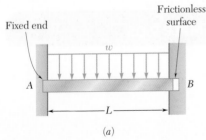

(a)

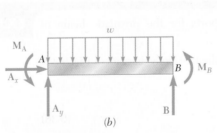

(b)

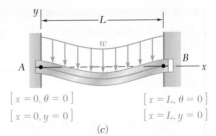

$$[x = 0, \theta = 0] \qquad\qquad [x = L, \theta = 0]$$
$$[x = 0, y = 0] \qquad\qquad [x = L, y = 0]$$

(c)

Fig. 9.26 Beam statically indeterminate to the second degree.

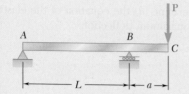

SAMPLE PROBLEM 9.1

The overhanging steel beam ABC carries a concentrated load $\mathbf{P}$ at end C. For portion AB of the beam, (*a*) derive the equation of the elastic curve, (*b*) determine the maximum deflection, (*c*) evaluate y_{max} for the following data:

$$W360 \times 101 \qquad I = 300 \times 10^6 \text{ mm}^4 \qquad E = 200 \text{ GPa}$$
$$P = 200 \text{ kN} \qquad L = 4.5 \text{ m} \qquad a = 1.2 \text{ m}$$

SOLUTION

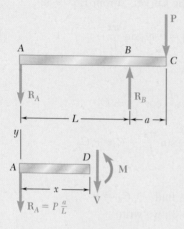

Free-Body Diagrams. Reactions: $\mathbf{R}_A = Pa/L \downarrow$ $\mathbf{R}_B = P(1 + a/L) \uparrow$
Using the free-body diagram of the portion of beam AD of length x, we find

$$M = -P\frac{a}{L}x \qquad (0 < x < L)$$

Differential Equation of the Elastic Curve. We use Eq. (9.4) and write

$$EI\frac{d^2y}{dx^2} = -P\frac{a}{L}x$$

Noting that the flexural rigidity EI is constant, we integrate twice and find

$$EI\frac{dy}{dx} = -\frac{1}{2}P\frac{a}{L}x^2 + C_1 \qquad (1)$$

$$EI\, y = -\frac{1}{6}P\frac{a}{L}x^3 + C_1x + C_2 \qquad (2)$$

Determination of Constants. For the boundary conditions shown, we have

$[x = 0, y = 0]$: From Eq. (2), we find $C_2 = 0$
$[x = L, y = 0]$: Again using Eq. (2), we write

$$EI(0) = -\frac{1}{6}P\frac{a}{L}L^3 + C_1L \qquad C_1 = +\frac{1}{6}PaL$$

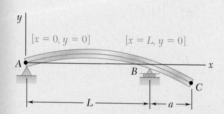

a. Equation of the Elastic Curve. Substituting for C_1 and C_2 into Eqs. (1) and (2), we have

$$EI\frac{dy}{dx} = -\frac{1}{2}P\frac{a}{L}x^2 + \frac{1}{6}PaL \qquad \frac{dy}{dx} = \frac{PaL}{6EI}\left[1 - 3\left(\frac{x}{L}\right)^2\right] \qquad (3)$$

$$EI\, y = -\frac{1}{6}P\frac{a}{L}x^3 + \frac{1}{6}PaLx \qquad y = \frac{PaL^2}{6EI}\left[\frac{x}{L} - \left(\frac{x}{L}\right)^3\right] \quad (4) \; \blacktriangleleft$$

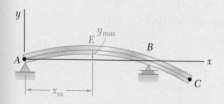

b. Maximum Deflection in Portion AB. The maximum deflection y_{max} occurs at point E where the slope of the elastic curve is zero. Setting $dy/dx = 0$ in Eq. (3), we determine the abscissa x_m of point E:

$$0 = \frac{PaL}{6EI}\left[1 - 3\left(\frac{x_m}{L}\right)^2\right] \qquad x_m = \frac{L}{\sqrt{3}} = 0.577L$$

We substitute $x_m/L = 0.577$ into Eq. (4) and have

$$y_{max} = \frac{PaL^2}{6EI}[(0.577) - (0.577)^3] \qquad y_{max} = 0.0642\frac{PaL^2}{EI} \; \blacktriangleleft$$

c. Evaluation of y_{max}. For the data given, the value of y_{max} is

$$y_{max} = 0.0642\frac{(200 \times 10^3 \text{ N})(1.2 \text{ m})(4.5 \text{ m})^2}{(200 \times 10^9 \text{ Pa})(300 \times 10^{-6} \text{ m}^4)} \qquad y_{max} = 5.2 \text{ mm} \; \blacktriangleleft$$

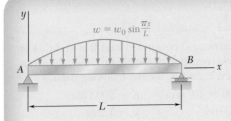

$$w = w_0 \sin \frac{\pi x}{L}$$

SAMPLE PROBLEM 9.2

For the beam and loading shown, determine (a) the equation of the elastic curve, (b) the slope at end A, (c) the maximum deflection.

SOLUTION

Differential Equation of the Elastic Curve. From Eq. (9.32),

$$EI \frac{d^4y}{dx^4} = -w(x) = -w_0 \sin \frac{\pi x}{L} \tag{1}$$

Integrate Eq. (1) twice:

$$EI \frac{d^3y}{dx^3} = V = +w_0 \frac{L}{\pi} \cos \frac{\pi x}{L} + C_1 \tag{2}$$

$$EI \frac{d^2y}{dx^2} = M = +w_0 \frac{L^2}{\pi^2} \sin \frac{\pi x}{L} + C_1 x + C_2 \tag{3}$$

Boundary Conditions:

[x = 0, M = 0]:   From Eq. (3), we find $C_2 = 0$

[x = L, M = 0]:   Again using Eq. (3), we write

$$0 = w_0 \frac{L^2}{\pi^2} \sin \pi + C_1 L \quad C_1 = 0$$

Thus:

$$EI \frac{d^2y}{dx^2} = +w_0 \frac{L^2}{\pi^2} \sin \frac{\pi x}{L} \tag{4}$$

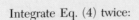

[x = 0, M = 0]   [x = L, M = 0]
[x = 0, y = 0]   [x = L, y = 0]

Integrate Eq. (4) twice:

$$EI \frac{dy}{dx} = EI \theta = -w_0 \frac{L^3}{\pi^3} \cos \frac{\pi x}{L} + C_3 \tag{5}$$

$$EI y = -w_0 \frac{L^4}{\pi^4} \sin \frac{\pi x}{L} + C_3 x + C_4 \tag{6}$$

Boundary Conditions:

[x = 0, y = 0]:   Using Eq. (6), we find $C_4 = 0$

[x = L, y = 0]:   Again using Eq. (6), we find $C_3 = 0$

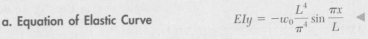

a. Equation of Elastic Curve $\quad EIy = -w_0 \dfrac{L^4}{\pi^4} \sin \dfrac{\pi x}{L}$ ◄

b. Slope at End A.   For x = 0, we have

$$EI \theta_A = -w_0 \frac{L^3}{\pi^3} \cos 0 \qquad \theta_A = \frac{w_0 L^3}{\pi^3 EI} \quad ◄$$

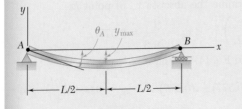

$\theta_A \quad y_{\text{max}}$

c. Maximum Deflection.   For $x = \frac{1}{2}L$

$$ELy_{\text{max}} = -w_0 \frac{L^4}{\pi^4} \sin \frac{\pi}{2} \qquad y_{\text{max}} = \frac{w_0 L^4}{\pi^4 EI} \downarrow \quad ◄$$

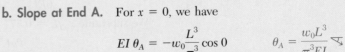

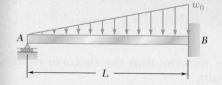

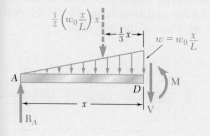

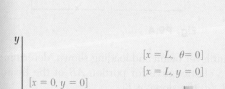

SAMPLE PROBLEM 9.3

For the uniform beam AB, (a) determine the reaction at A, (b) derive the equation of the elastic curve, (c) determine the slope at A. (Note that the beam is statically indeterminate to the first degree.)

SOLUTION

Bending Moment. Using the free body shown, we write

$$+\downarrow\Sigma M_D = 0: \qquad R_A x - \frac{1}{2}\left(\frac{w_0 x^2}{L}\right)\frac{x}{3} - M = 0 \qquad M = R_A x - \frac{w_0 x^3}{6L}$$

Differential Equation of the Elastic Curve. We use Eq. (9.4) and write

$$EI\frac{d^2 y}{dx^2} = R_A x - \frac{w_0 x^3}{6L}$$

Noting that the flexural rigidity EI is constant, we integrate twice and find

$$EI\frac{dy}{dx} = EI\,\theta = \frac{1}{2}R_A x^2 - \frac{w_0 x^4}{24L} + C_1 \qquad (1)$$

$$EI\,y = \frac{1}{6}R_A x^3 - \frac{w_0 x^5}{120L} + C_1 x + C_2 \qquad (2)$$

Boundary Conditions. The three boundary conditions that must be satisfied are shown on the sketch

$$[x = 0,\ y = 0]: \quad C_2 = 0 \qquad (3)$$

$$[x = L,\ \theta = 0] \quad \frac{1}{2}R_A L^2 - \frac{w_0 L^3}{24} + C_1 = 0 \qquad (4)$$

$$[x = L,\ y = 0] \quad \frac{1}{6}R_A L^3 - \frac{w_0 L^4}{120} + C_1 L + C_2 = 0 \qquad (5)$$

a. Reaction at A. Multiplying Eq. (4) by L, subtracting Eq. (5) member by member from the equation obtained, and noting that $C_2 = 0$, we have

$$\tfrac{1}{3}R_A L^3 - \tfrac{1}{30}w_0 L^4 = 0 \qquad \mathbf{R}_A = \tfrac{1}{10}w_0 L \uparrow \quad \blacktriangleleft$$

We note that the reaction is independent of E and I. Substituting $R_A = \frac{1}{10}w_0 L$ into Eq. (4), we have

$$\tfrac{1}{2}(\tfrac{1}{10}w_0 L)L^2 - \tfrac{1}{24}w_0 L^3 + C_1 = 0 \qquad C_1 = -\tfrac{1}{120}w_0 L^3$$

b. Equation of the Elastic Curve. Substituting for R_A, C_1, and C_2 into Eq. (2), we have

$$EI\,y = \frac{1}{6}\left(\frac{1}{10}w_0 L\right)x^3 - \frac{w_0 x^5}{120L} - \left(\frac{1}{120}w_0 L^3\right)x$$

$$y = \frac{w_0}{120EIL}(-x^5 + 2L^2 x^3 - L^4 x) \quad \blacktriangleleft$$

c. Slope at A. We differentiate the above equation with respect to x:

$$\theta = \frac{dy}{dx} = \frac{w_0}{120EIL}(-5x^4 + 6L^2 x^2 - L^4)$$

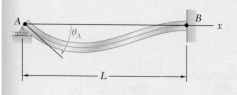

Making $x = 0$, we have $\qquad \theta_A = -\dfrac{w_0 L^3}{120EI} \qquad \theta_A = \dfrac{w_0 L^3}{120EI} \searrow \quad \blacktriangleleft$

PROBLEMS

In the following problems assume that the flexural rigidity *EI* of each beam is constant.

9.1 through 9.4 For the loading shown, determine (*a*) the equation of the elastic curve for the cantilever beam *AB*, (*b*) the deflection at the free end, (*c*) the slope at the free end.

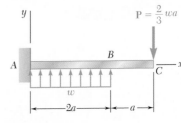

Fig. P9.1

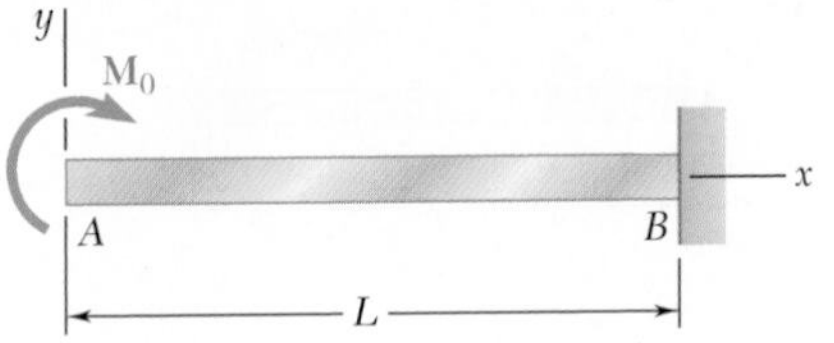

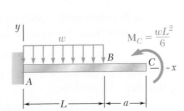

Fig. P9.2

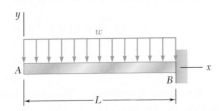

Fig. P9.3

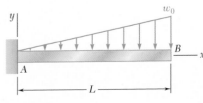

Fig. P9.4

9.5 and 9.6 For the cantilever beam and loading shown, determine (*a*) the equation of the elastic curve for portion *AB* of the beam, (*b*) the deflection at *B*, (*c*) the slope at *B*.

Fig. P9.5

Fig. P9.6

9.7 For the beam and loading shown, determine (*a*) the equation of the elastic curve for portion *BC* of the beam, (*b*) the deflection at midspan, (*c*) the slope at *B*.

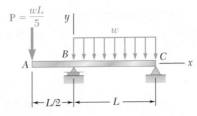

Fig. P9.7

9.8 For the beam and loading shown, determine (*a*) the equation of the elastic curve for portion *AB* of the beam, (*b*) the deflection at midspan, (*c*) the slope at *B*.

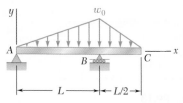

Fig. P9.8

9.9 Knowing that beam *AB* is a W130 × 23.8 rolled shape and that $P = 50$ kN, $L = 1.25$ m, and $E = 200$ GPa, determine (*a*) the slope at *A*, (*b*) the deflection at *C*.

9.10 Knowing that beam *AB* is an S200 × 27.4 rolled shape and that $w_0 = 60$ kN/m, $L = 2.7$ m, and $E = 200$ GPa, determine (*a*) the slope at *A*, (*b*) the deflection at *C*.

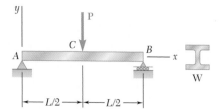

Fig. P9.9

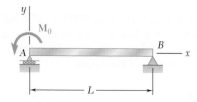

Fig. P9.10

9.11 (*a*) Determine the location and magnitude of the maximum deflection of beam *AB*. (*b*) Assuming that beam *AB* is a W360 × 64, $L = 3.5$ m, and $E = 200$ GPa, calculate the maximum allowable value of the applied moment $\mathbf{M_0}$ if the maximum deflection is not to exceed 1 mm.

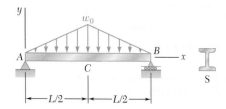

Fig. P9.11

9.12 For the beam and loading shown, (*a*) express the magnitude and location of the maximum deflection in terms of w_0, L, E, and I. (*b*) Calculate the value of the maximum deflection, assuming that beam *AB* is a W460 × 74 rolled shape and that $w_0 = 60$ kN/m, $L = 6$ m, and $E = 200$ GPa.

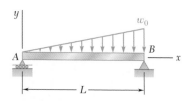

Fig. P9.12

9.13 For the beam and loading shown, determine the deflection at point C. Use $E = 200$ GPa.

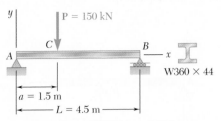

Fig. P9.13

9.14 For the beam and loading shown, knowing that $a = 2$ m, $w = 50$ kN/m, and $E = 200$ GPa, determine (a) the slope at support A, (b) the deflection at point C.

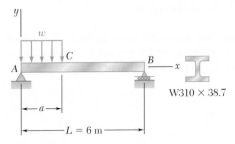

Fig. P9.14

9.15 For the beam and loading shown, determine the deflection at point C. Use $E = 200$ GPa.

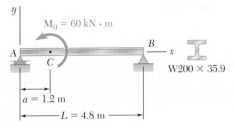

Fig. P9.15

9.16 Knowing that beam AE is an S200 × 27.4 rolled shape and that $P = 17.5$ kN, $L = 2.5$ m, $a = 0.8$ m, and $E = 200$ GPa, determine (a) the equation of the elastic curve for portion BD, (b) the deflection at the center C of the beam.

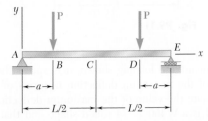

Fig. P9.16

9.17 For the beam and loading shown, determine (*a*) the equation of the elastic curve, (*b*) the deflection at the free end.

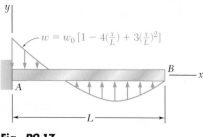

Fig. P9.17

9.18 For the beam and loading shown, determine (*a*) the equation of the elastic curve, (*b*) the slope at end *A*, (*c*) the deflection at the midpoint of the span.

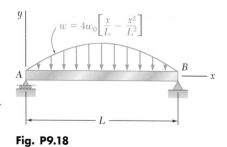

Fig. P9.18

9.19 through 9.22 For the beam and loading shown, determine the reaction at the roller support.

Fig. P9.19

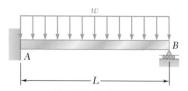

Fig. P9.20

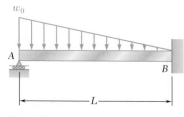

Fig. P9.21

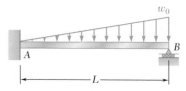

Fig. P9.22

9.23 For the beam shown, determine the reaction at the roller support when $w_0 = 15$ kN/m.

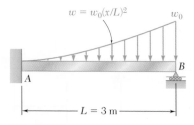

Fig. P9.23

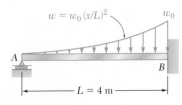

$w = w_0 (x/L)^2$

w_0

A

B

$L = 4$ m

Fig. P9.24

9.24 For the beam shown, determine the reaction at the roller support when $w_0 = 65$ kN/m.

9.25 through 9.28 Determine the reaction at the roller support and draw the bending moment diagram for the beam and loading shown.

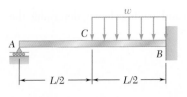

M_0

A

C

B

$L/2$

L

Fig. P9.25

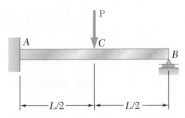

P

A

C

B

$L/2$

$L/2$

Fig. P9.26

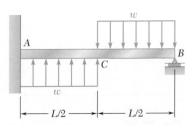

w

C

A

B

$L/2$

$L/2$

Fig. P9.27

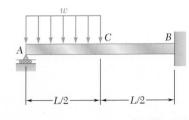

w_0

A

C

B

$\frac{1}{2}L$

L

Fig. P9.28

9.29 and 9.30 Determine the reaction at the roller support and the deflection at point C.

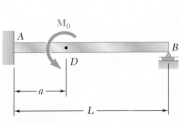

w

A

B

C

w

$L/2$

$L/2$

Fig. P9.29

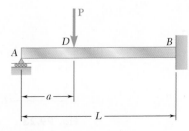

w

A

C

B

$L/2$

$L/2$

Fig. P9.30

9.31 and 9.32 Determine the reaction at the roller support and the deflection at point D if a is equal to $L/3$.

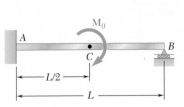

M_0

A

D

B

a

L

Fig. P9.31

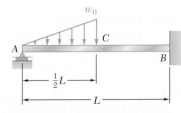

P

D

A

B

a

L

Fig. P9.32

9.33 and 9.34 Determine the reaction at A and draw the bending moment diagram for the beam and loading shown.

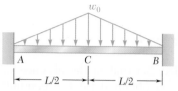

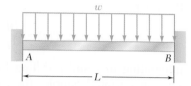

Fig. P9.33   **Fig. P9.34**

*9.6 USING SINGULARITY FUNCTIONS TO DETERMINE THE SLOPE AND DEFLECTION OF A BEAM

Reviewing the work done so far in this chapter, we note that the integration method provides a convenient and effective way of determining the slope and deflection at any point of a prismatic beam, *as long as the bending moment can be represented by a single analytical function $M(x)$*. However, when the loading of the beam is such that two different functions are needed to represent the bending moment over the entire length of the beam, as in Example 9.03 (Fig. 9.16), four constants of integration are required, and an equal number of equations, expressing continuity conditions at point D, as well as boundary conditions at the supports A and B, must be used to determine these constants. If three or more functions were needed to represent the bending moment, additional constants and a corresponding number of additional equations would be required, resulting in rather lengthy computations. Such would be the case for the beam shown in Photo 9.2. In this section these computations will be simplified through the use of the singularity functions discussed in Sec. 5.5.

Photo 9.2 In this roof structure, each of the joists applies a concentrated load to the beam that supports it.

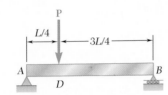

Fig. 9.16 (repeated)

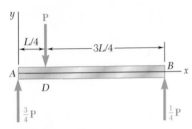

Fig. 9.27 Free-body diagram for beam of Fig. 9.16.

Fig. 9.28 Boundary conditions for beam of Fig. 9.16.

Let us consider again the beam and loading of Example 9.03 (Fig. 9.16) and draw the free-body diagram of that beam (Fig. 9.27). Using the appropriate singularity function, as explained in Sec. 5.5, to represent the contribution to the shear of the concentrated load **P,** we write

$$V(x) = \frac{3P}{4} - P\langle x - \tfrac{1}{4}L\rangle^0$$

Integrating in x and recalling from Sec. 5.5 that in the absence of any concentrated couple, the expression obtained for the bending moment will not contain any constant term, we have

$$M(x) = \frac{3P}{4}x - P\langle x - \tfrac{1}{4}L\rangle \tag{9.44}$$

Substituting for $M(x)$ from (9.44) into Eq. (9.4), we write

$$EI\frac{d^2y}{dx^2} = \frac{3P}{4}x - P\langle x - \tfrac{1}{4}L\rangle \tag{9.45}$$

and, integrating in x,

$$EI\,\theta = EI\frac{dy}{dx} = \frac{3}{8}Px^2 - \frac{1}{2}P\langle x - \tfrac{1}{4}L\rangle^2 + C_1 \tag{9.46}$$

$$EI\,y = \frac{1}{8}Px^3 - \frac{1}{6}P\langle x - \tfrac{1}{4}L\rangle^3 + C_1x + C_2 \tag{9.47}†$$

The constants C_1 and C_2 can be determined from the boundary conditions shown in Fig. 9.28. Letting $x = 0$, $y = 0$ in Eq. (9.47), we have

$$0 = 0 - \frac{1}{6}P\langle 0 - \tfrac{1}{4}L\rangle^3 + 0 + C_2$$

which reduces to $C_2 = 0$, since any bracket containing a negative quantity is equal to zero. Letting now $x = L$, $y = 0$, and $C_2 = 0$ in Eq. (9.47), we write

$$0 = \frac{1}{8}PL^3 - \frac{1}{6}P\langle\tfrac{3}{4}L\rangle^3 + C_1L$$

Since the quantity between brackets is positive, the brackets can be replaced by ordinary parentheses. Solving for C_1, we have

$$C_1 = -\frac{7PL^2}{128}$$

We check that the expressions obtained for the constants C_1 and C_2 are the same that were found earlier in Sec. 9.3. But the need for additional constants C_3 and C_4 has now been eliminated, and we do not have to write equations expressing that the slope and the deflection are continuous at point D.

---

†The continuity conditions for the slope and deflection at D are "built-in" in Eqs. (9.46) and (9.47). Indeed, the difference between the expressions for the slope θ_1 in AD and the slope θ_2 in DB is represented by the term $-\frac{1}{2}P\langle x - \frac{1}{4}L\rangle^2$ in Eq. (9.46), and this term is equal to zero at D. Similarly, the difference between the expressions for the deflection y_1 in AD and the deflection y_2 in DB is represented by the term $-\frac{1}{6}P\langle x - \frac{1}{4}L\rangle^3$ in Eq. (9.47), and this term is also equal to zero at D.

For the beam and loading shown (Fig. 9.29*a*) and using singularity functions, (*a*) express the slope and deflection as functions of the distance *x* from the support at *A*, (*b*) determine the deflection at the midpoint *D*. Use $E = 200$ GPa and $I = 6.87 \times 10^{-6}$ m⁴.

EXAMPLE 9.06

(*a*) We note that the beam is loaded and supported in the same manner as the beam of Example 5.05. Referring to that example, we recall that the given distributed loading was replaced by the two equivalent open-ended loadings shown in Fig. 9.29*b* and that the following expressions were obtained for the shear and bending moment:

$$V(x) = -1.5\langle x - 0.6\rangle^1 + 1.5\langle x - 1.8\rangle^1 + 2.6 - 1.2\langle x - 0.6\rangle^0$$

$$M(x) = -0.75\langle x - 0.6\rangle^2 + 0.75\langle x - 1.8\rangle^2$$
$$+ 2.6x - 1.2\langle x - 0.6\rangle^1 - 1.44\langle x - 2.6\rangle^0$$

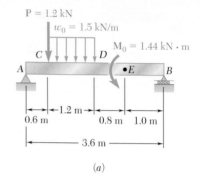

(*a*)

Integrating the last expression twice, we obtain

$$EI\theta = -0.25\langle x - 0.6\rangle^3 + 0.25\langle x - 1.8\rangle^3$$
$$+ 1.3x^2 - 0.6\langle x - 0.6\rangle^2 - 1.44\langle x - 2.6\rangle^1 + C_1 \qquad (9.48)$$

$$EIy = -0.0625\langle x - 0.6\rangle^4 + 0.0625\langle x - 1.8\rangle^4 + 0.4333x^3$$
$$- 0.2\langle x - 0.6\rangle^3 - 0.72\langle x - 2.6\rangle^2 + C_1x + C_2 \qquad (9.49)$$

The constants C_1 and C_2 can be determined from the boundary conditions shown in Fig. 9.30. Letting $x = 0$, $y = 0$ in Eq. (9.49) and noting that all the brackets contain negative quantities and, therefore, are equal to zero, we conclude that $C_2 = 0$. Letting now $x = 3.6$, $y = 0$, and $C_2 = 0$ in Eq. (9.49), we write

$$0 = -0.0625\langle 3.0\rangle^4 + 0.0625\langle 1.8\rangle^4$$
$$+ 0.4333(3.6)^3 - 0.2\langle 3.0\rangle^3 - 0.72\langle 1.0\rangle^2 + C_1(3.6) + 0$$

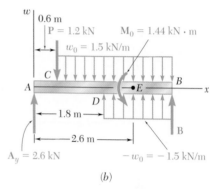

(*b*)

Fig. 9.29

Since all the quantities between brackets are positive, the brackets can be replaced by ordinary parentheses. Solving for C_1, we find $C_1 = -2.692$.

Fig. 9.30

(*b*) Substituting for C_1 and C_2 into Eq. (9.49) and making $x = x_D = 1.8$ m, we find that the deflection at point *D* is defined by the relation

$$EIy_D = -0.0625\langle 1.2\rangle^4 + 0.0625\langle 0\rangle^4$$
$$+ 0.4333(1.8)^3 - 0.2\langle 1.2\rangle^3 - 0.72\langle -0.8\rangle^2 - 2.692(1.8)$$

The last bracket contains a negative quantity and, therefore, is equal to zero. All the other brackets contain positive quantities and can be replaced by ordinary parentheses. We have

$$EIy_D = -0.0625(1.2)^4 + 0.0625(0)^4$$
$$+ 0.4333(1.8)^3 - 0.2(1.2)^3 - 0 - 2.692(1.8) = -2.794$$

Recalling the given numerical values of *E* and *I*, we write

$$(200 \text{ GPa})(6.87 \times 10^{-6} \text{ m}^4)y_D = -2.794 \text{ kN} \cdot \text{m}^3$$
$$y_D = -13.64 \times 10^{-3} \text{ m} = -2.03 \text{ mm}$$

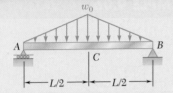

SAMPLE PROBLEM 9.4

For the prismatic beam and loading shown, determine (a) the equation of the elastic curve, (b) the slope at A, (c) the maximum deflection.

SOLUTION

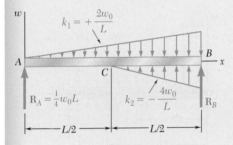

Bending Moment. The equation defining the bending moment of the beam was obtained in Sample Prob. 5.9. Using the modified loading diagram shown, we had [Eq. (3)]:

$$M(x) = -\frac{w_0}{3L}x^3 + \frac{2w_0}{3L}\langle x - \tfrac{1}{2}L\rangle^3 + \tfrac{1}{4}w_0Lx$$

a. Equation of the Elastic Curve. Using Eq. (9.4), we write

$$EI\frac{d^2y}{dx^2} = -\frac{w_0}{3L}x^3 + \frac{2w_0}{3L}\langle x - \tfrac{1}{2}L\rangle^3 + \tfrac{1}{4}w_0Lx \qquad (1)$$

and, integrating twice in x,

$$EI\,\theta = -\frac{w_0}{12L}x^4 + \frac{w_0}{6L}\langle x - \tfrac{1}{2}L\rangle^4 + \frac{w_0L}{8}x^2 + C_1 \qquad (2)$$

$$EI\,y = -\frac{w_0}{60L}x^5 + \frac{w_0}{30L}\langle x - \tfrac{1}{2}L\rangle^5 + \frac{w_0L}{24}x^3 + C_1x + C_2 \qquad (3)$$

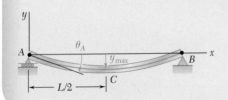

Boundary Conditions.
$[x = 0, y = 0]$: Using Eq. (3) and noting that each bracket $\langle\ \rangle$ contains a negative quantity and, thus, is equal to zero, we find $C_2 = 0$.

$[x = L, y = 0]$: Again using Eq. (3), we write

$$0 = -\frac{w_0L^4}{60} + \frac{w_0}{30L}\left(\frac{L}{2}\right)^5 + \frac{w_0L^4}{24} + C_1L \qquad C_1 = -\frac{5}{192}w_0L^3$$

Substituting C_1 and C_2 into Eqs. (2) and (3), we have

$$EI\,\theta = -\frac{w_0}{12L}x^4 + \frac{w_0}{6L}\langle x - \tfrac{1}{2}L\rangle^4 + \frac{w_0L}{8}x^2 - \frac{5}{192}w_0L^3 \qquad (4)$$

$$EI\,y = -\frac{w_0}{60L}x^5 + \frac{w_0}{30L}\langle x - \tfrac{1}{2}L\rangle^5 + \frac{w_0L}{24}x^3 - \frac{5}{192}w_0L^3x \qquad (5) \ \blacktriangleleft$$

b. Slope at A. Substituting $x = 0$ into Eq. (4), we find

$$EI\,\theta_A = -\frac{5}{192}w_0L^3 \qquad \theta_A = \frac{5w_0L^3}{192EI} \ \text{↖} \ \blacktriangleleft$$

c. Maximum Deflection. Because of the symmetry of the supports and loading, the maximum deflection occurs at point C, where $x = \tfrac{1}{2}L$. Substituting into Eq. (5), we obtain

$$EI\,y_{max} = w_0L^4\left[-\frac{1}{60(32)} + 0 + \frac{1}{24(8)} - \frac{5}{192(2)}\right] = -\frac{w_0L^4}{120}$$

$$y_{max} = \frac{w_0L^4}{120EI} \ \downarrow \ \blacktriangleleft$$

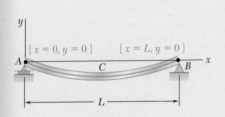

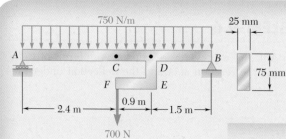

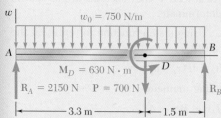

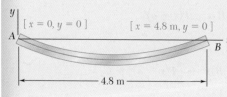

SAMPLE PROBLEM 9.5

The rigid bar DEF is welded at point D to the uniform steel beam AB. For the loading shown, determine (a) the equation of the elastic curve of the beam, (b) the deflection at the midpoint C of the beam. Use $E = 200$ GPa.

SOLUTION

Bending Moment. The equation defining the bending moment of the beam was obtained in Sample Prob. 5.10. Using the modified loading diagram shown and expressing x in metres, we had [Eq. (3)]:

$$M(x) = -375x^2 + 2150x - 700\langle x - 3.3\rangle^1 - 630\langle x - 3.3\rangle^0 \text{ N} \cdot \text{m}$$

a. Equation of the Elastic Curve. Using Eq. (8.4), we write

$$EI(d^2y/dx^2) = -375x^2 + 2150x - 700\langle x - 3.3\rangle^1 - 630\langle x - 3.3\rangle^0 \text{ N} \cdot \text{m} \quad (1)$$

and, integrating twice in x,

$$EI\,\theta = -125x^3 + 1075x^2 - 350\langle x - 3.3\rangle^2 - 630\langle x - 3.3\rangle^1 + C_1 \text{ N} \cdot \text{m}^2 \quad (2)$$

$$EI\,y = -31.25x^4 + 358.3x^3 - 116.7\langle x - 3.3\rangle^3 - 315\langle x - 3.3\rangle^2 + C_1x + C_2 \text{ N} \cdot \text{m}^3 \quad (3)$$

Boundary Conditions.

$[x = 0, y = 0]$: Using Eq. (3) and noting that each bracket $\langle\ \rangle$ contains a negative quantity and, thus, is equal to zero, we find $C_2 = 0$.

$[x = 4.8 \text{ m}, y = 0]$: Again using Eq. (3) and noting that each bracket contains a positive quantity and, thus, can be replaced by a parenthesis, we write

$$0 = -31.25(4.8)^4 + 358.3(4.8)^3 - 116.7(1.5)^3 - 315(1.5)^2 + C_1(4.8)$$

$$C_1 = -4570$$

Substituting the values found for C_1 and C_2 into Eq. (3), we have

$$EI\,y = -31.25x^4 + 358.3x^3 - 116.7\langle x - 3.3\rangle^3 - 315\langle x - 3.3\rangle^2 - 4570x \text{ N} \cdot \text{m}^3 \quad (3') \quad \blacktriangleleft$$

To determine EI, we recall that $E = 200$ GPa and compute

$$I = \tfrac{1}{12}bh^3 = \tfrac{1}{12}(25 \text{ mm})(75 \text{ mm})^3 = 0.879 \times 10^{-6} \text{ m}^4$$

$$EI = (200 \times 10^9 \text{ Pa})(0.879 \times 10^{-6} \text{ m}^4) = 175.8 \times 10^3 \text{ N} \cdot \text{m}^2$$

b. Deflection at Midpoint C. Making $x = 2.4$ m in Eq. (3'), we write

$$EI\,y_C = -31.25(2.4)^4 + 358.3(2.4)^3 - 116.7\langle -0.9\rangle^3 - 315\langle -0.9\rangle^2 - 4570(2.4)$$

Noting that each bracket $\langle\ \rangle$ is equal to zero and substituting for EI its numerical value, we have

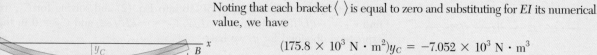

$$(175.8 \times 10^3 \text{ N} \cdot \text{m}^2)y_C = -7.052 \times 10^3 \text{ N} \cdot \text{m}^3$$

and, solving for y_C: $\qquad y_C = -0.04 \text{ m} \qquad y_C = -40 \text{ mm} \quad \blacktriangleleft$

Note that the deflection obtained *is not* the maximum deflection.

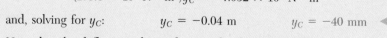

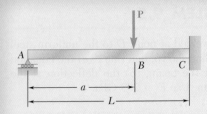

SAMPLE PROBLEM 9.6

For the uniform beam ABC, (a) express the reaction at A in terms of P, L, a, E, and I, (b) determine the reaction at A and the deflection under the load when $a = L/2$.

SOLUTION

Reactions. For the given vertical load $\mathbf{P}$ the reactions are as shown. We note that they are statically indeterminate.

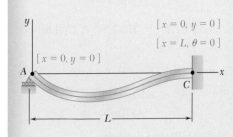

Shear and Bending Moment. Using a step function to represent the contribution of $\mathbf{P}$ to the shear, we write

$$V(x) = R_A - P\langle x - a\rangle^0$$

Integrating in x, we obtain the bending moment:

$$M(x) = R_A x - P\langle x - a\rangle^1$$

Equation of the Elastic Curve. Using Eq. (9.4), we write

$$EI\frac{d^2y}{dx^2} = R_A x - P\langle x - a\rangle^1$$

Integrating twice in x,

$$EI\frac{dy}{dx} = EI\,\theta = \frac{1}{2}R_A x^2 - \frac{1}{2}P\langle x - a\rangle^2 + C_1$$

$$EI\,y = \frac{1}{6}R_A x^3 - \frac{1}{6}P\langle x - a\rangle^3 + C_1 x + C_2$$

Boundary Conditions. Noting that the bracket $\langle x - a\rangle$ is equal to zero for $x = 0$, and to $(L - a)$ for $x = L$, we write

$$[x = 0, y = 0]: \quad C_2 = 0 \tag{1}$$

$$[x = L, \theta = 0]: \quad \tfrac{1}{2}R_A L^2 - \tfrac{1}{2}P(L - a)^2 + C_1 = 0 \tag{2}$$

$$[x = L, y = 0]: \quad \tfrac{1}{6}R_A L^3 - \tfrac{1}{6}P(L - a)^3 + C_1 L + C_2 = 0 \tag{3}$$

a. Reaction at A. Multiplying Eq. (2) by L, subtracting Eq. (3) member by member from the equation obtained, and noting that $C_2 = 0$, we have

$$\frac{1}{3}R_A L^3 - \frac{1}{6}P(L - a)^2[3L - (L - a)] = 0$$

$$\mathbf{R}_A = P\left(1 - \frac{a}{L}\right)^2\left(1 + \frac{a}{2L}\right)\uparrow \quad \blacktriangleleft$$

We note that the reaction is independent of E and I.

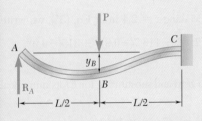

b. Reaction at A and Deflection at B when $a = \frac{1}{2}L$. Making $a = \frac{1}{2}L$ in the expression obtained for R_A, we have

$$R_A = P(1 - \tfrac{1}{2})^2(1 + \tfrac{1}{4}) = 5P/16 \qquad \mathbf{R}_A = \tfrac{5}{16}P\uparrow \quad \blacktriangleleft$$

Substituting $a = L/2$ and $R_A = 5P/16$ into Eq. (2) and solving for C_1, we find $C_1 = -PL^2/32$. Making $x = L/2$, $C_1 = -PL^2/32$, and $C_2 = 0$ in the expression obtained for y, we have

$$y_B = -\frac{7PL^3}{768EI} \qquad\qquad y_B = \frac{7PL^3}{768EI}\downarrow \quad \blacktriangleleft$$

Note that the deflection obtained is *not* the maximum deflection.

PROBLEMS

Use singularity functions to solve the following problems and assume that the flexural rigidity *EI* of each beam is constant.

9.35 and 9.36 For the beam and loading shown, determine (*a*) the equation of the elastic curve, (*b*) the slope at end *A*, (*c*) the deflection of point *C*.

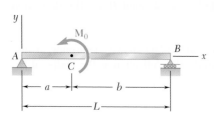

Fig. P9.35

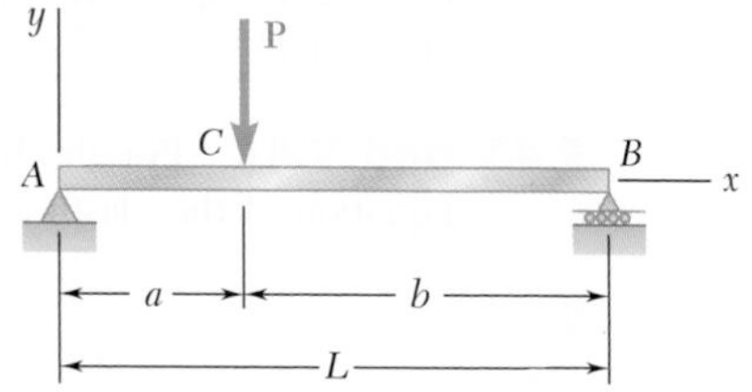

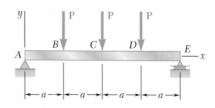

Fig. P9.36

9.37 and 9.38 For the beam and loading shown, determine the deflection at (*a*) point *B*, (*b*) point *C*, (*c*) point *D*.

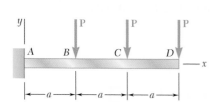

Fig. P9.37

Fig. P9.38

9.39 and 9.40 For the beam and loading shown, determine (*a*) the deflection at end *A*, (*b*) the deflection at point *C*, (*c*) the slope at end *D*.

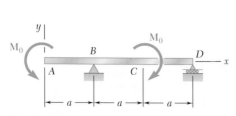

Fig. P9.39

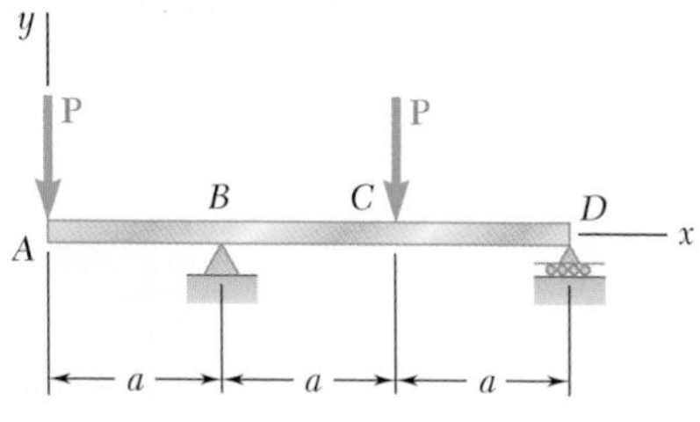

Fig. P9.40

9.41 For the beam and loading shown, determine (*a*) the equation of the elastic curve, (*b*) the deflection at the midpoint *C*.

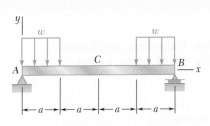

Fig. P9.41

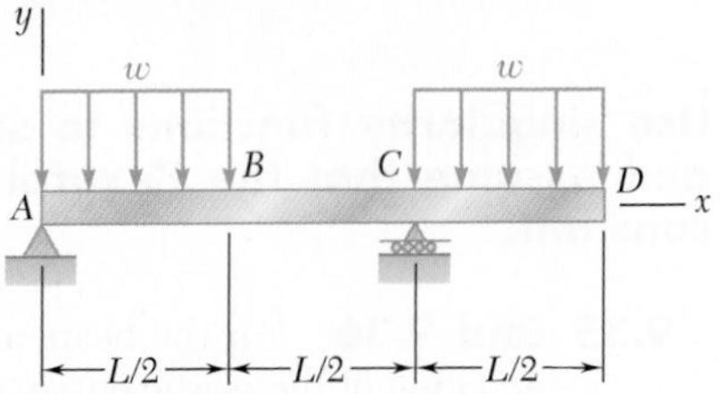

Fig. P9.42

9.42 For the beam and loading shown, determine (*a*) the equation of the elastic curve, (*b*) the deflection at point *B*, (*c*) the deflection at point *D*.

9.43 and 9.44 For the beam and loading shown, determine (*a*) the equation of the elastic curve, (*b*) the deflection at the midpoint *C*.

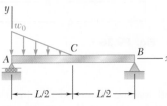

Fig. P9.43

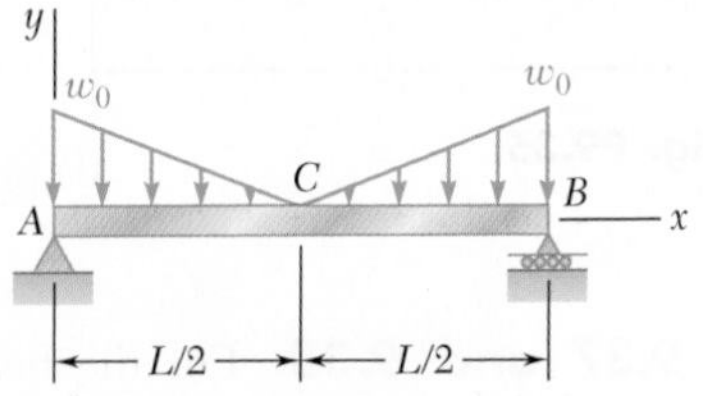

Fig. P9.44

9.45 For the beam and loading shown, determine (*a*) the slope at end *A*, (*b*) the deflection at point *C*. Use $E = 200$ GPa.

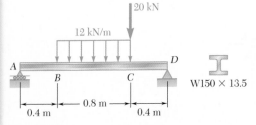

Fig. P9.45

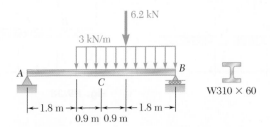

Fig. P9.46

9.46 For the beam and loading shown, determine (*a*) the slope at end *A*, (*b*) the deflection at the midpoint *C*. Use $E = 200$ GPa.

9.47 For the beam and loading shown, determine (*a*) the slope at end *A*, (*b*) the deflection at the midpoint *C*. Use $E = 200$ GPa.

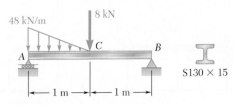

Fig. P9.47

9.48 For the timber beam and loading shown, determine (*a*) the slope at end *A*, (*b*) the deflection at the midpoint *C*. Use *E* = 12 GPa.

9.49 and 9.50 For the beam and loading shown, determine (*a*) the reaction at the roller support, (*b*) the deflection at point *C*.

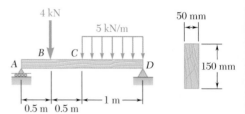

Fig. P9.48

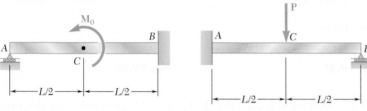

Fig. P9.49 **Fig. P9.50**

9.51 and 9.52 For the beam and loading shown, determine (*a*) the reaction at the roller support, (*b*) the deflection at point *B*.

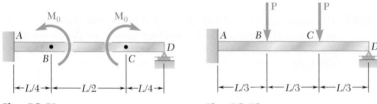

Fig. *P9.51* **Fig. P9.52**

9.53 For the beam and loading shown, determine (*a*) the reaction at point *C*, (*b*) the deflection at point *B*. Use *E* = 200 GPa.

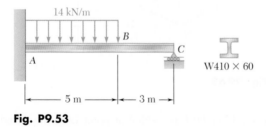

Fig. P9.53

9.54 and 9.55 For the beam and loading shown, determine (*a*) the reaction at point *A*, (*b*) the deflection at point *C*. Use *E* = 200 GPa.

9.56 For the beam shown and knowing that *P* = 40 kN, determine (*a*) the reaction at point *E*, (*b*) the deflection at point *C*. Use *E* = 200 GPa.

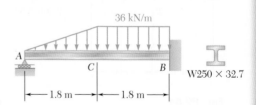

Fig. P9.54

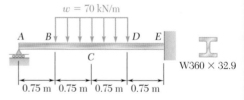

Fig. P9.55

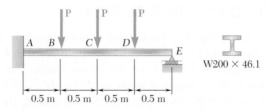

Fig. P9.56

9.57 and 9.58 For the beam and loading shown, determine (*a*) the reaction at point *A*, (*b*) the deflection at midpoint *C*.

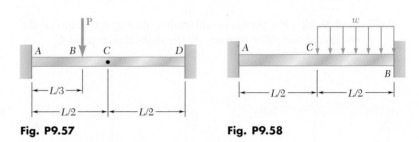

Fig. P9.57   **Fig. P9.58**

9.59 through 9.62 For the beam and loading indicated, determine the magnitude and location of the largest downward deflection.

    9.59 Beam and loading of Prob. 9.45.
    9.60 Beam and loading of Prob. 9.46.
    9.61 Beam and loading of Prob. 9.47.
    9.62 Beam and loading of Prob. 9.48.

9.63 The rigid bars *BF* and *DH* are welded to the rolled-steel beam *AE* as shown. Determine for the loading shown (*a*) the deflection at point *B*, (*b*) the deflection at midpoint *C* of the beam. Use $E = 200$ GPa.

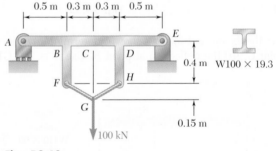

Fig. P9.63

9.64 The rigid bar *DEF* is welded at point *D* to the rolled-steel beam *AB*. For the loading shown, determine (*a*) the slope at point *A*, (*b*) the deflection at midpoint *C* of the beam. Use $E = 200$ GPa.

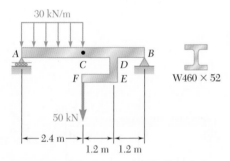

Fig. P9.64

9.7 METHOD OF SUPERPOSITION

When a beam is subjected to several concentrated or distributed loads, it is often found convenient to compute separately the slope and deflection caused by each of the given loads. The slope and deflection due to the combined loads are then obtained by applying the principle of superposition (Sec. 2.12) and adding the values of the slope or deflection corresponding to the various loads.

Determine the slope and deflection at D for the beam and loading shown (Fig. 9.31), knowing that the flexural rigidity of the beam is $EI = 100$ MN $\cdot$ m^2.

EXAMPLE 9.07

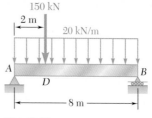

Fig. 9.31

The slope and deflection at any point of the beam can be obtained by superposing the slopes and deflections caused respectively by the concentrated load and by the distributed load (Fig. 9.32).

Since the concentrated load in Fig. 9.32b is applied at quarter span, we can use the results obtained for the beam and loading of Example 9.03 and write

$$(\theta_D)_P = -\frac{PL^2}{32EI} = -\frac{(150 \times 10^3)(8)^2}{32(100 \times 10^6)} = -3 \times 10^{-3} \text{ rad}$$

$$(y_D)_P = -\frac{3PL^3}{256EI} = -\frac{3(150 \times 10^3)(8)^3}{256(100 \times 10^6)} = -9 \times 10^{-3} \text{ m}$$

$$= -9 \text{ mm}$$

On the other hand, recalling the equation of the elastic curve obtained for a uniformly distributed load in Example 9.02, we express the deflection in Fig. 9.32c as

$$y = \frac{w}{24EI}(-x^4 + 2Lx^3 - L^3x) \qquad (9.50)$$

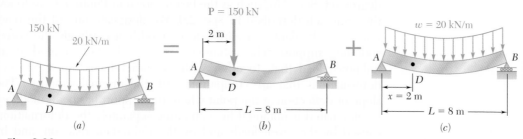

Fig. 9.32

and, differentiating with respect to x,

$$\theta = \frac{dy}{dx} = \frac{w}{24EI}(-4x^3 + 6Lx^2 - L^3) \qquad (9.51)$$

Making $w = 20$ kN/m, $x = 2$ m, and $L = 8$ m in Eqs. (9.51) and (9.50), we obtain

$$(\theta_D)_w = \frac{20 \times 10^3}{24(100 \times 10^6)}(-352) = -2.93 \times 10^{-3} \text{ rad}$$

$$(y_D)_w = \frac{20 \times 10^3}{24(100 \times 10^6)}(-912) = -7.60 \times 10^{-3} \text{ m}$$

$$= -7.60 \text{ mm}$$

Combining the slopes and deflections produced by the concentrated and the distributed loads, we have

$$\theta_D = (\theta_D)_P + (\theta_D)_w = -3 \times 10^{-3} - 2.93 \times 10^{-3}$$
$$= -5.93 \times 10^{-3} \text{ rad}$$

$$y_D = (y_D)_P + (y_D)_w = -9 \text{ mm} - 7.60 \text{ mm} = -16.60 \text{ mm}$$

To facilitate the task of practicing engineers, most structural and mechanical engineering handbooks include tables giving the deflections and slopes of beams for various loadings and types of support. Such a table will be found in Appendix D. We note that the slope and deflection of the beam of Fig. 9.31 could have been determined from that table. Indeed, using the information given under cases 5 and 6, we could have expressed the deflection of the beam for any value $x \leq L/4$. Taking the derivative of the expression obtained in this way would have yielded the slope of the beam over the same interval. We also note that the slope at both ends of the beam can be obtained by simply adding the corresponding values given in the table. However, the maximum deflection of the beam of Fig. 9.31 *cannot* be obtained by adding the maximum deflections of cases 5 and 6, since these deflections occur at different points of the beam.†

9.8 APPLICATION OF SUPERPOSITION TO STATICALLY INDETERMINATE BEAMS

We often find it convenient to use the method of superposition to determine the reactions at the supports of a statically indeterminate beam. Considering first the case of a beam indeterminate to the first degree (cf. Sec. 9.5), such as the beam shown in Photo 9.3, we follow the approach described in Sec. 2.9. We designate one of the reactions as redundant and eliminate or modify accordingly the corresponding support. The redundant reaction is then treated as an unknown load that, together with the other loads, must produce deformations that are compatible with the original supports. The slope or deflection at the point where the support has been modified or eliminated is obtained by computing separately the deformations caused by the given loads and by the redundant reaction, and by superposing the results obtained. Once the reactions at the supports have been found, the slope and deflection can be determined in the usual way at any other point of the beam.

Photo 9.3 The continuous beams supporting this highway overpass have three supports and are thus statically indeterminate.

---

†An approximate value of the maximum deflection of the beam can be obtained by plotting the values of y corresponding to various values of x. The determination of the exact location and magnitude of the maximum deflection would require setting equal to zero the expression obtained for the slope of the beam and solving this equation for x.

EXAMPLE 9.08

Determine the reactions at the supports for the prismatic beam and loading shown in Fig. 9.33. (This is the same beam and loading as in Example 9.05 of Sec. 9.5.)

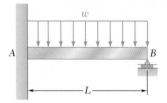

Fig. 9.33

We consider the reaction at B as redundant and release the beam from the support. The reaction $\mathbf{R}_B$ is now considered as an unknown load (Fig. 9.34a) and will be determined from the condition that the deflection of the beam at B must be zero. The solution is carried out by considering separately the deflection $(y_B)_w$ caused at B by the uniformly distributed load w (Fig. 9.34b) and the deflection $(y_B)_R$ produced at the same point by the redundant reaction $\mathbf{R}_B$ (Fig. 9.34c).

From the table of Appendix D (cases 2 and 1), we find that

$$(y_B)_w = -\frac{wL^4}{8EI} \qquad (y_B)_R = +\frac{R_B L^3}{3EI}$$

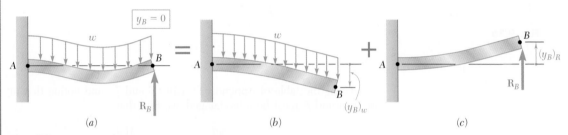

(a) (b) (c)

Fig. 9.34

Writing that the deflection at B is the sum of these two quantities and that it must be zero, we have

$$y_B = (y_B)_w + (y_B)_R = 0$$
$$y_B = -\frac{wL^4}{8EI} + \frac{R_B L^3}{3EI} = 0$$

and, solving for R_B, $\qquad R_B = \tfrac{3}{8}wL \qquad \mathbf{R}_B = \tfrac{3}{8}wL \uparrow$

Drawing the free-body diagram of the beam (Fig. 9.35) and writing the corresponding equilibrium equations, we have

$$+\uparrow \Sigma F_y = 0: \qquad R_A + R_B - wL = 0 \qquad\qquad (9.52)$$
$$R_A = wL - R_B = wL - \tfrac{3}{8}wL = \tfrac{5}{8}wL$$
$$\mathbf{R}_A = \tfrac{5}{8}wL \uparrow$$

$$+\curvearrowleft \Sigma M_A = 0: \quad M_A + R_B L - (wL)(\tfrac{1}{2}L) = 0 \qquad (9.53)$$
$$M_A = \tfrac{1}{2}wL^2 - R_B L = \tfrac{1}{2}wL^2 - \tfrac{3}{8}wL^2 = \tfrac{1}{8}wL^2$$
$$\mathbf{M}_A = \tfrac{1}{8}wL^2 \curvearrowright$$

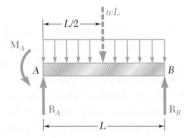

Fig. 9.35

605

Alternative Solution. We may consider the couple exerted at the fixed end A as redundant and replace the fixed end by a pin-and-bracket support. The couple $\mathbf{M}_A$ is now considered as an unknown load (Fig. 9.36a) and will be determined from the condition that the slope of the beam at A must be zero. The solution is carried out by considering separately the slope $(\theta_A)_w$ caused at A by the uniformly distributed load w (Fig. 9.36b) and the slope $(\theta_A)_M$ produced at the same point by the unknown couple $\mathbf{M}_A$ (Fig 9.36c).

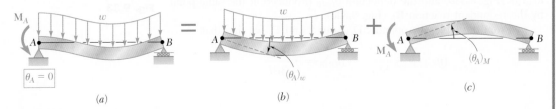

(a)   (b)   (c)

Fig. 9.36

Using the table of Appendix D (cases 6 and 7), and noting that in case 7, A and B must be interchanged, we find that

$$(\theta_A)_w = -\frac{wL^3}{24EI} \qquad (\theta_A)_M = \frac{M_A L}{3EI}$$

Writing that the slope at A is the sum of these two quantities and that it must be zero, we have

$$\theta_A = (\theta_A)_w + (\theta_A)_M = 0$$

$$\theta_A = -\frac{wL^3}{25EI} + \frac{M_A L}{3EI} = 0$$

and, solving for M_A,

$$M_A = \tfrac{1}{8}wL^2 \qquad \mathbf{M}_A = \tfrac{1}{8}wL^2 \text{\Large$\curvearrowright$}$$

The values of R_A and R_B may then be found from the equilibrium equations (9.52) and (9.53).

The beam considered in the preceding example was indeterminate to the first degree. In the case of a beam indeterminate to the second degree (cf. Sec. 9.5), two reactions must be designated as redundant, and the corresponding supports must be eliminated or modified accordingly. The redundant reactions are then treated as unknown loads which, simultaneously and together with the other loads, must produce deformations which are compatible with the original supports. (See Sample Prob. 9.9.)

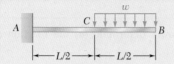

SAMPLE PROBLEM 9.7

For the beam and loading shown, determine the slope and deflection at point B.

SOLUTION

Principle of Superposition. The given loading can be obtained by superposing the loadings shown in the following "picture equation." The beam AB is, of course, the same in each part of the figure.

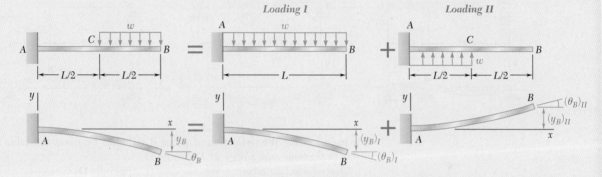

For each of the loadings I and II, we now determine the slope and deflection at B by using the table of *Beam Deflections and Slopes* in Appendix D.

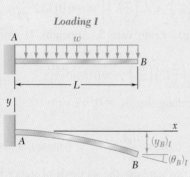

Loading I

$$(\theta_B)_I = -\frac{wL^3}{6EI} \qquad\qquad (y_B)_I = -\frac{wL^4}{8EI}$$

Loading II

$$(\theta_C)_{II} = +\frac{w(L/2)^3}{6EI} = +\frac{wL^3}{48EI} \qquad (y_C)_{II} = +\frac{w(L/2)^4}{8EI} = +\frac{wL^4}{128EI}$$

In portion CB, the bending moment for loading II is zero and thus the elastic curve is a straight line.

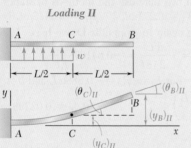

$$(\theta_B)_{II} = (\theta_C)_{II} = +\frac{wL^3}{48EI} \qquad (y_B)_{II} = (y_C)_{II} + (\theta_C)_{II}\left(\frac{L}{2}\right)$$

$$= \frac{wL^4}{128EI} + \frac{wL^3}{48EI}\left(\frac{L}{2}\right) = +\frac{7wL^4}{384EI}$$

Slope at Point B

$$\theta_B = (\theta_B)_I + (\theta_B)_{II} = -\frac{wL^3}{6EI} + \frac{wL^3}{48EI} = -\frac{7wL^3}{48EI} \qquad \theta_B = \frac{7wL^3}{48EI} \;\triangleleft$$

Deflection at B

$$y_B = (y_B)_I + (y_B)_{II} = -\frac{wL^4}{8EI} + \frac{7wL^4}{384EI} = -\frac{41wL^4}{384EI} \qquad y_B = \frac{41wL^4}{384EI} \downarrow \;\triangleleft$$

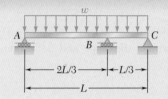

SAMPLE PROBLEM 9.8

For the uniform beam and loading shown, determine (a) the reaction at each support, (b) the slope at end A.

SOLUTION

Principle of Superposition. The reaction $\mathbf{R}_B$ is designated as redundant and considered as an unknown load. The deflections due to the distributed load and to the reaction $\mathbf{R}_B$ are considered separately as shown below.

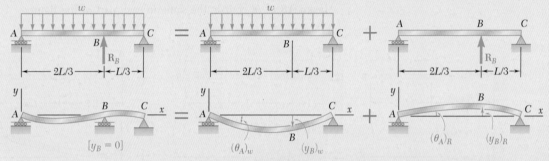

For each loading the deflection at point B is found by using the table of *Beam Deflections and Slopes* in Appendix D.

Distributed Loading. We use case 6, Appendix D

$$y = -\frac{w}{24EI}(x^4 - 2Lx^3 + L^3x)$$

At point B, $x = \frac{2}{3}L$:

$$(y_B)_w = -\frac{w}{24EI}\left[\left(\frac{2}{3}L\right)^4 - 2L\left(\frac{2}{3}L\right)^3 + L^3\left(\frac{2}{3}L\right)\right] = -0.01132\frac{wL^4}{EI}$$

Redundant Reaction Loading. From case 5, Appendix D, with $a = \frac{2}{3}L$ and $b = \frac{1}{3}L$, we have

$$(y_B)_R = -\frac{Pa^2b^2}{3EIL} = +\frac{R_B}{3EIL}\left(\frac{2}{3}L\right)^2\left(\frac{L}{3}\right)^2 = 0.01646\frac{R_BL^3}{EI}$$

a. Reactions at Supports. Recalling that $y_B = 0$, we write

$$y_B = (y_B)_w + (y_B)_R$$

$$0 = -0.01132\frac{wL^4}{EI} + 0.01646\frac{R_BL^3}{EI} \qquad R_B = 0.688wL \uparrow \quad \blacktriangleleft$$

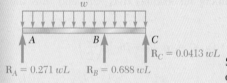

Since the reaction R_B is now known, we may use the methods of statics to determine the other reactions: $\quad R_A = 0.271wL \uparrow \quad R_C = 0.0413wL \uparrow \quad \blacktriangleleft$

b. Slope at End A. Referring again to Appendix D, we have

Distributed Loading. $(\theta_A)_w = -\dfrac{wL^3}{24EI} = -0.04167\dfrac{wL^3}{EI}$

Redundant Reaction Loading. For $P = -R_B = -0.688wL$ and $b = \frac{1}{3}L$

$$(\theta_A)_R = -\frac{Pb(L^2 - b^2)}{6EIL} = +\frac{0.688wL}{6EIL}\left(\frac{L}{3}\right)\left[L^2 - \left(\frac{L}{3}\right)^2\right] \qquad (\theta_A)_R = 0.03398\frac{wL^3}{EI}$$

Finally, $\theta_A = (\theta_A)_w + (\theta_A)_R$

$$\theta_A = -0.04167\frac{wL^3}{EI} + 0.03398\frac{wL^3}{EI} = -0.00769\frac{wL^3}{EI} \qquad \theta_A = 0.00769\frac{wL^3}{EI} \quad \blacktriangleleft$$

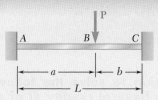

SAMPLE PROBLEM 9.9

For the beam and loading shown, determine the reaction at the fixed support C.

SOLUTION

Principle of Superposition. Assuming the axial force in the beam to be zero, the beam ABC is indeterminate to the second degree and we choose two reaction components as redundant, namely, the vertical force $\mathbf{R}_C$ and the couple $\mathbf{M}_C$. The deformations caused by the given load $\mathbf{P}$, the force $\mathbf{R}_C$, and the couple $\mathbf{M}_C$ will be considered separately as shown.

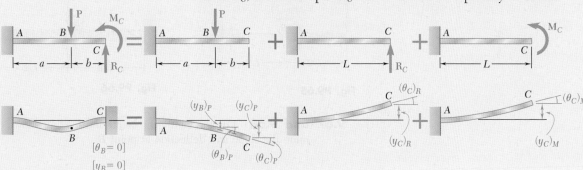

For each load, the slope and deflection at point C will be found by using the table of *Beam Deflections and Slopes* in Appendix D.

Load P. We note that, for this loading, portion BC of the beam is straight.

$$(\theta_C)_P = (\theta_B)_P = -\frac{Pa^2}{2EI} \qquad (y_C)_P = (y_B)_P + (\theta_B)_P b$$

$$= -\frac{Pa^3}{3EI} - \frac{Pa^2}{2EI}b = -\frac{Pa^2}{6EI}(2a + 3b)$$

Force R_C $\qquad (\theta_C)_R = +\frac{R_C L^2}{2EI} \qquad (y_C)_R = +\frac{R_C L^3}{3EI}$

Couple M_C $\quad (\theta_C)_M = +\frac{M_C L}{EI} \qquad (y_C)_M = +\frac{M_C L^2}{2EI}$

Boundary Conditions. At end C the slope and deflection must be zero:

$[x = L, \theta_C = 0]$: $\qquad \theta_C = (\theta_C)_P + (\theta_C)_R + (\theta_C)_M$

$$0 = -\frac{Pa^2}{2EI} + \frac{R_C L^2}{2EI} + \frac{M_C L}{EI} \qquad (1)$$

$[x = L, y_C = 0]$: $\qquad y_C = (y_C)_P + (y_C)_R + (y_C)_M$

$$0 = -\frac{Pa^2}{6EI}(2a + 3b) + \frac{R_C L^3}{3EI} + \frac{M_C L^2}{2EI} \qquad (2)$$

Reaction Components at C. Solving simultaneously Eqs. (1) and (2), we find after reductions

$$R_C = +\frac{Pa^2}{L^3}(a + 3b) \qquad \mathbf{R}_C = \frac{Pa^2}{L^3}(a + 3b) \uparrow \ \blacktriangleleft$$

$$M_C = -\frac{Pa^2 b}{L^2} \qquad \mathbf{M}_C = \frac{Pa^2 b}{L^2} \downarrow \ \blacktriangleleft$$

Using the methods of statics, we can now determine the reaction at A.

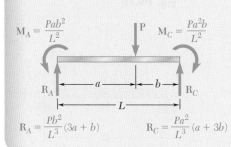

$$M_A = \frac{Pab^2}{L^2} \qquad M_C = \frac{Pa^2 b}{L^2}$$

$$R_A = \frac{Pb^2}{L^3}(3a + b) \qquad R_C = \frac{Pa^2}{L^3}(a + 3b)$$

PROBLEMS

Use the method of superposition to solve the following problems and assume that the flexural rigidity *EI* of each beam is constant.

9.65 and 9.66 For the cantilever beam and loading shown, determine the slope and deflection at the free end.

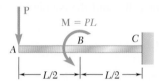

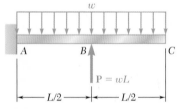

Fig. P9.65

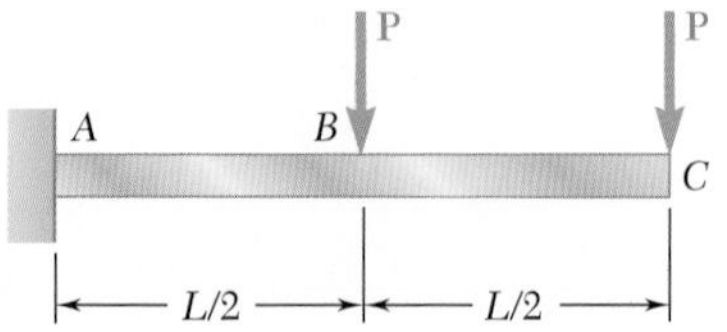

Fig. P9.66

9.67 and 9.68 For the cantilever beam and loading shown, determine the slope and deflection at point *B*.

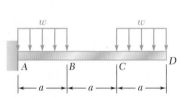

Fig. P9.67

Fig. P9.68

9.69 through 9.72 For the beam and loading shown, determine (*a*) the deflection at *C*, (*b*) the slope at end *A*.

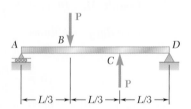

Fig. P9.69

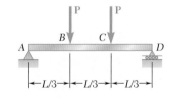

Fig. P9.70

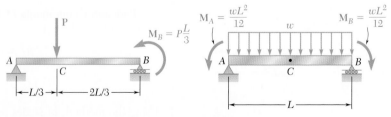

Fig. P9.71

Fig. P9.72

9.73 For the cantilever beam and loading shown, determine the slope and deflection at end *C*. Use $E = 200$ GPa.

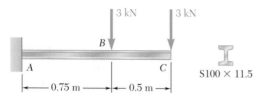

Fig. P9.73 and *P9.74*

9.74 For the cantilever beam and loading shown, determine the slope and deflection at point *B*. Use $E = 200$ GPa.

9.75 For the cantilever beam and loading shown, determine the slope and deflection at end *C*. Use $E = 200$ GPa.

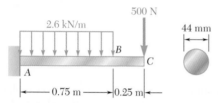

Fig. P9.75 and P9.76

9.76 For the cantilever beam and loading shown, determine the slope and deflection at point *B*. Use $E = 200$ GPa.

9.77 and 9.78 For the beam and loading shown, determine (*a*) the slope at end *A*, (*b*) the deflection at point *C*. Use $E = 200$ GPa.

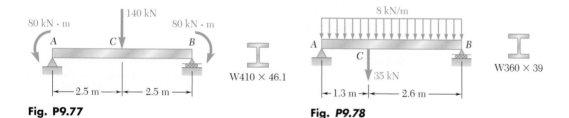

Fig. **P9.77**   Fig. **P9.78**

9.79 and 9.80 For the uniform beam shown, determine the reaction at each of the three supports.

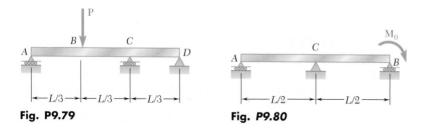

Fig. **P9.79**   Fig. **P9.80**

9.81 and 9.82 For the uniform beam shown, determine (*a*) the reaction at *A*, (*b*) the reaction at *B*.

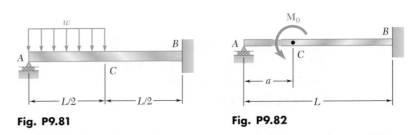

Fig. P9.81   **Fig. P9.82**

9.83 and 9.84 For the beam shown, determine the reaction at *B*.

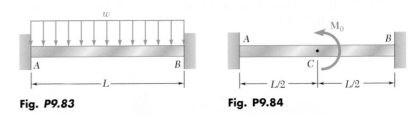

Fig. *P9.83*   **Fig. *P9.84***

9.85 A central beam *BD* is joined at hinges to two cantilever beams *AB* and *DE*. All beams have the cross section shown. For the loading shown, determine the largest *w* so that the deflection at *C* does not exceed 3 mm. Use *E* = 200 GPa.

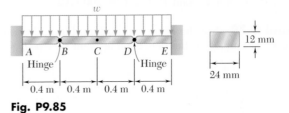

Fig. P9.85

9.86 The two beams shown have the same cross section and are joined by a hinge at *C*. For the loading shown, determine (*a*) the slope at point *A*, (*b*) the deflection at point *B*. Use *E* = 200 GPa.

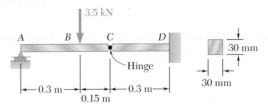

Fig. P9.86

9.87 Beam *DE* rests on the cantilever beam *AC* as shown. Knowing that a square rod of side 10 mm is used for each beam, determine the deflection at end *C* if the 25-N · m couple is applied (*a*) to end *E* of beam *DE*, (*b*) to end *C* of beam *AC*. Use *E* = 200 GPa.

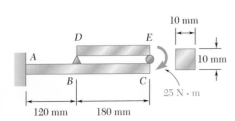

Fig. P9.87

9.88 Beam *AC* rests on the cantilever beam *DE* as shown. Knowing that a W410 × 38.8 rolled-steel shape is used for each beam, determine for the loading shown (*a*) the deflection at point *B*, (*b*) the deflection at point *D*. Use *E* = 200 GPa.

9.89 Before the 30-kN/m load is applied, a gap, δ_0 = 20 mm, exists between the W410 × 60 beam and the support at *C*. Knowing that *E* = 200 GPa, determine the reaction at each support after the uniformly distributed load is applied.

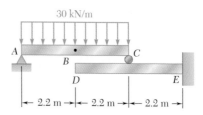

Fig. P9.88

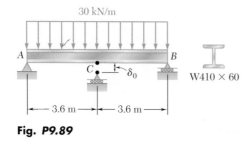

Fig. P9.89

9.90 The cantilever beam *BC* is attached to the steel cable *AB* as shown. Knowing that the cable is initially taut, determine the tension in the cable caused by the distributed load shown. Use *E* = 200 GPa.

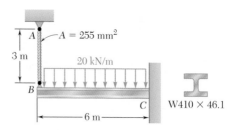

Fig. P9.90

9.91 Before the load **P** was applied, a gap, δ_0 = 0.5 mm, existed between the cantilever beam *AC* and the support at *B*. Knowing that *E* = 200 GPa, determine the magnitude of **P** for which the deflection at *C* is 1 mm.

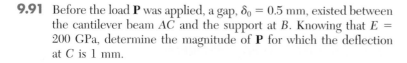

Fig. P9.91

9.92 For the loading shown, and knowing that beams *AB* and *DE* have the same flexural rigidity, determine the reaction (*a*) at *B*, (*b*) at *E*.

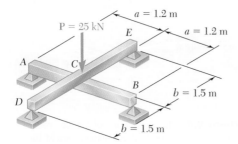

Fig. P9.92

9.93 A 22-mm-diameter rod BC is attached to the lever AB and to the fixed support at C. Lever AB has a uniform cross section 10 mm thick and 25 mm deep. For the loading shown, determine the deflection of point A. Use $E = 200$ GPa and $G = 77$ GPa.

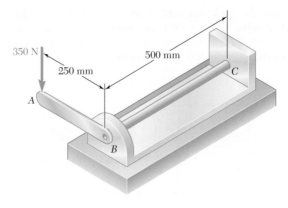

Fig. P9.93

9.94 A 16-mm-diameter rod has been bent into the shape shown. Determine the deflection of end C after the 200-N force is applied. Use $E = 200$ GPa and $G = 80$ GPa.

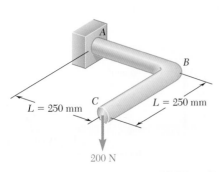

Fig. P9.94

*9.9 MOMENT-AREA THEOREMS

In Sec. 9.2 through Sec. 9.6 we used a mathematical method based on the integration of a differential equation to determine the deflection and slope of a beam at any given point. The bending moment was expressed as a function $M(x)$ of the distance x measured along the beam, and two successive integrations led to the functions $\theta(x)$ and $y(x)$ representing, respectively, the slope and deflection at any point of the beam. In this section you will see how geometric properties of the elastic curve can be used to determine the deflection and slope of a beam at a specific point (Photo 9.4).

Photo 9.4 The deflections of the beams supporting the floors of a building should be taken into account in the design process.

Consider a beam AB subjected to some arbitrary loading (Fig. 9.37a). We draw the diagram representing the variation along the beam of the quantity M/EI obtained by dividing the bending moment M by the flexural rigidity EI (Fig. 9.37b). We note that, except for a difference in the scales of ordinates, this diagram will be the same as the bending-moment diagram if the flexural rigidity of the beam is constant.

Recalling Eq. (9.4) of Sec. 9.3, and the fact that $dy/dx = \theta$, we write

$$\frac{d\theta}{dx} = \frac{d^2y}{dx^2} = \frac{M}{EI}$$

or

$$d\theta = \frac{M}{EI}\, dx \qquad (9.54)\dagger$$

Considering two arbitrary points C and D on the beam and integrating both members of Eq. (9.54) from C to D, we write

$$\int_{\theta_C}^{\theta_D} d\theta = \int_{x_C}^{x_D} \frac{M}{EI}\, dx$$

or

$$\theta_D - \theta_C = \int_{x_C}^{x_D} \frac{M}{EI}\, dx \qquad (9.55)$$

where θ_C and θ_D denote the slope at C and D, respectively (Fig. 9.37c). But the right-hand member of Eq. (9.55) represents the area under the (M/EI) diagram between C and D, and the left-hand member the angle between the tangents to the elastic curve at C and D (Fig. 9.37d). Denoting this angle by $\theta_{D/C}$, we have

$$\theta_{D/C} = \text{area under } (M/EI) \text{ diagram} \atop \text{between } C \text{ and } D \qquad (9.56)$$

This is the *first moment-area theorem*.

We note that the angle $\theta_{D/C}$ and the area under the (M/EI) diagram have the same sign. In other words, a positive area (i.e., an area located above the x axis) corresponds to a counterclockwise rotation of the tangent to the elastic curve as we move from C to D, and a negative area corresponds to a clockwise rotation.

---

†This relation can also be derived without referring to the results obtained in Sec. 9.3, by noting that the angle $d\theta$ formed by the tangents to the elastic curve at P and P' is also the angle formed by the corresponding normals to that curve (Fig. 9.38). We thus have $d\theta = ds/\rho$ where ds is the length of the arc PP' and ρ the radius of curvature at P. Substituting for $1/\rho$ from Eq. (4.21), and noting that, since the slope at P is very small, ds is equal in first approximation to the horizontal distance dx between P and P', we write

$$d\theta = \frac{M}{EI}\, dx \qquad (9.54)$$

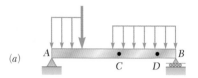

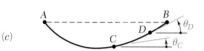

(a)

(b)

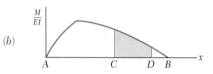

(c)

(d)

Fig. 9.37 First moment-area theorem.

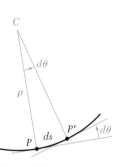

Fig. 9.38

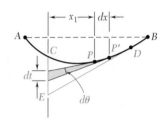

Fig. 9.39

Let us now consider two points P and P' located between C and D, and at a distance dx from each other (Fig. 9.39). The tangents to the elastic curve drawn at P and P' intercept a segment of length dt on the vertical through point C. Since the slope θ at P and the angle $d\theta$ formed by the tangents at P and P' are both small quantities, we can assume that dt is equal to the arc of circle of radius x_1 subtending the angle $d\theta$. We have, therefore,

$$dt = x_1 \, d\theta$$

or, substituting for $d\theta$ from Eq. (9.54),

$$dt = x_1 \frac{M}{EI} \, dx \tag{9.57}$$

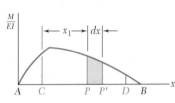

Fig. 9.40

We now integrate Eq. (9.57) from C to D. We note that, as point P describes the elastic curve from C to D, the tangent at P sweeps the vertical through C from C to E. The integral of the left-hand member is thus equal to the vertical distance from C to the tangent at D. This distance is denoted by $t_{C/D}$ and is called the *tangential deviation of C with respect to D*. We have, therefore,

$$t_{C/D} = \int_{x_C}^{x_D} x_1 \frac{M}{EI} \, dx \tag{9.58}$$

We now observe that $(M/EI) \, dx$ represents an element of area under the (M/EI) diagram, and $x_1 \, (M/EI) \, dx$ the first moment of that element with respect to a vertical axis through C (Fig. 9.40). The right-hand member in Eq. (9.58) thus represents the first moment with respect to that axis of the area located under the (M/EI) diagram between C and D.

We can, therefore, state the *second moment-area theorem* as follows: *The tangential deviation $t_{C/D}$ of C with respect to D is equal to the first moment with respect to a vertical axis through C of the area under the (M/EI) diagram between C and D.*

Recalling that the first moment of an area with respect to an axis is equal to the product of the area and of the distance from its centroid to that axis, we may also express the second moment-area theorem as follows:

$$t_{C/D} = (\text{area between } C \text{ and } D)\, \bar{x}_1 \tag{9.59}$$

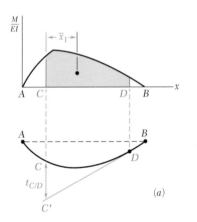

(a)

where the area refers to the area under the (M/EI) diagram, and where $\bar{x}_1$ is the distance from the centroid of the area to the vertical axis through C (Fig. 9.41a).

Care should be taken to distinguish between the tangential deviation of C with respect to D, denoted by $t_{C/D}$, and the tangential deviation of D with respect to C, which is denoted by $t_{D/C}$. The tangential deviation $t_{D/C}$ represents the vertical distance from D to the tangent to the elastic curve at C, and is obtained by multiplying the area under the (M/EI) diagram by the distance $\bar{x}_2$ from its centroid to the vertical axis through D (Fig. 9.41b):

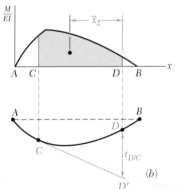

(b)

Fig. 9.41   Second moment-area theorem.

$$t_{D/C} = (\text{area between } C \text{ and } D)\, \bar{x}_2 \tag{9.60}$$

We note that, if an area under the (M/EI) diagram is located above the x axis, its first moment with respect to a vertical axis will be positive; if it is located below the x axis, its first moment will be negative. We check from Fig. 9.41, that a point with a *positive* tangential deviation is located *above* the corresponding tangent, while a point with a *negative* tangential deviation would be located *below* that tangent.

*9.10 APPLICATION TO CANTILEVER BEAMS AND BEAMS WITH SYMMETRIC LOADINGS

We recall that the first moment-area theorem derived in the preceding section defines the angle $\theta_{D/C}$ *between* the tangents at two points C and D of the elastic curve. Thus, the angle θ_D that the tangent at D forms with the horizontal, i.e., the slope at D, can be obtained only if the slope at C is known. Similarly, the second moment-area theorem defines the vertical distance of one point of the elastic curve from the tangent at another point. The tangential deviation $t_{D/C}$, therefore, will help us locate point D only if the tangent at C is known. We conclude that the two moment-area theorems can be applied effectively to the determination of slopes and deflections only if a certain *reference tangent* to the elastic curve has first been determined.

In the case of a *cantilever beam* (Fig. 9.42), the tangent to the elastic curve at the fixed end A is known and can be used as the reference tangent. Since $\theta_A = 0$, the slope of the beam at any point D is $\theta_D = \theta_{D/A}$ and can be obtained by the first moment-area theorem. On the other hand, the deflection y_D of point D is equal to the tangential deviation $t_{D/A}$ measured from the horizontal reference tangent at A and can be obtained by the second moment-area theorem.

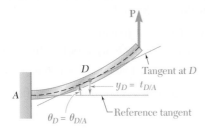

Fig. 9.42 Application of moment-area method to cantilever beams.

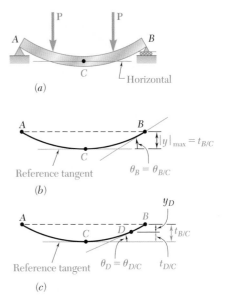

Fig. 9.43 Application of moment-area method to simply supported beams with symmetric loadings.

In the case of a simply supported beam AB with a *symmetric loading* (Fig. 9.43a) or in the case of an overhanging symmetric beam with a symmetric loading (see Sample Prob. 9.11), the tangent at the center C of the beam must be horizontal by reason of symmetry and can be used as the reference tangent (Fig. 9.43b). Since $\theta_C = 0$, the slope at the support B is $\theta_B = \theta_{B/C}$ and can be obtained by the first moment-area theorem. We also note that $|y|_{\max}$ is equal to the tangential deviation $t_{B/C}$ and can, therefore, be obtained by the second moment-area theorem. The slope at any other point D of the beam (Fig. 9.43c) is found in a similar fashion, and the deflection at D can be expressed as $y_D = t_{D/C} - t_{B/C}$.

EXAMPLE 9.09

Determine the slope and deflection at end B of the prismatic cantilever beam AB when it is loaded as shown (Fig. 9.44), knowing that the flexural rigidity of the beam is $EI = 10$ MN $\cdot$ m^2.

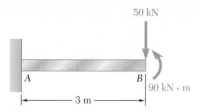

Fig. 9.44

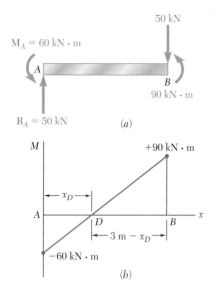

Fig. 9.45

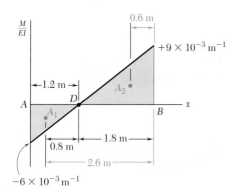

Fig. 9.46

We first draw the free-body diagram of the beam (Fig. 9.45a). Summing vertical components and moments about A, we find that the reaction at the fixed end A consists of a 50 kN upward vertical force $\mathbf{R}_A$ and a 60 kN $\cdot$ m counterclockwise couple $\mathbf{M}_A$. Next, we draw the bending-moment diagram (Fig. 9.45b) and determine from similar triangles the distance x_D from the end A to the point D of the beam where $M = 0$:

$$\frac{x_D}{60} = \frac{3 - x_D}{90} = \frac{3}{150} \qquad x_D = 1.2 \text{ m}$$

Dividing by the flexural rigidity EI the values obtained for M, we draw the (M/EI) diagram (Fig. 9.46) and compute the areas corresponding respectively to the segments AD and DB, assigning a positive sign to the area located above the x axis, and a negative sign to the area located below that axis. Using the first moment-area theorem, we write

$$
\begin{aligned}
\theta_{B/A} = \theta_B - \theta_A &= \text{area from } A \text{ to } B = A_1 + A_2 \\
&= -\tfrac{1}{2}(1.2 \text{ m})(6 \times 10^{-3} \text{ m}^{-1}) + \tfrac{1}{2}(1.8 \text{ m})(9 \times 10^{-3} \text{ m}^{-1}) \\
&= -3.6 \times 10^{-3} + 8.1 \times 10^{-3} \\
&= +4.5 \times 10^{-3} \text{ rad}
\end{aligned}
$$

and, since $\theta_A = 0$,

$$\theta_B = +4.5 \times 10^{-3} \text{ rad}$$

Using now the second moment-area theorem, we write that the tangential deviation $t_{B/A}$ is equal to the first moment about a vertical axis through B of the total area between A and B. Expressing the moment of each partial area as the product of that area and of the distance from its centroid to the axis through B, we have

$$
\begin{aligned}
t_{B/A} &= A_1(2.6 \text{ m}) + A_2(0.6 \text{ m}) \\
&= (-3.6 \times 10^{-3})(2.6 \text{ m}) + (8.1 \times 10^{-3})(0.6 \text{ m}) \\
&= -9.36 \text{ mm} + 4.86 \text{ mm} = -4.50 \text{ mm}
\end{aligned}
$$

Since the reference tangent at A is horizontal, the deflection at B is equal to $t_{B/A}$ and we have

$$y_B = t_{B/A} = -4.50 \text{ mm}$$

The deflected beam has been sketched in Fig. 9.47.

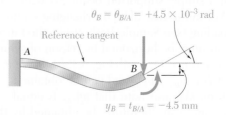

Fig. 9.47

*9.11 BENDING-MOMENT DIAGRAMS BY PARTS

*9.11 BENDING-MOMENT DIAGRAMS BY PARTS

In many applications the determination of the angle $\theta_{D/C}$ and of the tangential deviation $t_{D/C}$ is simplified if the effect of each load is evaluated independently. A separate (M/EI) diagram is drawn for each load, and the angle $\theta_{D/C}$ is obtained by adding algebraically the areas under the various diagrams. Similarly, the tangential deviation $t_{D/C}$ is obtained by adding the first moments of these areas about a vertical axis through D. A bending-moment or (M/EI) diagram plotted in this fashion is said to be *drawn by parts*.

When a bending-moment or (M/EI) diagram is drawn by parts, the various areas defined by the diagram consist of simple geometric shapes, such as rectangles, triangles, and parabolic spandrels. For convenience, the areas and centroids of these various shapes have been indicated in Fig. 9.48.

Shape		Area	c
Rectangle	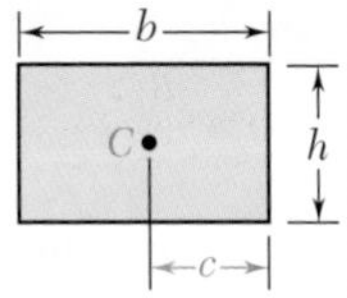	bh	$\dfrac{b}{2}$
Triangle	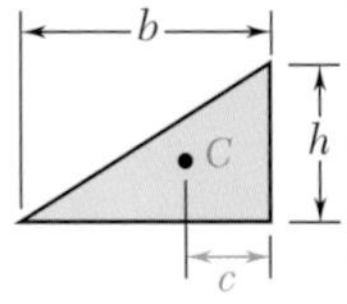	$\dfrac{bh}{2}$	$\dfrac{b}{3}$
Parabolic spandrel	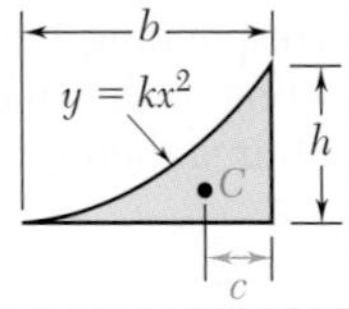 $y = kx^2$	$\dfrac{bh}{3}$	$\dfrac{b}{4}$
Cubic spandrel	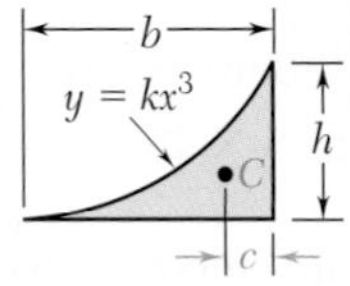 $y = kx^3$	$\dfrac{bh}{4}$	$\dfrac{b}{5}$
General spandrel	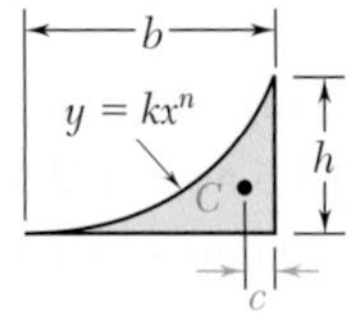 $y = kx^n$	$\dfrac{bh}{n+1}$	$\dfrac{b}{n+2}$

Fig. 9.48 Areas and centroids of common shapes.

EXAMPLE 9.10 Determine the slope and deflection at end B of the prismatic beam of Example 9.09, drawing the bending-moment diagram by parts.

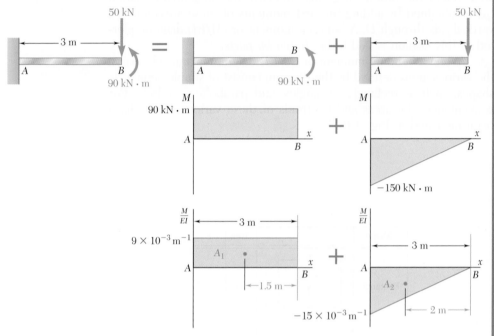

Fig. 9.49

We replace the given loading by the two equivalent loadings shown in Fig. 9.49, and draw the corresponding bending-moment and (M/EI) diagrams from right to left, starting at the free end B.

Applying the first moment-area theorem, and recalling that $\theta_A = 0$, we write

$$\theta_B = \theta_{B/A} = A_1 + A_2$$
$$= (9 \times 10^{-3}\,\text{m}^{-1})(3\,\text{m}) - \tfrac{1}{2}(15 \times 10^{-3}\,\text{m}^{-1})(3\,\text{m})$$
$$= 27 \times 10^{-3} - 22.5 \times 10^{-3} = 4.5 \times 10^{-3}\,\text{rad}$$

Applying the second moment-area theorem, we compute the first moment of each area about a vertical axis through B and write

$$y_B = t_{B/A} = A_1(1.5\,\text{m}) + A_2(2\,\text{m})$$
$$= (27 \times 10^{-3})(1.5\,\text{m}) - (22.5 \times 10^{-3})(2\,\text{m})$$
$$= 40.5\,\text{mm} - 45\,\text{mm} = -4.5\,\text{mm}$$

It is convenient, in practice, to group into a single drawing the two portions of the (M/EI) diagram (Fig. 9.50).

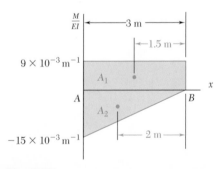

Fig. 9.50

For the prismatic beam AB and the loading shown (Fig. 9.51), determine the slope at a support and the maximum deflection.

EXAMPLE 9.11

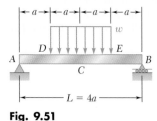

Fig. 9.51

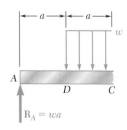

Fig. 9.52

We first sketch the deflected beam (Fig. 9.52). Since the tangent at the center C of the beam is horizontal, it will be used as the reference tangent, and we have $|y|_{max} = t_{A/C}$. On the other hand, since $\theta_C = 0$, we write

$$\theta_{C/A} = \theta_C - \theta_A = -\theta_A \qquad \text{or} \qquad \theta_A = -\theta_{C/A}$$

From the free-body diagram of the beam (Fig. 9.53), we find that

$$R_A = R_B = wa$$

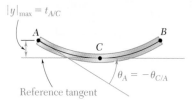

Fig. 9.53

Next, we draw the shear and bending-moment diagrams for the portion AC of the beam. We draw these diagrams by parts, considering separately the effects of the reaction $\mathbf{R}_A$ and of the distributed load. However, for convenience, the two parts of each diagram have been plotted together (Fig. 9.54). We recall from Sec. 5.3 that, the distributed load being uniform, the corresponding parts of the shear and bending-moment diagrams will be, respectively, linear and parabolic. The area and centroid of the triangle and of the parabolic spandrel can be obtained by referring to Fig. 9.48. The areas of the triangle and spandrel are found to be, respectively,

$$A_1 = \frac{1}{2}(2a)\left(\frac{2wa^2}{EI}\right) = \frac{2wa^3}{EI}$$

and

$$A_2 = -\frac{1}{3}(a)\left(\frac{wa^2}{2EI}\right) = -\frac{wa^3}{6EI}$$

Applying the first moment-area theorem, we write

$$\theta_{C/A} = A_1 + A_2 = \frac{2wa^3}{EI} - \frac{wa^3}{6EI} = \frac{11wa^3}{6EI}$$

Recalling from Figs. 9.51 and 9.52 that $a = \frac{1}{4}L$ and $\theta_A = -\theta_{C/A}$, we have

$$\theta_A = -\frac{11wa^3}{6EI} = -\frac{11wL^3}{384EI}$$

Applying now the second moment-area theorem, we write

$$t_{A/C} = A_1\frac{4a}{3} + A_2\frac{7a}{4} = \left(\frac{2wa^3}{EI}\right)\frac{4a}{3} + \left(-\frac{wa^3}{6EI}\right)\frac{7a}{4} = \frac{19wa^4}{8EI}$$

and

$$|y|_{max} = t_{A/C} = \frac{19wa^4}{8EI} = \frac{19wL^4}{2048EI}$$

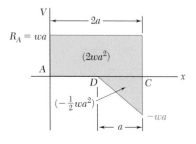

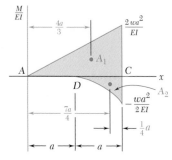

Fig. 9.54

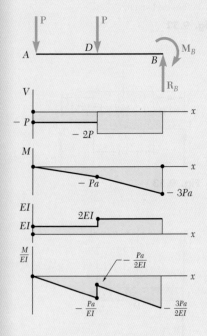

SAMPLE PROBLEM 9.10

The prismatic rods AD and DB are welded together to form the cantilever beam ADB. Knowing that the flexural rigidity is EI in portion AD of the beam and $2EI$ in portion DB, determine, for the loading shown, the slope and deflection at end A.

SOLUTION

(M/EI) Diagram. We first draw the bending-moment diagram for the beam and then obtain the (M/EI) diagram by dividing the value of M at each point of the beam by the corresponding value of the flexural rigidity.

Reference Tangent. We choose the horizontal tangent at the fixed end B as the reference tangent. Since $\theta_B = 0$ and $y_B = 0$, we note that

$$\theta_A = -\theta_{B/A} \qquad y_A = t_{A/B}$$

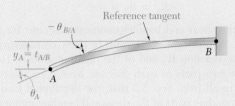

Slope at A. Dividing the (M/EI) diagram into the three triangular portions shown, we write

$$A_1 = -\frac{1}{2}\frac{Pa}{EI}a = -\frac{Pa^2}{2EI}$$

$$A_2 = -\frac{1}{2}\frac{Pa}{2EI}a = -\frac{Pa^2}{4EI}$$

$$A_3 = -\frac{1}{2}\frac{3Pa}{2EI}a = -\frac{3Pa^2}{4EI}$$

Using the first moment-area theorem, we have

$$\theta_{B/A} = A_1 + A_2 + A_3 = -\frac{Pa^2}{2EI} - \frac{Pa^2}{4EI} - \frac{3Pa^2}{4EI} = \frac{-3Pa^2}{2EI}$$

$$\theta_A = -\theta_{B/A} = +\frac{3Pa^2}{2EI} \qquad \theta_A = \frac{3Pa^2}{2EI} \measuredangle \ \blacktriangleleft$$

Deflection at A. Using the second moment-area theorem, we have

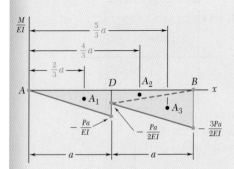

$$y_A = t_{A/B} = A_1\left(\frac{2}{3}a\right) + A_2\left(\frac{4}{3}a\right) + A_3\left(\frac{5}{3}a\right)$$

$$= \left(-\frac{Pa^2}{2EI}\right)\frac{2a}{3} + \left(-\frac{Pa^2}{4EI}\right)\frac{4a}{3} + \left(-\frac{3Pa^2}{4EI}\right)\frac{5a}{3}$$

$$y_A = -\frac{23Pa^3}{12EI} \qquad\qquad y_A = \frac{23Pa^3}{12EI} \downarrow \ \blacktriangleleft$$

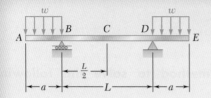

SAMPLE PROBLEM 9.11

For the prismatic beam and loading shown, determine the slope and deflection at end E.

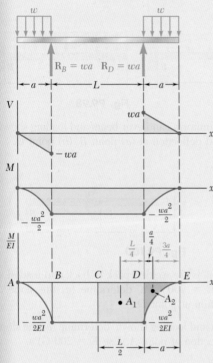

SOLUTION

(M/EI) Diagram. From a free-body diagram of the beam, we determine the reactions and then draw the shear and bending-moment diagrams. Since the flexural rigidity of the beam is constant, we divide each value of M by EI and obtain the (M/EI) diagram shown.

Reference Tangent. Since the beam and its loading are symmetric with respect to the midpoint C, the tangent at C is horizontal and is used as the reference tangent. Referring to the sketch, we observe that, since $\theta_C = 0$,

$$\theta_E = \theta_C + \theta_{E/C} = \theta_{E/C} \tag{1}$$

$$y_E = t_{E/C} - t_{D/C} \tag{2}$$

Slope at E. Referring to the (M/EI) diagram and using the first moment-area theorem, we write

$$A_1 = -\frac{wa^2}{2EI}\left(\frac{L}{2}\right) = -\frac{wa^2L}{4EI}$$

$$A_2 = -\frac{1}{3}\left(\frac{wa^2}{2EI}\right)(a) = -\frac{wa^3}{6EI}$$

Using Eq. (1), we have

$$\theta_E = \theta_{E/C} = A_1 + A_2 = -\frac{wa^2L}{4EI} - \frac{wa^3}{6EI}$$

$$\theta_E = -\frac{wa^2}{12EI}(3L + 2a) \qquad \theta_E = \frac{wa^2}{12EI}(3L + 2a) \;\text{⦢} \;\blacktriangleleft$$

Deflection at E. Using the second moment-area theorem, we write

$$t_{D/C} = A_1\frac{L}{4} = \left(-\frac{wa^2L}{4EI}\right)\frac{L}{4} = -\frac{wa^2L^2}{16EI}$$

$$t_{E/C} = A_1\left(a + \frac{L}{4}\right) + A_2\left(\frac{3a}{4}\right)$$

$$= \left(-\frac{wa^2L}{4EI}\right)\left(a + \frac{L}{4}\right) + \left(-\frac{wa^3}{6EI}\right)\left(\frac{3a}{4}\right)$$

$$= -\frac{wa^3L}{4EI} - \frac{wa^2L^2}{16EI} - \frac{wa^4}{8EI}$$

Using Eq. (2), we have

$$y_E = t_{E/C} - t_{D/C} = -\frac{wa^3L}{4EI} - \frac{wa^4}{8EI}$$

$$y_E = -\frac{wa^3}{8EI}(2L + a) \qquad y_E = \frac{wa^3}{8EI}(2L + a)\downarrow \;\blacktriangleleft$$

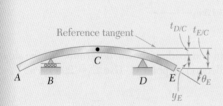

PROBLEMS

Use the moment-area method to solve the following problems.

9.95 through 9.98 For the uniform cantilever beam and loading shown, determine (a) the slope at the free end, (b) the deflection at the free end.

Fig. P9.95

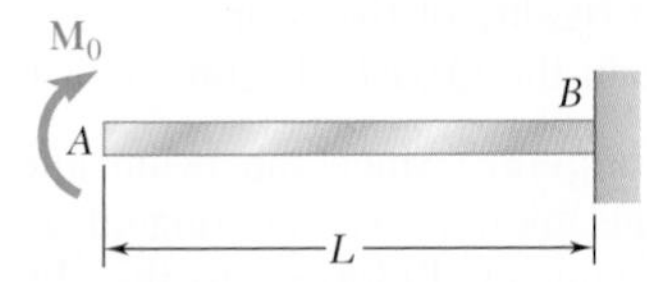

Fig. P9.96

Fig. P9.97

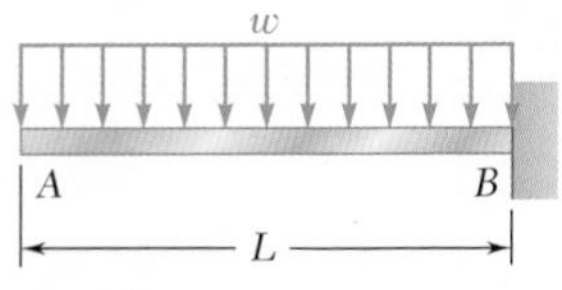

Fig. P9.98

9.99 and 9.100 For the uniform cantilever beam and loading shown, determine the slope and deflection at (a) point B, (b) point C.

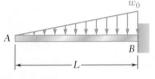

Fig. P9.99

Fig. P9.100

9.101 Two C150 × 12.2 channels are welded back to back and loaded as shown. Knowing that $E = 200$ GPa, determine (a) the slope at point D, (b) the deflection at point D.

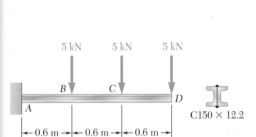

Fig. P9.101

9.102 For the cantilever beam and loading shown, determine (a) the slope at point A, (b) the deflection at point A. Use $E = 200$ GPa.

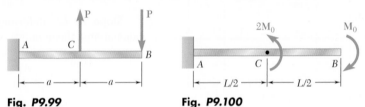

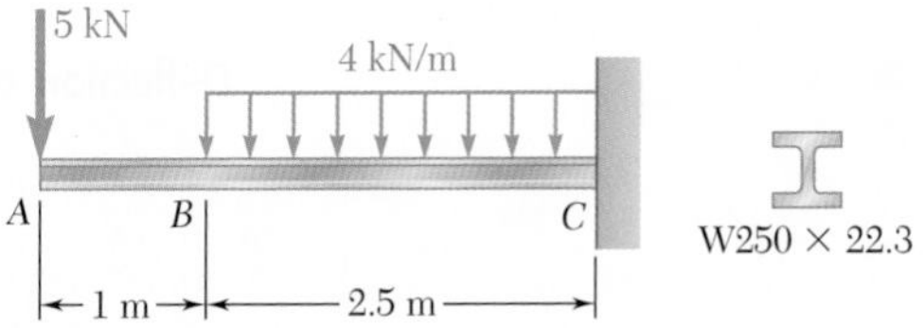

Fig. P9.102

9.103 For the cantilever beam and loading shown, determine (a) the slope at point C, (b) the deflection at point C. Use $E = 200$ GPa.

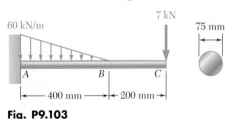

Fig. P9.103

9.104 For the cantilever beam and loading shown, determine (a) the slope at point A, (b) the deflection at point A. Use $E = 200$ GPa.

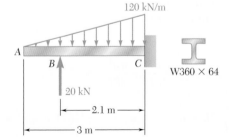

Fig. P9.104

9.105 For the cantilever beam and loading shown, determine (*a*) the slope at point *C*, (*b*) the deflection at point *C*.

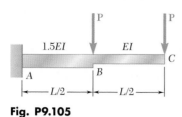

Fig. P9.105

9.106 For the cantilever beam and loading shown, determine the deflection and slope at end *A* caused by the moment $\mathbf{M_0}$.

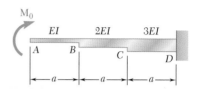

Fig. P9.106

9.107 Two cover plates are welded to the rolled-steel beam as shown. Using $E = 200$ GPa, determine the slope and deflection at end *C*.

9.108 Two cover plates are welded to the rolled-steel beam as shown. Using $E = 200$ GPa, determine (*a*) the slope at end *A*, (*b*) the deflection at end *A*.

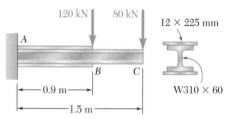

Fig. P9.107

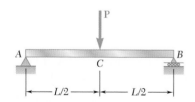

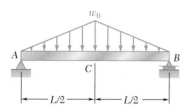

Fig. P9.108

9.109 through 9.114 For the prismatic beam and loading shown, determine (*a*) the slope at end *A*, (*b*) the deflection at the center *C* of the beam.

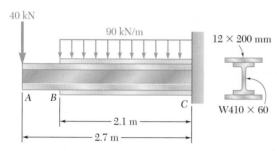

Fig. P9.109

Fig. P9.110

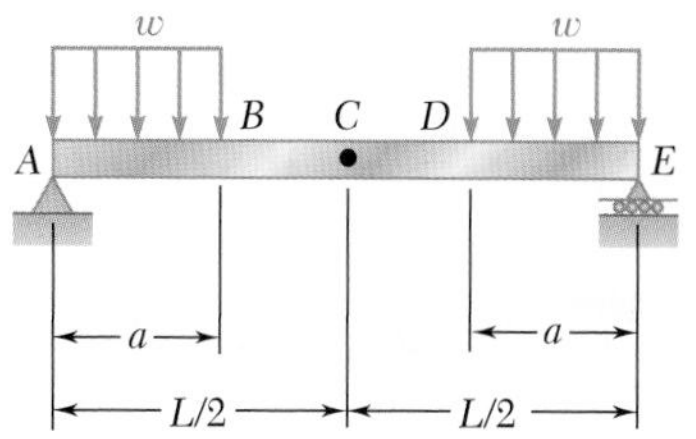

Fig. P9.111

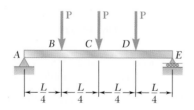

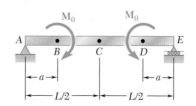

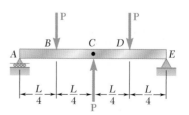

Fig. P9.112

Fig. P9.113

Fig. P9.114

9.115 and 9.116 For the beam and loading shown, determine (*a*) the slope at end *A*, (*b*) the deflection at the center *C* of the beam.

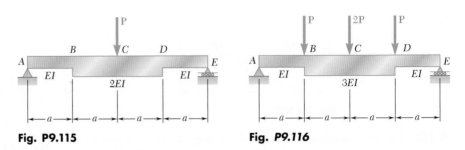

Fig. P9.115   **Fig. *P9.116***

9.117 For the beam and loading shown and knowing that *w* = 8 kN/m, determine (*a*) the slope at end *A*, (*b*) the deflection at midpoint *C*. Use *E* = 200 GPa.

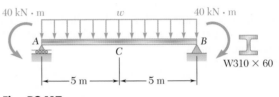

Fig. P9.117

9.118 and 9.119 For the beam and loading shown, determine (*a*) the slope at end *A*, (*b*) the deflection at the midpoint of the beam. Use *E* = 200 GPa.

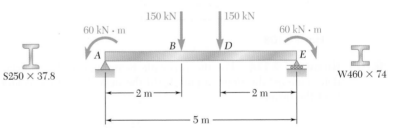

Fig. P9.118   **Fig. P9.119**

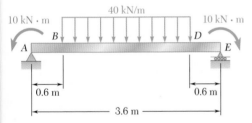

Fig. *P9.120*

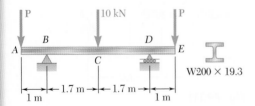

Fig. P9.123 and P9.124

9.120 Knowing that *P* = 8 kN, determine (*a*) the slope at end *A*, (*b*) the deflection at the midpoint *C* of the beam. Use *E* = 200 GPa.

9.121 For the beam and loading of Prob. 9.117, determine the value of *w* for which the deflection is zero at the midpoint *C* of the beam. Use *E* = 200 GPa.

9.122 For the beam and loading of Prob. 9.120, determine the magnitude of the forces **P** for which the deflection is zero at end *A* of the beam. Use *E* = 200 GPa.

*9.123 A uniform rod *AE* is to be supported at two points *B* and *D*. Determine the distance *a* for which the slope at ends *A* and *E* is zero.

*9.124 A uniform rod *AE* is to be supported at two points *B* and *D*. Determine the distance *a* from the ends of the rod to the points of support, if the downward deflections of points *A*, *C*, and *E* are to be equal.

*9.12 APPLICATION OF MOMENT-AREA THEOREMS TO BEAMS WITH UNSYMMETRIC LOADINGS

We saw in Sec. 9.10 that, when a simply supported or overhanging beam carries a symmetric load, the tangent at the center C of the beam is horizontal and can be used as the reference tangent. When a simply supported or overhanging beam carries an unsymmetric load, it is generally not possible to determine by inspection the point of the beam where the tangent is horizontal. Other means must then be found for locating a reference tangent, i.e., a tangent of known slope to be used in applying either of the two moment-area theorems.

It is usually most convenient to select the reference tangent at one of the beam supports. Considering, for example, the tangent at the support A of the simply supported beam AB (Fig. 9.55), we determine its slope by computing the tangential deviation $t_{B/A}$ of the support B with respect to A, and dividing $t_{B/A}$ by the distance L between the supports. Recalling that the tangential deviation of a point located above the tangent is positive, we write

$$\theta_A = -\frac{t_{B/A}}{L} \tag{9.61}$$

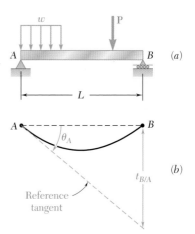

Fig. 9.55

Once the slope of the reference tangent has been found, the slope θ_D of the beam at any point D (Fig. 9.56) can be determined by using the first moment-area theorem to obtain $\theta_{D/A}$, and then writing

$$\theta_D = \theta_A + \theta_{D/A} \tag{9.62}$$

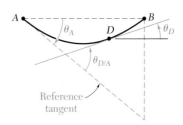

Fig. 9.56

The tangential deviation $t_{D/A}$ of D with respect to the support A can be obtained from the second moment-area theorem. We note that $t_{D/A}$ is equal to the segment ED (Fig. 9.57) and represents the vertical distance of D *from the reference tangent*. On the other hand, the deflection y_D of point D represents the vertical distance of D *from the horizontal line* AB (Fig. 9.58). Since y_D is equal in

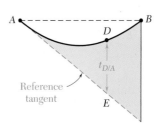

Fig. 9.57

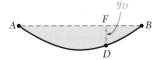

Fig. 9.58

magnitude to the segment FD, it can be expressed as the difference between EF and ED (Fig. 9.59). Observing from the similar triangles AFE and ABH that

$$\frac{EF}{x} = \frac{HB}{L} \qquad \text{or} \qquad EF = \frac{x}{L}t_{B/A}$$

and recalling the sign conventions used for deflections and tangential deviations, we write

$$y_D = ED - EF = t_{D/A} - \frac{x}{L}t_{B/A} \tag{9.63}$$

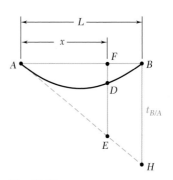

Fig. 9.59

EXAMPLE 9.12

For the prismatic beam and loading shown (Fig. 9.60), determine the slope and deflection at point D.

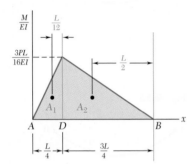

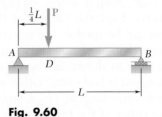

Fig. 9.60

Reference Tangent at Support A. We compute the reactions at the supports and draw the (M/EI) diagram (Fig. 9.61). We determine the tangential deviation $t_{B/A}$ of the support B with respect to the support A by applying the second moment-area theorem and computing the moments about a vertical axis through B of the areas A_1 and A_2. We have

$$A_1 = \frac{1}{2}\frac{L}{4}\frac{3PL}{16EI} = \frac{3PL^2}{128EI} \qquad A_2 = \frac{1}{2}\frac{3L}{4}\frac{3PL}{16EI} = \frac{9PL^2}{128EI}$$

$$t_{B/A} = A_1\left(\frac{L}{12} + \frac{3L}{4}\right) + A_2\left(\frac{L}{2}\right)$$

$$= \frac{3PL^2}{128EI}\frac{10L}{12} + \frac{9PL^2}{128EI}\frac{L}{2} = \frac{7PL^3}{128EI}$$

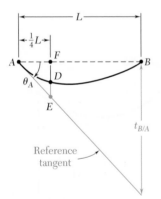

Fig. 9.61

The slope of the reference tangent at A (Fig. 9.62) is

$$\theta_A = -\frac{t_{B/A}}{L} = -\frac{7PL^2}{128EI}$$

Slope at D. Applying the first moment-area theorem from A to D, we write

$$\theta_{D/A} = A_1 = \frac{3PL^2}{128EI}$$

Thus, the slope at D is

$$\theta_D = \theta_A + \theta_{D/A} = -\frac{7PL^2}{128EI} + \frac{3PL^2}{128EI} = -\frac{PL^2}{32EI}$$

Fig. 9.62

Deflection at D. We first determine the tangential deviation $DE = t_{D/A}$ by computing the moment of the area A_1 about a vertical axis through D:

$$DE = t_{D/A} = A_1\left(\frac{L}{12}\right) = \frac{3PL^2}{128EI}\frac{L}{12} = \frac{PL^3}{512EI}$$

The deflection at D is equal to the difference between the segments DE and EF (Fig. 9.62). We have

$$y_D = DE - EF = t_{D/A} - \tfrac{1}{4}t_{B/A}$$

$$= \frac{PL^3}{512EI} - \frac{1}{4}\frac{7PL^3}{128EI}$$

$$y_D = -\frac{3PL^3}{256EI} = -0.01172PL^3/EI$$

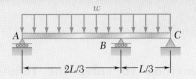

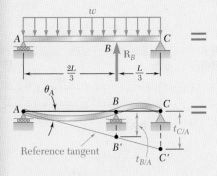

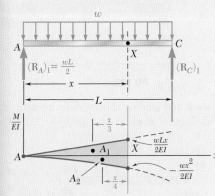

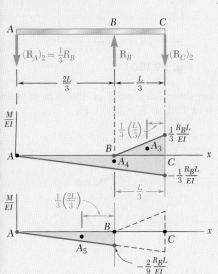

SAMPLE PROBLEM 9.14

For the uniform beam and loading shown, determine the reaction at B.

SOLUTION

The beam is indeterminate to the first degree. We choose the reaction $\mathbf{R}_B$ as redundant and consider separately the distributed loading and the redun-

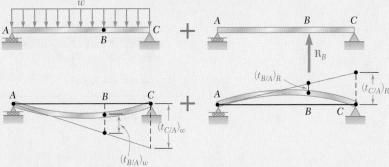

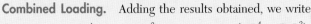

dant reaction loading. We next select the tangent at A as the reference tangent. From the similar triangles ABB' and ACC', we find that

$$\frac{t_{C/A}}{L} = \frac{t_{B/A}}{\frac{2}{3}L} \qquad t_{C/A} = \frac{3}{2}\,t_{B/A} \tag{1}$$

For each loading, we draw the (M/EI) diagram and then determine the tangential deviations of B and C with respect to A.

Distributed Loading. Considering the (M/EI) diagram from end A to an arbitrary point X, we write

$$(t_{X/A})_w = A_1 \frac{x}{3} + A_2 \frac{x}{4} = \left(\frac{1}{2}\frac{wLx}{2EI}x\right)\frac{x}{3} + \left(-\frac{1}{3}\frac{wx^2}{2EI}x\right)\frac{x}{4} = \frac{wx^3}{24EI}(2L - x)$$

Letting successively $x = L$ and $x = \frac{2}{3}L$, we have

$$(t_{C/A})_w = \frac{wL^4}{24EI} \qquad (t_{B/A})_w = \frac{4}{243}\frac{wL^4}{EI}$$

Redundant Reaction Loading

$$(t_{C/A})_R = A_3 \frac{L}{9} + A_4 \frac{L}{3} = \left(\frac{1}{2}\frac{R_B L}{3EI}\frac{L}{3}\right)\frac{L}{9} + \left(-\frac{1}{2}\frac{R_B L}{3EI}L\right)\frac{L}{3} = -\frac{4}{81}\frac{R_B L^3}{EI}$$

$$(t_{B/A})_R = A_5 \frac{2L}{9} = \left[-\frac{1}{2}\frac{2R_B L}{9EI}\left(\frac{2L}{3}\right)\right]\frac{2L}{9} = -\frac{4}{243}\frac{R_B L^3}{EI}$$

Combined Loading. Adding the results obtained, we write

$$t_{C/A} = \frac{wL^4}{24EI} - \frac{4}{81}\frac{R_B L^3}{EI} \qquad t_{B/A} = \frac{4}{243}\frac{(wL^4 - R_B L^3)}{EI}$$

Reaction at B. Substituting for $t_{C/A}$ and $t_{B/A}$ into Eq. (1), we have

$$\left(\frac{wL^4}{24EI} - \frac{4}{81}\frac{R_B L^3}{EI}\right) = \frac{3}{2}\left[\frac{4}{243}\frac{(wL^4 - R_B L^3)}{EI}\right]$$

$$R_B = 0.6875wL \qquad R_B = 0.688wL \uparrow \quad \blacktriangleleft$$

PROBLEMS

Use the moment-area method to solve the following problems.

9.125 through 9.128 For the prismatic beam and loading shown, determine (a) the deflection at point D, (b) the slope at end A.

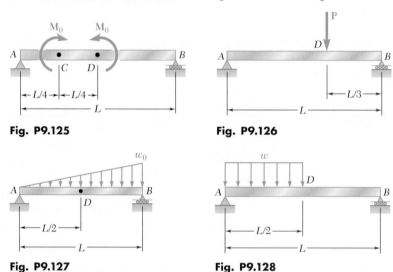

Fig. P9.125

Fig. P9.126

Fig. P9.127

Fig. P9.128

9.129 and 9.130 For the beam and loading shown, determine (a) the slope at end A, (b) the deflection at point D. Use $E = 200$ GPa.

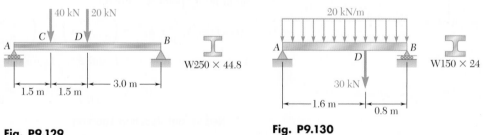

Fig. P9.129

Fig. P9.130

9.131 For the beam and loading shown, determine (a) the slope at point A, (b) the deflection at point E. Use $E = 200$ GPa.

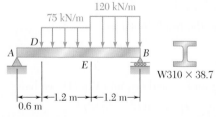

Fig. P9.131

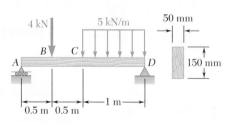

Fig. P9.132

9.132 For the timber beam and loading shown, determine (a) the slope at point A, (b) the deflection at point C. Use $E = 12$ GPa.

9.133 For the beam and loading shown, determine (*a*) the slope at point *A*, (*b*) the deflection at point *D*.

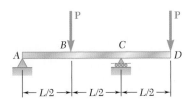

Fig. P9.133

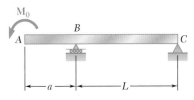

Fig. P9.134

9.134 For the beam and loading shown, determine (*a*) the slope at point *A*, (*b*) the deflection at point *A*.

9.135 For the beam and loading shown, determine (*a*) the slope at point *C*, (*b*) the deflection at point *D*. Use $E = 200$ GPa.

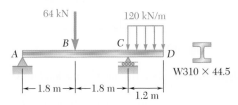

Fig. P9.135

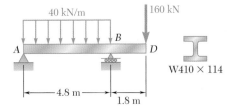

Fig. P9.136

9.136 For the beam and loading shown, determine (*a*) the slope at point *B*, (*b*) the deflection at point *D*. Use $E = 200$ GPa.

9.137 Knowing that the beam *AB* is made of a solid steel rod of diameter $d = 18$ mm, determine for the loading shown (*a*) the slope at point *D*, (*b*) the deflection at point *A*. Use $E = 200$ GPa.

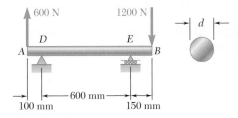

Fig. P9.137

9.138 Knowing that the beam *AD* is made of a solid steel bar, determine (*a*) the slope at point *B*, (*b*) the deflection at point *A*. Use $E = 200$ GPa.

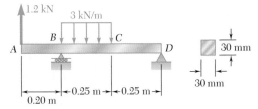

Fig. P9.138

9.139 For the beam and loading shown, determine the deflection (*a*) at point *D*, (*b*) at point *E*.

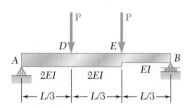

Fig. P9.139

9.140 For the beam and loading shown, determine (*a*) the slope at end *A*, (*b*) the slope at end *B*, (*c*) the deflection at the midpoint *C*.

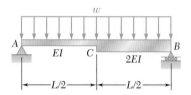

Fig. P9.140

9.141 through 9.144 For the beam and loading shown, determine the magnitude and location of the largest downward deflection.

9.141 Beam and loading of Prob. 9.126
9.142 Beam and loading of Prob. 9.127
9.143 Beam and loading of Prob. 9.129
9.144 Beam and loading of Prob. 9.131

9.145 For the beam and loading of Prob. 9.136, determine the largest upward deflection in span *AB*.

9.146 For the beam and loading of Prob. 9.137, determine the largest upward deflection in span *DE*.

9.147 through 9.150 For the beam and loading shown, determine the reaction at the roller support.

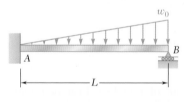

Fig. *P9.147*

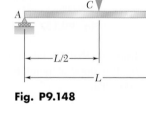

Fig. P9.148

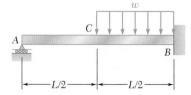

Fig. P9.149

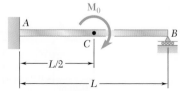

Fig. P9.150

REVIEW PROBLEMS

9.157 For the loading shown, determine (*a*) the equation of the elastic curve for the cantilever beam *AB*, (*b*) the deflection at the free end, (*c*) the slope at the free end.

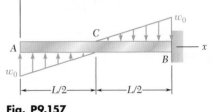

Fig. P9.157

9.158 (*a*) Determine the location and magnitude of the maximum absolute deflection in *AB* between *A* and the center of the beam. (*b*) Assuming that beam *AB* is a W460 × 113 rolled shape, $M_0 = 224$ kN · m, and $E = 200$ GPa, determine the maximum allowable length *L* so that the maximum deflection does not exceed 1.2 mm.

Fig. P9.158

9.159 For the beam and loading shown, determine (*a*) the equation of the elastic curve, (*b*) the slope at end *A*, (*c*) the deflection at the midpoint of the span.

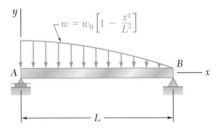

Fig. P9.159

9.160 Determine the reaction at *A* and draw the bending moment diagram for the beam and loading shown.

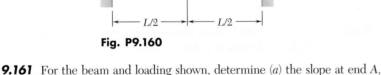

Fig. P9.160

9.161 For the beam and loading shown, determine (*a*) the slope at end *A*, (*b*) the deflection at point *B*. Use $E = 200$ GPa.

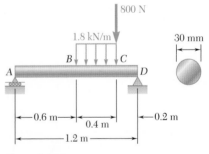

Fig. P9.161

9.162 The rigid bar *BDE* is welded at point *B* to the rolled-steel beam *AC*. For the loading shown, determine (*a*) the slope at point *A*, (*b*) the deflection at point *B*. Use $E = 200$ GPa.

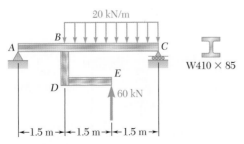

Fig. P9.162

647

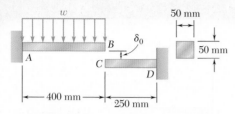

Fig. P9.163

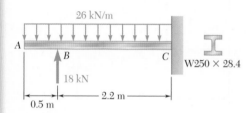

Fig. P9.165

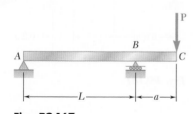

Fig. P9.167

9.163 Before the uniformly distributed load w is applied, a gap, $\delta_0 = 1.2$ mm, exists between the ends of the cantilever bars AB and CD. Knowing that $E = 105$ GPa and $w = 30$ kN/m, determine (a) the reaction at A, (b) the reaction at D.

9.164 A central beam BD is joined at hinges to two cantilever beams AB and DE. All beams have the cross section shown. For the loading shown, determine the largest w so that the deflection at C does not exceed 3 mm. Use $E = 200$ GPa.

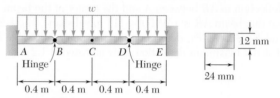

Fig. P9.164

9.165 For the cantilever beam and loading shown, determine (a) the slope at point A, (b) the deflection at point A. Use $E = 200$ GPa.

9.166 Knowing that the magnitude of the load **P** is 30 kN, determine (a) the slope at end A, (b) the deflection at end A, (c) the deflection at midpoint C of the beam. Use $E = 200$ GPa.

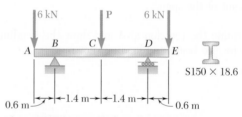

Fig. P9.166

9.167 For the beam and loading shown, determine (a) the slope at point C, (b) the deflection at point C.

9.168 A hydraulic jack can be used to raise point B of the cantilever beam ABC. The beam was originally straight, horizontal, and unloaded. A 20-kN load was then applied at point C, causing this point to move down. Determine (a) how much point B should be raised to return point C to its original position, (b) the final value of the reaction at B. Use $E = 200$ GPa.

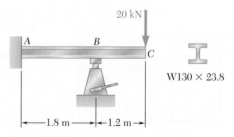

Fig. P9.168

COMPUTER PROBLEMS

The following problems are designed to be solved with a computer.

9.C1 Several concentrated loads can be applied to the cantilever beam AB. Write a computer program to calculate the slope and deflection of beam AB from $x = 0$ to $x = L$, using given increments Δx. Apply this program with increments $\Delta x = 50$ mm to the beam and loading of Prob. 9.73 and Prob. 9.74.

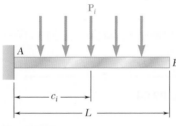

Fig. P9.C1

9.C2 The 6-m beam AB consists of a W530 × 92 rolled-steel shape and supports a 50-kN/m distributed load as shown. Write a computer program and use it to calculate for values of a from 0 to 6 m, using 0.3-m increments, (*a*) the slope and deflection at D, (*b*) the location and magnitude of the maximum deflection. Use $E = 200$ GPa.

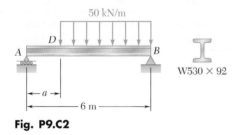

Fig. P9.C2

9.C3 The cantilever beam AB carries the distributed loads shown. Write a computer program to calculate the slope and deflection of beam AB from $x = 0$ to $x = L$ using given increments Δx. Apply this program with increments $\Delta x = 100$ mm, assuming that $L = 2.4$ m, $w = 36$ kN/m, and (*a*) $a = 0.6$ m, (*b*) $a = 1.2$ m, (*c*) $a = 1.8$ m. Use $E = 200$ GPa.

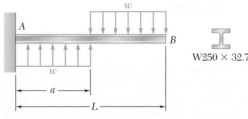

Fig. P9.C3

9.C4 The simple beam AB is of constant flexural rigidity EI and carries several concentrated loads as shown. Using the *Method of Integration*, write a computer program that can be used to calculate the slope and deflection at points along the beam from $x = 0$ to $x = L$ using given increments Δx. Apply this program to the beam and loading of (*a*) Prob. 9.13 with $\Delta x = 0.3$ m, (*b*) Prob. 9.16 with $\Delta x = 0.05$ m, (*c*) Prob. 9.129 with $\Delta x = 0.25$ m.

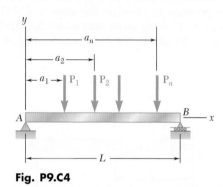

Fig. P9.C4

9.C5 The supports of beam AB consist of a fixed support at end A and a roller support located at point D. Write a computer program that can be used to calculate the slope and deflection at the free end of the beam for values of a from 0 to L using given increments Δa. Apply this program to calculate the slope and deflection at point B for each of the following cases:

	l	Δl	w	E	Shape
(a)	3.6 m	0.15 m	24 kN/m	200 GPa	W410 × 65
(b)	3 m	0.2 m	18 kN/m	200 GPa	W460 × 113

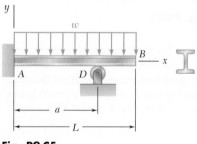

Fig. P9.C5

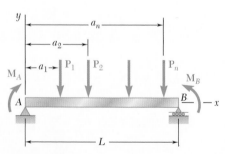

Fig. P9.C6

9.C6 For the beam and loading shown, use the *Moment-Area Method* to write a computer program to calculate the slope and deflection at points along the beam from $x = 0$ to $x = L$ using given increments Δx. Apply this program to calculate the slope and deflection at each concentrated load for the beam of (*a*) Prob. 9.77 with $\Delta x = 0.5$ m, (*b*) Prob. 9.119 with $\Delta x = 0.5$ m.

9.C7 Two 52-kN loads are maintained 2.5 m apart as they are moved slowly across beam AB. Write a computer program to calculate the deflection at the midpoint C of the beam for values of x from 0 to 9 m, using 0.5-m increments. Use $E = 200$ GPa.

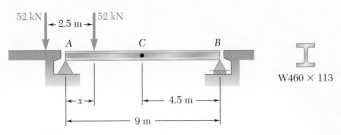

W460 × 113

Fig. P9.C7

9.C8 A uniformly distributed load w and several distributed loads P_i may be applied to beam AB. Write a computer program to determine the reaction at the roller support and apply this program to the beam and loading of (*a*) Prob. 9.53*a*, (*b*) Prob. 9.155.

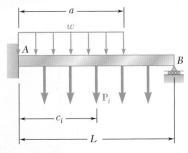

Fig. P9.C8

The curved pedestrian bridge is supported by a series of columns. The analysis and design of members supporting axial compressive loads will be discussed in this chapter.

652

CHAPTER 10

Columns

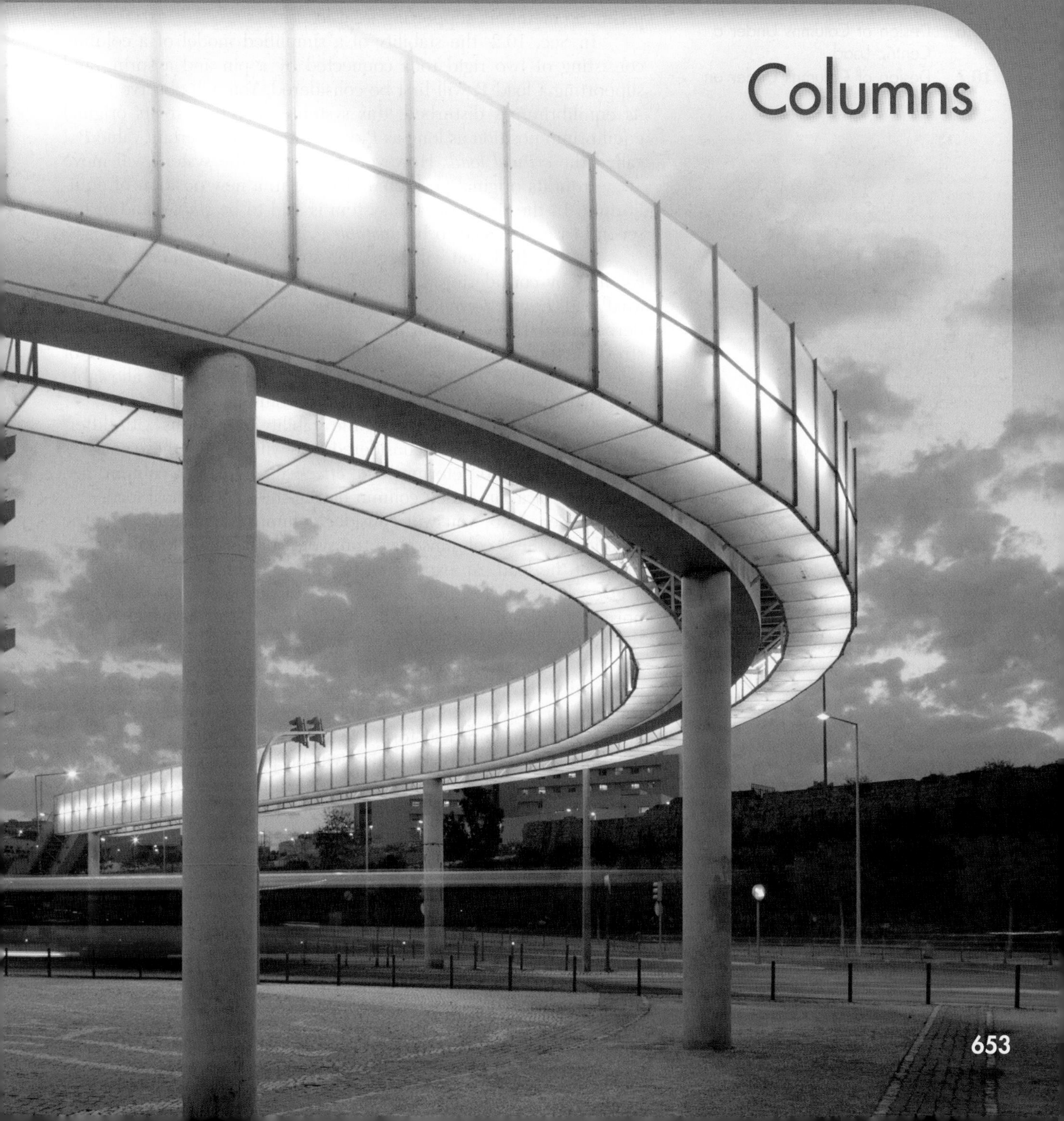

10.1   INTRODUCTION

In the preceding chapters, we had two primary concerns: (1) the strength of the structure, i.e., its ability to support a specified load without experiencing excessive stress; (2) the ability of the structure to support a specified load without undergoing unacceptable deformations. In this chapter, our concern will be with the stability of the structure, i.e., with its ability to support a given load without experiencing a sudden change in its configuration. Our discussion will relate chiefly to columns, i.e., to the analysis and design of vertical prismatic members supporting axial loads.

In Sec. 10.2, the stability of a simplified model of a column, consisting of two rigid rods connected by a pin and a spring and supporting a load **P,** will first be considered. You will observe that if its equilibrium is disturbed, this system will return to its original equilibrium position as long as P does not exceed a certain value P_{cr}, called the *critical load*. However, if $P > P_{cr}$, the system will move away from its original position and settle in a new position of equilibrium. In the first case, the system is said to be *stable,* and in the second case, it is said to be *unstable.*

In Sec. 10.3, you will begin the study of the *stability of elastic columns* by considering a pin-ended column subjected to a centric axial load. *Euler's formula* for the critical load of the column will be derived and from that formula the corresponding critical normal stress in the column will be determined. By applying a factor of safety to the critical load, you will be able to determine the allowable load that can be applied to a pin-ended column.

In Sec. 10.4, the analysis of the stability of columns with different end conditions will be considered. You will simplify these analyses by learning how to determine the *effective length* of a column, i.e., the length of a pin-ended column having the same critical load.

In Sec. 10.5, you will consider columns supporting eccentric axial loads; these columns have transverse deflections for all magnitudes of the load. An expression for the maximum deflection under a given load will be derived and used to determine the maximum normal stress in the column. Finally, the *secant formula,* which relates the average and maximum stresses in a column, will be developed.

In the first sections of the chapter, each column is initially assumed to be a straight homogeneous prism. In the last part of the chapter, you will consider real columns which are designed and analyzed using empirical formulas set forth by professional organizations. In Sec. 10.6, formulas will be presented for the allowable stress in columns made of steel, aluminum, or wood and subjected to a centric axial load. In the last section of the chapter (Sec. 10.7), the design of columns under an eccentric axial load will be considered.

10.2   STABILITY OF STRUCTURES

Suppose we are to design a column AB of length L to support a given load **P** (Fig. 10.1). The column will be pin-connected at both ends and we assume that **P** is a centric axial load. If the cross-sectional

Fig. 10.1   Column.

area A of the column is selected so that the value $\sigma = P/A$ of the stress on a transverse section is less than the allowable stress σ_{all} for the material used, and if the deformation $\delta = PL/AE$ falls within the given specifications, we might conclude that the column has been properly designed. However, it may happen that, as the load is applied, the column will *buckle*; instead of remaining straight, it will suddenly become sharply curved (Fig. 10.2). Photo 10.1 shows a column that has been loaded so that it is no longer straight; the column has buckled. Clearly, a column that buckles under the load it is to support is not properly designed.

Fig. 10.2 Buckled column.

Photo 10.1 Laboratory test showing a buckled column.

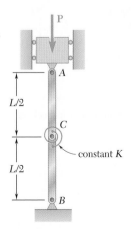

Fig. 10.3 Column model.

Before getting into the actual discussion of the stability of elastic columns, some insight will be gained on the problem by considering a simplified model consisting of two rigid rods AC and BC connected at C by a pin and a torsional spring of constant K (Fig. 10.3).

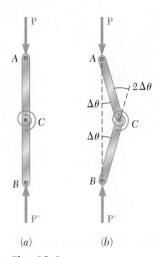

Fig. 10.4

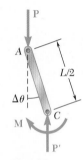

Fig. 10.5

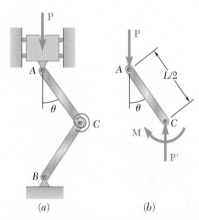

Fig. 10.6 Column model in buckled position.

If the two rods and the two forces **P** and **P′** are perfectly aligned, the system will remain in the position of equilibrium shown in Fig.10.4a as long as it is not disturbed. But suppose that we move C slightly to the right, so that each rod now forms a small angle $\Delta\theta$ with the vertical (Fig. 10.4b). Will the system return to its original equilibrium position, or will it move further away from that position? In the first case, the system is said to be *stable*, and in the second case, it is said to be *unstable*.

To determine whether the two-rod system is stable or unstable, we consider the forces acting on rod AC (Fig. 10.5). These forces consist of two couples, namely the couple formed by **P** and **P′**, of moment $P(L/2) \sin \Delta\theta$, which tends to move the rod away from the vertical, and the couple **M** exerted by the spring, which tends to bring the rod back into its original vertical position. Since the angle of deflection of the spring is $2\,\Delta\theta$, the moment of the couple **M** is $M = K(2\,\Delta\theta)$. If the moment of the second couple is larger than the moment of the first couple, the system tends to return to its original equilibrium position; the system is stable. If the moment of the first couple is larger than the moment of the second couple, the system tends to move away from its original equilibrium position; the system is unstable. The value of the load for which the two couples balance each other is called the *critical load* and is denoted by P_{cr}. We have

$$P_{cr}(L/2) \sin \Delta\theta = K(2\,\Delta\theta) \tag{10.1}$$

or, since $\sin \Delta\theta \approx \Delta\theta$,

$$P_{cr} = 4K/L \tag{10.2}$$

Clearly, the system is stable for $P < P_{cr}$, i.e., for values of the load smaller than the critical value, and unstable for $P > P_{cr}$.

Let us assume that a load $P > P_{cr}$ has been applied to the two rods of Fig. 10.3 and that the system has been disturbed. Since $P > P_{cr}$, the system will move further away from the vertical and, after some oscillations, will settle into a new equilibrium position (Fig. 10.6a). Considering the equilibrium of the free body AC (Fig. 10.6b), we obtain an equation similar to Eq. (10.1), but involving the finite angle θ, namely

$$P(L/2) \sin \theta = K(2\theta)$$

or

$$\frac{PL}{4K} = \frac{\theta}{\sin \theta} \tag{10.3}$$

The value of θ corresponding to the equilibrium position represented in Fig. 10.6 is obtained by solving Eq. (10.3) by trial and error. But we observe that, for any positive value of θ, we have $\sin \theta < \theta$. Thus, Eq. (10.3) yields a value of θ different from zero only when the left-hand member of the equation is larger than one. Recalling Eq. (10.2), we note that this is indeed the case here, since we have assumed $P > P_{cr}$. But, if we had assumed $P < P_{cr}$, the second equilibrium position shown in Fig. 10.6 would not exist and the only possible equilibrium position would be the position corresponding to $\theta = 0$. We thus check that, for $P < P_{cr}$, the position $\theta = 0$ must be stable.

This observation applies to structures and mechanical systems in general, and will be used in the next section, where the stability of elastic columns will be discussed.

10.3 EULER'S FORMULA FOR PIN-ENDED COLUMNS

Returning to the column AB considered in the preceding section (Fig. 10.1), we propose to determine the critical value of the load **P**, i.e., the value P_{cr} of the load for which the position shown in Fig. 10.1 ceases to be stable. If $P > P_{cr}$, the slightest misalignment or disturbance will cause the column to buckle, i.e., to assume a curved shape as shown in Fig. 10.2.

Our approach will be to determine the conditions under which the configuration of Fig. 10.2 is possible. Since a column can be considered as a beam placed in a vertical position and subjected to an axial load, we proceed as in Chap. 9 and denote by x the distance from end A of the column to a given point Q of its elastic curve, and by y the deflection of that point (Fig. 10.7a). It follows that the x axis will be vertical and directed downward, and the y axis horizontal and directed to the right. Considering the equilibrium of the free body AQ (Fig. 10.7b), we find that the bending moment at Q is $M = -Py$. Substituting this value for M in Eq. (9.4) of Sec. 9.3, we write

$$\frac{d^2y}{dx^2} = \frac{M}{EI} = -\frac{P}{EI}y \tag{10.4}$$

or, transposing the last term,

$$\frac{d^2y}{dx^2} + \frac{P}{EI}y = 0 \tag{10.5}$$

This equation is a linear, homogeneous differential equation of the second order with constant coefficients. Setting

$$p^2 = \frac{P}{EI} \tag{10.6}$$

we write Eq. (10.5) in the form

$$\frac{d^2y}{dx^2} + p^2y = 0 \tag{10.7}$$

which is the same as that of the differential equation for simple harmonic motion except, of course, that the independent variable is now the distance x instead of the time t. The general solution of Eq. (10.7) is

$$y = A \sin px + B \cos px \tag{10.8}$$

as we easily check by computing d^2y/dx^2 and substituting for y and d^2y/dx^2 into Eq. (10.7).

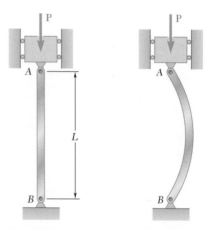

Fig. 10.1 Column (*repeated*) **Fig. 10.2** Buckled column (*repeated*)

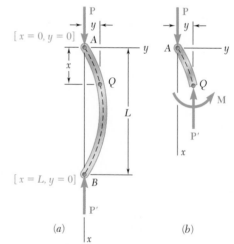

Fig. 10.7 Column in buckled position.

Recalling the boundary conditions that must be satisfied at ends A and B of the column (Fig. 10.7a), we first make $x = 0$, $y = 0$ in Eq. (10.8) and find that $B = 0$. Substituting next $x = L$, $y = 0$, we obtain

$$A \sin pL = 0 \tag{10.9}$$

This equation is satisfied either if $A = 0$, or if $\sin pL = 0$. If the first of these conditions is satisfied, Eq. (10.8) reduces to $y = 0$ and the column is straight (Fig. 10.1). For the second condition to be satisfied, we must have $pL = n\pi$ or, substituting for p from (10.6) and solving for P,

$$P = \frac{n^2 \pi^2 EI}{L^2} \tag{10.10}$$

The smallest of the values of P defined by Eq. (10.10) is that corresponding to $n = 1$. We thus have

$$P_{cr} = \frac{\pi^2 EI}{L^2} \tag{10.11}$$

The expression obtained is known as *Euler's formula,* after the Swiss mathematician Leonhard Euler (1707–1783). Substituting this expression for P into Eq. (10.6) and the value obtained for p into Eq. (10.8), and recalling that $B = 0$, we write

$$y = A \sin \frac{\pi x}{L} \tag{10.12}$$

which is the equation of the elastic curve after the column has buckled (Fig. 10.2). We note that the value of the maximum deflection, $y_m = A$, is indeterminate. This is due to the fact that the differential equation (10.5) is a linearized approximation of the actual governing differential equation for the elastic curve.†

If $P < P_{cr}$, the condition $\sin pL = 0$ cannot be satisfied, and the solution given by Eq. (10.12) does not exist. We must then have $A = 0$, and the only possible configuration for the column is a straight one. Thus, for $P < P_{cr}$ the straight configuration of Fig. 10.1 is stable.

In the case of a column with a circular or square cross section, the moment of inertia I of the cross section is the same about any centroidal axis, and the column is as likely to buckle in one plane as another, except for the restraints that can be imposed by the end connections. For other shapes of cross section, the critical load should be computed by making $I = I_{min}$ in Eq. (10.11); if buckling occurs, it will take place in a plane perpendicular to the corresponding principal axis of inertia.

The value of the stress corresponding to the critical load is called the *critical stress* and is denoted by σ_{cr}. Recalling Eq. (10.11)

---

†We recall that the equation $d^2y/dx^2 = M/EI$ was obtained in Sec. 9.3 by assuming that the slope dy/dx of the beam could be neglected and that the exact expression given in Eq. (9.3) for the curvature of the beam could be replaced by $1/\rho = d^2y/dx^2$.

and setting $I = Ar^2$, where A is the cross-sectional area and r its radius of gyration, we have

$$\sigma_{cr} = \frac{P_{cr}}{A} = \frac{\pi^2 EAr^2}{AL^2}$$

or

$$\sigma_{cr} = \frac{\pi^2 E}{(L/r)^2} \tag{10.13}$$

The quantity L/r is called the *slenderness ratio* of the column. It is clear, in view of the remark of the preceding paragraph, that the minimum value of the radius of gyration r should be used in computing the slenderness ratio and the critical stress in a column.

Equation (10.13) shows that the critical stress is proportional to the modulus of elasticity of the material, and inversely proportional to the square of the slenderness ratio of the column. The plot of σ_{cr} versus L/r is shown in Fig. 10.8 for structural steel, assuming $E = 200$ GPa and $\sigma_Y = 250$ MPa. We should keep in mind that no factor of safety has been used in plotting σ_{cr}. We also note that, if the value obtained for σ_{cr} from Eq. (10.13) or from the curve of Fig. 10.8 is larger than the yield strength σ_Y, this value is of no interest to us, since the column will yield in compression and cease to be elastic before it has a chance to buckle.

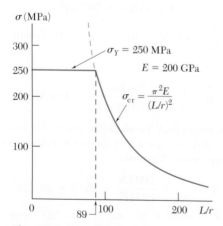

Fig. 10.8 Plot of critical stress.

Our analysis of the behavior of a column has been based so far on the assumption of a perfectly aligned centric load. In practice, this is seldom the case, and in Sec. 10.5 the effect of the eccentricity of the loading is taken into account. This approach will lead to a smoother transition from the buckling failure of long, slender columns to the compression failure of short, stubby columns. It will also provide us with a more realistic view of the relation between the slenderness ratio of a column and the load that causes it to fail.

EXAMPLE 10.01

A 2-m-long pin-ended column of square cross section is to be made of wood. Assuming $E = 13$ GPa, $\sigma_{all} = 12$ MPa, and using a factor of safety of 2.5 in computing Euler's critical load for buckling, determine the size of the cross section if the column is to safely support (*a*) a 100-kN load, (*b*) a 200-kN load.

(*a*) **For the 100-kN Load.** Using the given factor of safety, we make

$$P_{cr} = 2.5(100 \text{ kN}) = 250 \text{ kN} \qquad L = 2 \text{ m} \qquad E = 13 \text{ GPa}$$

in Euler's formula (10.11) and solve for I. We have

$$I = \frac{P_{cr}L^2}{\pi^2 E} = \frac{(250 \times 10^3 \text{ N})(2 \text{ m})^2}{\pi^2(13 \times 10^9 \text{ Pa})} = 7.794 \times 10^{-6} \text{ m}^4$$

Recalling that, for a square of side a, we have $I = a^4/12$, we write

$$\frac{a^4}{12} = 7.794 \times 10^{-6} \text{ m}^4 \qquad a = 98.3 \text{ mm} \approx 100 \text{ mm}$$

We check the value of the normal stress in the column:

$$\sigma = \frac{P}{A} = \frac{100 \text{ kN}}{(0.100 \text{ m})^2} = 10 \text{ MPa}$$

Since σ is smaller than the allowable stress, a 100×100-mm cross section is acceptable.

(*b*) **For the 200-kN Load.** Solving again Eq. (10.11) for I, but making now $P_{cr} = 2.5(200) = 500$ kN, we have

$$I = 15.588 \times 10^{-6} \text{ m}^4$$

$$\frac{a^4}{12} = 15.588 \times 10^{-6} \qquad a = 116.95 \text{ mm}$$

The value of the normal stress is

$$\sigma = \frac{P}{A} = \frac{200 \text{ kN}}{(0.11695 \text{ m})^2} = 14.62 \text{ MPa}$$

Since this value is larger than the allowable stress, the dimension obtained is not acceptable, and we must select the cross section on the basis of its resistance to compression. We write

$$A = \frac{P}{\sigma_{all}} = \frac{200 \text{ kN}}{12 \text{ MPa}} = 16.67 \times 10^{-3} \text{ m}^2$$

$$a^2 = 16.67 \times 10^{-3} \text{ m}^2 \qquad a = 129.1 \text{ mm}$$

A 130×130-mm cross section is acceptable.

10.4 EXTENSION OF EULER'S FORMULA TO COLUMNS WITH OTHER END CONDITIONS

Euler's formula (10.11) was derived in the preceding section for a column that was pin-connected at both ends. Now the critical load P_{cr} will be determined for columns with different end conditions.

In the case of a column with one free end A supporting a load **P** and one fixed end B (Fig. 10.9*a*), we observe that the column will behave as the upper half of a pin-connected column (Fig. 10.9*b*). The critical load for the column of Fig. 10.9*a* is thus the same as for the pin-ended column of Fig. 10.9*b* and can be obtained from Euler's

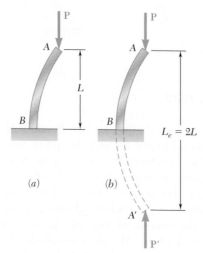

Fig. 10.9 Column with free end.

formula (10.11) by using a column length equal to twice the actual length L of the given column. We say that the *effective length* L_e of the column of Fig. 10.9 is equal to $2L$ and substitute $L_e = 2L$ in Euler's formula:

$$P_{cr} = \frac{\pi^2 EI}{L_e^2} \qquad (10.11')$$

The critical stress is found in a similar way from the formula

$$\sigma_{cr} = \frac{\pi^2 E}{(L_e/r)^2} \qquad (10.13')$$

The quantity L_e/r is referred to as the *effective slenderness ratio* of the column and, in the case considered here, is equal to $2L/r$.

Consider next a column with two fixed ends A and B supporting a load **P** (Fig. 10.10). The symmetry of the supports and of the loading about a horizontal axis through the midpoint C requires that the shear at C and the horizontal components of the reactions at A and B be zero (Fig. 10.11). It follows that the restraints imposed upon the upper half AC of the column by the support at A and by the

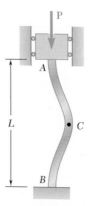

Fig. 10.10 Column with fixed ends.

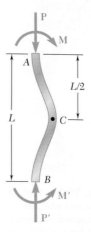

Fig. 10.11 Buckled shape of column with fixed ends.

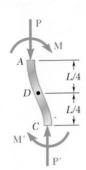

Fig. 10.12

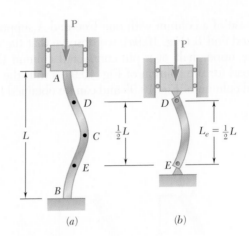

(a)

(b)

Fig. 10.13

lower half CB are identical (Fig. 10.12). Portion AC must thus be symmetric about its midpoint D, and this point must be a point of inflection, where the bending moment is zero. A similar reasoning shows that the bending moment at the midpoint E of the lower half of the column must also be zero (Fig. 10.13a). Since the bending moment at the ends of a pin-ended column is zero, it follows that the portion DE of the column of Fig. 10.13a must behave as a pin-ended column (Fig. 10.13b). We thus conclude that the effective length of a column with two fixed ends is $L_e = L/2$.

In the case of a column with one fixed end B and one pin-connected end A supporting a load **P** (Fig. 10.14), we must write and solve the differential equation of the elastic curve to determine the effective length of the column. From the free-body diagram of the entire column (Fig. 10.15), we first note that a transverse force **V** is exerted at end A, in addition to the axial load **P,** and that **V** is statically indeterminate. Considering now the free-body diagram of a portion AQ of the column (Fig. 10.16), we find that the bending moment at Q is

$$M = -Py - Vx$$

Fig. 10.14 Column with one end pin-connected and one end fixed.

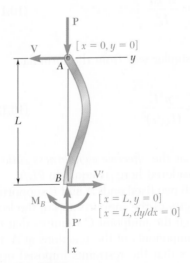

Fig. 10.15

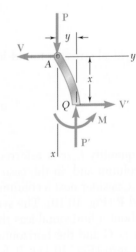

Fig. 10.16

Substituting this value into Eq. (9.4) of Sec. 9.3, we write

$$\frac{d^2y}{dx^2} = \frac{M}{EI} = -\frac{P}{EI}y - \frac{V}{EI}x$$

Transposing the term containing y and setting

$$p^2 = \frac{P}{EI} \tag{10.6}$$

as we did in Sec. 10.3, we write

$$\frac{d^2y}{dx^2} + p^2y = -\frac{V}{EI}x \tag{10.14}$$

This equation is a linear, nonhomogeneous differential equation of the second order with constant coefficients. Observing that the left-hand members of Eqs. (10.7) and (10.14) are identical, we conclude that the general solution of Eq. (10.14) can be obtained by adding a particular solution of Eq. (10.14) to the solution (10.8) obtained for Eq. (10.7). Such a particular solution is easily seen to be

$$y = -\frac{V}{p^2EI}x$$

or, recalling (10.6),

$$y = -\frac{V}{P}x \tag{10.15}$$

Adding the solutions (10.8) and (10.15), we write the general solution of Eq. (10.14) as

$$y = A \sin px + B \cos px - \frac{V}{P}x \tag{10.16}$$

The constants A and B, and the magnitude V of the unknown transverse force **V** are obtained from the boundary conditions indicated in Fig. 10.15. Making first $x = 0$, $y = 0$ in Eq. (10.16), we find that $B = 0$. Making next $x = L$, $y = 0$, we obtain

$$A \sin pL = \frac{V}{P}L \tag{10.17}$$

Finally, computing

$$\frac{dy}{dx} = Ap \cos px - \frac{V}{P}$$

and making $x = L$, $dy/dx = 0$, we have

$$Ap \cos pL = \frac{V}{P} \tag{10.18}$$

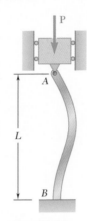

Fig. 10.14
(repeated)

Dividing (10.17) by (10.18) member by member, we conclude that a solution of the form (10.16) can exist only if

$$\tan pL = pL \tag{10.19}$$

Solving this equation by trial and error, we find that the smallest value of pL which satisfies (10.19) is

$$pL = 4.4934 \tag{10.20}$$

Carrying the value of p defined by Eq. (10.20) into Eq. (10.6) and solving for P, we obtain the critical load for the column of Fig. 10.14:

$$P_{cr} = \frac{20.19EI}{L^2} \tag{10.21}$$

The effective length of the column is obtained by equating the right-hand members of Eqs. (10.11′) and (10.21):

$$\frac{\pi^2 EI}{L_e^2} = \frac{20.19EI}{L^2}$$

Solving for L_e, we find that the effective length of a column with one fixed end and one pin-connected end is $L_e = 0.699L \approx 0.7L$.

The effective lengths corresponding to the various end conditions considered in this section are shown in Fig. 10.17.

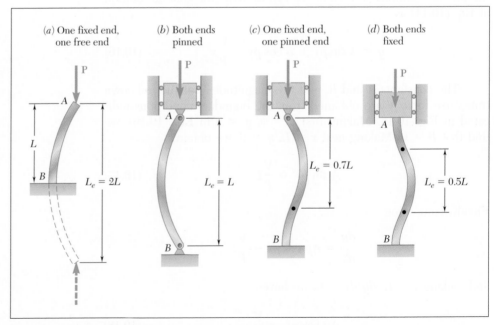

Fig. 10.17 Effective length of column for various end conditions.

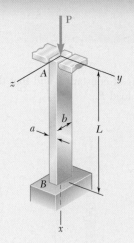

SAMPLE PROBLEM 10.1

An aluminum column of length L and rectangular cross section has a fixed end B and supports a centric load at A. Two smooth and rounded fixed plates restrain end A from moving in one of the vertical planes of symmetry of the column, but allow it to move in the other plane. (a) Determine the ratio a/b of the two sides of the cross section corresponding to the most efficient design against buckling. (b) Design the most efficient cross section for the column, knowing that $L = 0.5$ m, $E = 70$ GPa, $P = 20$ kN, and that a factor of safety of 2.5 is required.

SOLUTION

Buckling in xy Plane. Referring to Fig. 10.17, we note that the effective length of the column with respect to buckling in this plane is $L_e = 0.7L$. The radius of gyration r_z of the cross section is obtained by writing

$$I_x = \tfrac{1}{12}ba^3 \quad A = ab$$

and, since $I_z = Ar_z^2$, $\quad r_z^2 = \dfrac{I_z}{A} = \dfrac{\tfrac{1}{12}ba^3}{ab} = \dfrac{a^2}{12} \quad r_z = a/\sqrt{12}$

The effective slenderness ratio of the column with respect to buckling in the xy plane is

$$\frac{L_e}{r_z} = \frac{0.7L}{a/\sqrt{12}} \tag{1}$$

Buckling in xz Plane. The effective length of the column with respect to buckling in this plane is $L_e = 2L$, and the corresponding radius of gyration is $r_y = b/\sqrt{12}$. Thus,

$$\frac{L_e}{r_y} = \frac{2L}{b/\sqrt{12}} \tag{2}$$

a. Most Efficient Design. The most efficient design is that for which the critical stresses corresponding to the two possible modes of buckling are equal. Referring to Eq. (10.13'), we note that this will be the case if the two values obtained above for the effective slenderness ratio are equal. We write

$$\frac{0.7L}{a/\sqrt{12}} = \frac{2L}{b/\sqrt{12}}$$

and, solving for the ratio a/b, $\qquad \dfrac{a}{b} = \dfrac{0.7}{2} \qquad \dfrac{a}{b} = 0.35 \quad \blacktriangleleft$

b. Design for Given Data. Since $F.S. = 2.5$ is required,

$$P_{cr} = (F.S.)P = (2.5)(20 \text{ kN}) = 50 \text{ kN}$$

Using $a = 0.35b$, we have $A = ab = 0.35b^2$ and

$$\sigma_{cr} = \frac{P_{cr}}{A} = \frac{50\,000 \text{ N}}{0.35b^2}$$

Making $L = 0.5$ m in Eq. (2), we have $L_e/r_y = 3.464/b$. Substituting for E, L_e/r, and σ_{cr} into Eq. (10.13'), we write

$$\sigma_{cr} = \frac{\pi^2 E}{(L_e/r)^2} \qquad \frac{50\,000 \text{ N}}{0.35b^2} = \frac{\pi^2(70 \times 10^9 \text{ Pa})}{(3.464/b)^2}$$

$$b = 39.7 \text{ mm} \qquad a = 0.35b = 13.89 \text{ mm} \quad \blacktriangleleft$$

PROBLEMS

10.1 Knowing that the spring at A is of constant k and that the bar AB is rigid, determine the critical load P_{cr}.

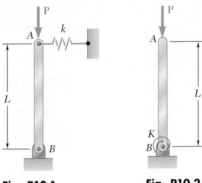

Fig. P10.1   **Fig. P10.2**

10.2 Knowing that the torsional spring at B is of constant K and that the bar AB is rigid, determine the critical load P_{cr}.

10.3 Two rigid bars AC and BC are connected by a pin at C as shown. Knowing that the torsional spring at B is of constant K, determine the critical load P_{cr} for the system.

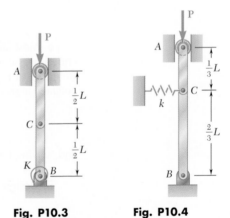

Fig. P10.3   **Fig. P10.4**

10.4 Two rigid bars AC and BC are connected as shown to a spring of constant k. Knowing that the spring can act in either tension or compression, determine the critical load P_{cr} for the system.

10.5 The rigid bar AD is attached to two springs of constant k and is in equilibrium in the position shown. Knowing that the equal and opposite loads P and P' *remain horizontal*, determine the magnitude P_{cr} of the critical load for the system.

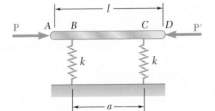

Fig. P10.5

666

10.6 The rigid rod AB is attached to a hinge at A and to two springs, each of constant k. If $h = 450$ mm, $d = 300$ mm, and $m = 200$ kg, determine the range of values of k for which the equilibrium of rod AB is stable in the position shown. Each spring can act in either tension or compression.

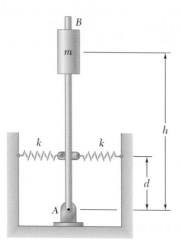

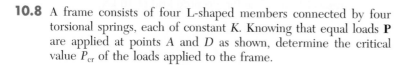

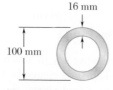

Fig. P10.6

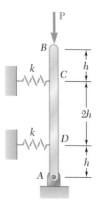

Fig. P10.7

10.7 The rigid rod AB is attached to a hinge at A and to two springs, each of constant $k = 30$ kN/m, that can act in either tension or compression. Knowing that $h = 0.6$ m, determine the critical load.

10.8 A frame consists of four L-shaped members connected by four torsional springs, each of constant K. Knowing that equal loads **P** are applied at points A and D as shown, determine the critical value P_{cr} of the loads applied to the frame.

10.9 Determine the critical load of a wooden meter stick that has a 7×24-mm rectangular cross section. Use $E = 12$ GPa.

10.10 Determine the critical load of a steel tube that is 5 m long and has a 100-mm outer diameter and a 16-mm wall thickness. Use $E = 200$ GPa.

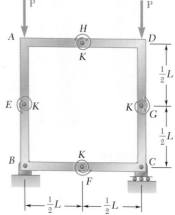

Fig. P10.8

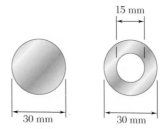

Fig. P10.10

10.11 A compression member of 1.5-m effective length consists of a solid 30-mm-diameter brass rod. In order to reduce the weight of the member by 25 percent, the solid rod is replaced by a hollow rod of the cross section shown. Determine (a) the percent reduction in the critical load, (b) the value of the critical load for the hollow rod. Use $E = 105$ GPa.

Fig. P10.11

10.12 Two brass rods used as compression members, each of 3-m effective length, have the cross sections shown. (a) Determine the wall thickness of the hollow square rod for which the rods have the same cross-sectional area. (b) Using $E = 105$ GPa, determine the critical load of each rod.

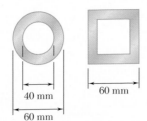

40 mm

60 mm

60 mm

Fig. P10.12

10.13 A column of effective length L can be made by gluing together identical planks in either of the arrangements shown. Determine the ratio of the critical load using the arrangement a to the critical load using the arrangement b.

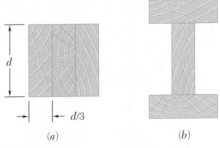

d

$d/3$

(a)

(b)

Fig. P10.13

10.14 Determine the radius of the round strut so that the round and square struts have the same cross-sectional area, and compute the critical load of each strut. Use $E = 200$ GPa.

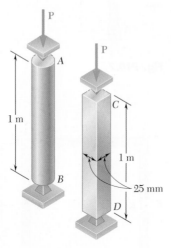

P

A

P

C

1 m

1 m

B

25 mm

D

Fig. P10.14

10.15 A column of 6-m effective length is to be made from three plates as shown. Using $E = 200$ GPa, determine the factor of safety with respect to buckling for a centric load of 16 kN.

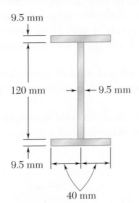

9.5 mm

120 mm

9.5 mm

9.5 mm

40 mm

Fig. P10.15

10.16 A column of 3-m effective length is to be made by welding together two C130 $\times$ 13 rolled-steel channels. Using $E = 200$ GPa, determine for each arrangement shown the allowable centric load if a factor of safety of 2.4 is required.

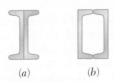

(a)

(b)

Fig. P10.16

10.17 A single compression member of 8.2-m effective length is obtained by connecting two C200 × 17.1 steel channels with lacing bars as shown. Knowing that the factor of safety is 1.85, determine the allowable centric load for the member. Use E = 200 GPa and d = 100 mm.

10.18 A column of 7-m effective length is made by welding two 228 × 12-mm plates to a W200 × 52 as shown. Determine the allowable centric load if a factor of safety of 2.3 is required. Use E = 200 GPa.

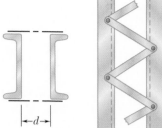

Fig. P10.17

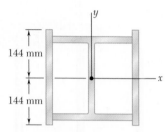

Fig. P10.18

10.19 Member AB consists of a single C130 × 10.4 steel channel of length 2.5 m. Knowing that the pins A and B pass through the centroid of the cross section of the channel, determine the factor of safety for the load shown with respect to buckling in the plane of the figure when θ = 30°. Use E = 200 GPa.

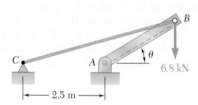

Fig. P10.19

10.20 Knowing that a factor of safety of 2.6 is required, determine the largest load **P** that can be applied to the structure shown. Use E = 200 GPa and consider only buckling in the plane of the structure.

10.21 A rigid block of mass m can be supported in each of the four ways shown. Each column consists of an aluminum tube that has a 44-mm outer diameter and a 4-mm wall thickness. Using E = 70 GPa and a factor of safety of 2.8, determine the allowable mass for each support condition.

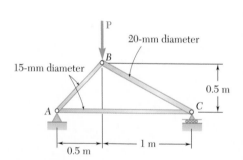

Fig. P10.20

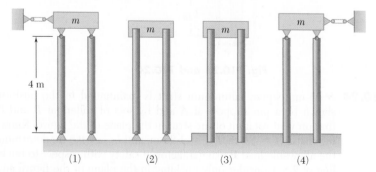

Fig. P10.21

10.22 Each of the five struts shown consists of a solid steel rod. (*a*) Knowing that the strut of Fig. (1) is of a 20-mm diameter, determine the factor of safety with respect to buckling for the loading shown. (*b*) Determine the diameter of each of the other struts for which the factor of safety is the same as the factor of safety obtained in part *a*. Use $E = 200$ GPa.

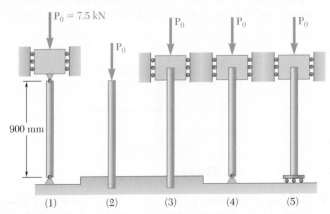

Fig. P10.22

10.23 A 32-mm-square aluminum strut is maintained in the position shown by a pin support at *A* and by sets of rollers at *B* and *C* that prevent rotation of the strut in the plane of the figure. Knowing that $L_{AB} = 1.4$ m, determine (*a*) the largest values of L_{BC} and L_{CD} that can be used if the allowable load **P** is to be as large as possible, (*b*) the magnitude of the corresponding allowable load if the factor of safety is 2.8. Consider only buckling in the plane of the figure and use $E = 72$ GPa.

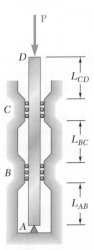

Fig. P10.23 and *P10.24*

10.24 A 25-mm-square aluminum strut is maintained in the position shown by a pin support at *A* and by sets of rollers at *B* and *C* that prevent rotation of the strut in the plane of the figure. Knowing that $L_{AB} = 1.0$ m, $L_{BC} = 1.25$ m, and $L_{CD} = 0.5$ m, determine the allowable load **P** using a factor of safety with respect to buckling of 2.8. Consider only buckling in the plane of the figure and use $E = 75$ GPa.

10.25 Column AB carries a centric load **P** of magnitude 60 kN. Cables BC and BD are taut and prevent motion of point B in the xz plane. Using Euler's formula and a factor of safety of 2.2, and neglecting the tension in the cables, determine the maximum allowable length L. Use $E = 200$ GPa.

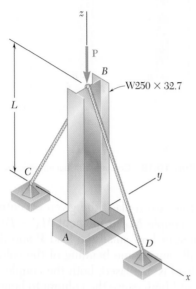

Fig. P10.25

10.26 A W200 × 31.3 rolled-steel shape is used with the support and cable arrangement shown in Prob. 10.25. Knowing that $L = 7$ m, determine the allowable centric load **P** if a factor of safety of 2.2 is required. Use $E = 200$ GPa.

10.27 Column ABC has a uniform rectangular cross section with $b = 12$ mm and $d = 22$ mm. The column is braced in the xz plane at its midpoint C and carries a centric load **P** of magnitude 3.8 kN. Knowing that a factor of safety of 3.2 is required, determine the largest allowable length L. Use $E = 200$ GPa.

10.28 Column ABC has a uniform rectangular cross section and is braced in the xz plane at its midpoint C. (*a*) Determine the ratio b/d for which the factor of safety is the same with respect to buckling in the xz and yz planes. (*b*) Using the ratio found in part a, design the cross section of the column so that the factor of safety will be 3.0 when $P = 4.4$ kN, $L = 1$ m, and $E = 200$ GPa.

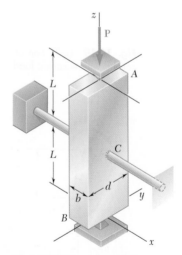

Fig. P10.27 and P10.28

*10.5 ECCENTRIC LOADING; THE SECANT FORMULA

In this section the problem of column buckling will be approached in a different way, by observing that the load **P** applied to a column is never perfectly centric. Denoting by e the eccentricity of the load, i.e., the distance between the line of action **P** and the axis of the

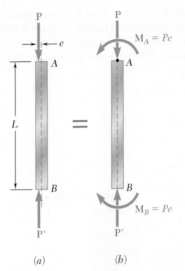

(a)   (b)

Fig. 10.18 Column with eccentric load.

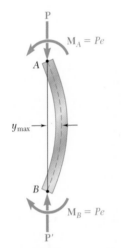

Fig. 10.19 Deflection of column with eccentric load.

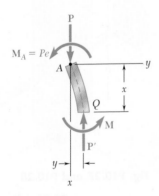

Fig. 10.20

column (Fig. 10.18a), we replace the given eccentric load by a centric force **P** and a couple $\mathbf{M}_A$ of moment $M_A = Pe$ (Fig. 10.18b). It is clear that, no matter how small the load **P** and the eccentricity e, the couple $\mathbf{M}_A$ will cause some bending of the column (Fig. 10.19). As the eccentric load is increased, both the couple $\mathbf{M}_A$ and the axial force **P** increase, and both cause the column to bend further. Viewed in this way, the problem of buckling is not a question of determining how long the column can remain straight and stable under an increasing load, but rather how much the column can be permitted to bend under the increasing load, if the allowable stress is not to be exceeded and if the deflection y_{max} is not to become excessive.

We first write and solve the differential equation of the elastic curve, proceeding in the same manner as we did earlier in Secs. 10.3 and 10.4. Drawing the free-body diagram of a portion AQ of the column and choosing the coordinate axes as shown (Fig. 10.20), we find that the bending moment at Q is

$$M = -Py - M_A = -Py - Pe \qquad (10.22)$$

Substituting the value of M into Eq. (9.4) of Sec. 9.3, we write

$$\frac{d^2y}{dx^2} = \frac{M}{EI} = -\frac{P}{EI}y - \frac{Pe}{EI}$$

Transposing the term containing y and setting

$$p^2 = \frac{P}{EI} \qquad (10.6)$$

as done earlier, we write

$$\frac{d^2y}{dx^2} + p^2y = -p^2e \qquad (10.23)$$

Since the left-hand member of this equation is the same as that of Eq. (10.7), which was solved in Sec. 10.3, we write the general solution of Eq. (10.23) as

$$y = A \sin px + B \cos px - e \qquad (10.24)$$

where the last term is a particular solution of Eq. (10.23).

The constants A and B are obtained from the boundary conditions shown in Fig. 10.21. Making first $x = 0$, $y = 0$ in Eq. (10.24), we have

$$B = e$$

Making next $x = L$, $y = 0$, we write

$$A \sin pL = e(1 - \cos pL) \tag{10.25}$$

Recalling that

$$\sin pL = 2 \sin \frac{pL}{2} \cos \frac{pL}{2}$$

and

$$1 - \cos pL = 2 \sin^2 \frac{pL}{2}$$

and substituting into Eq. (10.25), we obtain, after reductions,

$$A = e \tan \frac{pL}{2}$$

Substituting for A and B into Eq. (10.24), we write the equation of the elastic curve:

$$y = e\left(\tan \frac{pL}{2} \sin px + \cos px - 1 \right) \tag{10.26}$$

The value of the maximum deflection is obtained by setting $x = L/2$ in Eq. (10.26). We have

$$y_{max} = e\left(\tan \frac{pL}{2} \sin \frac{pL}{2} + \cos \frac{pL}{2} - 1 \right)$$

$$= e\left(\frac{\sin^2 \dfrac{pL}{2} + \cos^2 \dfrac{pL}{2}}{\cos \dfrac{pL}{2}} - 1 \right)$$

$$y_{max} = e\left(\sec \frac{pL}{2} - 1 \right) \tag{10.27}$$

Recalling Eq. (10.6), we write

$$y_{max} = e\left[\sec\left(\sqrt{\frac{P}{EI}} \frac{L}{2} \right) - 1 \right] \tag{10.28}$$

We note from the expression obtained that y_{max} becomes infinite when

$$\sqrt{\frac{P}{EI}} \frac{L}{2} = \frac{\pi}{2} \tag{10.29}$$

While the deflection does not actually become infinite, it nevertheless becomes unacceptably large, and P should not be allowed to reach the critical value which satisfies Eq. (10.29). Solving (10.29) for P, we have

$$P_{cr} = \frac{\pi^2 EI}{L^2} \tag{10.30}$$

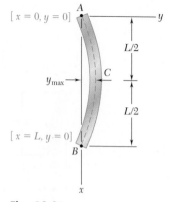

$[x = 0, y = 0]$

A

y

$L/2$

y_{max}

C

$L/2$

$[x = L, y = 0]$

B

x

Fig. 10.21

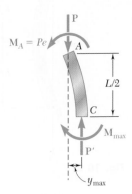

Fig. 10.22

which is the value that we obtained in Sec. 10.3 for a column under a centric load. Solving (10.30) for EI and substituting into (10.28), we can express the maximum deflection in the alternative form

$$y_{max} = e\left(\sec \frac{\pi}{2}\sqrt{\frac{P}{P_{cr}}} - 1 \right) \tag{10.31}$$

The maximum stress σ_{max} occurs in the section of the column where the bending moment is maximum, i.e., in the transverse section through the midpoint C, and can be obtained by adding the normal stresses due, respectively, to the axial force and the bending couple exerted on that section (cf. Sec. 4.12). We have

$$\sigma_{max} = \frac{P}{A} + \frac{M_{max}c}{I} \tag{10.32}$$

From the free-body diagram of the portion AC of the column (Fig. 10.22), we find that

$$M_{max} = Py_{max} + M_A = P(y_{max} + e)$$

Substituting this value into (10.32) and recalling that $I = Ar^2$, we write

$$\sigma_{max} = \frac{P}{A}\left[1 + \frac{(y_{max} + e)c}{r^2} \right] \tag{10.33}$$

Substituting for y_{max} the value obtained in (10.28), we write

$$\sigma_{max} = \frac{P}{A}\left[1 + \frac{ec}{r^2} \sec\left(\sqrt{\frac{P}{EI}}\frac{L}{2} \right) \right] \tag{10.34}$$

An alternative form for σ_{max} is obtained by substituting for y_{max} from (10.31) into (10.33). We have

$$\sigma_{max} = \frac{P}{A}\left(1 + \frac{ec}{r^2} \sec \frac{\pi}{2}\sqrt{\frac{P}{P_{cr}}} \right) \tag{10.35}$$

The equation obtained can be used with any end conditions, as long as the appropriate value is used for the critical load (cf. Sec. 10.4).

We note that, since σ_{max} does not vary linearly with the load P, the principle of superposition does not apply to the determination of the stress due to the simultaneous application of several loads; the resultant load must first be computed, and then Eq. (10.34) or Eq. (10.35) can be used to determine the corresponding stress. For the same reason, any given factor of safety should be applied to the load, and not to the stress.

Making $I = Ar^2$ in Eq. (10.34) and solving for the ratio P/A in front of the bracket, we write

$$\frac{P}{A} = \frac{\sigma_{max}}{1 + \dfrac{ec}{r^2} \sec\left(\dfrac{1}{2}\sqrt{\dfrac{P}{EA}}\dfrac{L_e}{r} \right)} \tag{10.36}$$

where the effective length is used to make the formula applicable to various end conditions. This formula is referred to as *the secant formula*; it defines the force per unit area, P/A, that causes a specified maximum stress σ_{max} in a column of given effective slenderness ratio, L_e/r, for a given value of the ratio ec/r^2, where e is the eccentricity of the applied load. We note that, since P/A appears in both members, it is necessary to solve a transcendental equation by trial and error to obtain the value of P/A corresponding to a given column and loading condition.

Equation (10.36) was used to draw the curves shown in Fig. 10.23 for a steel column, assuming the values of E and σ_Y shown in the figure. These curves make it possible to determine the load per unit area P/A, which causes the column to yield for given values of the ratios L_e/r and ec/r^2.

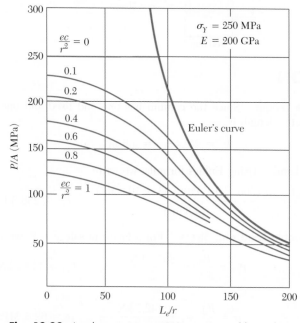

Fig. 10.23 Load per unit area, P/A, causing yield in column.

We note that, for small values of L_e/r, the secant is almost equal to 1 in Eq. (10.36), and P/A can be assumed equal to

$$\frac{P}{A} = \frac{\sigma_{max}}{1 + \dfrac{ec}{r^2}} \qquad (10.37)$$

a value that could be obtained by neglecting the effect of the lateral deflection of the column and using the method of Sec. 4.12. On the other hand, we note from Fig. 10.23 that, for large values of L_e/r, the curves corresponding to the various values of the ratio ec/r^2 get very close to Euler's curve defined by Eq. (10.13′), and thus that the effect of the eccentricity of the loading on the value of P/A becomes negligible. The secant formula is chiefly useful for intermediate values of L_e/r. However, to use it effectively, we should know the value of the eccentricity e of the loading, and this quantity, unfortunately, is seldom known with any degree of precision.

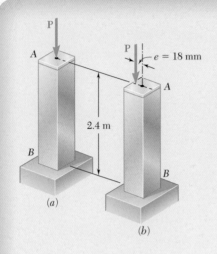

(a)

(b)

SAMPLE PROBLEM 10.2

The uniform column AB consists of a 2.4-m section of structural tubing having the cross section shown. (a) Using Euler's formula and a factor of safety of 2.0, determine the allowable centric load for the column and the corresponding normal stress. (b) Assuming that the allowable load, found in part a, is applied as shown at a point 18 mm from the geometric axis of the column, determine the horizontal deflection of the top of the column and the maximum normal stress in the column. Use $E = 200$ GPa.

$$A = 2284 \text{ mm}^2$$
$$I = 3.33 \times 10^6 \text{ mm}^4$$
$$r = 38 \text{ mm}$$
$$c = 50 \text{ mm}$$

SOLUTION

Effective Length. Since the column has one end fixed and one end free, its effective length is

$$L_e = 2(2.4 \text{ m}) = 4.8 \text{ m}$$

Critical Load. Using Euler's formula, we write

$$P_{cr} = \frac{\pi^2 EI}{L_e^2} = \frac{\pi^2(200 \times 10^9 \text{ Pa})(3.33 \times 10^{-6} \text{ m}^4)}{(4.8 \text{ m}^2)} \qquad P_{cr} = 285.3 \text{ kN}$$

a. Allowable Load and Stress. For a factor of safety of 2, we find

$$P_{all} = \frac{P_{cr}}{F.S.} = \frac{285.3 \text{ kN}}{2} \qquad P_{all} = 142.7 \text{ kN} \blacktriangleleft$$

and

$$\sigma = \frac{P_{all}}{A} = \frac{142.7 \text{ kN}}{2284 \times 10^{-6} \text{ m}^2} \qquad \sigma = 62.5 \text{ MPa} \blacktriangleleft$$

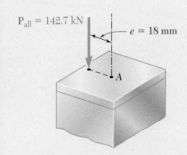

b. Eccentric Load. We observe that column AB and its loading are identical to the upper half of the column of Fig. 10.19 which was used in the derivation of the secant formulas; we conclude that the formulas of Sec. 10.5 apply directly to the case considered here. Recalling that $P_{all}/P_{cr} = \frac{1}{2}$ and using Eq. (10.31), we compute the horizontal deflection of point A:

$$y_m = e\left[\sec\left(\frac{\pi}{2}\sqrt{\frac{P}{P_{cr}}}\right) - 1\right] = (18 \text{ mm})\left[\sec\left(\frac{\pi}{2\sqrt{2}}\right) - 1\right]$$

$$= (18 \text{ mm})(2.252 - 1) \qquad y_m = 22.5 \text{ mm} \blacktriangleleft$$

The maximum normal stress is obtained from Eq. (10.35):

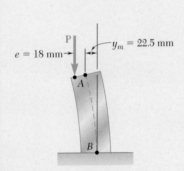

$$\sigma_m = \frac{P}{A}\left[1 + \frac{ec}{r^2}\sec\left(\frac{\pi}{2}\sqrt{\frac{P}{P_{cr}}}\right)\right]$$

$$= \frac{142.7 \times 10^3 \text{ N}}{2284 \times 10^{-6} \text{ m}^2}\left[1 + \frac{(0.018 \text{ m})(0.05 \text{ m})}{(0.038 \text{ m})^2}\sec\left(\frac{\pi}{2\sqrt{2}}\right)\right]$$

$$= (62.48 \text{ MPa})[1 + 0.6233(2.252)] \qquad \sigma_m = 150.2 \text{ MPa} \blacktriangleleft$$

PROBLEMS

10.29 An axial load **P** is applied to the 32-mm-diameter steel rod *AB* as shown. For $P = 37$ kN and $e = 1.2$ mm, determine (*a*) the deflection at the midpoint *C* of the rod, (*b*) the maximum stress in the rod. Use $E = 200$ GPa.

Fig. P10.29

10.30 An axial load **P** = 15 kN is applied at point *D* that is 4 mm from the geometric axis of the square aluminum bar *BC*. Using $E = 70$ GPa, determine (*a*) the horizontal deflection of end *C*, (*b*) the maximum stress in the column.

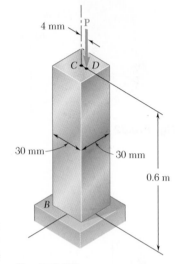

Fig. P10.30

10.31 The line of action of the 300-kN axial load is parallel to the geometric axis of the column *AB* and intersects the *x* axis at $x = 15$ mm. Using $E = 200$ GPa, determine (*a*) the horizontal deflection of the midpoint *C* of the column, (*b*) the maximum stress in the column.

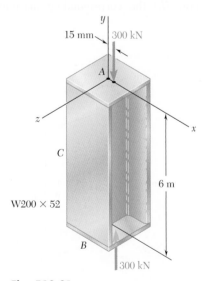

Fig. P10.31

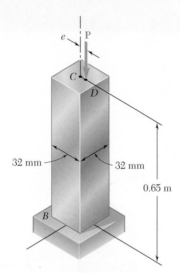

Fig. P10.32

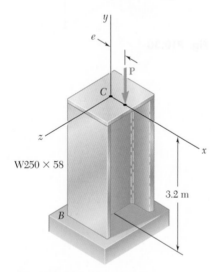

Fig. P10.34

10.32 An axial load **P** is applied to the 32-mm-square aluminum bar *BC* as shown. When $P = 24$ kN, the horizontal deflection at end *C* is 4 mm. Using $E = 70$ GPa, determine (*a*) the eccentricity *e* of the load, (*b*) the maximum stress in the bar.

10.33 An axial load **P** is applied to the 32-mm-diameter steel rod *AB* as shown. For $P = 37$ kN and $e = 1.2$ mm, determine (*a*) the deflection of the midpoint *C* of the rod, (*b*) the maximum stress in the rod. Use $E = 200$ GPa.

Fig. P10.33

10.34 The axial load **P** is applied at a point located on the *x* axis at a distance *e* from the geometric axis of the rolled-steel column *BC*. When $P = 350$ kN, the horizontal deflection of the top of the column is 5 mm. Using $E = 200$ GPa, determine (*a*) the eccentricity *e* of the load, (*b*) the maximum stress in the column.

10.35 An axial load **P** is applied at point *D* that is 6 mm from the geometric axis of the square aluminum bar *BC*. Using $E = 70$ GPa, determine (*a*) the load **P** for which the horizontal deflection of end *C* is 12 mm, (*b*) the corresponding maximum stress in the column.

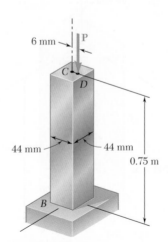

Fig. P10.35

10.36 An axial load **P** is applied at a point located on the x axis at a distance $e = 12$ mm from the geometric axis of the W310 × 60 rolled-steel column BC. Assuming that $L = 3.5$ m and using $E = 200$ GPa, determine (a) the load **P** for which the horizontal deflection at end C is 15 mm, (b) the corresponding maximum stress in the column.

10.37 Solve Prob. 10.36, assuming that $L = 4.5$ m.

10.38 The line of action of an axial load **P** is parallel to the geometric axis of the column AB and intersects the x axis at $x = 20$ mm. Using $E = 200$ GPa, determine (a) the load **P** for which the horizontal deflection of the midpoint C of the column is 12 mm, (b) the corresponding maximum stress in the column.

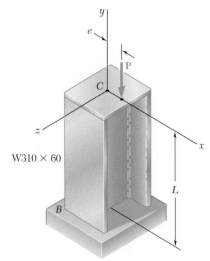

Fig. P10.36

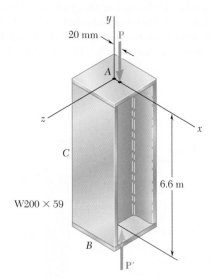

Fig. P10.38

10.39 A brass pipe having the cross section shown has an axial load **P** applied 5 mm from its geometric axis. Using $E = 120$ GPa, determine (a) the load **P** for which the horizontal deflection at the midpoint C is 5 mm, (b) the corresponding maximum stress in the column.

10.40 Solve Prob. 10.39, assuming that the axial load **P** is applied 10 mm from the geometric axis of the column.

10.41 The steel bar AB has a 10 × 10-mm square cross section and is held by pins that are a fixed distance apart and are located at a distance $e = 0.9$ mm from the geometric axis of the bar. Knowing that at temperature T_0 the pins are in contact with the bar and that the force in the bar is zero, determine the increase in temperature for which the bar will just make contact with point C if $d = 0.3$ mm. Use $E = 200$ GPa and a coefficient of thermal expansion $\alpha = 11.7 \times 10^{-6}/°C$.

10.42 For the bar of Prob. 10.41, determine the required distance d for which the bar will just make contact with point C when the temperature increases by 60°C.

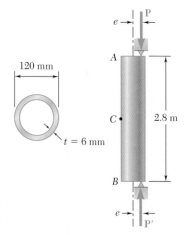

Fig. P10.39

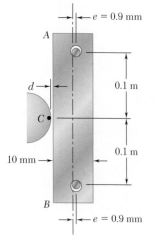

Fig. P10.41

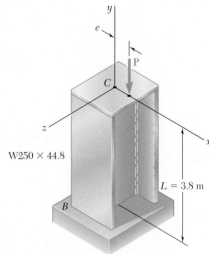

W250 × 44.8

$L = 3.8$ m

Fig. P10.43

10.43 An axial load **P** is applied to the W250 × 44.8 rolled-steel column BC that is free at its top C and fixed at its base B. Knowing that the eccentricity of the load is $e = 12$ mm and that for the grade of steel used $\sigma_Y = 250$ MPa and $E = 200$ GPa, determine (a) the magnitude of P of the allowable load when a factor of safety of 2.4 with respect to permanent deformation is required, (b) the ratio of the load found in part a to the magnitude of the allowable centric load for the column. (*Hint:* Since the factor of safety must be applied to the load **P**, not to the stress, use Fig. 10.23 to determine P_Y).

10.44 Solve Prob. 10.43, assuming that the length of the column is reduced to 1.5 m.

10.45 A 3.5-m-long steel tube having the cross section and properties shown is used as a column. For the grade of steel used $\sigma_Y = 250$ MPa and $E = 200$ GPa. Knowing that a factor of safety of 2.6 with respect to permanent deformation is required, determine the allowable load **P** when the eccentricity e is (a) 15 mm, (b) 7.5 mm. (See hint of Prob. 10.43).

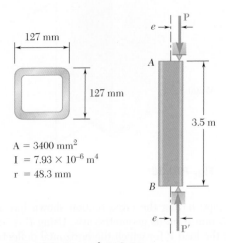

127 mm

127 mm

$A = 3400$ mm^2
$I = 7.93 \times 10^{-6}$ m^4
$r = 48.3$ mm

A

3.5 m

B

Fig. P10.45 and P10.46

10.46 Solve Prob. 10.45, assuming that the length of the tube is increased to 5 m.

10.47 A 250-kN axial load **P** is applied to a W200 × 35.9 rolled-steel column BC that is free at its top C and fixed at its base B. Knowing that the eccentricity of the load is $e = 6$ mm, determine the largest permissible length L if the allowable stress in the column is 80 MPa. Use $E = 200$ GPa.

10.48 A 100-kN axial load **P** is applied to the W150 × 18 rolled-steel column BC that is free at its top C and fixed at its base B. Knowing that the eccentricity of the load is $e = 6$ mm, determine the largest permissible length L if the allowable stress in the column is 80 MPa. Use $E = 200$ GPa.

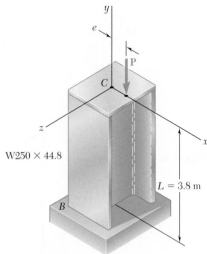

L

Fig. P10.47 and P10.48

10.49 Axial loads of magnitude $P = 84$ kN are applied parallel to the geometric axis of the W200 × 22.5 rolled-steel column AB and intersect the x axis at a distance e from the geometric axis. Knowing that $\sigma_{all} = 75$ MPa and $E = 200$ GPa, determine the largest permissible length L when (a) $e = 5$ mm, (b) $e = 12$ mm.

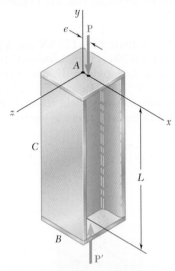

Fig. P10.49 and *P10.50*

10.50 Axial loads of magnitude $P = 580$ kN are applied parallel to the geometric axis of the W250 × 80 rolled-steel column AB and intersect the x axis at a distance e from the geometric axis. Knowing that $\sigma_{all} = 75$ MPa and $E = 200$ GPa, determine the largest permissible length L when (a) $e = 5$ mm, (b) $e = 10$ mm.

10.51 A 54-kN axial load is applied with an eccentricity $e = 10$ mm to the circular steel rod BC that is free at its top C and fixed at its base B. Knowing that the stock of rods available for use have diameters in increments of 4 mm from 44 mm to 72 mm, determine the lightest rod that can be used if $\sigma_{all} = 110$ MPa. Use $E = 200$ GPa.

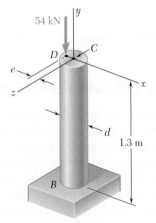

Fig. P10.51

10.52 Solve Prob. 10.51, assuming that the 54-kN axial load will be applied to the rod with an eccentricity $e = \frac{1}{2}d$.

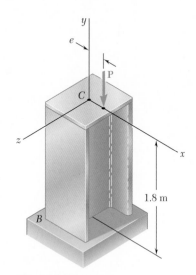

Fig. P10.53

10.53 An axial load of magnitude $P = 220$ kN is applied at a point located on the x axis at a distance $e = 6$ mm from the geometric axis of the wide-flange column BC. Knowing that $E = 200$ GPa, choose the lightest W200 shape that can be used if $\sigma_{all} = 120$ MPa.

10.54 Solve Prob. 10.53, assuming that the magnitude of the axial load is $P = 345$ kN.

10.55 Axial loads of magnitude $P = 175$ kN are applied parallel to the geometric axis of a W250 × 44.8 rolled-steel column AB and intersect the axis at a distance $e = 12$ mm from its geometric axis. Knowing that $\sigma_Y = 250$ MPa and $E = 200$ GPa, determine the factor of safety with respect to yield. (*Hint:* Since the factor of safety must be applied to the load **P**, not to the stresses, use Fig. 10.23 to determine P_Y.)

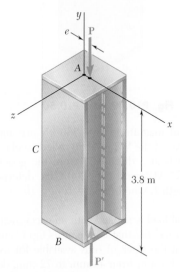

Fig. P10.55

10.56 Solve Prob. 10.55, assuming that $e = 0.16$ mm and $P = 155$ kN.

10.6 DESIGN OF COLUMNS UNDER A CENTRIC LOAD

In the preceding sections, we have determined the critical load of a column by using Euler's formula, and we have investigated the deformations and stresses in eccentrically loaded columns by using the secant formula. In each case we assumed that all stresses remained below the proportional limit and that the column was initially a straight homogeneous prism. Real columns fall short of such an idealization, and in practice the design of columns is based on empirical formulas that reflect the results of numerous laboratory tests.

Over the last century, many steel columns have been tested by applying to them a centric axial load and increasing the load until failure occurred. The results of such tests are represented in

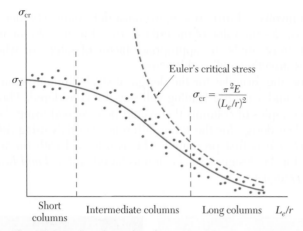

Fig. 10.24 Plot of test data for steel columns.

Fig. 10.24 where, for each of many tests, a point has been plotted with its ordinate equal to the normal stress σ_{cr} at failure, and its abscissa equal to the corresponding value of the effective slenderness ratio, L_e/r. Although there is considerable scatter in the test results, regions corresponding to three types of failure can be observed. For long columns, where L_e/r is large, failure is closely predicted by Euler's formula, and the value of σ_{cr} is observed to depend on the modulus of elasticity E of the steel used, but not on its yield strength σ_Y. For very short columns and compression blocks, failure occurs essentially as a result of yield, and we have $\sigma_{cr} \approx \sigma_Y$. Columns of intermediate length comprise those cases where failure is dependent on both σ_Y and E. In this range, column failure is an extremely complex phenomenon, and test data have been used extensively to guide the development of specifications and design formulas.

Empirical formulas that express an allowable stress or critical stress in terms of the effective slenderness ratio were first introduced over a century ago, and since then have undergone a continuous process of refinement and improvement. Typical empirical formulas previously used to approximate test data are shown in Fig. 10.25. It is not always feasible to use a single formula for all values of L_e/r. Most design specifications use different formulas, each with a definite range of applicability. In each case we must check that the formula we propose to use is applicable for the value of L_e/r for the

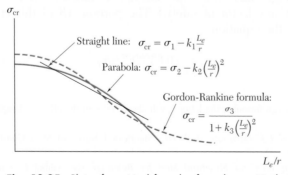

Fig. 10.25 Plots of empirical formulas for column critical stress.

column involved. Furthermore, we must determine whether the formula provides the value of the critical stress for the column, in which case we must apply the appropriate factor of safety, or whether it provides directly an allowable stress.

Specific formulas for the design of steel, aluminum, and wood columns under centric loading will now be considered. Photo 10.2 shows examples of columns that would be designed using these formulas. The design for the three different materials using *Allowable Stress Design* is first presented. This is followed with the formulas needed for the design of steel columns based on *Load and Resistance Factor Design.*†

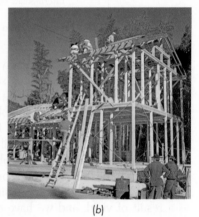

(a)   (b)

Photo 10.2 The water tank in (a) is supported by steel columns and the building under construction in (b) is framed with wood columns.

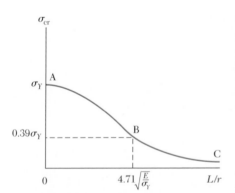

Fig. 10.26 Steel column design.

Structural Steel—Allowable Stress Design. The formulas most widely used for the allowable stress design of steel columns under a centric load are found in the Specification for Structural Steel Buildings of the American Institute of Steel Construction.‡ As we shall see, an exponential expression is used to predict σ_{all} for columns of short and intermediate lengths, and an Euler-based relation is used for long columns. The design relations are developed in two steps:

1. First a curve representing the variation of σ_{cr} with L/r is obtained (Fig. 10.26). It is important to note that this curve does not incorporate any factor of safety.§ The portion AB of this curve is defined by the equation

$$\sigma_{\text{cr}} = \left[0.658^{(\sigma_Y/\sigma_e)} \right] \sigma_Y \qquad (10.38)$$

---

†In specific design formulas, the letter L will always refer to the effective length of the column.

‡*Manual of Steel Construction*, 13th ed., American Institute of Steel Construction, Chicago, 2005.

§In the *Specification for Structural Steel for Buildings*, the symbol F is used for stresses.

where

$$\sigma_e = \frac{\pi^2 E}{(L/r)^2} \tag{10.39}$$

The portion BC is defined by the equation

$$\sigma_{cr} = 0.877\sigma_e \tag{10.40}$$

We note that when $L/r = 0$, $\sigma_{cr} = \sigma_Y$ in Eq. (10.38). At point B, Eq. (10.38) joins Eq. (10.40). The value of slenderness L/r at the junction between the two equations is

$$\frac{L}{r} = 4.71 \sqrt{\frac{E}{\sigma_Y}} \tag{10.41}$$

If L/r is smaller than the value in Eq. (10.41), σ_{cr} is determined from Eq. (10.38), and if L/r is greater, σ_{cr} is determined from Eq. (10.40). At the value of the slenderness L/r specified in Eq. (10.41), the stress $\sigma_e = 0.44\,\sigma_Y$. Using Eq. (10.40), $\sigma_{cr} = 0.877\,(0.44\,\sigma_Y) = 0.39\,\sigma_Y$.

2. A factor of safety must be introduced to obtain the final AISC design formulas. The factor of safety specified by the specification is 1.67. Thus

$$\sigma_{all} = \frac{\sigma_{cr}}{1.67} \tag{10.42}$$

The formulas obtained can be used with SI units.

We observe that, by using Eqs. (10.38), (10.40), (10.41), and (10.42), we can determine the allowable axial stress for a given grade of steel and any given value of L/r. The procedure is to first compute the value of L/r at the intersection between the two equations from Eq. (10.41). For given values of L/r smaller than that in Eq. (10.41), we use Eqs. (10.38) and (10.42) to calculate σ_{all}, and for values greater than that in Eq. (10.41), we use Eqs. (10.40) and (10.42) to calculate σ_{all}. Figure 10.27 provides a general illustration of how σ_e varies as a function of L/r for different grades of structural steel.

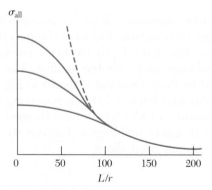

Fig. 10.27 Steel column design for different grades of steel.

EXAMPLE 10.02

P = 60 kN

A

L

B

Fig. 10.28

Determine the longest unsupported length L for which the S100 × 11.5 rolled-steel compression member AB can safely carry the centric load shown (Fig. 10.28). Assume $\sigma_Y = 250$ MPa and $E = 200$ GPa.

From Appendix C we find that, for an S100 × 11.5 shape,

$$A = 1460 \text{ mm}^2 \qquad r_x = 41.7 \text{ mm} \qquad r_y = 14.6 \text{ mm}$$

If the 60-kN load is to be safely supported, we must have

$$\sigma_{\text{all}} = \frac{P}{A} = \frac{60 \times 10^3 \text{ N}}{1460 \times 10^{-6} \text{ m}^2} = 41.1 \times 10^6 \text{ Pa}$$

We must compute the critical stress σ_{cr}. Assuming L/r is larger than the slenderness specified by Eq. (10.41), we use Eq. (10.40) with (10.39) and write

$$\sigma_{\text{cr}} = 0.877 \, \sigma_e = 0.877 \frac{\pi^2 E}{(L/r)^2}$$

$$= 0.877 \frac{\pi^2 (200 \times 10^9 \text{ Pa})}{(L/r)^2} = \frac{1.731 \times 10^{12} \text{ Pa}}{(L/r)^2}$$

Using this expression in Eq. (10.42) for σ_{all}, we write

$$\sigma_{\text{all}} = \frac{\sigma_{\text{cr}}}{1.67} = \frac{1.037 \times 10^{12} \text{ Pa}}{(L/r)^2}$$

Equating this expression to the required value of σ_{all}, we write

$$\frac{1.037 \times 10^{12} \text{ Pa}}{(L/r)^2} = 1.41 \times 10^6 \text{ Pa} \qquad L/r = 158.8$$

The slenderness ratio from Eq. (10.41) is

$$\frac{L}{r} = 4.71 \sqrt{\frac{200 \times 10^9}{250 \times 10^6}} = 133.2$$

Our assumption that L/r is greater than this slenderness ratio was correct. Choosing the smaller of the two radii of gyration, we have

$$\frac{L}{r_y} = \frac{L}{14.6 \times 10^{-3} \text{ m}} = 158.8 \qquad L = 2.32 \text{ m}$$

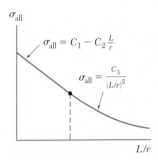

σ_{all}

$\sigma_{\text{all}} = C_1 - C_2 \dfrac{L}{r}$

$\sigma_{\text{all}} = \dfrac{C_3}{(L/r)^2}$

L/r

Fig. 10.29 Aluminum column design.

Aluminum. Many aluminum alloys are available for use in structural and machine construction. For most columns the specifications of the Aluminum Association† provide two formulas for the allowable stress in columns under centric loading. The variation of σ_{all} with L/r defined by these formulas is shown in Fig. 10.29. We note that for short columns a linear relation between σ_{all} with L/r is used and for long columns an Euler-type formula is used. Specific formulas for use in the design of buildings and similar structures are given here for two commonly used alloys.

†*Specifications for Aluminum Structures,* Aluminum Association, Inc., Washington, D.C., 2010.

$P = 60$ kN

A

L

d

B

c

d

SAMPLE PROBLEM 10.4

Using the aluminum alloy 2014-T6, determine the smallest diameter rod that can be used to support the centric load $P = 60$ kN if (a) $L = 750$ mm, (b) $L = 300$ mm.

SOLUTION

For the cross section of a solid circular rod, we have

$$I = \frac{\pi}{4}c^4 \qquad A = \pi c^2 \qquad r = \sqrt{\frac{I}{A}} = \sqrt{\frac{\pi c^4/4}{\pi c^2}} = \frac{c}{2}$$

a. Length of 750 mm. Since the diameter of the rod is not known, a value of L/r must be assumed; we *assume* that $L/r > 55$ and use Eq. (10.46). For the centric load **P**, we have $\sigma = P/A$ and write

$$\frac{P}{A} = \sigma_{all} = \frac{382 \times 10^3 \text{ MPa}}{(L/r)^2}$$

$$\frac{60 \times 10^3 \text{ N}}{\pi c^2} = \frac{382 \times 10^9 \text{ Pa}}{\left(\dfrac{0.750 \text{ m}}{c/2}\right)^2}$$

$$c^4 = 112.5 \times 10^{-9} \text{ m}^4 \qquad c = 18.31 \text{ mm}$$

For $c = 18.44$ mm, the slenderness ratio is

$$\frac{L}{r} = \frac{L}{c/2} = \frac{750 \text{ mm}}{(18.31 \text{ mm})/2} = 81.9 > 55$$

Our assumption is correct, and for $L = 750$ mm, the required diameter is

$$d = 2c = 2(18.31 \text{ mm}) \qquad d = 36.6 \text{ mm} \blacktriangleleft$$

b. Length of 300 mm. We again *assume* that $L/r > 55$. Using Eq. (10.46), and following the procedure used in part a, we find that $c = 11.58$ mm and $L/r = 51.8$. Since L/r is less than 55, our assumption is wrong; we now assume that $L/r < 55$ and use Eq. (10.45′) for the design of this rod.

$$\frac{P}{A} = \sigma_{all} = \left[213 - 1.577\left(\frac{L}{r}\right)\right] \text{MPa}$$

$$\frac{60 \times 10^3 \text{ N}}{\pi c^2} = \left[213 - 1.577\left(\frac{0.3 \text{ m}}{c/2}\right)\right] 10^6 \text{ Pa}$$

$$c = 11.95 \text{ mm}$$

For $c = 11.95$ mm, the slenderness ratio is

$$\frac{L}{r} = \frac{L}{c/2} = \frac{300 \text{ mm}}{(11.95 \text{ mm})/2} = 50.2$$

Our second assumption that $L/r < 55$ is correct. For $L = 300$ mm, the required diameter is

$$d = 2c = 2(11.95 \text{ mm}) \qquad d = 23.9 \text{ mm} \blacktriangleleft$$

10.57 Using allowable stress design, determine the allowable centric load for a column of 6-m effective length that is made from the following rolled-steel shape: (*a*) W200 × 35.9, (*b*) W200 × 86. Use $\sigma_Y = 250$ MPa and $E = 200$ GPa.

10.58 A W200 × 46.1 rolled-steel shape is used for a column of 6-m effective length. Using allowable stress design, determine the allowable centric load if the yield strength of the grade of steel used is (*a*) $\sigma_Y = 250$ MPa, (*b*) $\sigma_Y = 345$ MPa. Use $E = 200$ GPa.

10.59 A steel pipe having the cross section shown is used as a column. Using the allowable stress design determine the allowable centric load if the effective length of the column is (*a*) 6 m, (*b*) 4 m. Use $\sigma_Y = 250$ MPa and $E = 200$ GPa.

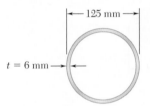

|← 125 mm →|

$t = 6$ mm

Fig. P10.59

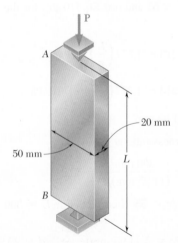

y

C •———— x

$A = 13.75 \times 10^3$ mm^2
$I_x = 26.0 \times 10^6$ mm^4
$I_y = 141.0 \times 10^6$ mm^4

Fig. P10.60

10.60 A column is made from half of a W360 × 216 rolled-steel shape, with the geometric properties as shown. Using allowable stress design, determine the allowable centric load if the effective length of the column is (*a*) 4.0 m, (*b*) 6.5 m. Use $\sigma_Y = 345$ MPa and $E = 200$ GPa.

10.61 A column with the cross section shown has a 4-m effective length. Using allowable stress design, determine the largest centric load that can be applied to the column. Use $\sigma_Y = 250$ MPa and $E = 200$ GPa.

12 mm

6 mm →|← 250 mm

150 mm

12 mm

Fig. P10.61

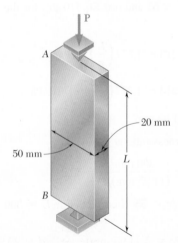

P

A

20 mm

50 mm

L

B

Fig. P10.62

10.62 Using the aluminum alloy 2014-T6, determine the largest allowable length of the aluminum bar *AB* for a centric load **P** of magnitude (*a*) 150 kN, (*b*) 90 kN, (*c*) 25 kN.

10.63 A sawn lumber column with a 190 × 140-mm cross section has a 5.5-m effective length. Knowing that for the grade of wood used the adjusted allowable stress for compression parallel to the grain is σ_C = 8.5 MPa and that the adjusted modulus E = 9 GPa, determine the maximum allowable centric load for the column.

10.64 A column having a 3.5-m effective length is made of sawn lumber with a 114 × 140-mm cross section. Knowing that for the grade of wood used the adjusted allowable stress for compression parallel to the grain is σ_C = 7.6 MPa and the adjusted modulus E = 2.8 GPa, determine the maximum allowable centric load for the column.

10.65 A compression member of 2.3-m effective length is obtained by bolting together two L127 × 76 × 12.7-mm steel angles as shown. Using allowable stress design, determine the allowable centric load for the column. Use σ_Y = 250 MPa and E = 200 GPa.

Fig. P10.65

10.66 and 10.67 A compression member of 9-m effective length is obtained by welding two 10-mm-thick steel plates to a W250 × 80 rolled-steel shape as shown. Knowing that σ_Y = 345 MPa and E = 200 GPa and using allowable stress design, determine the allowable centric load for the compression member.

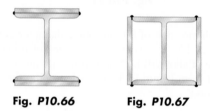

Fig. P10.66   **Fig. P10.67**

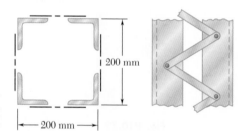

200 mm

200 mm

Fig. P10.68

10.68 A column of 6.4-m effective length is obtained by connecting four L89 × 89 × 9.5-mm steel angles with lacing bars as shown. Using allowable stress design, determine the allowable centric load for the column. Use σ_Y = 345 MPa and E = 200 GPa.

10.69 An aluminum structural tube is reinforced by bolting two plates to it as shown for use as a column of 1.7-m effective length. Knowing that all material is aluminum alloy 2014-T6, determine the maximum allowable centric load.

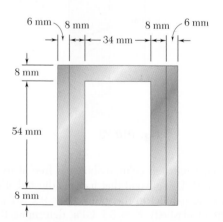

6 mm   8 mm   8 mm   6 mm

34 mm

8 mm

54 mm

8 mm

Fig. P10.69

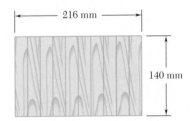

Fig. P10.70

Fig. P10.72

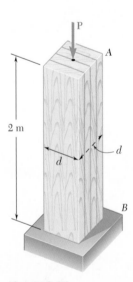

Fig. P10.74

10.70 A rectangular column with a 4.4-m effective length is made of glued laminated wood. Knowing that for the grade of wood used the adjusted allowable stress for compression parallel to the grain is $\sigma_C = 8.3$ MPa and the adjusted modulus $E = 4.6$ GPa, determine the maximum allowable centric load for the column.

10.71 A 280-kN centric load is applied to the column shown, which is free at its top A and fixed at its base B. Using aluminum alloy 2014-T6, select the smallest square section that can be used.

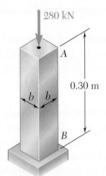

Fig. P10.71

10.72 An aluminum tube of 90-mm outer diameter is to carry a centric load of 120 kN. Knowing that the stock of tubes available for use are made of alloy 2014-T6 and with wall thicknesses in increments of 3 mm from 6 mm to 15 mm, determine the lightest tube that can be used.

10.73 A 72-kN centric load must be supported by an aluminum column as shown. Using the aluminum alloy 6061-T6, determine the minimum dimension b that can be used.

Fig. P10.73

10.74 The glued laminated column shown is free at its top A and fixed at its base B. Using wood that has an adjusted allowable stress for compression parallel to the grain $\sigma_C = 9.2$ MPa and an adjusted modulus of elasticity $E = 5.7$ GPa, determine the smallest cross section that can support a centric load of 62 kN.

10.75 A 75-kN centric load is applied to a rectangular sawn lumber column of 6.5-m effective length. Using sawn lumber for which the adjusted allowable stress for compression parallel to the grain is $\sigma_C = 7.3$ MPa and the adjusted modulus is $E = 10$ GPa, determine the smallest cross section that can be used. Use $b = 2d$.

10.76 A glue laminated column of 3-m effective length is to be made from boards of 24×100-mm cross section. Knowing that for the grade of wood used, $E = 11$ GPa and the adjusted allowable stress for compression parallel to the grain is $\sigma_C = 9$ MPa, determine the number of boards that must be used to support the centric load shown when (a) $P = 34$ kN, (b) $P = 17$ kN.

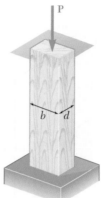

Fig. P10.75

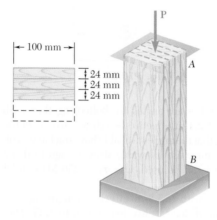

Fig. P10.76

10.77 A column of 4.5-m effective length must carry a centric load of 900 kN. Knowing that $\sigma_Y = 345$ MPa and $E = 200$ GPa, use allowable stress design to select the wide-flange shape of 250-mm nominal depth that should be used.

10.78 A column of 4.6-m effective length must carry a centric load of 525 kN. Knowing that $\sigma_Y = 345$ MPa and $E = 200$ GPa, use allowable stress design to select the wide-flange shape of 200-mm nominal depth that should be used.

10.79 A column of 6.8-m effective length must carry a centric load of 1200 kN. Using allowable stress design, select the wide-flange shape of 350-mm nominal depth that should be used. Use $\sigma_Y = 350$ MPa and $E = 200$ GPa.

10.80 A square steel tube having the cross section shown is used as a column of 7.8-m effective length to carry a centric load of 260 kN. Knowing that the tubes available for use are made with wall thicknesses ranging from 6 mm to 18 mm in increments of 1.5 mm, use allowable stress design to determine the lightest tube that can be used. Use $\sigma_Y = 250$ MPa and $E = 200$ GPa.

10.81 A compression member has the cross section shown and an effective length of 1.5 m. Knowing that the aluminum alloy used is 2014-T6, determine the allowable centric load.

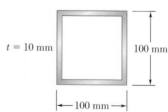

Fig. P10.80

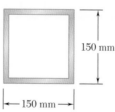

Fig. P10.81

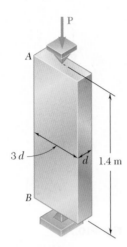

Fig. P10.82

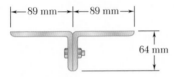

Fig. P10.84

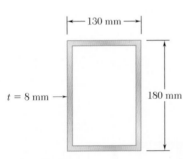

Fig. P10.86

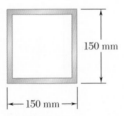

Fig. P10.88

10.82 A centric load **P** must be supported by the steel bar *AB*. Using allowable stress design, determine the smallest dimension *d* of the cross section that can be used when (*a*) *P* = 108 kN, (*b*) *P* = 166 kN. Use σ_Y = 250 MPa and *E* = 200 GPa.

10.83 Two 89 × 64-mm angles are bolted together as shown for use as a column of 2.4-m effective length to carry a centric load of 325 kN. Knowing that the angles available have thicknesses of 6.4 mm, 9.5 mm, and 12.7 mm, use allowable stress design to determine the lightest angles that can be used. Use σ_Y = 250 MPa and *E* = 200 GPa.

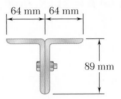

Fig. P10.83

10.84 Two 89 × 64-mm angles are bolted together as shown for use as a column of 2.4-m effective length to carry a centric load of 180 kN. Knowing that the angles available have thicknesses of 6.4 mm, 9.5 mm, and 12.7 mm, use allowable stress design to determine the lightest angles that can be used. Use σ_Y = 250 MPa and *E* = 200 GPa.

*10.85 A column with a 5.8-m effective length supports a centric load, with ratio of dead to live load equal to 1.35. The dead load factor is γ_D = 1.2, the live load factor γ_L = 1.6, and the resistance factor ϕ = 0.90. Use load and resistance factor design to determine the allowable centric dead and live loads if the column is made of the following rolled-steel shape: (*a*) W250 × 67, (*b*) W360 × 101. Use σ_Y = 345 MPa and *E* = 200 GPa.

*10.86 A rectangular steel tube having the cross section shown is used as a column of 4.4-m effective length. Knowing that σ_Y = 250 MPa and *E* = 200 GPa, use load and resistance factor design to determine the largest centric live load that can be applied if the centric dead load is 220 kN. Use a dead load factor γ_D = 1.2, a live load factor γ_L = 1.6, and the resistance factor ϕ = 0.85.

*10.87 A steel column of 5.5-m effective length must carry a centric dead load of 310 kN and a centric live load of 375 kN. Knowing that σ_Y = 250 MPa and *E* = 200 GPa, use load and resistance factor design to select the wide-flange shape of 310-mm nominal depth that should be used. The dead load factor γ_D = 1.2, the live load factor γ_L = 1.6, and the resistance factor ϕ = 0.90.

*10.88 The steel tube having the cross section shown is used as a column of 4.5-m effective length to carry a centric dead load of 210 kN and a centric live load of 240 kN. Knowing that the tubes available for use are made with wall thicknesses in increments of 2 mm from 4 mm to 10 mm, use load and resistance factor design to determine the lightest tube that can be used. Use σ_Y = 250 MPa and *E* = 200 GPa. The dead load factor γ_D = 1.2, the live load factor γ_L = 1.6, and the resistance factor ϕ = 0.85.

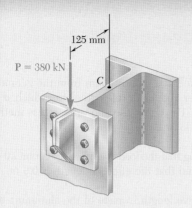

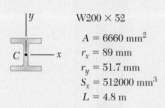

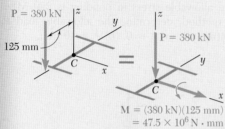

W200 × 52

$A = 6660 \text{ mm}^2$
$r_x = 89 \text{ mm}$
$r_y = 51.7 \text{ mm}$
$S_x = 512000 \text{ mm}^3$
$L = 4.8 \text{ m}$

W200 × 71

$A = 9100 \text{ mm}^2$
$r_x = 91.7 \text{ mm}$
$r_y = 52.8 \text{ mm}$
$S_x = 709000 \text{ mm}^3$
$L = 4.8 \text{ m}$

W200 × 59

$A = 7560 \text{ mm}^2$
$r_x = 89.9 \text{ mm}$
$r_y = 51.9 \text{ mm}$
$S_x = 582000 \text{ mm}^3$
$L = 4.8 \text{ m}$

SAMPLE PROBLEM 10.7

A steel column having an effective length of 4.8 m is loaded eccentrically as shown. Using the interaction method, select the wide-flange shape of 200-mm nominal depth that should be used. Assume $E = 200$ GPa and $\sigma_Y = 250$ MPa, and use an allowable stress in bending of 150 MPa.

SOLUTION

So that we can select a trial section, we use the allowable-stress method with $\sigma_{all} = 150$ MPa and write

$$\sigma_{all} = \frac{P}{A} + \frac{Mc}{I_x} = \frac{P}{A} + \frac{Mc}{Ar_x^2} \qquad (1)$$

From Appendix C we observe for shapes of 200-mm nominal depth that $c \approx 100$ mm and $r_x \approx 89$ mm. Substituting into Eq. (1), we have

$$150 \text{ N/mm}^2 = \frac{380 \text{ kN}}{A} + \frac{(47.5 \times 10^6 \text{ N} \cdot \text{mm})(100 \text{ mm})}{A(89 \text{ mm})^2} \quad A \approx 6.53 \times 10^3 \text{ mm}^2$$

We select for a first trial shape: W200 × 52.

Trial 1: W200 × 52. The allowable stresses are
Allowable Bending Stress: (see data) $(\sigma_{all})_{bending} = 150$ MPa
Allowable Centric Stress: The largest slenderness ratio is $L/r_y = (4.8 \text{ m})/(0.0517 \text{ m}) = 92.8$. For $E = 200$ GPa and $\sigma_Y = 250$ MPa, we find that the value of slenderness L/r at the junction between the two equations is 133.2 (Eq. 10.41). Thus, we use Eqs. (10.38) and (10.39) to find that $\sigma_{cr} = 158.4$ MPa. Using Eq. (10.42) the allowable stress is

$$(\sigma_{all})_{centric} = \frac{158.4 \text{ MPa}}{1.67} = 94.85 \text{ MPa}$$

For the W200 × 52 trial shape, we have

$$\frac{P}{A} = \frac{380 \text{ kN}}{6660 \text{ mm}^2} = 57 \text{ MPa} \qquad \frac{Mc}{I} = \frac{M}{S_x} = \frac{47.5 \times 10^6 \text{ N} \cdot \text{mm}}{512000 \text{ mm}} = 92.8 \text{ MPa}$$

With this data we find that the left-hand member of Eq. (10.60) is

$$\frac{P/A}{(\sigma_{all})_{centric}} + \frac{Mc/I}{(\sigma_{all})_{bending}} = \frac{57 \text{ MPa}}{94.85 \text{ MPa}} + \frac{92.8 \text{ MPa}}{150 \text{ MPa}} = 1.22$$

Since $1.22 > 1.000$, the requirement expressed by the interaction formula is not satisfied; we must select a larger trial shape.

Trial 2: W200 × 71. Following the procedure used in trial 1, we write

$$\frac{L}{r_y} = \frac{4.8 \text{ m}}{0.0528 \text{ m}} = 90.9 \qquad (\sigma_{all})_{centric} = 96.6 \text{ MPa}$$

$$\frac{P}{A} = \frac{380 \text{ kN}}{9100 \text{ mm}^2} = 41.8 \text{ MPa} \qquad \frac{Mc}{I} = \frac{M}{S_x} = \frac{47.5 \times 10^6 \text{ N} \cdot \text{mm}}{709000 \text{ mm}^3} = 67 \text{ MPa}$$

Substituting into Eq. (10.60) gives

$$\frac{P/A}{(\sigma_{all})_{centric}} + \frac{Mc/I}{(\sigma_{all})_{bending}} = \frac{41.8 \text{ MPa}}{96.6} + \frac{67 \text{ MPa}}{150 \text{ MPa}} = 0.879 < 1.000$$

The W200 × 71 shape is satisfactory but may be unnecessarily large.

Trial 3: W200 × 59. Following again the same procedure, we find that the interaction formula is not satisfied.

Selection of Shape. The shape to be used is   W200 × 71 ◀

PROBLEMS

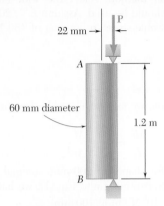

Fig. P10.89

10.89 An eccentric load is applied at a point 22 mm from the geometric axis of a 60-mm-diameter rod made of a steel for which $\sigma_Y = 250$ MPa and $E = 200$ GPa. Using the allowable-stress method, determine the allowable load **P**.

10.90 Solve Prob. 10.89, assuming that the load is applied at a point 40 mm from the geometric axis and that the effective length is 0.9 m.

10.91 A column of 5.5-m effective length is made of the aluminum alloy 2014-T6, for which the allowable stress in bending is 220 MPa. Using the interaction method, determine the allowable load **P**, knowing that the eccentricity is (a) $e = 0$, (b) $e = 40$ mm.

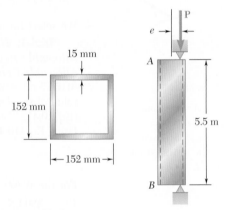

Fig. P10.91

10.92 Solve Prob. 10.91, assuming that the effective length of the column is 3.0 m.

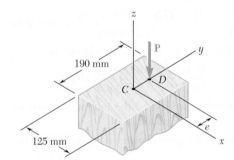

Fig. P10.93

10.93 A sawn-lumber column of 125×190-mm cross section has an effective length of 2.5 m. The grade of wood used has an adjusted allowable stress for compression parallel to the grain $\sigma_C = 8$ MPa and a modulus $E = 8.3$ GPa. Using the allowable-stress method, determine the largest eccentric load **P** that can be applied when (a) $e = 12$ mm, (b) $e = 24$ mm.

10.94 Solve Prob. 10.93 using the interaction method and an allowable stress in bending of 9 MPa.

10.95 A steel compression member of 2.75-m effective length supports an eccentric load as shown. Using the allowable-stress method and assuming $e = 40$ mm, determine the maximum allowable load **P.** Use $\sigma_Y = 250$ MPa and $E = 200$ GPa.

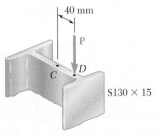

Fig. P10.95

10.96 Solve Prob. 10.95, using $e = 60$ mm.

10.97 The compression member AB is made of a steel for which $\sigma_Y =$ 250 MPa and $E = 200$ GPa. It is free at its top A and fixed at its base B. Using the allowable-stress method, determine the largest allowable eccentricity e_x, knowing that (a) $e_y = 0$, (b) $e_y = 8$ mm.

10.98 The compression member AB is made of a steel for which $\sigma_Y =$ 250 MPa and $E = 200$ GPa. It is free at its top A and fixed at its base B. Using the interaction method with an allowable bending stress equal to 120 MPa and knowing that the eccentricities e_x and e_y are equal, determine their largest allowable common value.

10.99 An eccentric load $P = 48$ kN is applied at a point 20 mm from the geometric axis of a 50-mm-diameter rod made of the aluminum alloy 6061-T6. Using the interaction method and an allowable stress in bending of 145 MPa, determine the largest allowable effective length L that can be used.

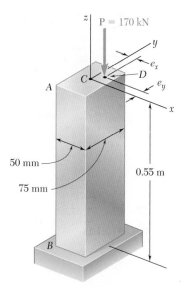

Fig. P10.97 and P10.98

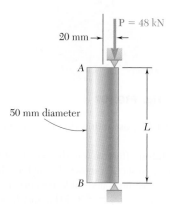

Fig. P10.99

10.100 Solve Prob. 10.99, assuming that the aluminum alloy used is 2014-T6 and that the allowable stress in bending is 180 MPa.

10.101 A rectangular column is made of a grade of sawn wood that has an adjusted allowable stress for compression parallel to the grain $\sigma_C = 8.3$ MPa and an adjusted modulus of elasticity $E = 11.1$ GPa. Using the allowable-stress method, determine the largest allowable effective length L that can be used.

10.102 Solve Prob. 10.101, assuming that $P = 105$ kN.

10.103 An 32-kN vertical load **P** is applied at the midpoint of one edge of the square cross section of the steel compression member AB, which is free at its top A and fixed at its base B. Knowing that the alloy used is 6061-T6 and using the allowable-stress method, determine the smallest allowable dimension d.

10.104 Solve Prob. 10.103, assuming that the vertical load **P** is applied at the corner of the cross section.

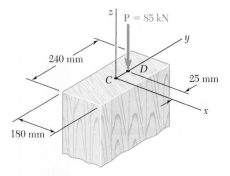

Fig. P10.101

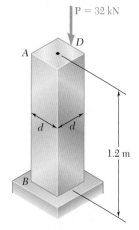

Fig. P10.103

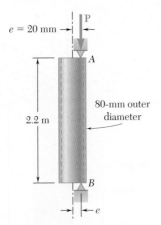

$e = 20$ mm

P

A

80-mm outer
diameter

2.2 m

B

e

Fig. P10.105

10.105 A steel tube of 80-mm outer diameter is to carry a 93-kN load **P** with an eccentricity of 20 mm. The tubes available for use are made with wall thicknesses in increments of 3 mm from 6 mm to 15 mm. Using the allowable-stress method, determine the lightest tube that can be used. Assume $E = 200$ GPa and $\sigma_Y = 250$ MPa.

10.106 Solve Prob. 10.105, using the interaction method with $P = 165$ kN, $e = 15$ mm, and an allowable stress in bending of 150 MPa.

10.107 A compression member of rectangular cross section has an effective length of 0.9 m and is made of the aluminum alloy 2014-T6 for which the allowable stress in bending is 160 MPa. Using the interaction method, determine the smallest dimension d of the cross section that can be used when $e = 10$ mm.

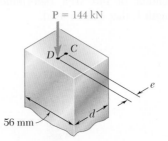

$P = 144$ kN

D C

e

56 mm

d

Fig. P10.107

10.108 Solve Prob. 10.107, assuming that $e = 5$ mm.

10.109 A column of 4.2-m effective length consists of a section of steel tubing having the cross section shown. Using the allowable-stress method, determine the maximum allowable eccentricity e if (a) $P = 220$ kN, (b) $P = 140$ kN. Use $\sigma_Y = 250$ MPa and $E = 200$ GPa.

10.110 Solve Prob. 10.109, assuming that the effective length of the column is increased to 5.4 m and that (a) $P = 110$ kN, (b) $P = 70$ kN.

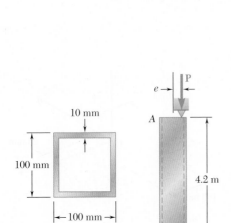

10 mm

100 mm

100 mm

e

P

A

4.2 m

B

Fig. P10.109

10.111 A sawn lumber column of rectangular cross section has a 2.2-m effective length and supports a 41-kN load as shown. The sizes available for use have b equal to 90 mm, 140 mm, 190 mm, and 240 mm. The grade of wood has an adjusted allowable stress for compression parallel to the grain $\sigma_C = 8.1$ MPa and an adjusted modulus $E = 8.3$ GPa. Using the allowable-stress method, determine the lightest section that can be used.

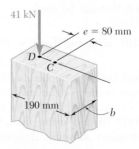

41 kN

$e = 80$ mm

D

C

190 mm

b

Fig. P10.111

10.112 Solve Prob. 10.111, assuming that $e = 40$ mm.

10.113 A steel column having a 7.2-m effective length is loaded eccentrically as shown. Using the allowable-stress method, select the wide-flange shape of 350-mm nominal depth that should be used. Use $\sigma_Y = 250$ MPa and $E = 200$ GPa.

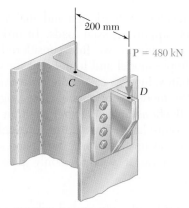

200 mm

P = 480 kN

C

D

Fig. P10.113

10.114 Solve Prob. 10.113 using the interaction method, assuming that $\sigma_Y = 350$ MPa and the allowable stress in bending is 200 MPa.

10.115 A steel column of 7.2-m effective length is to support an 83-kN eccentric load **P** at a point D, located on the x axis as shown. Using the allowable-stress method, select the wide-flange shape of 250-mm nominal depth that should be used. Use $E = 200$ GPa and $\sigma_Y = 250$ MPa.

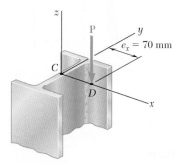

z

P

y

$e_x = 70$ mm

C

D

x

Fig. P10.115

10.116 A steel compression member of 5.8-m effective length is to support a 296-kN eccentric load **P.** Using the interaction method, select the wide-flange shape of 200-mm nominal depth that should be used. Use $E = 200$ GPa, $\sigma_Y = 250$ MPa, and $\sigma_{\text{all}} = 150$ MPa in bending.

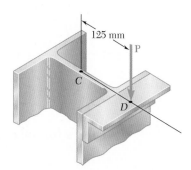

125 mm

P

C

D

Fig. P10.116

This chapter was devoted to the design and analysis of columns, i.e., prismatic members supporting axial loads. In order to gain insight into the behavior of columns, we first considered in Sec. 10.2 the equilibrium of a simple model and found that for values of the load **Critical load** P exceeding a certain value P_{cr}, called the *critical load*, two equilibrium positions of the model were possible: the original position with zero transverse deflections and a second position involving deflections that could be quite large. This led us to conclude that the first equilibrium position was unstable for $P > P_{cr}$, and stable for $P < P_{cr}$, since in the latter case it was the only possible equilibrium position.

In Sec. 10.3, we considered a pin-ended column of length L and of constant flexural rigidity EI subjected to an axial centric load P. Assuming that the column had buckled (Fig. 10.36), we noted that the bending moment at point Q was equal to $-Py$ and wrote

$$\frac{d^2y}{dx^2} = \frac{M}{EI} = -\frac{P}{EI}y \qquad (10.4)$$

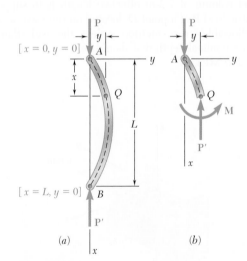

Fig. 10.36

Solving this differential equation, subject to the boundary conditions corresponding to a pin-ended column, we determined the smallest load P for which buckling can take place. This load, known as the **Euler's formula** *critical load* and denoted by P_{cr}, is given by *Euler's formula:*

$$P_{cr} = \frac{\pi^2 EI}{L^2} \qquad (10.11)$$

where L is the length of the column. For this load or any larger load, the equilibrium of the column is unstable and transverse deflections will occur.

Denoting the cross-sectional area of the column by A and its radius of gyration by r, we determined the critical stress σ_{cr} corresponding to the critical load P_{cr}:

$$\sigma_{cr} = \frac{\pi^2 E}{(L/r)^2} \tag{10.13}$$

The quantity L/r is called the *slenderness ratio* and we plotted σ_{cr} as a function of L/r (Fig. 10.37). Since our analysis was based on stresses remaining below the yield strength of the material, we noted that the column would fail by yielding when $\sigma_{cr} > \sigma_Y$.

In Sec. 10.4, we discussed the critical load of columns with various end conditions and wrote

$$P_{cr} = \frac{\pi^2 EI}{L_e^2} \tag{10.11'}$$

where L_e is the *effective length* of the column, i.e., the length of an equivalent pin-ended column. The effective lengths of several columns with various end conditions were calculated and shown in Fig. 10.17 on page 664.

In Sec. 10.5, we considered columns supporting an *eccentric axial load*. For a pin-ended column subjected to a load P applied with an eccentricity e, we replaced the load by a centric axial load and a couple of moment $M_A = Pe$ (Figs. 10.38 and 10.39) and derived the following expression for the maximum transverse deflection:

$$y_{max} = e\left[\sec\left(\sqrt{\frac{P}{EI}}\frac{L}{2}\right) - 1\right] \tag{10.28}$$

Slenderness ratio

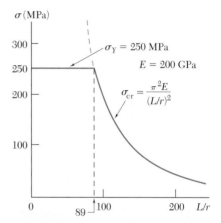

Fig. 10.37

Effective length

Eccentric axial load.

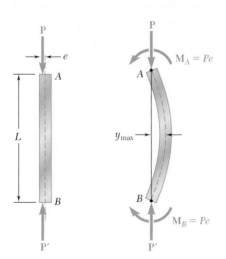

Fig. 10.38   **Fig. 10.39**

Secant formula

We then determined the maximum stress in the column, and from the expression obtained for that stress, we derived the *secant formula:*

$$\frac{P}{A} = \frac{\sigma_{max}}{1 + \dfrac{ec}{r^2} \sec\left(\dfrac{1}{2}\sqrt{\dfrac{P}{EA}}\dfrac{L_e}{r}\right)} \tag{10.36}$$

This equation can be solved for the force per unit area, P/A, that causes a specified maximum stress σ_{max} in a pin-ended column or any other column of effective slenderness ratio L_e/r.

Design of real columns

In the first part of the chapter we considered each column as a straight homogeneous prism. Since imperfections exist in all real columns, the *design of real columns* is done by using empirical formulas based on laboratory tests and set forth in specifications and codes issued by professional organizations. In Sec. 10.6, we discussed the design of *centrically loaded columns* made of steel, aluminum, or wood. For each material, the design of the column was based on formulas expressing the allowable stress as a function of the slenderness ratio L/r of the column. For structural steel, we also discussed the alternative method of *Load and Resistance Factor Design.*

Centrically loaded columns

Eccentrically loaded columns

In the last section of the chapter [Sec. 10.7], we studied two methods used for the design of columns under an *eccentric* load. The first method was the *allowable-stress method,* a conservative method in which it is assumed that the allowable stress is the same as if the column were centrically loaded. The allowble-stress method requires that the following inequality be satisfied:

Allowable-stress method

$$\frac{P}{A} + \frac{Mc}{I} \le \sigma_{all} \tag{10.53}$$

Interaction method

The second method was the *interaction method,* a method used in most modern specifications. In this method the allowable stress for a centrically loaded column is used for the portion of the total stress due to the axial load and the allowable stress in bending for the stress due to bending. Thus, the inequality to be satisfied is

$$\frac{P/A}{(\sigma_{all})_{centric}} + \frac{Mc/I}{(\sigma_{all})_{bending}} \le 1 \tag{10.55}$$

REVIEW PROBLEMS

10.117 The rigid bar AD is attached to two springs of constant k and is in equilibrium in the position shown. Knowing that the equal and opposite loads **P** and **P′** *remain vertical*, determine the magnitude P_{cr} of the critical load for the system. Each spring can act in either tension or compression.

10.118 The steel rod BC is attached to the rigid bar AB and to the fixed support at C. Knowing that $G = 77$ GPa, determine the diameter of rod BC for which the critical load P_{cr} of the system is 350 N.

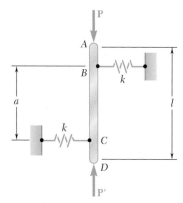

Fig. P10.117

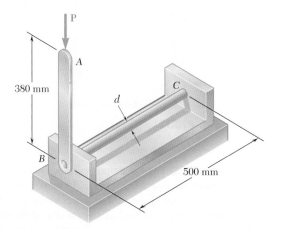

Fig. P10.118

10.119 Determine (*a*) the critical load for the steel strut, (*b*) the dimension d for which the aluminum strut will have the same critical load. (*c*) Express the weight of the aluminum strut as a percent of the weight of the steel strut.

10.120 Supports A and B of the pin-ended column shown are at a fixed distance L from each other. Knowing that at a temperature T_0 the force in the column is zero and that buckling occurs when the temperature is $T_1 = T_0 + \Delta T$, express ΔT in terms of b, L and the coefficient of thermal expansion α.

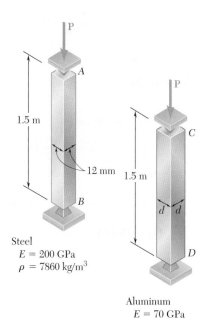

Steel
$E = 200$ GPa
$\rho = 7860$ kg/m³

Aluminum
$E = 70$ GPa
$\rho = 2800$ kg/m³

Fig. P10.119

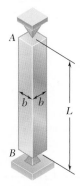

Fig. P10.120

10.121 Members AB and CD are 30-mm-diameter steel rods, and members BC and AD are 22-mm-diameter steel rods. When the turnbuckle is tightened, the diagonal member AC is put in tension. Knowing that a factor of safety with respect to buckling of 2.75 is required, determine the largest allowable tension in AC. Use $E = 200$ GPa and consider only buckling in the plane of the structure.

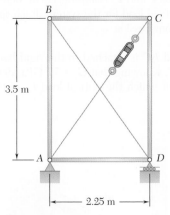

Fig. P10.121

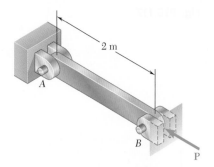

Fig. P10.122

10.122 The uniform aluminum bar AB has a 20 × 36-mm rectangular cross section and is supported by pins and brackets as shown. Each end of the bar may rotate freely about a horizontal axis through the pin, but rotation about a vertical axis is prevented by the brackets. Using $E = 70$ GPa, determine the allowable centric load **P** if a factor of safety of 2.5 is required.

10.123 A compression member has the cross section shown and an effective length of 1.5 m. Knowing that the aluminum alloy used is 6061-T6, determine the allowable centric load.

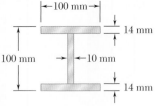

Fig. P10.123

10.124 Knowing that $P = 5.2$ kN, determine the factor of safety for the structure shown. Use $E = 200$ GPa and consider only buckling in the plane of the structure.

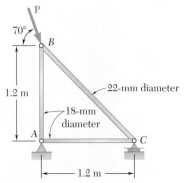

Fig. P10.124

10.125 An axial load **P** of magnitude 560 kN is applied at a point on the *x* axis at a distance *e* = 6 mm from the geometric axis of the W200 × 46.1 rolled-steel column *BC*. Using *E* = 200 GPa, determine (*a*) the horizontal deflection of end *C*, (*b*) the maximum stress in the column.

10.126 A column of 5-m effective length must carry a centric load of 950 kN. Using allowable stress design, select the wide-flange shape of 250 mm nominal depth that should be used. Use σ_Y = 250 MPa and *E* = 200 GPa.

10.127 Bar *AB* is free at its end *A* and fixed at its base *B*. Determine the allowable centric load **P** if the aluminum alloy is (*a*) 6061-T6, (*b*) 2014-T6.

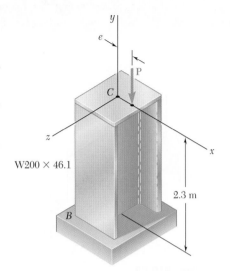

W200 × 46.1

Fig. P10.125

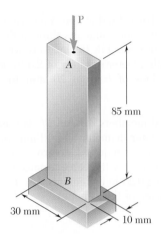

Fig. *P10.127*

10.128 A 180-kN axial load **P** is applied to the rolled-steel column *BC* at a point on the *x* axis at a distance *e* = 62 mm from the geometric axis of the column. Using the allowable-stress method, select the wide-flange shape of 200-mm nominal depth that should be used. Use *E* = 200 GPa and σ_Y = 250 MPa.

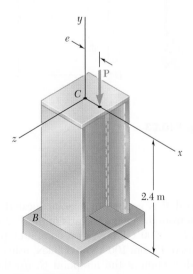

Fig. P10.128

COMPUTER PROBLEMS

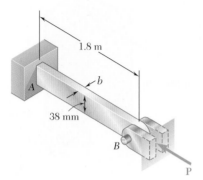

Fig. P10.C2

The following problems are designed to be solved with a computer.

10.C1 A solid steel rod having an effective length of 500 mm is to be used as a compression strut to carry a centric load **P.** For the grade of steel used, $E = 200$ GPa and $\sigma_Y = 245$ MPa. Knowing that a factor of safety of 2.8 is required and using Euler's formula, write a computer program and use it to calculate the allowable centric load P_{all} for values of the radius of the rod from 6 mm to 24 mm, using 2-mm increments.

10.C2 An aluminum bar is fixed at end A and supported at end B so that it is free to rotate about a horizontal axis through the pin. Rotation about a vertical axis at end B is prevented by the brackets. Knowing that $E = 70$ GPa, use Euler's formula with a factor of safety of 2.5 to determine the allowable centric load **P** for values of b from 20 mm to 38 mm, using 3-mm increments.

10.C3 The pin-ended members AB and BC consist of sections of aluminum pipe of 120-mm outer diameter and 10-mm wall thickness. Knowing that a factor of safety of 3.5 is required, determine the mass m of the largest block that can be supported by the cable arrangement shown for values of h from 4 m to 8 m, using 0.25-m increments. Use $E = 70$ GPa and consider only buckling in the plane of the structure.

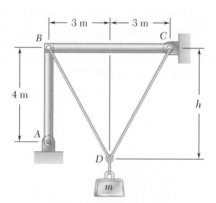

Fig. P10.C3

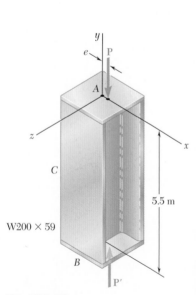

Fig. P10.C4

10.C4 An axial load **P** is applied at a point located on the x axis at a distance $e = 12$ mm from the geometric axis of the W200 × 59 rolled-steel column AB. Using $E = 200$ GPa, write a computer program and use it to calculate for values of P from 100 kN to 300 kN, using 20-kN increments, (*a*) the horizontal deflection at the midpoint C, (*b*) the maximum stress in the column.

10.C5 A column of effective length L is made from a rolled-steel shape and carries a centric axial load **P.** The yield strength for the grade of steel used is denoted by σ_Y, the modulus of elasticity by E, the cross-sectional area of the selected shape by A, and its smallest radius of gyration by r. Using the AISC design formulas for allowable stress design, write a computer program to determine the allowable load **P.** Use this program to solve (*a*) Prob. 10.57, (*b*) Prob. 10.58, (*c*) Prob. 10.60.

10.C6 A column of effective length L is made from a rolled-steel shape and is loaded eccentrically as shown. The yield strength of the grade of steel used is denoted by σ_Y, the allowable stress in bending by σ_{all}, the modulus of elasticity by E, the cross-sectional area of the selected shape by A, and its smallest radius of gyration by r. Write a computer program to determine the allowable load **P,** using either the allowable-stress method or the interaction method. Use this program to check the given answer for (*a*) Prob. 10.113, (*b*) Prob. 10.114.

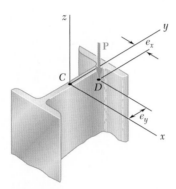

Fig. P10.C6

As the diver comes down on the diving board the potential energy due to his elevation above the board will be converted into strain energy due to the bending of the board. The normal and shearing stresses resulting from energy loadings will be determined in this chapter.

Energy Methods

11.1 INTRODUCTION

In the previous chapter we were concerned with the relations existing between forces and deformations under various loading conditions. Our analysis was based on two fundamental concepts, the concept of stress (Chap. 1) and the concept of strain (Chap. 2). A third important concept, the concept of *strain energy*, will now be introduced.

In Sec. 11.2, the *strain energy* of a member will be defined as the increase in energy associated with the deformation of the member. You will see that the strain energy is equal to the work done by a slowly increasing load applied to the member. The *strain-energy density* of a material will be defined as the strain energy per unit volume; it will be seen that it is equal to the area under the stress-strain diagram of the material (Sec. 11.3). From the stress-strain diagram of a material two additional properties will be defined, namely, the *modulus of toughness* and the *modulus of resilience* of the material.

In Sec. 11.4 the elastic strain energy associated with *normal stresses* will be discussed, first in members under axial loading and then in members in bending. Later you will consider the elastic strain energy associated with shearing stresses such as occur in torsional loadings of shafts and in transverse loadings of beams (Sec. 11.5). Strain energy for a *general state of stress* will be considered in Sec. 11.6, where the *maximum-distortion-energy criterion* for yielding will be derived.

The effect of *impact loading* on members will be considered in Sec. 11.7. You will learn to calculate both the *maximum stress* and the *maximum deflection* caused by a moving mass impacting on a member. Properties that increase the ability of a structure to withstand impact loads effectively will be discussed in Sec. 11.8.

In Sec. 11.9 the elastic strain of a member subjected to a *single concentrated load* will be calculated, and in Sec. 11.10 the deflection at the point of application of a single load will be determined.

The last portion of the chapter will be devoted to the determination of the strain energy of structures subjected to *several loads* (Sec. 11.11). *Castigliano's theorem* will be derived in Sec. 11.12 and used in Sec. 11.13 to determine the deflection at a given point of a structure subjected to several loads. In the last section Castigliano's theorem will be applied to the analysis of indeterminate structures (Sec. 11.14).

11.2 STRAIN ENERGY

Consider a rod BC of length L and uniform cross-sectional area A, which is attached at B to a fixed support, and subjected at C to a *slowly increasing* axial load $\mathbf{P}$ (Fig. 11.1). As we noted in Sec. 2.2, by plotting the magnitude P of the load against the deformation x of the rod, we obtain a certain load-deformation diagram (Fig. 11.2) that is characteristic of the rod BC.

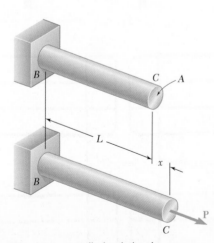

Fig. 11.1 Axially loaded rod.

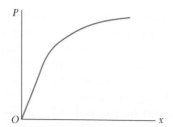

Fig. 11.2 Load-deformation diagram.

Let us now consider the work dU done by the load **P** as the rod elongates by a small amount dx. This *elementary work* is equal to the product of the magnitude P of the load and of the small elongation dx. We write

$$dU = P \, dx \qquad (11.1)$$

and note that the expression obtained is equal to the element of area of width dx located under the load-deformation diagram (Fig. 11.3). The *total work* U done by the load as the rod undergoes a deformation x_1 is thus

$$U = \int_0^{x_1} P \, dx$$

and is equal to the area under the load-deformation diagram between $x = 0$ and $x = x_1$.

The work done by the load **P** as it is slowly applied to the rod must result in the increase of some energy associated with the deformation of the rod. This energy is referred to as the *strain energy* of the rod. We have, by definition,

$$\text{Strain energy} = U = \int_0^{x_1} P \, dx \qquad (11.2)$$

We recall that work and energy should be expressed in units obtained by multiplying units of length by units of force. Thus, if SI metric units are used, work and energy are expressed in N · m; this unit is called a *joule* (J).

In the case of a linear and elastic deformation, the portion of the load-deformation diagram involved can be represented by a straight line of equation $P = kx$ (Fig. 11.4). Substituting for P in Eq. (11.2), we have

$$U = \int_0^{x_1} kx \, dx = \tfrac{1}{2}kx_1^2$$

or

$$U = \tfrac{1}{2}P_1 x_1 \qquad (11.3)$$

where P_1 is the value of the load corresponding to the deformation x_1.

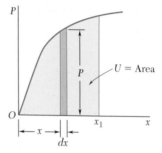

Fig. 11.3 Work due to load **P.**

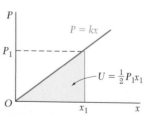

Fig. 11.4 Work due to linear, elastic deformation.

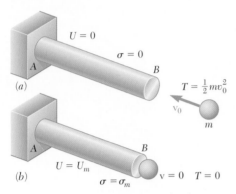

Fig. 11.5 Rod subject to impact loading.

The concept of strain energy is particularly useful in the determination of the effects of impact loadings on structures or machine components. Consider, for example, a body of mass m moving with a velocity $\mathbf{v}_0$ which strikes the end B of a rod AB (Fig. 11.5a). Neglecting the inertia of the elements of the rod, and assuming no dissipation of energy during the impact, we find that the maximum strain energy U_m acquired by the rod (Fig. 11.5b) is equal to the original kinetic energy $T = \frac{1}{2}mv_0^2$ of the moving body. We then determine the value P_m of the static load which would have produced the same strain energy in the rod, and obtain the value σ_m of the largest stress occurring in the rod by dividing P_m by the cross-sectional area of the rod.

11.3 STRAIN-ENERGY DENSITY

As we noted in Sec. 2.2, the load-deformation diagram for a rod BC depends upon the length L and the cross-sectional area A of the rod. The strain energy U defined by Eq. (11.2), therefore, will also depend upon the dimensions of the rod. In order to eliminate the effect of size from our discussion and direct our attention to the properties of the material, the strain energy per unit volume will be considered. Dividing the strain energy U by the volume $V = AL$ of the rod (Fig. 11.1), and using Eq. (11.2), we have

$$\frac{U}{V} = \int_0^{x_1} \frac{P}{A}\frac{dx}{L}$$

Recalling that P/A represents the normal stress σ_x in the rod, and x/L the normal strain ϵ_x, we write

$$\frac{U}{V} = \int_0^{\epsilon_1} \sigma_x \, d\epsilon_x$$

where ϵ_1 denotes the value of the strain corresponding to the elongation x_1. The strain energy per unit volume, U/V, is referred to as the *strain-energy density* and will be denoted by the letter u. We have, therefore,

$$\text{Strain-energy density} = u = \int_0^{\epsilon_1} \sigma_x \, d\epsilon_x \qquad (11.4)$$

The strain-energy density u is expressed in units obtained by dividing units of energy by units of volume. Thus, if SI metric units are used, the strain-energy density is expressed in J/m^3 or its multiples kJ/m^3 and MJ/m^3.†

---

†We note that 1 J/m^3 and 1 Pa are both equal to 1 N/m^2. Thus, strain-energy density and stress are dimensionally equal and could be expressed in the same units.

Referring to Fig. 11.6, we note that the strain-energy density u is equal to the area under the stress-strain curve, measured from $\epsilon_x = 0$ to $\epsilon_x = \epsilon_1$. If the material is unloaded, the stress returns to zero, but there is a permanent deformation represented by the strain ϵ_p, and only the portion of the strain energy per unit volume corresponding to the triangular area is recovered. The remainder of the energy spent in deforming the material is dissipated in the form of heat.

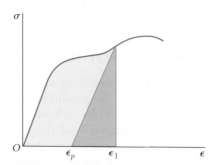

Fig. 11.6 Strain energy.

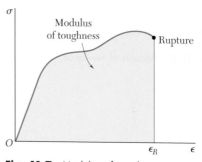

Fig. 11.7 Modulus of toughness.

The value of the strain-energy density obtained by setting $\epsilon_1 = \epsilon_R$ in Eq. (11.4), where ϵ_R is the strain at rupture, is known as the *modulus of toughness* of the material. It is equal to the area under the entire stress-strain diagram (Fig. 11.7) and represents the energy per unit volume required to cause the material to rupture. It is clear that the toughness of a material is related to its ductility as well as to its ultimate strength (Sec. 2.3), and that the capacity of a structure to withstand an impact load depends upon the toughness of the material used (Photo 11.1).

If the stress σ_x remains within the proportional limit of the material, Hooke's law applies and we write

$$\sigma_x = E\epsilon_x \tag{11.5}$$

Substituting for σ_x from (11.5) into (11.4), we have

$$u = \int_0^{\epsilon_1} E\epsilon_x \, d\epsilon_x = \frac{E\epsilon_1^2}{2} \tag{11.6}$$

or, using Eq. (11.5) to express ϵ_1 in terms of the corresponding stress σ_1,

$$u = \frac{\sigma_1^2}{2E} \tag{11.7}$$

The value u_Y of the strain-energy density obtained by setting $\sigma_1 = \sigma_Y$ in Eq. (11.7), where σ_Y is the yield strength, is called the *modulus of resilience* of the material. We have

$$u_Y = \frac{\sigma_Y^2}{2E} \tag{11.8}$$

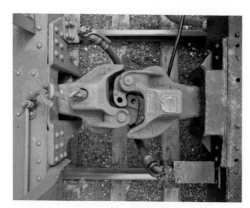

Photo 11.1 The railroad coupler is made of a ductile steel that has a large modulus of toughness.

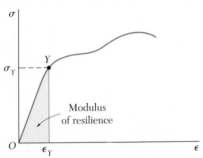

σ

σ_Y ---- Y

Modulus of resilience

O ϵ_Y ϵ

Fig. 11.8 Modulus of resilience.

The modulus of resilience is equal to the area under the straight-line portion OY of the stress-strain diagram (Fig. 11.8) and represents the energy per unit volume that the material can absorb without yielding. The capacity of a structure to withstand an impact load without being permanently deformed clearly depends upon the resilience of the material used.

Since the modulus of toughness and the modulus of resilience represent characteristic values of the strain-energy density of the material considered, they are both expressed in J/m^3 or its multiples if SI units are used.†

11.4 ELASTIC STRAIN ENERGY FOR NORMAL STRESSES

Since the rod considered in the preceding section was subjected to uniformly distributed stresses σ_x, the strain-energy density was constant throughout the rod and could be defined as the ratio U/V of the strain energy U and the volume V of the rod. In a structural element or machine part with a nonuniform stress distribution, the strain-energy density u can be defined by considering the strain energy of a small element of material of volume ΔV and writing

$$u = \lim_{\Delta V \to 0} \frac{\Delta U}{\Delta V}$$

or

$$u = \frac{dU}{dV} \qquad (11.9)$$

The expression obtained for u in Sec. 11.3 in terms of σ_x and ϵ_x remains valid, i.e., we still have

$$u = \int_0^{\epsilon_x} \sigma_x \, d\epsilon_x \qquad (11.10)$$

but the stress σ_x, the strain ϵ_x, and the strain-energy density u will generally vary from point to point.

For values of σ_x within the proportional limit, we may set $\sigma_x = E\epsilon_x$ in Eq. (11.10) and write

$$u = \frac{1}{2}E\epsilon_x^2 = \frac{1}{2}\sigma_x \epsilon_x = \frac{1}{2}\frac{\sigma_x^2}{E} \qquad (11.11)$$

The value of the strain energy U of a body subjected to uniaxial normal stresses can be obtained by substituting for u from Eq. (11.11) into Eq. (11.9) and integrating both members. We have

$$U = \int \frac{\sigma_x^2}{2E} \, dV \qquad (11.12)$$

The expression obtained is valid only for elastic deformations and is referred to as the *elastic strain energy* of the body.

†However, referring to the footnote on page 718, we note that the modulus of toughness and the modulus of resilience could be expressed in the same units as stress.

Strain Energy Under Axial Loading. We recall from Sec. 2.17 that, when a rod is subjected to a centric axial loading, the normal stresses σ_x can be assumed uniformly distributed in any given transverse section. Denoting by A the area of the section located at a distance x from the end B of the rod (Fig. 11.9), and by P the internal force in that section, we write $\sigma_x = P/A$. Substituting for σ_x into Eq. (11.12), we have

$$U = \int \frac{P^2}{2EA^2} \, dV$$

or, setting $dV = A \, dx$,

$$U = \int_0^L \frac{P^2}{2AE} \, dx \tag{11.13}$$

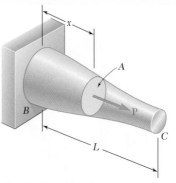

Fig. 11.9 Rod with centric axial load.

In the case of a rod of uniform cross section subjected at its ends to equal and opposite forces of magnitude P (Fig. 11.10), Eq. (11.13) yields

$$U = \frac{P^2 L}{2AE} \tag{11.14}$$

Fig. 11.10

A rod consists of two portions BC and CD of the same material and same length, but of different cross sections (Fig. 11.11). Determine the strain energy of the rod when it is subjected to a centric axial load **P**, expressing the result in terms of P, L, E, the cross-sectional area A of portion CD, and the ratio n of the two diameters.

EXAMPLE 11.01

We use Eq. (11.14) to compute the strain energy of each of the two portions, and add the expressions obtained:

$$U_n = \frac{P^2(\frac{1}{2}L)}{2AE} + \frac{P^2(\frac{1}{2}L)}{2(n^2 A)E} = \frac{P^2 L}{4AE}\left(1 + \frac{1}{n^2}\right)$$

or

$$U_n = \frac{1 + n^2}{2n^2} \frac{P^2 L}{2AE} \tag{11.15}$$

We check that, for $n = 1$, we have

$$U_1 = \frac{P^2 L}{2AE}$$

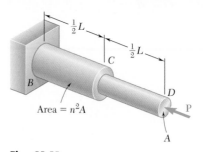

Fig. 11.11

which is the expression given in Eq. (11.14) for a rod of length L and uniform cross section of area A. We also note that, for $n > 1$, we have $U_n < U_1$; for example, when $n = 2$, we have $U_2 = (\frac{5}{8})U_1$. Since the maximum stress occurs in portion CD of the rod and is equal to $\sigma_{\max} = P/A$, it follows that, for a given allowable stress, increasing the diameter of portion BC of the rod results in a *decrease* of the overall energy-absorbing capacity of the rod. Unnecessary changes in cross-sectional area should therefore be avoided in the design of members that may be subjected to loadings, such as impact loadings, where the energy-absorbing capacity of the member is critical.

EXAMPLE 11.02

A load **P** is supported at B by two rods of the same material and of the same uniform cross section of area A (Fig. 11.12). Determine the strain energy of the system.

Denoting by F_{BC} and F_{BD}, respectively, the forces in members BC and BD, and recalling Eq. (11.14), we express the strain energy of the system as

$$U = \frac{F_{BC}^2(BC)}{2AE} + \frac{F_{BD}^2(BD)}{2AE} \qquad (11.16)$$

But we note from Fig. 11.12 that

$$BC = 0.6l \qquad BD = 0.8l$$

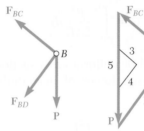

Fig. 11.13

and from the free-body diagram of pin B and the corresponding force triangle (Fig. 11.13) that

$$F_{BC} = +0.6P \qquad F_{BD} = -0.8P$$

Substituting into Eq. (11.16), we have

$$U = \frac{P^2l[(0.6)^3 + (0.8)^3]}{2AE} = 0.364 \frac{P^2l}{AE}$$

Fig. 11.12

Strain Energy in Bending. Consider a beam AB subjected to a given loading (Fig. 11.14), and let M be the bending moment at a distance x from end A. Neglecting for the time being the effect of shear, and taking into account only the normal stresses $\sigma_x = My/I$, we substitute this expression into Eq. (11.12) and write

$$U = \int \frac{\sigma_x^2}{2E} dV = \int \frac{M^2y^2}{2EI^2} dV$$

Setting $dV = dA\, dx$, where dA represents an element of the cross-sectional area, and recalling that $M^2/2EI^2$ is a function of x alone, we have

$$U = \int_0^L \frac{M^2}{2EI^2} \left(\int y^2\, dA \right) dx$$

Recalling that the integral within the parentheses represents the moment of inertia I of the cross section about its neutral axis, we write

$$U = \int_0^L \frac{M^2}{2EI} dx \qquad (11.17)$$

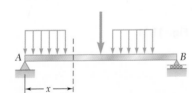

Fig. 11.14 Beam subject to transverse loads.

EXAMPLE 11.03

Determine the strain energy of the prismatic cantilever beam AB (Fig. 11.15), taking into account only the effect of the normal stresses.

The bending moment at a distance x from end A is $M = -Px$. Substituting this expression into Eq. (11.17), we write

$$U = \int_0^L \frac{P^2x^2}{2EI}\, dx = \frac{P^2L^3}{6EI}$$

Fig. 11.15

11.5 ELASTIC STRAIN ENERGY FOR SHEARING STRESSES

When a material is subjected to plane shearing stresses τ_{xy}, the strain-energy density at a given point can be expressed as

$$u = \int_0^{\gamma_{xy}} \tau_{xy}\, d\gamma_{xy} \tag{11.18}$$

where γ_{xy} is the shearing strain corresponding to τ_{xy} (Fig. 11.16a). We note that the strain-energy density u is equal to the area under the shearing-stress-strain diagram (Fig. 11.16b).

For values of τ_{xy} within the proportional limit, we have $\tau_{xy} = G\gamma_{xy}$, where G is the modulus of rigidity of the material. Substituting for τ_{xy} into Eq. (11.18) and performing the integration, we write

$$u = \frac{1}{2}G\gamma_{xy}^2 = \frac{1}{2}\tau_{xy}\gamma_{xy} = \frac{\tau_{xy}^2}{2G} \tag{11.19}$$

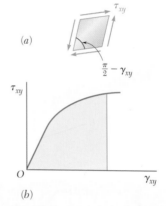

Fig. 11.16 Strain energy due to shear.

The value of the strain energy U of a body subjected to plane shearing stresses can be obtained by recalling from Sec. 11.4 that

$$u = \frac{dU}{dV} \tag{11.9}$$

Substituting for u from Eq. (11.19) into Eq. (11.9) and integrating both members, we have

$$U = \int \frac{\tau_{xy}^2}{2G}\, dV \tag{11.20}$$

This expression defines the elastic strain associated with the shear deformations of the body. Like the similar expression obtained in Sec. 11.4 for uniaxial normal stresses, it is valid only for elastic deformations.

Strain Energy in Torsion. Consider a shaft BC of length L subjected to one or several twisting couples. Denoting by J the polar moment of inertia of the cross section located at a distance x from B (Fig. 11.17), and by T the internal torque in that section, we recall

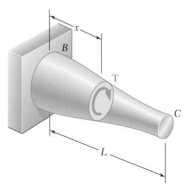

Fig. 11.17 Shaft subject to torque.

that the shearing stresses in the section are $\tau_{xy} = T\rho/J$. Substituting for τ_{xy} into Eq. (11.20), we have

$$U = \int \frac{\tau_{xy}^2}{2G} dV = \int \frac{T^2\rho^2}{2GJ^2} dV$$

Setting $dV = dA\, dx$, where dA represents an element of the cross-sectional area, and observing that $T^2/2GJ^2$ is a function of x alone, we write

$$U = \int_0^L \frac{T^2}{2GJ^2} \left(\int \rho^2\, dA \right) dx$$

Recalling that the integral within the parentheses represents the polar moment of inertia J of the cross section, we have

$$U = \int_0^L \frac{T^2}{2GJ} dx \qquad (11.21)$$

In the case of a shaft of uniform cross section subjected at its ends to equal and opposite couples of magnitude T (Fig. 11.18), Eq. (11.21) yields

$$U = \frac{T^2 L}{2GJ} \qquad (11.22)$$

Fig. 11.18

EXAMPLE 11.04

A circular shaft consists of two portions BC and CD of the same material and same length, but of different cross sections (Fig. 11.19). Determine the strain energy of the shaft when it is subjected to a twisting couple **T** at end D, expressing the result in terms of T, L, G, the polar moment of inertia J of the smaller cross section, and the ratio n of the two diameters.

We use Eq. (11.22) to compute the strain energy of each of the two portions of shaft, and add the expressions obtained. Noting that the polar moment of inertia of portion BC is equal to $n^4 J$, we write

$$U_n = \frac{T^2(\frac{1}{2}L)}{2GJ} + \frac{T^2(\frac{1}{2}L)}{2G(n^4 J)} = \frac{T^2 L}{4GJ}\left(1 + \frac{1}{n^4}\right)$$

or

$$U_n = \frac{1 + n^4}{2n^4}\frac{T^2 L}{2GJ} \qquad (11.23)$$

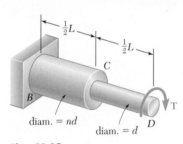

Fig. 11.19

We check that, for $n = 1$, we have

$$U_1 = \frac{T^2 L}{2GJ}$$

which is the expression given in Eq. (11.22) for a shaft of length L and uniform cross section. We also note that, for $n > 1$, we have $U_n < U_1$; for example, when $n = 2$, we have $U_2 = (\frac{17}{32})U_1$. Since the maximum shearing stress occurs in the portion CD of the shaft and is proportional to the torque T, we note as we did earlier in the case of the axial loading of a rod that, for a given allowable stress, increasing the diameter of portion BC of the shaft results in a *decrease* of the overall energy-absorbing capacity of the shaft.

Strain Energy Under Transverse Loading. In Sec. 11.4 we obtained an expression for the strain energy of a beam subjected to a transverse loading. However, in deriving that expression we took into account only the effect of the normal stresses due to bending and neglected the effect of the shearing stresses. In Example 11.05 both types of stresses will be taken into account.

EXAMPLE 11.05

Determine the strain energy of the rectangular cantilever beam AB (Fig. 11.20), taking into account the effect of both normal and shearing stresses.

We first recall from Example 11.03 that the strain energy due to the normal stresses σ_x is

$$U_\sigma = \frac{P^2 L^3}{6EI}$$

Fig. 11.20

To determine the strain energy U_τ due to the shearing stresses τ_{xy}, we recall Eq. (6.9) of Sec. 6.4 and find that, for a beam with a rectangular cross section of width b and depth h,

$$\tau_{xy} = \frac{3}{2} \frac{V}{A} \left(1 - \frac{y^2}{c^2} \right) = \frac{3}{2} \frac{P}{bh} \left(1 - \frac{y^2}{c^2} \right)$$

Substituting for τ_{xy} into Eq. (11.20), we write

$$U_\tau = \frac{1}{2G} \left(\frac{3}{2} \frac{P}{bh} \right)^2 \int \left(1 - \frac{y^2}{c^2} \right)^2 dV$$

or, setting $dV = b \, dy \, dx$, and after reductions,

$$U_\tau = \frac{9P^2}{8Gbh^2} \int_{-c}^{c} \left(1 - 2\frac{y^2}{c^2} + \frac{y^4}{c^4} \right) dy \int_0^L dx$$

Performing the integrations, and recalling that $c = h/2$, we have

$$U_\tau = \frac{9P^2 L}{8Gbh^2} \left[y - \frac{2}{3}\frac{y^3}{c^2} + \frac{1}{5}\frac{y^5}{c^4} \right]_{-c}^{+c} = \frac{3P^2 L}{5Gbh} = \frac{3P^2 L}{5GA}$$

The total strain energy of the beam is thus

$$U = U_\sigma + U_\tau = \frac{P^2 L^3}{6EI} + \frac{3P^2 L}{5GA}$$

or, noting that $I/A = h^2/12$ and factoring the expression for U_σ,

$$U = \frac{P^2 L^3}{6EI} \left(1 + \frac{3Eh^2}{10GL^2} \right) = U_\sigma \left(1 + \frac{3Eh^2}{10GL^2} \right) \qquad (11.24)$$

Recalling from Sec. 2.14 that $G \geq E/3$, we conclude that the parenthesis in the expression obtained is less than $1 + 0.9(h/L)^2$ and, thus, that the relative error is less than $0.9(h/L)^2$ when the effect of shear is neglected. For a beam with a ratio h/L less than $\frac{1}{10}$, the percentage error is less than 0.9 percent. It is therefore customary in engineering practice to neglect the effect of shear in computing the strain energy of slender beams.

11.6 STRAIN ENERGY FOR A GENERAL STATE OF STRESS

In the preceding sections, we determined the strain energy of a body in a state of uniaxial stress (Sec. 11.4) and in a state of plane shearing stress (Sec. 11.5). In the case of a body in a general state of stress characterized by the six stress components σ_x, σ_y, σ_z, τ_{xy}, τ_{yz}, and τ_{zx}, the strain-energy density can be obtained by adding the expressions given in Eqs. (11.10) and (11.18), as well as the four other expressions obtained through a permutation of the subscripts.

In the case of the elastic deformation of an isotropic body, each of the six stress-strain relations involved is linear, and the strain-energy density can be expressed as

$$u = \tfrac{1}{2}(\sigma_x\epsilon_x + \sigma_y\epsilon_y + \sigma_z\epsilon_z + \tau_{xy}\gamma_{xy} + \tau_{yz}\gamma_{yz} + \tau_{zx}\gamma_{zx}) \quad (11.25)$$

Recalling the relations (2.38) obtained in Sec. 2.14, and substituting for the strain components into (11.25), we have, for the most general state of stress at a given point of an elastic isotropic body,

$$u = \frac{1}{2E}\left[\sigma_x^2 + \sigma_y^2 + \sigma_z^2 - 2\nu(\sigma_x\sigma_y + \sigma_y\sigma_z + \sigma_z\sigma_x)\right]$$

$$+ \frac{1}{2G}(\tau_{xy}^2 + \tau_{yz}^2 + \tau_{zx}^2) \quad (11.26)$$

If the principal axes at the given point are used as coordinate axes, the shearing stresses become zero and Eq. (11.26) reduces to

$$u = \frac{1}{2E}\left[\sigma_a^2 + \sigma_b^2 + \sigma_c^2 - 2\nu(\sigma_a\sigma_b + \sigma_b\sigma_c + \sigma_c\sigma_a)\right] \quad (11.27)$$

where σ_a, σ_b, and σ_c are the principal stresses at the given point.

We now recall from Sec. 7.7 that one of the criteria used to predict whether a given state of stress will cause a ductile material to yield, namely, the maximum-distortion-energy criterion, is based on the determination of the energy per unit volume associated with the distortion, or change in shape, of that material. Let us, therefore, attempt to separate the strain-energy density u at a given point into two parts, a part u_v associated with a change in volume of the material at that point, and a part u_d associated with a distortion, or change in shape, of the material at the same point. We write

$$u = u_v + u_d \quad (11.28)$$

In order to determine u_v and u_d, we introduce the *average value* $\overline{\sigma}$ of the principal stresses at the point considered,

$$\overline{\sigma} = \frac{\sigma_a + \sigma_b + \sigma_c}{3} \quad (11.29)$$

and set

$$\sigma_a = \overline{\sigma} + \sigma_a' \qquad \sigma_b = \overline{\sigma} + \sigma_b' \qquad \sigma_c = \overline{\sigma} + \sigma_c' \quad (11.30)$$

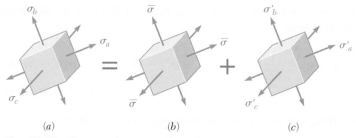

Fig. 11.21 Element subject to multiaxial stress.

Thus, the given state of stress (Fig. 11.21a) can be obtained by superposing the states of stress shown in Fig. 11.21b and c. We note that the state of stress described in Fig. 11.21b tends to change the volume of the element of material, but not its shape, since all the faces of the element are subjected to the same stress $\bar{\sigma}$. On the other hand, it follows from Eqs. (11.29) and (11.30) that

$$\sigma'_a + \sigma'_b + \sigma'_c = 0 \qquad (11.31)$$

which indicates that some of the stresses shown in Fig. 11.21c are tensile and others compressive. Thus, this state of stress tends to change the shape of the element. However, it does not tend to change its volume. Indeed, recalling Eq. (2.31) of Sec. 2.13, we note that the dilatation e (i.e., the change in volume per unit volume) caused by this state of stress is

$$e = \frac{1 - 2\nu}{E}(\sigma'_a + \sigma'_b + \sigma'_c)$$

or $e = 0$, in view of Eq. (11.31). We conclude from these observations that the portion u_v of the strain-energy density must be associated with the state of stress shown in Fig. 11.21b, while the portion u_d must be associated with the state of stress shown in Fig. 11.21c.

It follows that the portion u_v of the strain-energy density corresponding to a change in volume of the element can be obtained by substituting $\bar{\sigma}$ for each of the principal stresses in Eq. (11.27). We have

$$u_v = \frac{1}{2E}\left[3\bar{\sigma}^2 - 2\nu(3\bar{\sigma}^2)\right] = \frac{3(1 - 2\nu)}{2E}\bar{\sigma}^2$$

or, recalling Eq. (11.29),

$$u_v = \frac{1 - 2\nu}{6E}(\sigma_a + \sigma_b + \sigma_c)^2 \qquad (11.32)$$

The portion of the strain-energy density corresponding to the distortion of the element is obtained by solving Eq. (11.28) for u_d

and substituting for u and u_v from Eqs. (11.27) and (11.32), respectively. We write

$$u_d = u - u_v = \frac{1}{6E} \left[3(\sigma_a^2 + \sigma_b^2 + \sigma_c^2) - 6\nu(\sigma_a\sigma_b + \sigma_b\sigma_c + \sigma_c\sigma_a) \right.$$
$$\left. - (1 - 2\nu)(\sigma_a + \sigma_b + \sigma_c)^2 \right]$$

Expanding the square and rearranging terms, we have

$$u_d = \frac{1 + \nu}{6E} \left[(\sigma_a^2 - 2\sigma_a\sigma_b + \sigma_b^2) + (\sigma_b^2 - 2\sigma_b\sigma_c + \sigma_c^2) \right.$$
$$\left. + (\sigma_c^2 - 2\sigma_c\sigma_a + \sigma_a^2) \right]$$

Noting that each of the parentheses inside the bracket is a perfect square, and recalling from Eq. (2.43) of Sec. 2.15 that the coefficient in front of the bracket is equal to $1/12G$, we obtain the following expression for the portion u_d of the strain-energy density, i.e., for the distortion energy per unit volume,

$$u_d = \frac{1}{12G} \left[(\sigma_a - \sigma_b)^2 + (\sigma_b - \sigma_c)^2 + (\sigma_c - \sigma_a)^2 \right] \qquad (11.33)$$

In the case of *plane stress*, and assuming that the c axis is perpendicular to the plane of stress, we have $\sigma_c = 0$ and Eq. (11.33) reduces to

$$u_d = \frac{1}{6G} (\sigma_a^2 - \sigma_a\sigma_b + \sigma_b^2) \qquad (11.34)$$

Considering the particular case of a tensile-test specimen, we note that, at yield, we have $\sigma_a = \sigma_Y$, $\sigma_b = 0$, and thus $(u_d)_Y = \sigma_Y^2/6G$. The maximum-distortion-energy criterion for plane stress indicates that a given state of stress is safe as long as $u_d < (u_d)_Y$ or, substituting for u_d from Eq. (11.34), as long as

$$\sigma_a^2 - \sigma_a\sigma_b + \sigma_b^2 < \sigma_Y^2 \qquad (7.26)$$

which is the condition stated in Sec. 7.7 and represented graphically by the ellipse of Fig. 7.39. In the case of a general state of stress, the expression (11.33) obtained for u_d should be used. The maximum-distortion-energy criterion is then expressed by the condition.

$$(\sigma_a - \sigma_b)^2 + (\sigma_b - \sigma_c)^2 + (\sigma_c - \sigma_a)^2 < 2\sigma_Y^2 \qquad (11.35)$$

which indicates that a given state of stress is safe if the point of coordinates σ_a, σ_b, σ_c is located within the surface defined by the equation

$$(\sigma_a - \sigma_b)^2 + (\sigma_b - \sigma_c)^2 + (\sigma_c - \sigma_a)^2 = 2\sigma_Y^2 \qquad (11.36)$$

This surface is a circular cylinder of radius $\sqrt{2/3}\,\sigma_Y$ with an axis of symmetry forming equal angles with the three principal axes of stress.

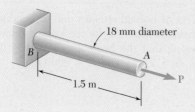

SAMPLE PROBLEM 11.1

During a routine manufacturing operation, rod AB must acquire an elastic strain energy of 13.6 N · m. Using $E = 200$ GPa, determine the required yield strength of the steel if the factor of safety with respect to permanent deformation is to be five.

SOLUTION

Factor of Safety. Since a factor of safety of five is required, the rod should be designed for a strain energy of

$$U = 5(13.6 \text{ N} \cdot \text{m}) = 68 \text{ N} \cdot \text{m}$$

Strain-Energy Density. The volume of the rod is

$$V = AL = \frac{\pi}{4}(18 \text{ mm})^2(1500 \text{ mm}) = 381\ 700 \text{ mm}^3$$

Since the rod is of uniform cross section, the required strain-energy density is

$$u = \frac{U}{V} = \frac{68 \text{ N} \cdot \text{m}}{(381\ 700 \times 10^{-9} \text{ m}^3)} = 0.1782 \times 10^6 \text{ N/m}^2$$

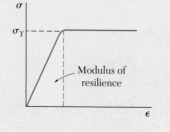

Yield Strength. We recall that the modulus of resilience is equal to the strain-energy density when the maximum stress is equal to σ_Y. Using Eq. (11.8), we write

$$u = \frac{\sigma_Y^2}{2E}$$

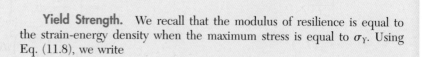

$$0.1782 \times 10^6 \text{ N/m}^2 = \frac{\sigma_Y^2}{2(200 \times 10^9 \text{ N/m}^2)} \qquad\qquad \sigma_Y = 267 \text{ MPa} \blacktriangleleft$$

Comment. It is important to note that, since energy loads are not linearly related to the stresses they produce, factors of safety associated with energy loads should be applied to the energy loads and not to the stresses.

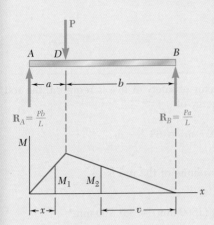

SAMPLE PROBLEM 11.2

(a) Taking into account only the effect of normal stresses due to bending, determine the strain energy of the prismatic beam AB for the loading shown. (b) Evaluate the strain energy, knowing that the beam is a W250 × 67, $P = 160$ kN, $L = 3.6$ m, $a = 0.9$ m, $b = 2.7$ m, and $E = 200$ GPa.

SOLUTION

Bending Moment. Using the free-body diagram of the entire beam, we determine the reactions

$$\mathbf{R}_A = \frac{Pb}{L} \uparrow \qquad \mathbf{R}_B = \frac{Pa}{L} \uparrow$$

For portion AD of the beam, the bending moment is

$$M_1 = \frac{Pb}{L} x$$

For portion DB, the bending moment at a distance v from end B is

$$M_2 = \frac{Pa}{L} v$$

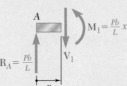

From A to D:

a. Strain Energy. Since strain energy is a scalar quantity, we add the strain energy of portion AD to that of portion DB to obtain the total strain energy of the beam. Using Eq. (11.17), we write

$$U = U_{AD} + U_{DB}$$
$$= \int_0^a \frac{M_1^2}{2EI} dx + \int_0^b \frac{M_2^2}{2EI} dv$$
$$= \frac{1}{2EI} \int_0^a \left(\frac{Pb}{L} x\right)^2 dx + \frac{1}{2EI} \int_0^b \left(\frac{Pa}{L} v\right)^2 dv$$
$$= \frac{1}{2EI} \frac{P^2}{L^2} \left(\frac{b^2 a^3}{3} + \frac{a^2 b^3}{3}\right) = \frac{P^2 a^2 b^2}{6EIL^2}(a + b)$$

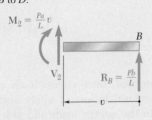

From B to D:

or, since $(a + b) = L$, $\qquad\qquad U = \dfrac{P^2 a^2 b^2}{6EIL}$ ◀

b. Evaluation of the Strain Energy. The moment of inertia of a W250 × 67 rolled-steel shape is obtained from Appendix C.

$$\begin{aligned} P &= 160 \text{ kN} & L &= 3.6 \text{ m} \\ a &= 0.9 \text{ m} & b &= 2.7 \text{ m} \\ E &= 200 \text{ GPa} & I &= 104 \times 10^6 \text{ mm}^4 \end{aligned}$$

Substituting into the expression for U, we have

$$U = \frac{(160 \times 10^3 \text{ N})^2 (0.9 \text{ m})^2 (2.7 \text{ m})^2}{6(200 \times 10^9 \text{ Pa})(104 \times 10^{-6} \text{ m}^4)(3.6 \text{ m})} \qquad U = 336 \text{ N} \cdot \text{m} \blacktriangleleft$$

PROBLEMS

11.1 Determine the modulus of resilience for each of the following metals:
- (a) Stainless steel
  AISI 302 (annealed): $E = 190$ GPa $\sigma_Y = 260$ MPa
- (b) Stainless steel 2014-T6
  AISI 302 (cold-rolled): $E = 190$ GPa $\sigma_Y = 520$ MPa
- (c) Malleable cast iron: $E = 165$ GPa $\sigma_Y = 230$ MPa

11.2 Determine the modulus of resilience for each of the following alloys:
- (a) Titanium: $E = 115$ GPa $\sigma_Y = 875$ MPa
- (b) Magnesium: $E = 45$ GPa $\sigma_Y = 200$ MPa
- (c) Cupronickel (annealed) $E = 140$ GPa $\sigma_Y = 125$ MPa

11.3 Determine the modulus of resilience for each of the following grades of structural steel:
- (a) ASTM A709 Grade 50: $\sigma_Y = 350$ MPa
- (b) ASTM A913 Grade 65: $\sigma_Y = 450$ MPa
- (c) ASTM A709 Grade 100: $\sigma_Y = 700$ MPa

11.4 Determine the modulus of resilience for each of the following aluminum alloys:
- (a) 1100-H14: $E = 70$ GPa $\sigma_Y = 55$ MPa
- (b) 2014-T6 $E = 72$ GPa: $\sigma_Y = 220$ MPa
- (c) 6061-T6 $E = 69$ GPa: $\sigma_Y = 150$ MPa

11.5 The stress-strain diagram shown has been drawn from data obtained during a tensile test of an aluminum alloy. Using $E = 72$ GPa, determine (a) the modulus of resilience of the alloy, (b) the modulus of toughness of the alloy.

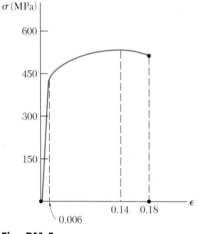

Fig. P11.5

11.6 The stress-strain diagram shown has been drawn from data obtained during a tensile test of a specimen of structural steel. Using $E = 200$ GPa, determine (a) the modulus of resilience of the steel, (b) the modulus of toughness of the steel.

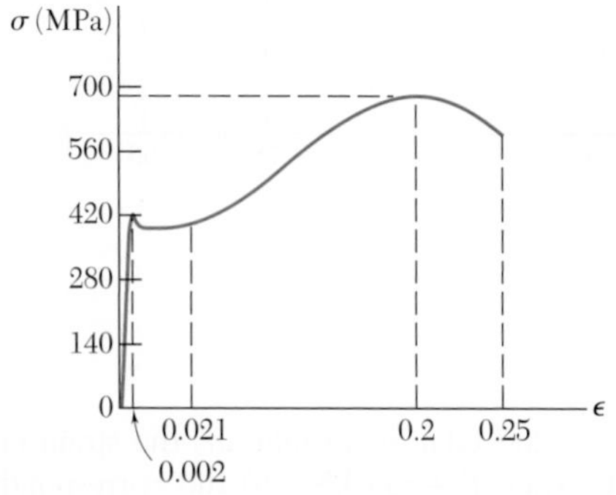

Fig. P11.6

11.7 The load-deformation diagram shown has been drawn from data obtained during a tensile test of a 20-mm-diameter rod of an aluminum alloy. Knowing that the deformation was measured using a 400-mm gage length, determine (*a*) the modulus of resilience of the alloy, (*b*) the modulus of toughness of the alloy.

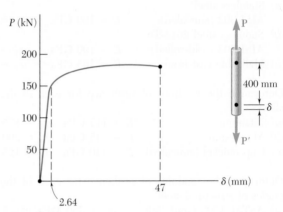

Fig. P11.7

11.8 The load-deformation diagram shown has been drawn from data obtained during a tensile test of structural steel. Knowing that the cross-sectional area of the specimen is 250 mm² and that the deformation was measured using a 500-mm gage length, determine (*a*) the modulus of resilience of the steel, (*b*) the modulus of toughness of the steel.

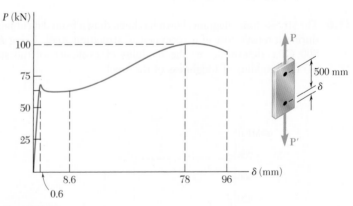

Fig. P11.8

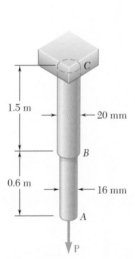

Fig. P11.9

11.9 Using $E = 200$ GPa, determine (*a*) the strain energy of the steel rod *ABC* when $P = 35$ kN, (*b*) the corresponding strain energy density in portions *AB* and *BC* of the rod.

11.10 Rods *AB* and *BC* are made of a steel for which the yield strength is $\sigma_Y = 300$ MPa and the modulus of elasticity is $E = 200$ GPa. Determine the maximum strain energy that can be acquired by the assembly without causing permanent deformation when the length *a* of rod *AB* is (*a*) 2 m, (*b*) 4 m.

11.11 A 0.75-m length of aluminum pipe of cross-sectional area 1190 mm² is welded to a fixed support *A* and to a rigid cap *B*. The steel rod *EF*, of 18-mm diameter, is welded to cap *B*. Knowing that the modulus of elasticity is 200 GPa for the steel and 73 GPa for the aluminum, determine (*a*) the total strain energy of the system when P = 40 kN, (*b*) the corresponding strain-energy density of the pipe *CD* and in the rod *EF*.

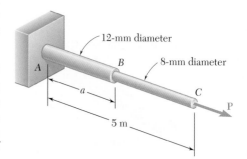

Fig. P11.10

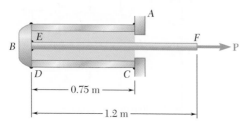

Fig. P11.11

11.12 Rod *AB* is made of a steel for which the yield strength is $\sigma_Y = 450$ MPa and $E = 200$ GPa; rod *BC* is made of an aluminum alloy for which $\sigma_Y = 280$ MPa and $E = 73$ GPa. Determine the maximum strain energy that can be acquired by the composite rod *ABC* without causing any permanent deformations.

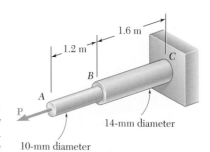

Fig. P11.12

11.13 A single 6-mm-diameter steel pin *B* is used to connect the steel strip *DE* to two aluminum strips, each of 20-mm width and 5-mm thickness. The modulus of elasticity is 200 GPa for the steel and 70 GPa for the aluminum. Knowing that for the pin at *B* the allowable shearing stress is $\tau_{all} = 85$ MPa, determine, for the loading shown, the maximum strain energy that can be acquired by the assembled strips.

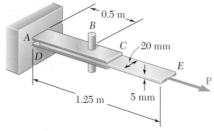

Fig. P11.13

11.14 Rod *BC* is made of a steel for which the yield strength is $\sigma_Y = 300$ MPa and the modulus of elasticity is $E = 200$ GPa. Knowing that a strain energy of 10 J must be acquired by the rod when the axial load **P** is applied, determine the diameter of the rod for which the factor of safety with respect to permanent deformation is six.

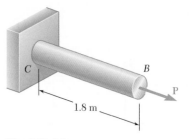

Fig. P11.14

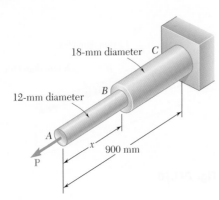

Fig. P11.15

11.15 The assembly ABC is made of a steel for which $E = 200$ GPa and $\sigma_Y = 320$ MPa. Knowing that a strain energy of 5 J must be acquired by the assembly as the axial load $\mathbf{P}$ is applied, determine the factor of safety with respect to permanent deformation when (a) $x = 300$ mm, (b) $x = 600$ mm.

11.16 Using $E = 74$ GPa, determine by approximate means the maximum strain energy that can be acquired by the aluminum rod shown if the allowable normal stress is $\sigma_{all} = 154$ MPa.

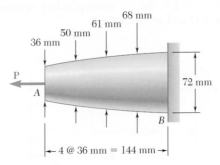

Fig. P11.16

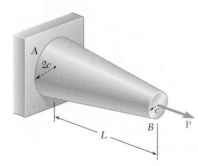

Fig. P11.17

11.17 Show by integration that the strain energy of the tapered rod AB is

$$U = \frac{1}{4}\frac{P^2L}{EA_{min}}$$

where A_{min} is the cross-sectional area at end B.

11.18 through 11.21 In the truss shown, all members are made of the same material and have the uniform cross-sectional area indicated. Determine the strain energy of the truss when the load $\mathbf{P}$ is applied.

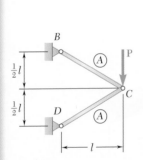

Fig. P11.18

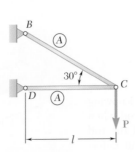

Fig. P11.19

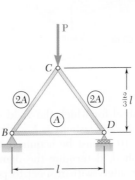

Fig. P11.20

Fig. P11.21

11.22 In the truss shown, all members are made of aluminum and have the uniform cross-sectional area shown. Using $E = 72$ GPa, determine the strain energy of the truss for the loading shown.

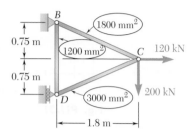

Fig. P11.22 and P11.23

11.23 Solve Prob. 11.22, assuming that the 120-kN load is removed.

11.24 through 11.27 Taking into account only the effect of normal stresses, determine the strain energy of the prismatic beam AB for the loading shown.

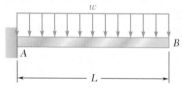

Fig. P11.24

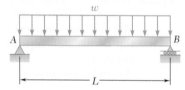

Fig. P11.25

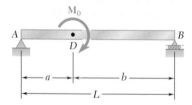

Fig. P11.26

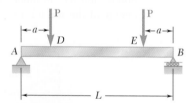

Fig. P11.27

11.28 and 11.29 Using $E = 200$ GPa, determine the strain energy due to bending for the steel beam and loading shown. (Ignore the effect of shearing stresses.)

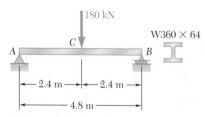

Fig. P11.28

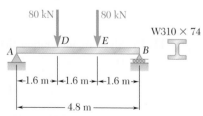

Fig. P11.29

11.30 and 11.31 Using $E = 200$ GPa, determine the strain energy due to bending for the steel beam and loading shown.

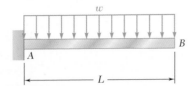

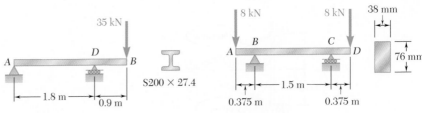

Fig. P11.30

Fig. P11.31

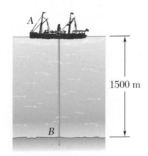

Fig. P11.32

11.32 Assuming that the prismatic beam AB has a rectangular cross section, show that for the given loading the maximum value of the strain-energy density in the beam is

$$u_{max} = 15\frac{U}{V}$$

where U is the strain energy of the beam and V is its volume.

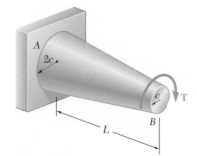

Fig. P11.33

11.33 The ship at A has just started to drill for oil on the ocean floor at a depth of 1500 m. The steel drill pipe has an outer diameter of 200 mm and a uniform wall thickness of 12 mm. Knowing that the top of the drill pipe rotates through two complete revolutions before the drill bit at B starts to operate and using $G = 77$ GPa, determine the maximum strain energy acquired by the drill pipe.

11.34 Rod AC is made of aluminum and is subjected to a torque $\mathbf{T}$ applied at $\mathbf{C}$. Knowing that $G = 73$ GPa and that portion BC of the rod is hollow and has an inner diameter of 16 mm, determine the strain energy of the rod for a maximum shearing stress of 120 MPa.

Fig. P11.34

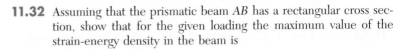

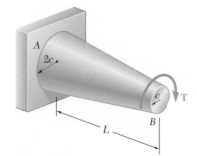

Fig. P11.35

11.35 Show by integration that the strain energy in the tapered rod AB is

$$U = \frac{7}{48}\frac{T^2L}{GJ_{min}}$$

where J_{min} is the polar moment of inertia of the rod at end B.

11.36 The state of stress shown occurs in a machine component made of a grade of steel for which $\sigma_Y = 450$ MPa. Using the maximum-distortion-energy criterion, determine the factor of safety associated with the yield strength when (a) $\sigma_y = +110$ MPa, (b) $\sigma_y = -110$ MPa.

11.37 The state of stress shown occurs in a machine component made of a grade of steel for which $\sigma_Y = 450$ MPa. Using the maximum-distortion-energy criterion, determine the range of values of σ_y for which the factor of safety associated with the yield strength is equal to or larger than 2.2.

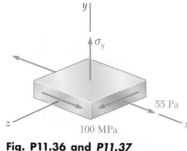

Fig. P11.36 and P11.37

11.38 The state of stress shown occurs in a machine component made of a brass for which $\sigma_Y = 160$ MPa. Using the maximum-distortion-energy criterion, determine the range of values of σ_z for which yield does not occur.

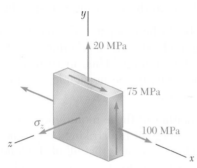

Fig. P11.38 and P11.39

11.39 The state of stress shown occurs in a machine component made of a brass for which $\sigma_Y = 160$ MPa. Using the maximum-distortion-energy criterion, determine whether yield occurs when (a) $\sigma_z = +45$ MPa, (b) $\sigma_z = -45$ MPa.

11.40 Determine the strain energy of the prismatic beam AB, taking into account the effect of both normal and shearing stresses.

*11.41 A vibration isolation support is made by bonding a rod A, of radius R_1, and a tube B, of inner radius R_2, to a hollow rubber cylinder. Denoting by G the modulus of rigidity of the rubber, determine the strain energy of the hollow rubber cylinder for the loading shown.

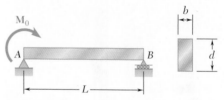

Fig. P11.40

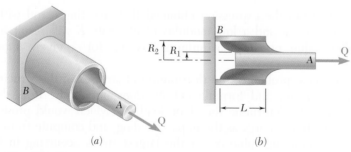

Fig. P11.41

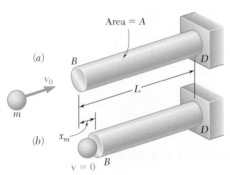

Area = A

(a)

B D

v_0

L

m

(b) x_m

D

$v = 0$ B

Fig. 11.22 Rod subject to impact loading.

Photo 11.2 Steam alternately lifts a weight inside the pile driver and then propels it downward. This delivers a large impact load to the pile that is being driven into the ground.

11.7 IMPACT LOADING

Consider a rod BD of uniform cross section which is hit at its end B by a body of mass m moving with a velocity $\mathbf{v}_0$ (Fig. 11.22a). As the rod deforms under the impact (Fig. 11.22b), stresses develop within the rod and reach a maximum value σ_m. After vibrating for a while, the rod will come to rest, and all stresses will disappear. Such a sequence of events is referred to as an *impact loading* (Photo 11.2).

In order to determine the maximum value σ_m of the stress occurring at a given point of a structure subjected to an impact loading, we are going to make several simplifying assumptions.

First, we assume that the kinetic energy $T = \frac{1}{2}mv_0^2$ of the striking body is transferred entirely to the structure and, thus, that the strain energy U_m corresponding to the maximum deformation x_m is

$$U_m = \tfrac{1}{2}mv_0^2 \tag{11.37}$$

This assumption leads to the following two specific requirements:

1. No energy should be dissipated during the impact.
2. The striking body should not bounce off the structure and retain part of its energy. This, in turn, necessitates that the inertia of the structure be negligible, compared to the inertia of the striking body.

In practice, neither of these requirements is satisfied, and only part of the kinetic energy of the striking body is actually transferred to the structure. Thus, assuming that all of the kinetic energy of the striking body is transferred to the structure leads to a conservative design of that structure.

We further assume that the stress-strain diagram obtained from a static test of the material is also valid under impact loading. Thus, for an elastic deformation of the structure, we can express the maximum value of the strain energy as

$$U_m = \int \frac{\sigma_m^2}{2E}\,dV \tag{11.38}$$

In the case of the uniform rod of Fig. 11.22, the maximum stress σ_m has the same value throughout the rod, and we write $U_m = \sigma_m^2\,V/2E$. Solving for σ_m and substituting for U_m from Eq. (11.37), we write

$$\sigma_m = \sqrt{\frac{2U_m E}{V}} = \sqrt{\frac{mv_0^2 E}{V}} \tag{11.39}$$

We note from the expression obtained that selecting a rod with a large volume V and a low modulus of elasticity E will result in a smaller value of the maximum stress σ_m for a given impact loading.

In most problems, the distribution of stresses in the structure is not uniform, and formula (11.39) does not apply. It is then convenient to determine the static load $\mathbf{P}_m$, which would produce the same strain energy as the impact loading, and compute from P_m the corresponding value σ_m of the largest stress occurring in the structure.

EXAMPLE 11.06

A body of mass m moving with a velocity $\mathbf{v}_0$ hits the end B of the non-uniform rod BCD (Fig. 11.23). Knowing that the diameter of portion BC is twice the diameter of portion CD, determine the maximum value σ_m of the stress in the rod.

Making $n = 2$ in the expression (11.15) obtained in Example 11.01, we find that when rod BCD is subjected to a static load P_m, its strain energy is

$$U_m = \frac{5P_m^2 L}{16AE} \tag{11.40}$$

where A is the cross-sectional area of portion CD of the rod. Solving Eq. (11.40) for P_m, we find that the static load that produces in the rod the same strain energy as the given impact loading is

$$P_m = \sqrt{\frac{16}{5}\frac{U_m AE}{L}}$$

where U_m is given by Eq. (11.37). The largest stress occurs in portion CD of the rod. Dividing P_m by the area A of that portion, we have

$$\sigma_m = \frac{P_m}{A} = \sqrt{\frac{16}{5}\frac{U_m E}{AL}} \tag{11.41}$$

or, substituting for U_m from Eq. (11.37),

$$\sigma_m = \sqrt{\frac{8}{5}\frac{mv_0^2 E}{AL}} = 1.265\sqrt{\frac{mv_0^2 E}{AL}}$$

Comparing this value with the value obtained for σ_m in the case of the uniform rod of Fig. 11.22 and making $V = AL$ in Eq. (11.39), we note that the maximum stress in the rod of variable cross section is 26.5 percent larger than in the lighter uniform rod. Thus, as we observed earlier in our discussion of Example 11.01, increasing the diameter of portion BC of the rod results in a *decrease* of the energy-absorbing capacity of the rod.

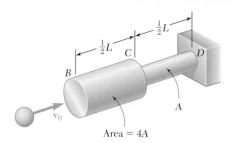

Fig. 11.23

Area = 4A

EXAMPLE 11.07

A block of weight W is dropped from a height h onto the free end of the cantilever beam AB (Fig. 11.24). Determine the maximum value of the stress in the beam.

As it falls through the distance h, the potential energy Wh of the block is transformed into kinetic energy. As a result of the impact, the kinetic energy in turn is transformed into strain energy. We have, therefore,†

$$U_m = Wh \tag{11.42}$$

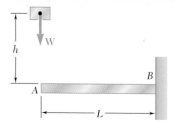

Fig. 11.24

†The total distance through which the block drops is actually $h + y_m$, where y_m is the maximum deflection of the end of the beam. Thus, a more accurate expression for U_m (see Sample Prob. 11.3) is

$$U_m = W(h + y_m) \tag{11.42'}$$

However, when $h \gg y_m$, we may neglect y_m and use Eq. (11.42).

Recalling the expression obtained for the strain energy of the cantilever beam AB in Example 11.03 and neglecting the effect of shear, we write

$$U_m = \frac{P_m^2 L^3}{6EI}$$

Solving this equation for P_m, we find that the static force that produces in the beam the same strain energy is

$$P_m = \sqrt{\frac{6U_m EI}{L^3}} \qquad (11.43)$$

The maximum stress σ_m occurs at the fixed end B and is

$$\sigma_m = \frac{|M|c}{I} = \frac{P_m Lc}{I}$$

Substituting for P_m from (11.43), we write

$$\sigma_m = \sqrt{\frac{6U_m E}{L(I/c^2)}} \qquad (11.44)$$

or, recalling (11.42),

$$\sigma_m = \sqrt{\frac{6WhE}{L(I/c^2)}}$$

11.8 DESIGN FOR IMPACT LOADS

Let us now compare the values obtained in the preceding section for the maximum stress σ_m (a) in the rod of uniform cross section of Fig. 11.22, (b) in the rod of variable cross section of Example 11.06, and (c) in the cantilever beam of Example 11.07, assuming that the last has a circular cross section of radius c.

(a) We first recall from Eq. (11.39) that, if U_m denotes the amount of energy transferred to the rod as a result of the impact loading, the maximum stress in the rod of uniform cross section is

$$\sigma_m = \sqrt{\frac{2U_m E}{V}} \qquad (11.45a)$$

where V is the volume of the rod.

(b) Considering next the rod of Example 11.06 and observing that the volume of the rod is

$$V = 4A(L/2) + A(L/2) = 5AL/2$$

we substitute $AL = 2V/5$ into Eq. (11.41) and write

$$\sigma_m = \sqrt{\frac{8U_m E}{V}} \qquad (11.45b)$$

(c) Finally, recalling that $I = \frac{1}{4}\pi c^4$ for a beam of circular cross section, we note that

$$L(I/c^2) = L(\tfrac{1}{4}\pi c^4/c^2) = \tfrac{1}{4}(\pi c^2 L) = \tfrac{1}{4}V$$

Eq. (11.44), we express the maximum stress in the cantilever beam
of Example 11.07 as

$$\sigma_m = \sqrt{\frac{24U_m E}{V}} \qquad (11.45c)$$

We note that, in each case, the maximum stress σ_m is proportional to the square root of the modulus of elasticity of the material and inversely proportional to the square root of the volume of the member. Assuming all three members to have the same volume and to be of the same material, we also note that, for a given value of the absorbed energy, the uniform rod will experience the lowest maximum stress, and the cantilever beam the highest one.

This observation can be explained by the fact that, the distribution of stresses being uniform in case a, the strain energy will be uniformly distributed throughout the rod. In case b, on the other hand, the stresses in portion BC of the rod are only 25 percent as large as the stresses in portion CD. This uneven distribution of the stresses and of the strain energy results in a maximum stress σ_m twice as large as the corresponding stress in the uniform rod. Finally, in case c, where the cantilever beam is subjected to a transverse impact loading, the stresses vary linearly along the beam as well as across a transverse section. The very uneven resulting distribution of strain energy causes the maximum stress σ_m to be 3.46 times larger than if the same member had been loaded axially as in case a.

The properties noted in the three specific cases discussed in this section are quite general and can be observed in all types of structures and impact loadings. We thus conclude that a structure designed to withstand effectively an impact load should

1. Have a large volume
2. Be made of a material with a low modulus of elasticity and a high yield strength
3. Be shaped so that the stresses are distributed as evenly as possible throughout the structure

11.9 WORK AND ENERGY UNDER A SINGLE LOAD

When we first introduced the concept of strain energy at the beginning of this chapter, we considered the work done by an axial load **P** applied to the end of a rod of uniform cross section (Fig. 11.1). We defined the strain energy of the rod for an elongation x_1 as the work of the load **P** as it is slowly increased from 0 to the value P_1 corresponding to x_1. We wrote

$$\text{Strain energy} = U = \int_0^{x_1} P \, dx \qquad (11.2)$$

In the case of an elastic deformation, the work of the load **P**, and thus the strain energy of the rod, were expressed as

$$U = \tfrac{1}{2} P_1 x_1 \qquad (11.3)$$

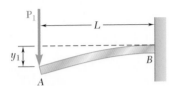

Fig. 11.25 Cantilever beam with load $\mathbf{P}_1$.

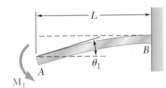

Fig. 11.26 Cantilever beam with couple $\mathbf{M}_1$.

Later, in Secs. 11.4 and 11.5, we computed the strain energy of structural members under various loading conditions by determining the strain-energy density u at every point of the member and integrating u over the entire member.

However, when a structure or member is subjected to a *single concentrated load*, it is possible to use Eq. (11.3) to evaluate its elastic strain energy, provided, of course, that the relation between the load and the resulting deformation is known. For instance, in the case of the cantilever beam of Example 11.03 (Fig. 11.25), we write

$$U = \tfrac{1}{2}P_1 y_1$$

and, substituting for y_1 the value obtained from the table of *Beam Deflections and Slopes* of Appendix D,

$$U = \frac{1}{2}P_1\left(\frac{P_1 L^3}{3EI}\right) = \frac{P_1^2 L^3}{6EI} \tag{11.46}$$

A similar approach can be used to determine the strain energy of a structure or member subjected to a *single couple*. Recalling that the elementary work of a couple of moment M is $M\,d\theta$, where $d\theta$ is a small angle, we find, since M and θ are linearly related, that the elastic strain energy of a cantilever beam AB subjected to a single couple $\mathbf{M}_1$ at its end A (Fig. 11.26) can be expressed as

$$U = \int_0^{\theta_1} M\,d\theta = \tfrac{1}{2}M_1\theta_1 \tag{11.47}$$

where θ_1 is the slope of the beam at A. Substituting for θ_1 the value obtained from Appendix D, we write

$$U = \frac{1}{2}M_1\left(\frac{M_1 L}{EI}\right) = \frac{M_1^2 L}{2EI} \tag{11.48}$$

In a similar way, the elastic strain energy of a uniform circular shaft AB of length L subjected at its end B to a single torque $\mathbf{T}_1$ (Fig. 11.27) can be expressed as

$$U = \int_0^{\phi_1} T\,d\phi = \tfrac{1}{2}T_1\phi_1 \tag{11.49}$$

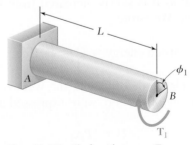

Fig. 11.27 Shaft with torque $\mathbf{T}_1$.

Substituting for the angle of twist ϕ_1 from Eq. (3.16), we verify that

$$U = \frac{1}{2}T_1\left(\frac{T_1L}{JG}\right) = \frac{T_1^2L}{2JG}$$

as previously obtained in Sec. 11.5.

The method presented in this section may simplify the solution of many impact-loading problems. In Example 11.08, the crash of an automobile into a barrier (Photo 11.3) is considered by using a simplified model consisting of a block and a simple beam.

Photo 11.3 As the automobile crashed into the barrier, considerable energy was dissipated as heat during the permanent deformation of the automobile and the barrier. *Source:* Crash test photo courtesy of Sec-Envel and L.I.E.R., France.

A block of mass m moving with a velocity $\mathbf{v}_0$ hits squarely the prismatic member AB at its midpoint C (Fig. 11.28). Determine (*a*) the equivalent static load P_m, (*b*) the maximum stress σ_m in the member, and (*c*) the maximum deflection x_m at point C.

EXAMPLE 11.08

(a) Equivalent Static Load. The maximum strain energy of the member is equal to the kinetic energy of the block before impact. We have

$$U_m = \tfrac{1}{2}mv_0^2 \tag{11.50}$$

On the other hand, expressing U_m as the work of the equivalent horizontal static load as it is slowly applied at the midpoint C of the member, we write

$$U_m = \tfrac{1}{2}P_m x_m \tag{11.51}$$

where x_m is the deflection of C corresponding to the static load P_m. From the table of *Beam Deflections and Slopes* of Appendix D, we find that

$$x_m = \frac{P_m L^3}{48EI} \tag{11.52}$$

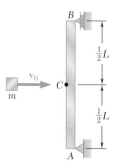

Fig. 11.28

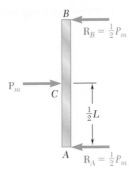

Fig. 11.29

Substituting for x_m from (11.52) into (11.51), we write

$$U_m = \frac{1}{2}\frac{P_m^2 L^3}{48EI}$$

Solving for P_m and recalling Eq. (11.50), we find that the static load equivalent to the given impact loading is

$$P_m = \sqrt{\frac{96U_mEI}{L^3}} = \sqrt{\frac{48mv_0^2\,EI}{L^3}} \qquad (11.53)$$

(b) Maximum Stress. Drawing the free-body diagram of the member (Fig. 11.29), we find that the maximum value of the bending moment occurs at C and is $M_{\max} = P_mL/4$. The maximum stress, therefore, occurs in a transverse section through C and is equal to

$$\sigma_m = \frac{M_{\max}c}{I} = \frac{P_mLc}{4I}$$

Substituting for P_m from (11.53), we write

$$\sigma_m = \sqrt{\frac{3mv_0^2\,EI}{L(I/c)^2}}$$

(c) Maximum Deflection. Substituting into Eq. (11.52) the expression obtained for P_m in (11.53), we have

$$x_m = \frac{L^3}{48EI}\sqrt{\frac{48mv_0^2\,EI}{L^3}} = \sqrt{\frac{mv_0^2\,L^3}{48EI}}$$

11.10 DEFLECTION UNDER A SINGLE LOAD BY THE WORK-ENERGY METHOD

We saw in the preceding section that, if the deflection x_1 of a structure or member under a single concentrated load $\mathbf{P}_1$ is known, the corresponding strain energy U is obtained by writing

$$U = \tfrac{1}{2}P_1x_1 \qquad (11.3)$$

A similar expression for the strain energy of a structural member under a single couple $\mathbf{M}_1$ is:

$$U = \tfrac{1}{2}M_1\theta_1 \qquad (11.47)$$

Conversely, if the strain energy U of a structure or member subjected to a single concentrated load $\mathbf{P}_1$ or couple $\mathbf{M}_1$ is known, Eq. (11.3) or (11.47) can be used to determine the corresponding deflection x_1 or angle θ_1. In order to determine the deflection under a single load applied to a structure consisting of several component parts, it is easier, rather than use one of the methods of Chap. 9, to first compute the strain energy of the structure by integrating the strain-energy density over its various parts, as was done in Secs. 11.4 and 11.5, and then use either Eq. (11.3) or Eq. (11.47) to obtain the desired deflection. Similarly, the angle of twist ϕ_1 of a composite shaft can be obtained by integrating the strain-energy density over the various parts of the shaft and solving Eq. (11.49) for ϕ_1.

It should be kept in mind that the method presented in this section can be used *only if the given structure is subjected to a single concentrated load or couple.* The strain energy of a structure subjected to several loads *cannot* be determined by computing the work of each load as if it were applied independently to the structure (see Sec. 11.11). We can also observe that, even if it were possible to compute the strain energy of the structure in this manner, only one equation would be available to determine the deflections corresponding to the various loads. In Secs. 11.12 and 11.13, another method based on the concept of strain energy is presented, one that can be used to determine the deflection or slope at a given point of a structure, even when that structure is subjected simultaneously to several concentrated loads, distributed loads, or couples.

EXAMPLE 11.09

A load **P** is supported at B by two uniform rods of the same cross-sectional area A (Fig. 11.30). Determine the vertical deflection of point B.

The strain energy of the system under the given load was determined in Example 11.02. Equating the expression obtained for U to the work of the load, we write

$$U = 0.364 \frac{P^2 l}{AE} = \frac{1}{2} P y_B$$

and, solving for the vertical deflection of B,

$$y_B = 0.728 \frac{Pl}{AE}$$

Remark. We should note that, once the forces in the two rods have been obtained (see Example 11.02), the deformations $\delta_{B/C}$ and $\delta_{B/D}$ of the rods could be obtained by the method of Chap. 2. Determining the vertical deflection of point B from these deformations, however, would require a careful geometric analysis of the various displacements involved. The strain-energy method used here makes such an analysis unnecessary.

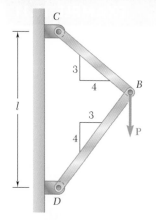

Fig. 11.30

EXAMPLE 11.10

Determine the deflection of end A of the cantilever beam AB (Fig. 11.31), taking into account the effect of (*a*) the normal stresses only, (*b*) both the normal and shearing stresses.

(**a**) **Effect of Normal Stresses.** The work of the force **P** as it is slowly applied to A is

$$U = \tfrac{1}{2} P y_A$$

Substituting for U the expression obtained for the strain energy of the beam in Example 11.03, where only the effect of the normal stresses was considered, we write

$$\frac{P^2 L^3}{6EI} = \frac{1}{2} P y_A$$

Fig. 11.31

and, solving for y_A,

$$y_A = \frac{PL^3}{3EI}$$

(b) Effect of Normal and Shearing Stresses. We now substitute for U the expression (11.24) obtained in Example 11.05, where the effects of both the normal and shearing stresses were taken into account. We have

$$\frac{P^2L^3}{6EI}\left(1 + \frac{3Eh^2}{10GL^2}\right) = \frac{1}{2}Py_A$$

and, solving for y_A,

$$y_A = \frac{PL^3}{3EI}\left(1 + \frac{3Eh^2}{10GL^2}\right)$$

We note that the relative error when the effect of shear is neglected is the same that was obtained in Example 11.05, i.e., less than $0.9(h/L)^2$. As we indicated then, this is less than 0.9 percent for a beam with a ratio h/L less than $\frac{1}{10}$.

EXAMPLE 11.11

A torque **T** is applied at the end D of shaft BCD (Fig. 11.32). Knowing that both portions of the shaft are of the same material and same length, but that the diameter of BC is twice the diameter of CD, determine the angle of twist for the entire shaft.

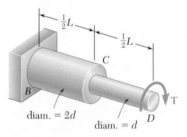

Fig. 11.32

The strain energy of a similar shaft was determined in Example 11.04 by breaking the shaft into its component parts BC and CD. Making $n = 2$ in Eq. (11.23), we have

$$U = \frac{17}{32}\frac{T^2L}{2GJ}$$

where G is the modulus of rigidity of the material and J the polar moment of inertia of portion CD of the shaft. Setting U equal to the work of the torque as it is slowly applied to end D, and recalling Eq. (11.49), we write

$$\frac{17}{32}\frac{T^2L}{2GJ} = \frac{1}{2}T\phi_{D/B}$$

and, solving for the angle of twist $\phi_{D/B}$,

$$\phi_{D/B} = \frac{17TL}{32GJ}$$

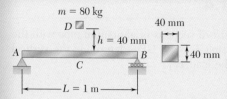

SAMPLE PROBLEM 11.3

The block D of mass m is released from rest and falls a distance h before it strikes the midpoint C of the aluminum beam AB. Using $E = 73$ GPa, determine (a) the maximum deflection of point C, (b) the maximum stress that occurs in the beam.

SOLUTION

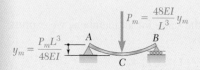

Position 1 y_m Position 2

Principle of Work and Energy. Since the block is released from rest, we note that in position 1 both the kinetic energy and the strain energy are zero. In position 2, where the maximum deflection y_m occurs, the kinetic energy is again zero. Referring to the table of *Beam Deflections and Slopes* of Appendix D, we find the expression for y_m shown. The strain energy of the beam in position 2 is

$$U_2 = \frac{1}{2}P_m y_m = \frac{1}{2}\frac{48EI}{L^3}y_m^2 \qquad U_2 = \frac{24EI}{L^3}y_m^2$$

We observe that the work done by the weight **W** of the block is $W(h + y_m)$. Equating the strain energy of the beam to the work done by **W,** we have

$$\frac{24EI}{L^3}y_m^2 = W(h + y_m) \tag{1}$$

From Appendix D

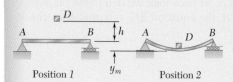

$$y_m = \frac{P_m L^3}{48EI} \qquad P_m = \frac{48EI}{L^3}y_m$$

a. Maximum Deflection of Point C. From the given data we have

$$EI = (73 \times 10^9 \text{ Pa})\tfrac{1}{12}(0.04 \text{ m})^4 = 15.573 \times 10^3 \text{ N} \cdot \text{m}^2$$

$$L = 1 \text{ m} \qquad h = 0.040 \text{ m} \qquad W = mg = (80 \text{ kg})(9.81 \text{ m/s}^2) = 784.8 \text{ N}$$

Substituting into Eq. (1), we obtain and solve the quadratic equation

$$(373.8 \times 10^3)y_m^2 - 784.8y_m - 31.39 = 0 \qquad y_m = 10.27 \text{ mm} \blacktriangleleft$$

b. Maximum Stress. The value of P_m is

$$P_m = \frac{48EI}{L^3}y_m = \frac{48(15.573 \times 10^3 \text{ N} \cdot \text{m})}{(1 \text{ m})^3}(0.01027 \text{ m}) \qquad P_m = 7677 \text{ N}$$

Recalling that $\sigma_m = M_{\text{max}}c/I$ and $M_{\text{max}} = \frac{1}{4}P_m L$, we write

$$\sigma_m = \frac{(\frac{1}{4}P_m L)c}{I} = \frac{\frac{1}{4}(7677 \text{ N})(1 \text{ m})(0.020 \text{ m})}{\frac{1}{12}(0.040 \text{ m})^4} \qquad \sigma_m = 179.9 \text{ MPa} \blacktriangleleft$$

An approximation for the work done by the weight of the block can be obtained by omitting y_m from the expression for the work and from the right-hand member of Eq. (1), as was done in Example 11.07. If this approximation is used here, we find $y_m = 9.16$ mm; the error is 10.8 percent. However, if an 8-kg block is dropped from a height of 400 mm, producing the same value of Wh, omitting y_m from the right-hand member of Eq. (1) results in an error of only 1.2 percent. A further discussion of this approximation is given in Prob. 11.70.

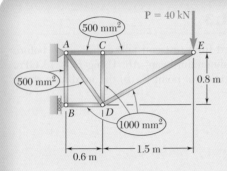

SAMPLE PROBLEM 11.4

Members of the truss shown consist of sections of aluminum pipe with the cross-sectional areas indicated. Using $E = 73$ GPa, determine the vertical deflection of point E caused by the load **P**.

SOLUTION

Axial Forces in Truss Members. The reactions are found by using the free-body diagram of the entire truss. We then consider in sequence the equilibrium of joints, E, C, D, and B. At each joint we determine the forces indicated by dashed lines. At joint B, the equation $\Sigma F_x = 0$ provides a check of our computations.

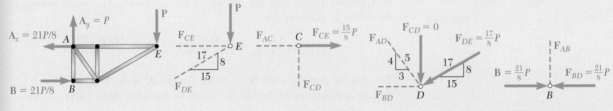

$$\Sigma F_y = 0: F_{DE} = -\tfrac{17}{8}P \qquad \Sigma F_x = 0: F_{AC} = +\tfrac{15}{8}P \qquad \Sigma F_y = 0: F_{AD} = +\tfrac{5}{4}P \qquad \Sigma F_y = 0: F_{AB} = 0$$
$$\Sigma F_x = 0: F_{CE} = +\tfrac{15}{8}P \qquad \Sigma F_y = 0: F_{CD} = 0 \qquad \Sigma F_x = 0: F_{BD} = -\tfrac{21}{8}P \qquad \Sigma F_x = 0: \text{(Checks)}$$

Strain Energy. Noting that E is the same for all members, we express the strain energy of the truss as follows

$$U = \sum \frac{F_i^2 L_i}{2A_i E} = \frac{1}{2E} \sum \frac{F_i^2 L_i}{A_i} \qquad (1)$$

where F_i is the force in a given member as indicated in the following table and where the summation is extended over all members of the truss.

Member	F_i	L_i, m	A_i, m^2	$\dfrac{F_i^2 L_i}{A_i}$
AB	0	0.8	500×10^{-6}	0
AC	$+15P/8$	0.6	500×10^{-6}	$4\ 219P^2$
AD	$+5P/4$	1.0	500×10^{-6}	$3\ 125P^2$
BD	$-21P/8$	0.6	1000×10^{-6}	$4\ 134P^2$
CD	0	0.8	1000×10^{-6}	0
CE	$+15P/8$	1.5	500×10^{-6}	$10\ 547P^2$
DE	$-17P/8$	1.7	1000×10^{-6}	$7\ 677P^2$

$$\sum \frac{F_i^2 L_i}{A_i} = 29\ 700P^2$$

Returning to Eq. (1), we have

$$U = (1/2E)(29.7 \times 10^3 P^2).$$

Principle of Work-Energy. We recall that the work done by the load **P** as it is gradually applied is $\tfrac{1}{2}Py_E$. Equating the work done by **P** to the strain energy U and recalling that $E = 73$ GPa and $P = 40$ kN, we have

$$\frac{1}{2}Py_E = U \qquad \frac{1}{2}Py_E = \frac{1}{2E}(29.7 \times 10^3 P^2)$$

$$y_E = \frac{1}{E}(29.7 \times 10^3 P) = \frac{(29.7 \times 10^3)(40 \times 10^3)}{73 \times 10^9}$$

$$y_E = 16.27 \times 10^{-3} \text{ m} \qquad\qquad y_E = 16.27 \text{ mm} \downarrow \quad \blacktriangleleft$$

PROBLEMS

11.42 A 6-kg collar has a speed $v_0 = 4.5$ m/s when it strikes a small plate attached to end A of the 20-mm-diameter rod AB. Using $E = 200$ GPa, determine (a) the equivalent static load, (b) the maximum stress in the rod, (c) the maximum deflection of end A.

11.43 A 5-kg collar D moves along the uniform rod AB and has a speed $v_0 = 6$ m/s when it strikes a small plate attached to end A of the rod. Using $E = 200$ GPa and knowing that the allowable stress in the rod is 250 MPa, determine the smallest diameter that can be used for the rod.

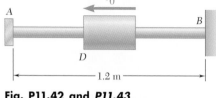

Fig. P11.42 and P11.43

11.44 Collar D is released from rest in the position shown and is stopped by a small plate attached at end C of the vertical rod ABC. Determine the mass of the collar for which the maximum normal stress in portion BC is 125 MPa.

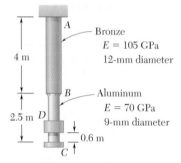

Bronze
$E = 105$ GPa
12-mm diameter

Aluminum
$E = 70$ GPa
9-mm diameter

4 m

2.5 m

0.6 m

Fig. P11.44

11.45 Solve Prob. 11.44, assuming that both portions of rod ABC are made of aluminum.

11.46 The 48-kg collar G is released from rest in the position shown and is stopped by plate BDF that is attached to the 20-mm-diameter steel rod CD and to the 15-mm-diameter steel rods AB and EF. Knowing that for the grade of steel used $\sigma_{all} = 180$ MPa and $E = 200$ GPa, determine the largest allowable distance h.

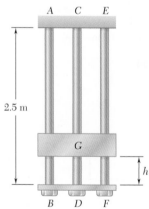

Fig. P11.46

11.47 Solve Prob. 11.46, assuming that the 20-mm-diameter steel rod CD is replaced by a 20-mm-diameter rod made of an aluminum alloy for which $\sigma_{all} = 150$ MPa and $E = 75$ GPa.

11.48 The steel beam AB is struck squarely at its midpoint C by a 45-kg block moving horizontally with a speed $v_0 = 2$ m/s. Using $E = 200$ GPa, determine (a) the equivalent static load, (b) the maximum normal stress in the beam, (c) the maximum deflection of the midpoint C of the beam.

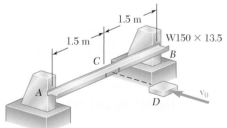

11.49 Solve Prob. 11.48, assuming that the W150 × 13.5 rolled-steel beam is rotated by 90° about its longitudinal axis so that its web is vertical.

Fig. P11.48

749

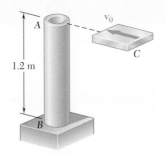

Fig. P11.50

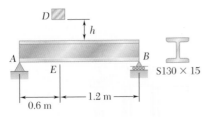

Fig. P11.54

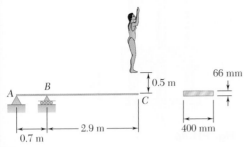

Fig. P11.55

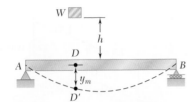

Fig. P11.56 and P11.57

11.50 The post AB consists of a steel pipe of 80-mm diameter and 6-mm wall thickness. A 6-kg block C moving horizontally with a velocity $\mathbf{v}_0$ hits the post squarely at A. Using $E = 200$ GPa, determine the largest speed v_0 for which the maximum normal stress in the pipe does not exceed 180 MPa.

11.51 Solve Prob. 11.50, assuming that the post AB consists of a solid steel rod of 80-mm outer diameter.

11.52 and 11.53 The 2-kg block D is dropped from the position shown onto the end of a 16-mm-diameter rod. Knowing that $E = 200$ GPa, determine (a) the maximum deflection of end A, (b) the maximum bending moment in the rod, (c) the maximum normal stress in the rod.

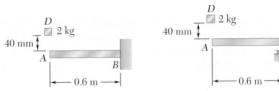

Fig. P11.52   **Fig. P11.53**

11.54 The 20-kg block D is dropped from a height $h = 180$ mm onto the steel beam AB. Knowing that $E = 200$ GPa, determine (a) the maximum deflection at point E, (b) the maximum normal stress in the beam.

11.55 A 72-kg diver jumps from a height of 0.5 m onto end C of a diving board having the uniform cross section shown. Assuming that the diver's legs remain rigid and using $E = 12$ GPa, determine (a) the maximum deflection at point C, (b) the maximum normal stress in the board, (c) the equivalent static load.

11.56 A block of weight W is dropped from a height h onto the horizontal beam AB and hits it at point D. (a) Show that the maximum deflection y_m at point D can be expressed as

$$y_m = y_{st}\left(1 + \sqrt{1 + \frac{2h}{y_{st}}}\right)$$

where y_{st} represents the deflection at D caused by a static load W applied at that point and where the quantity in parenthesis is referred to as the *impact factor*. (b) Compute the impact factor for the beam and the impact of Prob. 11.52.

11.57 A block of weight W is dropped from a height h onto the horizontal beam AB and hits point D. (a) Denoting by y_m the exact value of the maximum deflection at D and by y'_m the value obtained by neglecting the effect of this deflection on the change in potential energy of the block, show that the absolute value of the relative error is $(y'_m - y_m)/y_m$, never exceeding $y'_m/2h$. (b) Check the result obtained in part a by solving part a of Prob. 11.52 without taking y_m into account when determining the change in potential energy of the load, and comparing the answer obtained in this way with the exact answer to that problem.

11.58 and 11.59 Using the method of work and energy, determine the deflection at point D caused by the load **P.**

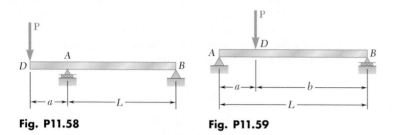

Fig. P11.58   **Fig. P11.59**

11.60 and 11.61 Using the method of work and energy, determine the slope at point D caused by the couple M_0.

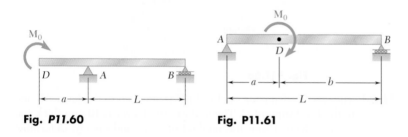

Fig. *P11.60*   **Fig. P11.61**

11.62 and 11.63 Using the method of work and energy, determine the deflection at point C caused by the load **P.**

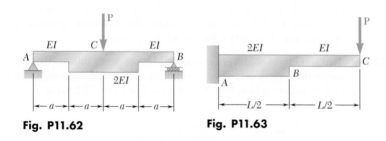

Fig. P11.62   **Fig. P11.63**

11.64 Using the method of work and energy, determine the slope at point A caused by the couple M_0.

11.65 Using the method of work and energy, determine the slope at point D caused by the couple M_0.

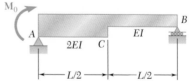

Fig. *P11.64*

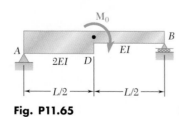

Fig. P11.65

11.66 The 20-mm-diameter steel rod CD is welded to the 20-mm-diameter steel shaft AB as shown. End C of rod CD is touching the rigid surface shown when a couple $\mathbf{T}_B$ applied to a disk attached to shaft AB. Knowing that the bearings are self aligning and exert no couples on the shaft, determine the angle of rotation of the disk when $T_B = 400$ N $\cdot$ m. Use $E = 200$ GPa and $G = 77.2$ GPa. (Consider the strain energy due to both bending and twisting in shaft AB and to bending in arm CD.)

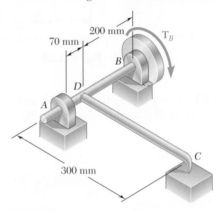

Fig. P11.66

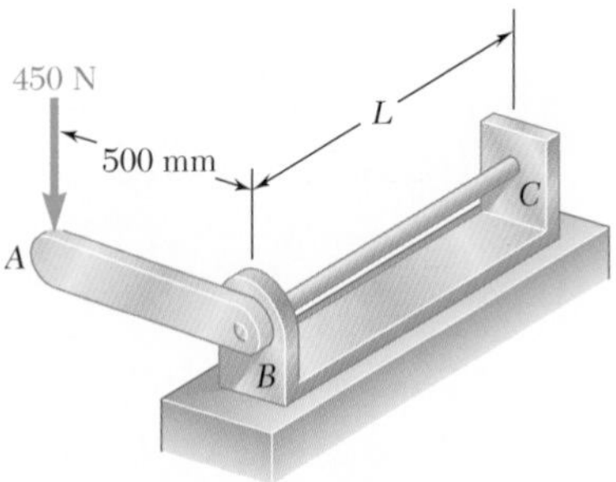

11.67 The 20-mm-diameter steel rod BC is attached to the lever AB and to the fixed support C. The uniform steel lever is 10 mm thick and 30 mm deep. Using the method of work and energy, determine the deflection of point A when $L = 600$ mm. Use $E = 200$ GPa and $G = 77.2$ GPa.

11.68 The 20-mm-diameter steel rod BC is attached to the lever AB and to the fixed support C. The uniform steel lever is 10 mm thick and 30 mm deep. Using the method of work and energy, determine the length L of the rod BC for which the deflection at point A is 40 mm. Use $E = 200$ GPa and $G = 77.2$ GPa.

Fig. P11.67 and P11.68

11.69 Two solid steel shafts are connected by the gears shown. Using the method of work and energy, determine the angle through which end D rotates when $T = 820$ N $\cdot$ m. Use $G = 77.2$ GPa.

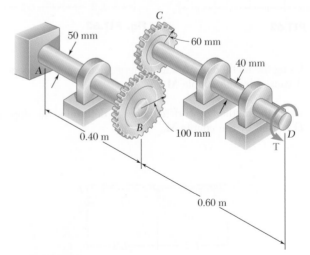

Fig. P11.69

11.70 The thin-walled hollow cylindrical member *AB* has a noncircular cross section of nonuniform thickness. Using the expression given in Eq. (3.53) of Sec. 3.13, and the expression for the strain-energy density given in Eq. (11.19), show that the angle of twist of member *AB* is

$$\phi = \frac{TL}{4\alpha^2 G} \oint \frac{ds}{t}$$

where *ds* is an element of the center line of the wall cross section and *α* is the area enclosed by that center line.

Fig. P11.70

11.71 Each member of the truss shown has a uniform cross-sectional area *A*. Using the method of work and energy, determine the vertical deflection of the point of application of the load **P.**

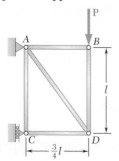

Fig. P11.71

11.72 Each member of the truss shown is made of steel and has a cross-sectional area of 400 mm². Using *E* = 200 GPa, determine the deflection of point *D* caused by the 16-kN load.

11.73 Each member of the truss shown is made of steel and has a uniform cross-sectional area of 1945 mm². Using *E* = 200 GPa, determine the vertical deflection of joint *A* caused by the application of the 100-kN load.

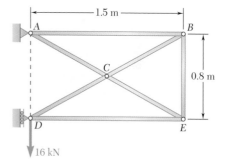

Fig. P11.72

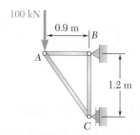

Fig. P11.73

11.74 Each member of the truss shown is made of steel and has a uniform cross-sectional area of 3220 mm². Using *E* = 200 GPa, determine the vertical deflection of joint *C* caused by the application of the 60-kN load.

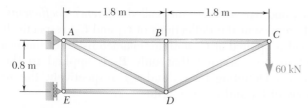

Fig. P11.74

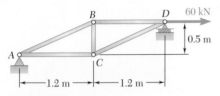

Fig. P11.75

11.75 Each member of the truss shown is made of steel; the cross-sectional area of member BC is 800 mm^2 and for all other members the cross-sectional area is 400 mm^2. Using $E = 200$ GPa, determine the deflection of point D caused by the 60-kN load.

11.76 The steel rod BC has a 24-mm diameter and the steel cable $ABDCA$ has a 12-mm diameter. Using $E = 200$ GPa, determine the deflection of joint D caused by the 12-kN load.

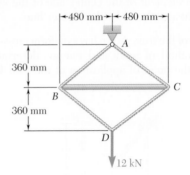

Fig. P11.76

*11.11 WORK AND ENERGY UNDER SEVERAL LOADS

In this section, the strain energy of a structure subjected to several loads will be considered and will be expressed in terms of the loads and the resulting deflections.

Consider an elastic beam AB subjected to two concentrated loads $\mathbf{P}_1$ and $\mathbf{P}_2$. The strain energy of the beam is equal to the work of $\mathbf{P}_1$ and $\mathbf{P}_2$ as they are slowly applied to the beam at C_1 and C_2, respectively (Fig. 11.33). However, in order to evaluate this work, we must first express the deflections x_1 and x_2 in terms of the loads $\mathbf{P}_1$ and $\mathbf{P}_2$.

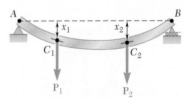

Fig. 11.33 Beam with multiple loads.

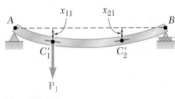

Fig. 11.34

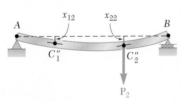

Fig. 11.35

Let us assume that only $\mathbf{P}_1$ is applied to the beam (Fig. 11.34). We note that both C_1 and C_2 are deflected and that their deflections are proportional to the load P_1. Denoting these deflections by x_{11} and x_{21}, respectively, we write

$$x_{11} = \alpha_{11}P_1 \qquad x_{21} = \alpha_{21}P_1 \qquad (11.54)$$

where α_{11} and α_{21} are constants called *influence coefficients*. These constants represent the deflections of C_1 and C_2, respectively, when a unit load is applied at C_1 and are characteristics of the beam AB.

Let us now assume that only $\mathbf{P}_2$ is applied to the beam (Fig. 11.35). Denoting by x_{12} and x_{22}, respectively, the resulting deflections of C_1 and C_2, we write

$$x_{12} = \alpha_{12}P_2 \qquad x_{22} = \alpha_{22}P_2 \qquad (11.55)$$

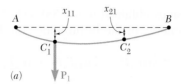

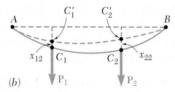

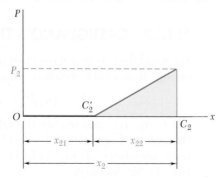

Fig. 11.36

where α_{12} and α_{22} are the influence coefficients representing the deflections of C_1 and C_2, respectively, when a unit load is applied at C_2. Applying the principle of superposition, we express the deflections x_1 and x_2 of C_1 and C_2 when both loads are applied (Fig. 11.33) as

$$x_1 = x_{11} + x_{12} = \alpha_{11}P_1 + \alpha_{12}P_2 \tag{11.56}$$
$$x_2 = x_{21} + x_{22} = \alpha_{21}P_1 + \alpha_{22}P_2 \tag{11.57}$$

To compute the work done by $\mathbf{P}_1$ and $\mathbf{P}_2$, and thus the strain energy of the beam, it is convenient to assume that $\mathbf{P}_1$ is first applied slowly at C_1 (Fig. 11.36a). Recalling the first of Eqs. (11.54), we express the work of $\mathbf{P}_1$ as

$$\tfrac{1}{2}P_1x_{11} = \tfrac{1}{2}P_1(\alpha_{11}P_1) = \tfrac{1}{2}\alpha_{11}P_1^2 \tag{11.58}$$

and note that $\mathbf{P}_2$ does no work while C_2 moves through x_{21}, since it has not yet been applied to the beam.

Now we slowly apply $\mathbf{P}_2$ at C_2 (Fig. 11.36b); recalling the second of Eqs. (11.55), we express the work of $\mathbf{P}_2$ as

$$\tfrac{1}{2}P_2x_{22} = \tfrac{1}{2}P_2(\alpha_{22}P_2) = \tfrac{1}{2}\alpha_{22}P_2^2 \tag{11.59}$$

But, as $\mathbf{P}_2$ is slowly applied at C_2, the point of application of $\mathbf{P}_1$ moves through x_{12} from C_1' to C_1, and the load $\mathbf{P}_1$ does work. Since $\mathbf{P}_1$ is *fully applied* during this displacement (Fig. 11.37), its work is equal to P_1x_{12} or, recalling the first of Eqs. (11.55),

$$P_1x_{12} = P_1(\alpha_{12}P_2) = \alpha_{12}P_1P_2 \tag{11.60}$$

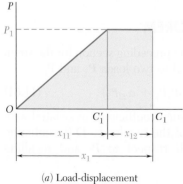

(a) Load-displacement diagram for C_1

(b) Load-displacement diagram for C_2

Fig. 11.37 Load-displacement diagrams.

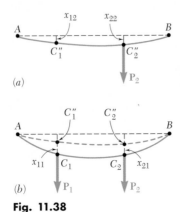

(a)

(b)

Fig. 11.38

Adding the expressions obtained in (11.58), (11.59), and (11.60), we express the strain energy of the beam under the loads $\mathbf{P}_1$ and $\mathbf{P}_2$ as

$$U = \tfrac{1}{2}(\alpha_{11}P_1^2 + 2\alpha_{12}P_1P_2 + \alpha_{22}P_2^2) \qquad (11.61)$$

If the load $\mathbf{P}_2$ had first been applied to the beam (Fig. 11.38a), and then the load $\mathbf{P}_1$ (Fig. 11.38b), the work done by each load would have been as shown in Fig. 11.39. Calculations similar to those we have just carried out would lead to the following alternative expression for the strain energy of the beam:

$$U = \tfrac{1}{2}(\alpha_{22}P_2^2 + 2\alpha_{21}P_2P_1 + \alpha_{11}P_1^2) \qquad (11.62)$$

Equating the right-hand members of Eqs. (11.61) and (11.62), we find that $\alpha_{12} = \alpha_{21}$, and thus conclude that the deflection produced at C_1 by a unit load applied at C_2 is equal to the deflection produced at C_2 by a unit load applied at C_1. This is known as *Maxwell's reciprocal theorem*, after the British physicist James Clerk Maxwell (1831–1879).

While we are now able to express the strain energy U of a structure subjected to several loads as a function of these loads, we cannot use the method of Sec. 11.10 to determine the deflection of such a structure. Indeed, computing the strain energy U by integrating the strain-energy density u over the structure and substituting the expression obtained into (11.61) would yield only one equation, which clearly could not be solved for the various coefficients α.

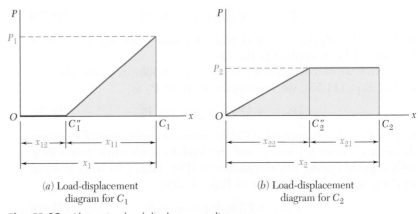

(a) Load-displacement diagram for C_1

(b) Load-displacement diagram for C_2

Fig. 11.39 Alternative load-displacement diagrams.

*11.12 CASTIGLIANO'S THEOREM

We recall the expression obtained in the preceding section for the strain energy of an elastic structure subjected to two loads $\mathbf{P}_1$ and $\mathbf{P}_2$:

$$U = \tfrac{1}{2}(\alpha_{11}P_1^2 + 2\alpha_{12}P_1P_2 + \alpha_{22}P_2^2) \qquad (11.61)$$

where α_{11}, α_{12}, and α_{22} are the influence coefficients associated with the points of application C_1 and C_2 of the two loads. Differentiating both members of Eq. (11.61) with respect to P_1 and recalling Eq. (11.56), we write

$$\frac{\partial U}{\partial P_1} = \alpha_{11}P_1 + \alpha_{12}P_2 = x_1 \qquad (11.63)$$

Differentiating both members of Eq. (11.61) with respect to P_2, recalling Eq. (11.57), and keeping in mind that $\alpha_{12} = \alpha_{21}$, we have

$$\frac{\partial U}{\partial P_2} = \alpha_{12}P_1 + \alpha_{22}P_2 = x_2 \qquad (11.64)$$

More generally, if an elastic structure is subjected to n loads $\mathbf{P}_1, \mathbf{P}_2, \ldots, \mathbf{P}_n$, the deflection x_j of the point of application of $\mathbf{P}_j$, measured along the line of action of $\mathbf{P}_j$, can be expressed as the partial derivative of the strain energy of the structure with respect to the load $\mathbf{P}_j$. We write

$$x_j = \frac{\partial U}{\partial P_j} \qquad (11.65)$$

This is *Castigliano's theorem,* named after the Italian engineer Alberto Castigliano (1847–1884) who first stated it.†

Recalling that the work of a couple $\mathbf{M}$ is $\frac{1}{2}M\theta$, where θ is the angle of rotation at the point where the couple is slowly applied, we note that Castigliano's theorem may be used to determine the slope of a beam at the point of application of a couple $\mathbf{M}_j$. We have

$$\theta_j = \frac{\partial U}{\partial M_j} \qquad (11.68)$$

Similarly, the angle of twist ϕ_j in a section of a shaft where a torque $\mathbf{T}_j$ is slowly applied is obtained by differentiating the strain energy of the shaft with respect to T_j:

$$\phi_j = \frac{\partial U}{\partial T_j} \qquad (11.69)$$

---

†In the case of an elastic structure subjected to n loads $\mathbf{P}_1, \mathbf{P}_2, \ldots, \mathbf{P}_n$, the deflection of the point of application of $\mathbf{P}_j$, measured along the line of action of $\mathbf{P}_j$, can be expressed as

$$x_j = \sum_k \alpha_{jk}P_k \qquad (11.66)$$

and the strain energy of the structure is found to be

$$U = \tfrac{1}{2} \sum_i \sum_k \alpha_{ik}P_iP_k \qquad (11.67)$$

Differentiating U with respect to P_j, and observing that P_j is found in terms corresponding to either $i = j$ or $k = j$, we write

$$\frac{\partial U}{\partial P_j} = \frac{1}{2}\sum_k \alpha_{jk}P_k + \frac{1}{2}\sum_i \alpha_{ij}P_i$$

or, since $\alpha_{ij} = \alpha_{ji}$,

$$\frac{\partial U}{\partial P_j} = \frac{1}{2}\sum_k \alpha_{jk}P_k + \frac{1}{2}\sum_i \alpha_{ji}P_i = \sum_k \alpha_{jk}P_k$$

Recalling Eq. (11.66), we verify that

$$x_j = \frac{\partial U}{\partial P_j} \qquad (11.65)$$

*11.13 DEFLECTIONS BY CASTIGLIANO'S THEOREM

We saw in the preceding section that the deflection x_j of a structure at the point of application of a load $\mathbf{P}_j$ can be determined by computing the partial derivative $\partial U/\partial P_j$ of the strain energy U of the structure. As we recall from Secs. 11.4 and 11.5, the strain energy U is obtained by integrating or summing over the structure the strain energy of each element of the structure. The calculation by Castigliano's theorem of the deflection x_j is simplified if the differentiation with respect to the load P_j is carried out before the integration or summation.

In the case of a beam, for example, we recall from Sec. 11.4 that

$$U = \int_0^L \frac{M^2}{2EI}\,dx \tag{11.17}$$

and determine the deflection x_j of the point of application of the load $\mathbf{P}_j$ by writing

$$x_j = \frac{\partial U}{\partial P_j} = \int_0^L \frac{M}{EI}\frac{\partial M}{\partial P_j}\,dx \tag{11.70}$$

In the case of a truss consisting of n uniform members of length L_i, cross-sectional area A_i, and internal force F_i, we recall Eq. (11.14) and express the strain energy U of the truss as

$$U = \sum_{i=1}^n \frac{F_i^2 L_i}{2A_i E} \tag{11.71}$$

The deflection x_j of the point of application of the load $\mathbf{P}_j$ is obtained by differentiating with respect to P_j each term of the sum. We write

$$x_j = \frac{\partial U}{\partial P_j} = \sum_{i=1}^n \frac{F_i L_i}{A_i E}\frac{\partial F_i}{\partial P_j} \tag{11.72}$$

EXAMPLE 11.12

The cantilever beam AB supports a uniformly distributed load w and a concentrated load $\mathbf{P}$ as shown (Fig. 11.40). Knowing that $L = 2$ m, $w = 4$ kN/m, $P = 6$ kN, and $EI = 5$ MN $\cdot$ m^2, determine the deflection at A.

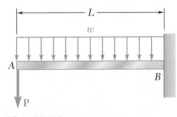

Fig. 11.40

The deflection y_A of the point A where the load $\mathbf{P}$ is applied is obtained from Eq. (11.70). Since $\mathbf{P}$ is vertical and directed downward, y_A represents a vertical deflection and is positive downward. We have

$$y_A = \frac{\partial U}{\partial P} = \int_0^L \frac{M}{EI} \frac{\partial M}{\partial P} dx \tag{11.73}$$

The bending moment M at a distance x from A is

$$M = -(Px + \tfrac{1}{2}wx^2) \tag{11.74}$$

and its derivative with respect to P is

$$\frac{\partial M}{\partial P} = -x$$

Substituting for M and $\partial M/\partial P$ into Eq. (11.73), we write

$$y_A = \frac{1}{EI} \int_0^L \left(Px^2 + \frac{1}{2}wx^3 \right) dx$$
$$y_A = \frac{1}{EI} \left(\frac{PL^3}{3} + \frac{wL^4}{8} \right) \tag{11.75}$$

Substituting the given data, we have

$$y_A = \frac{1}{5 \times 10^6 \text{ N} \cdot \text{m}^2} \left[\frac{(6 \times 10^3 \text{ N})(2 \text{ m})^3}{3} + \frac{(4 \times 10^3 \text{ N/m})(2 \text{ m})^4}{8} \right]$$
$$y_A = 4.8 \times 10^{-3} \text{ m} \qquad y_A = 4.8 \text{ mm} \downarrow$$

We note that the computation of the partial derivative $\partial M/\partial P$ *could not have been carried out* if the numerical value of P had been substituted for P in the expression (11.74) for the bending moment.

We can observe that the deflection x_j of a structure at a given point C_j can be obtained by the direct application of Castigliano's theorem only if a load $\mathbf{P}_j$ happens to be applied at C_j in the direction in which x_j is to be determined. When no load is applied at C_j, or when a load is applied in a direction other than the desired one, we can still obtain the deflection x_j by Castigliano's theorem if we use the following procedure: We apply a fictitious or "dummy" load $\mathbf{Q}_j$ at C_j in the direction in which the deflection x_j is to be determined and use Castigliano's theorem to obtain the deflection

$$x_j = \frac{\partial U}{\partial Q_j} \tag{11.76}$$

due to $\mathbf{Q}_j$ and the actual loads. Making $Q_j = 0$ in Eq. (11.76) yields the deflection at C_j in the desired direction under the given loading.

The slope θ_j of a beam at a point C_j can be determined in a similar manner by applying a fictitious couple $\mathbf{M}_j$ at C_j, computing the partial derivative $\partial U/\partial M_j$, and making $M_j = 0$ in the expression obtained.

EXAMPLE 11.13

The cantilever beam AB supports a uniformly distributed load w (Fig. 11.41). Determine the deflection and slope at A.

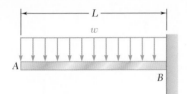

Fig. 11.41

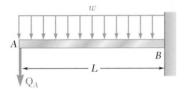

Fig. 11.42

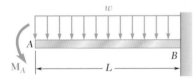

Fig. 11.43

Deflection at A. We apply a dummy downward load $\mathbf{Q}_A$ at A (Fig. 11.42) and write

$$y_A = \frac{\partial U}{\partial Q_A} = \int_0^L \frac{M}{EI} \frac{\partial M}{\partial Q_A} dx \qquad (11.77)$$

The bending moment M at a distance x from A is

$$M = -Q_A x - \tfrac{1}{2} w x^2 \qquad (11.78)$$

and its derivative with respect to Q_A is

$$\frac{\partial M}{\partial Q_A} = -x \qquad (11.79)$$

Substituting for M and $\partial M/\partial Q_A$ from (11.78) and (11.79) into (11.77), and making $Q_A = 0$, we obtain the deflection at A for the given loading:

$$y_A = \frac{1}{EI} \int_0^L (-\tfrac{1}{2} w x^2)(-x)\,dx = +\frac{wL^4}{8EI}$$

Since the dummy load was directed downward, the positive sign indicates that

$$y_A = \frac{wL^4}{8EI} \downarrow$$

Slope at A. We apply a dummy counterclockwise couple $\mathbf{M}_A$ at A (Fig. 11.43) and write

$$\theta_A = \frac{\partial U}{\partial M_A}$$

Recalling Eq. (11.17), we have

$$\theta_A = \frac{\partial}{\partial M_A} \int_0^L \frac{M^2}{2EI}\,dx = \int_0^L \frac{M}{EI} \frac{\partial M}{\partial M_A}\,dx \qquad (11.80)$$

The bending moment M at a distance x from A is

$$M = -M_A - \tfrac{1}{2} w x^2 \qquad (11.81)$$

and its derivative with respect to M_A is

$$\frac{\partial M}{\partial M_A} = -1 \qquad (11.82)$$

Substituting for M and $\partial M/\partial M_A$ from (11.81) and (11.82) into (11.80), and making $M_A = 0$, we obtain the slope at A for the given loading:

$$\theta_A = \frac{1}{EI} \int_0^L (-\tfrac{1}{2} w x^2)(-1)\,dx = +\frac{wL^3}{6EI}$$

Since the dummy couple was counterclockwise, the positive sign indicates that the angle θ_A is also counterclockwise:

$$\theta_A = \frac{wL^3}{6EI} \;\measuredangle$$

A load **P** is supported at B by two rods of the same material and of the same cross-sectional area A (Fig. 11.44). Determine the horizontal and vertical deflection of point B.

EXAMPLE 11.14

We apply a dummy horizontal load **Q** at B (Fig. 11.45). From Castigliano's theorem we have

$$x_B = \frac{\partial U}{\partial Q} \qquad y_B = \frac{\partial U}{\partial P}$$

Recalling from Sec. 11.4 the expression (11.14) for the strain energy of a rod, we write

$$U = \frac{F_{BC}^2 \,(BC)}{2AE} + \frac{F_{BD}^2 \,(BD)}{2AE}$$

where F_{BC} and F_{BD} represent the forces in BC and BD, respectively. We have, therefore,

$$x_B = \frac{\partial U}{\partial Q} = \frac{F_{BC}\,(BC)}{AE}\frac{\partial F_{BC}}{\partial Q} + \frac{F_{BD}\,(BD)}{AE}\frac{\partial F_{BD}}{\partial Q} \qquad (11.83)$$

and

$$y_B = \frac{\partial U}{\partial P} = \frac{F_{BC}\,(BC)}{AE}\frac{\partial F_{BC}}{\partial P} + \frac{F_{BD}\,(BD)}{AE}\frac{\partial F_{BD}}{\partial P} \qquad (11.84)$$

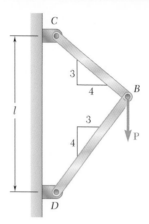

Fig. 11.44

From the free-body diagram of pin B (Fig. 11.46), we obtain

$$F_{BC} = 0.6P + 0.8Q \qquad F_{BD} = -0.8P + 0.6Q \qquad (11.85)$$

Differentiating these expressions with respect to Q and P, we write

$$\frac{\partial F_{BC}}{\partial Q} = 0.8 \qquad \frac{\partial F_{BD}}{\partial Q} = 0.6$$

$$\frac{\partial F_{BC}}{\partial P} = 0.6 \qquad \frac{\partial F_{BD}}{\partial P} = -0.8 \qquad (11.86)$$

Substituting from (11.85) and (11.86) into both (11.83) and (11.84), making $Q = 0$, and noting that $BC = 0.6l$ and $BD = 0.8l$, we obtain the horizontal and vertical deflections of point B under the given load **P**:

$$x_B = \frac{(0.6P)(0.6l)}{AE}(0.8) + \frac{(-0.8P)(0.8l)}{AE}(0.6)$$

$$= -0.096\frac{Pl}{AE}$$

$$y_B = \frac{(0.6P)(0.6l)}{AE}(0.6) + \frac{(-0.8P)(0.8l)}{AE}(-0.8)$$

$$= +0.728\frac{Pl}{AE}$$

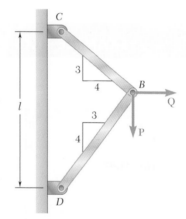

Fig. 11.45

Referring to the directions of the loads **Q** and **P,** we conclude that

$$x_B = 0.096\frac{Pl}{AE} \leftarrow \qquad y_B = 0.728\frac{Pl}{AE} \downarrow$$

We check that the expression obtained for the vertical deflection of B is the same that was found in Example 11.09.

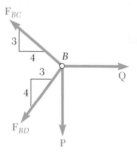

Fig. 11.46

*11.14   STATICALLY INDETERMINATE STRUCTURES

The reactions at the supports of a statically indeterminate elastic structure can be determined by Castigliano's theorem. In the case of a structure indeterminate to the first degree, for example, we designate one of the reactions as redundant and eliminate or modify accordingly the corresponding support. The redundant reaction is then treated as an unknown load that, together with the other loads, must produce deformations that are compatible with the original supports. We first calculate the strain energy U of the structure due to the combined action of the given loads and the redundant reaction. Observing that the partial derivative of U with respect to the redundant reaction represents the deflection (or slope) at the support that has been eliminated or modified, we then set this derivative equal to zero and solve the equation obtained for the redundant reaction.† The remaining reactions can be obtained from the equations of statics.

†This is in the case of a rigid support allowing no deflection. For other types of support, the partial derivative of U should be set equal to the allowed deflection.

EXAMPLE 11.15   Determine the reactions at the supports for the prismatic beam and loading shown (Fig. 11.47).

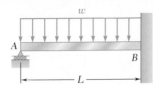

Fig. 11.47

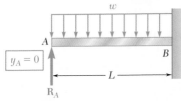

Fig. 11.48

The beam is statically indeterminate to the first degree. We consider the reaction at A as redundant and release the beam from that support. The reaction $\mathbf{R}_A$ is now considered as an unknown load (Fig. 11.48) and will be determined from the condition that the deflection y_A at A must be zero. By Castigliano's theorem $y_A = \partial U/\partial R_A$, where U is the strain energy of the beam under the distributed load and the redundant reaction. Recalling Eq. (11.70), we write

$$y_A = \frac{\partial U}{\partial R_A} = \int_0^L \frac{M}{EI}\frac{\partial M}{\partial R_A}\,dx \qquad (11.87)$$

We now express the bending moment M for the loading of Fig. 11.48. The bending moment at a distance x from A is

$$M = R_A x - \tfrac{1}{2}wx^2 \qquad (11.88)$$

and its derivative with respect to R_A is

$$\frac{\partial M}{\partial R_A} = x \qquad (11.89)$$

Substituting for M and $\partial M/\partial R_A$ from (11.88) and (11.89) into (11.87), we write

$$y_A = \frac{1}{EI}\int_0^L \left(R_A x^2 - \frac{1}{2}wx^3\right)dx = \frac{1}{EI}\left(\frac{R_A L^3}{3} - \frac{wL^4}{8}\right)$$

Setting $y_A = 0$ and solving for R_A, we have

$$R_A = \tfrac{3}{8}wL \qquad \mathbf{R}_A = \tfrac{3}{8}wL \uparrow$$

From the conditions of equilibrium for the beam, we find that the reaction at B consists of the following force and couple:

$$\mathbf{R}_B = \tfrac{5}{8}wL \uparrow \qquad \mathbf{M}_B = \tfrac{1}{8}wL^2 \downarrow$$

A load **P** is supported at B by three rods of the same material and the same cross-sectional area A (Fig. 11.49). Determine the force in each rod.

EXAMPLE 11.16

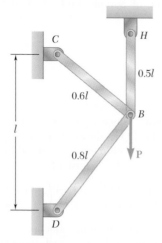

Fig. 11.49

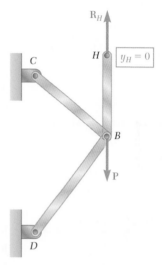

Fig. 11.50

The structure is statically indeterminate to the first degree. We consider the reaction at H as redundant and release rod BH from its support at H. The reaction $\mathbf{R}_H$ is now considered as an unknown load (Fig. 11.50) and will be determined from the condition that the deflection y_H of point H must be zero. By Castigliano's theorem $y_H = \partial U/\partial R_H$, where U is the strain energy of the three-rod system under the load **P** and the redundant reaction $\mathbf{R}_H$. Recalling Eq. (11.72), we write

$$y_H = \frac{F_{BC}\,(BC)}{AE}\frac{\partial F_{BC}}{\partial R_H} + \frac{F_{BD}\,(BD)}{AE}\frac{\partial F_{BD}}{\partial R_H} + \frac{F_{BH}\,(BH)}{AE}\frac{\partial F_{BH}}{\partial R_H} \quad (11.90)$$

We note that the force in rod BH is equal to R_H and write

$$F_{BH} = R_H \quad (11.91)$$

Then, from the free-body diagram of pin B (Fig. 11.51), we obtain

$$F_{BC} = 0.6P - 0.6R_H \qquad F_{BD} = 0.8R_H - 0.8P \quad (11.92)$$

Differentiating with respect to R_H the force in each rod, we write

$$\frac{\partial F_{BC}}{\partial R_H} = -0.6 \qquad \frac{\partial F_{BD}}{\partial R_H} = 0.8 \qquad \frac{\partial F_{BH}}{\partial R_H} = 1 \quad (11.93)$$

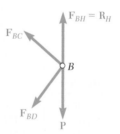

Fig. 11.51

Substituting from (11.91), (11.92), and (11.93) into (11.90), and noting that the lengths BC, BD, and BH are, respectively, equal to $0.6l$, $0.8l$, and $0.5l$, we write

$$y_H = \frac{1}{AE}\left[(0.6P - 0.6R_H)(0.6l)(-0.6)\right.$$
$$\left. + (0.8R_H - 0.8P)(0.8l)(0.8) + R_H\,(0.5l)(1)\right]$$

Setting $y_H = 0$, we obtain

$$1.228R_H - 0.728P = 0$$

and, solving for R_H,

$$R_H = 0.593P$$

Carrying this value into Eqs. (11.91) and (11.92), we obtain the forces in the three rods:

$$F_{BC} = +0.244P \qquad F_{BD} = -0.326P \qquad F_{BH} = +0.593P$$

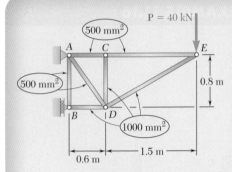

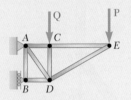

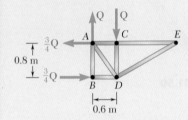

SAMPLE PROBLEM 11.5

For the truss and loading of Sample Prob. 11.4, determine the vertical deflection of joint C.

SOLUTION

Castigliano's Theorem. Since no vertical load is applied at joint C, we introduce the dummy load $\mathbf{Q}$ as shown. Using Castigliano's theorem, and denoting by F_i the force in a given member i caused by the combined loading of $\mathbf{P}$ and $\mathbf{Q}$, we have, since $E = $ constant,

$$y_C = \sum \left(\frac{F_i L_i}{A_i E}\right)\frac{\partial F_i}{\partial Q} = \frac{1}{E}\sum \left(\frac{F_i L_i}{A_i}\right)\frac{\partial F_i}{\partial Q} \tag{1}$$

Force in Members. Considering in sequence the equilibrium of joints E, C, B, and D, we determine the force in each member caused by load $\mathbf{Q}$.

Joint E: $F_{CE} = F_{DE} = 0$

Joint C: $F_{AC} = 0$; $F_{CD} = -Q$

Joint B: $F_{AB} = 0$; $F_{BD} = -\frac{3}{4}Q$

Joint D

F_{AD} $F_{CD} = Q$

$F_{BD} = \frac{3}{4}Q$ D

Force triangle

$F_{CD} = Q$ $F_{AD} = \frac{5}{4}Q$

$F_{BD} = \frac{3}{4}Q$

The force in each member caused by the load $\mathbf{P}$ was previously found in Sample Prob. 11.4. The total force in each member under the combined action of $\mathbf{Q}$ and $\mathbf{P}$ is shown in the following table. Forming $\partial F_i/\partial Q$ for each member, we then compute $(F_i L_i/A_i)(\partial F_i/\partial Q)$ as indicated in the table.

Member	F_i	$\partial F_i/\partial Q$	L_i, m	A_i, m^2	$\left(\dfrac{F_i L_i}{A_i}\right)\dfrac{\partial F_i}{\partial Q}$
AB	0	0	0.8	500×10^{-6}	0
AC	$+15P/8$	0	0.6	500×10^{-6}	0
AD	$+5P/4 + 5Q/4$	$\frac{5}{4}$	1.0	500×10^{-6}	$+3125P + 3125Q$
BD	$-21P/8 - 3Q/4$	$-\frac{3}{4}$	0.6	1000×10^{-6}	$+1181P + 338Q$
CD	$-Q$	-1	0.8	1000×10^{-6}	$+ 800Q$
CE	$+15P/8$	0	1.5	500×10^{-6}	0
DE	$-17P/8$	0	1.7	1000×10^{-6}	0

$$\sum \left(\frac{F_i L_i}{A_i}\right)\frac{\partial F_i}{\partial Q} = 4306P + 4263Q$$

Deflection of C. Substituting into Eq. (1), we have

$$y_C = \frac{1}{E}\sum \left(\frac{F_i L_i}{A_i}\right)\frac{\partial F_i}{\partial Q} = \frac{1}{E}(4306P + 4263Q)$$

Since the load $\mathbf{Q}$ is not part of the original loading, we set $Q = 0$. Substituting the given data, $P = 40$ kN and $E = 73$ GPa, we find

$$y_C = \frac{4306\,(40 \times 10^3\ \text{N})}{73 \times 10^9\ \text{Pa}} = 2.36 \times 10^{-3}\ \text{m} \qquad y_C = 2.36\ \text{mm}\downarrow \quad \blacktriangleleft$$

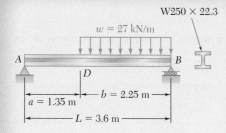

W250 × 22.3

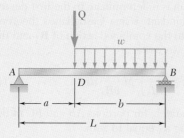

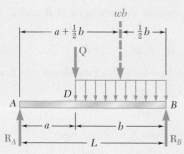

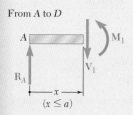

From A to D

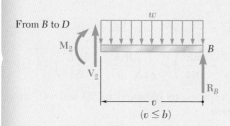

From B to D

SAMPLE PROBLEM 11.6

For the beam and loading shown, determine the deflection at point D. Use $E = 200$ GPa.

SOLUTION

Castigliano's Theorem. Since the given loading does not include a vertical load at point D, we introduce the dummy load $\mathbf{Q}$ as shown. Using Castigliano's theorem and noting that EI is constant, we write

$$y_D = \int \frac{M}{EI}\left(\frac{\partial M}{\partial Q}\right)dx = \frac{1}{EI}\int M\left(\frac{\partial M}{\partial Q}\right)dx \tag{1}$$

The integration will be performed separately for portions AD and DB.

Reactions. Using the free-body diagram of the entire beam, we find

$$\mathbf{R}_A = \frac{wb^2}{2L} + Q\frac{b}{L}\uparrow \qquad \mathbf{R}_B = \frac{wb(a + \tfrac{1}{2}b)}{L} + Q\frac{a}{L}\uparrow$$

Portion AD of Beam. Using the free body shown, we find

$$M_1 = R_A x = \left(\frac{wb^2}{2L} + Q\frac{b}{L}\right)x \qquad \frac{\partial M_1}{\partial Q} = +\frac{bx}{L}$$

Substituting into Eq. (1) and integrating from A to D gives

$$\frac{1}{EI}\int M_1\frac{\partial M_1}{\partial Q}dx = \frac{1}{EI}\int_0^a R_A x\left(\frac{bx}{L}\right)dx = \frac{R_A a^3 b}{3EIL}$$

We substitute for R_A and then set the dummy load Q equal to zero.

$$\frac{1}{EI}\int M_1\frac{\partial M_1}{\partial Q}dx = \frac{wa^3 b^3}{6EIL^2} \tag{2}$$

Portion DB of Beam. Using the free body shown, we find that the bending moment at a distance v from end B is

$$M_2 = R_B v - \frac{wv^2}{2} = \left[\frac{wb(a + \tfrac{1}{2}b)}{L} + Q\frac{a}{L}\right]v - \frac{wv^2}{2} \qquad \frac{\partial M_2}{\partial Q} = +\frac{av}{L}$$

Substituting into Eq. (1) and integrating from point B where $v = 0$, to point D where $v = b$, we write

$$\frac{1}{EI}\int M_2\frac{\partial M_2}{\partial Q}dv = \frac{1}{EI}\int_0^b\left(R_B v - \frac{wv^2}{2}\right)\left(\frac{av}{L}\right)dv = \frac{R_B ab^3}{3EIL} - \frac{wab^4}{8EIL}$$

Substituting for R_B and setting $Q = 0$,

$$\frac{1}{EI}\int M_2\frac{\partial M_2}{\partial Q}dv = \left[\frac{wb(a + \tfrac{1}{2}b)}{L}\right]\frac{ab^3}{3EIL} - \frac{wab^4}{8EIL} = \frac{5a^2 b^4 + ab^5}{24EIL^2}w \tag{3}$$

Deflection at Point D. Recalling Eqs. (1), (2), and (3), we have

$$y_D = \frac{wab^3}{24EIL^2}(4a^2 + 5ab + b^2) = \frac{wab^3}{24EIL^2}(4a + b)(a + b) = \frac{wab^3}{24EIL}(4a + b)$$

From Appendix C we find that $I = 28.9 \times 10^6$ mm^4 for a W250 × 22.3. Substituting for I, w, a, b, and L their numerical values, we obtain

$$y_D = 6.36 \text{ mm} \downarrow \quad \blacktriangleleft$$

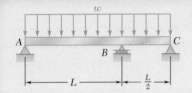

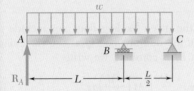

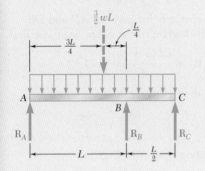

From A to B

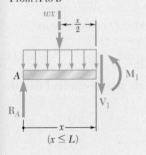

From C to B

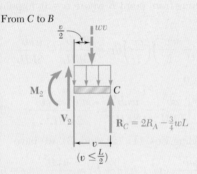

SAMPLE PROBLEM 11.7

For the uniform beam and loading shown, determine the reactions at the supports.

SOLUTION

Castigliano's Theorem. The beam is indeterminate to the first degree and we choose the reaction $\mathbf{R}_A$ as redundant. Using Castigliano's theorem, we determine the deflection at A due to the combined action of $\mathbf{R}_A$ and the distributed load. Since EI is constant, we write

$$y_A = \int \frac{M}{EI}\left(\frac{\partial M}{\partial R_A}\right) dx = \frac{1}{EI}\int M \frac{\partial M}{\partial R_A} dx \tag{1}$$

The integration will be performed separately for portions AB and BC of the beam. Finally, $\mathbf{R}_A$ is obtained by setting y_A equal to zero.

Free Body: Entire Beam. We express the reactions at B and C in terms of R_A and the distributed load

$$R_B = \tfrac{9}{4}wL - 3R_A \qquad R_C = 2R_A - \tfrac{3}{4}wL \tag{2}$$

Portion AB of Beam. Using the free-body diagram shown, we find

$$M_1 = R_A x - \frac{wx^2}{2} \qquad \frac{\partial M_1}{\partial R_A} = x$$

Substituting into Eq. (1) and integrating from A to B, we have

$$\frac{1}{EI}\int M_1 \frac{\partial M}{\partial R_A} dx = \frac{1}{EI}\int_0^L \left(R_A x^2 - \frac{wx^3}{2}\right) dx = \frac{1}{EI}\left(\frac{R_A L^3}{3} - \frac{wL^4}{8}\right) \tag{3}$$

Portion BC of Beam. We have

$$M_2 = \left(2R_A - \frac{3}{4}wL\right)v - \frac{wv^2}{2} \qquad \frac{\partial M_2}{\partial R_A} = 2v$$

Substituting into Eq. (1) and integrating from C, where $v = 0$, to B, where $v = \tfrac{1}{2}L$, we have

$$\frac{1}{EI}\int M_2 \frac{\partial M_2}{\partial R_A} dv = \frac{1}{EI}\int_0^{L/2}\left(4R_A v^2 - \frac{3}{2}wLv^2 - wv^3\right) dv$$

$$= \frac{1}{EI}\left(\frac{R_A L^3}{6} - \frac{wL^4}{16} - \frac{wL^4}{64}\right) = \frac{1}{EI}\left(\frac{R_A L^3}{6} - \frac{5wL^4}{64}\right) \tag{4}$$

Reaction at A. Adding the expressions obtained in (3) and (4), we determine y_A and set it equal to zero

$$y_A = \frac{1}{EI}\left(\frac{R_A L^3}{3} - \frac{wL^4}{8}\right) + \frac{1}{EI}\left(\frac{R_A L^3}{6} - \frac{5wL^4}{64}\right) = 0$$

Solving for R_A, $\qquad\qquad R_A = \dfrac{13}{32}wL \qquad\qquad R_A = \dfrac{13}{32}wL \uparrow$ ◀

Reactions at B and C. Substituting for R_A into Eqs. (2), we obtain

$$R_B = \frac{33}{32}wL \uparrow \qquad R_C = \frac{wL}{16} \uparrow \quad ◀$$

PROBLEMS

11.77 through 11.79 Using the information in Appendix D, compute the work of the loads as they are applied to the beam (*a*) if the load **P** is applied first, (*b*) if the couple **M** is applied first.

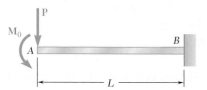

Fig. P11.77

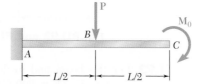

Fig. P11.78

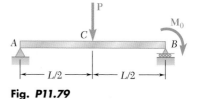

Fig. P11.79

11.80 through 11.82 For the beam and loading shown, (*a*) compute the work of the loads as they are applied successively to the beam, using the information provided in Appendix D, (*b*) compute the strain energy of the beam by the method of Sec. 11.4 and show that it is equal to the work obtained in part *a*.

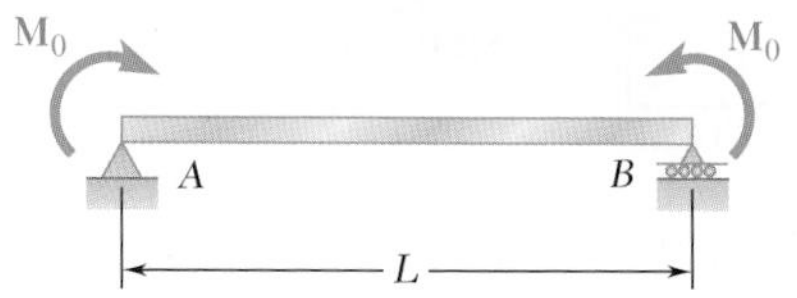

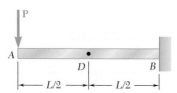

Fig. P11.80

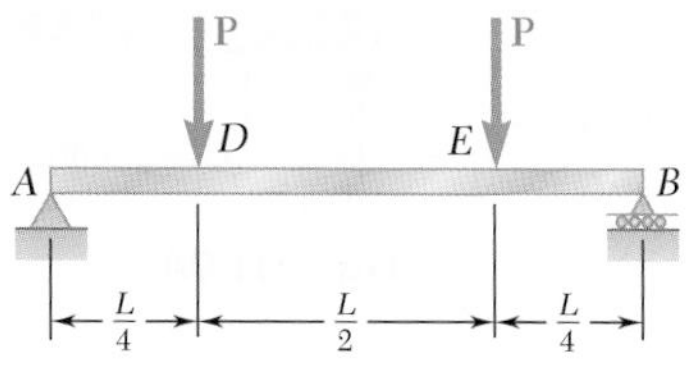

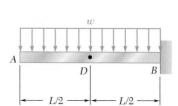

Fig. P11.81

Fig. P11.82

11.83 and 11.84 For the prismatic beam shown, determine the deflection of point *D*.

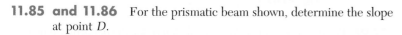

Fig. P11.83 and P11.85

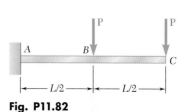

Fig. P11.84 and P11.86

11.85 and 11.86 For the prismatic beam shown, determine the slope at point *D*.

11.87 and 11.88 For the prismatic beam shown, determine the deflection at point *D*.

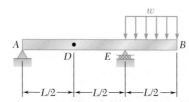

Fig. P11.87 and P11.89

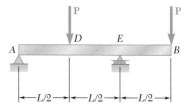

Fig. P11.88 and P11.90

11.89 and 11.90 For the prismatic beam shown, determine the slope at point *D*.

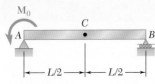

Fig. P11.91

11.91 For the prismatic beam shown, determine the slope at point *B*.

11.92 For the beam and loading shown, determine the deflection of point *A*. Use $E = 200$ GPa.

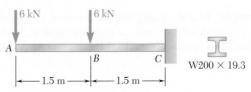

Fig. P11.92 and *P11.93*

11.93 For the beam and loading shown, determine the deflection of point *B*. Use $E = 200$ GPa.

11.94 For the beam and loading shown, determine the deflection at point *D*. Use $E = 200$ GPa.

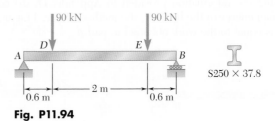

Fig. P11.94

11.95 and 11.96 For the beam and loading shown, determine the deflection at point *B*. Use $E = 200$ GPa.

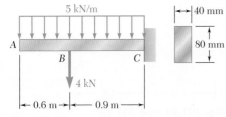

Fig. P11.95

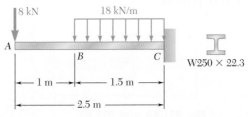

Fig. P11.96

11.97 For the beam and loading shown, determine the deflection at point *C*. Use $E = 200$ GPa.

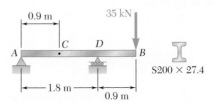

Fig. *P11.97* and P11.98

11.98 For the beam and loading shown, determine the slope at end *A*. Use $E = 200$ GPa.

11.99 and 11.100   For the truss and loading shown, determine the horizontal and vertical deflection of joint C.

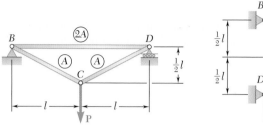

Fig. P11.99

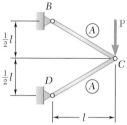

Fig. P11.100

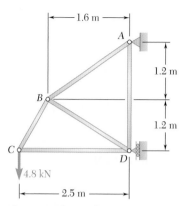

Fig. *P11.101 and P11.102*

11.101 and 11.102   Each member of the truss shown is made of steel and has a cross-sectional area of 500 mm^2. Using $E = 200$ GPa, determine the deflection indicated.

      11.101   Vertical deflection of joint B

      11.102   Horizontal deflection of joint B

11.103 and 11.104   Each member of the truss shown is made of steel and has the cross-sectional area shown. Using $E = 200$ GPa, determine the deflection indicated.

      11.103   Vertical deflection of joint C

      11.104   Horizontal deflection of joint C

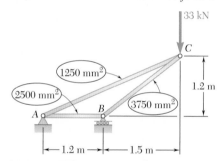

Fig. P11.103 and P11.104

11.105   Two rods AB and BC of the same flexural rigidity EI are welded together at B. For the loading shown, determine (a) the deflection of point C, (b) the slope of member BC at point C.

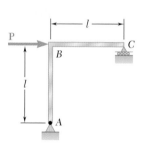

Fig. P11.105

11.106   A uniform rod of flexural rigidity EI is bent and loaded as shown. Determine (a) the horizontal deflection of point D, (b) the slope at point D.

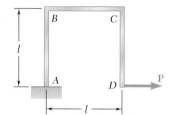

Fig. P11.106 and P11.107

11.107   A uniform rod of flexural rigidity EI is bent and loaded as shown. Determine (a) the vertical deflection of point D, (b) the slope of BC at point C.

11.108 A uniform rod of flexural rigidity EI is bent and loaded as shown. Determine (a) the vertical deflection of point A, (b) the horizontal deflection of point A.

11.109 For the beam and loading shown and using Castigliano's theorem, determine (a) the horizontal deflection of point B, (b) the vertical deflection of point B.

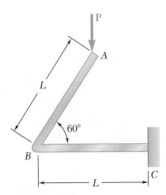

Fig. P11.108

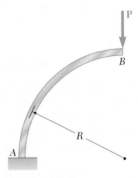

Fig. P11.109

Fig. P11.110

11.110 For the uniform rod and loading shown and using Castigliano's theorem, determine the deflection of point B.

11.111 through 11.114 Determine the reaction at the roller support and draw the bending-moment diagram for the beam and loading shown.

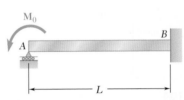

Fig. P11.111

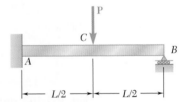

Fig. P11.112

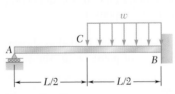

Fig. P11.113

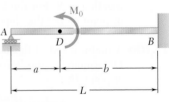

Fig. P11.114

11.115 For the uniform beam and loading shown, determine the reaction at each support.

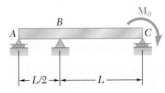

Fig. P11.115

11.116 Determine the reaction at the roller support and draw the bending-moment diagram for the beam and loading shown.

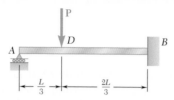

Fig. P11.116

11.117 through 11.120 Three members of the same material and same cross-sectional area are used to support the loading **P.** Determine the force in member *BC.*

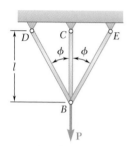

Fig. P11.117

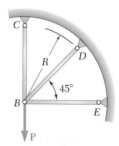

Fig. P11.118

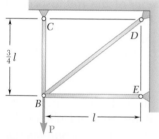

Fig. P11.119

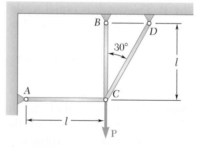

Fig. P11.120

11.121 and 11.122 Knowing that the eight members of the indeterminate truss shown have the same uniform cross-sectional area, determine the force in member *AB.*

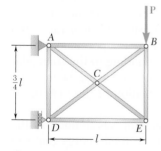

Fig. P11.121

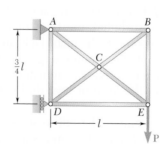

Fig. P11.122

REVIEW AND SUMMARY

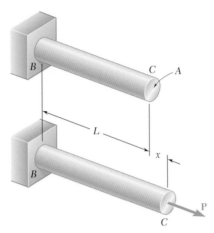

Fig. 11.52

This chapter was devoted to the study of strain energy and to the ways in which it can be used to determine the stresses and deformations in structures subjected to both static and impact loadings.

In Sec. 11.2 we considered a uniform rod subjected to a slowly increasing axial load **P** (Fig. 11.52). We noted that the area under

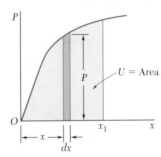

Fig. 11.53

Strain energy

the load-deformation diagram (Fig. 11.53) represents the work done by **P**. This work is equal to the *strain energy* of the rod associated with the deformation caused by the load **P**:

$$\text{Strain energy} = U = \int_0^{x_1} P\,dx \qquad (11.2)$$

Strain-energy density

Since the stress is uniform throughout the rod, we were able to divide the strain energy by the volume of the rod and obtain the strain energy per unit volume, which we defined as the *strain-energy density* of the material [Sec. 11.3]. We found that

$$\text{Strain-energy density} = u = \int_0^{\epsilon_1} \sigma_x\,d\epsilon_x \qquad (11.4)$$

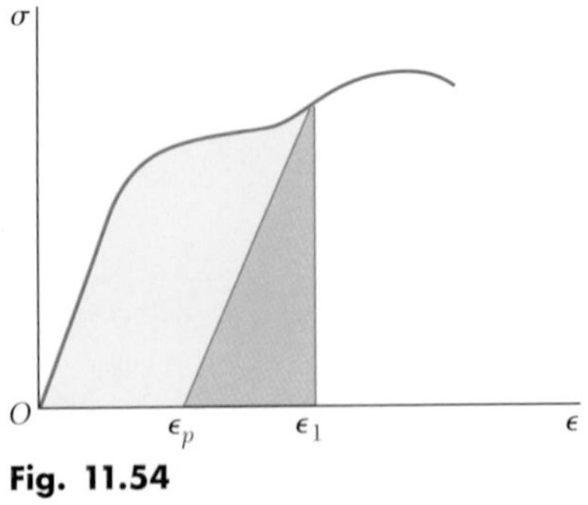

and noted that the strain-energy density is equal to the area under the stress-strain diagram of the material (Fig. 11.54). As we saw in Sec. 11.4, Eq. (11.4) remains valid when the stresses are not uniformly distributed, but the strain-energy density will then vary from point to point. If the material is unloaded, there is a permanent strain ϵ_p and only the strain-energy density corresponding to the triangular area is recovered, the remainder of the energy having been dissipated in the form of heat during the deformation of the material.

Fig. 11.54

Modulus of toughness

The area under the entire stress-strain diagram was defined as the *modulus of toughness* and is a measure of the total energy that can be acquired by the material.

If the normal stress σ remains within the proportional limit of the material, the strain-energy density u is expressed as

$$u = \frac{\sigma^2}{2E}$$

The area under the stress-strain curve from zero strain to the strain ϵ_Y at yield (Fig. 11.55) is referred to as the *modulus of resilience* of the material and represents the energy per unit volume that the material can absorb without yielding. We wrote

$$u_Y = \frac{\sigma_Y^2}{2E} \tag{11.8}$$

Modulus of resilience

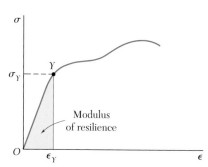

Fig. 11.55

In Sec. 11.4 we considered the strain energy associated with *normal stresses*. We saw that if a rod of length L and *variable cross-sectional area* A is subjected at its end to a centric axial load **P**, the strain energy of the rod is

$$U = \int_0^L \frac{P^2}{2AE}\,dx \tag{11.13}$$

If the rod is of *uniform cross section* of area A, the strain energy is

$$U = \frac{P^2L}{2AE} \tag{11.14}$$

Strain energy under axial load

We saw that for a beam subjected to transverse loads (Fig. 11.56) the strain energy associated with the normal stresses is

$$U = \int_0^L \frac{M^2}{2EI}\,dx \tag{11.17}$$

where M is the bending moment and EI the flexural rigidity of the beam.

Strain energy due to bending

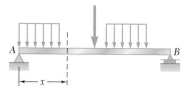

Fig. 11.56

The strain energy associated with *shearing stresses* was considered in Sec. 11.5. We found that the strain-energy density for a material in pure shear is

$$u = \frac{\tau_{xy}^2}{2G} \tag{11.19}$$

where τ_{xy} is the shearing stress and G the modulus of rigidity of the material.

Strain energy due to shearing stresses

For a shaft of length L and uniform cross section subjected at its ends to couples of magnitude T (Fig. 11.57) the strain energy was found to be

$$U = \frac{T^2L}{2GJ} \tag{11.22}$$

where J is the polar moment of inertia of the cross-sectional area of the shaft.

Strain energy due to torsion

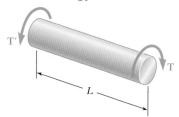

Fig. 11.57

General state of stress

In Sec. 11.6 we considered the strain energy of an elastic isotropic material under a general state of stress and expressed the strain-energy density at a given point in terms of the principal stresses σ_a, σ_b, and σ_c at that point:

$$u = \frac{1}{2E} [\sigma_a^2 + \sigma_b^2 + \sigma_c^2 - 2\nu(\sigma_a\sigma_b + \sigma_b\sigma_c + \sigma_c\sigma_a)] \quad (11.27)$$

The strain-energy density at a given point was divided into two parts: u_v, associated with a change in volume of the material at that point, and u_d, associated with a distortion of the material at the same point. We wrote $u = u_v + u_d$, where

$$u_v = \frac{1 - 2\nu}{6E} (\sigma_a + \sigma_b + \sigma_c)^2 \quad (11.32)$$

and

$$u_d = \frac{1}{12G} [(\sigma_a - \sigma_b)^2 + (\sigma_b - \sigma_c)^2 + (\sigma_c - \sigma_a)^2] \quad (11.33)$$

Using the expression obtained for u_d, we derived the maximum-distortion-energy criterion, which was used in Sec. 7.7 to predict whether a ductile material would yield under a given state of plane stress.

Impact loading

Equivalent static load

In Sec. 11.7 we considered the *impact loading* of an elastic structure being hit by a mass moving with a given velocity. We assumed that the kinetic energy of the mass is transferred entirely to the structure and defined the *equivalent static load* as the load that would cause the same deformations and stresses as are caused by the impact loading.

After discussing several examples, we noted that a structure designed to withstand effectively an impact load should be shaped in such a way that stresses are evenly distributed throughout the structure, and that the material used should have a low modulus of elasticity and a high yield strength [Sec. 11.8].

Members subjected to a single load

The strain energy of structural members subjected to a *single load* was considered in Sec. 11.9. In the case of the beam and loading of Fig. 11.58 we found that the strain energy of the beam is

$$U = \frac{P_1^2 L^3}{6EI} \quad (11.46)$$

Observing that the work done by the load $\mathbf{P}$ is equal to $\frac{1}{2}P_1y_1$, we equated the work of the load and the strain energy of the beam and determined the deflection y_1 at the point of application of the load [Sec. 11.10 and Example 11.10].

The method just described is of limited value, since it is restricted to structures subjected to a single concentrated load and to the determination of the deflection at the point of application of that load. In the remaining sections of the chapter, we presented a

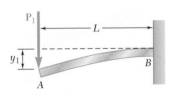

Fig. 11.58

more general method, which can be used to determine deflections at various points of structures subjected to several loads.

In Sec. 11.11 we discussed the strain energy of a structure subjected to several loads, and in Sec. 11.12 introduced *Castigliano's theorem*, which states that the deflection x_j, of the point of application of a load $\mathbf{P}_j$ measured along the line of action of $\mathbf{P}_j$ is equal to the partial derivative of the strain energy of the structure with respect to the load $\mathbf{P}_j$. We wrote

Castigliano's theorem

$$x_j = \frac{\partial U}{\partial P_j} \tag{11.65}$$

We also found that we could use Castigliano's theorem to determine the *slope* of a beam at the point of application of a couple $\mathbf{M}_j$ by writing

$$\theta_j = \frac{\partial U}{\partial M_j} \tag{11.68}$$

and the *angle of twist* in a section of a shaft where a torque $\mathbf{T}_j$ is applied by writing

$$\phi_j = \frac{\partial U}{\partial T_j} \tag{11.69}$$

In Sec. 11.13, Castigliano's theorem was applied to the determination of deflections and slopes at various points of a given structure. The use of "dummy" loads enabled us to include points where no actual load was applied. We also observed that the calculation of a deflection x_j was simplified if the differentiation with respect to the load P_j was carried out before the integration. In the case of a beam, recalling Eq. (11.17), we wrote

$$x_j = \frac{\partial U}{\partial P_j} = \int_0^L \frac{M}{EI} \frac{\partial M}{\partial P_j} dx \tag{11.70}$$

Similarly, for a truss consisting of n members, the deflection x_j at the point of application of the load $\mathbf{P}_j$ was found by writing

$$x_j = \frac{\partial U}{\partial P_j} = \sum_{i=1}^n \frac{F_i L_i}{A_i E} \frac{\partial F_i}{\partial P_j} \tag{11.72}$$

The chapter concluded [Sec. 11.14] with the application of Castigliano's theorem to the analysis of *statically indeterminate structures* [Sample Prob. 11.7, Examples 11.15 and 11.16].

Statically indeterminate structures

REVIEW PROBLEMS

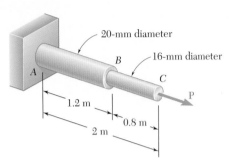

Fig. P11.123

11.123 Using $E = 200$ GPa, determine (*a*) the strain energy of the steel rod *ABC* when $P = 25$ kN, (*b*) the corresponding strain-energy density in portions *AB* and *BC* of the rod.

11.124 Assuming that the prismatic beam *AB* has a rectangular cross section, show that for the given loading the maximum value of the strain-energy density in the beam is

$$u_{max} = \frac{45}{8} \frac{U}{V}$$

where *U* is the strain energy of the beam and *V* is its volume.

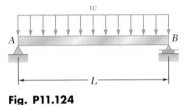

Fig. P11.124

11.125 The cylindrical block *E* has a speed $v_0 = 4.8$ m/s when it strikes squarely the yoke *BD* that is attached to the 22-mm-diameter rods *AB* and *CD*. Knowing that the rods are made of a steel for which $\sigma_Y = 350$ MPa and $E = 200$ GPa, determine the weight of block *E* for which the factor of safety is five with respect to permanent deformation of the rods.

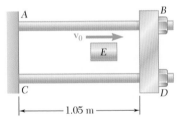

Fig. P11.125

11.126 Each member of the truss shown is made of aluminum and has the cross-sectional area shown. Using $E = 72$ GPa, determine the strain energy of the truss for the loading shown.

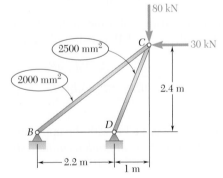

Fig. P11.26

11.127 A block of weight *W* is placed in contact with a beam at some given point *D* and released. Show that the resulting maximum deflection at point *D* is twice as large as the deflection due to a static load *W* applied at *D*.

11.128 The 12-mm-diameter steel rod ABC has been bent into the shape shown. Knowing that $E = 200$ GPa and $G = 77.2$ GPa, determine the deflection of end C caused by the 150-N force.

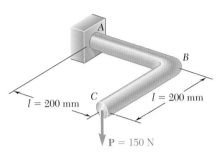

Fig. P11.128

11.129 Two steel shafts, each of 22-mm diameter, are connected by the gears shown. Knowing that $G = 77$ GPa and that shaft DF is fixed at F, determine the angle through which end A rotates when a 135 N $\cdot$ m torque is applied at A. (Ignore the strain energy due to the bending of the shafts.)

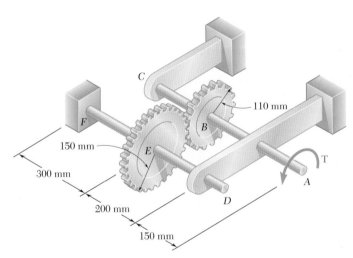

Fig. P11.129

11.130 Each member of the truss shown is made of steel; the cross-sectional area of member BC is 800 mm² and for all other members the cross-sectional area is 400 mm². Using $E = 200$ GPa, determine the deflection of point D caused by the 60-kN load.

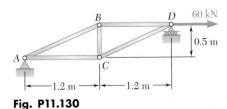

Fig. P11.130

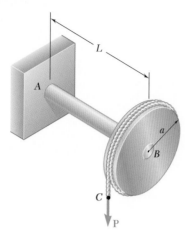

Fig. P11.131

11.131 A disk of radius a has been welded to end B of the solid steel shaft AB. A cable is then wrapped around the disk and a vertical force **P** is applied to end C of the cable. Knowing that the radius of the shaft is r and neglecting the deformations of the disk and of the cable, show that the deflection of point C caused by the application of **P** is

$$\delta_C = \frac{PL^3}{3EI}\left(1 + 1.5\frac{Er^2}{GL^2}\right)$$

11.132 Three rods, each of the same flexural rigidity EI, are welded to form the frame $ABCD$. For the loading shown, determine the angle formed by the frame at point D.

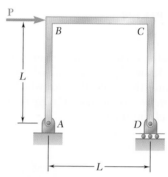

Fig. P11.132

11.133 The steel bar ABC has a square cross section of side 20 mm and is subjected to a 220-N load **P**. Using $E = 200$ GPa for rod BD and the bar, determine the deflection of point C.

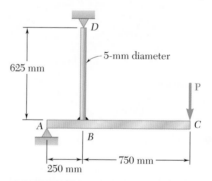

Fig. P11.133

11.134 For the uniform beam and loading shown, determine the reaction at each support.

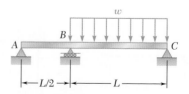

Fig. P11.134

COMPUTER PROBLEMS

The following problems are designed to be solved with a computer.

11.C1 A rod consisting of n elements, each of which is homogeneous and of uniform cross section, is subjected to a load **P** applied at its free end. The length of element i is denoted by L_i and its diameter by d_i. (*a*) Denoting by E the modulus of elasticity of the material used in the rod, write a computer program that can be used to determine the strain energy acquired by the rod and the deformation measured at its free end. (*b*) Use this program to determine the strain energy and deformation for the rods of Probs. 11.9 and 11.123.

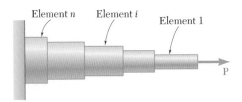

Fig. P11.C1

11.C2 Two 18×150-mm cover plates are welded to a W200 $\times$ 26.6 rolled-steel beam as shown. The 670-kg block is to be dropped from a height $h = 50$ mm onto the beam. (*a*) Write a computer program to calculate the maximum normal stress on transverse sections just to the left of D and at the center of the beam for values of a from 0 to 1.5 m using 125-mm increments. (*b*) From the values considered in part a, select the distance a for which the maximum normal stress is as small as possible. Use $E = 200$ GPa.

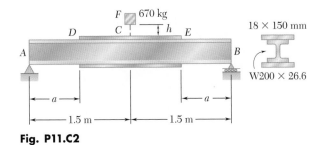

Fig. P11.C2

11.C3 The 16-kg block D is dropped from a height h onto the free end of the steel bar AB. For the steel used $\sigma_{\text{all}} = 120$ MPa and $E = 200$ GPa. (*a*) Write a computer program to calculate the maximum allowable height h for values of the length L from 100 mm to 1.2 m, using 100-mm increments. (*b*) From the values considered in part a, select the length corresponding to the largest allowable height.

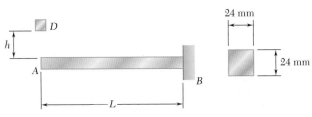

Fig. P11.C3

779

11.C4 The block D of mass $m = 8$ kg is dropped from a height $h = 750$ mm onto the rolled-steel beam AB. Knowing that $E = 200$ GPa, write a computer program to calculate the maximum deflection of point E and the maximum normal stress in the beam for values of a from 100 to 900 m, using 100-mm increments.

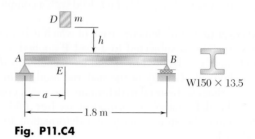

Fig. P11.C4

11.C5 The steel rods AB and BC are made of a steel for which $\sigma_Y = 300$ MPa and $E = 200$ GPa. (*a*) Write a computer program to calculate for values of a from 0 to 6 m, using 1-m increments, the maximum strain energy that can be acquired by the assembly without causing any permanent deformation. (*b*) For each value of a considered, calculate the diameter of a uniform rod of length 6 m and of the same mass as the original assembly, and the maximum strain energy that could be acquired by this uniform rod without causing permanent deformation.

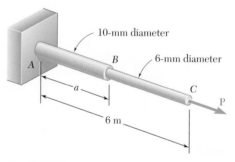

Fig. P11.C5

11.C6 A 72-kg diver jumps from a height of 0.5 m onto end C of a diving board having the uniform cross section shown. Write a computer program to calculate for values of a from 250 to 1250 mm, using 250-mm increments, (*a*) the maximum deflection of point C, (*b*) the maximum bending moment in the board, (*c*) the equivalent static load. Assume that the diver's legs remain rigid and use $E = 12$ GPa.

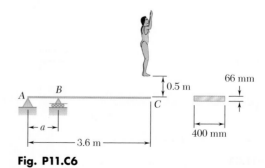

Fig. P11.C6

Appendices

†Courtesy of the American Institute of Steel Construction, Chicago, Illinois.

Appendix A

Moments of Areas

A.1 FIRST MOMENT OF AN AREA; CENTROID OF AN AREA

Consider an area A located in the xy plane (Fig. A.1). Denoting by x and y the coordinates of an element of area dA, we define the *first moment of the area A with respect to the x axis* as the integral

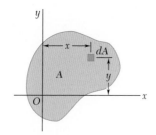

Fig. A.1

$$Q_x = \int_A y \, dA \tag{A.1}$$

Similarly, the *first moment of the area A with respect to the y axis* is defined as the integral

$$Q_y = \int_A x \, dA \tag{A.2}$$

We note that each of these integrals may be positive, negative, or zero, depending on the position of the coordinate axes. If SI units are used, the first moments Q_x and Q_y are expressed in m^3 or mm^3.

The *centroid of the area A* is defined as the point C of coordinates $\bar{x}$ and $\bar{y}$ (Fig. A.2), which satisfy the relations

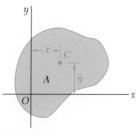

Fig. A.2

$$\int_A x \, dA = A\bar{x} \qquad \int_A y \, dA = A\bar{y} \tag{A.3}$$

Comparing Eqs. (A.1) and (A.2) with Eqs. (A.3), we note that the first moments of the area A can be expressed as the products of the area and of the coordinates of its centroid:

$$Q_x = A\bar{y} \qquad Q_y = A\bar{x} \tag{A.4}$$

When an area possesses an *axis of symmetry*, the first moment of the area with respect to that axis is zero. Indeed, considering the area A of Fig. A.3, which is symmetric with respect to the y axis, we observe that to every element of area dA of abscissa x corresponds an element of area dA' of abscissa −x. It follows that the integral in Eq. (A.2) is zero and, thus, that $Q_y = 0$. It also follows from the first of the relations (A.3) that $\bar{x} = 0$. Thus, if an area A possesses an axis of symmetry, its centroid C is located on that axis.

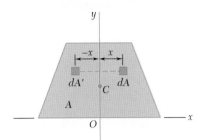

Fig. A.3

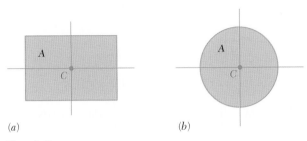

(a)   (b)

Fig. A.4

Since a rectangle possesses two axes of symmetry (Fig. A.4a), the centroid C of a rectangular area coincides with its geometric center. Similarly, the centroid of a circular area coincides with the center of the circle (Fig. A.4b).

When an area possesses a *center of symmetry* O, the first moment of the area about any axis through O is zero. Indeed, considering the area A of Fig. A.5, we observe that to every element of area dA of coordinates x and y corresponds an element of area dA' of coordinates −x and −y. It follows that the integrals in Eqs. (A.1) and (A.2) are both zero, and that $Q_x = Q_y = 0$. It also follows from Eqs. (A.3) that $\bar{x} = \bar{y} = 0$, i.e., the centroid of the area coincides with its center of symmetry.

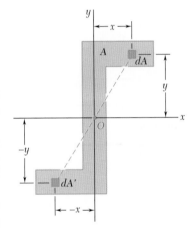

Fig. A.5

When the centroid C of an area can be located by symmetry, the first moment of that area with respect to any given axis can be readily obtained from Eqs. (A.4). For example, in the case of the rectangular area of Fig. A.6, we have

$$Q_x = A\bar{y} = (bh)(\tfrac{1}{2}h) = \tfrac{1}{2}bh^2$$

and

$$Q_y = A\bar{x} = (bh)(\tfrac{1}{2}b) = \tfrac{1}{2}b^2h$$

In most cases, however, it is necessary to perform the integrations indicated in Eqs. (A.1) through (A.3) to determine the first moments and the centroid of a given area. While each of the integrals involved is actually a double integral, it is possible in many applications to select elements of area dA in the shape of thin horizontal or vertical strips, and thus to reduce the computations to integrations in a single variable. This is illustrated in Example A.01.

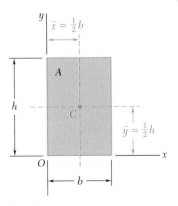

Fig. A.6

For the triangular area of Fig. A.7, determine (*a*) the first moment Q_x of the area with respect to the *x* axis, (*b*) the ordinate $\bar{y}$ of the centroid of the area.

(a) First Moment Q_x. We select as an element of area a horizontal strip of length *u* and thickness *dy*, and note that all the points within the element are at the same distance *y* from the *x* axis (Fig. A.8). From similar triangles, we have

$$\frac{u}{b} = \frac{h-y}{h} \qquad u = b\frac{h-y}{h}$$

and

$$dA = u \, dy = b\frac{h-y}{h} \, dy$$

The first moment of the area with respect to the *x* axis is

$$Q_x = \int_A y \, dA = \int_0^h yb\frac{h-y}{h} \, dy = \frac{b}{h}\int_0^h (hy - y^2) \, dy$$

$$= \frac{b}{h}\left[h\frac{y^2}{2} - \frac{y^3}{3}\right]_0^h \qquad Q_x = \tfrac{1}{6}bh^2$$

(b) Ordinate of Centroid. Recalling the first of Eqs. (A.4) and observing that $A = \tfrac{1}{2}bh$, we have

$$Q_x = A\bar{y} \qquad \tfrac{1}{6}bh^2 = (\tfrac{1}{2}bh)\bar{y}$$

$$\bar{y} = \tfrac{1}{3}h$$

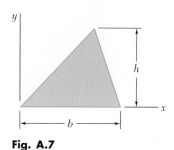

Fig. A.7

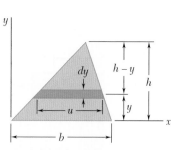

Fig. A.8

A.2 DETERMINATION OF THE FIRST MOMENT AND CENTROID OF A COMPOSITE AREA

Consider an area *A*, such as the trapezoidal area shown in Fig. A.9, which may be divided into simple geometric shapes. As we saw in the preceding section, the first moment Q_x of the area with respect to the *x* axis is represented by the integral $\int y \, dA$, which extends over the entire area *A*. Dividing *A* into its component parts A_1, A_2, A_3, we write

$$Q_x = \int_A y \, dA = \int_{A_1} y \, dA + \int_{A_2} y \, dA + \int_{A_3} y \, dA$$

or, recalling the second of Eqs. (A.3),

$$Q_x = A_1\bar{y}_1 + A_2\bar{y}_2 + A_3\bar{y}_3$$

where $\bar{y}_1$, $\bar{y}_2$, and $\bar{y}_3$ represent the ordinates of the centroids of the component areas. Extending this result to an arbitrary number of component areas, and noting that a similar expression may be obtained for Q_y, we write

$$Q_x = \sum A_i\bar{y}_i \qquad Q_y = \sum A_i\bar{x}_i \qquad \text{(A.5)}$$

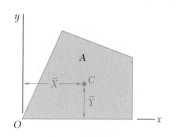

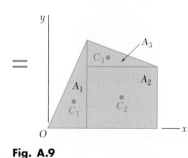

Fig. A.9

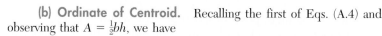

To obtain the coordinates $\overline{X}$ and $\overline{Y}$ of the centroid C of the composite area A, we substitute $Q_x = A\overline{Y}$ and $Q_y = A\overline{X}$ into Eqs. (A.5). We have

$$A\overline{Y} = \sum_i A_i \overline{y}_i \quad A\overline{X} = \sum_i A_i \overline{x}_i$$

Solving for $\overline{X}$ and $\overline{Y}$ and recalling that the area A is the sum of the component areas A_i, we write

$$\overline{X} = \frac{\sum_i A_i \overline{x}_i}{\sum_i A_i} \quad \overline{Y} = \frac{\sum_i A_i \overline{y}_i}{\sum_i A_i} \qquad (A.6)$$

EXAMPLE A.02

Locate the centroid C of the area A shown in Fig. A.10.

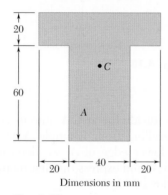

Dimensions in mm

Fig. A.10

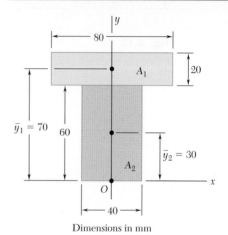

Dimensions in mm

Fig. A.11

Selecting the coordinate axes shown in Fig. A.11, we note that the centroid C must be located on the y axis, since this axis is an axis of symmetry; thus, $\overline{X} = 0$.

Dividing A into its component parts A_1 and A_2, we use the second of Eqs. (A.6) to determine the ordinate $\overline{Y}$ of the centroid. The actual computation is best carried out in tabular form.

	Area, mm^2	$\overline{y}_i$, mm	$A_i\overline{y}_i$, mm^3
A_1	$(20)(80) = 1600$	70	112×10^3
A_2	$(40)(60) = 2400$	30	72×10^3
	$\sum_i A_i = 4000$		$\sum_i A_i\overline{y}_i = 184 \times 10^3$

$$\overline{Y} = \frac{\sum_i A_i \overline{y}_i}{\sum_i A_i} = \frac{184 \times 10^3 \text{ mm}^3}{4 \times 10^3 \text{ mm}^2} = 46 \text{ mm}$$

Referring to the area A of Example A.02, we consider the horizontal x' axis through its centroid C. (Such an axis is called a *centroidal axis.*) Denoting by A' the portion of A located above that axis (Fig. A.12), determine the first moment of A' with respect to the x' axis.

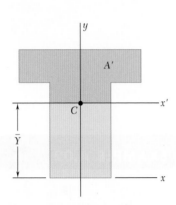

Fig. A.12

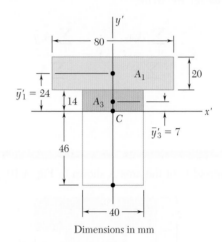

Dimensions in mm

Fig. A.13

Solution. We divide the area A' into its components A_1 and A_3 (Fig. A.13). Recalling from Example A.02 that C is located 46 mm above the lower edge of A, we determine the ordinates $\bar{y}_1'$ and $\bar{y}_3'$ of A_1 and A_3 and express the first moment $Q_{x'}'$ of A' with respect to x' as follows:

$$Q_{x'}' = A_1\bar{y}_1' + A_3\bar{y}_3'$$
$$= (20 \times 80)(24) + (14 \times 40)(7) = 42.3 \times 10^3 \text{ mm}^3$$

Alternative Solution. We first note that since the centroid C of A is located on the x' axis, the first moment $Q_{x'}$ of the *entire area* A with respect to that axis is zero:

$$Q_{x'} = A\bar{y}' = A(0) = 0$$

Denoting by A'' the portion of A located below the x' axis and by $Q_{x'}''$ its first moment with respect to that axis, we have therefore

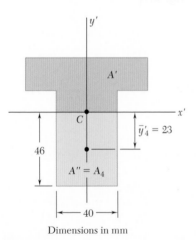

Dimensions in mm

Fig. A.14

$$Q_{x'} = Q_{x'}' + Q_{x'}'' = 0 \quad \text{or} \quad Q_{x'}' = -Q_{x'}''$$

which shows that the first moments of A' and A'' have the same magnitude and opposite signs. Referring to Fig. A.14, we write

$$Q_{x'}'' = A_4\bar{y}_4' = (40 \times 46)(-23) = -42.3 \times 10^3 \text{ mm}^3$$

and

$$Q_{x'}' = -Q_{x'}'' = +42.3 \times 10^3 \text{ mm}^3$$

A.3 SECOND MOMENT, OR MOMENT OF INERTIA, OF AN AREA; RADIUS OF GYRATION

Consider again an area A located in the xy plane (Fig. A.1) and the element of area dA of coordinates x and y. The *second moment*, or *moment of inertia*, of the area A with respect to the x axis, and the second moment, or moment of inertia, of A with respect to the y axis are defined, respectively, as

$$I_x = \int_A y^2 \, dA \quad I_y = \int_A x^2 \, dA \tag{A.7}$$

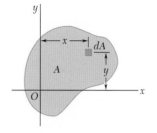

Fig. A.1 (repeated)

These integrals are referred to as *rectangular moments of inertia*, since they are computed from the rectangular coordinates of the element dA. While each integral is actually a double integral, it is possible in many applications to select elements of area dA in the shape of thin horizontal or vertical strips, and thus reduce the computations to integrations in a single variable. This is illustrated in Example A.04.

We now define the *polar moment of inertia* of the area A with respect to point O (Fig. A.15) as the integral

$$J_O = \int_A \rho^2 \, dA \tag{A.8}$$

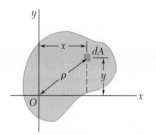

Fig. A.15

where ρ is the distance from O to the element dA. While this integral is again a double integral, it is possible in the case of a circular area to select elements of area dA in the shape of thin circular rings, and thus reduce the computation of J_O to a single integration (see Example A.05).

We note from Eqs. (A.7) and (A.8) that the moments of inertia of an area are positive quantities. If SI units are used, moments of inertia are expressed in m^4 or mm^4.

An important relation may be established between the polar moment of inertia J_O of a given area and the rectangular moments of inertia I_x and I_y of the same area. Noting that $\rho^2 = x^2 + y^2$, we write

$$J_O = \int_A \rho^2 \, dA = \int_A (x^2 + y^2) \, dA = \int_A y^2 \, dA + \int_A x^2 \, dA$$

or

$$J_O = I_x + I_y \tag{A.9}$$

The *radius of gyration* of an area A with respect to the x axis is defined as the quantity r_x, that satisfies the relation

$$I_x = r_x^2 A \tag{A.10}$$

where I_x is the moment of inertia of A with respect to the x axis. Solving Eq. (A.10) for r_x, we have

$$r_x = \sqrt{\frac{I_x}{A}} \tag{A.11}$$

In a similar way, we define the radii of gyration with respect to the y axis and the origin O. We write

$$I_y = r_y^2 A \qquad r_y = \sqrt{\frac{I_y}{A}} \tag{A.12}$$

$$J_O = r_O^2 A \qquad r_O = \sqrt{\frac{J_O}{A}} \tag{A.13}$$

Substituting for J_O, I_x, and I_y in terms of the corresponding radii of gyration in Eq. (A.9), we observe that

$$r_O^2 = r_x^2 + r_y^2 \tag{A.14}$$

EXAMPLE A.04

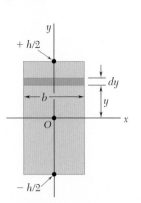

Fig. A.16

Fig. A.17

For the rectangular area of Fig. A.16, determine (a) the moment of inertia I_x of the area with respect to the centroidal x axis, (b) the corresponding radius of gyration r_x.

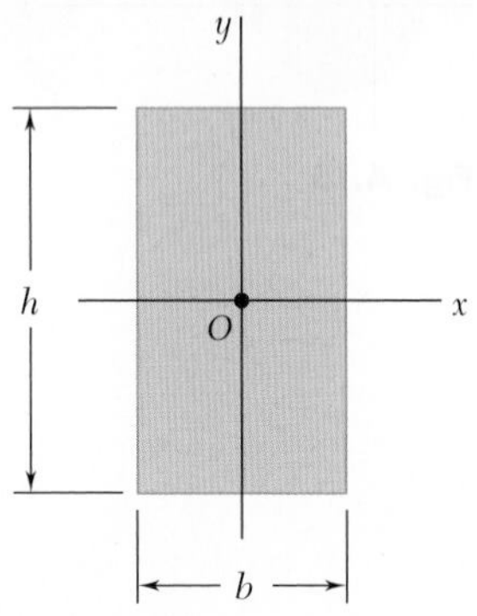

(a) Moment of Inertia I_x. We select as an element of area a horizontal strip of length b and thickness dy (Fig. A.17). Since all the points within the strip are at the same distance y from the x axis, the moment of inertia of the strip with respect to that axis is

$$dI_x = y^2\, dA = y^2(b\, dy)$$

Integrating from $y = -h/2$ to $y = +h/2$, we write

$$I_x = \int_A y^2\, dA = \int_{-h/2}^{+h/2} y^2(b\, dy) = \tfrac{1}{3}b[y^3]_{-h/2}^{+h/2}$$

$$= \tfrac{1}{3}b\left(\frac{h^3}{8} + \frac{h^3}{8}\right)$$

or

$$I_x = \tfrac{1}{12}bh^3$$

(b) Radius of Gyration r_x. From Eq. (A.10), we have

$$I_x = r_x^2 A \qquad \tfrac{1}{12}bh^3 = r_x^2(bh)$$

and, solving for r_x,

$$r_x = h/\sqrt{12}$$

For the circular area of Fig. A.18, determine (*a*) the polar moment of inertia J_O, (*b*) the rectangular moments of inertia I_x and I_y.

(a) Polar Moment of Inertia. We select as an element of area a ring of radius ρ and thickness $d\rho$ (Fig. A.19). Since all the points within the ring are at the same distance ρ from the origin O, the polar moment of inertia of the ring is

$$dJ_O = \rho^2 \, dA = \rho^2 (2\pi\rho \, d\rho)$$

Integrating in ρ from 0 to c, we write

$$J_O = \int_A \rho^2 \, dA = \int_0^c \rho^2 (2\pi\rho \, d\rho) = 2\pi \int_0^c \rho^3 \, d\rho$$

$$J_O = \tfrac{1}{2}\pi c^4$$

(b) Rectangular Moments of Inertia. Because of the symmetry of the circular area, we have $I_x = I_y$. Recalling Eq. (A.9), we write

$$J_O = I_x + I_y = 2I_x \quad \tfrac{1}{2}\pi c^4 = 2I_x$$

and, thus,

$$I_x = I_y = \tfrac{1}{4}\pi c^4$$

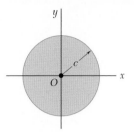

Fig. A.18

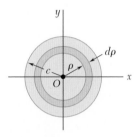

Fig. A.19

A.4 PARALLEL-AXIS THEOREM

Consider the moment of inertia I_x of an area A with respect to an arbitrary x axis (Fig. A.20). Denoting by y the distance from an element of area dA to that axis, we recall from Sec. A.3 that

$$I_x = \int_A y^2 \, dA$$

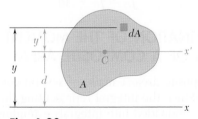

Fig. A.20

Let us now draw the *centroidal x′ axis*, i.e., the axis parallel to the x axis which passes through the centroid C of the area. Denoting by y' the distance from the element dA to that axis, we write $y = y' + d$,

A9

where d is the distance between the two axes. Substituting for y in the integral representing I_x, we write

$$I_x = \int_A y^2 \, dA = \int_A (y' + d)^2 dA$$

$$I_x = \int_A y'^2 \, dA + 2d \int_A y' \, dA + d^2 \int_A dA \qquad \text{(A.15)}$$

The first integral in Eq. (A.15) represents the moment of inertia $\bar{I}_{x'}$ of the area with respect to the centroidal x' axis. The second integral represents the first moment $Q_{x'}$ of the area with respect to the x' axis and is equal to zero, since the centroid C of the area is located on that axis. Indeed, we recall from Sec. A.1 that

$$Q_{x'} = A\bar{y}' = A(0) = 0$$

Finally, we observe that the last integral in Eq. (A.15) is equal to the total area A. We have, therefore,

$$I_x = \bar{I}_{x'} + Ad^2 \qquad \text{(A.16)}$$

This formula expresses that the moment of inertia I_x of an area with respect to an arbitrary x axis is equal to the moment of inertia $\bar{I}_{x'}$ of the area with respect to the centroidal x' axis parallel to the x axis, *plus* the product Ad^2 of the area A and of the square of the distance d between the two axes. This result is known as the *parallel-axis theorem*. It makes it possible to determine the moment of inertia of an area with respect to a given axis, when its moment of inertia with respect to a centroidal axis of the same direction is known. Conversely, it makes it possible to determine the moment of inertia $\bar{I}_{x'}$ of an area A with respect to a centroidal axis x', when the moment of inertia I_x of A with respect to a parallel axis is known, by *subtracting* from I_x the product Ad^2. We should note that the parallel-axis theorem may be used *only if one of the two axes involved is a centroidal axis*.

A similar formula may be derived, which relates the polar moment of inertia J_O of an area with respect to an arbitrary point O and the polar moment of inertia $\bar{J}_C$ of the same area with respect to its centroid C. Denoting by d the distance between O and C, we write

$$J_O = \bar{J}_C + Ad^2 \qquad \text{(A.17)}$$

A.5 DETERMINATION OF THE MOMENT OF INERTIA OF A COMPOSITE AREA

Consider a composite area A made of several component parts A_1, A_2 and so forth. Since the integral representing the moment of inertia of A may be subdivided into integrals extending over A_1, A_2 and so forth, the moment of inertia of A with respect to a given axis will be obtained by adding the moments of inertia of the areas A_1, A_2, and so forth, with respect to the same axis. Before adding the moments of inertia of the component areas, however, the parallel-axis theorem should be used to transfer each moment of inertia to the desired axis. This is shown in Example A.06.

Determine the moment of inertia $\bar{I}_x$ of the area shown with respect to the centroidal x axis (Fig. A.21).

Location of Centroid. The centroid C of the area must first be located. However, this has already been done in Example A.02 for the given area. We recall from that example that C is located 46 mm above the lower edge of the area A.

Computation of Moment of Inertia. We divide the area A into the two rectangular areas A_1 and A_2 (Fig. A.22), and compute the moment of inertia of each area with respect to the x axis.

Rectangular Area A_1. To obtain the moment of inertia $(I_x)_1$ of A_1 with respect to the x axis, we first compute the moment of inertia of A_1 *with respect to its own centroidal axis* x'. Recalling the formula derived in part *a* of Example A.04 for the centroidal moment of inertia of a rectangular area, we have

$$(\bar{I}_{x'})_1 = \tfrac{1}{12}bh^3 = \tfrac{1}{12}(80 \text{ mm})(20 \text{ mm})^3 = 53.3 \times 10^3 \text{ mm}^4$$

Using the parallel-axis theorem, we transfer the moment of inertia of A_1 from its centroidal axis x' to the parallel axis x:

$$(I_x)_1 = (\bar{I}_{x'})_1 + A_1 d_1^2 = 53.3 \times 10^3 + (80 \times 20)(24)^2$$
$$= 975 \times 10^3 \text{ mm}^4$$

Rectangular Area A_2. Computing the moment of inertia of A_2 with respect to its centroidal axis x'', and using the parallel-axis theorem to transfer it to the x axis, we have

$$(\bar{I}_{x''})_2 = \tfrac{1}{12}bh^3 = \tfrac{1}{12}(40)(60)^3 = 720 \times 10^3 \text{ mm}^4$$
$$(I_x)_2 = (\bar{I}_{x''})_2 + A_2 d_2^2 = 720 \times 10^3 + (40 \times 60)(16)^2$$
$$= 1334 \times 10^3 \text{ mm}^4$$

Entire Area A. Adding the values computed for the moments of inertia of A_1 and A_2 with respect to the x axis, we obtain the moment of inertia $\bar{I}_x$ of the entire area:

$$\bar{I}_x = (I_x)_1 + (I_x)_2 = 975 \times 10^3 + 1334 \times 10^3$$
$$\bar{I}_x = 2.31 \times 10^6 \text{ mm}^4$$

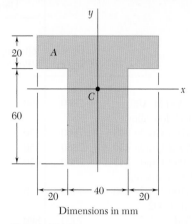

Dimensions in mm

Fig. A.21

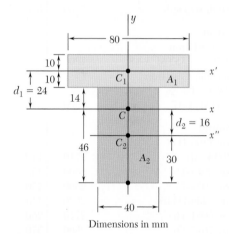

Dimensions in mm

Fig. A.22

Material	Density kg/m³	Ultimate Strength Tension, MPa	Compression,[2] MPa	Shear, MPa	Yield Strength[3] Tension, MPa	Shear, MPa	Modulus of Elasticity, GPa	Modulus of Rigidity, GPa	Coefficient of Thermal Expansion, 10⁻⁶/°C	Ductility, Percent Elongation in 50 mm
Steel										
Structural (ASTM-A36)	7860	400			250	145	200	77.2	11.7	21
High-strength-low-alloy										
ASTM-A709 Grade 345	7860	450			345		200	77.2	11.7	21
ASTM-A913 Grade 450	7860	550			450		200	77.2	11.7	17
ASTM-A992 Grade 345	7860	450			345		200	77.2	11.7	21
Quenched & tempered										
ASTM-A709 Grade 690	7860	760			690		200	77.2	11.7	18
Stainless, AISI 302										
Cold-rolled	7920	860			520		190	75	17.3	12
Annealed	7920	655			260	150	190	75	17.3	50
Reinforcing Steel										
Medium strength	7860	480			275		200	77	11.7	
High strength	7860	620			415		200	77	11.7	
Cast Iron										
Gray Cast Iron										
4.5% C, ASTM A-48	7200	170	655	240			69	28	12.1	0.5
Malleable Cast Iron										
2% C, 1% Si,										
ASTM A-47	7300	345	620	330	230		165	65	12.1	10
Aluminum										
Alloy 1100-H14										
(99% Al)	2710	110		70	95	55	70	26	23.6	9
Alloy 2014-T6	2800	455		275	400	230	75	27	23.0	13
Alloy-2024-T4	2800	470		280	325		73		23.2	19
Alloy-5456-H116	2630	315		185	230	130	72		23.9	16
Alloy 6061-T6	2710	260		165	240	140	70	26	23.6	17
Alloy 7075-T6	2800	570		330	500		72	28	23.6	11
Copper										
Oxygen-free copper										
(99.9% Cu)										
Annealed	8910	220		150	70		120	44	16.9	45
Hard-drawn	8910	390		200	265		120	44	16.9	4
Yellow-Brass										
(65% Cu, 35% Zn)										
Cold-rolled	8470	510		300	410	250	105	39	20.9	8
Annealed	8470	320		220	100	60	105	39	20.9	65
Red Brass										
(85% Cu, 15% Zn)										
Cold-rolled	8740	585		320	435		120	44	18.7	3
Annealed	8740	270		210	70		120	44	18.7	48
Tin bronze	8800	310			145		95		18.0	30
(88 Cu, 8Sn, 4Zn)										
Manganese bronze	8360	655			330		105		21.6	20
(63 Cu, 25 Zn, 6 Al, 3 Mn, 3 Fe)										
Aluminum bronze	8330	620	900		275		110	42	16.2	6
(81 Cu, 4 Ni, 4 Fe, 11 Al)										

(Table continued on page A13)

Continued from page A12

Material	Density kg/m³	Ultimate Strength Tension, MPa	Compression,[2] MPa	Shear, MPa	Yield Strength[3] Tension, MPa	Shear, MPa	Modulus of Elasticity, GPa	Modulus of Rigidity, GPa	Coefficient of Thermal Expansion, 10⁻⁶/°C	Ductility, Percent Elongation in 50 mm
Magnesium Alloys										
Alloy AZ80 (Forging)	1800	345		160	250		45	16	25.2	6
Alloy AZ31 (Extrusion)	1770	255		130	200		45	16	25.2	12
Titanium										
Alloy (6% Al, 4% V)	4730	900			830		115		9.5	10
Monel Alloy 400(Ni-Cu)										
Cold-worked	8830	675			585	345	180		13.9	22
Annealed	8830	550			220	125	180		13.9	46
Cupronickel										
(90% Cu, 10% Ni)										
Annealed	8940	365			110		140	52	17.1	35
Cold-worked	8940	585			545		140	52	17.1	3
Timber, air dry										
Douglas fir	470	100	50	7.6			13	0.7	Varies	
Spruce, Sitka	415	60	39	7.6			10	0.5	3.0 to 4.5	
Shortleaf pine	500		50	9.7			12			
Western white pine	390		34	7.0			10			
Ponderosa pine	415	55	36	7.6			9			
White oak	690		51	13.8			12			
Red oak	660		47	12.4			12			
Western hemlock	440	90	50	10.0			11			
Shagbark hickory	720		63	16.5			15			
Redwood	415	65	42	6.2			9			
Concrete										
Medium strength	2320		28				25		9.9	
High strength	2320		40				30		9.9	
Plastics										
Nylon, type 6/6, (molding compound)	1140	75	95		45		2.8		144	50
Polycarbonate	1200	65	85		35		2.4		122	110
Polyester, PBT (thermoplastic)	1340	55	75		55		2.4		135	150
Polyester elastomer	1200	45		40			0.2			500
Polystyrene	1030	55	90		55		3.1		125	2
Vinyl, rigid PVC	1440	40	70		45		3.1		135	40
Rubber	910	15							162	600
Granite (Avg. values)	2770	20	240	35			70	4	7.2	
Marble (Avg. values)	2770	15	125	28			55	3	10.8	
Sandstone (Avg. values)	2300	7	85	14			40	2	9.0	
Glass, 98% silica	2190		50				65	4.1	80	

[1]Properties of metals vary widely as a result of variations in composition, heat treatment, and mechanical working.
[2]For ductile metals the compression strength is generally assumed to be equal to the tension strength.
[3]Offset of 0.2 percent.
[4]Timber properties are for loading parallel to the grain.
[5]See also *Marks' Mechanical Engineering Handbook*, 10th ed., McGraw-Hill, New York, 1996; *Annual Book of ASTM*, American Society for Testing Materials, Philadelphia, Pa.; *Metals Handbook*, American Society of Metals, Metals Park, Ohio; and *Aluminum Design Manual*, The Aluminum Association, Washington, DC.

APPENDIX C Properties of Rolled-Steel Shapes
(SI Units)

W Shapes
(Wide-Flange Shapes)

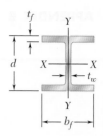

Designation†	Area A, mm²	Depth d, mm	Flange		Web Thickness t_w, mm	Axis X-X			Axis Y-Y		
			Width b_f, mm	Thickness t_f, mm		I_x 10^6 mm⁴	S_x 10^3 mm³	r_x mm	I_y 10^6 mm⁴	S_y 10^3 mm³	r_y mm
W920 × 449	57300	947	424	42.7	24.0	8780	18500	391	541	2560	97.0
201	25600	904	305	20.1	15.2	3250	7190	356	93.7	618	60.5
W840 × 299	38200	856	399	29.2	18.2	4830	11200	356	312	1560	90.4
176	22400	836	292	18.8	14.0	2460	5880	330	77.8	534	58.9
W760 × 257	32900	772	381	27.2	16.6	3430	8870	323	249	1310	86.9
147	18800	754	267	17.0	13.2	1660	4410	297	53.3	401	53.3
W690 × 217	27800	696	356	24.8	15.4	2360	6780	292	184	1040	81.3
125	16000	678	254	16.3	11.7	1190	3490	272	44.1	347	52.6
W610 × 155	19700	612	325	19.1	12.7	1290	4230	257	108	667	73.9
101	13000	602	228	14.9	10.5	762	2520	243	29.3	257	47.5
W530 × 150	19200	544	312	20.3	12.7	1010	3720	229	103	660	73.4
92	11800	533	209	15.6	10.2	554	2080	217	23.9	229	45.0
66	8390	526	165	11.4	8.89	351	1340	205	8.62	104	32.0
W460 × 158	20100	475	284	23.9	15.0	795	3340	199	91.6	646	67.6
113	14400	462	279	17.3	10.8	554	2390	196	63.3	452	66.3
74	9480	457	191	14.5	9.02	333	1460	187	16.7	175	41.9
52	6650	450	152	10.8	7.62	212	944	179	6.37	83.9	31.0
W410 × 114	14600	419	262	19.3	11.6	462	2200	178	57.4	441	62.7
85	10800	417	181	18.2	10.9	316	1510	171	17.9	198	40.6
60	7610	406	178	12.8	7.75	216	1060	168	12.0	135	39.9
46.1	5890	404	140	11.2	6.99	156	773	163	5.16	73.6	29.7
38.8	4950	399	140	8.76	6.35	125	629	159	3.99	57.2	28.4
W360 × 551	70300	455	419	67.6	42.2	2260	9950	180	828	3950	108
216	27500	376	394	27.7	17.3	712	3800	161	282	1430	101
122	15500	363	257	21.7	13.0	367	2020	154	61.6	480	63.0
101	12900	356	254	18.3	10.5	301	1690	153	50.4	397	62.5
79	10100	353	205	16.8	9.40	225	1270	150	24.0	234	48.8
64	8130	348	203	13.5	7.75	178	1030	148	18.8	185	48.0
57.8	7230	358	172	13.1	7.87	160	895	149	11.1	129	39.4
44	5710	351	171	9.78	6.86	121	688	146	8.16	95.4	37.8
39	4960	353	128	10.7	6.48	102	578	144	3.71	58.2	27.4
32.9	4190	348	127	8.51	5.84	82.8	475	141	2.91	45.9	26.4

†A wide-flange shape is designated by the letter W followed by the nominal depth in millimeters and the mass in kilograms per meter.

(Table continued on page A15)

APPENDIX C Properties of Rolled-Steel Shapes
(SI Units)
Continued from page A14

W Shapes
(Wide-Flange Shapes)

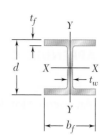

Designation†	Area A, mm²	Depth d, mm	Flange Width b_f, mm	Flange Thickness t_f, mm	Web Thickness t_w, mm	Axis X-X I_x 10^6 mm⁴	Axis X-X S_x 10^3 mm³	Axis X-X r_x mm	Axis Y-Y I_y 10^6 mm⁴	Axis Y-Y S_y 10^3 mm³	Axis Y-Y r_y mm
W310 × 143	18200	323	310	22.9	14.0	347	2150	138	112	728	78.5
107	13600	312	305	17.0	10.9	248	1600	135	81.2	531	77.2
74	9420	310	205	16.3	9.40	163	1050	132	23.4	228	49.8
60	7550	302	203	13.1	7.49	128	844	130	18.4	180	49.3
52	6650	318	167	13.2	7.62	119	747	133	10.2	122	39.1
44.5	5670	312	166	11.2	6.60	99.1	633	132	8.45	102	38.6
38.7	4940	310	165	9.65	5.84	84.9	547	131	7.20	87.5	38.4
32.7	4180	312	102	10.8	6.60	64.9	416	125	1.94	37.9	21.5
23.8	3040	305	101	6.73	5.59	42.9	280	119	1.17	23.1	19.6
W250 × 167	21200	290	264	31.8	19.2	298	2060	118	98.2	742	68.1
101	12900	264	257	19.6	11.9	164	1240	113	55.8	433	65.8
80	10200	257	254	15.6	9.4	126	983	111	42.9	338	65.0
67	8580	257	204	15.7	8.89	103	805	110	22.2	218	51.1
58	7420	252	203	13.5	8.00	87.0	690	108	18.7	185	50.3
49.1	6260	247	202	11.0	7.37	71.2	574	106	15.2	151	49.3
44.8	5700	267	148	13.0	7.62	70.8	531	111	6.95	94.2	34.8
32.7	4190	259	146	9.14	6.10	49.1	380	108	4.75	65.1	33.8
28.4	3630	259	102	10.0	6.35	40.1	308	105	1.79	35.1	22.2
22.3	2850	254	102	6.86	5.84	28.7	226	100	1.20	23.8	20.6
W200 × 86	11000	222	209	20.6	13.0	94.9	852	92.7	31.3	300	53.3
71	9100	216	206	17.4	10.2	76.6	708	91.7	25.3	246	52.8
59	7550	210	205	14.2	9.14	60.8	582	89.7	20.4	200	51.8
52	6650	206	204	12.6	7.87	52.9	511	89.2	17.7	174	51.6
46.1	5880	203	203	11.0	7.24	45.8	451	88.1	15.4	152	51.3
41.7	5320	205	166	11.8	7.24	40.8	398	87.6	9.03	109	41.1
35.9	4570	201	165	10.2	6.22	34.4	342	86.9	7.62	92.3	40.9
31.3	3970	210	134	10.2	6.35	31.3	298	88.6	4.07	60.8	32.0
26.6	3390	207	133	8.38	5.84	25.8	249	87.1	3.32	49.8	31.2
22.5	2860	206	102	8.00	6.22	20.0	193	83.6	1.42	27.9	22.3
19.3	2480	203	102	6.48	5.84	16.5	162	81.5	1.14	22.5	21.4
W150 × 37.1	4740	162	154	11.6	8.13	22.2	274	68.6	7.12	91.9	38.6
29.8	3790	157	153	9.27	6.60	17.2	220	67.6	5.54	72.3	38.1
24	3060	160	102	10.3	6.60	13.4	167	66.0	1.84	36.1	24.6
18	2290	153	102	7.11	5.84	9.20	120	63.2	1.24	24.6	23.3
13.5	1730	150	100	5.46	4.32	6.83	91.1	62.7	0.916	18.2	23.0
W130 × 28.1	3590	131	128	10.9	6.86	10.9	167	55.1	3.80	59.5	32.5
23.8	3040	127	127	9.14	6.10	8.91	140	54.1	3.13	49.2	32.0
W100 × 19.3	2470	106	103	8.76	7.11	4.70	89.5	43.7	1.61	31.1	25.4

†A wide-flange shape is designated by the letter W followed by the nominal depth in millimeters and the mass in kilograms per meter.

APPENDIX C Properties of Rolled-Steel Shapes
(SI Units)

S Shapes
(American Standard Shapes)

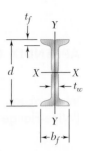

Designation†	Area A, mm²	Depth d, mm	Flange Width b_f, mm	Flange Thickness t_f, mm	Web Thickness t_w, mm	Axis X-X I_x 10⁶ mm⁴	Axis X-X S_x 10³ mm³	Axis X-X r_x mm	Axis Y-Y I_y 10⁶ mm⁴	Axis Y-Y S_y 10³ mm³	Axis Y-Y r_y mm
S610 × 180	22900	622	204	27.7	20.3	1320	4230	240	34.5	338	38.9
158	20100	622	200	27.7	15.7	1220	3930	247	32.0	320	39.9
149	18900	610	184	22.1	18.9	991	3260	229	19.7	215	32.3
134	17100	610	181	22.1	15.9	937	3060	234	18.6	205	33.0
119	15200	610	178	22.1	12.7	874	2870	241	17.5	197	34.0
S510 × 143	18200	516	183	23.4	20.3	695	2700	196	20.8	228	33.8
128	16300	516	179	23.4	16.8	653	2540	200	19.4	216	34.5
112	14200	508	162	20.2	16.1	533	2100	194	12.3	152	29.5
98.2	12500	508	159	20.2	12.8	495	1950	199	11.4	144	30.2
S460 × 104	13200	457	159	17.6	18.1	384	1690	170	10.0	126	27.4
81.4	10300	457	152	17.6	11.7	333	1460	180	8.62	113	29.0
S380 × 74	9480	381	143	15.8	14.0	202	1060	146	6.49	90.6	26.2
64	8130	381	140	15.8	10.4	186	973	151	5.95	85.0	26.9
S310 × 74	9420	305	139	16.7	17.4	126	829	116	6.49	93.2	26.2
60.7	7680	305	133	16.7	11.7	112	739	121	5.62	84.1	26.9
52	6580	305	129	13.8	10.9	94.9	624	120	4.10	63.6	24.9
47.3	6010	305	127	13.8	8.89	90.3	593	123	3.88	61.1	25.4
S250 × 52	6650	254	125	12.5	15.1	61.2	482	96.0	3.45	55.1	22.8
37.8	4810	254	118	12.5	7.90	51.2	403	103	2.80	47.4	24.1
S200 × 34	4360	203	106	10.8	11.2	26.9	265	78.5	1.78	33.6	20.2
27.4	3480	203	102	10.8	6.88	23.9	236	82.8	1.54	30.2	21.0
S150 × 25.7	3260	152	90.7	9.12	11.8	10.9	143	57.9	0.953	21.0	17.1
18.6	2360	152	84.6	9.12	5.89	9.16	120	62.2	0.749	17.7	17.8
S130 × 15	1890	127	76.2	8.28	5.44	5.12	80.3	52.1	0.495	13.0	16.2
S100 × 14.1	1800	102	71.1	7.44	8.28	2.81	55.4	39.6	0.369	10.4	14.3
11.5	1460	102	67.6	7.44	4.90	2.52	49.7	41.7	0.311	9.21	14.6
S75 × 11.2	1420	76.2	63.8	6.60	8.86	1.21	31.8	29.2	0.241	7.55	13.0
8.5	1070	76.2	59.2	6.60	4.32	1.04	27.4	31.2	0.186	6.28	13.2

†An American Standard Beam is designated by the letter S followed by the nominal depth in millimeters and the mass in kilograms per meter.

APPENDIX C   Properties of Rolled-Steel Shapes
(SI Units)

C Shapes
(American Standard Channels)

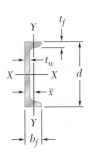

Designation†	Area A, mm²	Depth d, mm	Flange		Web Thickness t_w, mm	Axis X-X			Axis Y-Y			
			Width b_f, mm	Thickness t_f, mm		I_x 10^6 mm⁴	S_x 10^3 mm³	r_x mm	I_y 10^6 mm⁴	S_y 10^3 mm³	r_y mm	x mm
C380 × 74	9480	381	94.5	16.5	18.2	168	882	133	4.58	61.8	22.0	20.3
60	7610	381	89.4	16.5	13.2	145	762	138	3.82	54.7	22.4	19.8
50.4	6450	381	86.4	16.5	10.2	131	688	143	3.36	50.6	22.9	20.0
C310 × 45	5680	305	80.5	12.7	13.0	67.4	442	109	2.13	33.6	19.4	17.1
37	4740	305	77.5	12.7	9.83	59.9	393	113	1.85	30.6	19.8	17.1
30.8	3920	305	74.7	12.7	7.16	53.7	352	117	1.61	28.2	20.2	17.7
C250 × 45	5680	254	77.0	11.1	17.1	42.9	339	86.9	1.64	27.0	17.0	16.5
37	4740	254	73.4	11.1	13.4	37.9	298	89.4	1.39	24.1	17.1	15.7
30	3790	254	69.6	11.1	9.63	32.8	259	93.0	1.17	21.5	17.5	15.4
22.8	2890	254	66.0	11.1	6.10	28.0	221	98.3	0.945	18.8	18.1	16.1
C230 × 30	3790	229	67.3	10.5	11.4	25.3	221	81.8	1.00	19.2	16.3	14.8
22	2850	229	63.2	10.5	7.24	21.2	185	86.4	0.795	16.6	16.7	14.9
19.9	2540	229	61.7	10.5	5.92	19.9	174	88.6	0.728	15.6	16.9	15.3
C200 × 27.9	3550	203	64.3	9.91	12.4	18.3	180	71.6	0.820	16.6	15.2	14.4
20.5	2610	203	59.4	9.91	7.70	15.0	148	75.9	0.633	13.9	15.6	14.1
17.1	2170	203	57.4	9.91	5.59	13.5	133	79.0	0.545	12.7	15.8	14.5
C180 × 18.2	2320	178	55.6	9.30	7.98	10.1	113	66.0	0.483	11.4	14.4	13.3
14.6	1850	178	53.1	9.30	5.33	8.82	100	69.1	0.398	10.1	14.7	13.7
C150 × 19.3	2460	152	54.9	8.71	11.1	7.20	94.7	54.1	0.437	10.5	13.3	13.1
15.6	1990	152	51.6	8.71	7.98	6.29	82.6	56.4	0.358	9.19	13.4	12.7
12.2	1540	152	48.8	8.71	5.08	5.45	71.3	59.4	0.286	8.00	13.6	13.0
C130 × 13	1700	127	48.0	8.13	8.26	3.70	58.3	46.5	0.260	7.28	12.3	12.1
10.4	1270	127	44.5	8.13	4.83	3.11	49.0	49.5	0.196	6.10	12.4	12.3
C100 × 10.8	1370	102	43.7	7.52	8.15	1.91	37.5	37.3	0.177	5.52	11.4	11.7
8	1020	102	40.1	7.52	4.67	1.60	31.5	39.6	0.130	4.54	11.3	11.6
C75 × 8.9	1140	76.2	40.6	6.93	9.04	0.862	22.6	27.4	0.125	4.31	10.5	11.6
7.4	948	76.2	38.1	6.93	6.55	0.770	20.2	28.4	0.100	3.74	10.3	11.2
6.1	774	76.2	35.8	6.93	4.32	0.687	18.0	29.7	0.0795	3.21	10.1	11.1

†An American Standard Channel is designated by the letter C followed by the nominal depth in millimeters and the mass in kilograms per meter.

APPENDIX C Properties of Rolled-Steel Shapes
(SI Units)

Angles
Equal Legs

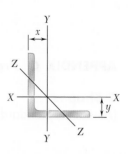

Size and Thickness, mm	Mass per Meter, kg/m	Area, mm²	Axis X-X				Axis Z-Z
			I 10^6 mm⁴	S 10^3 mm³	r mm	x or y mm	r_z mm
L203 × 203 × 25.4	75.9	9680	37.1	259	61.7	59.9	39.6
19	57.9	7350	29.1	200	62.5	57.4	39.9
12.7	39.3	5000	20.3	137	63.2	55.1	40.4
L152 × 152 × 25.4	55.7	7100	14.7	140	45.5	47.2	29.7
19	42.7	5460	11.7	109	46.2	45.0	29.7
15.9	36.0	4600	10.0	92.4	46.7	43.7	29.7
12.7	29.2	3720	8.28	75.2	47.2	42.4	30.0
9.5	22.2	2830	6.41	57.5	47.5	41.1	30.2
L127 × 127 × 19	35.1	4480	6.53	74.1	38.1	38.6	24.7
15.9	29.8	3780	5.66	63.1	38.6	37.3	24.8
12.7	24.1	3060	4.70	51.6	38.9	36.1	24.9
9.5	18.3	2330	3.65	39.5	39.4	34.8	25.0
L102 × 102 × 19	27.5	3510	3.17	45.7	30.0	32.3	19.7
15.9	23.4	2970	2.76	39.0	30.5	31.0	19.7
12.7	19.0	2420	2.30	32.1	30.7	30.0	19.7
9.5	14.6	1850	1.80	24.6	31.2	28.7	19.8
6.4	9.80	1250	1.25	16.9	31.8	27.4	19.9
L89 × 89 × 12.7	16.5	2100	1.51	24.3	26.7	26.7	17.2
9.5	12.6	1600	1.19	18.8	27.2	25.4	17.3
6.4	8.60	1090	0.832	12.9	27.7	24.2	17.5
L76 × 76 × 12.7	14.0	1770	0.916	17.4	22.7	23.6	14.7
9.5	10.7	1360	0.728	13.5	23.1	22.5	14.8
6.4	7.30	929	0.512	9.32	23.5	21.2	14.9
L64 × 64 × 12.7	11.4	1450	0.508	11.7	18.7	20.4	12.2
9.5	8.70	1120	0.405	9.14	19.0	19.3	12.2
6.4	6.10	768	0.288	6.34	19.4	18.1	12.2
4.8	4.60	581	0.223	4.83	19.6	17.4	12.2
L51 × 51 × 9.5	7.00	877	0.198	5.70	15.0	16.1	9.80
6.4	4.70	605	0.144	4.00	15.4	14.9	9.83
3.2	2.40	312	0.0787	2.11	15.7	13.6	9.93

APPENDIX C Properties of Rolled-Steel Shapes
(SI Units)

Angles
Unequal Legs

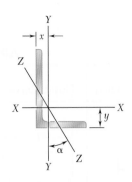

Size and Thickness, mm	Mass per Meter kg/m	Area mm²	Axis X-X				Axis Y-Y				Axis Z-Z	
			I_x 10⁶ mm⁴	S_x 10³ mm³	r_x mm	y mm	I_y 10⁶ mm⁴	S_y 10³ mm³	r_y mm	x mm	r_z mm	tan α
L203 × 152 × 25.4	65.5	8390	33.7	247	63.2	67.3	16.1	146	43.7	41.9	32.5	0.542
19	50.1	6410	26.4	192	64.0	64.8	12.8	113	44.5	39.6	32.8	0.550
12.7	34.1	4350	18.5	131	64.8	62.5	9.03	78.5	45.5	37.1	33.0	0.557
L152 × 102 × 19	35.0	4480	10.2	102	47.8	52.6	3.59	48.3	28.4	27.2	21.7	0.428
12.7	24.0	3060	7.20	70.6	48.5	50.3	2.59	33.8	29.0	24.9	21.9	0.440
9.5	18.2	2330	5.58	54.1	49.0	49.0	2.02	25.9	29.5	23.7	22.1	0.446
L127 × 76 × 12.7	19.0	2420	3.93	47.4	40.1	44.2	1.06	18.5	20.9	18.9	16.3	0.357
9.5	14.5	1850	3.06	36.4	40.6	42.9	0.837	14.3	21.3	17.7	16.4	0.364
6.4	9.80	1250	2.12	24.7	41.1	41.7	0.587	9.83	21.7	16.5	16.6	0.371
L102 × 76 × 12.7	16.4	2100	2.09	30.6	31.5	33.5	0.999	18.0	21.8	20.9	16.1	0.542
9.5	12.6	1600	1.64	23.6	32.0	32.3	0.787	13.9	22.2	19.7	16.2	0.551
6.4	8.60	1090	1.14	16.2	32.3	31.0	0.554	9.59	22.5	18.4	16.2	0.558
L89 × 64 × 12.7	13.9	1770	1.35	23.1	27.4	30.5	0.566	12.4	17.8	17.8	13.5	0.485
9.5	10.7	1360	1.07	17.9	27.9	29.2	0.454	9.65	18.2	16.6	13.6	0.495
6.4	7.30	929	0.753	12.3	28.4	27.9	0.323	6.72	18.6	15.4	13.7	0.504
L76 × 51 × 12.7	11.5	1450	0.799	16.4	23.4	27.4	0.278	7.70	13.8	14.7	10.8	0.413
9.5	8.80	1120	0.641	12.8	23.8	26.2	0.224	6.03	14.1	13.6	10.8	0.426
6.4	6.10	768	0.454	8.87	24.2	24.9	0.162	4.23	14.5	12.4	10.9	0.437
L64 × 51 × 9.5	7.90	1000	0.380	8.95	19.5	21.0	0.214	5.92	14.6	14.7	10.6	0.612
6.4	5.40	684	0.273	6.24	19.9	19.8	0.155	4.15	15.0	13.5	10.7	0.624

APPENDIX D Beam Deflections and Slopes

Beam and Loading	Elastic Curve	Maximum Deflection	Slope at End	Equation of Elastic Curve
1		$-\dfrac{PL^3}{3EI}$	$-\dfrac{PL^2}{2EI}$	$y = \dfrac{P}{6EI}(x^3 - 3Lx^2)$
2		$-\dfrac{wL^4}{8EI}$	$-\dfrac{wL^3}{6EI}$	$y = -\dfrac{w}{24EI}(x^4 - 4Lx^3 + 6L^2x^2)$
3		$-\dfrac{ML^2}{2EI}$	$-\dfrac{ML}{EI}$	$y = -\dfrac{M}{2EI}x^2$
4		$-\dfrac{PL^3}{48EI}$	$\pm\dfrac{PL^2}{16EI}$	For $x \leq \frac{1}{2}L$: $y = \dfrac{P}{48EI}(4x^3 - 3L^2x)$
5		For $a > b$: $-\dfrac{Pb(L^2 - b^2)^{3/2}}{9\sqrt{3}EIL}$ at $x_m = \sqrt{\dfrac{L^2 - b^2}{3}}$	$\theta_A = -\dfrac{Pb(L^2 - b^2)}{6EIL}$ $\theta_B = +\dfrac{Pa(L^2 - a^2)}{6EIL}$	For $x < a$: $y = \dfrac{Pb}{6EIL}[x^3 - (L^2 - b^2)x]$ For $x = a$: $y = -\dfrac{Pa^2b^2}{3EIL}$
6		$-\dfrac{5wL^4}{384EI}$	$\pm\dfrac{wL^3}{24EI}$	$y = -\dfrac{w}{24EI}(x^4 - 2Lx^3 + L^3x)$
7		$\dfrac{ML^2}{9\sqrt{3}EI}$	$\theta_A = +\dfrac{ML}{6EI}$ $\theta_B = -\dfrac{ML}{3EI}$	$y = -\dfrac{M}{6EIL}(x^3 - L^2x)$

Appendix E

Fundamentals of Engineering Examination

Engineers are required to be licensed when their work directly affects the public health, safety, and welfare. The intent is to ensure that engineers have met minimum qualifications, involving competence, ability, experience, and character. The licensing process involves an initial exam, called the *Fundamentals of Engineering Examination*, professional experience, and a second exam, called the *Principles and Practice of Engineering*. Those who successfully complete these requirements are licensed as a *Professional Engineer*. The exams are developed under the auspices of the *National Council of Examiners for Engineering and Surveying*.

The first exam, the *Fundamentals of Engineering Examination*, can be taken just before or after graduation from a four-year accredited engineering program. The exam stresses subject material in a typical undergraduate engineering program, including *Mechanics of Materials*. The topics included in the exam cover much of the material in this book. The following is a list of the main topic areas, with references to the appropriate sections in this book. Also included are problems that can be solved to review this material.

Stresses (1.3–1.8; 1.11–1.12)
   Problems: 1.1, 1.7, 1.31, 1.37

Strains (2.2–2.3; 2.5–2.6; 2.8–2.11; 2.14–2.15)
   Problems: 2.7, 2.19, 2.41, 2.49, 2.63, 2.68

Torsion (3.2–3.6; 3.13)
   Problems: 3.6, 3.28, 3.35, 3.51, 3.132, 3.138

Bending (4.2–4.6; 4.12)
   Problems: 4.11, 4.23, 4.34, 4.47, 4.104, 4.109

Shear and Bending-Moment Diagrams (5.2–5.3)
Problems: 5.6, 5.9, 5.42, 5.48

Normal Stresses in Beams (5.1–5.3)
Problems: 5.18, 5.21, 5.55, 5.61

Shear (6.2–6.4; 6.6–6.7)
Problems: 6.2, 6.12, 6.32, 6.36

Transformation of Stresses and Strains (7.2–7.4; 7.7–7.9)
Problems: 7.6, 7.13, 7.33, 7.41, 7.81, 7.87, 7.102, 7.109

Deflection of Beams (9.2–9.4; 9.7)
Problems: 9.6, 9.10, 9.72, 9.75

Columns (10.2–10.4)
Problems: 10.11, 10.21, 10.28

Strain Energy (11.2–11.4)
Problems: 11.10, 11.14, 11.19

Photo Credits

CHAPTER 1
Opener: © Construction Photography/CORBIS RF; 1.1: © Vince Streano/CORBIS; 1.2: © John DeWolf.

CHAPTER 2
Opener: © Construction Photography/CORBIS; 2.1: © John DeWolf; 2.2: Courtesy of Tinius Olsen Testing Machine Co., Inc.; 2.3, 2.4, 2.5: © John DeWolf.

CHAPTER 3
Opener: © Brownie Harris; 3.1: © 2008 Ford Motor Company; 3.2: © John DeWolf; 3.3: Courtesy of Tinius Olsen Testing Machine Co., Inc.

CHAPTER 4
Opener: © Lawrence Manning/CORBIS; 4.1: Courtesy of Flexifoil; 4.2: © Tony Freeman/Photo Edit; 4.3: © Hisham Ibrahim/Getty Images RF; 4.4: © Kevin R. Morris/CORBIS; 4.5: © Tony Freeman/Photo Edit; 4.6: © John DeWolf.

CHAPTER 5
Opener: © Mark Segal/Digital Vision/Getty Images RF; 5.1: © David Papazian/CORBIS RF; 5.2: © Godden Collection, National Information Service for Earthquake Engineering, University of California, Berkeley.

CHAPTER 6
Opener: © Godden Collection, National Information Service for Earthquake Engineering, University of California, Berkeley; 6.1: © John DeWolf; 6.2: © Jake Wyman/Getty Images; 6.3: © Rodho/shutterstock.com.

CHAPTER 7
Opener: NASA; 7.1: © Radlund & Associates/Getty Images RF; 7.2: © Spencer C. Grant/Photo Edit; 7.3: © Clair Dunn/Alamy; 7.4: © Spencer C. Grant/Photo Edit.

CHAPTER 8
Opener: © Mark Read.

CHAPTER 9
Opener: © Construction Photography/CORBIS; 9.1: Royalty-Free/CORBIS; 9.2 and 9.3: © John DeWolf; 9.4: Courtesy of Aztec Galvenizing Services; 9.5: Royalty-Free/CORBIS.

CHAPTER 10
Opener: © Jose Manuel/Photographer's Choice RF/Getty Images; 10.1: © Courtesy of Fritz Engineering Laboratory, Lehigh University; 10.2a: © Godden Collection, National Information Service for Earthquake Engineering, University of California, Berkeley; 10.2b: © Peter Marlow/Magnum Photos.

CHAPTER 11
Opener: © Corbis Super RF/Alamy; 11.1: © Daniel Schwen; 11.2: © Tony Freeman/Photo Edit Inc.; 11.3: Courtesy of L.I.E.R. and Sec Envel.

Photo Credits

CHAPTER 1

Opener: © Construction Photography/CORBIS; 1.1: © J. Shaw; Starter: CORBIS; 1.2: © John DeWolf.

CHAPTER 2

Opener: © Geostock/Photographer's/CORBIS; 2.1: © Digital Stock; 2.2: Courtesy of Tony Olsson/Index Machine Co., Inc.; 2.3: © John DeWolf.

CHAPTER 3

Opener: © Bettmann/Illinois; 3.4: © UPI; Final figure: Courtesy of © John DeWolf; 3.4: Courtesy of Tony Olsson/Index Machine Co., Inc.

CHAPTER 4

Opener: © Lawrence Manning/CORBIS; 4.1: Courtesy of Pearson; 4.2: © Roy Ziegmun/Photo Edit; 4.3: © Digital Stock/Corbis; Images 10.3: 4.5: © Ernst H. Weiss/CORBIS; 4.6: © Tony Freeman; Photo Edit; 4.7: © John DeWolf.

CHAPTER 5

Opener: © Stock.xpd/Royalty/Getty Images; IPC 5.1: © David Equena/CORBIS; 5.3, 5.4: © Lockheed Corporation; Starter: National Information Service for Earthquake Engineering, University of California, Berkeley.

CHAPTER 6

Opener: © Geostock; Starter: National Information Service for Earthquake Engineering, University of California, Berkeley.

© John DeWolf; 6.1: Los Warningale; Images 6.5: © Purdue; Shutterstock.com.

CHAPTER 7

Opener: NASA; 7.1: © Bullard & Associates/Getty Images; IPC 7.4: © Sprock; 7.5: © Image Bank; Starter: 7.8: © Getty; IPC; 7.10: © Sprock; Starter: © ChuckShow Fund.

CHAPTER 8

Opener: © Matt Rush.

CHAPTER 9

Opener: © Construction Photography/CORBIS; 9.1: J. Barnett; IPC 9.2: Digital Stock; 9.3: Laura J. Wolfe; 9.4: Courtesy of Auto Club racing. See bags; 9.5: Digital Stock/CORBIS.

CHAPTER 10

Opener: Jose Manuel/Photographer's Choice/Getty Images; 10.1: Courtesy of UPI; Engineering Laboratory, College Cambridge 10.2; Courtesy University; National Information Service for Earthquake Engineering, University of California, Berkeley; IPC 10.6: Plus.3; Digital Stock.

CHAPTER 11

Opener: © Corbis Images; IPC Starter: LLC © Digital Stock; 11.2; Ray: Freeman/Photo Edit Inc.; IPC: Courtesy of T.J. Clark and © Stock.

Index